HVAC

Second Edition

EUGENE SILBERSTEIN

Australia • Brazil • Japan • Korea • Mexico • Singapore • Spain • United Kingdom • United States

Residential Construction Academy HVAC, Second Edition
Eugene Silberstein

Vice President, Career and Professional Editorial: Dave Garza

Director of Learning Solutions: Sandy Clark

Senior Acquisitions Editor: Jim DeVoe

Managing Editor: Larry Main

Product Manager: Brooke Wilson, Ohlinger Publishing Services

Editorial Assistant: Cris Savino

Vice President, Career and Professional Marketing: Jennifer Baker

Marketing Director: Deborah Yarnell

Marketing Manager: Katie Hall

Marketing Coordinator: Mark Pierro

Production Director: Wendy Troeger

Production Manager: Mark Bernard

Content Project Manager: David E. Plagenza

Art Director: Casey Kirchmayer

Technology Project Manager: Joe Pliss

For product information and technology assistance, contact us at
Cengage Learning Customer & Sales Support, 1-800-354-9706
For permission to use material from this text or product,
submit all requests online at **www.cengage.com/permissions**.
Further permissions questions can be e-mailed to
permissionrequest@cengage.com

Library of Congress Control Number: 2010932977

ISBN-13: 978-1-4390-5634-9

ISBN-10: 1-4390-5634-X

Delmar
5 Maxwell Drive
Clifton Park, NY 12065-2919
USA

Cengage Learning is a leading provider of customized learning solutions with office locations around the globe, including Singapore, the United Kingdom, Australia, Mexico, Brazil, and Japan. Locate your local office at:
international.cengage.com/region

Cengage Learning products are represented in Canada by Nelson Education, Ltd.

To learn more about Delmar, visit **www.cengage.com/delmar**

Purchase any of our products at your local college store or at our preferred online store **www.ichapters.com**

Notice to the Reader

Publisher does not warrant or guarantee any of the products described herein or perform any independent analysis in connection with any of the product information contained herein. Publisher does not assume, and expressly disclaims, any obligation to obtain and include information other than that provided to it by the manufacturer. The reader is expressly warned to consider and adopt all safety precautions that might be indicated by the activities described herein and to avoid all potential hazards. By following the instructions contained herein, the reader willingly assumes all risks in connection with such instructions. The publisher makes no representations or warranties of any kind, including but not limited to, the warranties of fitness for particular purpose or merchantability, nor are any such representations implied with respect to the material set forth herein, and the publisher takes no responsibility with respect to such material. The publisher shall not be liable for any special, consequential, or exemplary damages resulting, in whole or part, from the readers' use of, or reliance upon, this material.

Printed in the United States of America
2 3 4 5 6 7   13 12 11

Table of Contents

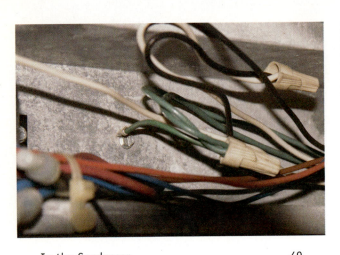

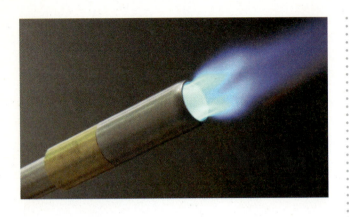

CHAPTER 6

Tubing and Piping Tools 134

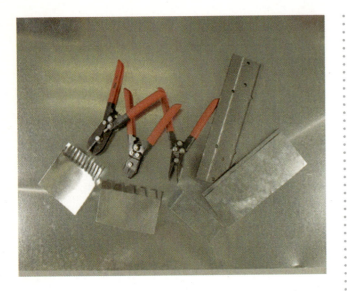

SECTION 4

Electricity for HVAC 333

CHAPTER 14

Electricity Fundamentals 334

CHAPTER 15

Electric Motors 373

CHAPTER 16

Electric Controls 407

CHAPTER 17

Wiring Diagrams 440

CHAPTER 18

Electric Codes............... 456

SECTION 5

System Commissioning........ 479

CHAPTER 19

Testing, Adjusting, and Balancing........................... 480

CHAPTER 20

Indoor Air Quality 508

CHAPTER 21

Mechanical Troubleshooting 529

Electrical Troubleshooting.... 556

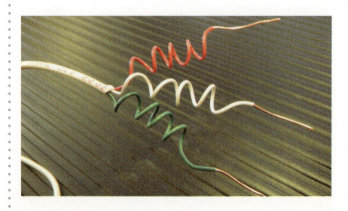

SECTION 6

Heating Systems 589

CHAPTER 23

Electric Heat 590

CHAPTER 24

Gas Heat 611

CHAPTER 25

Oil Heat 654

CHAPTER 26

Hydronic Heat 688

Preface

HOME BUILDERS INSTITUTE RESIDENTIAL CONSTRUCTION ACADEMY: HVAC

About the Residential Construction Academy Series

One of the most pressing problems confronting the building industry today is the shortage of skilled labor. The construction industry must recruit an estimated 200,000 to 250,000 new craft workers each year to meet future needs. This shortage is expected to continue well into the next decade because of projected job growth and a decline in the number of available workers. At the same time, the training of available labor is becoming an increasing concern throughout the country. This lack of training opportunities has resulted in a shortage of 65,000 to 80,000 skilled workers per year. This challenge is affecting all construction trades and is threatening the ability of builders to construct quality homes.

These challenges led to the creation of the innovative *Residential Construction Academy Series*. The *Residential Construction Academy Series* is the perfect way to introduce people of all ages to the building trades while guiding them in the development of essential workplace skills, including carpentry, electrical wiring, HVAC, plumbing, masonry, and facilities maintenance. The products and services offered through the *Residential Construction Academy* are the result of cooperative planning and rigorous joint efforts between industry and education. The program was originally conceived by the National Association of Home Builders (NAHB)—the premier association of more than 200,000 member groups in the residential construction industry—and its workforce development arm, the Home Builders Institute (HBI).

For the first time, construction professionals and educators created national skills standards for the construction trades. In the summer of 2001, NAHB, through the HBI, began the process of developing residential craft standards in six trades: carpentry, electrical wiring, HVAC, plumbing, masonry, and facilities maintenance. Groups of employers from across the country met with an independent research and measurement organization to begin the development of new craft training standards. Care was taken to assure representation of builders and remodelers, residential and light commercial, custom single family and high production or volume builders. The guidelines from the National Skills Standards Board were followed in developing the new standards. In addition, the process met or exceeded American Psychological Association standards for occupational credentialing.

Next, through a partnership between HBI and Cengage/Delmar Learning, learning materials—textbooks, videos, and instructor's curriculum and teaching tools—were created to teach these standards effectively. A foundational tenet of this series is that students *learn by doing*. Integrated into this colorful, highly illustrated text are Procedure sections designed to help students apply information through hands-on, active application. A constant focus of the *Residential Construction Academy Series* is teaching the skills needed to be successful in the construction industry and constantly applying the learning to real-world applications.

The newest programming component to the Residential Construction Academy Series is industry Program Credentialing and Certification for both instructors and students by the Home Builders Institute. National Instructor Certification ensures consistency in instructor teaching/training methodologies and knowledge competency when teaching to the industry's national skills standards. Student Certification is offered for each trade area of the Residential Construction Academy Series in the form of rigorous testing. Student Certification is tied to a national database that will provide an opportunity for easy access for potential employers to verify skills and competencies. Instructor and Student Certification serve the basis for Program Credentialing offered by HBI. For more information on HBI Program Credentialing and Instructor and Student Certification, please go to **www.hbi.org/certification**.

About This Book

A home is an essential part of life. It provides protection, security, and privacy to its occupants. It is often viewed as the single most important thing a family can own. This book is written for students who want to learn how to properly install and service the air conditioning and heating equipment that provides comfort for those who reside within the home.

The 2nd edition of HVAC covers the basics of air conditioning and heating theory as well as the processes and skills needed to safely install and service residential HVAC equipment. These required skills are discussed and presented in a manner that not only explains what needs to be done but also shows how to accomplish these tasks. General and task-specific safety issues are addressed throughout the book.

This textbook provides a valuable resource for the areas of heating, ventilation, and air conditioning that are required of an entry-level HVAC service technician, although those actively involved in the industry will also benefit from the material covered. The basic "hands-on" skills as well as the procedures outlined in this book will help individuals gain proficiency in this ever-changing trade.

In addition to topics such as heat and refrigeration theory, electrical theory, motors, and automatic controls, this book covers a wide range of topics including the installation of piping circuits, air distribution systems, system installation, startup and, of course, system servicing. The concepts of mechanical refrigeration, heat pumps, fossil fuel heating furnaces, and hydronic systems are also covered in great detail. The format of this material is intended to be easy to read and easy to teach.

ORGANIZATION

This textbook is organized in a manner so that those new to the industry as well as those already working in the field can gain maximum benefit from its content. The six main sections of the book cover the major aspects of the HVAC industry as they affect residential construction:

- *Section 1: Refrigeration Fundamentals* discusses matter, energy, heat basics, the refrigeration process, general safety practices, and refrigerant management.

- *Section 2: Tools* covers hand tools, tubing and piping tools, and specialized HVAC/R tools.

- *Section 3: Installation* introduces unit location, duct systems, system connections as well as system leak testing, system evacuation and system start-up, and charging.

- *Section 4: Electricity for HVAC* addresses electric fundamentals and motors, electric controls, wiring diagrams, and electric codes.

- *Section 5: System Commissioning* discusses the concepts of testing, adjusting and balancing, indoor air quality as well as mechanical and electrical troubleshooting. This section also includes step-by-step service calls that take the reader through situations commonly encountered in the field.

- *Section 6: Heating* introduces various methods of residential heating, including electric heating, gas heating, oil heating, hydronics, heat pumps, and geothermal systems. This section also includes step-by-step service calls that take the reader through situations commonly encountered in the field.

NEW TO THIS EDITION

The second edition of HVAC has undergone some very exciting changes. In addition to six completely new chapters, the existing installation chapters have been split into smaller chapters, making it easier to teach and evaluate individual installation tasks. Here are the highlights:

NEW CONTENT

CHAPTER 5

- New chapter on hand tools. This chapter provides information regarding the basic hand tools that are used by most individuals in the building trades.

CHAPTER 6

- New chapter on tubing and piping tools. This chapter is devoted to the tools that are used to connect piping circuits.

CHAPTER 7

- New chapter on specialized HVAC/R tools. This chapter discusses the tools that are commonly used by the HVAC/R system installer and technician to aid in the installation, evaluation and troubleshooting of HVAC/R equipment.

CHAPTER 8

- System location selection is now its own chapter.

CHAPTER 9

- Duct systems is now its own chapter.

CHAPTER 10

- System connections is now a separate chapter.

CHAPTER 11

- System leak testing is now a separate chapter.

CHAPTER 12

- System evacuation is now a separate chapter.

CHAPTER 13

- System startup and charging is now a separate chapter.

CHAPTER 17

- New chapter on wiring diagrams. This chapter is devoted to wiring diagram symbols and the reading of wiring diagrams.

CHAPTER 18

- New chapter covering national electric codes. This chapter provides the reader with important information from the National Electric Code® that directly affects the HVAC/R installer and technician.

CHAPTER 19

- New chapter on system testing, adjusting, and balancing. This chapter addresses the importance of proper system balancing. Detailed methods are discussed for ensuring proper airflow through the air distribution system.

In addition to the changes noted above, additional content will be found throughout the text. These changes include:

- Expanded discussion of R-410A and the inclusion of R-410A in air conditioning examples.
- Expanded discussion regarding new generation replacement refrigerants.
- New coverage of ductless split systems.
- New coverage of oxy-acetylene torch kits.
- Expanded material on ladder safety.
- Expanded content on system evacuation.
- New coverage of ultraviolet (UV) air cleaners.

NEW FEATURES

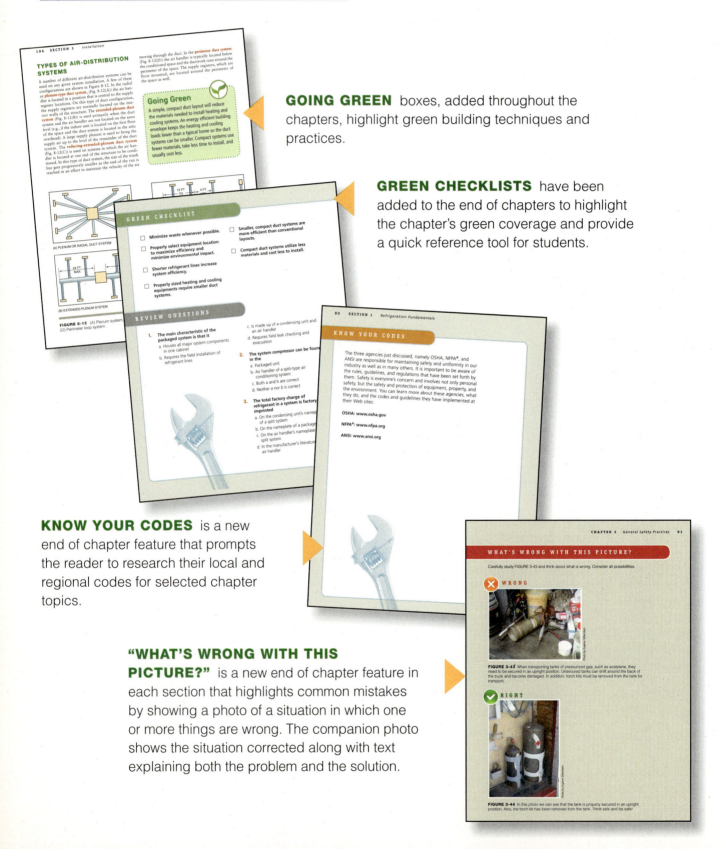

GOING GREEN boxes, added throughout the chapters, highlight green building techniques and practices.

GREEN CHECKLISTS have been added to the end of chapters to highlight the chapter's green coverage and provide a quick reference tool for students.

KNOW YOUR CODES is a new end of chapter feature that prompts the reader to research their local and regional codes for selected chapter topics.

"WHAT'S WRONG WITH THIS PICTURE?" is a new end of chapter feature in each section that highlights common mistakes by showing a photo of a situation in which one or more things are wrong. The companion photo shows the situation corrected along with text explaining both the problem and the solution.

KEY FEATURES

This innovative series was designed with input from educators and industry and informed by the curriculum and training objectives established by the Standards Committee. The following features aid in the learning process:

LEARNING FEATURES such as the Introduction, Objectives, and Glossary set the stage for the coming body of knowledge and help the learner identify key concepts and information. These learning features serve as a road map for the chapter. The learner also may use them as a reference later.

GOING GREEN boxes, added throughout the chapters, highlight green building techniques and practices.

ACTIVE LEARNING is a core concept of the *Residential Construction Academy Series.* Information is heavily illustrated to provide a visual of new tools and tasks encountered by the learner. In the Procedures, various tasks used in HVAC installation and service are grouped in a step-by-step approach. The overall effect is a clear view of the task, making learning easier.

FIGURE 4-11 Color-coded refrigerant cylinders.

FROM EXPERIENCE provides tricks of the trade and mentoring wisdom that make a particular task a little easier for the novice to accomplish.

SAFETY is featured throughout the text to instill safety as an "attitude" among learners. Safe jobsite practices by all workers is essential; if one person acts in an unsafe manner, then all workers on the job are at risk of being injured, too. Learners will come to appreciate that safety is a blend of ability, skill, and knowledge that should be continuously applied to all they do in the Construction industry.

CAUTION features highlight safety issues and urgent safety reminders for the trade.

TECH TOOL BELT exposes the learner to specialized tools that are required to accomplish the tasks that are necessary of installation and repair personnel.

EXAMPLES of applied mathematical formulas and calculations are provided to give the student more exposure to the calculations and their use.

SERVICE CALLS describe typical service procedures that are followed in the field to familiarize learners with real life applications. The service call begins with the customer complaint, followed by step-by-step sequences of the actions taken by the field technician, until the system problem has been located and repaired. The service call is followed by a brief discussion, highlighting and commenting on the troubleshooting techniques used by the technician in that particular service call.

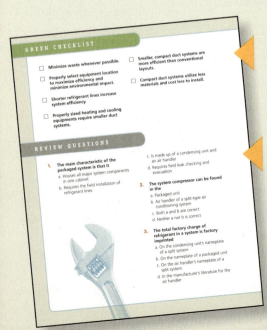

NEW GREEN CHECKLISTS have been added to the end of chapters to highlight the chapter's green coverage and provide a quick reference tool for students.

REVIEW QUESTIONS complete each chapter. These are designed to reinforce the information learned in the chapter as well as give learners the opportunity to think about what has been taught and what they have accomplished.

NEW **KNOW YOUR CODES** is a new end of chapter feature that prompts the reader to research their local and regional codes for selected chapter topics.

NEW **"WHAT'S WRONG WITH THIS PICTURE?"** is a new end of chapter feature in each section that highlights common mistakes by showing a photo of a situation in which one or more things are wrong. The companion photo shows the situation corrected along with text explaining both the problem and the solution.

TURNKEY CURRICULUM AND TEACHING MATERIAL PACKAGE

We understand that a text is only one part of a complete, turnkey educational system. We also understand that instructors want to spend their time teaching, not preparing to teach. The *Residential Construction Academy Series* is committed to providing thorough curriculum and preparatory materials to aid instructors and alleviate some of their heavy preparation commitments. An integrated teaching solution is provided with the text, including the Instructor's Resource CD, a printed Instructor's Resource Guide, and Workbook.

Workbook

Designed to accompany *Residential Construction Academy HVAC*, 2nd edition, the Workbook is an extension of the core text, and provides additional review questions and problems designed to challenge and reinforce the student's comprehension of the content presented in the core text.

Instructor Resources

The **Instructor Resources** CD contains lecture outlines, notes to instructors with teaching hints, cautions, and answers to review questions and other aids for the Instructor using this *Series.* These features are available for each chapter of the book, and are easily customizable in Microsoft Word. Designed as a complete and integrated package, the Instructor is also provided with suggestions for when and how to use the accompanying **PowerPoint, Computerized Test Bank, Video,** and **CD Courseware** package components. There are also print and pdf versions of the **Instructor's Resource Guide** available, as well as other aids for the Instructor using this *Series.*

The **Computerized Testbank** in examview makes generating tests and quizzes a snap. With hundreds of questions and different styles to choose from, you can create customized assessments for your students with the click of a button. Add your own unique questions and print rationales for easy class preparation.

Customizable **PowerPoint® Presentations** focus on key points for each chapter through lecture outlines that can be used to teach the course. Instructors may teach from this outline or make changes to suit individual classroom needs.

Use the hundreds of images from the **Image Library** to enhance your PowerPoint® Presentations, create test questions or add visuals wherever you need them. These valuable images are pulled from the accompanying textbook, are organized by chapter, and are easily searchable.

The **HVAC Video Series** is an integrated part of the *Residential Construction Academy HVAC* package. The series contains a set of four 20-minute videos that provide step-by-step HVAC instructions. All the essential information is covered in this series, beginning with the science behind HVAC principles to ensuring optimum operation of the system for customer satisfaction. Installation, troubleshooting, and air quality control are also addressed. Need-to-know Technician's Tips and Safety Tips offer practical advice from the experts.

The complete set includes the following: Video #1: The Science of HVAC, Video #2: New System Installation, Video #3: Basic Electricity and Troubleshooting, Video #4: Introduction to Heating and Air Quality Control. An Instructor's Guide is included as well.

Online Companion

The Online Companion is an excellent supplement for students. It features many useful resources to support the HVAC book. Linked from the student materials section of www.residentialacademy.com, the Online Companion includes chapter quizzes, an online glossary, product updates, related links, and more.

About the Author

The author of this book, Eugene Silberstein, has been involved in all aspects of the HVAC/R industry from field technician and system designer to company owner, teacher, administrator, consultant, and author over the past 30+ years. He recently had the title of Certified Master HVAC/R Educator, CMHE, bestowed upon him by HVAC Excellence. Mr. Silberstein is presently the Director and Professor of the HVAC/R Associates of Applied Sciences program on the Michael J. Grant Campus of Suffolk County Community College in Brentwood, New York, and has been involved with that program since its inception in 2003. In addition to teaching at the post-secondary level, Mr. Silberstein has taught air conditioning, heating, and plumbing at the high school level at Nassau BOCES in Westbury, New York. Mr. Silberstein was also an instructor, refrigeration department chair, and educational supervisor at Apex Technical School in New York City.

Eugene Silberstein is an active member of the American Society of Heating, Refrigerating and Air-Conditioning Engineers, ASHRAE, and the Refrigeration Service Engineers Society, RSES. Mr. Silberstein graduated with dual degrees from The City College of New York in New York City. Mr. Silberstein's book credits include *Heat Pumps* (Cengage Delmar Learning), *Refrigeration and Air Conditioning Technology*, 6th and 7th Editions (Cengage Delmar Learning), and *Pressure Enthalpy Without Tears* (The HVAC Prof, Inc.). Other writing credits include over 12 instructional videos that accompany *Electricity for Refrigeration, Heating and Air Conditioning*, 7th Edition, and 24 instructional videos that accompany *Refrigeration and Air Conditioning Technology*, 6th Edition.

Acknowledgments

HVAC National Skill Standards

The NAHB and the HBI would like to thank the many individual members and companies that participated in the creation of the HVAC National Skills Standards. Special thanks are extended to the following individuals and companies:

Janie Ade, Arlington Fuel & Oil

David Gannon, Tempo Mechanical Services

Rick Harper, Florida Comfort

James Henderson, North Texas Job Corps Center

Fred Humphreys, Home Builders Institute

Mark Huth, Delmar Cengage Learning

Joe Jurewicz, SCI Coal Township

Gerry Kennedy, Plumbing-Heating-Cooling-Contractors-National Association (PHCC)

Warren Lupson, Lupson and Associates LLC

Tom Moore, Climate Control

Mark Newey, Southface Energy Institute

Gene Porter, Refrigeration Service Engineers Society

Terry Reeves, Climate Masters, Inc.

Ronald Rogers, Wasdyke Associates

Leslie Sandler, Air-Conditioning & Refrigeration Institute

Eugene Silberstein, Suffolk County Community College

Ray Wasdyke, Wasdyke Associates

In addition to the standards committee, many other people contributed their time and expertise to the project. They have spent hours attending focus groups, reviewing, and contributing to the work. Delmar Learning and the author extend our sincere gratitude to:

Richard Anderson, Mississippi Job Corps Center

Barry Burkan, Apex Technical School

William J. Connelly, Orleans Technical Institute

Michael D. Frank, Quentin Burdick Job Corps

Kevin Fry, Excelsior Spring Job Corps Center

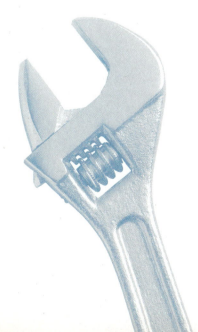

Elwin Hunt, San Joaquin Valley College

Jim Johnson, Technical Training Associates

John Levey, Oil Heat Associates

Joseph P. Moravek, Lee College

Greg Skudlarek, Minneapolis Community and Technical College

A special thank you goes out to the following individuals:

Kevin Standiford for his work on the Instructors Guide that accompanies this text.

My developmental editor on this project, Brooke Wilson, for her devotion to this work, her attention to detail, and her creative input. You are absolutely great!

Monica Ohlinger, an amazing and energetic individual, for always putting out high-quality, premium products.

John Levey of Oilheat Associates, Inc. for his assistance with the oil heat chapter and the images contained therein.

Dave Boyd of Appion and Dave Foster of Uniweld for providing some of the high quality images that appear in the book.

Dawnmarie Martino, Richard Zimmerman and Barrington White for their assistance with the Career Profile portion of the text.

The author would also like to give thanks to a very special friend, Dan Holohan of Dan Holohan Associates in Bethpage, New York, and of Heatinghelp.com (www.heatinghelp.com) for his assistance in creating an awesome hydronics chapter. Dan's unique insight to the world of hydronics has proven invaluable to those in the trade and to anyone who has had the opportunity to cross his path. It is to all of these individuals that this book is dedicated.

Introduction

ORGANIZATION OF THE INDUSTRY

The residential construction industry is one of the biggest sectors of the American economy. According to the U.S. Department of Labor, construction is one of the Nation's largest industries, with 7.2 million wage and salary jobs and 1.8 million self-employed and unpaid family workers in 2008. About 64% of wage and salary jobs in construction were in the specialty trade contractors sector, primarily plumbing, heating and air conditioning, electrical, and masonry. The National Association of Home Builders (NAHB) reports that home building traditionally accounts for 50%–55% of the construction industry. Opportunities are available for people to work at all levels in the construction industry, from those who handle the tools and materials on the job site to the senior engineers and architects who spend most of their time in offices. Few people spend their entire lives in a single occupation, and even fewer spend their lives working for one employer. You should be aware of all the opportunities in the construction industry so that you can make career decisions in the future, even if you are sure of what you want to do at this time.

CONSTRUCTION PERSONNEL

The occupations in the construction industry can be divided into four categories:

- Unskilled or semiskilled labor
- Skilled trades or crafts
- Technicians
- Design and management

UNSKILLED OR SEMISKILLED LABOR

Construction is labor intensive. That means it requires a lot of labor to produce the same dollar value of end products by comparison with other industries, where labor may be a smaller part of the picture. Construction workers with limited skills are called *laborers*. Laborers are sometimes assigned the tasks of moving materials, running errands, and working under the close supervision of a skilled worker. Their work is strenuous, and so construction laborers must be in excellent physical condition.

Construction laborers are construction workers who have not reached a high level of skill in a particular trade and are not registered in an apprenticeship program. These laborers often specialize in working with a particular trade, such as mason's tenders or carpenter's helpers (Fig. I-1). Although the mason's tender may not have the skill of a bricklayer, the mason's tender knows how to mix mortar for particular conditions, can erect scaffolding, and is familiar with the bricklayer's tools. Many laborers go on to acquire skills and become skilled workers. Laborers who specialize in a particular trade are often paid slightly more than completely unskilled laborers.

SKILLED TRADES

A *craft* or *skilled trade* is an occupation working with tools and materials and building structures. The building trades are the crafts that deal most directly with building construction (see Fig. I-2).

The building trades are among the highest paying of all skilled occupations. However, work in the building trades can involve working in cold conditions in winter or blistering sun in the summer. Also, job opportunities will be best in an area where a lot of construction is being done. The construction industry is growing at a high rate nationwide. Generally, plenty of work is available to provide a comfortable living for a good worker.

APPRENTICESHIP

The skill needed to be employed in the building trades is often learned in an apprentice program. Apprenticeships are usually offered by trade unions, trade associations,

Delmar/Cengage Learning.

FIGURE I-1 This construction laborer is a mason's tender.

technical colleges, and large employers. *Apprentices* attend class a few hours a week to learn the necessary theory. The rest of the week they work on a job site under the supervision of a *journeyman* (a skilled worker who has completed the apprenticeship and has experience on the job). The term "journeyman" is a gender neutral term that has been used for decades. It is worth noting that many highly skilled building trades workers are women. Apprentices receive a much lower salary than journeymen, often about 50% of what a journeyman receives. The apprentice wage usually increases as stages of the apprenticeship are successfully completed. By the time the apprenticeship is completed, the apprentice can be earning as much as 95% of what a journeyman earns. Many apprentices receive college credit for their training. Some journeymen receive their training through school or community college and on-the-job training. In one way or another, some classroom training and some on-the-job supervised experience are usually necessary to reach journeyman status. Not all apprentice programs are the same, but a typical apprenticeship lasts 4 or 5 years and

Carpenter
>   Framing carpenter
>   Finish carpenter
>   Cabinetmaker

Plumber
>   New construction
>   Maintenance and repair

Roofer

Electrician
>   Construction electrician
>   Maintenance electrician

Mason
>   Bricklayer (also lays concrete blocks)
>   Cement finisher

HVAC technician

Plasterer
>   Finish plaster
>   Stucco plaster

Tile setter

Equipment operator

Drywall installer
>   Installer
>   Taper

Painter

FIGURE I-2 Building trades.

requires between 100 and 200 hours per year of class-room training along with 1,200–1,500 hours per year of supervised work experience.

TECHNICIANS

Technicians provide a link between the skilled trades and the professions. Technicians often work in offices, but their work also takes them to construction sites. Technicians use mathematics, computer skills, specialized equipment, and knowledge of construction to perform a variety of jobs. Figure I-3 lists several technical occupations.

Most technicians have some type of college education, often combined with on-the-job experience, to prepare them for their technical jobs. Community colleges often have programs aimed at preparing people to work at the technician level in construction. Some community college programs are intended especially for preparing workers for the building trades, while others have more of a construction management focus. Construction management courses give the graduate a good overview of the business of construction. The starting salary for a construction technician is about the same as for a skilled trade, but the technician can

Technical Career	Some Common Jobs
Surveyor	Measures land, draws maps, lays out building lines, and lays out roadways
Estimator	Calculates time and materials necessary for project
Drafter	Draws plans and construction details in conjunction with architects and engineers
Expeditor	Ensures that labor and materials are scheduled properly
Superintendent	Supervises all activities at one or more job sites
Inspector	Inspects project for compliance with local building codes at various stages of completion
Planner	Plans for best land and community development

FIGURE I-3 Technicians.

be more certain of regular work and will have better opportunities for advancement.

DESIGN AND MANAGEMENT

Architecture, engineering, and contracting are the design and management professions. The *professions* are those occupations that require more than 4 years of college and a license to practice. Many contractors have less than 4 years of college, but they often operate at a very high level of business, influencing millions of dollars, and so they are included with the professions here. These construction professionals spend most of their time in offices and are not frequently seen on the job site.

Architects usually have a strong background in art, so they are well prepared to design attractive, functional buildings. A typical architect's education includes a 4-year degree in fine art, followed by a master's degree in architecture. Most of their construction education comes during the final years of work on the architecture degree.

Engineers generally have more background in math and science, so they are prepared to analyze conditions and calculate structural characteristics. There are many specialties within engineering, but civil engineers are the ones most commonly found in construction. Some civil engineers work mostly in road layout and building. Other civil engineers work mostly with structures in buildings. They are sometimes referred to as structural engineers.

Contractors are the owners of the businesses that do most of the building. In larger construction firms, the principal (the owner) may be more concerned with running the business than with supervising construction. Some contractors are referred to as general contractors and others as *subcontractors* (Fig. I-4). The general contractor is the principal construction company hired by the owner to construct the building. A general contractor might have only a skeleton crew, relying on subcontractors for most of the actual construction. The general contractor's superintendent coordinates the work of all the subcontractors.

It is quite common for a successful journeyman to start his or her own business as a contractor, specializing

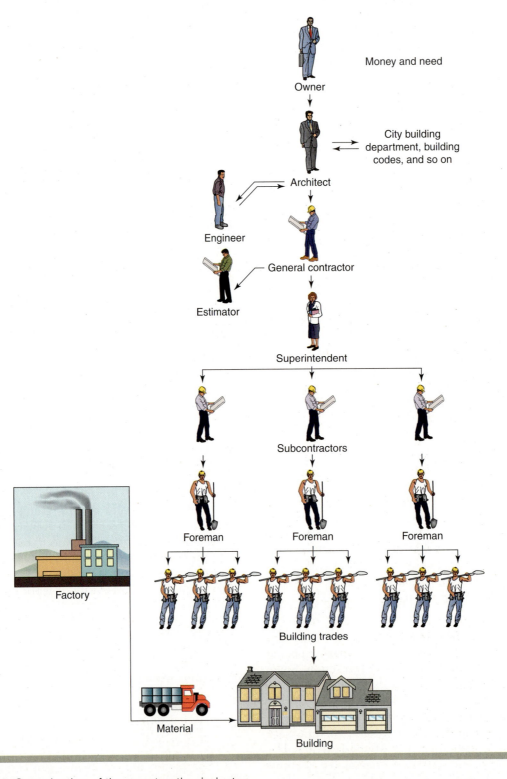

FIGURE I-4 Organization of the construction industry.

in the field in which he or she was a journeyman. These are the subcontractors that sign on to do a specific part of the construction, such as framing or plumbing. As the contractor's company grows and the company works on several projects at one time, the skilled workers with the best ability to lead others may become foremen. A foreman is a working supervisor of a small crew of workers in a specific trade. All contractors have to be concerned with business management. For this reason, many successful contractors attend college and get a degree in construction management. Most states require contractors to have a license to do contracting in their state. Requirements vary from state to state, but a contractor's license usually requires several years of experience in the trade and a test on both trade information and the contracting business.

AN OVERALL VIEW OF DESIGN AND CONSTRUCTION

To understand the relationships between some of the design and construction occupations, we shall look at a scenario for a typical housing development. The first people to be involved are the community planners and the real estate *developer*. The real estate developer has identified a 300-acre tract on which he would like to build nearly 1,000 homes, which he will later sell at a good profit. The developer must work with the city planners to ensure that the use he has planned is acceptable to the city. The city planner is responsible for ensuring that all building in the city fits the city's development plan and zoning ordinances. On a project this big, the developer might even bring in a planner of his own to help decide where parks and community buildings should be located and how much parking space they will need.

As the plans for development begin to take shape, it becomes necessary to plan streets and to start designing houses to be built throughout the development. A civil engineer is hired to plan and design the streets. The civil engineer will first work with the developer and planners to lay out the locations of the streets, their widths, and drainage provisions to get rid of storm water. The civil engineer also considers soil conditions and expected traffic to design the foundation for the roadway.

An architectural firm, or perhaps a single architect, will design the houses. Typically, several stock plans are used throughout a development, but many homeowners wish to pay extra to have a custom home designed and built. In a custom home, everything is designed for that particular house. Usually the homeowner, who will eventually live in the house, works with the architect to specify the sizes, shapes, and locations of rooms, interior and exterior trim, type of roof, built-in cabinets

and appliances, use of outdoor spaces, and other special features. Architects specialize in use of space, aesthetics (attractive appearance), and livability features. Most architectural features do not involve special structural considerations, but when they do, a structural engineer is employed to analyze the structural requirements and help ensure that the structure will adequately support the architectural features.

One part of construction that almost always involves an engineer is the design of roof trusses. Roof trusses are the assemblies that make up the frame of the roof Trusses are made up of the top chords, bottom chords, web members, and gussets. The engineer considers the weight of the framing materials, the weight of the roof covering, the anticipated weight of any snow that will fall on the roof in winter, and the span (the distance between supports) of the truss to design trusses for a particular purpose. The architect usually hires the engineer for this work, and so the end product is one set of construction drawings that includes all the architectural and engineering specifications for the building. Even though the drawings are sometimes referred to as architectural drawings, they include work done by architects, engineers, and their technicians. Building codes require an architect's seal on the drawings before work can begin. The architect will require an engineer to certify certain aspects of the drawings before putting the architect's seal on them.

FORMS OF OWNERSHIP

Construction companies vary in size from small, one-person companies to very large international organizations that do many kinds of construction. However, the size of the company does not necessarily indicate the form of ownership. Three types of ownership and the advantages and disadvantages of each are shown in Figure I-5.

UNIONS AND CONTRACTORS' ASSOCIATIONS

The construction industry contains thousands of organizations of people with common interests and goals. Whole directories of these organizations are available in libraries and on the Internet. Two categories of

Forms of Ownership	What it Means	Advantages	Disadvantages
Sole Proprietorship	A sole proprietorship is a business whose owner and operator are the same person.	The owner has complete control over the business and there is a minimum of government regulation. If the company is successful, the owner receives high profits.	If the business goes into debt the owner is responsible for that debt. The owner can be sued for the company, and the owner suffers all the losses of the company.
Partnership (*General* and *Limited Liability Partnership* (LLP))	A partnership is similar to a sole proprietorship, but there are two or more owners. *General:* In a general partnership, each partner shares the profits and losses of the company in proportion to the partner's share of investment in the company. *LLP:* A limited liability partner is one who invests in the business, receives a proportional share of the profit or loss, but has limited liability.	*General Partnership:* The advantage is that the partners share the expense of starting the business and partnerships are not controlled by extensive government regulations. *LLP:* A limited liability partner can only lose his or her investment.	*General Partnership:* Each partner can be held responsible for all the debts of the company. *LLP:* Every LLP must have one or more general partners who run the business. The general partners in an LLP have unlimited liability and they can be personally sued for any debts of the company.
Corporation	In a corporation a group of people own the company. Another, usually smaller, group of people manage the business. The owners buy shares of stock. A share of stock is a share or a part of the business. The value of each share increases or decreases according to the success of the company.	In a corporation, no person has unlimited liability. The owners can only lose the amount of money they invested in stock. The owners of a corporation are not responsible for the debts of the corporation. The corporation itself is the legal body and is responsible for its own debts.	The government has stricter regulations for corporations than for the other forms of ownership. Also, corporations are more expensive to form and to operate than are proprietorships and partnerships.

FIGURE I-5 Three forms of business ownership.

construction organizations are of particular importance to construction students: craft unions and contractors' associations.

UNIONS

A *craft union*, usually just called a "union," is an organization of workers in a particular building trade. Workers' unions were first formed in the 1800s when factory workers were being forced to work extreme hours under unsafe conditions—and for very low wages. Although working conditions in both factories and construction have improved dramatically, unions continue to serve a valuable role in the construction industry. Figure I-6 lists several national construction craft unions.

Union members pay dues to be members of the union. Dues money pays for the benefits the union provides for its members. Most unions have an apprenticeship program that includes both classroom instruction and on-the-job supervised work experience. Some of the members' dues pay for instructors, classroom space, and training supplies. Unions usually provide a pension for members who have worked in the trade. Because they represent a large block of members, unions can be a powerful force in influencing government to do such things as pass worker safety laws, encourage more construction, and support technology that is good for construction. Unions negotiate with employers (contractors) to establish both a pay rate and working conditions for their members. It is quite typical to find that union members enjoy a higher hourly pay rate than nonunion workers in the same trade.

CONTRACTORS' ASSOCIATIONS

Associations of contractors include just about every imaginable type of construction contractor. Figure I-7 lists only a small number of the largest associations that have apprenticeship programs. Some contractors' associations are formed to represent only nonunion contractors; a few represent only union contractors; and others represent both. Many associations of nonunion contractors were originally formed because the contractor members felt a need to work together to provide some of the benefits that union contractors receive—such as apprentice training and a lobbying voice in Washington, D.C.

International Association of Bridge, Structural, Ornamental and Reinforcing Iron Workers (www.ironworkers.org/)

International Association of Heat and Frost Insulators and Asbestos Workers (www.insulators.org/)

International Brotherhood of Boilermakers, Iron Ship Builders, Blacksmiths, Forgers and Helpers (www.boilermakers.org/)

International Brotherhood of Electrical Workers (www.ibew.org/)

International Brotherhood of Teamsters (www.teamster.org/)

International Union of Bricklayers and Allied Craftworkers (www.bacweb.org/)

International Union of Elevator Constructors (www.iuec.org/)

International Union of Operating Engineers (www.iuoe.org/)

International Union of Painters and Allied Trades (www.iupat.org/)

Laborers' International Union of North America (www.liuna.org/)

Operative Plasterers' and Cement Masons' International Association of the United States and Canada (www.opcmia.org/)

Sheet Metal Workers' International Association (www.smwia.org/)

United Association of Journeymen and Apprentices of the Plumbing and Pipefitting Industry of the United States and Canada (www.ua.org/)

United Brotherhood of Carpenters and Joiners of America (www.carpenters.org/)

United Union of Roofers, Waterproofers and Allied Workers (www.unionroofers.com/)

Utility Workers Union of America (www.uwua.org/)

FIGURE I-6 Construction craft unions.

Air Conditioning Contractors of America (http://www.acca.org)

Air Conditioning Heating and Refrigeration Institute (http://www.ahrinet.org/)

Associated Builders and Contractors (http://www.abc.org)

National Association of Home Builders (http://www.nahb.org)

Home Builder's Institute (http://www.hbi.org)

Independent Electrical Contractors Association (http://www.ieci.org)

National Electrical Contractors Association (http://www.necanet.org)

National Utility Contractors Association (http://www.nuca.com)

Plumbing-Heating-Cooling Contractors Association (http://www.phccweb.org)

The Associated General Contractors (AGC) of America (http://www.agc.org)

FIGURE I-7 These are only a few of the largest construction associations.

BUILDING CODES

Most towns, cities, and counties have building codes. A *building code* is a set of regulations (usually in the form of a book) that ensure that all buildings in that jurisdiction (area covered by a certain government agency) are of safe construction. Building codes specify such things as minimum size and spacing of lumber for wall framing, steepness of stairs, and fire rating of critical components. The local building department enforces the local building codes. States usually have their own building codes, and state codes often require local building codes to be at least as strict as the state code. Most small cities and counties adopt the state code as their own, meaning that the state building code is the one enforced by the local building department.

Until recently, three major model codes were published by independent organizations. (A model code is a suggested building code that is intended to be adopted as is or with revisions to become a government's official code.) Each model code was widely used in a different region of the United States. By themselves, model codes have no authority. They are simply a model that a government agency can choose to adopt as their own or modify as they see fit. In 2009, the International Code Council published a new model code called the *International Building Code*. They also published the *International Residential Code* to cover home construction (Fig. I-8).

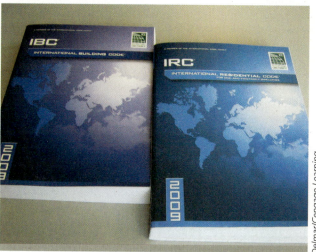

FIGURE I-8 *International Building Code* and *International Residential Code*.

Since publication of the first *International Building Code,* states have increasingly adopted it as their building code.

Other than the building code, many codes govern the safe construction of buildings: plumbing codes, fire protection codes, and electrical codes. Most workers on the job site do not need to refer to the codes much during construction. It is the architects and engineers who design the buildings that usually see that the code requirements are covered by their designs. Plumbers and electricians do, however, need to refer to their respective codes frequently. Especially in residential construction, it is common for the plans to indicate where fixtures and outlets are to be located, but the plumbers and electricians must calculate loads and plan their work so it meets the requirements of their codes. The electrical and plumbing codes are updated frequently, so the workers in those trades spend a certain amount of their time learning what is new in their codes.

WORKING IN THE INDUSTRY

Often success in a career depends more on how people act or how they present themselves to the world than it does on how skilled they are at their job. Most employers would prefer to have a person with modest skills but a great work ethic than a person with great skills but a weak ethic.

ETHICS

Ethics are principles of conduct that determine which behaviors are right and wrong. The two aspects of ethics are values and actions. *Values* have to do with what we believe to be right or wrong. We can have a very strong sense of values, knowing the difference between right and wrong, but not act on those values. If we know what is right but we act otherwise, we lack ethics. To be ethical, we must have good values and act accordingly.

We often hear that someone has a great work ethic. That simply means that the person has good ethics in matters pertaining to work. Work ethic is the quality of putting your full effort into your job and striving to do the best job you can. A person with a strong work ethic has the qualities listed in Figure I-9. Good work ethics become habits, and the easiest way to develop good work ethics is to consciously practice them.

COMMON RATIONALIZATIONS

We judge ourselves by our best intentions and our best actions. Others judge us by our last worst act. Conscientious people who want to do their jobs well

A person with a strong work ethic:
• Shows up to work a few minutes early instead of a few minutes late.
• Looks for a job to do as soon as the previous one is done. (This person is sometimes described as a self-starter.)
• Does every job as well as possible.
• Stays with a task until it is completely finished.
• Looks for opportunities to learn more about the job.
• Cooperates with others on the job.
• Is honest with the employer's materials, time, and resources.

FIGURE I-9 Characteristics of a good work ethic.

often fail to consider their behavior at work. They tend to compartmentalize ethics into two parts: private and occupational. As a result, sometimes good people think it is okay to do things at work that they know would be wrong outside of work. They forget that everyone's first job is to be a good person. People can easily fall prey to rationalizations when they are trying to support a good cause. "It is all for a good cause" is an attractive rationale that changes how we see deception, concealment, conflicts of interest, favoritism, and violations of established rules and procedures. In making tough decisions, do not be distracted by rationalizations.

Good work ethics yield great benefits. As little children, most of us learned the difference between right and wrong. As adults, when we do what we know is right, we feel good about ourselves and what we are doing. On the other hand, doing what we know is wrong is depressing. We lose respect for ourselves, knowing that what we have done is not something we would want others to do to us. Employers recognize people with a good work ethic. They are the people who are always doing something productive, their work turns out better, and they seem cheerful most of the time. Which person do you think an employer will give the most opportunities to: a person who is always busy and whose work is usually well done or a person who seems glum and must always be told what to do next?

WORKING ON A TEAM

Constructing a building is not a job for one person acting alone (Fig. I-10). The work at the site requires cooperative effort by carpenters, masons, plumbers, painters, electricians, and others. Usually, several workers from each of these trades collaborate. A construction project without teamwork would have lots of problems. For example, one carpenter's work might not match up with another carpenter's work. There could be too much of some materials and not enough of others. Walls may be enclosed before the electrician runs the wiring in them.

Delmar/Cengage Learning.

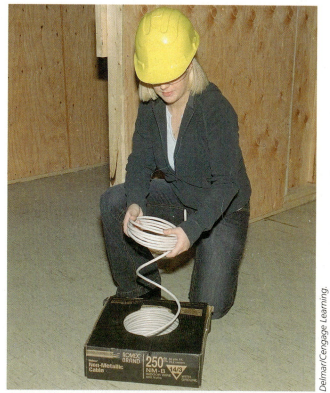

Delmar/Cengage Learning.

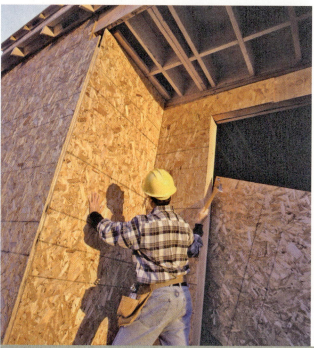

Courtesy of Lousiana Pacific Corporation.

Delmar/Cengage Learning.

FIGURE I-10 Work on the job requires cooperative effort by individuals from different trade areas.

Teamwork is very important on a construction site, but what does being a team player on a construction team mean? Effective team members have the best interests of the whole team at heart. Each team member has to carry his or her own load, but it goes beyond that. Sometimes, a team member might have to carry more than his or her own load, just because that is what is best for the team. If you are installing electrical boxes and the plumber says one of your boxes is in the way of a pipe, it might be in the best interests of the project to move the electrical box. That would mean you would have to undo work you had just completed and then redo it. It is, after all, a lot easier to relocate an outlet box than to reroute a sink drain.

The following are six traits of an effective team:

- *Listening.* Team members listen to one another's ideas. They build on teammates' ideas.
- *Questioning.* Team members ask one another sincere questions.
- *Respect.* Team members respect one another's opinions. They encourage and support the ideas of others.
- *Helping.* Team members help one another.
- *Sharing.* Team members offer ideas to one another and tell one another what they have learned.
- *Participation.* Team members contribute ideas, discuss them, and play an active role together in projects.

COMMUNICATION

How could members function as a team without communication? Good communication is one of the most important skills for success in any career. Employers want workers who can communicate effectively; but more importantly, you must be able to communicate with others to do your job well and to be a good team member. How many of the six traits of an effective team require communication?

Many forms of communication exist, but the most basic ones are speaking, listening, writing, reading, and body language. If you master those five forms of communication, you will probably succeed in your career.

SPEAKING

To communicate well through speech, you need a reasonably good vocabulary. It is not necessary, or even desirable, to fill your speech with a lot of flowery words that do not say much or that you do not really understand. What is necessary is to know the words that convey what you want the listener to hear, and it is equally necessary to use good enough grammar so those words can be communicated properly. Using the wrong word or using it improperly can cause two serious problems: For one thing, if you use the wrong word, you will not be saying what you intended to say. This is also often true if you use a great word wrong since you still might not be saying what you thought you were saying. For another thing (the second serious problem), using a poor choice of words or using bad grammar gives the listener the impression that you are poorly educated or that maybe you just do not care about good communication skills. As a businessperson, you will find that communicating is critical to earning respect as a professional as well as to gaining people's business. Three important steps of effective communication are:

- looking your listeners in the eye.
- asking yourself if you think they understand what you are saying. If it is important, ask them if they understand.
- trying a different approach if they do not understand.

LISTENING

Good listening is an important skill. Have you ever had people say something to you, and after they were finished and gone, you wondered what they said or you missed some of the details? Perhaps they were giving you directions or telling you about a school assignment. If only you could listen to them again! If possible, try paraphrasing. Paraphrasing means to repeat what they said but in different words. If someone gives you directions, wait until the person is finished. Then repeat the directions to person, so he or she can tell you if you are correct. Look at the speaker and form a mental picture of what the speaker is saying. Make what the speaker is saying important to you. Good listening can mean hearing and acting on a detail of a job that will result in giving a competitive edge in bidding.

WRITING

Writing is a lot like speaking, except you do not have the advantage of seeing if the person seems to understand or of asking if the person understands. That means you really have to consider your reader. If you are giving instructions, keep them as simple as possible. If you are reporting something to a supervisor, make your report complete, but do not take up his or her time with unrelated trivia. Penmanship, spelling, and grammar count. Always use good grammar to ensure that you are saying what you intend and that your reader will take you seriously. Use standard penmanship, and make it as neat as possible. Do not invent new ways of forming letters, and do not try to make your penmanship ornate. You will only make it harder to read. If you are unsure of how to spell a word, look it up in a dictionary. Next time, you will know the word and will not have to look it up. After you write something—read it, thinking about how your intended reader will take it. Make changes if necessary. Your writing is important! Sole proprietors have to demonstrate good writing skills in proposals and contracts. If either of these is poorly written, it can cost the business a lot of money.

READING

You will have to read at work. That is a fact no matter what your occupation. You will have to read building specifications, instructions for use of materials and tools, safety notices, and notes from the boss (Fig. I-11). To develop reading skills, find something you are interested in and spend at least 10 or 15 minutes every day reading it. You might read the sports section of the newspaper, books about your hobby, hunting and fishing magazines, or anything else that is interesting to you. What is important is that you read. Practicing reading will make you a better reader. It will also make you a better writer and a better speaker. When you come across a word you do not know how to pronounce or you do not know the meaning of, look it up or ask someone for help. You will find that you learn pronunciation and meaning very quickly, and your communication skills will improve faster than you expect. In practically no time, you will not need help very often.

BODY LANGUAGE

Body language is an important form of communication. How you position your body and what you do with your hands, face, and eyes all convey a lot of information to the person you are communicating with. Whole books are written about how body language is used to communicate and how to read body language. We will only discuss a couple of key points here.

FIGURE I-11 Copies of Material Data Safety Sheets are one example of relevant materials.

Delmar/Cengage Learning.

When you look happy and confident, the message you convey is that you are honest (you have nothing to hide or to worry about) and you probably know what you are talking about. If you look unhappy, unsure of yourself, or uninterested, your body language tells the other person to be wary of what you are saying—something is wrong. The following are a few rules for body language that will help you convey a favorable message:

- Look the other person in the eye. Looking toward the floor makes you look untrustworthy. Looking off in space makes you seem uninterested in the other person.

- Keep your hands out of your pockets, and do not wring your hands. Just let your hands rest at your sides or in your lap if you are sitting. An occasional hand gesture is okay, but do not overdo it.

- Dress neatly. Even if you are wearing work clothes, you can be neat. Faddish clothes, extra baggy or extra tight fitting clothes, and T-shirts with offensive messages on them all distract from the real you.

- Speak up. How loudly you speak might not seem like body language, but it has a lot to do with how people react to you. If they have to strain to hear what you are saying, they will think that either you are not confident in what you are saying or you are angry and not to be trusted. If you see your listeners straining to hear you or if they frequently ask you to repeat what you are saying, speak a little louder.

CUSTOMER SERVICE

In any industry, you will only be as successful as you are good at building your reputation for doing quality work and for the degree to which your customers are happy with you and your job. On the job site, your customer might be a crew chief, a foreman, a subcontractor, or a contractor. If you are the contractor or subcontractor, the customer will be whoever hired you. It doesn't actually matter who hired you, though—your role will always be to do the very best job you can for whomever it is that you are working.

Good customer service also includes providing a good value for your fees, being honest, communicating clearly, being cooperative, and looking to provide the best possible experience your customer can have in working with you. Just as when you practice good ethics, when you provide great customer service you will enjoy your job much more. You will be proud of your work, others will want to hire you more often, and your career will be much easier to build. Think about how you like being treated when you are a customer—and always try to treat your customers at least as well.

LIFELONG LEARNING

Lifelong learning refers to the idea that we all need to continue to learn throughout our entire lives. We have greater opportunities to learn and greater opportunities to move up a career ladder today. Our lives are filled with technology, innovative new materials, and new opportunities. People change not only jobs, but entire careers several times during their working life. Those workers who do not understand the new technology in the workplace, along with those who do not keep up with the changes in how their company is managed, are destined to fall behind economically. There is little room in a fast-paced company of this century for a person whose knowledge and skills are not growing as fast as the company. To keep up with new information and to develop new skills for the changing workplace, everyone must continue to learn throughout life.

CONSTRUCTION TRENDS

Every industry has innovations, and construction is no exception. As a construction professional, it is important to be aware of new technologies, new methods, and new ways of thinking about your work. This is as important for a worker's future employment as being aware of safety and ethical business practices. Some of the key technological trends include disaster mitigation, maintenance, building modeling, and green building.

DISASTER MITIGATION

Both new and existing buildings need to be strengthened and improved to deal with earthquakes, floods, hurricanes, and tornados. Actions like improving wall bracing or preparing moisture management reduces damage and improves safety when these events occur. These actions are increasingly required by building regulations (especially in disaster-prone areas) and requested by property owners and insurers.

MAINTENANCE

Unlike single natural disasters, preventing long-term wear and tear is also an important industry trend. Property owners are more concerned about the costs, effort, and time required to repair and to maintain their homes and buildings. So, there has been significant research into materials that are more durable, construction assemblies that manage moisture, air and elements better, and overall higher quality construction work.

BUILDING MODELING

One of the biggest new trends in construction technology doesn't include construction materials at all: it includes being able to design, simulate, and manage buildings with the use of computer and information technology. Some of these tools, like Computer-Aided Drafting (CAD) and Computer-Aided Manufacturing (CAM), have been around for decades. Others, like energy modeling and simulation software or project management tools, are being used more and more. Still others, like Building Information Modeling (BIM) are gathering many of these previous tools into single computing platforms. In all cases, the ability to use computers and professional software is becoming mandatory among workers.

GREEN BUILDING

Probably the biggest trend in the construction industry over the last decade has been *green building*—that is, planning, design, construction, and maintenance practices that try to minimize a building's impact on the environment throughout its use. Although a set definition of green building is still evolving, everyone agrees on a few key concepts that are important and that in themselves are also major construction trends.

OCCUPANT HEALTH AND SAFETY

The quality of indoor air is influenced by the kinds of surface paints and sealants that are used as well as the management of moisture in plumbing lines, HVAC equipment, and fixtures. Long-term maintenance and care by homeowners and remodelers also can shape the prevalence of pests, damage, and mold. Builders and remodelers are becoming more aware of the products and assemblies they use that could have an effect on indoor environments.

WATER CONSERVATION AND EFFICIENCY

Many builders and property owners are attempting to collect, efficiently use, and reuse water in ways that all save the overall amount being used. From using collectors of rainwater to irrigate lawns, to installing low-flow toilets and water-conserving appliances, to feeding used "greywater" from sinks and showers into secondary nonoccupant water needs, water efficiency is a trend in all green building but especially where water shortages or droughts are prevalent.

LOW-IMPACT DEVELOPMENT

Builders concerned with the effect of the construction site on the land, soils, and water underneath are incorporating storm water techniques, foundation and pavement treatments, and landscaping preservation methods to minimize disturbances to the land and surrounding natural environments.

MATERIAL EFFICIENCY

Builders are becoming more aware of the amount of waste coming from construction sites, and inefficiency in the amount of materials (like structural members) that they install in buildings. Many of the materials that are used in construction also do not come from naturally renewable sources or from recycled content materials. Using materials from preferred sources, using them wisely, and then appropriately recycling what is left is a big industry trend (Fig. I-12).

ENERGY EFFICIENCY
AND RENEWABLE ENERGY SOURCES

The most widely known of all green building trends involves the kind and amount of energy that buildings use. Oftentimes, builders can incorporate the use of renewable energy sources (like solar photovoltaics) or passive solar orientation into their designs. Then, the combination of good building envelope construction and efficient equipment and appliances can all reduce utility costs for property owners, much like the maintenance trend reduces repair costs (Fig. I-13).

There are many ways to keep track of the latest trends in the construction industry. Trade or company's journals, online resources and blogs, and the latest research coming out of government and university laboratories are several ways to keep informed and up-to-date on the latest industry trends.

FIGURE I-12 Recycling construction waste is becoming widely practiced in the industry.

FIGURE I-13 Duct insulation increases system energy efficiency.

JOB OPPORTUNITIES

Heating, Ventilating, Air Conditioning, and Refrigeration, (HVAC/R), are needed, for the most part, by everyone every single day. This puts HVAC installers and technicians in demand no matter where they are in the country, or even the world for that matter! The HVAC industry is largely unaffected by the economy although portions of the industry may experience temporary swings in growth depending on economic and governmental changes. For example, if the new home building industry experiences large growth, the installation end of the HVAC industry grows while the service end of the industry typically slows. On the other hand, if the new home building industry experiences a slowdown, the installation side of the industry slows, but the service end typically experiences significant growth.

For many, HVAC/R is not a luxury or an option—it's a necessity. Consider the HVAC/R needs of hospitals, restaurants, movie theaters, supermarkets and other retail and commercial establishments. If the HVAC/R equipment fails in one of these locations, the results can be disastrous. Those with breathing disorders rely on air conditioning to live. Restaurants rely on HVAC/R to keep their customers comfortable, their food fresh and the air clean. Everyone relies on HVAC/R every day. It would be impossible for you to go through a single day without reaping the benefits that HVAC/R provide.

HVAC/R CAREERS

People who enter the HVAC/R industry find themselves doing many different things since this industry encompasses so many of the traditional building trades.

HVAC/R installers and technicians often perform tasks that involve plumbing, electricity, masonry, and carpentry. In addition, there are many different aspects to the industry and individuals who are trained in HVAC/R can enter the industry in a number of different ways. Here is a brief list of jobs that are available to individuals in the HVAC industry:

- Installation Technician
- Service Technician
- Maintenance Personnel
- HVAC/R Supply House Counterperson
- Blueprint Reader
- Field Supervisor
- Project Coordinator
- Project Supervisor
- Service Manager
- Dispatcher
- Sheet Metal Worker
- Testing, Adjusting, and Balancing Technician
- Indoor Air Quality Specialist

HVAC/R: A UNIQUE JOB EXPERIENCE

The HVAC/R industry has a lot to offer to the right people! If you enjoy going to different places and seeing different things everyday, this is probably a great industry for you!

As an HVAC system installer or technician, everyday is a new experience. You have the opportunity to work on many different types of equipment and learn something new every single day. Because of the diversity associated with this industry, your work schedule may also be somewhat unconventional. Depending on where you have to work and what must be done, you might have to work into the late hours of the night or very early in the morning. If you are on call and a customer calls because she has no heat, the HVAC technician might be called out in the middle of the night to fix the system. The HVAC/R industry is very unique, very exciting and very rewarding as well.

FINANCIAL REWARDS

According to the United States Bureau of Labor Statistics, as of May 2009 (the most recent statistics available at the time of this printing) the average hourly wage for an HVAC/R installer or mechanic is $19.76. The middle 50% of installers and technicians earn between $15.42 and $25.80. The top 10% of installers and technicians earn over $31.53 per hour and the bottom 10% of installers and technicians earn less than $12.38 per hour. It is quite obvious from these statistics that those with more experience and those who strive for better careers will earn more money.

It should be noted that the figures just mentioned are base hourly rates and do not reflect overtime. It should also be noted that these numbers are national averages and will vary based on location.

Barrington Richard White

HVAC TECHNICIAN

EDUCATION

Barrington attended the HVAC/R program at Suffolk Community College in Brentwood, NY and earned his Associates of Applied Science (A.A.S.) degree. His industry certifications include EPA Section 608 Universal, National Oilheat Research Alliance (NORA) Gold oil burner certification, and NATE certifications in technical areas including air conditioning, heat pumps, gas furnaces, oil furnaces, and light commercial refrigeration.

HISTORY

Barrington chose the HVAC/R industry because his skills would be in demand all year round—everybody needs heat, air conditioning, and refrigeration, even when economic times may get tough. Since graduating from school, Barrington has worked continuously and feels a sense of job security. Even when it was time to change positions, he found it very easy to make the transition from one company to another.

ON THE JOB

Barrington's work day usually starts at 8:00 am. His day often consists of a mixture of service calls and preventive maintenance tasks. At work, it is his responsibility to not only perform the tasks at hand, but also to obtain all of the filters, parts, and supplies that are needed to get the job done.

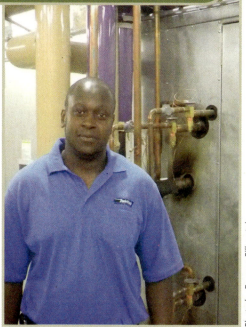

Photo by Eugene Silberstein.

CHALLENGES

One of the biggest challenges that Barrington faces in the field is coming across a piece of equipment or a situation that he has not encountered before. Even though he might not initially know what that particular component did, taking some time to evaluate a system goes a long way.

IMPORTANCE OF EDUCATION

Barrington feels that his formal education and training have given him a substantial head start in this field. His 5-year history in the industry, including his training and on-the-job experience, helped him develop a solid understanding of the HVAC/R trade.

FUTURE OPPORTUNITIES

Barrington plans on going back to school to earn a Bachelor's degree in Design Engineering Technology. He is hoping to eventually go into the design aspect of the industry.

(CONTINUED)

WORDS OF ADVICE

If you put in the time and hard work, you should be able to achieve anything in this field. Although you might not know everything, you will need to know how to obtain the knowledge when you need it. To do this, you need to develop the skill of networking. To move up quickly in this field, you will need to make friends with other technicians, fellow students, and people associated in this field. This would be your resource if you have any questions on troubleshooting or could be your way to scoring that next job.

Dawnmarie Martino

SERVICE TECHNICIAN (EMCOR SERVICES, NY/NJ)

EDUCATION

Dawnmarie attended the HVAC/R program at Suffolk Community College in Brentwood, NY and was a member of the first graduating class of that program. At Suffolk, she earned her Associates of Applied Science (A.A.S.) degree. She holds EPA Section 608 Universal certification as well as number of other industry credentials including R-410A Safety and Training, National Oilheat Research Alliance (NORA) Gold oil burner certification, and NATE certifications. Dawnmarie also makes every effort to continue her education by taking various advanced technician training courses across the country such as Direct Digital Controls (DDC) and courses offered by major equipment manufacturers.

HISTORY

Dawnmarie comes from a food service family. Her brother owns a restaurant on Long Island, New York and Dawnmarie worked with him for a period of time and then took her own place in the industry as a caterer. As a caterer, she quickly learned the importance of air conditioning, and refrigeration. She was bitten by the HVAC/R bug, left the food service industry and began training to become a service technician. Her never-ending quest for knowledge keeps her on her toes and in the classroom. She

Photo by Eugene Silberstein.

enjoys learning and, as a result, has become a "go to" person at her company.

ON THE JOB

Her typical workday starts when she clocks in at about 8:00 a.m., but her day actually starts much earlier. She is required to be on site at her first job at that time and, depending on how far that is from home, she might be on the road at 5:30 am. Before she arrives at the job, Dawnmarie has to make certain that she has all the equipment and supplies she needs to accomplish the tasks at hand.

Although performing repair work is routine for Dawnmarie, she understands that system failure is often a major inconvenience for the equipment owner. Types of service work that she often performs ranges from routine preventive maintenance to compressor changeouts and system troubleshooting.

Dawnmarie has learned that being on time for work is essential and that she is working within the schedules of others. Her day does not end until her last

(CONTINUED)

service call has been completed. This can be stressful at times, but the personal satisfaction in knowing that she has played an integral role in ensuring the comfort of others outweighs any negatives.

CHALLENGES

Dawnmarie found that her gender was one of her biggest challenges. When entering the industry, it took some time to prove to others that she is a valuable and productive member of the HVAC/R industry. "Now, before they see me as a woman, they see me as a skilled craftsperson."

IMPORTANCE OF EDUCATION

Dawnmarie's formal HVAC/R training and education provided her with a solid knowledge of industry basics. The laws of physics do not change, so understanding how refrigerants, fossil fuels, electricity, and other elements react under different conditions makes it much easier when called on to evaluate the "tough" stuff. Quite often, what appears to be a complicated problem turns out to be straightforward when you look at the big picture.

FUTURE OPPORTUNITIES

Dawnmarie recognizes the industry is moving toward Green practices and an emphasis on conservation and sustainability. She plans to continue her Green building practices education by attending continuing education classes and possibly obtaining Green-related technician certifications.

WORDS OF ADVICE

"Your success depends on you and your own ambitions. Your employer will only take you so far; the rest of the trip depends on you. You will only get out as much as you put in."

Richard H. Zimmerman

Photo by Eugene Silberstein.

SERVICE MANAGER (AIR STREAM AIR CONDITIONING, CORP., PLAINVIEW, NY)

EDUCATION

Richard graduated from Farmingdale Senior High School, Farmingdale, NY in 1976 with a Regents Diploma and received his A.A.S. in HVAC/R from Suffolk County Community College, Grant Campus Brentwood, NY in May 2006 as a returning adult student. He is currently preparing for the L.E.E.D. G.A. exam (Leadership in Energy and Environmental Design [Green Associate]) given by the United States Green Building Council.

HISTORY

Richard began his career in the construction industry by working construction on the weekends while he worked a "suit and tie" job during the week. While working construction, he met an HVAC technician and was introduced to "the world of HVAC/R." Richard spent his first 8 years in the HVAC industry training as a stationary engineer in residential/ light commercial as a nonunion technician before eventually becoming a union technician in Steamfitters Local #638b. During his tenure there, he received additional training and joined the Refrigeration Service Engineer's Society.

After years in the field, Richard decided to return to college and earn his Associates Degree in Applied Science. Once he received his degree, he began teaching at a vocational school at night while working as an HVAC/R technician and service manager by day. Richard has come full circle, moving on from teaching at a vocational school to teaching part-time at the college where he earned his degree.

ON THE JOB

A typical day at the office for Richard starts at 7:00 a.m. and ends at 5:00 p.m. His day includes being a liaison with customers, providing technical support for technicians, providing engineering services for our customers and, by request, preparation of quotes for the repair and/or replacement of equipment.

CHALLENGES

Richard explains that the largest challenge he finds in the industry is that the technology of today continues to advance, but the majority of the technicians in the field are rapidly falling behind. He thinks it is important for technicians to remember that the responsibility of continuing their education and exposure to new

(CONTINUED)

technology is that of the technician, not the responsibility of the employer.

IMPORTANCE OF EDUCATION

Richard believes that "education is a win for everyone." Building your knowledge base increases the variety of skills and services you can offer an employer. The employer can then offer more services, in turn making you a very valuable asset. Richard states it simply as "the more you know the more income you can generate."

FUTURE OPPORTUNITIES

Currently, Richard is a member of the U.S. Green building council known as ASHRAE and pursuing his L.E.E.D. accreditation. Richard takes continuing education

seriously and plans to always continue to evolve along with technology, for as long as he remains active in the trade.

WORDS OF ADVICE

Be prepared! This is a continually evolving trade. While the basics remain the same, the technology continues to progress and technicians today must keep themselves fully informed to remain competitive. To continually advance one's training and skill level is the technician's responsibility and no one else's. Knowledge is the only way to remain at the front of the race. Don't be afraid to ask questions if you do not understand a particular topic, either in the classroom or out in the field. It is much easier to stay current than to catch up. In the long run, you'll be a much better technician.

LOOKING FORWARD

As many HVAC/R technicians and installers retire from the industry, many opportunities present themselves for qualified, educated, and energetic individuals. The HVAC industry has undergone more changes in the last 10 years than in the previous 50, so it is the newly trained workforce that will become the face of the future in the years to come. From environmental concerns and technological changes to new governmental laws and guidelines and the need for more efficient equipment, the HVAC/R industry has much to offer to those who are willing to step up and become a part of it! The pages that follow will provide the reader with an in depth look at our industry and the valuable information you will need to be successful as an HVAC installer or technician. Enjoy!

SECTION **1**

Refrigeration Fundamentals

Matter, Energy, and Heat Basics

OBJECTIVES

Upon completion of this chapter, the student should be able to

- Define the terms *solid*, *liquid*, and *gas*.

- Explain the three states of matter.

- Explain the law of conservation of energy.

- Explain the three types of heat transfer: conduction, convection, and radiation.

- Relate the concepts of conduction, convection, and radiation to real-life situations.

- Explain the differences between sensible heat and latent heat.

- Relate the concepts of sensible and latent heat to real-life situations.

- Explain the difference between heat content and heat level.

- Convert Fahrenheit temperature readings to the Celsius scale and vice versa.

- Explain the differences between gauge pressure and absolute pressure.

- Convert absolute pressures to their equivalent gauge pressures.

For an air conditioning system to be properly installed, the members of the installation crew should possess at least a basic understanding of matter, energy, and heat theory. Having this knowledge tends to improve the quality of the initial installation, which will lead to an air conditioning system with fewer operational problems in the future. Matter can exist in nature as a solid, liquid, or vapor, and is any substance that occupies space and has mass. Energy is often defined as the potential to do work. Common forms of energy are heat energy, electrical energy, and mechanical energy. Heat is the term used to describe the motion of molecules. If the molecules of a substance are not moving, there is zero heat content. In our study of heat theory, we will also discuss two factors that affect the operation of air conditioning systems: temperature and pressure.

The purpose of an air conditioning system is to transfer heat from inside the occupied space to a remote location where this heat is not objectionable. To clarify this concept, in this chapter we will study heat transfer as well as the concepts just described. For heat transfer to take place between two substances, there must be a temperature difference between the substances. In the air conditioning industry, the two substances we primarily concern ourselves with are the temperature of the occupied space and the temperature of the outside air. By maintaining desired pressures within an air conditioning system, we can maintain and control the rate and direction of heat transfer.

GLOSSARY OF TERMS

Absolute pressure The pressure scale that takes atmospheric pressure into account

Atmospheric pressure The weight of the gases that exert a force on the Earth's surface

British thermal unit (Btu) The amount of heat required to raise the temperature of 1 pound of water 1°F

Conduction The method of heat transfer by which heat is transferred from molecule to molecule within a substance

Convection The method of heat transfer that is facilitated by the flow of a fluid—typically air or water

Energy The ability to do work

Gas laws Laws of physics that govern the behavior of gases or vapors

Gauge pressure The pressure scale that does not take atmospheric pressure into account; at sea level, the gauge pressure will be 0 psig

Heat Energy that increases molecular movement within a substance

Horsepower Unit of power equal to 33,000 ft lb/min

Inches of mercury vacuum When reading gauge pressure, a reading below atmospheric pressure

Latent heat Heat energy that results in a change in state of a substance while maintaining a constant temperature

Matter Any substance that has weight and mass and occupies space

Power The rate at which work is done; work per unit time

Pressure Force per unit area; common units are pounds per square inch (psi)

psia Pounds per square inch absolute; takes into account the pressure of the atmosphere and is approximately equal to the gauge pressure plus 15

psig Pounds per square inch gauge; ignores the pressure of the atmosphere; used to measure the pressure in sealed vessels such as car tires and air conditioning systems

Radiation Method of heat transfer by which heat travels through the air and heats the first object the rays come in contact with

Sensible heat Heat energy that results in the change of temperature of a substance

Temperature Term used to describe the level of heat intensity

Thermal kinetic energy See **sensible heat**

Thermal potential energy See **latent heat**

Work Force exerted on an object times the distance the object is moved, measured in foot-pounds (ft lb)

MATTER

Matter, by definition, is anything that has weight and mass and occupies space. When in the form of only one of the naturally occurring substances, it is called an element. The smallest particle of an element is an atom, and it too is considered to be matter. Matter can exist in any one of three physical states: solid, liquid, or gas.

SOLIDS

A solid is a substance that has definite volume and sufficient mechanical strength to maintain a constant shape. The molecules in a solid have a strong attraction for each other so they are able to hold together. Examples of solids include a block of ice and a piece of lumber (Fig. 1-1).

Solids exert all of their forces in the downward direction either toward the surface on which they are resting or toward the earth (Fig. 1-2). The weight of a solid is a combination of the mass of the object and the gravitational force acted on it by the earth. Mass can be referred to as the object itself without taking into account the earth's gravitational force.

LIQUIDS

Liquids, like solids, have definite volumes, but they do not have definite shapes. The attraction between liquid molecules is weaker than between those of a solid, so liquids need to be stored in containers. The shape a liquid takes depends on the shape of the container that holds it. Because a liquid has a definite volume, it cannot be compressed into a smaller space.

If pressure is exerted on a liquid in a container, the force will be exerted toward the sides and bottom of the container because the weight of the liquid pushes down and to the sides. Pascal's law states that, when a liquid is involved, pressure is transmitted equally in all directions irrespective of the area over which the pressure is applied. The physical concept on which Pascal's law is based is the foundation for hydraulic systems.

Consider an example where a liquid such as oil is contained within a system (Fig. 1-3). The cross-sectional area of the smaller tube is 1 square inch and the cross-sectional area of the larger tube is 10 square inches. Assume that pistons are fitted into each of the tubes and a 1-pound force is exerted downward on the smaller piston. This downward pressure is equal to 1 pound per square inch (psi). Since, following Pascal's

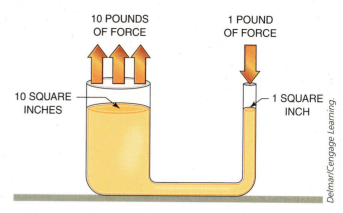

FIGURE 1-3 Graphical representation of Pascal's law.

law, the pressure is transmitted equally across the entire system, the larger tube will be able to support a weight of 10 pounds because the area is ten times the area of the smaller tube. If the cross-sectional area of the larger tube were 20 square inches, it would be able to support a weight of 20 pounds.

GASES

Gases have neither definite volume nor definite shape. Gas molecules have little attraction for each other, and gas will completely fill any vessel that contains it. Gas exerts pressure in all directions against the walls of the container that holds it.

For example, when a balloon is blown up, the balloon expands equally in all directions (Fig. 1-4). However, if 50% of the air is released from the balloon, the remaining air will expand to completely fill the balloon. The

FIGURE 1-1 Wood and blocks of ice are two examples of matter in the solid form.

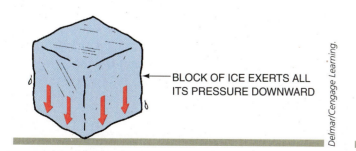

FIGURE 1-2 Solids exert all of their pressure downward.

FIGURE 1-4 When a gas fills a vessel, the pressure is exerted in all directions.

pressure in the balloon with only 50% of the air in it will be lower than when the balloon contained more air.

Three laws of physics, referred to as **gas laws,** govern the behavior of gases:

- Boyle's Law
- Charles' Law
- Dalton's Law

BOYLE'S LAW

Boyle's law states that there is a definite relationship between the pressure and volume of a gas if the temperature is kept constant. Boyle's law states that the pressure and volume of a gas are inversely related, meaning that if the volume is decreased, the pressure will increase. Conversely, if the volume is increased, the pressure will decrease. Mathematically, this relationship is

$$P_1 V_1 = P_2 V_2$$

where

P_1 = The original pressure of the gas

V_1 = The original volume of the gas

P_2 = The new pressure of the gas

V_2 = The new volume of the gas

Note that the pressures used in the above example are denoted psia, which is pounds per square inch absolute. Absolute pressures must be used when performing calculations using Boyle's law. Refer to the section in this chapter on absolute pressure for more information on this topic.

CHARLES' LAW

Charles' law states that there is a definite relationship between the volume and the temperature of a gas if the pressure is kept constant. In addition, this law states that there is a direct relationship between the pressure and the temperature of a gas, provided that the volume is kept constant. Mathematically, these two relationships can be expressed as

$$P_1 T_2 = P_2 T_1$$
$$V_1 T_2 = V_2 T_1$$

EXAMPLE 1

Assume the volume of a vessel that contains a gas is decreased from 4 ft³ to 3 ft³. The original pressure of the gas was 60 psia. What is the new pressure of the gas if the temperature remains constant?

Solution

From the information provided, we get

P_1 = 60 psia
V_1 = 4 ft³
V_2 = 3 ft³
P_2 = Unknown quantity

Substituting into the original formula, we get

(60 psia) × (4 ft³) = P_2 × (3 ft³)

P_2 = (60 psia) × (4 ft³)/(3 ft³)

P_2 = 80 psia

EXAMPLE 2

Assume the pressure of a fixed volume of gas is increased from 20 psia to 40 psia. The original temperature of the gas was 500°R. What is the new temperature of the gas?

Solution

From the information provided, we get

P_1 = 20 psia
T_1 = 500°R
P_2 = 40 psia
T_2 = Unknown quantity

Substituting into the first formula for Charles' law, we get

(20 psia) × T_2 = (40 psia) × (500°R)

T_2 = (40 psia) × (500°R)/(20 psia)

T_2 = 1000°R

EXAMPLE 3

Assume the volume of gas is increased from 4 ft³ to 6 ft³. The original temperature of the gas was 400°R. What is the new temperature of the gas if the pressure is unchanged?

Solution

From the information provided, we get

$$V_1 = 4 \text{ ft}^3$$

$$V_2 = 6 \text{ ft}^3$$

$$T_1 = 400°R$$

$$T_2 = \text{Unknown quantity}$$

Substituting into the second formula for Charles' law, we get

$$(4 \text{ ft}^3) \times T_2 = (6 \text{ ft}^3) \times (400°R)$$

$$T_2 = (6 \text{ ft}^3) \times (400°R)/(4 \text{ ft}^3)$$

$$T_2 = 600°R$$

DALTON'S LAW

Dalton's law states that the total pressure of a confined mixture of different gases is equal to the sum of the pressures of each gas in the mixture. This is also referred to as the law of partial pressures. For example, if a vessel of oxygen at 30 psig is combined with a vessel of nitrogen at 40 psig, the total pressure of the mixture will be 70 psig (Fig. 1-5). This law assumes that the volumes of each gas as well as the volume of the mixture are the same. For example, if one cubic foot of gas 1 is combined with one cubic foot of gas 2, then the result will be the sum of the pressures of gas 1 and that of gas 2 provided that the mixture is contained in a one cubic foot vessel.

EXAMPLE 4

Assume that a one cubic foot sample of a gas at 50 psig is combined with a one cubic foot sample of another gas at 80 psig. If the resulting mixture is contained in a *one cubic foot container*, what is its pressure?

Solution

Dalton's Law states that the final pressure is equal to the sum of the individual pressures, so the final pressure will be the sum of 50 psig and 80 psig, or 130 psig.

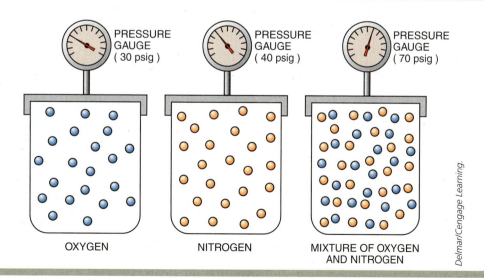

PRESSURE GAUGE (30 psig)

PRESSURE GAUGE (40 psig)

PRESSURE GAUGE (70 psig)

OXYGEN

NITROGEN

MIXTURE OF OXYGEN AND NITROGEN

Delmar/Cengage Learning.

FIGURE 1-5 Dalton's law of partial pressures. The total pressure of a gaseous mixture is the sum of the pressures of each individual gas.

EXAMPLE 5

Assume that a one cubic foot sample of a gas at 50 psig is combined with a one cubic foot sample of another gas at 80 psig. If the resulting mixture is contained in a *two cubic foot container*, what is its pressure?

Solution

This example differs from the first, since the volume of the mixture is now 2 ft³ instead of 1 ft³. In this case, the final pressure will be the average of the two pressures. We take the average by adding the 50 psig and the 80 psig and then dividing by two. The result is then (50 psig + 80 psig)/2, which gives us 65 psig.

ENERGY

Energy is most accurately defined as the ability or capacity to do work. Energy can exist in a number of different forms, including heat or thermal energy, mechanical energy, electrical energy, and chemical energy. Energy is often converted from one form to another.

Consider the example of a power plant that uses a steam turbine. Water is heated, causing it to boil into a vapor: steam. This steam is then used to turn a turbine. In this example, the heat energy (steam) is converted into mechanical energy (the turning of the turbine). The turning of the turbine, in turn, is used to generate electrical energy. Thus, the mechanical energy is now being converted into electrical energy. As the turbine operates, however, internal friction causes the turbine to get hot. This is an example of mechanical energy being converted back into heat energy. For the sake of completeness, let us assume that some of the electrical energy that was generated is being used to power a battery charger. The electrical energy has now been converted to chemical energy inside the battery. If the batteries are used to power other circuits, the chemical energy is being converted back to electrical energy.

The air conditioning industry is concerned mostly with the transfer of heat from one place to another. An air conditioning system is designed to transfer heat from

inside a structure, where it is not desired, to the outside, where it makes little or no difference. This concept can be referred to as the motion of heat or, more scientifically, thermodynamics, where *thermo* refers to heat and *dynamics* refers to motion. Regarding energy, two laws of thermodynamics apply directly to energy in general and, more specifically, heat energy.

FIRST LAW OF THERMODYNAMICS

The first law of thermodynamics states that energy cannot be created or destroyed, but it can be converted from one form to another. Simply stated, when one form of energy is generated, it is done so at the expense of another energy form, as mentioned in the previous section. When electrical energy is converted into mechanical energy, as in the example of an electric motor, a certain amount of the electrical energy is also converted into heat energy that causes the motor to heat up. This brings us to the second law of thermodynamics.

SECOND LAW OF THERMODYNAMICS

The second law of thermodynamics states that energy tends to degrade into low-level heat energy. In other words, heat energy is a by-product of energy conversion that flows from a warmer substance to a cooler substance. Therefore, in order for heat energy to travel, there must be a temperature difference between the materials. When converting electrical energy into mechanical energy, heat energy is generated as well. The same holds true when electrical energy is converted into chemical energy and when mechanical energy is converted to electrical energy.

When we speak of low-level heat energy, the fact that heat is a by-product of all energy transfers is being addressed. Heat can never be completely converted into other forms of energy, but other forms of energy can be completely converted into heat energy.

WORK

As discussed in the previous section, energy is the ability to do work. **Work**, in turn, can be defined as the force exerted on an object times the distance the object is moved, measured in foot-pounds (ft lb). Work occurs when a force moves a mass a certain distance. Mathematically

$$\text{Work} = \text{Force} \times \text{Distance}$$

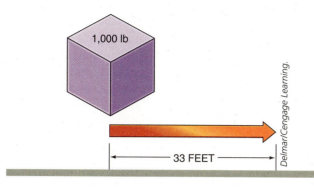

FIGURE 1-6 When a 1,000-lb weight is moved a distance of 33 feet, 33,000 ft lb of work is being done.

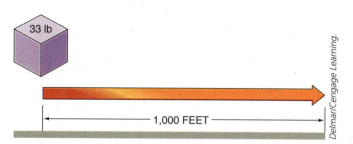

FIGURE 1-7 When a 33-lb weight is moved a distance of 1,000 feet, 33,000 ft lb of work is being done.

If a 1,000-pound weight is moved a distance of 33 feet (Fig. 1-6), the amount of work that is being done is

Work = 1,000 pounds × 33 feet = 33,000 ft lb

Similarly, if a 33-pound weight is moved a distance of 1,000 feet (Fig. 1-7), the amount of work is

Work = 33 pounds × 1,000 feet = 33,000 ft lb

Notice that the amount of work in these two examples is the same. This indicates that moving a lighter object further is the same as moving a heavier object a shorter distance, provided that the product (multiplication) of the forces and distances are equal.

POWER

As energy is the ability to do work, **power** is the rate of doing work, which is the amount of work per unit time. Saying that a man can move a 100-pound weight 50 feet is not a measurement of power, because the factor of time is not included. If, however, that same man can move a 100-pound weight 50 feet in *1 minute*, power units are being addressed.

The term **horsepower** (hp) is used to measure units of power. Long ago it was estimated that a horse could

lift 33,000 pounds to a height of 1 foot in one minute. Mathematically speaking

1 hp = 33,000 ft lb/min

EXAMPLE 6

If a weight of 33 pounds is lifted to a height of 1,000 feet in 1 minute, what is the power, or rate of work, being done?

Solution

**(33 pounds) × (1,000 feet)/1 min
= 33,000 ft lb/min**

EXAMPLE 7

If the previous example is changed to lifting 66 pounds to a height of 1,000 feet in 1 minute, what is the rate of work being done?

Solution

**(66 pounds) × (1,000 feet)/1 min
= 66,000 ft lb/min**

The amount of power in example 6 is 1 hp because the total is 33,000 ft lb/min, which is the amount of work per unit time that equals one horsepower.

In example 7, the amount of power is equal to 2 hp, since the amount of work per unit time is twice that of 1 hp.

HEAT

In our discussion of energy, it was determined that there are many types of energy including electrical, mechanical, chemical, and heat or thermal. Energy is not considered to be matter because it does not occupy space and does not have mass. The air conditioning industry concerns itself primarily with the transfer of heat energy from one area to another, namely from inside a structure to the outside of the structure being cooled.

The amount of **heat energy** present in a substance is related to the rate of vibration of the molecules within a substance. If the molecules move faster, the heat content will be higher. Conversely, if the molecules move slower, the heat content will be reduced.

Heat content is measured in **British thermal units**, or Btu. The **Btu** is defined as the amount of heat required to raise the temperature of 1 pound of water 1°F at sea level (Fig. 1-8). For example, if the temperature of 1 pound of water is raised from 68°F to 69°F, 1 Btu of heat has been added to the water. Similarly, if the temperature of 1 pound of water is increased from 50°F to 60°F, 10 Btu of heat have been added to the water.

EXAMPLE 8

How many Btu are required to increase the temperature of 5 pounds of water from 85°F to 100°F?

Solution

Since 1 Btu is the amount of heat required to raise the temperature of 1 pound of water 1°F, the number of Btu required to raise 1 pound of water from 85°F to 100°F is 15 Btu (100°F − 85°F = 15°F). Since there are 5 pounds of water in this example, the number of Btu required will be 5 times 15, or 75 Btu.

HEAT AND POWER RELATIONSHIP

Since air conditioning systems are powered by electricity and the main purpose of air conditioning systems is to transfer heat, a relationship between heat content and electric power is useful. In electrical terms, power is expressed in watts where

$$1 \text{ Watt} = 3.413 \text{ Btu}$$

and

$$746 \text{ Watts} = 1 \text{ Horsepower}$$

Going Green

The amount of power consumed is directly related to the amount of work done. If we can reduce the amount of work required to heat or cool a structure, we will save power.

TEMPERATURE AND TEMPERATURE SCALES

Heat content, as defined in terms of Btu, should not be confused with the level of heat. The level, or intensity of heat, is defined as the **temperature** of a substance. Temperature, as we know it, is measured with a thermometer. Consider two vessels that contain water. One vessel contains 1 pound of water; the other contains

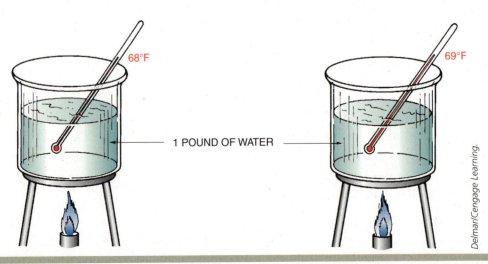

68°F   69°F

— 1 POUND OF WATER —

Delmar/Cengage Learning.

FIGURE 1-8 One Btu is the amount of heat energy required to increase the temperature of 1 pound of water 1°F.

10 pounds of water (Fig. 1-9). The temperature of the water in each container is 50°F. If both vessels are heated until the water reaches 100°F, more heat in Btu would have to be added to the container with 10 pounds of water. It would take 50 Btu to raise the temperature of 1 pound of water, but 500 Btu to raise the temperature of 10 pounds of water (Fig. 1-10). So, even though the level of heat, or temperature, in both containers is the same, the heat content is not.

Four temperature scales are used to measure the level of heat intensity. These are the Fahrenheit, Celsius, Rankine, and Kelvin scales. The Fahrenheit and Celsius scales are the most common for the average individual, while the Rankine and Kelvin scales are used primarily for scientific purposes.

On the Fahrenheit temperature scale, water boils at 212°F and freezes at 32°F. Absolute zero, the temperature at which all molecular movement stops, is −460°F. On the Rankine scale, absolute zero is at 0°R. In addition, water freezes at 492°R and boils at 672°R. The relationship between the Fahrenheit and Rankine temperature scales is shown in Figure 1-11.

On the Celsius temperature scale, water boils at 100°C and freezes at 0°C. Absolute zero is at −273°C. On the Kelvin scale, water boils at 373°K and freezes at 273°K. Absolute zero is at 0°K. The relationship between the Celsius and Kelvin temperature scales is shown in Figure 1-12.

TEMPERATURE CONVERSIONS

On occasion, it is necessary to convert from one temperature scale to another. Most often, conversions between the Celsius and Fahrenheit scales will have to

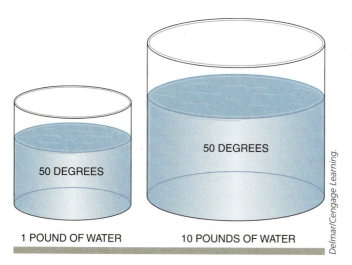

FIGURE 1-9 Two vessels of water at the same temperature. The vessel with more water has a higher heat content.

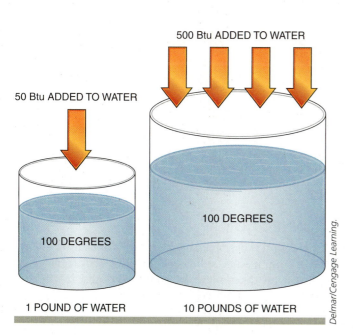

FIGURE 1-10 More heat is required to raise the temperature of the vessel containing more water than the vessel containing less.

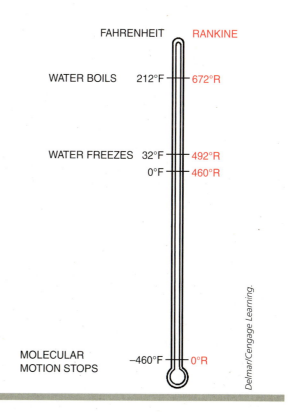

FIGURE 1-11 Relationship between the Rankine and Fahrenheit scales shown on a thermometer.

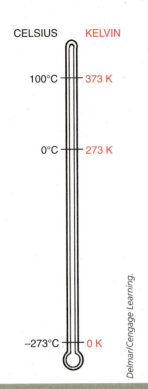

FIGURE 1-12 Relationship between the Kelvin and Celsius scales shown on a thermometer.

be done. The conversions between these two scales are as follows:

$$°F = (1.8°C) + 32$$

$$°C = \frac{(°F - 32)}{1.8}$$

The relationship between the Kelvin scale and the Celsius scale is as follows:

$$°K = °C + 273$$

The relationship between the Rankine scale and the Fahrenheit scale is as follows:

$$°R = °F + 460$$

HEAT TRANSFER

When heat energy is applied to a substance, the molecules of the substance begin to move more rapidly. The molecules collide with each other more readily and the friction of the colliding molecules causes the temperature of the substance to rise. Conversely, when heat energy is removed from a substance, the molecules slow down and the amount of friction decreases. The temperature of the substance decreases as well. When two substances of different temperatures come in contact with each other,

heat energy is transferred between the substances from the warmer substance to the cooler. Heat is transferred by three common methods:

- Conduction
- Convection
- Radiation

CONDUCTION

When heat is transferred by **conduction**, the energy travels from one molecule to another within the substance. For example, if one end of a metal rod is heated by a flame, the end of the rod in the flame will get hot very quickly while the other end may remain cool. After a short time, however, the cooler end of the rod will get hot as well (Fig. 1-13). Water in a pot is heated by conduction as the heat from the pot is transferred to the water.

The rate of heat transfer due to conduction is different in different substances. For example, heat does not travel as well through wood as it does through an iron rod. Typically, substances that are good conductors of electricity are also good conductors of heat.

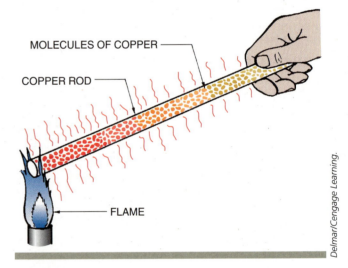

MOLECULES OF COPPER

COPPER ROD

FLAME

FIGURE 1-13 Heat applied to one end of a metal rod transmits through the rod to the other end by conduction.

CONVECTION

Convection is described as the transfer of heat by a flowing medium. Since solids do not flow, heat transfer by convection takes place in liquids and gases, typically water and air. One common example of heat transfer by convection is baseboard heating. As air next to the baseboard heater is warmed, the molecules expand, making them lighter. This heated air rises and is replaced by cooler air which, in turn, is heated as well. The heated air rises and the cooler air falls, creating air currents. These

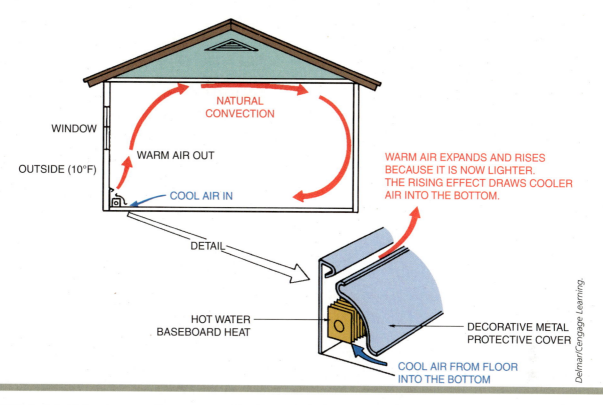

FIGURE 1-14 Natural convection occurs when heated air rises and is replaced by cooler air.

currents are called convection currents and, in the case of the baseboard heating example, are natural convection currents (Fig. 1-14).

In the typical air conditioning system, a fan or blower moves air over a heat transfer surface to remove heat from the air in the space being cooled. In this case, instead of relying on natural convection currents, as in the baseboard heating example, the blower speeds up the air to increase the volume of air treated per unit time. When using a blower to move air or a pump to move water to facilitate the transfer of heat via convection, the process is referred to as forced convection (Fig. 1-15).

RADIATION

Radiation refers to heat that moves in the form of straight rays. These rays produce heat when they come in contact with a surface without heating anything in their path. The sun is an example of heating by radiation. The sun's rays travel until they come in contact with the surface of the Earth and other celestial bodies. The Earth is not as hot as the sun because, as the distance between the two objects increases by a factor of two, the heat intensity decreases by a factor of four. For example, if the distance between the sun and an object is doubled, the amount of heat sensed by the object will be one-fourth of the heat originally sensed (Fig. 1-16).

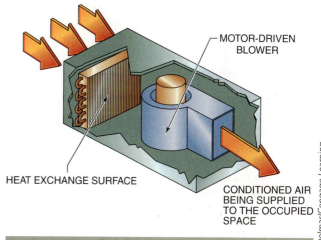

FIGURE 1-15 Convection currents are speeded up with a blower or fan.

In geographic regions of the world that have four distinct seasons, the distance between the sun and those regions plays a major role in the temperature differences that exist between the summer and winter months. In locations near the equator, the sun is at a constant distance from the Earth, so those areas experience warm weather year round.

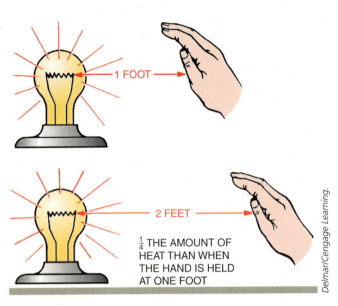

Delmar/Cengage Learning.

FIGURE 1-16 As the distance from a radiant heat source is doubled, the heat intensity is reduced by a factor of four.

SENSIBLE HEAT

As water or another substance is heated, say, from 50°F to 100°F, the difference in temperature can be readily measured with a thermometer. When we turn on the news to get a weather report, the meteorologist expresses weather in terms of the present air temperature as well as the temperature that is likely to be encountered in the days to follow. These changes in temperature are examples of **thermal kinetic energy**, which result from the motion of the molecules within a substance and can be sensed with thermometers. Since these heat transfers can

Going Green

As HVAC technicians, installers, and service personnel, we are responsible for moving heat from one place to another. In the heating season we add heat to the structure, while in the cooling season we remove heat from the structure. By moving heat energy in the most efficient manner possible, we are saving energy and increasing the quality of the air inside the structure.

be *sensed*, they are referred to as **sensible heat** transfers. Examples of sensible heat transfers include increasing the temperature of water from 40°F to 50°F, or decreasing the temperature of air from 80°F to 65°F. Changing the temperature of ice or steam are also examples of sensible heat transfers. When changing the temperature of ice or steam, however, only 0.5 Btu/lb is required as opposed to the 1 Btu/lb for water.

LATENT HEAT

When water boils, it changes from the liquid state to the vapor state. When water reaches 212°F at sea level, it begins to vaporize. The temperature of the steam that is

---

EXAMPLE 9

Assume that we want to change a 1-pound block of ice at 0°F to steam at 392°F. The steps involved in making this transition are as follows:

1. The ice at 0°F must first change to ice at 32°F

2. The ice at 32°F must change to water at 32°F

3. The water at 32°F must change to water at 212°F

4. The water at 212°F must change to steam at 212°F

5. The steam at 212°F must change to steam at 392°F

Solution

The total Btu requirement can be outlined as follows:

Step 1. 0.5 Btu/lb (32°F − 0°F)
= 0.5 Btu/lb (32°F) = 16 Btu

Step 2. 144 Btu

Step 3. 1.0 Btu/lb (212°F − 32°F) = 180 Btu

Step 4. 970.3 Btu

Step 5. 0.5 Btu/lb (392°F − 212°F)
= 0.5 Btu/lb (180°F) = 90 Btu

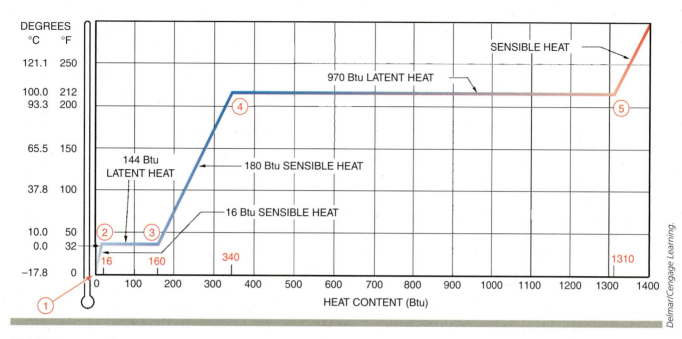

FIGURE 1-17 Graphical representation showing how the heat content of ice, water, and steam changes as heat energy is added.

created is also 212°F. The level of heat, or temperature, has not changed, but the state of the fluid has. Obviously, though, a transfer of heat has taken place. This type of heat transfer is referred to as **thermal potential energy**, which is the heat associated with the forces of attraction within a substance. It cannot be measured with a thermometer and is therefore considered to be **latent heat**, which comes from the Greek word for hidden. Another example of latent heat is when ice at 32°F changes to water at 32°F.

When latent heat transfers take place, the transfer is much greater than during sensible heat transfers. For example, when ice at 32°F changes to water at 32°F the heat transfer is 144 Btu/lb, but when water at 212°F changes to steam at 212°F, the heat transfer is approximately 970.3 Btu/lb. Systems such as evaporative coolers rely on latent heat transfers to provide cooling in geographic areas where the humidity is low.

The process is shown graphically in Figure 1-17. Adding up the required Btu for each of the previous steps, we conclude that a total of 1400.3 Btu are needed to change the 1 pound of ice at 0°F to steam at 392°F.

PRESSURE

When fluids, gases, or liquids are contained in a vessel, they exert a force on the walls of the container that holds them. As in the earlier example of the balloon, as more air is introduced to the balloon, the force that is exerted on the walls of the balloon will increase as well. This can continue until the force inside becomes greater than the force that holds the rubber together, and the balloon bursts. In addition, if instead of adding more air to the balloon, the balloon is heated, the air molecules in the balloon will move faster resulting in weaker attraction forces. This causes the air to expand, increasing the force on the walls of the balloon.

Similarly, if the balloon is cooled, the forces of attraction of the air molecules will increase, reducing the amount of force exerted on the balloon walls. This force is referred to as pressure. **Pressure** is therefore defined as the force that is exerted on the walls of a vessel and is measured in force per unit area. In the air conditioning industry, the units of force are psi. Earlier discussions of the gas laws provide relationships between the pressure, volume, and temperature of gases.

The weight of a fluid also contributes to the amount of pressure that is exerted. Consider the example of a deep sea diver. As the diver swims further and further below the surface of the ocean, the pressure exerted on her body increases as the weight of the water directly above her increases.

ATMOSPHERIC PRESSURE

Our atmosphere is made up mostly of oxygen, nitrogen, and water vapor, with small percentages of other gases. The vapor pressure of each gas exerts a force on the surface of the earth. To determine the pressure in psi that is exerted on the earth by these gases, a 1-square-inch column of air is extended from sea level up through the atmosphere to a maximum height of about 600 miles. This column of air, called **atmospheric pressure**, weighs approximately 14.7 pounds and exerts this pressure over the 1-square-inch area (Fig. 1-18). It can be said then that a pressure of 14.7 pounds presses against each 1-square-inch space on the surface of the earth at sea level. As the altitude increases, this pressure decreases because there is less air above the ground at higher altitudes. For example, at an altitude at 5,000 feet above sea level the pressure of the air is approximately 12 psi. In the air conditioning industry, two pressure readings are used to evaluate system operation:

- Gauge pressure
- Absolute pressure

GAUGE PRESSURE, PSIG

One of the most useful pieces of information that an air conditioning service technician uses to evaluate a system is the pressure within the piping circuit at various points in the system. If the system is open to the atmosphere, a pressure gauge will read 0 psi (Fig. 1-19). This **gauge pressure** reading does not take into account the 1-square-inch

column of air that extends into the atmosphere. This pressure does have an effect on the system, but for evaluation purposes, it is ignored. When inflating a car or bicycle tire, for example, the air pressure in the tire is gauge pressure and does not take atmospheric pressure into account.

To avoid confusion, when gauge pressure is used, the pressure reading is followed by **psig**, pounds per square inch gauge, to indicate that atmospheric pressure has not been accounted for in the reading.

ABSOLUTE PRESSURE, PSIA

When making certain calculations regarding system operation or when calculating pressures, volumes, and temperatures using the gas laws discussed earlier, the pressure of the atmosphere must be considered. When the pressure of the atmosphere is taken into account, the pressure is referred to as **absolute pressure**, and the pressure reading is followed by **psia**, pounds per square inch absolute.

Since the difference between the gauge pressure and the absolute pressure is 14.7 pounds per square inch at sea level, the relationship between the two pressures can be expressed mathematically as

$$\text{psia} = \text{psig} + 14.7$$

From this, we can see that 0 psig = 14.7 psia

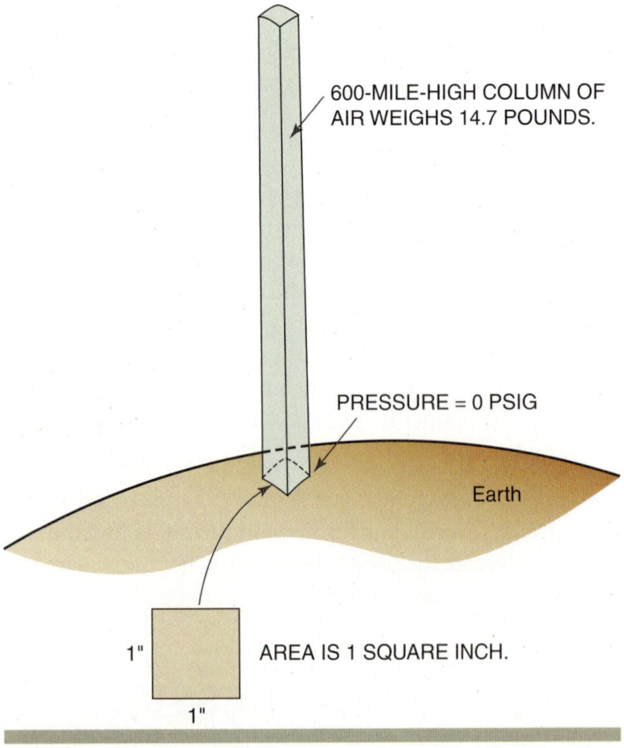

600-MILE-HIGH COLUMN OF AIR WEIGHS 14.7 POUNDS.

PRESSURE = 0 PSIG

Earth

1"   AREA IS 1 SQUARE INCH.

1"

FIGURE 1-18 A 1-square-inch column of air over the earth weighs 14.7 pounds at sea level.

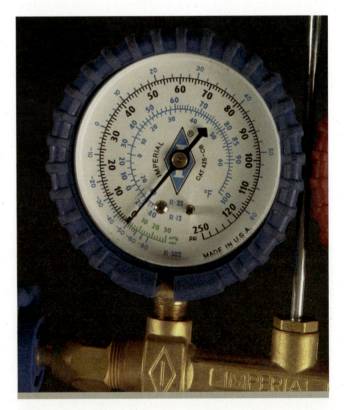

FIGURE 1-19 This gauge reading of 0 psig does not take atmospheric pressure into account.

VACUUM PRESSURES

When the pressure is equal to or greater than 0 psig, the conversion between psia and psig is straightforward, as outlined in the previous section. However, when the gauge pressure in a vessel is below atmospheric pressure, the absolute pressure will be somewhere below 14.7 psia.

Atmospheric pressure can be read with a barometer, which is a mercury-filled device that is open to the atmosphere (Fig. 1-20). A tube, approximately 36″ in length and filled with mercury (Hg) is inverted into a reservoir that is also filled with mercury. A vacuum will be pulled in the top portion of the tube holding the mercury in the tube. The height of this mercury column will be 29.92″; it is referred to as 29.92″ of Hg column and 0″ of Hg vacuum. This is the condition at 0 psig or 14.7 psia. As the pressure outside the tube is reduced, the column of mercury will fall as the amount of pressure holding the mercury in place is reduced (Fig. 1-21). The mercury column measurement will fall from 29.92″ toward zero and the mercury vacuum measurement will increase from zero toward a perfect vacuum reading of 29.92″. Further reductions in the pressure outside the tube will result in further reductions in the mercury level until, at 0 psia, the level of mercury in the tube will be at the same level as the reservoir itself. At this point, there will be measurements of 0″ Hg column and a 29.92″ Hg vacuum (Fig. 1-22).

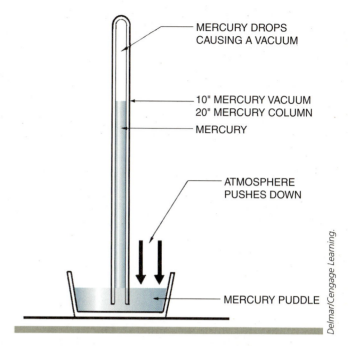

FIGURE 1-21 As atmospheric pressure falls, the mercury column falls, indicating a partial vacuum.

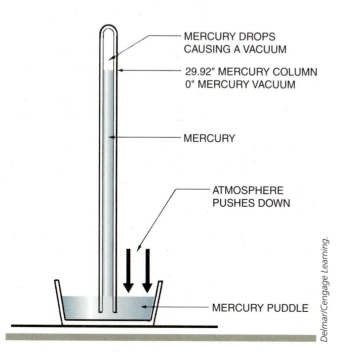

FIGURE 1-20 Mercury barometer indicating atmospheric pressure.

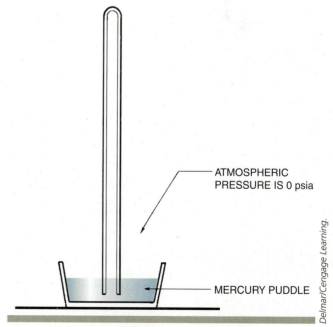

FIGURE 1-22 In the absence of atmospheric pressure, the mercury will be at the same height as the reservoir, indicating a perfect 29.92″ Hg vacuum.

VACUUM PRESSURE CALCULATIONS AND CONVERSIONS

Quite often, the pressure in an air conditioning system will be below atmospheric pressure, either when the system is malfunctioning or when the system is being prepared to be placed into operation. The gauges used by the service technician can read gauge pressures below atmospheric pressure (Fig. 1-23). These pressures are read in terms of **inches of mercury** ("Hg). In order to convert the gauge pressure to absolute pressure, the following formula is used:

$$\text{psia} = (30'' \text{ Hg} - \text{gauge reading in } '' \text{Hg})/2$$

For calculation purposes, the perfect vacuum reading of 29.92" Hg is often rounded off to 30" Hg. For example, if the gauge reading was 10" Hg vacuum, the psia equivalent would be

$$\text{psia} = (30'' \text{ Hg} - 10'' \text{ Hg})/2 = 20/2 = 10 \text{ psia}$$

If a perfect vacuum of 30" Hg was achieved in an air conditioning system, the absolute pressure would be

$$\text{psia} = (30'' \text{ Hg} - 30'' \text{ Hg})/2 = 0/2 = 0 \text{ psia}$$

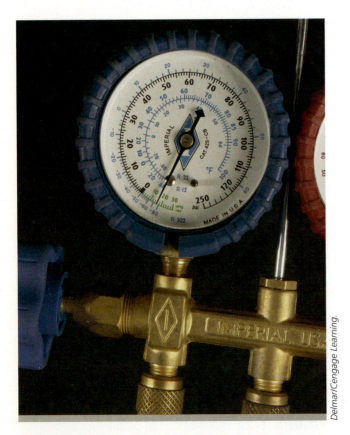

Delmar/Cengage Learning.

FIGURE 1-23 This pressure gauge is reading a pressure below atmospheric pressure. The pressure is below 0 psig.

SUMMARY

- Matter is any substance that has weight and mass and occupies space.
- Matter can be in the form of a solid, liquid, or gas.
- Gas behavior is governed by Boyle's law, Charles' law, and Dalton's law.
- Energy is the capacity to do work.
- Energy can be in the form of heat energy, mechanical energy, electrical energy, or chemical energy.
- Energy cannot be created or destroyed, but can be converted from one form to another.
- Work is defined as the force exerted on an object times the distance it is moved.
- Power is the rate of doing work.
- The heat present in a substance is related to the vibration of the molecules within the substance.
- The Btu is the heat required to raise the temperature of 1 pound of water 1°F.
- The watt is the unit of power and is equal to 3.413 Btu.
- Temperature is the measure of heat level or intensity.
- Heat can be transferred by conduction, convection, and radiation.
- Sensible heat transfers cause the temperature of a substance to change.
- Latent heat transfers result in a change of state while maintaining a constant temperature.
- Pressure is the force that is exerted on the walls of a vessel and is measured in force per unit area.
- Atmospheric pressure is the pressure of the atmosphere above the surface of the earth.
- Gauge pressure provides pressure readings without taking the pressure of the atmosphere into account.
- Absolute pressure provides pressure readings that take atmospheric pressure into account.
- Atmospheric pressure can be read with a barometer.

GREEN CHECKLIST

☐ **The amount of power consumed is directly related to the amount of work done.**

☐ **Power consumption can be reduced by reducing the amount of work done.**

☐ **Effective heat transfer saves energy.**

☐ **Ineffective heat transfer can have a negative effect on indoor air quality.**

REVIEW QUESTIONS

1. **The three states of matter are**
   a. Solid, vapor, and gas
   b. Solid, fluid, and liquid
   c. Solid, liquid, and gas
   d. Liquid, gas, and fluid

2. **The gas law that states the total pressure of a combined mixture is equal to the sum of the individual pressures is**
   a. Charles' law
   b. Pascal's law
   c. Boyle's law
   d. Dalton's law

3. **At constant temperature,**
   a. The pressure of a vapor will increase as the volume is decreased
   b. The pressure of a vapor will decrease as the volume is increased
   c. The pressure of a vapor will decrease as the volume is decreased
   d. Both a and b are correct

4. **When electrical energy is converted to mechanical energy,**
   a. 100% of the electrical energy is converted to mechanical energy
   b. Mechanical energy is generated, but so is heat energy

   c. Mechanical energy in generated, but so is chemical energy
   d. Electrical energy cannot be converted into mechanical energy

5. **What must be done to destroy energy?**
   a. The energy must be heated
   b. The energy must be cooled
   c. Both a and b are correct
   d. Neither a nor b is correct

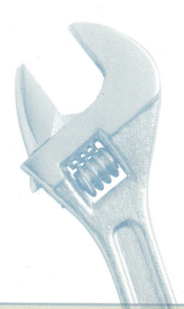

6. **If a 500-pound object is moved 200 feet, how much work is being done?**
    a. 200 ft lb
    b. 500 ft lb
    c. 1,000 ft lb
    d. 100,000 ft lb

7. **Which of the following is true regarding power?**
    a. Power is the rate of doing work
    b. Power is measured in horsepower
    c. Electrical power is measured in watts
    d. All of the above are correct

8. **Vibration of molecules within a substance directly affects the substance's**
    a. Mechanical energy
    b. Heat energy
    c. Electrical energy
    d. Chemical energy

9. **Raising the temperature of 1 pound of water from 50°F to 100°F will require**
    a. 50 Btu to be added to the water
    b. 50 Btu to be removed from the water
    c. 100 Btu to be added to the water
    d. 100 Btu to be removed from the water

10. **Quantity of heat is measured**
    a. With a thermometer
    b. In degrees
    c. In Btu
    d. All of the above are correct

11. **Absolute zero, the temperature at which molecular motion stops, occurs at**
    a. −460°F
    b. 0°R
    c. 0°K
    d. All of the above are correct

12. **The method of heat transfer that occurs when heat energy travels from one molecule to another within a substance is called**
    a. Conduction
    b. Convention
    c. Convection
    d. Radiation

13. **Good conductors of electricity are said to be**
    a. Poor conductors of heat
    b. Good conductors of heat
    c. Made of nonmetallic substances
    d. Both a and c are correct

14. **As the distance from a glowing heat source such as a light bulb is doubled,**
    a. The amount of heat sensed will also double
    b. The amount of heat sensed will be one-half of the original heat sensed
    c. The amount of heat sensed will be one-fourth of the original heat sensed
    d. The amount of heat sensed will be twice the original amount of heat sensed

15. **Which of the following is an example of sensible heat?**

    a. Water changing temperature from 40°F to 100°F

    b. Ice at 32°F changing to water at 32°F

    c. Steam at 212°F changing to water at 212°F

    d. All of the above are correct

16. **When one pound of ice at 32°F changes to water at 32°F,**

    a. Zero Btu are absorbed by the ice

    b. 144 Btu are absorbed by the ice

    c. 970.3 Btu are absorbed by the ice

    d. 72 Btu are absorbed by the ice

17. **The barometer is typically used to measure**

    a. Atmospheric pressure

    b. Gauge pressure

    c. Absolute pressure

    d. All of the above are correct

18. **At sea level, atmospheric pressure is**

    a. 14.7 psi

    b. 14.7 psig

    c. 0 psia

    d. 30" Hg

19. **If the gauge pressure in a system is equal to 30 psig, what is the absolute pressure?**

    a. 0 psia

    b. 15.3 psia

    c. 44.7 psia

    d. None of the above is correct

20. **If the gauge pressure in a system is 6" Hg vacuum, what is the absolute pressure?**

    a. 0 psia

    b. 12 psia

    c. 20.7 psia

    d. 30 psia

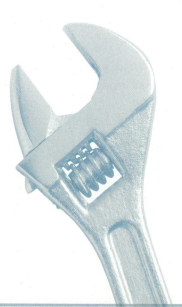

The Refrigeration Process

OBJECTIVES   *Upon completion of this chapter, the student should be able to*

- Describe the basic vapor-compression refrigeration cycle.

- List the basic components that make up a vapor-compression refrigeration system.

- Describe the function of a compressor.

- List various types of compressors.

- Describe the function of the condenser.

- Describe the function of the metering device.

- List three commonly used metering devices.

- Describe the function of the evaporator.

In today's society, it is the knowledgeable, well-trained technician who helps ensure the continued satisfactory operation of air conditioning equipment. To do this, the HVAC technician needs to fully understand the fundamental concepts of the vapor-compression refrigeration system and the four key components that, together, allow the system to operate effectively and efficiently: the compressor, the condenser, the metering device, and the evaporator.

Having studied heat theory, temperature, and pressure in the previous chapter, as well as the relationship that exists among them, we can now begin our discussion of the basic refrigeration cycle on which these concepts are based. One important factor that must be considered is that the vapor-compression refrigeration process is a repeating cycle. This means that when one cycle has been completed, the system and its refrigerant are in the proper configuration and state to begin another cycle. This repeating process will continue indefinitely as long as the system is in operation. The components in the refrigeration system are intended to facilitate this process and the supply of refrigerant contained within the system will not deplete as long as the system remains leak free.

Automatic expansion valve (AEV) Metering device that maintains a constant evaporator pressure

Bubble point For blended refrigerants, the temperature used to calculate condenser subcooling

Bulb pressure Pressure that facilitates the opening of the thermostatic expansion valve

Capillary tube A fixed bore metering device

Compression The portion of the compression process in which the refrigerant is compressed

Compressor The component of an air conditioning system that pumps refrigerant through the system by increasing the pressure of the vapor refrigerant

Condenser A heat transfer surface in an air conditioning system that rejects heat

Condenser saturation temperature The temperature at which system refrigerant will condense in the condenser; this temperature corresponds to the high side pressure in the system on a pressure/temperature chart

GLOSSARY OF TERMS (CONT'D)

Condensing medium The medium, usually air or water, that absorbs the heat that is rejected by the system condenser

Critically charged system A system that requires an exact quantity of refrigerant; all of the refrigerant is moving through the system at all times

Cylinder The component part of a reciprocating compressor that houses the piston

Dehumidifying The process of removing humidity from the air

Desuperheating The process by which the discharge refrigerant from the compressor is cooled down to the condenser saturation temperature

Dew point The temperature at which air reaches 100% humidity; for blended refrigerants, the temperature used to calculate evaporator superheat

Discharge The portion of the compression process when the refrigerant is discharged from the compressor

Discharge line The refrigerant line that carries the discharge refrigerant from the compressor to the condenser

Discharge pressure The pressure of the refrigerant in the high-pressure side of the air conditioning system; also referred to as the high-side or head pressure

Discharge valve The component part of a reciprocating compressor that opens to discharge refrigerant from the compressor into the discharge line

Dry-type evaporator An evaporator that is designed to have all of the refrigerant boil off into a vapor before leaving the coil

Evaporator The component part of an air conditioning system that is responsible for absorbing heat from the space to be cooled

Evaporator pressure The pressure that corresponds to the temperature at which refrigerant vaporizes in the evaporator on a pressure/temperature chart

Externally equalized TEV A thermostatic expansion valve that measures the evaporator pressure at the outlet of the coil

Fixed-bore metering device A metering device that does not open or close in response to changes in the load on the system; examples of the fixed bore metering device are the capillary tube and the piston

Flash gas The process by which some of the liquid refrigerant instantly vaporizes upon entering the evaporator

Internally equalized TEV A thermostatic expansion valve that measures the evaporator pressure at the inlet of the coil

Liquid line The refrigerant-carrying line that connects the condenser to the metering device

Metering device The component of an air conditioning system that controls the flow of refrigerant to the evaporator

Outside ambient temperature The temperature of the air that surrounds the outdoor coil of an air conditioning system

Piston The component part of the reciprocating compressor that moves back and forth within the cylinder to compress or expand the refrigerant in the cylinder

Pressure/temperature chart Shows the relationship that exists between the temperatures and pressures of saturated refrigerants

Psig Pounds per square inch gauge; pressure readings taken from a gauge manifold

Reexpansion In a reciprocating compressor, the process by which the refrigerant trapped in the cylinder at the end of a cycle is expanded to reduce the pressure in the cylinder

Repeating cycle A process that can repeat itself indefinitely without depleting resources

Return air Air that is brought from the conditioned space to the evaporator

Saturated Refrigerant that is a mixture of liquid and

vapor; saturated refrigerants follow a pressure/temperature relationship

Spring pressure One of the closing pressures on the thermostatic expansion valve; also the opening pressure on the automatic expansion valve

Subcooling The process by which the refrigerant in the condenser is cooled below the condenser saturation temperature

Suction In a reciprocating compressor, the process by which refrigerant in the suction line is pulled into the compression chamber prior to being compressed

Suction line The refrigerant-carrying line that connects the outlet of the evaporator to the compressor

Suction pressure The pressure of the low side of the system; also called low-side or back pressure

Suction valve The component part of a reciprocating compressor that opens to pull refrigerant from the evaporator into the compressor for compression

Superheat The process by which vapor refrigerant is heated above its evaporator saturation temperature

Supply air The air that is returned to the conditioned space after it has passed through the evaporator

Thermal bulb TEV component that senses the temperature at the outlet of the evaporator and converts this temperature to the valve's opening pressure

Thermostatic expansion valve (TEV) Metering device designed to maintain a constant evaporator superheat

Total system charge The amount of refrigerant a system contains

Transmission line The small diameter tube that connects the thermal bulb to the body of a thermostatic expansion valve

THE REFRIGERATION PROCESS

The refrigeration process can best be described as the transferring of heat from an objectionable location to a place where the addition of this heat has little or no effect on the surrounding conditions. The process of air conditioning, therefore, involves removing heat from inside a structure or residence and depositing the heat outside. In the case of domestic refrigeration, the heat is transferred from the inside of the refrigerator or freezer to the air surrounding the appliance. The basic air conditioning system is made up of four major components that enable this heat transfer to take place. These components are:

- Evaporator
- Compressor
- Condenser
- Metering device

These four components are connected by a piping arrangement that carries a fluid called refrigerant. Figure 2-1 shows a piping diagram of the basic refrigeration system. Refrigerant has the ability to absorb heat at a low pressure and temperature and reject heat at a higher temperature and pressure. The heat that is absorbed by the system is the heat that is present in the space to be conditioned.

One of the main characteristics of a refrigerant is that when it is **saturated**, a definite relationship between its pressure and temperature is created. A saturated refrigerant is a fluid that exists simultaneously as both a vapor and a liquid. The relationship that exists between the pressure and the temperature of the refrigerant is referred to simply as a pressure/temperature relationship.

PRESSURE/TEMPERATURE RELATIONSHIP

A *saturated* refrigerant behaves in a predictable manner with respect to the pressure and temperature of the substance. As the pressure of the refrigerant changes, so does the temperature. An increase in the pressure results in an increase in temperature, and a decrease in the pressure results in a decrease in temperature. For this reason, it is possible for a refrigeration system to absorb heat at a lower temperature and reject heat at a higher temperature by controlling the pressure of the refrigerant at various stages in the refrigeration cycle. A **pressure/ temperature chart** provides the relationship between pressure and temperature for some commonly used refrigerants (Fig. 2-2).

From this chart, we can see that the temperature of saturated R-22 at 68.5 **psig** is 40°F, while the temperature of saturated R-410A at 118 psig is 40°F. It can be readily seen from the pressure/temperature chart that different refrigerants have different pressure/temperature relationships. The pressure/ temperature chart is a useful tool for the service technician in troubleshooting and evaluating air conditioning systems.

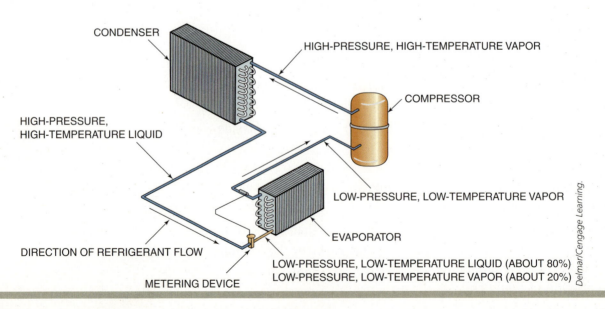

FIGURE 2-1 Pictorial diagram of the basic refrigeration cycle.

TEMPERATURE °F	12	22	134a	502	404A	410A
−60	19.0	12.0		7.2	6.6	0.3
−55	17.3	9.2		3.8	3.1	2.6
−50	15.4	6.2		0.2	0.8	5.0
−45	13.3	2.7		1.9	2.5	7.8
−40	11.0	0.5	14.7	4.1	4.8	9.8
−35	8.4	2.6	12.4	6.5	7.4	14.2
−30	5.5	4.9	9.7	9.2	10.2	17.9
−25	2.3	7.4	6.8	12.1	13.3	21.9
−20	0.6	10.1	3.6	15.3	16.7	26.4
−18	1.3	11.3	2.2	16.7	18.2	28.2
−16	2.0	12.5	0.7	18.1	19.6	30.2
−14	2.8	13.8	0.3	19.5	21.1	32.2
−12	3.6	15.1	1.2	21.0	22.7	34.3
−10	4.5	16.5	2.0	22.6	24.3	36.4
−8	5.4	17.9	2.8	24.2	26.0	38.7
−6	6.3	19.3	3.7	25.8	27.8	40.9
−4	7.2	20.8	4.6	27.5	30.0	42.3
−2	8.2	22.4	5.5	29.3	31.4	45.8
0	9.2	24.0	6.5	31.1	33.3	48.3
1	9.7	24.8	7.0	32.0	34.3	49.6
2	10.2	25.6	7.5	32.9	35.3	50.9
3	10.7	26.4	8.0	33.9	36.4	52.3
4	11.2	27.3	8.6	34.9	37.4	53.6
5	11.8	28.2	9.1	35.8	38.4	55.0
6	12.3	29.1	9.7	36.8	39.5	56.4
7	12.9	30.0	10.2	37.9	40.6	57.8
8	13.5	30.9	10.8	38.9	41.7	59.3
9	14.0	31.8	11.4	39.9	42.8	60.7
10	14.6	32.8	11.9	41.0	43.9	62.2
11	15.2	33.7	12.5	42.1	45.0	63.7
12	15.8	34.7	13.2	43.2	46.2	65.3
13	16.4	35.7	13.8	44.3	47.4	66.8
14	17.1	36.7	14.4	45.4	48.6	68.4
15	17.7	37.7	15.1	46.5	49.8	70.0
16	18.4	38.7	15.7	47.7	51.0	71.6
17	19.0	39.8	16.4	48.8	52.3	73.2
18	19.7	40.8	17.1	50.0	53.5	75.0
19	20.4	41.9	17.7	51.2	54.8	76.7
20	21.0	43.0	18.4	52.4	56.1	78.4
21	21.7	44.1	19.2	53.7	57.4	80.1
22	22.4	45.3	19.9	54.9	58.8	81.9
23	23.2	46.4	20.6	56.2	60.1	83.7
24	23.9	47.6	21.4	57.5	61.5	85.5
25	24.6	48.8	22.0	58.8	62.9	87.3
26	25.4	49.9	22.9	60.1	64.3	90.2
27	26.1	51.2	23.7	61.5	65.8	91.1
28	26.9	52.4	24.5	62.8	67.2	93.0
29	27.7	53.6	25.3	64.2	68.7	95.0
30	28.4	54.9	26.1	65.6	70.2	97.0
31	29.2	56.2	26.9	67.0	71.7	99.0
32	30.1	57.5	27.8	68.4	73.2	101.0
33	30.9	58.8	28.7	69.9	74.8	103.1
34	31.7	60.1	29.5	71.3	76.4	105.1
35	32.6	61.5	30.4	72.8	78.0	107.3
36	33.4	62.8	31.3	74.3	79.6	108.4
37	34.3	64.2	32.2	75.8	81.2	111.6
38	35.2	65.6	33.2	77.4	82.9	113.8
39	36.1	67.1	34.1	79.0	84.6	116.0
40	37.0	68.5	35.1	80.5	86.3	118.3
41	37.9	70.0	36.0	82.1	88.0	120.5
42	38.8	71.4	37.0	83.8	89.7	122.9
43	39.8	73.0	38.0	85.4	91.5	125.2
44	40.7	74.5	39.0	87.0	93.3	127.6
45	41.7	76.0	40.1	88.7	95.1	130.0
46	42.6	77.6	41.1	90.4	97.0	132.4
47	43.6	79.2	42.2	92.1	98.8	134.9
48	44.6	80.8	43.3	93.9	100.7	136.4
49	45.7	82.4	44.4	95.6	102.6	139.9
50	46.7	84.0	45.5	97.4	104.5	142.5
55	52.0	92.6	51.3	106.6	114.6	156.0
60	57.7	101.6	57.3	116.4	125.2	170.0
65	63.8	111.2	64.1	126.7	136.5	185.0
70	70.2	121.4	71.2	137.6	148.5	200.8
75	77.0	132.2	78.7	149.1	161.1	217.6
80	84.2	143.6	86.8	161.2	174.5	235.4
85	91.8	155.7	95.3	174.0	188.6	254.2
90	99.8	168.4	104.4	187.4	203.5	274.1
95	108.2	181.8	114.0	201.4	219.2	295.0
100	117.2	195.9	124.2	216.2	235.7	317.1
105	126.6	210.8	135.0	231.7	253.1	340.3
110	136.4	226.4	146.4	247.9	271.4	364.8
115	146.8	242.7	158.5	264.9	290.6	390.5
120	157.6	259.9	171.2	282.7	310.7	417.4
125	169.1	277.9	184.6	301.4	331.8	445.8
130	181.0	296.8	198.7	320.8	354.0	475.4
135	193.5	316.6	213.5	341.2	377.1	506.5
140	206.6	337.2	229.1	362.6	401.4	539.1
145	220.3	358.9	245.5	385.9	426.8	573.2
150	234.6	381.5	262.7	408.4	453.3	608.9
155	249.5	405.1	280.7	432.9	479.8	616.2

VACUUM (in. Hg) – RED FIGURES
GAGE PRESSURE (psig) – BOLD FIGURES

Delmar/Cengage Learning.

FIGURE 2-2 Pressure/temperature relationship chart for some common refrigerants.

from experience...

It is important to note that only saturated refrigerant follows the pressure/temperature relationship. When heat is added to 100% vapor refrigerant or removed from 100% liquid refrigerant, the pressure/temperature relationship does not hold.

Courtesy of Carrier.

FIGURE 2-3 Typical evaporator coil in the air handler of a residential central air conditioning system.

EVAPORATORS

The **evaporator** (Fig. 2-3) is the system component that is responsible for performing the actual *cooling* or *refrigerating* of the occupied space. The evaporator, while reducing the temperature of a given space, does not actually make the space cold. *Cold* can be defined as the absence of heat. Therefore, to cool down a room one must not add cold to it but, instead, heat must be removed from it.

The purpose of the evaporator, therefore, is to absorb heat from the space that is to be cooled.

The normal direction of heat transfer is from a warmer substance to a cooler substance so, for heat to flow to the refrigerant in the evaporator, the evaporator must be cooler than the space temperature. Consider the following example. We pour warm juice into a glass of ice to make the juice cold, but in reality, the juice is making the ice warm. In this instance, the heat travels from the juice to the ice, removing heat from the juice. This is why the ice melts so fast. The operation of the evaporator is basically the same. However, in the juice example, once the ice melts, there is no more "cooling." An evaporator cools continuously because there is a constant flow of refrigerant through the coil that does not melt or deteriorate.

The evaporator is also responsible for **dehumidifying**, or removing humidity from the air. The evaporator operates at a temperature below the **dew point** temperature, so it causes moisture in the air—humidity—to condense out of the air. The dew point temperature is defined as the temperature at which air reaches 100% humidity, implying that condensation can occur.

The refrigerant in the evaporator is able to maintain a low temperature because its pressure is greatly reduced by the combined effects of the compressor and the metering device. Remember, the refrigerant follows a pressure/temperature relationship that determines a refrigerant's temperature at a given pressure as long as the refrigerant

is saturated. When the refrigerant enters the evaporator, it is saturated as approximately 80% liquid and 20% vapor.

As the refrigerant flows through the evaporator, heat is transferred from the air passing over the coil to the refrigerant flowing inside the coil. As the heat is absorbed by the refrigerant, more and more of the liquid refrigerant boils, or evaporates, into a vapor. Latent heat, also called hidden heat because it cannot be measured with a thermometer, causes the refrigerant to boil. This heat transfer is latent because the temperature of the refrigerant is not changing, but the state of the refrigerant is changing from a liquid to a vapor. This process will continue until all the liquid refrigerant boils into a vapor, at which point it begins to **superheat**. Superheat is defined as the amount of heat added to a vapor that results in an increase in temperature. Heat transfers that result in the superheating of the refrigerant are sensible heat transfers, which, unlike latent heat transfers, *can* be measured with a thermometer. Nevertheless, the refrigerant will continue to absorb heat.

The refrigerant superheats because it is now 100% vapor and is still much cooler than the medium passing over the coil so it continues to absorb heat. Figure 2-4 illustrates a typical evaporator on an air conditioning system operating with R-22 as its refrigerant. Evaporators on residential systems are **dry-type evaporators**, meaning that all the refrigerant is boiled off into a vapor before leaving the coil; hence, the refrigerant is "dry."

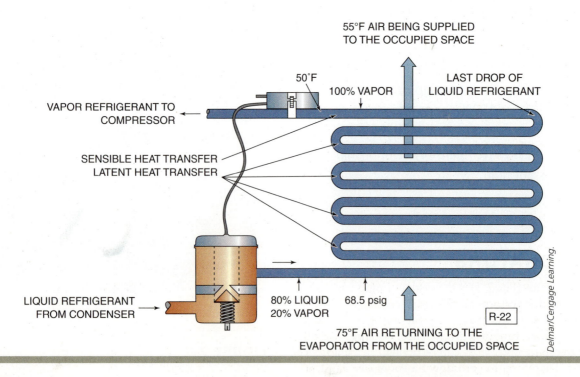

FIGURE 2-4 Evaporator at design conditions for an R-22 air conditioning application.

COMPRESSORS

The **compressor** (Fig. 2-5) is the component in an air conditioning system that is responsible for raising the pressure and temperature of the vapor refrigerant that leaves the evaporator. This is necessary for two reasons. First, increasing the pressure of the vapor creates the pressure difference needed for refrigerant flow to occur in the system. A higher-pressure fluid will flow toward a lower-pressure fluid. Since pressure and temperature are related, increasing the pressure will increase the temperature as well. Second, increasing the temperature of the refrigerant enables it to easily reject system heat, because the direction of heat transfer is from a higher-temperature substance to a lower-temperature substance. So, if the heat from the system is to be rejected to the **condensing medium**, it must be at a higher temperature than the medium. The condensing medium, which is typically air for residential systems, is the substance that absorbs the heat that the system must reject to remain functional.

The compressor is considered a vapor pump as it is not intended to receive liquid. Liquids are noncompressible fluids. Liquid entering the compressor will result in reduced efficiency and capacity and, in many cases, mechanical damage to the compressor's internal components. Compressors found on residential systems are often hermetically sealed. Hermetically sealed compressors cannot be serviced internally since they are welded closed and are normally replaced when internal failure occurs. Large hermetic compressors can be rebuilt by companies specializing in this type of work, as long as doing so is economically feasible. This is not common on compressors used in residential applications. Some commonly used types of compressors are reciprocating, rotary, and scroll.

RECIPROCATING COMPRESSORS

A reciprocating compressor uses pistons, cylinders, and valves to accomplish the compression of the refrigerant. The **piston** (Fig. 2-6) moves in a back-and-forth, or reciprocating, path inside the **cylinder** (Fig. 2-7). Reciprocating compressors vary in size and capacity depending on the requirements of the system.

The compressor is a dividing point between the high- and low-pressure sides of the system and houses other integral components including suction and discharge valves. The **suction valve** connects the compressor to the low-pressure side of the system via the **suction line** that

Courtesy of Bristol Compressors, Inc.

FIGURE 2-5 Welded hermetic compressor. The suction line usually is piped directly into the shell and is open to the crankcase. The discharge line is piped from the compressor inside the shell to the outside of the shell. The compressor shell is typically thought of as a low-side component.

Courtesy of Carrier.

FIGURE 2-6 Piston in a reciprocating compressor.

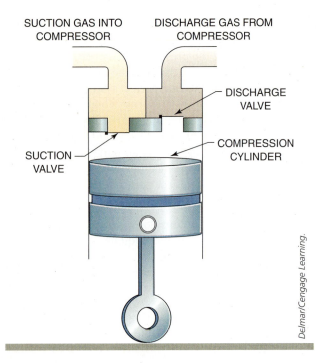

FIGURE 2-7 The piston moves back and forth within the cylinder to expand and compress the refrigerant.

carries refrigerant into the compressor. The **discharge valve** connects the compressor to the high-pressure side of the system via the **discharge line** that carries refrigerant from the compressor after it has been compressed. The suction and discharge valves open and close depending on the pressure difference across them and allow the vapor refrigerant to enter and leave the compression chamber at the proper point in time.

Reciprocating Compressor Operation. For ease in explanation, the compression process can be divided into four distinct "miniprocesses":

* Reexpansion
* Suction or intake
* Compression
* Discharge

As a reference point, we will start our discussion about the compression process where the piston is at the highest possible position within the cylinder. This position is referred to as *top dead center*. At top dead center, both the suction and the discharge valves are in the closed position and the refrigerant in the compression chamber is equal to the discharge pressure [Fig. 2-8(A)].

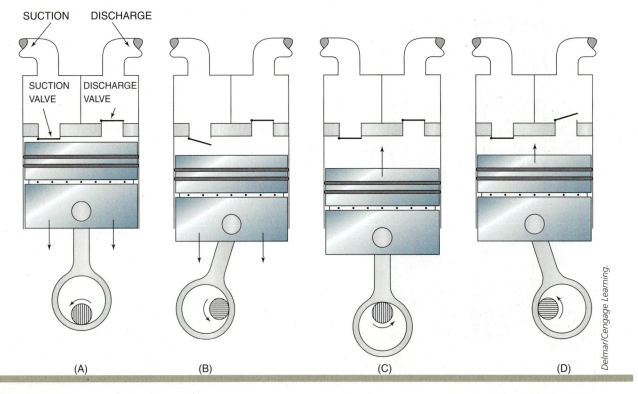

FIGURE 2-8 (A) The piston is at top dead center and is beginning to move downward in the cylinder. Both the suction and discharge valves are closed. (B) The suction valve opens when the cylinder pressure drops below the suction pressure. Refrigerant from the suction line now enters the compression cylinder. (C) The piston is at bottom dead center and is beginning to move up in the cylinder. The suction valve is pushed closed and compression begins. (D) When the pressure in the cylinder is greater than the discharge pressure, the discharge valve is pushed open. The high-pressure, high-temperature vapor is now discharged to the condenser.

Since both valves are closed, no refrigerant can enter or leave the compression chamber. As the compressor motor continues to turn, the piston starts to move down in the cylinder. This increases the volume of the cylinder. The pressure of the refrigerant in the cylinder begins to decrease, because the same amount of refrigerant is now occupying a larger space. The refrigerant in the cylinder goes through a process of expansion. This part of the cycle is referred to as **reexpansion** because the refrigerant that occupies the space above the piston at top dead center is expanded during the first part of every downward stroke.

The pressure of the refrigerant in the cylinder will continue to drop until it reaches a point just below the suction pressure of the system. **Suction pressure** is the term used to describe the low-side pressure of the system. At this pressure, the suction pressure will now be greater than the compression chamber pressure and the suction valve will open [Fig. 2-8(B)]. At this point, the pressure in the suction line is equal to the pressure in the compression chamber. As the piston continues to move downward, suction gas is drawn into the compression chamber. This portion of the cycle is referred to as **suction** or intake. Suction will continue until the piston stops moving in the downward direction. When the piston reaches its lowest position in the cylinder, *bottom dead center*, the suction portion of the cycle ends.

As the compressor continues to operate, the piston starts to move upward in the cylinder. This upward motion of the piston pushes the suction valve closed, trapping refrigerant in the cylinder. The piston continues to move upward, reducing the volume of the cylinder and increasing the pressure of the refrigerant. This part of the cycle is referred to as **compression** [Fig. 2-8(C)]. Compression will continue until the pressure of the refrigerant inside the cylinder is slightly greater than the pressure of the refrigerant in the discharge line.

When the cylinder pressure is greater than the **discharge pressure**, the discharge valve will be pushed open, allowing the high-pressure refrigerant to be pushed out of the cylinder into the discharge line as the piston continues to move upward. This part of the cycle is referred to as **discharge**, [Fig. 2-8(D)]. Discharge will continue until the piston reaches top dead center, where the discharge refrigerant will push the discharge valve closed as the piston again starts to move downward. The cycle then repeats itself as long as the compressor is energized.

ROTARY COMPRESSORS

A rotary compressor can be identified by its cylindrical shape and the fact that the discharge line is located at the very top center of the compressor (Fig. 2-9). This type of compressor operates with an eccentric roller connected directly to the shaft of the motor. This roller rotates at the

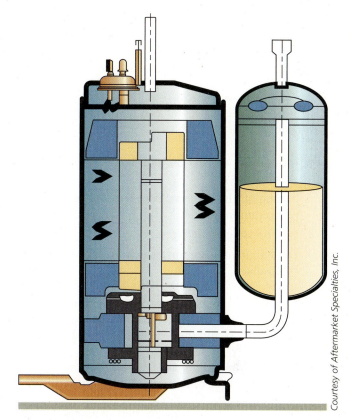

FIGURE 2-9 Rotary compressor.

Courtesy of Aftermarket Specialties, Inc.

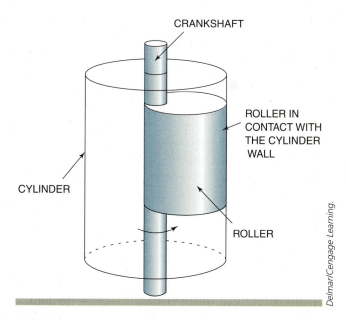

FIGURE 2-10 Roller scraping against the interior surface of the cylinder.

Delmar/Cengage Learning.

same speed as the motor within the shell of the compressor, rubs against the inside of the compression chamber (Fig. 2-10), and forms a compartment in which refrigerant vapor is trapped. This movable vane creates a seal with the roller due to a spring pressure that constantly pushes the vane up against the roller as it rotates. Like the reciprocating compressor, refrigerant is introduced into the compressor via the suction line and discharged from the compressor through the discharge line.

Rotary Compressor Operation. The compression process in a rotary compressor can be divided into four intermediate steps:

1. Introducing suction gas into the compression chamber
2. Sealing off the suction chamber, trapping refrigerant inside
3. Compressing the refrigerant
4. Discharging the high-pressure refrigerant from the compressor

To illustrate the compression process in a rotary compressor, we start where the roller is blocking off the discharge port and the refrigerant from the suction port is permitted to enter the cylinder [Fig. 2-11(A)]. As the roller rotates in a counterclockwise direction, the suction or intake port is sealed off from the refrigerant in the cylinder. The refrigerant is now trapped in the cylinder by the vane, the discharge valve, and the roller's contact with the cylinder wall [Fig. 2-11(B)]. As the roller continues to turn, the space the refrigerant occupies decreases and the pressure and temperature of the refrigerant increase [Fig. 2-11(C)]. Once the pressure of the refrigerant in the cylinder rises above the pressure of the refrigerant in the discharge line, the discharge valve will open and the high-pressure, high-temperature refrigerant vapor will be pushed out of the compressor [Fig. 2-11(D)].

While the refrigerant is being compressed, the other side of the cylinder is open to the suction side, so new suction gas is entering the chamber in preparation for the next compression cycle. When all of the high-temperature, high-pressure vapor has been discharged from the cylinder, the suction gas that has entered will be compressed and the cycle will repeat itself.

SCROLL COMPRESSORS

Another type of compressor that is rapidly growing in popularity is the scroll compressor (Fig. 2-12). The concept used in this type of compressor has been known for quite some time, but it was ahead of the technology needed to manufacture it. The successful and efficient operation of the scroll is reliant on the production of two perfectly machined spirals or scrolls (Fig. 2-13). One of the scrolls is stationary, while the other scroll vibrates or wobbles. As this scroll vibrates, refrigerant vapor is pushed and compressed toward the center of the compressor where it is discharged from the device.

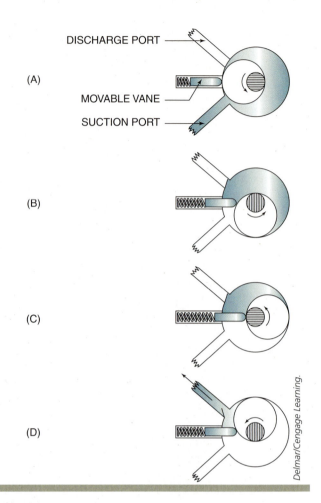

FIGURE 2-11 Compression process in a rotary compressor: (A) Vapor refrigerant enters the chamber through the suction port. (B) The refrigerant is trapped in the chamber. (C) The refrigerant is being compressed. (D) The high-pressure refrigerant is discharged.

Scroll Compressor Operation. The compression process of the scroll compressor is illustrated in Figure 2-14. The configuration of the scrolls forms multiple chambers, each of which is at a different stage of compression, allowing the compressor to operate smoothly and continuously. Scroll compressors are becoming a popular choice for replacement, because they are better able to handle any liquid that may enter the device.

Because one of the scrolls is stationary and the other floats, the scrolls have some *play*. This play will allow the floating scroll to move if liquid, which is not compressible, should enter the compressor, thereby preventing damage to the device. Damage to the compressor is averted since the scroll's movement increases the volume of the chamber, thereby accommodating the volume of the liquid. The pistons on a reciprocating compressor move in a well-defined path and the introduction of

Courtesy of Copeland Corporation.

FIGURE 2-12 Scroll compressor.

Courtesy of Copeland Corporation.

FIGURE 2-13 Two identical scrolls that form crescent-shaped pockets when nested together.

1. GAS ENTERS AN OUTER OPENING AS ONE SCROLL ORBITS THE OTHER.

2. THE OPEN PASSAGE IS SEALED AS GAS IS DRAWN INTO THE COMPRESSION CHAMBER.

3. AS ONE SCROLL CONTINUES ORBITING, THE GAS IS COMPRESSED INTO AN INCREASINGLY SMALLER "POCKET."

4. GAS IS CONTINUALLY COMPRESSED TO THE CENTER OF THE SCROLLS, WHERE IT IS DISCHARGED THROUGH PRECISELY MACHINED PORTS AND RETURNED TO THE SYSTEM.

5. DURING ACTUAL OPERATION, ALL PASSAGES ARE IN VARIOUS STAGES OF COMPRESSION AT ALL TIMES, RESULTING IN NEAR-CONTINUOUS INTAKE AND DISCHARGE.

Courtesy of Copeland Corporation.

FIGURE 2-14 Compression process in a scroll compressor. Compression is created by the interaction of an orbiting spiral and a stationary spiral. Gas enters the outer openings as one of the spirals orbits. As the spiral orbits, gas is compressed into an increasingly smaller pocket, reaching discharge pressure at the center port. Actually, during operation, all six gas passages are in various stages of compression at all times, resulting in nearly continuous suction and discharge.

liquid could result in major component failure. Liquid entering the scroll compressor will, however, reduce the capacity of the system.

Another benefit of scroll compressors is that they do not use suction and discharge valves. The scrolls, as they age, tend to "wear in," making them more efficient, as opposed to valves, which wear out with time. Because there are no suction or discharge valves, scroll compressors are equipped with a low-mass, disc-type check valve at the discharge port that prevents the high-pressure refrigerant from traveling back through the compressor during the off cycle.

Going Green

Scroll compressors are a popular choice for use on high-efficiency air conditioning equipment. The operation of the scroll compressor allows for all the compressed refrigerant to be discharged to the condenser, unlike the operation of the reciprocating compressor. Reciprocating compressors allow for some of the compressed vapor to remain in the compression cylinder even after the piston has reached its highest point in the cylinder.

CONDENSERS

The **condenser** (Fig. 2-15) is the system component responsible for rejecting heat from the system. The condenser is a heat-exchange surface that permits the transfer of heat from the system refrigerant to the condensing medium. This heat comes from the space that is being cooled and from the heat generated during the compression process. The condensing medium, which is usually air for residential systems, must be at a lower temperature than the refrigerant to allow heat transfer to occur. This rejection of heat causes the heat-laden vapor refrigerant coming from the compressor to condense. The temperature at which the refrigerant will condense is called the **condenser saturation temperature**.

THE CONDENSING PROCESS

The refrigerant leaving the compressor is superheated and well above the saturation temperature, so it must first be cooled before it can condense. The process of

Courtesy of Carrier.

FIGURE 2-15 Condenser coil of a typical split-type central air conditioning system.

removing the superheat from the discharge refrigerant is referred to as **desuperheating**. The desuperheating process removes superheat and is a sensible heat exchange since this change can be measured with a thermometer. This is the first function of the condenser. For example, if an R-22 air conditioning system is operating as in Figure 2-16, the superheated refrigerant is leaving the compressor at 220°F and must be cooled down to 110°F before it will begin to condense. Although the temperature of the refrigerant is changing, the pressure will remain constant. The condenser saturation temperature of 110°F corresponds to the high-side pressure of 226 psig on a pressure/temperature chart. Once the refrigerant's temperature falls to 110°F, it begins to change state. This is the second function of the condenser. The condensing process, which is a latent heat transfer, will continue until all the refrigerant has become a liquid. This transfer of heat causes the 110°F vapor to change into a 110°F liquid. Since the 110°F liquid is still warmer than the condensing medium, it will continue to give up heat to the medium and begin to subcool. **Subcooling** is the third process that takes place in the condenser. One degree of subcooling is equivalent to one degree below the condenser saturation temperature. In this example, the refrigerant is condensing at 110°F and leaving the condenser at 90°F, so the condenser is operating with 20 degrees of subcooling. The overall purpose of the condenser can be summed up in the chart shown in Figure 2-17.

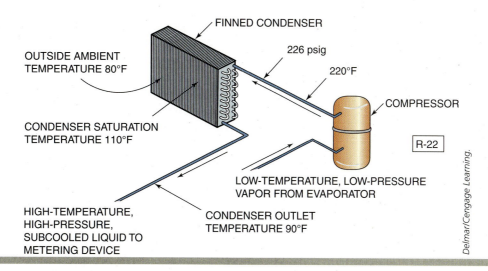

FIGURE 2-16 The superheated discharge gas from the compressor must desuperheat from 220°F to 110°F before it can begin to condense.

Condenser Process	Type of Heat Transfer	Purpose	Location in Condenser
Desuperheating	Sensible	Reduces refrigerant to condenser saturation temperature	Discharge line and top portion of the condenser
Condensing (change of state)	Latent	Allows refrigerant to change from a vapor to a liquid	Middle portion of the condenser
Subcooling	Sensible	Cools refrigerant to a temperature below the condenser saturation temperature	Usually the last 10% of the condenser

FIGURE 2-17 Description of the basic condenser functions.

AIR-COOLED CONDENSERS

In a condenser, heat is transferred from one location to another. Air-cooled condensers transfer heat into the air that passes over the condenser coil. Residential air-cooled condensers are located outdoors because, if they were located in the occupied space, the heat rejected by the system would raise the space temperature. Air-cooled condensers use fans to increase the amount of air that passes over and through the condenser coil.

The condenser in Figure 2-16 shows outside, or ambient, air passing over the coil at 80°F. For a standard-efficiency condenser, the condenser saturation temperature is roughly 30 to 35 degrees higher than the outside ambient temperature. In this example, the condenser saturation temperature is 110°F, indicating a 30-degree difference. Air must be allowed to flow freely through the coil to ensure maximum efficiency. If the coil becomes dirty or a fan motor burns out, the system's efficiency will be greatly reduced and it may stop operating altogether.

Higher-efficiency condensers, which are typically larger than standard-efficiency models, can operate with saturation temperatures as low as 20 degrees higher than outside ambient temperature. The larger coils increase the effective surface area and increase the rate of heat transfer. They also allow for more subcooling, which further reduces the temperature of the refrigerant in the liquid line, leading to more efficient system operation.

Whether the condenser is standard or high efficiency, the condenser saturation temperature is still determined by the outside ambient temperature. When it is hot, the condenser saturation temperature is higher and the system efficiency lower. On cooler days, the condenser saturation temperature is lower and the system efficiency and effectiveness higher.

Going Green

High-efficiency air conditioning systems are manufactured with condenser coils that are larger than those on standard efficiency systems. This leads to increased heat transfer from the system, lower head pressures, increased subcooling, and more efficient operation.

METERING DEVICES

As previously discussed, the refrigeration system uses a closed cycle that repeats itself. For this to occur, whatever is done must later be "undone" at another point in the cycle. If pressure is increased, it must be decreased. If heat is absorbed, it must be rejected. Just as the compressor increased the temperature and pressure of the refrigerant entering it, another component is responsible for reducing the pressure and temperature of the refrigerant in the system. This component is the **metering device**, also referred to as the expansion device.

The metering device is responsible for feeding the proper amount of refrigerant to the evaporator. The refrigerant that enters the metering device is a high-temperature, high-pressure, subcooled liquid that leaves the device as a low-temperature, low-pressure, saturated liquid. The saturated liquid is roughly 80% liquid and 20% vapor, and, since the refrigerant is saturated, there is a pressure/temperature relationship for the refrigerant. For an air conditioning system operating with R-22 as its refrigerant, the pressure of the refrigerant entering the metering device is approximately the same as the high-side pressure, which for an ideal air conditioning system is about 226 psig, assuming an air-cooled condenser is used. The temperature of the refrigerant entering the metering device will be roughly 95°F, assuming that the condenser is operating with 15 degrees of subcooling. The pressure of the refrigerant as it leaves the metering device, at design conditions, should be roughly 68.5 psig and its boiling temperature 40°F (refer to the pressure/temperature chart for R-22) (Fig. 2-18). The metering device creates a restriction to flow just like a car accident restricts the flow of traffic (Fig. 2-19). The reduction in pressure after

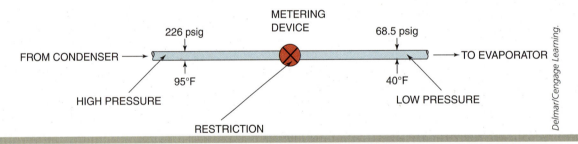

FIGURE 2-18 The metering device restricts normal refrigerant flow and separates the high- and low-pressure sides of the system.

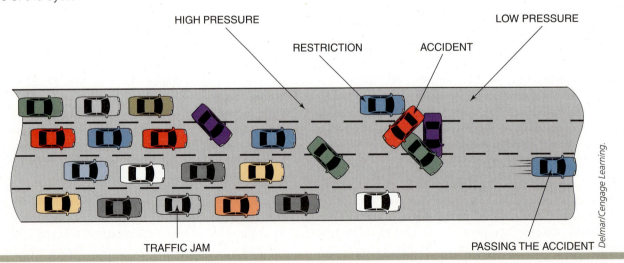

FIGURE 2-19 A car accident causes a restriction of traffic flow. The cars build up before the accident and pressure is reduced after the accident.

refrigerant passes the expansion device is created by the combined effects of the metering device and the compressor's suction.

Three types of expansion devices are mentioned in this text and, although the basic purpose of each is similar, their operation is very different. These devices are the capillary tube, the automatic expansion valve, and the thermostatic expansion valve. The capillary tube is the least complicated.

CAPILLARY TUBE

The **capillary tube** is a **fixed-bore** device, meaning that the opening the refrigerant flows through does not change in size (Fig. 2-20). Capillary tubes are found on **critically charged** systems which, by definition, have the **total system charge** moving through the system whenever it is operating. Total system charge is the total amount of refrigerant that the system holds.

> ## *from experience...*
>
> Many residential central air conditioning systems are not critically charged, so the capillary tube is not commonly found on them. The capillary tube is most commonly found on through-the-wall or window air conditioners.

Capillary-tube systems operate differently from other systems that store refrigerant until it is needed by the evaporator, so the amount of refrigerant in the system must correspond precisely to the manufacturer's specifications. During operation, as the high-temperature, high-pressure, subcooled liquid refrigerant from the liquid line enters the capillary tube, its flow is restricted through the device. As the refrigerant flows through the capillary

tube, its pressure begins to drop and some of the liquid boils into a vapor. When the refrigerant reaches the end of the capillary tube, the pressure of the refrigerant has dropped significantly. When it leaves the capillary tube as a saturated liquid, it expands because the tubing size of the evaporator is larger, and the pressure drops down to the desired evaporator saturation pressure.

More liquid refrigerant boils into a vapor upon leaving the capillary tube and helps cool the remaining vapor by absorbing heat from it. This boiling refrigerant is referred to as **flash gas** because it immediately flashes to a vapor. This flashing of the liquid to a vapor occurs within the capillary tube as well as its outlet. Although flash gas helps the system operate efficiently, too much is an indication of a system problem.

AUTOMATIC EXPANSION VALVE

The **automatic expansion valve** (AEV) modulates the flow of refrigerant into the evaporator to keep the evaporator pressure constant (Fig. 2-21). Unlike the capillary tube, which cannot adjust the flow of refrigerant, the automatic expansion valve opens and closes to either increase or decrease the amount of refrigerant feeding into the evaporator in response to the pressure of the refrigerant in the evaporator.

The valve operates on a "needle-and-seat" mechanism that modulates the amount of refrigerant that is permitted to pass through the valve. The position of the needle is determined by the difference between two pressures:

- The spring pressure
- The evaporator pressure

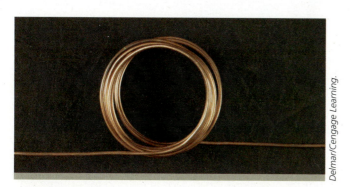

Delmar/Cengage Learning.

FIGURE 2-20 Capillary tube metering device.

Courtesy of Parker Hannefin Corporation.

FIGURE 2-21 Automatic expansion valve.

The **spring pressure** is the pressure that opens the valve. This pressure is adjustable and is set at the desired evaporator pressure. The higher the spring pressure, the higher the evaporator pressure will be. The **evaporator pressure** is the pressure that closes the valve. The spring and evaporator pressures push in opposite directions, and the valve position changes accordingly [Fig. 2-22(A)]. For example, if the spring pressure is 70 psig and the evaporator pressure is only 60 psig, the spring pressure, which is greater than the evaporator pressure, will cause the valve to open in order to feed more refrigerant to the evaporator [Fig. 2-22(B)]. This is done to increase the evaporator pressure. When the evaporator pressure reaches the desired level, the valve

will maintain the proper flow to maintain that pressure. When the spring pressure is equal to the evaporator pressure, the valve is in equilibrium (Fig. 2-22A). If, on the other hand, the spring pressure is set at 70 psig and the evaporator pressure is 80 psig, the evaporator pressure, which is greater than the spring pressure, will close the valve [Fig. 2-22(C)].

The automatic expansion valve operates in a manner that is opposite to what one might expect. If the cooling requirement of the system increases, one would assume that the metering device would open to introduce more refrigerant to the evaporator. Such is not the case. Consider the following example of a small refrigerator equipped

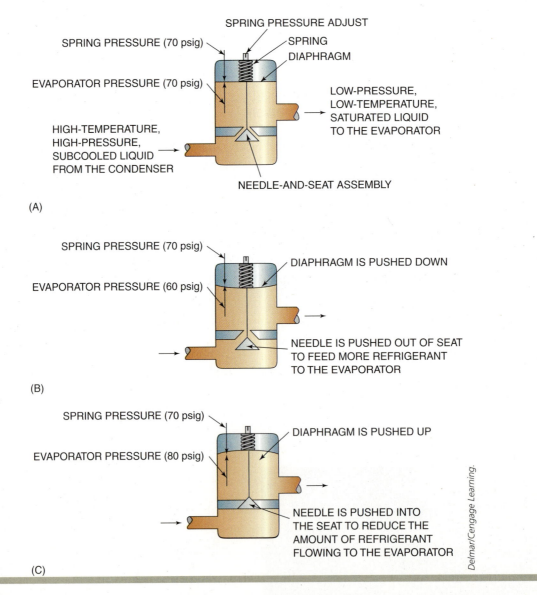

FIGURE 2-22 (A) The spring and evaporator pressures push in opposite directions. This valve is in equilibrium. (B) The evaporator pressure is lower than desired, so the valve is opening to allow more refrigerant to flow to the evaporator. (C) The evaporator pressure is higher than desired. This valve is closing to reduce refrigerant flow to the evaporator.

with an automatic expansion valve. The refrigerator was empty when a large amount of room-temperature food was placed inside the box. The temperature of the product is far above the desired box temperature, so the temperature of the air in the box will increase. This increase in box temperature results in an increase in the temperature of the air moving over the evaporator coil. This higher temperature results in an increase of the pressure in the evaporator. As mentioned, an increase in evaporator pressure pushes the valve toward the closed position, feeding less refrigerant to the evaporator. On the other hand, when the heat load is low, or when the product has reached the desired temperature, the evaporator pressure decreases, causing the automatic expansion valve to open, feeding more refrigerant to the evaporator. For this reason, the automatic expansion valve is ideal for systems that experience a constant heat load as opposed to systems that operate under a wide range of conditions.

THERMOSTATIC EXPANSION VALVE

The **thermostatic expansion valve (TEV)** (Fig. 2-23) is a modulating valve that opens and closes in order to feed the proper amount of refrigerant to the evaporator.

Courtesy of Emerson Climate Technologies Flow Controls.

FIGURE 2-23 Thermostatic expansion valve.

Unlike the capillary tube, which does not modulate fluid flow, and the automatic expansion valve, which is designed to maintain a constant pressure in the evaporator, the thermostatic expansion valve is designed to maintain a constant evaporator superheat. Modulating devices, such as the automatic expansion valve and the thermostatic expansion valve, can open and close in response to external conditions to regulate refrigerant flow through the devices. The thermostatic expansion valve operates on a needle-and-seat concept that is very similar to the automatic expansion valve. The main difference between the TEV and the AEV is that the TEV closes as the system load is reduced, while the AEV closes as the system load is increased.

To understand how the thermostatic expansion valve operates, it is important to understand how evaporator superheat is measured. Superheat, as mentioned earlier, is the amount of sensible heat that is added to the refrigerant after it has all boiled off into a vapor. Consider the R-22 evaporator in Figure 2-24. The evaporator saturation pressure is 68.5 psig and the temperature at the outlet of the evaporator is 50°F. Superheat is calculated by subtracting the evaporator saturation temperature from the evaporator outlet temperature. In this case, the saturation temperature is 40°F, from the pressure temperature chart, and the outlet temperature of the evaporator is 50°F. The superheat in this evaporator is the difference between these two temperatures, or 10 degrees. An evaporator superheat between 8 and 15 degrees is acceptable for most high-temperature air conditioning applications.

Going Green

The thermostatic expansion valve is a popular choice for use on high-efficiency air conditioning systems. During periods of high heat loads, when the evaporator can often be underfed and operates with high superheat, the TEV opens to allow for a higher net refrigeration effect and greater evaporator capacity, leading to increased system efficiency.

As stated in the section on compressors, the refrigerant entering the compressor needs to be 100% vapor to prevent component damage. The thermostatic

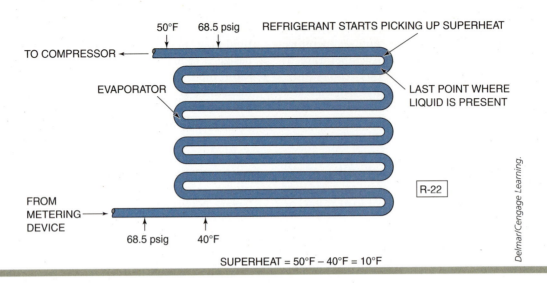

FIGURE 2-24 Evaporator superheat calculation.

expansion valve helps ensure that this is indeed the case. The amount of superheat that will be maintained by the valve depends on the setting of the superheat spring and the size of the valve itself. A cutaway view of a thermostatic expansion valve is shown in Figure 2-25.

The thermostatic expansion valve's needle-and-seat assembly is controlled by three pressures that position the needle properly to feed the correct amount of refrigerant to the evaporator. The pressures push on a diaphragm, which is a thin and very flexible piece of steel, whose position determines the position of the needle in the seat. The three pressures are:

• Evaporator pressure
• Spring pressure
• Bulb pressure

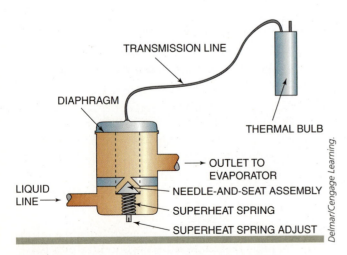

FIGURE 2-25 Cutaway illustration of a TEV.

Evaporator Pressure. The evaporator pressure is one of the pressures that helps close the valve. It attempts to push the needle into the seat to reduce the flow of refrigerant into the evaporator. The evaporator pressure can be taken from either the inlet or the outlet of the coil. If the pressure drop (difference between the inlet and outlet pressures) across the evaporator coil is small, the inlet and outlet pressures are relatively close to each other and the pressure reading can be taken from the inlet of the coil. A thermostatic expansion valve that senses pressure at the inlet of the evaporator is called an **internally equalized** valve (Fig. 2-26A); a thermostatic expansion valve that senses the evaporator outlet pressure is called an **externally equalized** valve (Fig. 2-26B). An example illustrating the difference in operation of internally and externally equalized valves can be found at the end of the section.

Spring Pressure. The spring pressure, also known as the superheat spring pressure, determines how much superheat the evaporator will operate with. The higher the spring pressure, the higher the amount of superheat. The spring comes factory set and should only be adjusted by trained professionals, because improperly adjusted superheat springs can cause major system damage, including compressor failure. The spring pressure is the other pressure that closes the valve, reducing the amount of refrigerant flowing into the evaporator. The evaporator pressure added to the spring pressure provides the total *closing pressure* for the valve. Figure 2-27 shows the superheat spring adjustment on the thermostatic expansion valve.

Bulb Pressure. The **bulb pressure** is the only pressure that opens the valve. This pressure is generated inside a thermal bulb that is mounted at the outlet of the evaporator (Fig. 2-26). The line that connects the thermal bulb

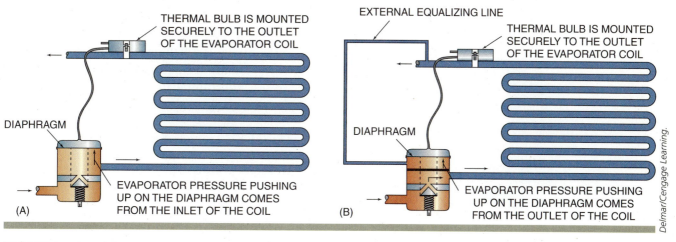

FIGURE 2-26 (A) An internally equalized TEV. The valve senses the evaporator pressure from the inlet of the coil. (B) An externally equalized TEV. The valve senses the evaporator pressure from the outlet of the coil.

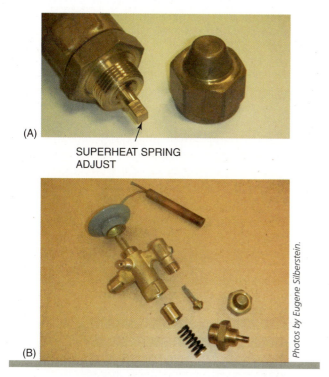

FIGURE 2-27 (A) Superheat spring adjustment screw on a TXV. (B) Exploded view of TXV components.

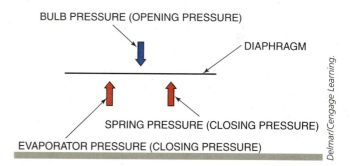

FIGURE 2-28 The bulb pressure pushes to open the TEV. The spring and the evaporator pressures push to close the valve.

to the thermostatic expansion valve is called the **transmission line**. The **thermal bulb** is refrigerant filled and, for the most part, follows a pressure/temperature relationship. It is the thermal bulb that senses the evaporator outlet temperature. The refrigerant in the thermal bulb is isolated from that of the system so no mixing takes place. The refrigerant at this temperature exerts a specific amount of pressure (pressure/temperature relationship) that pushes down on the diaphragm, opposing the evaporator and spring pressures (Fig. 2-28).

Proper operation of the thermostatic expansion valve relies on the proper mounting of the thermal bulb. Figure 2-29 shows the thermal bulb secured to the outlet of the evaporator. The following guidelines should be followed:

1. Make certain the bulb is strapped securely to the suction line, as close to the outlet of the evaporator as possible.
2. Make certain the bulb is secured to a clean, straight section of hard-drawn pipe.
3. Do not mount the thermal bulb on a fitting or other component that will prevent good thermal contact.
4. Whenever possible, use the strapping material that is supplied with the valve.
5. Never solder the bulb to the suction line.
6. Do not use glues, tapes, string, or other adhesive materials to secure the bulb.

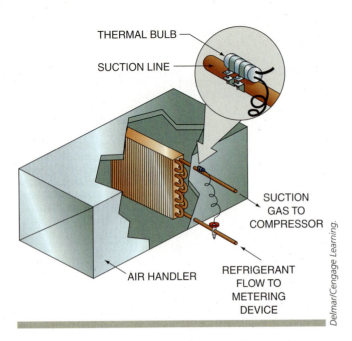

THERMAL BULB

SUCTION LINE

SUCTION GAS TO COMPRESSOR

AIR HANDLER

REFRIGERANT FLOW TO METERING DEVICE

Delmar/Cengage Learning.

FIGURE 2-29 Thermal bulb mounted securely to the outlet of the evaporator.

7. Always wrap the bulb with a good-quality insulation tape to ensure accurate TEV operation.
8. On residential systems, the thermal bulb should be strapped to the top of the suction line.
9. Always follow manufacturer's instructions to ensure proper installation of the specific valve being used.

The valve in example 1 is being pushed closed since the closing pressure is larger than the opening pressure. If the refrigerant in the system and the bulb is R-22, you can determine the amount of superheat in the evaporator by converting the pressures into temperatures using the pressure/temperature relationship for R-22. If the evaporator pressure is 70 psig, the evaporator saturation temperature is 41 degrees. If the bulb pressure is 73 psig, this means that the temperature the bulb is sensing is 43 degrees at the outlet of the evaporator. Therefore, the superheat in the evaporator is 2 degrees, which is the result of subtracting the evaporator saturation temperature from the evaporator outlet temperature (43°F − 41°F).

EXAMPLE 1

In the system illustrated in Figure 2-30, you are given the following information:

- Evaporator pressure is 70 psig.
- Spring pressure is 10 psig.
- Bulb pressure is 73 psig.

What is the opening pressure and the total closing pressure?

Solution

You can conclude the following:

- Opening pressure = 73 psig (bulb pressure)
- Total closing pressure = 80 psig (evaporator pressure + spring pressure)

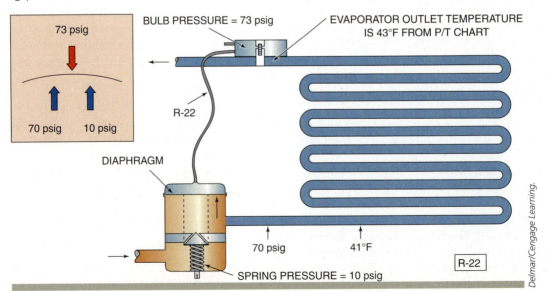

73 psig

70 psig   10 psig

BULB PRESSURE = 73 psig

EVAPORATOR OUTLET TEMPERATURE IS 43°F FROM P/T CHART

R-22

DIAPHRAGM

70 psig   41°F

R-22

SPRING PRESSURE = 10 psig

Delmar/Cengage Learning.

FIGURE 2-30 This valve is closing to increase the superheat in the evaporator.

EXAMPLE 2

In the system illustrated in Figure 2-31, you are given the following:

- Evaporator pressure is 70 psig.
- Spring pressure is 10 psig.
- Bulb pressure is 84 psig.

What is the opening pressure and the total closing pressure?

Solution

You can conclude the following:

- Opening pressure = 84 psig (bulb pressure)
- Total closing pressure = 80 psig (evaporator pressure + spring pressure)

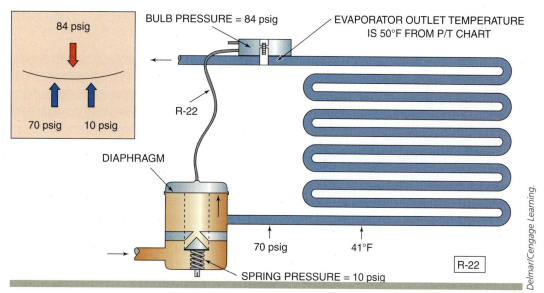

FIGURE 2-31 This valve is opening to reduce the superheat in the coil.

The valve in example 2 is being pushed open since the opening pressure is larger than the closing pressure. If the refrigerant in the system and the bulb is R-22, you can determine the amount of superheat in the evaporator by converting the pressures into temperatures using the pressure/temperature relationship for R-22. If the evaporator pressure is 70 psig, the evaporator saturation temperature is 41 degrees. If the bulb pressure is 84 psig, this means that the temperature the bulb is sensing is 50 degrees at the outlet of the evaporator. Therefore, the superheat in the evaporator is 9 degrees, which is the result of subtracting the evaporator saturation temperature from the evaporator outlet temperature (50°F − 41°F).

The valve in example 3 is in equilibrium since the closing pressure and the opening pressure are the same. If the refrigerant in the system and the bulb is R-22, you can determine the amount of superheat in the evaporator by converting the pressures into temperatures using the pressure/temperature relationship for R-22. If the evaporator pressure is 70 psig, the evaporator saturation temperature is 41 degrees.

If the bulb pressure is 80 psig, the temperature the bulb is sensing is 48 degrees at the outlet of the evaporator. Therefore, the superheat in the evaporator is 7 degrees, which is the result of subtracting the evaporator saturation temperature from the evaporator outlet temperature (48°F − 41°F).

EVAPORATOR PRESSURE DROPS AND THE TEV

Now we will examine a situation in which the evaporator pressure drop is small, less than 2 psig, and compare with an evaporator that has a large, more than 2 psig, pressure drop. Since the piping offers resistance to flow, the pressure of the refrigerant will decrease as it flows through the coil. For this reason, larger coils typically have larger pressure drops than smaller coils. This will illustrate when an internally equalized valve is desirable and when an externally equalized valve is desirable.

Consider an ideal air conditioning evaporator that has no pressure drop, although a pressure drop of less

EXAMPLE 3

In the system illustrated in Figure 2-32 you are given the following:

- Evaporator pressure is 70 psig.
- Spring pressure is 10 psig.
- Bulb pressure is 80 psig.

What is the opening pressure and the total closing pressure?

Solution

You can conclude the following:

- Opening pressure = 80 psig (bulb pressure)
- Total closing pressure = 80 psig (evaporator pressure + spring pressure)

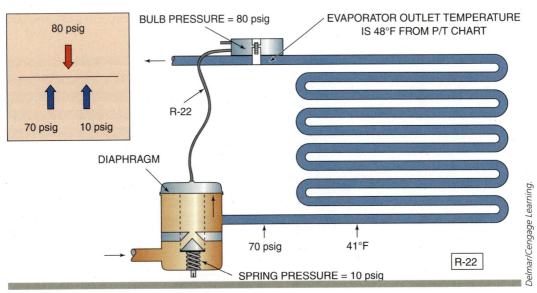

FIGURE 2-32 This valve is in equilibrium. It is feeding the proper amount of refrigerant to maintain 7°F of superheat in the evaporator coil.

than 2 psig through the coil is considered negligible. The evaporator pressure remains constant at 68.5 psig through the entire coil operating with R-22 as its refrigerant. The thermostatic expansion valve operates with R-22 in its bulb and has a spring pressure of 15.5 psig and a bulb pressure of 84 psig. The evaporator saturation temperature is 40°F, and the evaporator outlet temperature is 50°F (from the pressure/temperature chart). The evaporator superheat is 10°F and the valve is in equilibrium since the opening pressure (84 psig) is equal to the closing pressure (15.5 psig + 68.5 psig = 84 psig) (Fig. 2-33).

Now consider the same evaporator, but this time assume that the evaporator has a pressure drop across it of 11 psig. This indicates that the pressure at the inlet of the evaporator is 68.5 psig and the pressure at the outlet of the evaporator is 57.5 psig. Assume that an internally equalized valve is being used so you are reading the evaporator pressure at the inlet of the coil. Since the refrigerant is saturated and follows a pressure/temperature relationship, as the pressure of the refrigerant drops, the boiling temperature of the refrigerant also drops. Assume that the last of the refrigerant boils to a vapor at 32°F (57.5 psig). At this point, the refrigerant will start picking up superheat. If the valve is to reach equilibrium, assuming the same spring pressure of 15.5 psig, the bulb pressure will have to reach 84 psig (68.5 psig + 15.5 psig), which corresponds to a 50°F evaporator outlet temperature (Fig. 2-34). This evaporator is now operating with 18 degrees of superheat, compared with the 10 degrees of superheat in the previous evaporator with no pressure drop.

Now look at the same coil again, but this time assume that an externally equalized valve is being used. Instead of measuring the pressure at the inlet of the evaporator coil, now measure the evaporator pressure at the outlet of the coil. This evaporator pressure is 57.5 psig. The spring pressure is still 15.5 psig, so the total closing pressure is 73 psig (57.5 psig + 15.5 psig). For the valve to reach equilibrium, the opening pressure will have to be 73 psig. This pressure corresponds to about 43°F, so the temperature at the outlet of the evaporator coil is 43°F.

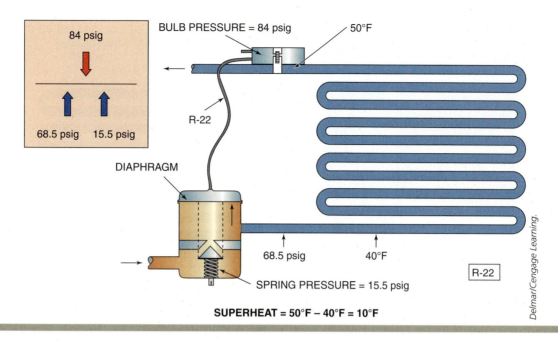

FIGURE 2-33 An ideal evaporator coil with no pressure drop across it. The inlet pressure and outlet pressure are both 68.5 psig.

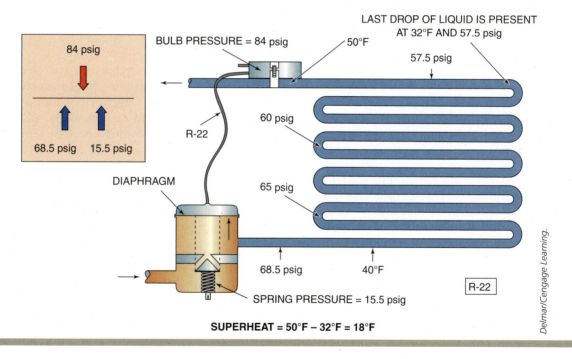

FIGURE 2-34 This evaporator has a large pressure drop across it. The evaporator must operate with excess superheat in order to reach equilibrium.

The superheat in the evaporator will therefore be 11°F (43°F − 32°F), which will allow the evaporator to operate more efficiently (Figure 2-35).

You can see from the preceding examples that if the pressure drop across the evaporator coil is large, an externally equalized thermostatic expansion valve should be used, because it compensates for the pressure drop through the coil. When the pressure drop across the coil is small, either an internally equalized or an externally equalized valve can be used. The internally equalized valve is usually preferred when the pressure drop is small, less than 2 psig, because there are two fewer piping connections. This reduces the possibility of future refrigerant leaks.

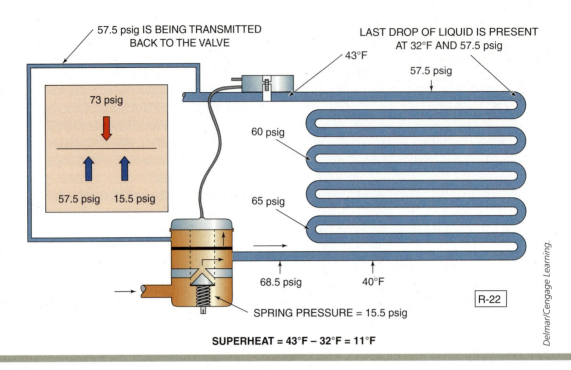

FIGURE 2-35 Using an externally equalized TEV on evaporators with large pressure drops helps to reduce the amount of superheat in the coil.

REFRIGERANTS IN THE RESIDENTIAL SYSTEM

As mentioned throughout our discussion of the refrigeration components in this chapter, a refrigerant is a substance that facilitates the transfer of heat by absorbing heat at a lower temperature and pressure (evaporating) and rejecting heat at a higher temperature and pressure (condensing). Although many different substances can act as refrigerants, water being the best, system configuration and design constraints determine the best refrigerant for the job. Refrigerants are rated by their Ozone Depletion Potential, ODP, which determines the substance's ability to deplete the ozone layer and their Global Warming Potential, GWP, which indicates their contribution to global warming. GWP is measured using a relative scale, which compares the mass of the selected gas to carbon dioxide, whose GWP is 1.

R-22

R-22 is a very popular refrigerant and is found in many residential air conditioning systems. Although it is no longer used in the manufacturing of new air conditioning systems, the number of existing systems using this

refrigerant is large, so technicians will be encountering R-22 for many years to come. R-22 is an HCFC refrigerant, meaning that its compound is made up of hydrogen, chlorine, fluorine, and carbon. HCFC-22 is therefore classified as a hydrochlorofluorocarbon refrigerant. R-22 has an ODP of 0.055 on a scale from 0 to 1 and a GWP of 0.34. R-22 is a chemical compound, so it cannot break down into its component elements. For this reason, R-22 can be added to an air conditioning system in either the vapor or liquid form without fractionating or breaking down into its component parts.

Government regulations set forth by the Environmental Protection Agency under the Montreal Protocol Act mandate that R-22 be phased out over the next 10 years. As of January 1, 2010, it became illegal for equipment manufacturers to use R-22 in newly produced systems and the total phase-out of R-22 is slated for 2020. At the present time, it seems that HFC-410A will be the accepted replacement for R-22 in new equipment, while R-407C is a popular choice for retrofitting existing systems.

R-410A

Over the past few years, R-410A has become the refrigerant of choice for newly manufactured air conditioning systems. Since it became illegal to use R-22 in new equipment, R-410A has become a popular replacement. Also known as Puron and AZ-20, HFC-410A is a hydrofluorocarbon blended refrigerant with an ODP of 0 and a GWP of 0.42. R-410A has its benefits as well as its drawbacks. One of the main drawbacks is that the operating pressures of R-410A systems are about 40% to 70% higher than the operating pressures on equivalent R-22 systems. For this reason, systems that are presently operating with R-22 cannot be switched over or retrofitted to operate with the new refrigerant. These systems will ultimately have to be replaced upon failure. The compressors that are designed to operate with R-410A are constructed differently with thicker shells as well as other mechanical differences. In addition, other system components, such as filter driers and refrigerant flow controls, must be rated at much higher pressures. Evaporator coils, for example, must meet the UL design pressure rating of 235 psig in order to be used with R-410A. It should also be noted that the metering devices specified for use in conjunction with R-22 are not compatible with R-410A. In addition, the alkyl-benzene and mineral oils used in R-22 systems are not compatible with the ester-based lubricants used in R-410A systems.

R-410A is a binary refrigerant blend made up of R-32 and R-125. Since these component refrigerants do not follow the same pressure/temperature relationship, R-410A will evaporate and condense at a number of different temperatures for a given pressure. This range in temperatures is referred to as the temperature glide. Temperature glide typically ranges from 0.2 to 16 degrees, and the temperature glide for R-410A is 0.3. Typically, when dealing with a blended refrigerant, it must be added to an air conditioning system in the liquid form to keep the component refrigerants in a homogeneous mixture. Adding a blended refrigerant to a system in the vapor form will cause the refrigerant to separate into its component refrigerants. Because the temperature glide of R-410A is low, however, it is often disregarded.

The benefits of this new refrigerant include increased system efficiency and zero ozone depletion potential. R-410A exhibits increases in efficiency in the 5% range.

R-407C

Although R-410A is the popular choice for newly manufactured air conditioning equipment, it is not a very good choice when a system is to be retrofitted for use with a new, nonchlorine-containing refrigerant. When the existing system is to remain, for the most part, intact, a good refrigerant option is R-407C. R-407C is a ternary HFC refrigerant blend, made up of R-32, R-125, and R-134a. As with R-410A, R-407C contains no chlorine and is used with ester-based lubricants.

The operating pressures of R-407C are very similar to those of R-22, so the pressure ratings of the existing system components are not a concern. The main cause for concern is the incompatibility between the alkyl-benzene or mineral oils used in R-22 systems and the ester-based lubricants that are required for use with R-407C. Because of this incompatibility, the changeover to R-407C is most commonly performed when an R-22 compressor failure has occurred. This is because most of the refrigerant oil, which is located in the compressor itself, will be removed from the system when the defective compressor is removed.

Although R-407C has a slightly lower efficiency than R-22, it has an ODP of 0 and a GWP of 0.34. Unlike the low-temperature glide of R-410A, R-407C has a large temperature glide, about 10, indicating a high potential for fractionation. Because of this, R-407C should always leave its storage tank as a liquid to ensure that the correct proportion of its component refrigerants is maintained. When evaluating systems that are charged with R-407C, it is important to keep the temperature glide in mind, as there are two temperatures that correspond to a given pressure and vice versa. The pressure/temperature chart for R-407C looks a little different than that of R-22 and R-410A in the sense there are **bubble point** values and **dew point** values (Fig. 2-36).

The bubble point temperature represents the temperature at which the first bubble of vapor will appear in a high-pressure, high-temperature liquid refrigerant. The bubble point temperature is used to calculate and determine the condenser subcooling. The dew point temperature represents the temperature at which the first drop of liquid refrigerant will appear in the low-pressure, low-temperature vapor refrigerant. The dew point is used to calculate and determine evaporator superheat.

THE VAPOR-COMPRESSION REFRIGERATION CYCLE—PUTTING IT TOGETHER

Having discussed the component parts of the basic air conditioning system, we will now put the pieces together and evaluate a sample, properly operating, R-22 air conditioning system. We will examine the system shown in Figure 2-37 and will assume that it is a standard efficiency model.

R-407C							
psig	°F DP/BP	psig	°F DP/BP	psig	°F DP/BP	psig	°F DP/BP
0	-34/--	52	32/--	105	64/54	200	--/92
4	-25/--	54	33/--	110	67/56	205	--/94
8	-17/--	56	35/--	115	69/59	210	--/95
12	-11/--	58	36/--	120	71/61	220	--/98
16	-5/--	60	38/--	125	--/63	230	--/101
20	1/--	62	39/--	130	--/65	240	--/104
24	6/--	64	41/29	135	--/68	250	--/107
28	10/--	66	42/31	140	--/70	260	--/110
30	12/--	68	43/32	145	--/72	275	--/114
32	14/--	70	45/34	150	--/74	290	--/118
34	16/--	72	46/35	155	--/76	305	--/122
36	18/--	74	47/36	160	--/78	320	--/125
38	20/--	76	48/37	165	--/80	335	--/129
40	22/--	78	50/39	170	--/82	350	--/132
42	24/--	80	51/40	175	--/83	365	--/135
44	25/--	85	54/43	180	--/85	380	--/139
46	27/--	90	56/46	185	--/87	400	--/143
48	29/--	95	59/48	190	--/89	420	--/147
50	30/--	100	62/51	195	--/90	440	--/150

Delmar/Cengage Learning.

FIGURE 2-36 Pressure/temperature chart for R-407C.

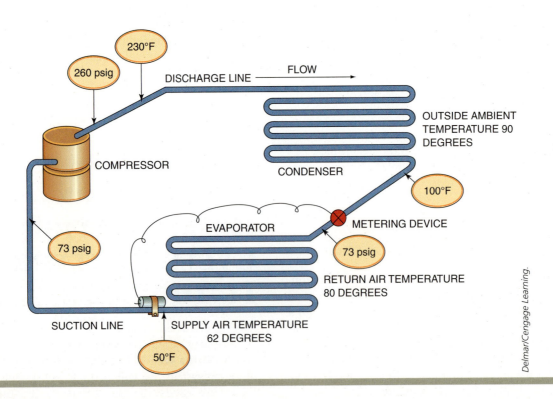

Delmar/Cengage Learning.

FIGURE 2-37 Sample air conditioning system.

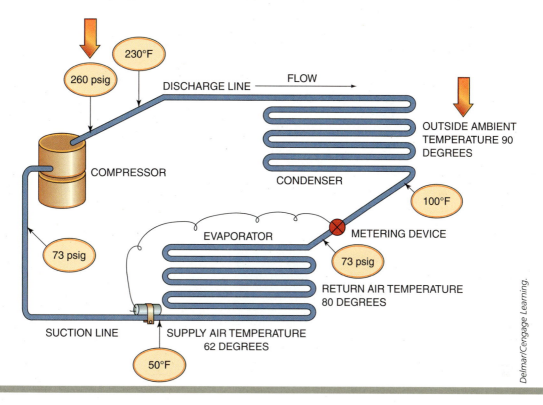

FIGURE 2-38 The outside ambient temperature is 90°F. The compressor discharge temperature of 260 psig corresponds to a condenser saturation temperature of 120°F, which corresponds to the outside ambient temperature plus 30 degrees.

OUTSIDE AIR CONDITIONS

The outside air temperature, referred to as the **outside ambient temperature**, is 90°F (Fig. 2-38), so the condenser saturation temperature should be approximately 120°F. Remember that on standard efficiency systems, the condenser saturation temperature is about 30 higher than the outside ambient temperature (90°F + 30°F = 120°F). From the figure we can see that the high-side pressure in this system is 260 psig, which corresponds to a saturation temperature of 120°F from the pressure/temperature chart for R-22.

> ### from experience...
>
> Higher efficiency systems are designed to operate with lower condenser saturation temperatures than standard efficiency systems. At an outside ambient temperature of 90°F, the condenser saturation temperature might very well be in the range from 110°F to 115°F.

FROM THE COMPRESSOR

The refrigerant leaving the compressor is at a temperature of 230°F (Fig. 2-39). This is well above the 120°F saturation temperature of the refrigerant, as indicated by the high-side pressure of 260 psig. The refrigerant is therefore *superheated*, and must cool down to 120°F before it can begin to condense. As the refrigerant flows from the compressor via the discharge line, it begins to cool, a *sensible heat transfer*. This process was referred to as *desuperheating*.

IN THE CONDENSER

In the first portion of the condenser, the refrigerant is still giving up sensible heat in order to cool down to the saturation temperature of 120°F, where it can begin to condense. The condensing process is a *latent heat transfer*, as the refrigerant is changing state without changing temperature. At this point, the refrigerant is following the *pressure/temperature relationship*.

Once the refrigerant has completely condensed into a liquid, it will continue to give up heat. This is because the liquid refrigerant is at a temperature of 120°F and the outside ambient temperature is 90°F. The direction of natural heat transfer is from a higher temperature

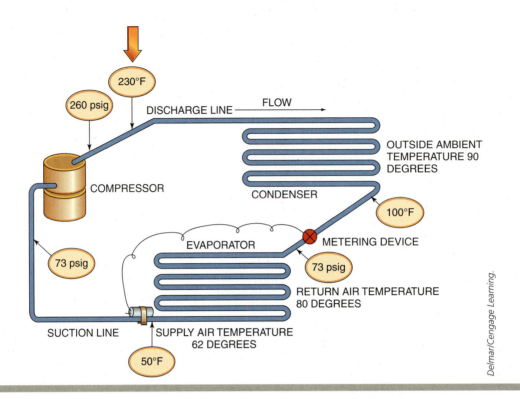

FIGURE 2-39 The temperature of the refrigerant leaving the compressor is 230°F.

substance to a lower temperature substance, so the refrigerant will continue to reject heat to the surrounding air.

The outlet of the condenser is at a temperature of 100°F (Fig. 2-40) indicating that the refrigerant has dropped to a temperature 20 degrees lower than the condenser saturation temperature. This indicates that the condenser is operating with 20 degrees of *subcooling*. Remember that subcooling is equal to the condenser saturation temperature minus the condenser outlet temperature. Since the refrigerant leaves the condenser as a liquid, the line leaving the condenser is called the **liquid line**. Since the refrigerant is now 100% liquid and cooler than the condenser saturation temperature, it no longer follows the pressure/temperature relationship.

> *from experience...*
>
> Only saturated refrigerant follows the pressure/temperature relationships!

THROUGH THE METERING DEVICE

The pressure at the inlet of the metering device is the same as the high pressure of the system which, in this case, is 260 psig. The refrigerant flows through the metering device as a high-temperature, high-pressure liquid and leaves the metering device as a low-temperature, low-pressure saturated liquid. Remember, the refrigerant now follows the pressure/temperature relationship because the refrigerant is *saturated*. The pressure of the refrigerant at the outlet of the metering device is 73 psig (Fig. 2-41), which corresponds to an evaporator saturation temperature of 43°F. This means that the refrigerant is boiling into a vapor at a temperature of 43°F.

INDOOR AIR CONDITIONS

In our example, the temperature of the space to be conditioned is 80°F. This is the temperature of the air that is passing through the evaporator. The air coming from the space is referred to as **return air**. As the air passes through the coil, heat is transferred to the cooler surface of the evaporator. This removes heat from the air, which is returned to the occupied space at a temperature of 62°F (Fig. 2-42). The air being sent back to the space is called **supply air**. The temperature difference between the return air and the supply air is the temperature differential. In this case, the temperature differential is 18 degrees. Under normal operating conditions and humidity (in the 50% range), the temperature differential is expected to be in the 17 to 20 degree range. When the humidity is higher, the temperature differential will be lower. When the humidity is lower, the temperature differential will be higher.

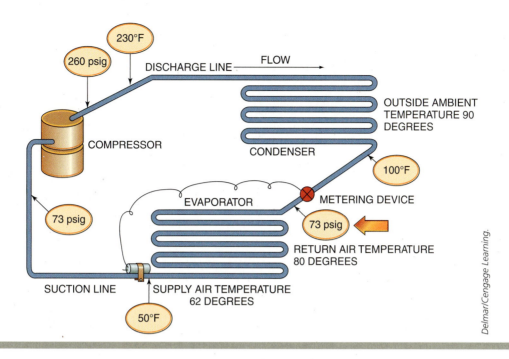

FIGURE 2-40 The temperature of the refrigerant at the outlet of the condenser is 100°F.

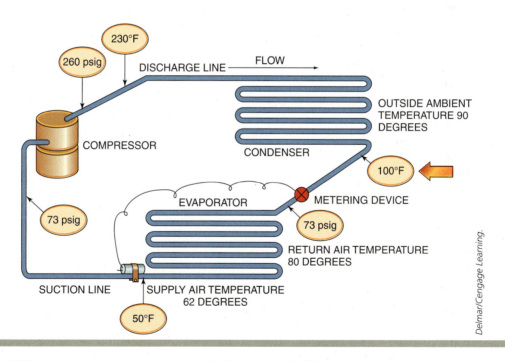

FIGURE 2-41 The low-side or suction pressure in this system is 73 psig.

ABSORBING HEAT IN THE EVAPORATOR

As the cool refrigerant flowing through the evaporator comes in contact with the air from the occupied space, the refrigerant absorbs heat from the air. This addition of heat causes the refrigerant to boil off into a vapor. This is a *latent*

heat transfer as the refrigerant remains at a constant temperature while changing from a liquid to a vapor. At some point toward the outlet of the evaporator, the refrigerant will be 100% vapor at 43°F. Since the vapor is still cooler than the air passing over the coil, the vapor will continue to absorb heat from the air, thereby increasing the temperature of the refrigerant. At the outlet of the evaporator,

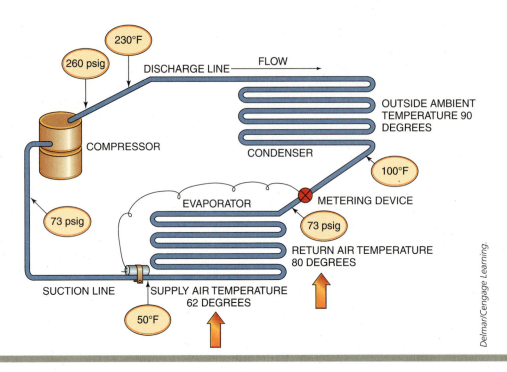

FIGURE 2-42 The air temperature in the occupied space is 80°F. The air being supplied to the space has been cooled 18 degrees to a temperature of 62°F.

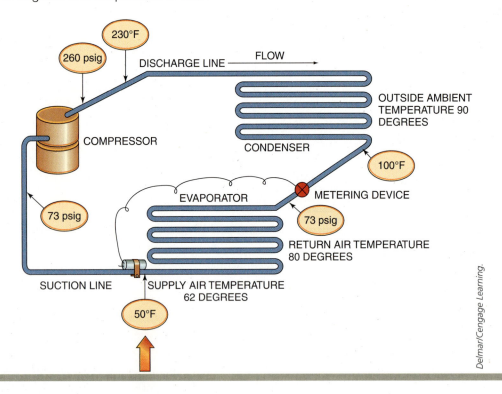

FIGURE 2-43 The temperature of the refrigerant at the outlet of the evaporator is 50°F.

the temperature of the refrigerant is 50°F (Fig. 2-43). This refrigerant is *superheated*. The amount of superheat is the difference between the evaporator saturation temperature and the evaporator outlet temperature. In this case, the superheat in the evaporator is 10 degrees (50°F − 40°F).

ON TO THE COMPRESSOR

Once the refrigerant leaves the evaporator, the refrigerant travels to the compressor as a low-pressure, low-temperature vapor through the suction line (Fig. 2-44). The suction line should be well insulated upon system

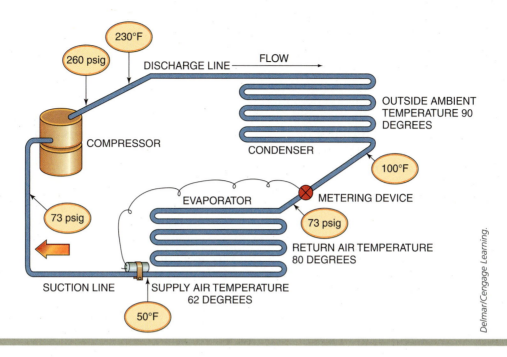

DISCHARGE LINE → FLOW

230°F

260 psig

COMPRESSOR

OUTSIDE AMBIENT TEMPERATURE 90 DEGREES

CONDENSER

100°F

METERING DEVICE

EVAPORATOR

73 psig

73 psig

RETURN AIR TEMPERATURE 80 DEGREES

SUCTION LINE

SUPPLY AIR TEMPERATURE 62 DEGREES

50°F

Delmar/Cengage Learning.

FIGURE 2-44 The suction line connects the evaporator to the inlet of the compressor.

installation to prevent the system from picking up any unwanted heat.

IN THE COMPRESSOR

After entering the compressor, the refrigerant is compressed, increasing the pressure of the refrigerant. During the compression process, the temperature of the refrigerant is increased as well. In this system, the refrigerant enters the compressor as a low-temperature (50°F), low-pressure (73 psig) vapor and is discharged from the compressor as a high-temperature (230°F), high-pressure (260 psig) vapor. The discharge refrigerant from the compressor then flows to the condenser and the cycle repeats itself.

EVAPORATIVE COOLING

The vapor-compression refrigeration cycle is not the only method by which comfort cooling can be obtained. In dry climates, evaporative cooling is a popular alternative. The evaporative cooler, also referred to as a swamp cooler, uses a wetted fiber pad to provide cooling. The unit is located outside, close to the structure and outdoor air is passed over the wet fiber pad by means of a blower. As the water from the pad evaporates, it absorbs heat from the air passing through the cooler. This cooler air is then introduced to the conditioned space. A plumbing line must be run to the unit to provide the water needed to wet the pad. A float valve and a pump maintain the desired water level in the unit and keep the pad wet.

SUMMARY

- The vapor-compression refrigeration cycle is a repeating cycle, consisting of a compressor, a condenser, a metering device, and an evaporator.
- The compressor changes the refrigerant from a low-temperature, low-pressure, superheated vapor to a high-temperature, high-pressure, superheated vapor.
- Three common types of compressors are the rotary, the reciprocating, and the scroll.
- The condenser rejects heat from the system and changes the refrigerant from a high-temperature, high-pressure vapor into a high-temperature, high-pressure liquid.
- Condensers used on residential systems are typically air cooled
- Latent-heat transfers cannot be measured with a thermometer.
- When the temperature of a substance changes, a sensible-heat transfer takes place.
- The metering device controls the flow of refrigerant to the evaporator. It changes high-temperature, high-pressure liquid into a low-temperature, low-pressure liquid.
- Three common metering devices are the capillary tube, the automatic expansion valve, and the thermostatic expansion valve.
- The refrigerant in the evaporator absorbs heat into the refrigeration system.

GREEN CHECKLIST

☐ Larger evaporator coils are used to increase system efficiency by increasing the heat transfer rate from the space to the system.

☐ Larger evaporator coils provide for more effective dehumidification of the occupied space.

☐ Scroll compressors are a popular choice for use on high-efficiency systems.

☐ Larger condenser coils allow an air conditioning system to reject heat more effectively and efficiently.

☐ Larger condenser coils allow for lower head pressures and increased subcooling.

☐ Thermostatic expansion valves are desired for use on high-efficiency air conditioning systems.

☐ R-22 and other HCFC refrigerants are in the process of being phased out.

☐ Popular replacements for HCFC refrigerants include HFC-410A and HFC-407C.

REVIEW QUESTIONS

1. **Ideally, refrigerant enters the compressor as a**
   a. Low-temperature vapor
   b. Low-temperature liquid
   c. High-temperature vapor
   d. Low-pressure liquid

2. **The piping between the compressor and the condenser is called the**
   a. Suction line
   b. Discharge line
   c. Expansion line
   d. Liquid line

3. **Under normal operating conditions, the refrigerant leaving the condenser is a**
   a. Subcooled liquid
   b. Saturated liquid
   c. Superheated vapor
   d. Saturated vapor

4. **Refrigerant enters the metering device as a**
   a. Low-pressure liquid
   b. High-pressure liquid
   c. Low-pressure vapor
   d. High-pressure vapor

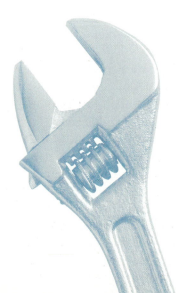

5. **The refrigerant in the evaporator**
   a. Absorbs heat from the air passing over it
   b. Removes moisture from the air passing over the coil
   c. Removes both latent heat and sensible heat from the air passing over the coil
   d. All of the above are correct

6. **The evaporator saturation temperature for an R-22 evaporator operating at 68.5 psig is approximately**
   a. 68.5°F
   b. 40°F
   c. 22°F
   d. 40°C

7. **If the thermal bulb on a TEV comes loose from its mounting at the outlet of the evaporator, which of the following will occur?**
   a. The valve will open, feeding more refrigerant to the evaporator
   b. The valve position will remain unchanged, and the pressure in the bulb will remain the same
   c. The valve will close, starving the evaporator
   d. The suction pressure will drop

8. **Evaporator superheat is defined as**
   a. The evaporator outlet temperature minus the condenser saturation temperature
   b. The compressor outlet temperature minus the evaporator outlet temperature
   c. The evaporator outlet temperature minus the evaporator saturation temperature
   d. The evaporator outlet temperature minus the evaporator saturation pressure

9. **Which of the following is an example of a latent heat transfer?**
   a. Refrigerant changing from a vapor at 50 degrees to a vapor at 200 degrees
   b. Refrigerant changing from a liquid at 90 degrees to a liquid at 80 degrees
   c. Refrigerant changing from a liquid at 40 degrees to a vapor at 40 degrees
   d. All of the above are correct

10. **If the condenser saturation pressure for an R-22 system is 226.4 psig and the condenser outlet temperature is 90 degrees, what is the condenser subcooling?**
    a. 110 degrees
    b. 90 degrees
    c. 20 degrees
    d. 10 degrees

11. **When measuring the condenser subcooling on an R-407C system,**
    a. The low side pressures and temperatures should be used
    b. The bubble point temperature should be used
    c. The dew point temperature should be used
    d. The subcooling is read directly from the gauge manifold

General Safety Practices

OBJECTIVES   *Upon completion of this chapter, the student should be able to*

- Describe acceptable dress for air conditioning system installers and technicians.

- Dress appropriately to work in the field.

- Explain acceptable tool and equipment safety practices.

- Describe the factors that affect the severity of electric shock.

- Properly ground tools and equipment for use in the field.

- Explain and demonstrate proper fire extinguisher use.

- Explain the importance of knowing proper first aid procedures.

- List the agencies that have an effect on safety related issues.

Air conditioning system installers and technicians are faced with a number of possible hazards on a daily basis. Working with open flames, electricity, and pressurized vessels creates the potential for serious injury and equipment damage. Quite often, service technicians work in confined spaces, which increases the chances of sustaining injuries through inhaling fumes and gases from refrigerants, adhesives, acetylene, and other materials commonly used when servicing and installing air conditioning systems. In addition, installers and technicians often work under hot conditions, which leads to perspiration that, in turn, can increase the severity of electric shocks. Technicians must also protect themselves from other potential dangers such as working in excessively cold environments, working on ladders and working in dimly lit attics and basements.

Because of these potential hazards, service personnel should do everything possible to reduce risks. Most accidents result from carelessness. Being aware of the immediate surroundings and potential safety hazards is the best way for individuals to help ensure their safety. Knowing how to use a fire extinguisher and how to administer first aid and cardiopulmonary resuscitation (CPR) are skills that have proven invaluable in times of emergency. This chapter will address a number of safety issues and practices that can help reduce the chances of sustaining injuries while working in the field.

American National Standards Institute (ANSI) Organization that coordinates the voluntary formation of standards that ensures the uniformity of products, processes, and systems

Asphyxiation Loss of consciousness that is caused by a lack of oxygen or excessive carbon dioxide in the blood

Carbon dioxide (CO_2) extinguishers Fire extinguisher that uses vaporizing liquid carbon dioxide to remove heat from fire

Cardiopulmonary resuscitation (CPR) Emergency first aid procedure to maintain circulation of blood to the brain

Class A fire extinguishers Fire extinguishers intended for use on fires that result from burning wood, paper, or other ordinary combustibles

Class B fire extinguishers Fire extinguishers intended for use on fires that involve flammable liquids such as grease, gasoline, or oil

Class C fire extinguishers Fire extinguishers intended for use on electrically energized fires

Class D fire extinguishers Fire extinguisher typically used on flammable metals

Dry chemical extinguisher Fire extinguisher that contains an extinguishing agent and a compressed, nonflammable gas, which is used as the propellant

Frostbite Injury to the skin resulting from prolonged exposure to freezing temperatures

Frostnip The first stage of frostbite

Ground Term used to describe an electrical connection between the casing of equipment or tools and the earth

Ground fault circuit interrupter (GFCI) Electrical device designed to sense small current leaks to ground and de-energize the circuit before injury can result

Halon extinguisher Fire extinguisher that contains a gas that interrupts the chemical reaction when fuel burns

Material Safety Data Sheets (MSDS) Forms that provide storage, transport, and first aid information regarding chemicals used in the field

National Fire Protection Agency® (NFPA) Agency that provides codes, standards, research, training, and education regarding safety and fire-related issues

Occupational Safety and Health Administration (OSHA) Branch of the U.S. Department of Labor that strives to reduce injuries and deaths in the work place

PASS Acronym used to describe fire extinguisher use (Pull, Aim, Squeeze, Sweep)

Service wrench Tool designed to turn square stems on valves

PERSONAL SAFETY

While working in the field, the single most important thing a worker can do is protect his or her body from personal injury. Mechanical equipment and personal property can be replaced if they become damaged, but eyes, fingers, and other body parts cannot. Permanent back injury can result from the improper lifting of equipment. The choice of clothing, equipment, tools, and safety protection items can help ensure that the technician remains safe while working in the field. By addressing safety issues on the job, employers save money and add value to the company. In addition, if the workforce remains safe, the company will operate more efficiently.

PERSONAL BEHAVIOR

When working on a job site, it is important to behave in a safe and professional manner. Running in the workplace can result in slip and fall accidents that might lead to severe personal injury. Quite often, water, oils, and other slippery materials are spilled on the floor and running can increase the chances of slipping. Therefore, all spills should be cleaned up immediately and walking should be the only means of getting from one place to another. Wearing rubber-soled shoes also helps reduce the chance of falling.

Accidents on the job often result from carelessness. It is very important for workers to be aware of their surroundings at all times and to evaluate the immediate area for possible safety hazards. Being aware of potential dangers, including anything that holds back pressure, conducts heat or electricity, is rough or sharp, or can be dropped, will help reduce the chance of an accident occurring.

CLOTHING

Dressing for work is an extremely important first step a technician can take to remain safe even before arriving at the job site. Baggy clothing should not be worn as it can get caught in machinery. Long-sleeved shirts should be worn, tucked into the pants; short-sleeve shirts do not protect the individual from burns or sharp objects. Technicians should never wear short pants when working in the field, even though the outside temperature may make this a very tempting option. Long hair should be tied back and/or placed inside the shirt to prevent it from getting caught in rotating parts or getting burned while soldering or brazing. Quite often, insurance companies will reject claims for injuries that were sustained as a result of improper dress.

JEWELRY

Jewelry, if at all possible, should not be worn while working. Metallic watches, rings, and necklaces are good conductors of electricity and can result in electric shock. If a watch must be worn, make certain that it is made of plastic or other nonconductive material and has a leather or plastic strap.

If necklaces are worn, make certain they are tucked inside the shirt, as they can get caught in a piece of machinery and pull the wearer into the equipment. The best decision, of course, is to remove them altogether. As well as being good conductors of electricity, metal rings can also get caught on nails or other objects that may protrude from a work surface. This can lead to a severed finger should the ring get caught. If an individual does not want to take off a ring, such as a wedding band, a piece of electrical tape wrapped around the ring will reduce the possibility of electric shock.

Large, hanging earrings should also be avoided as they can conduct electricity and get caught in equipment. A good rule of thumb is to avoid wearing anything metal while working. This applies to other items as well, such as wallets with chains and bracelets.

SAFETY GLASSES AND GOGGLES

One of the most important pieces of safety equipment worn on the job are safety glasses and goggles. When installing or servicing air conditioning equipment,

particles can easily become airborne and get lodged in the eyes, resulting in eye injury or blindness. When installing duct systems, for example, metal filings are created when drilling into the metal duct sections. These filings are very sharp and can easily scratch the cornea of the eye. When soldering or brazing copper pipes, the molten solder can drip or splatter, burning the eye. In addition, when working with pressurized gases, an untimely release can result in injury to the eyes. In order to prevent personal injury, the proper eye protection should be selected.

A number of factors must be considered when selecting eye protection. Safety glasses, goggles, and face shields are all intended to provide different degrees of protection and should be selected based on the job that is going to be performed. For example, some safety glasses provide protection only from objects directed toward the front of the glasses, while others provide protection from objects directed toward the side of the glasses as well. The style or type of eye protection that is used must be able to provide sufficient protection from the potential hazards. For example, if the technician is going to be performing tasks such as chipping, grinding, sawing, drilling, or chiseling, there is the danger of becoming injured by flying fragments, chips, filings, sand, or dirt. In this case, the individual should be wearing eye protection that is equipped with side protection (Fig. 3-1). For severe exposure, face shields should be worn.

If a technician is using an acetylene torch, the dangers associated with molten metal and flying particles are faced. In this case, welding goggles are a good choice for eye protection, especially those equipped with tight-fitting eye cups. For grinding or spot welding, flexible fitting goggles are acceptable. This type of goggle has a plastic front

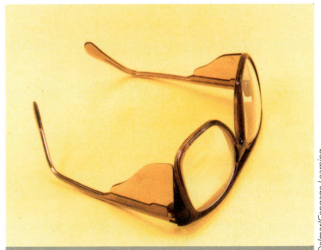

Delmar/Cengage Learning.

FIGURE 3-1 Safety glasses are used to protect the eyes from airborne particulate matter and released gases.

Photo by Eugene Silberstein.

FIGURE 3-2 Goggles that fit close to the face provide more protection.

and a molded rubber side section that fits tight against the face around the eyes. They are held in place by a rubber or elastic strap that fits around the head (Fig. 3-2).

The American National Standards Institute, ANSI, provides detailed information regarding the specific type of eye protection that should be used for various tasks in Standard Z87. You can visit the ANSI Web site at www.ansi.org. ANSI is an organization that coordinates the voluntary formation of standards that ensures the uniformity of products, processes, and systems.

The second factor to consider is comfort. The goggles just described fit tight against the face and provide very good eye protection, but are often uncomfortable and can cause the wearer to sweat as the air inside becomes stagnant. Since many people find them uncomfortable, they are often left inside the toolbox and not worn as they should be. As a compromise, there are more comfortable options that are worn as regular glasses. These glasses often have the side shields but they do not fit tight against the face. Most technicians feel that some degree of protection from airborne particulate matter is better than none, so the eyeglass type of eye protection is very popular.

WORK BOOTS

To protect the feet from falling objects, proper footwear should be worn. Rubber-soled work boots made of heavy leather (Fig. 3-3) provide ample protection from electric shock and from falling objects that are not very heavy. In addition, heavy leather boots also help protect the wearer from sharp objects such as nails. When working around heavy equipment, work boots with steel toes are often desirable as they prevent the toes from getting crushed should a heavy object fall on them. Steel-toed shoes are typically much more expensive than those without steel

Courtesy of Northern Safety Co., Inc., Frankfort, NY.

FIGURE 3-3 Rubber-soled work boots.

toes but they are well worth the extra expense when the added protection they provide is factored in. Under no circumstances should sneakers be worn in the field. Since air conditioning technicians often work with oils and grease, it is a good idea to make certain that the soles of the boots are oil resistant. Also, make certain that the laces are tied snugly and that any loose shoelace ends are not long enough to drag on the floor.

EAR PROTECTION

Damage to the ears can result from working around loud tools and machinery. Even though the tools used to install and service residential air conditioning systems typically do not generate unsafe noise levels, the use of these tools in confined areas magnifies the noise levels generated. When working with tools that can generate potentially dangerous noise levels, field personnel should wear ear protection. The most popular type of ear protection device is the self-expanding ear plug, Figure 3-4. These ear plugs are squeezed into small balls and inserted into the ear. Once inserted, they slowly expand and shape themselves to fit perfectly into the ear.

Other common types of ear protection are premolded earplugs and earmuffs. In order for any type of ear protection to be effective, they must be properly inserted or worn, so it is important to follow the manufacturer's directions carefully. Premolded ear plugs should be fully inserted into the ear. Earmuffs should fit snugly around the ears, making certain that nothing prevents them from fitting tightly against the skin.

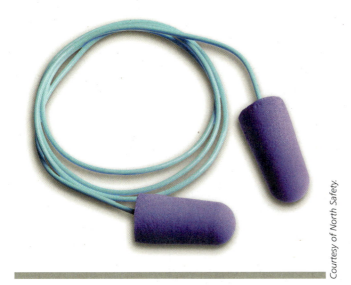

Courtesy of North Safety.

FIGURE 3-4 Expanding ear plugs.

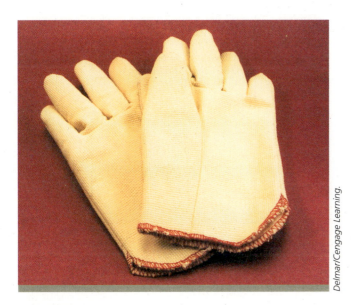

Delmar/Cengage Learning.

FIGURE 3-5 Heavy-duty work gloves.

GLOVES

Heavy duty work gloves (Fig. 3-5) provide protection for the hands under a number of conditions and situations. When soldering or brazing, gloves can provide burn protection from both the torch flame and from the pipes that remain hot for a period of time even after the torch has been extinguished. In addition, the discharge line on air conditioning systems can be in the range of 200°F or more and should never be touched.

When working with sharp objects such as metal duct sections, gloves can help prevent cuts. Also, when lifting objects, gloves help give a better grasp on the object,

reducing the chances of dropping it. Finally, when working with refrigerants, gloves can prevent frostbite that can result when released refrigerant comes in contact with the skin.

LIFTING OBJECTS AND BACK SUPPORT BELTS

Some of the most common injuries sustained by those working in the field are to the back. These injuries result from improper lifting of objects, even though they may not be very heavy. Objects should be lifted with the legs, not the back. Do not bend over with straight legs to lift the object. Always bend at the knees, and once the object is supported, straighten the legs to complete the lift (Fig. 3-6). To provide additional back support, a back brace should be worn (Fig. 3-7).

KNEE PADS

Quite often, service technicians find themselves on their knees, either in an attic or kneeling at a condensing unit. Both installation and service technicians are found in this often painful position, yet many do not utilize knee pads. The main purpose of knee pads is to provide a cushioned surface between the hard surface of ground or attic and the patella, which is the triangular bone that covers the knee joint. Knee pads (Fig 3-8) come in a wide variety of styles and typically have caps made from gels, plastic, or rubber.

ELECTRICAL SAFETY

A great deal of the work done by air conditioning system installers and technicians involves electricity. Approximately 85% of all system problems are electrical in nature. When installing air conditioning systems, line-voltage electrical connections must be made at both the air handler and condensing unit locations. These connections must be made in accordance with all local building and electric codes to ensure the safety of the equipment, the structure, and those in and around it.

Those performing electric work must be well aware of the potential hazards that exist and the precautions that can be taken to reduce their effects. In addition to dressing properly, the proper use of power tools, extension cords, and other electric devices can also reduce the chance of a mishap. Proper grounding and circuit protection helps to further ensure the continued safe operation of the equipment.

ELECTRIC SHOCK

Electric shock occurs when a portion of the body becomes part of an active, or energized, electric circuit. The severity of the electric shock depends on the amount of current

1	2	3	4
APPROACH THE LOAD AND SIZE IT UP AS TO WEIGHT, SIZE, AND SHAPE. CONSIDER YOUR PHYSICAL ABILITY TO HANDLE THE LOAD.	PLACE FEET CLOSE TO THE OBJECT TO BE LIFTED AND 8 TO 12 INCHES APART FOR GOOD BALANCE.	BEND THE KNEES TO THE DEGREE THAT IS COMFORT-ABLE AND GET A HANDHOLD. THEN USING BOTH LEG AND BACK MUSCLES . . .	LIFT THE LOAD STRAIGHT UP, SMOOTHLY AND EVENLY. PUSH WITH YOUR LEGS AND KEEP THE LOAD CLOSE TO YOUR BODY.

5	6	7
LIFT THE OBJECT INTO CARRYING POSITION, MAKING NO TURNING OR TWISTING MOVEMENTS UNTIL THE LIFT IS COMPLETED.	TURN YOUR BODY WITH CHANGES OF FOOT POSITION AFTER LOOKING OVER YOUR PATH OF TRAVEL, MAKING SURE IT IS CLEAR.	SETTING THE LOAD DOWN IS JUST AS IMPORTANT AS PICKING IT UP. USING LEG AND BACK MUSCLES, COMFORTABLY LOWER LOAD BY BENDING YOUR KNEES. WHEN LOAD IS SECURELY POSITIONED, RELEASE YOUR GRIP.

Delmar/Cengage Learning.

FIGURE 3-6 How to lift safely.

flowing in the circuit, the path the current takes through the body, the amount of moisture in the immediate surroundings, the amount of time the shock is received, and the general health of the individual. The amount of current flow depends on the voltage supplied to the circuit and the resistance of the circuit. The resistance of the body is greatly reduced when perspiration forms on the skin, which is common while working in the field. As the resistance of the body decreases, the amount of current flow increases. Very small amounts of current can cause serious injuries or death. Figure 3-9 shows the relationship between electric current and the likely effects these currents will have on the human body. It is important to note that the difference between a current that causes a minor shock and one that can cause death is only 100 mA, or 0.10 Ampere. Refer to the chapter on electricity for more information regarding electrical theory and the relationships among voltage, current, and resistance.

Electric shocks can range in severity from minor discomfort and tingling to electrocution. The best way to avoid receiving an electric shock is to work on circuits that are de-energized. This is not always possible, but wearing rubber-soled shoes, avoiding metallic jewelry, and being aware that the possibility of shock exists will help reduce the risk of receiving a shock. If working on

CAUTION

CAUTION: Never work on electric circuits while standing on a wet floor or when not wearing rubber-soled boots. Shocks received when standing in a wet location are quite often deadly as the current passes through the heart, causing it to stop pumping. Water and electricity do not mix! Stay dry and stay safe.

Courtesy of Ergodyne.

FIGURE 3-7 Brace for the lower back provides extra support.

Photo by Eugene Silberstein.

FIGURE 3-8 Knee pads.

CURRENT IN MILLIAMPERES, mA	LIKELY EFFECTS ON HUMAN BODY
1 mA	Slight tingling sensation.
5 mA	Slight shock. Disturbing but not painful.
6 to 30 mA	Painful shock. Loss of muscle control.
50 to 150 mA	Extreme pain. Respiratory arrest. Sever muscle contractions.
1,000 to 4,300 mA	Muscular contractions and nerve damage. Erratic heart pumping (ventricular fibrillation). Possible death.
10,000 mA	Cardiac arrest, severe burns. Death is likely.

Delmar/Cengage Learning.

FIGURE 3-9 Relationship between electric current and its likely effects on the human body.

from experience...

When working on electric circuits, treat *all* circuits as though they are live, even when you know that they are de-energized. By doing this, good work habits are formed and you will likely prevent a shock should you find yourself working on a live circuit that you thought was de-energized.

CAUTION

CAUTION: When working on electric circuits, it is recommended that the circuit be de-energized at the main distribution panel or disconnect switch and locked with a padlock to prevent anyone from accidentally energizing the circuit while you are working on it.

live circuits is absolutely necessary, be sure to proceed cautiously to avoid becoming part of the circuit.

To reduce the risk of receiving an electric shock, make certain that you do not touch metallic surfaces while working on electric circuits or while using power tools. Should your thumb and fingers of the same hand come in contact with, and become part of, an electric circuit,

the current will pass through the hand without flowing through other parts of the body (Fig. 3-10). However, if one hand is in contact with a metallic surface while the other hand touches a live wire, the current will flow through the body from one hand to the other through the heart (Fig. 3-11). This shock will most likely result in a more severe injury. Similarly, if the worker is standing on a wet location and one or both hands come in

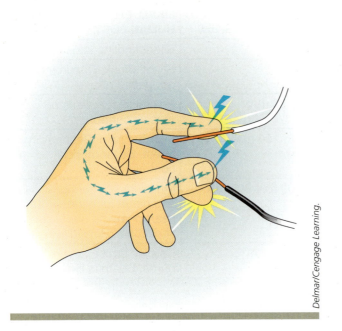

Delmar/Cengage Learning.

FIGURE 3-10 When the thumb and fingers of the same hand become part of an electric circuit, current will flow through only the hand.

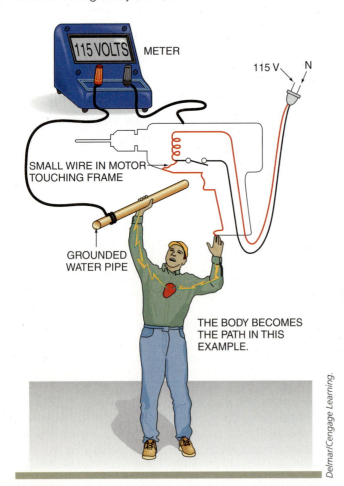

Delmar/Cengage Learning.

FIGURE 3-11 Electric current flows from one hand to the other, flowing through the heart.

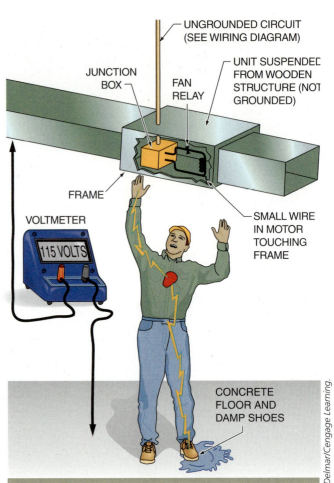

Delmar/Cengage Learning.

FIGURE 3-12 When current flows through the body from one hand to the leg, severe electric shock can result.

contact with a live wire, the current will flow from the hand through the body to the leg. This shock often results in severe injury also as the current flows through the heart (Fig. 3-12).

⚠ **CAUTION**

CAUTION: Should a wire come loose from inside an air conditioning system and come in contact with the casing of the equipment, electric shock can result by simply touching the surface of the unit.

GROUND WIRES

To reduce the risk of receiving an electric shock from equipment or tools, all equipment should be properly grounded. The term **ground** is used to describe an electrical connection between the casing of equipment or tools and the earth. The ground wire can be easily identified because it is typically green in color, but it can also be a bare copper wire (Fig. 3-13). The ground wire should NEVER be disconnected from the equipment.

> **CAUTION**
>
> **CAUTION:** Should a compressor become grounded, the copper pipes that carry refrigerant to and away from the compressor can become *live* and, if touched, can cause a shock. Shocks received in this manner can be quite dangerous and unexpected, since they can be received when the technician is not anywhere near the system control panel or electric service disconnect switch.

The ground wire acts as a bypass if the casing of a piece of equipment comes in contact with a hot or energized power lead. If the grounded equipment is energized, the electric current will flow through the ground wire

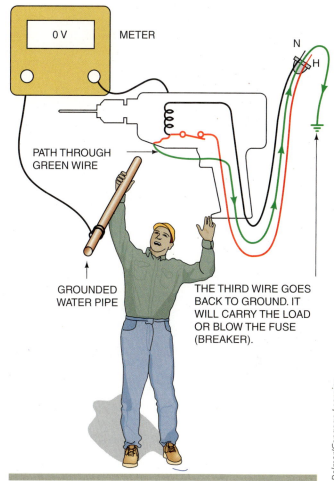

FIGURE 3-14 The ground wire protects from electric shock by allowing current to bypass the body.

and carry the current flow, or trip the circuit breaker or fuse. Since the human body has more resistance than the ground wire, current will take the path of least resistance and flow through the ground wire instead of the body. This reduces the chance of electric shock (Fig. 3-14). The ground wire is never to be used for any purpose other than grounding.

> **CAUTION**
>
> **CAUTION:** Never cut or remove the ground wire from a piece of equipment or tool.

EXTENSION CORDS AND GROUND PRONGS

Power tools and extension cords are often manufactured with three-pronged plugs (Fig. 3-15). Two of these prongs provide power to the tool, while the rounded

FIGURE 3-13 Ground wire in an air conditioning system.

FIGURE 3-15 Three-pronged plug.

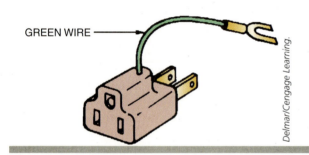

FIGURE 3-16 Adapter used to connect a three-pronged plug to a receptacle that does not have a ground connection.

prong at the bottom of the plug is the ground prong. This prong should never be cut off as doing so will eliminate the protection that the ground is designed to provide. In the event that a receptacle or outlet has only two slots, an approved adapter must be used. The adapter is manufactured with a green wire that extends from the device that must be screwed to a ground for the circuit to be protected (Fig. 3-16). Extension cords that are used on the jobsite in conjunction with power tools are subject to a great deal of wear and tear. Damage to an extension cord may not be immediately visible but insulation damage can lead to electric shock. Proper grounding will not protect the user from shock in the event of a damaged extension cord, so a ground fault circuit interrupter is needed.

GROUND FAULT CIRCUIT INTERRUPTERS

An additional safety often used to protect the technician is the **ground fault circuit interrupter** (GFCI) or simply GFI. The GFI (Fig. 3-17) is designed to sense small current leaks to ground and de-energize the circuit before injury can result. In residential and commercial installations as well, electric codes require the installation of a GFI on any circuit that is located close to a wet location. For example, outlets in bathrooms are required to

FIGURE 3-17 Ground fault circuit interrupter.

have GFI protection. When working in the field, power tools should be protected by a GFI to reduce the risk of electric shock.

LOCKOUT/TAGOUT

When inspecting, repairing, or servicing electrical equipment, the power supply to the equipment should be de-energized to reduce the risk of electric shock. The Occupational Safety and Health Administration, OSHA, standards require the lockout/tagout of equipment prior to servicing. This, of course assumes that the equipment does not need to be energized for servicing.

Lockout devices are devices that, when in place, prevent the accidental energizing of equipment (Fig. 3-18). The device is secured to the switch on a power source and locked in place so it is impossible to energize the circuits being worked on. Some lockout devices provide for multiple users, each with his/her own lock, when more than one individual is working on the system. Tagouts are clearly visible warning devices, such as tags, that are securely fastened to a device or piece of equipment

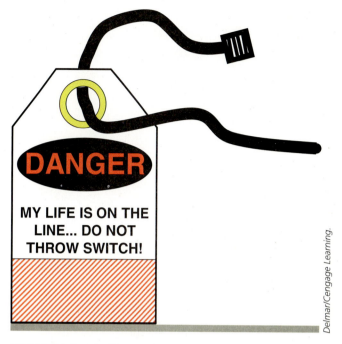

FIGURE 3-19 Tagout device.

(Fig. 3-19). The tags indicate that the equipment may not be operated until the tagout device is removed. The ties that are used to secure the tag in place must be self-locking and nonreusable.

FIRE SAFETY

To ensure the safety of the technician, the home owner's property and equipment, and those around it, proper precautions must be taken to reduce the effects of a fire should one start. A fire extinguisher (Fig. 3-20) should be carried on the truck and brought along as a tool to the work location. Running to the truck to retrieve a fire extinguisher once a fire has started will cost valuable time and can result in severe property damage or personal injury.

There are several different types of fire extinguishers based on the fires they are intended to combat. For instance, one would not use water to extinguish an electrical fire as water is an excellent conductor of electricity. Fire extinguishers are rated as class A, class B, class C, and class D, based on the method and materials used to extinguish the fire.

CLASS A FIRE EXTINGUISHERS

Class A fire extinguishers are designed to be used on fires that result from burning wood, paper, or other ordinary combustibles. Extinguishers used solely for class A fires contain water and compressed gas. Such extinguishers

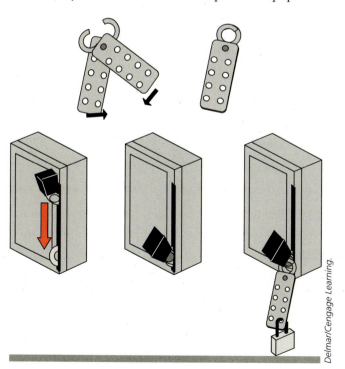

FIGURE 3-18 Lockout devices.

Courtesy of Kidde Safety Products.

FIGURE 3-20 Fire extinguisher.

Delmar/Cengage Learning.

FIGURE 3-21 Class A fire extinguishers can be labeled with a picture of burning combustibles.

are labeled with either a picture of burning combustibles, or a green triangle in which the letter A is clearly visible (Fig. 3-21).

CLASS B FIRE EXTINGUISHERS

Class B fire extinguishers are designed for use on fires that involve flammable liquids such as grease, gasoline, or oil. Extinguishers that can be used on class B fires are labeled with either a picture of burning fuel, or the letter B inside a red square (Fig. 3-22).

Extinguishers used on class B fires can be of the dry chemical, halon, or carbon dioxide type. **Dry chemical extinguishers** contain an extinguishing agent and a compressed, nonflammable gas, which is used as the propellant. **Halon extinguishers** contain a gas that interrupts

Delmar/Cengage Learning.

FIGURE 3-22 Class B fire extinguishers can be labeled with a picture of burning fuel.

the chemical reaction when fuel burns. In **carbon dioxide (CO_2) extinguishers**, the CO_2 is stored as a compressed liquid in the extinguisher. As it is released from the cylinder of the extinguisher, it expands, vaporizes, and cools the air surrounding the fire.

CLASS C FIRE EXTINGUISHERS

Class C fire extinguishers are designed for use on electrically energized fires and contain an extinguishing agent that is nonconductive to prevent the fire from spreading. These extinguishers are labeled with either a picture of a burning outlet, or the letter C inside a blue circle (Fig. 3-23). Carbon dioxide extinguishers are typically the best choice for class C fires.

CLASS D FIRE EXTINGUISHERS

Class D fire extinguishers are typically used on flammable metals and are designed for specific metals and materials. They are labeled with the letter D inside a yellow star, (Fig. 3-24). This type of extinguisher is typically not found in the residential air conditioning setting.

Delmar/Cengage Learning.

FIGURE 3-23 Class C fire extinguishers can be labeled with a picture of a burning receptacle.

Delmar/Cengage Learning.

FIGURE 3-24 Label for class D fire extinguishers.

MULTIPLE PURPOSE FIRE EXTINGUISHERS

Many extinguishers used by air conditioning technicians and installers are rated for multiple uses and can be used on different types of fires. For example, if a fire extinguisher can be used on class A and class B fires, it will be labeled with both the green triangle and the red square. Some extinguishers are designed to be used on A-B fires, B-C fires, or A-B-C fires. The label in Figure 3-25 indicates that the fire extinguisher can be used on class A, class B, and class C fires, and the label in Figure 3-26 indicates that the extinguisher can be used on class A and class B fires, but not class C fires. Since air conditioning installers or technicians typically work around ordinary combustibles as well as electricity and fuel products, the multi-purpose fire extinguisher is the way to go.

FIRE EXTINGUISHER USE

Although there are several different types of fire extinguishers, they all work in more or less the same way. The acronym **PASS** is associated with fire extinguisher operation and use.

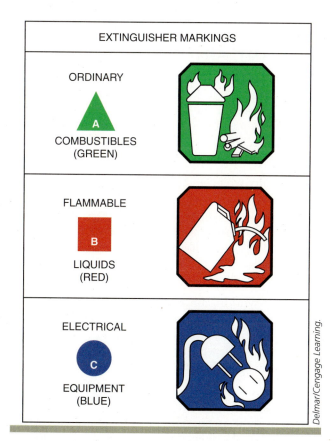

FIGURE 3-25 This label indicates that the fire extinguisher can be used on class A, B, or C fires.

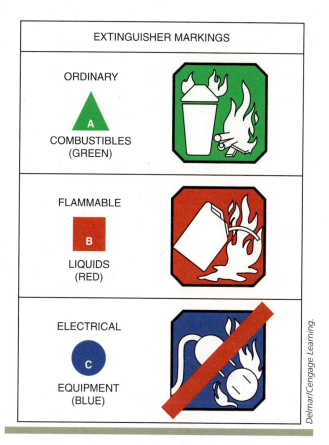

FIGURE 3-26 This label indicates that the fire extinguisher can be used on class A or class B fires.

P—Pull the pin (Fig. 3-27) at the top of the extinguisher to allow the handle to be squeezed

A—Aim the nozzle of the fire extinguisher at the base of the fire

S—Squeeze the handle to discharge the extinguisher from a distance of 6 to 8 feet from the fire

S—Sweep the nozzle back and forth at the base of the fire until it is put out

FIRE EXTINGUISHER MAINTENANCE

Do not wait until there is a fire to look for or inspect the extinguisher. Fire extinguishers should be inspected periodically to ensure that the indicator arrow is in the green, or safe, region of the dial (Fig. 3-28). If the extinguisher has been discharged, it should be recharged immediately.

Fire extinguishers should be stored in an easily accessible location and should never be locked away. Trying to find a key to unlock the fire extinguisher will cost valuable time and can result in property damage or personal injury.

PIN

Courtesy of Kidde Safety Products.

FIGURE 3-27 Pin on a fire extinguisher.

Courtesy of Kidde Safety Products.

FIGURE 3-28 Arrow pointing to the safe range on a fire extinguisher.

TOOL, EQUIPMENT, AND MATERIAL SAFETY

To ensure that individuals remain safe while working in the field, care must be taken in working with tools and around equipment. Improper tool usage can result

in personal injury. Tools are rated for certain jobs and have specific limitations. Screwdrivers, for example, should not be used as chisels, and wrenches should not be used as hammers. Exceeding the limitations of the tool can result in tool failure and, more importantly, severe injury.

Here are some general safety tips regarding tool use:

- Keep all tools in good condition.
- Replace all tools with fractured or damaged handles.
- Make certain the tool handle fits securely into the head of the tool.
- Make certain all tools have the correct handles.
- Use proper personal protection equipment when using tools.
- Make certain that all tool cutting edges are kept sharp to prevent binding.
- Keep all tools clean and free from oils.
- Make certain all guards and safeties are in place and operational
- Use the correct size and type tool for the job being performed.

WRENCHES

When using a wrench, it is important to pull on the wrench, not push it. Pushing a wrench while exerting excessive force on the tool can result in the tool slipping, throwing the technician off balance. This can cause the technician to fall into the work and onto a sharp object.

Specialty wrenches are designed to perform specific tasks and should be used accordingly. The acetylene tank, used in the soldering and brazing processes, is opened and closed by turning a square-shaped stem on the tank (Fig. 3-29) with a service wrench (Fig. 3-30). The **service wrench** is designed to turn square stems on tanks and refrigeration valves.

When working on piping systems, properly sized pipe wrenches (Fig. 3-31) should be used. Using a wrench that is too small can result in having to exert excessive force on the tool, which can result in injury should the wrench slip from the pipe.

When using an adjustable, or crescent, wrench (Fig. 3-32), be sure to tighten the wrench around the object as tight as possible to prevent slippage. Once again, pull on the wrench, do not push it. It is important to tighten nuts until snug but overtightening the nuts can result in breakage and injury.

FIGURE 3-29 Square stem on the acetylene tank valve.

FIGURE 3-30 Refrigeration service wrench.

SCREWDRIVERS

Screwdrivers (Fig. 3-33) are intended to be used to, well, drive screws. They are not to be used as chisels, as crowbars, as pry bars, or for any other nonscrew task. Using the screwdriver as a chisel can result in breaking the tool, causing shards of metal to fly into the air. Commonly used screwdrivers are the straight slot screwdriver (Fig. 3-34) and the Phillips screwdriver (Fig. 3-35). Straight slot screwdrivers should not be used to tighten or loosen Phillips-head screws.

FIGURE 3-31 Pipe wrenches.

FIGURE 3-32 Adjustable wench.

FIGURE 3-33 Assorted screwdrivers.

FIGURE 3-34 Straight slot screwdriver.

FIGURE 3-35 Phillips head screwdriver.

Courtesy of Louisville Ladder.

FIGURE 3-36 Fiberglass and wood ladders.

LADDERS

It is recommended that ladders made of nonconducting material be used when working in the field. Commonly used materials in nonconducting ladders are fiberglass and wood (Fig. 3-36). Although wood ladders are heavier than metal ladders, they perform exactly the same functions and can protect the technician in the event the ladder comes in contact with a live electric wire.

Avoid standing on the top rung of a ladder (Fig. 3-37). If a technician is unable to reach the desired height without standing on the top rung, a longer ladder should be used. Avoid working overhead whenever possible; falling objects can lead to injury. Reaching while on a ladder is also not desirable as the worker can lose his balance and fall. Make sure the ladder is used on a level surface to help prevent it from falling over.

> ### CAUTION
>
> **CAUTION:** Never leave tools or other materials on the top platform of the ladder. If the ladder is moved by another individual the object can fall, causing injury.

INCORRECT

Delmar/Cengage Learning.

FIGURE 3-37 Standing on the top rung of the ladder can cause an individual to lose his balance.

FIGURE 3-38 Metal spreaders must be in place and not damaged.

Here are some important things to keep in mind when using ladders:

- Never use buckets, boxes, chairs, or other items in place of a ladder!
- Be careful when carrying ladders. You don't want to injure others or damage property.
- Do not use damaged ladders.
- Make certain that the metal spreaders/locking devices are in place and not damaged (Fig. 3-38).
- Keep ladders free from oil, grease, and other substances that can cause slipping.
- Observe the weight limitations of the ladder.
- Make certain that ladders are only used on level surfaces.
- Never move or adjust a ladder while someone is on it.
- Never use the top of the ladder as a step.
- Always face the ladder! This applies to going up as well as going down.
- Do not carry items that can cause you to lose your balance.
- Always have at least one free hand to hold onto the ladder.

SOLDERING AND BRAZING EQUIPMENT

When using acetylene torches to solder or braze refrigerant lines, make certain there is ample ventilation. Raw acetylene and the fumes from the torch can cause respiratory soreness, dizziness, and, depending on the amount of vapor inhaled, unconsciousness.

When using the torch, secure the tank in the vertical position to prevent it from tipping over. Should the tank fall over, the valve could break off, creating a potentially explosive situation.

CAUTION

CAUTION: Never direct the torch flame toward the tank!

CAUTION

CAUTION: Always leak-check the hoses, regulators, and connections on the torch kit before lighting the torch. Refer to the chapter on soldering and brazing for more on acetylene tank safety.

PRESSURIZED GAS TANKS AND CYLINDERS

Nitrogen is often used to pressurize air conditioning systems for leak-checking purposes. Nitrogen, as well as other gases including refrigerant, is often supplied in 125-pound cylinders containing high-pressure fluid. The cylinders (Fig. 3-39) are designed with screw-on caps

FIGURE 3-39 Nitrogen tank.

that protect the valve arrangement at the top of the cylinder. They should not be moved unless the cap is screwed securely in place. Large tanks should be chained securely to a cart to be rolled from place to place.

CAUTION: Refrigerant and nitrogen tanks should be stored and transported in the upright position.

CAUTION: If the tank is equipped with a protective cap, do not move the tank unless the cap is in place.

CHEMICAL AND MATERIAL SAFETY

The air conditioning and refrigeration technician often works with a number of chemical solutions, including cleaners, adhesives, joint compounds, soldering flux, and oils. When handling such chemicals, follow the manufacturer's directions for use and heed the warnings on the product labels (Fig. 3-40). Many of these chemicals can cause skin irritation or burns if they come in contact with the bare skin. The vapors from these chemicals can also

FIGURE 3-40 Chemical containers are labeled with safety and usage directions.

cause respiratory inflammation, unconsciousness, irregular heartbeats, or death.

When purchasing chemicals, obtain **Material Safety Data Sheets** (MSDS) (Fig. 3-41) for each item used. These forms provide chemical information about the item, as

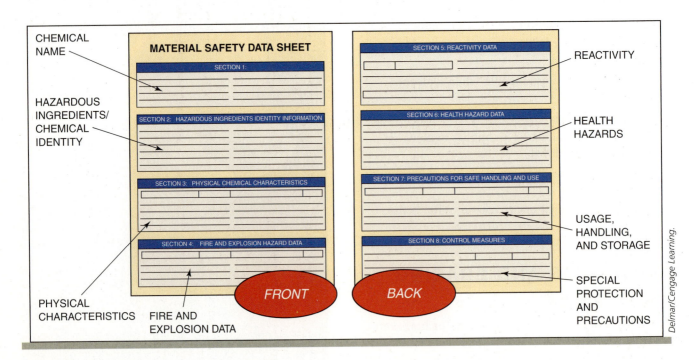

FIGURE 3-41 Material safety data sheets.

CAUTION

CAUTION: Always follow the manufacturer's recommendations and instructions when using chemicals of any sort.

well as information regarding safe storage, transporting, and first aid should the product be ingested, spilled, or permitted to contact the skin.

FIRST AID

No matter how careful service technicians are, there are bound to be accidents and mishaps that require immediate medical attention. Although not all injuries require a visit to the doctor or hospital, some treatment is necessary to prevent further injury or infection. All service vehicles should be equipped with a first aid kit (Fig. 3-42) that contains the basic medical supplies, such as burn cream, bandages, alcohol pads, eye wash, eye pads, tweezers, antiseptic spray, gauze bandages, and CPR shields. Note that the following sections are not intended to provide medical advice, but to provide basic information regarding immediate treatment for a number of situations commonly encountered in the field.

Courtesy of Custom Kits Company, Inc.

FIGURE 3-42 Typical first aid kit.

FROSTBITE

Air conditioning technicians work around refrigerants on a daily basis. These refrigerants, when released to the atmosphere, vaporize at temperatures well below 0°F—some as low as −40°F. Should refrigerant come in contact with the skin, frostbite can result. Frostbite, which can occur any time there is prolonged exposure to freezing temperatures, occurs in three stages. The first stage, called **frostnip**, causes whitening of the skin, itching, tingling, and loss of feeling. The second stage, called **frostbite**, results in the skin turning purple and blisters forming on the skin. The third stage is very rare. It results in gangrene and requires amputation of the affected area.

CAUTION

CAUTION: Exposure to released refrigerant can result in immediate frostbite.

Frostbite can best be treated by covering the area with something warm and dry and then obtaining professional medical attention. Never rub, massage, poke, or squeeze the affected area as this can result in tissue damage. A warm bottle of water can be placed gently against the affected area to warm it slightly.

BLEEDING

In the event that a cut results in bleeding, place a clean folded cloth over the area and apply firm pressure. If blood soaks through the cloth, do not remove it. Simply cover the cloth with another and continue to apply pressure until the bleeding stops. If at all possible, elevate the cut area to a level above the heart to help stop the bleeding. If the cut is relatively small, the injury can be washed with soap and warm water and then bandaged.

ASPHYXIATION

Asphyxiation is loss of consciousness caused by a lack of oxygen or excessive carbon dioxide in the blood. An oxygen level below 19% may result in unconsciousness. As a result of electric shock or inhalation of refrigerant, the victim may stop breathing. When a victim's respiratory system fails, the flow of oxygen through the body may stop within a matter of minutes. If the victim stops breathing, **cardiopulmonary resuscitation (CPR)** should be administered.

CAUTION

CAUTION: Only properly trained individuals should administer CPR. Do not attempt CPR if you have not received professional training.

CHEMICAL BURNS

In the event of a chemical burn, remove the clothing on or near the burn area. Under no circumstances should clothing be pulled-off over the head, as residual chemical may enter the eyes. If necessary, cut the clothing from the body. Wash the affected area with low-pressure water for at least twenty minutes and apply a clean bandage to the wound. Seek medical attention immediately.

ELECTRIC SHOCK

The most immediate response to an electric shock may be the most dangerous. DO NOT TOUCH THE SHOCK VICTIM! First disconnect the power at the main disconnect switch to ensure that the circuit is not energized. Once the power has been turned off, use a nonconducting object, such as a wooden stick, to move the wires from the victim or, if possible, move the victim from the area.

If the victim is conscious, place a pillow or rolled jacket under his head and await medical attention. If the victim is not breathing, CPR should be administered by someone who has been properly trained.

AGENCIES

Many agencies, both governmental and private, provide input and/or act as regulatory agencies to ensure that work environments, tools, and equipment are maintained and designed to function under safe conditions. These agencies set and enforce standards that help ensure the safety of workers as well as their surroundings.

OCCUPATIONAL SAFETY AND HEALTH ADMINISTRATION

OSHA was formed as a result of the Occupational Safety and Health Act of 1970 and is responsible for protecting the health of America's workers. OSHA, a government agency, has more than 200 offices nationwide and employs engineers, physicians, educators, and other professionals to ultimately save lives and prevent personal injury. OSHA identifies potentially dangerous situations in the workplace by conducting inspections.

The top priority of OSHA is to investigate reports of immediate danger. When an accident occurs on a job site that results in the hospitalization of at least three workers, an OSHA investigation is conducted. In addition, OSHA will investigate safety-related complaints from employees of a particular company. Based on the agency's findings, fines are levied against the employer and typically range from $0 to $70,000 based on the accident potential of the violation.

NATIONAL FIRE PROTECTION AGENCY

The **National Fire Protection Agency® (NFPA)** is responsible for reducing the burdens of fire and other hazards by providing codes, standards, research, training, and education regarding safety-related issues. The NFPA® is an international agency that has more than 75,000 members and more than 300 codes and standards that influence building processes, designs, and installations. Some of the codes that have the most immediate effect on air conditioning and refrigeration personnel are the Fire Protection Code, the National Fuel Gas Code, the *National Electrical Code®*, and the Building Construction and Safety Code.

AMERICAN NATIONAL STANDARDS INSTITUTE

The **American National Standards Institute (ANSI)** is an organization that coordinates the voluntary formation of standards to ensure the uniformity of products, processes, and systems. ANSI provides a neutral forum for the development of these standards. It also accredits organizations that, in turn, draft American National Standards, which are accepted and followed by the participating companies and businesses, including manufacturers, professional societies, and trade associations.

SUMMARY

- Most accidents result from carelessness.
- Always be aware of your surroundings and potential hazards.
- Dress properly for work wearing long pants, long-sleeved shirts, and work boots.
- Remove metallic jewelry, as it is a good conductor of heat and electricity.
- Safety glasses, ear plugs, and gloves provide additional protection from dangerous conditions on the job site.
- Power tools and equipment should be grounded to protect against electric shock.
- Electric shock occurs when the body becomes part of an electric circuit.
- Always de-energize electric circuits before working on them.
- Ground wires and prongs should never be cut or disconnected.
- The GFI de-energizes a circuit when a current leak to ground is sensed.
- Fire extinguishers are classified by the types of fires they are designed to be used on.
- Fire extinguisher use: Pull, Aim, Squeeze, Sweep (PASS).
- Always use tools for the tasks they are intended to perform.
- Handle and use chemicals according to the manufacturer's directions.
- Be prepared for injuries on the job and have a first aid kit handy.
- OSHA, NFPA®, and ANSI help ensure safety in the work place.

REVIEW QUESTIONS

1. **Water on the floor of the work area can**
   a. Increase the chance of slipping
   b. Increase the chance of receiving an electric shock
   c. Both a and b are correct
   d. Neither a nor b is correct

2. **Injuries on the job are often the result of**
   a. Electric shock
   b. Carelessness
   c. Slips and falls
   d. Cuts and burns

3. **Why should metal rings not be worn while working on air conditioning equipment?**
   a. They are poor conductors of electricity
   b. They are poor conductors of heat
   c. They can get caught on nails or other objects protruding from the work surface
   d. All of the above are correct

4. **Safety glasses protect the eyes from**
   a. Airborne particulate matter
   b. Splattering solder
   c. Pressurized gases
   d. All of the above are correct

5. **Rubber soled, steel-toed work boots help prevent injuries involving**
   a. Falling objects
   b. Electric shock
   c. Bitter cold weather
   d. Both a and b are correct

6. **Noise levels can be increased by**
   a. Working in confined areas
   b. Wearing ear plugs
   c. Using lower speeds on power tools
   d. All of the above are correct

7. **When lifting heavy objects from the floor**
   a. Lift with the back, not the legs
   b. Lift with straight legs and a bent back
   c. Lift with the legs, not the back
   d. Lift with straight legs and bent arms

8. **An electric shock occurs when**
   a. The body becomes part of an energized electric circuit
   b. The electric current bypasses the body and flows through the ground wire
   c. There is no current flow in the circuit
   d. None of the above is correct

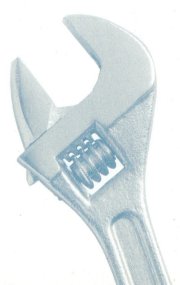

9. The severity of an electric shock is increased by
a. Increasing the voltage supplied to the circuit
b. Reducing the resistance of the body
c. Increasing the amount of current flow in the circuit
d. All of the above

10. The ground prong on an extension cord can be cut off when
a. The outlet to be used does not have a ground connection
b. The tool is double insulated
c. The extension cord is too long
d. The ground prong should never be cut off

11. A fire that was started as a result of a match being thrown into a garbage pail is classified as a
a. Class A fire
b. Class B fire
c. Class C fire
d. Class D fire

12. Burning fuel can be extinguished with a
a. Halon extinguisher
b. Carbon dioxide extinguisher
c. Dry chemical extinguisher
d. All of the above are correct

13. When referring to fire extinguisher use, the acronym PASS stands for
a. Push, Aim, Stand, Squeeze
b. Pull, Aim, Stand, Squeeze
c. Pull, Aim, Squeeze, Sweep
d. Pin, Aim, Sweep, Squeeze

14. Which of the following tools should be used if a chisel is not available?
a. Wrench
b. Screwdriver
c. Pry bar
d. None of the above

15. Which agency investigates employee complaints regarding unsafe conditions on the job site?
a. OSHA
b. NFPA®
c. EPA
d. ANSI

KNOW YOUR CODES

The three agencies just discussed, namely OSHA, NFPA®, and ANSI are responsible for maintaining safety and uniformity in our industry as well as in many others. It is important to be aware of the rules, guidelines, and regulations that have been set forth by them. Safety is everyone's concern and involves not only personal safety, but the safety and protection of equipment, property, and the environment. You can learn more about these agencies, what they do, and the codes and guidelines they have implemented at their Web sites:

OSHA: www.osha.gov

NFPA®: www.nfpa.org

ANSI: www.ansi.org

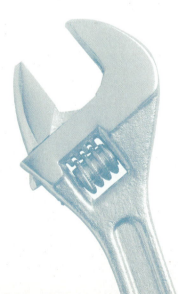

WHAT'S WRONG WITH THIS PICTURE?

Carefully study FIGURE 3-43 and think about what is wrong. Consider all possibilities.

 WRONG

Photo by Eugene Silberstein.

FIGURE 3-43 When transporting tanks of pressurized gas, such as acetylene, they need to be secured in an upright position. Unsecured tanks can shift around the back of the truck and become damaged. In addition, torch kits must be removed from the tank for transport.

 **RIGHT**

Photo by Eugene Silberstein.

FIGURE 3-44 In this photo we can see that the tank is properly secured in an upright position. Also, the torch kit has been removed from the tank. Think safe and be safe!

Refrigerant Management

OBJECTIVES *Upon completion of this chapter, the student should be able to*

- List the desired properties for refrigerants.
- Explain the characteristics of hydrocarbons.
- Explain how a hydrocarbon becomes halogenated.
- Explain the chemical elements that make up CFC, HC, HCFC, and HFC refrigerants.
- Differentiate between azeotropic, near-azeotropic, and zeotropic refrigerant blends.
- Explain how ozone protects the earth.
- Explain the difference between good and bad ozone.
- Explain how ozone is destroyed.
- Explain the concept of global warming.
- Interpret ozone depletion potentials (ODP) and global warming potentials (GWP).
- Determine the proper oil to use with various types of refrigerant.
- Differentiate between reusable and disposable refrigerant cylinders.
- Obey laws regarding certification and refrigerant handling.
- Differentiate between Type I, Type II, Type III, and universal certification.
- Differentiate between refrigerant recovering, recycling, and reclaiming.
- Demonstrate passive and active recovery methods.
- Demonstrate methods to recycle refrigerant.

To maintain a safe work environment, the field technician must be aware of the various refrigerants that are used in the industry and be able to work with them according to the guidelines and laws that are in place. There are a large number of refrigerant types, such as chlorofluorocarbons, (CFCs), hydrochlorofluorocarbons (HCFCs) hydrofluorocarbons (HFCs), and hydrocarbons (HCs) that technicians and air conditioning installers must deal with on a daily basis. Refrigerants can be of the single compound variety or can be blended. Blended refrigerants are formed by mixing two or more single compound refrigerants together in specific proportions to form new refrigerants.

In an effort to protect the ozone layer and the environment as a whole, Federal laws have been put into effect that govern the safe handling, storage, and transporting of refrigerants as well as the certification of field technicians. Refrigerants cannot be released to the atmosphere! Refrigerants must be recovered, which is the process of removing the refrigerant from the system and storing it in an external container. Once recovered, the refrigerant can be recycled; that is, filtered and replaced in the system. Heavily contaminated refrigerants must be reclaimed; that is, sent to a chemical plant where the refrigerant is reprocessed to meet new refrigerant standards. In this chapter, we will discuss the various types of refrigerants available, as well as their chemical make-ups and the legal aspects of safe handling, transport, usage, recovery, and storage.

Active recovery recovery method that uses a self-contained recovery unit

Azeotropes blended refrigerants that behave as a single refrigerant with one saturation temperature for a given pressure

Blended refrigerants refrigerants that are created by mixing two or more other refrigerants

Ethane a hydrocarbon made up of two carbon atoms surrounded by six hydrogen atoms

Fractionation the process by which blended refrigerants separate into their component refrigerants

Fully halogenated a hydrocarbon-based refrigerant that has all of its hydrogen atoms replaced with chlorine or fluorine atoms

Global warming the result when the atmosphere traps the heat radiated from the earth

Global warming potential (GWP) the index used to measure a chemical substance's effect on global warming as compared with the effects of carbon dioxide

Greenhouse effect see global warming

GLOSSARY OF TERMS (CONT'D)

Halogen chemical elements such as chlorine and fluorine that replace hydrogen atoms to create halogenated hydrocarbons. Other halogens include iodine and bromine

Hydrocarbons molecules that are comprised of only hydrogen and carbon atoms; methane and ethane are examples of hydrocarbons

Low loss fittings fittings attached to refrigerant hoses to prevent the release of refrigerant from them when removed from a system

Methane a hydrocarbon made up of one carbon atom surrounded by four hydrogen atoms

Montreal Protocol Act 1987 legislation resulting from a meeting of 23 countries that jump started the program that slowed the production of ozone depleting substances

Near-azeotropes blended refrigerants that behave similar to azeotropes, but that operate with temperature glides and are subject to fractionation

Ozone depletion potential (ODP) the index used to measure a chemical substance's effect on ozone depletion as compared to the effects of CFC-11; the ODP ranges from 0 to 1

Ozone layer refers to the layer of good ozone that exists in the stratosphere, which is 7 to 30 miles above the Earth's surface

Partially halogenated a hydrocarbon based refrigerant that has some of its hydrogen atoms replaced with chlorine or fluorine atoms

Passive recovery recovery method that uses the system compressor to remove refrigerant from the system

Personal protection equipment (PPE) any equipment that will provide protection from potential injury

Stratosphere the atmospheric shell around the earth that is between 7 and 30 miles above the surface of the earth

Temperature glide the range of temperatures in which a blended refrigerant will vaporize or condense at a given pressure

Troposphere the atmospheric shell around the earth that is between 0 and 7 miles above the surface of the earth

Zeotropes blended refrigerants that behave similar to azeotropes, but operate with temperature glides and are subject to fractionation; these refrigerants experience larger temperature glides than near azeotropic refrigerant blends

REFRIGERANT TYPES

Just as there are many different cooling applications, there are many different refrigerants available in the industry, because it would be impossible for one refrigerant to satisfy the needs of all systems. Refrigerants should have a series of desired properties to ensure that the systems are safe and can be serviced relatively easily. Some of these desired properties include being:

- Environmentally friendly
- Nontoxic
- Nonflammable
- Chemically stable
- Recyclable
- Relatively low cost
- Detectable at low concentrations

Most refrigerants in use today originate from one of two base molecules: ethane and methane. Methane and ethane are referred to as pure **hydrocarbons** because they contain only hydrogen and carbon. **Methane** [Fig. 4-1(A)] contains one carbon atom and four hydrogen atoms and has the chemical formula CH_4. **Ethane** [Fig. 4-1(B)] is made up of two carbon atoms and six hydrogen atoms and has the chemical formula CH_3CH_3. Both methane and ethane are considered to be very good refrigerants but their use has been greatly reduced because of their high flammability.

To create safer refrigerants that are not quite as flammable, some or all of the hydrogen atoms are removed from either the ethane or methane molecules and replaced with other elements, such as fluorine and/or chlorine. For example, when hydrogen atoms are replaced with fluorine, the new molecule is said to be fluorinated. When the hydrogen atoms are replaced with chlorine, the molecule is said to be chlorinated. Hydrogen atoms can also be replaced with a combination of chlorine and fluorine atoms. When some of the hydrogen atoms of a hydrocarbon are replaced with either chlorine or fluorine, the result becomes a **partially halogenated** hydrocarbon, or simply a **halogen** refrigerant. When all of the hydrogen atoms have been replaced with chlorine, fluorine, or a combination of both, the refrigerant is said to be **fully halogenated**. Four common classes of refrigerants in use today are:

- Hydrocarbons (HC)
- Hydrochlorofluorocarbons (HCFC)
- Chlorofluorocarbons (CFC)
- Hydrofluorocarbons (HFC)

HYDROCARBON REFRIGERANTS

As mentioned in the first portion of this section, hydrocarbons (HC) are comprised of only hydrogen and carbon atoms. None of the hydrogen atoms have been replaced with chlorine or fluorine. Two common hydrocarbon refrigerants are methane and ethane. Methane, R-50, has a boiling point of $-259°F$ at atmospheric pressure, while ethane, R-170, has a boiling point of $-127.5°F$ at atmospheric pressure. Other common hydrocarbons include butane (R-600), commonly found in cigarette lighters, and propane (R-290), which is often used for cooking and heating. Hydrocarbons are typically not found in air conditioning or refrigeration systems because of their high flammability.

HYDROCHLOROFLUOROCARBON (HCFC) REFRIGERANTS

As the name implies, hydrochlorofluorocarbon (HCFC) refrigerants contain hydrogen, chlorine, fluorine, and carbon. For residential applications, the most commonly used HCFC refrigerant is R-22, which is made up of one carbon atom, one chlorine atom, one hydrogen atom, and two fluorine atoms. The chemical name of R-22 is monochlorodifluoromethane or chlorodifluoromethane and its chemical formula is $CHClF_2$. A molecule of R-22 is shown in Figure 4-2. Other commonly used HCFC refrigerants are R-21, R-123, and R-124.

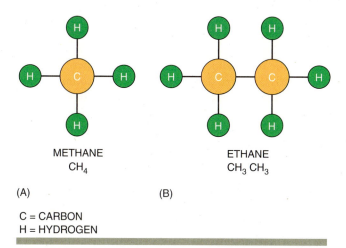

METHANE
CH_4

ETHANE
$CH_3 CH_3$

(A)   (B)

C = CARBON
H = HYDROGEN

FIGURE 4-1 (A) Methane molecule. (B) Ethane molecule.

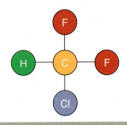

FIGURE 4-2 Molecule of R-22.

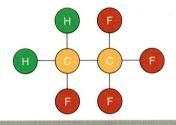

FIGURE 4-4 Molecule of R-134a.

HCFC refrigerants still contain hydrogen and are therefore referred to as partially halogenated refrigerants. These refrigerants have shorter lives in the atmosphere and are therefore less dangerous to the environment than are the CFC refrigerants that will be discussed next. The HVAC/R industry is turning away from HCFC refrigerants because of the negative effects they have on the environment. The phaseout of HCFC refrigerants began in the year 2004, with the total phaseout scheduled for 2030. The total phaseout for R-22 is scheduled for the year 2020.

CHLOROFLUOROCARBON (CFC) REFRIGERANTS

Chlorofluorocarbon (CFC) refrigerants contain no hydrogen as all of the hydrogen atoms have been replaced by either chlorine or fluorine and have become fully halogenated. CFCs have a very long life in the atmosphere, more than 75 years, and are dangerous to the environment, as will be discussed in the next section of this text. Because of the danger CFCs pose to the environment, they have not been produced since December 1995. Previously, domestic refrigerators, freezers, and automotive air conditioners commonly used R-12, a CFC, as their refrigerant. It is made up of one carbon atom, two chlorine atoms, and two fluorine atoms (Fig. 4-3). Since the phaseout of CFCs, domestic appliances and automotive air conditioners are now manufactured with alternative refrigerants such as R-134a, an HFC refrigerant. Prior to their phaseout, CFCs were commonly found in

our everyday lives and were used as propellants in spray cans and asthma inhalers.

HYDROFLUOROCARBON (HFC) REFRIGERANTS

Hydrofluorocarbon (HFC) refrigerants do not contain any chlorine and are safer for the environment than are CFC and HCFC refrigerants. These refrigerants are halogenated, but only by fluorine atoms. The most popular HFC refrigerant is R-134a, which was originally intended to replace R-12 in domestic refrigerators and automotive air conditioning systems. One major problem that arose with R-134a is that it is not compatible with the oil and gaskets used in R-12 systems. To use the new refrigerant, the gaskets needed to be changed and all of the old oil removed from the system. A molecule of R-134a can be seen in Figure 4-4.

BLENDED REFRIGERANTS

Increased attention is being paid to the safety of the environment, namely ozone depletion and global warming. As will be seen, some of the refrigerant types discussed previously are very dangerous to the environment, and much research has been done to locate alternative refrigerants that will serve the same purposes while posing a lesser risk to the environment. These new refrigerants, called blended refrigerants, are mixtures of two or more HCFC or HFC refrigerants that, when properly combined, perform in a manner similar to those refrigerants that are more harmful to the environment.

Some of these mixtures behave as a single refrigerant with only one saturation temperature at a given pressure. Others, which are capable of separating into their component refrigerants, exhibit a temperature glide, which means that, at a given pressure, the refrigerant will vaporize or condense at a number of different

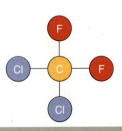

FIGURE 4-3 Molecule of R-12.

temperatures. For this reason, blended refrigerants are classified as azeotropes, near-azeotropes, or zeotropes.

Azeotropes are blended refrigerants that behave as a single refrigerant with a single vaporization or condensation temperature for a given pressure. Refrigerants such as R-500 (a blend of R-152A and R-12) and R-502 (a blend of R-22 and R-115) are azeotropic blends and have been used successfully in the air conditioning industry for many years. It should be noted that both R-500 and R-502 contain CFC refrigerants (R-500 contains R-12 and R-502 contains R-115) and have been phased out as of January 1, 1996. Near-azeotropic refrigerant blends, called **near-azeotropes**, experience a temperature glide. The refrigerants in a near-azeotropic blend can separate into the component refrigerants as the individual molecules of the different refrigerants do not combine chemically with each other. Each refrigerant in the mixture will boil or vaporize at different temperatures at a given pressure. This is called **fractionation**. **Zeotropes** are refrigerant mixtures that also experience temperature glides and fractionation. The near azeotropic blends, however, perform more like azeotropes than do zeotropes. This means that the zeotropes experience larger temperature glides and are more susceptible to fractionation than are near-azeotropic blends.

NEW GENERATION BLENDED REFRIGERANTS

As our industry has moved away from using CFC refrigerants and is in the process of moving away from HCFC refrigerants, more and more air conditioning and refrigeration applications are utilizing HFC refrigerants and HFC refrigerant blends. For residential air conditioning systems, HFC blended refrigerants R-410A and R-407C are popular choices for replacing R-22. R-410A, a binary blend of R-32 and R-125, is the popular choice for newly manufactured equipment. R-410A operates with a very low temperature glide of about 0.3°F, which is often ignored as the chances of fractionation are very low. R-407C, a ternary blend of R-32, R-125, and R-134a, is the popular choice for retrofitting existing R-22 systems. Unlike R-410A, R-407C operates with a large temperature glide of about 10°F. For this reason, special care must be taken when charging R-407C into any system. Please refer back to Chapter 2 for more information on these two refrigerants.

Another new generation refrigerant that is available for R-22 replacement is R-438A. This refrigerant is a blend of R-32, R-125, R-134a, R-600, and R-601a. This refrigerant blend has the same component refrigerants as R-407C, but two pure hydrocarbons, butane (R-600) and isopentane (R-601a), have been added in small quantities. The main benefit of this refrigerant is that it can be used in conjunction with the mineral oils that are presently used in conjunction with R-22. The presence of the hydrocarbons in R-438A facilitates oil return to the compressor by reducing the viscosity, or thickness, of the oil. R-438A is an HFC blend, so its ODP is zero. The operating pressures of R-438A are similar to those of R-22, but its capacity is slightly lower than that of R-22.

> ### from experience...
> Whenever retrofitting any air conditioning or refrigeration system, be sure to follow all manufacturers' instructions as well as all industry-accepted standards and procedures.

OZONE, OZONE DEPLETION, AND GLOBAL WARMING

In recent years, a great deal of attention has been paid to the protection of the ozone layer. Ozone is a form of oxygen that is made up of three oxygen atoms [Fig. 4-5(A)] as opposed to the oxygen we breathe, which contains only two oxygen atoms [Fig. 4-5(B)]. The **ozone layer**, the blanket of ozone gas that surrounds the earth, is responsible for filtering and absorbing ultraviolet rays, thereby

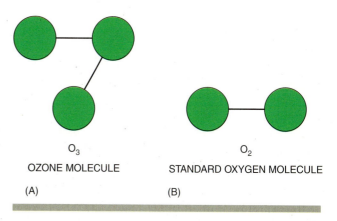

FIGURE 4-5 (A) Ozone molecule. (B) Oxygen molecule.

reducing the amount of ultraviolet radiation that reaches Earth. Ultraviolet radiation can cause:

- Skin cancer
- Eye-related issues including cataracts
- Decreased plant growth rates
- Compromised immune systems

TROPOSPHERIC AND STRATOSPHERIC OZONE

Ozone can be found in both the troposphere and the stratosphere. The troposphere is the region that exists from the earth's surface up to a distance of 7 miles above the Earth's surface. The stratosphere exists from 7 to 30 miles above the Earth (Fig. 4-6). The ozone in the tropospheric region is referred to as bad ozone, as it is responsible for the formation of smog when the sun's rays react with air contaminants. The ozone in the stratospheric region is referred to as good ozone, because it protects us from ultraviolet radiation.

HOW IS OZONE DESTROYED?

Ozone is destroyed when an ozone molecule comes in contact with a chlorine atom. Chlorine is a component part of CFC refrigerants, which is why the release of CFC refrigerants to the atmosphere has come under such close scrutiny over the past decade. A chlorofluorocarbon can exist in the atmosphere for more than 50 years before making its way to the stratosphere. When it does, the sun reacts with it, causing a chlorine atom to break

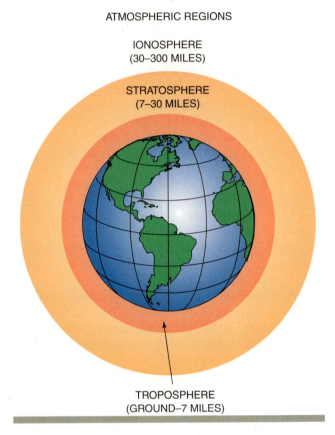

ATMOSPHERIC REGIONS

IONOSPHERE
(30–300 MILES)

STRATOSPHERE
(7–30 MILES)

TROPOSPHERE
(GROUND–7 MILES)

FIGURE 4-6 Atmospheric regions.

away from the CFC (Fig. 4-7). This chlorine atom reacts with an ozone molecule, breaking the molecule into an oxygen molecule and a chlorine monoxide molecule (Fig. 4-8). Any loose oxygen molecule can, in turn, break up the chlorine monoxide molecule. This will cause the

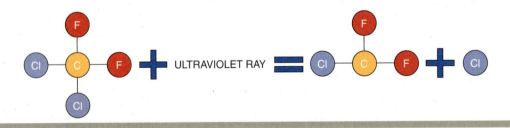

FIGURE 4-7 When ultraviolet radiation reacts with a CFC molecule, a chlorine atom is separated from the CFC molecule.

FIGURE 4-8 When the chlorine atom comes in contact with an ozone molecule, the chlorine atom bonds with one oxygen atom, creating an oxygen molecule and a chlorine monoxide molecule.

FIGURE 4-9 A loose oxygen atom will bond with the oxygen atom in the chlorine monoxide molecule, releasing the chlorine atom to destroy another ozone molecule.

formation of another oxygen molecule, freeing the chlorine atom to destroy another ozone molecule (Fig. 4-9). It is estimated that one chlorine atom can destroy up to 100,000 ozone molecules.

Going Green

A single chlorine atom can destroy up to 100,000 ozone molecules! Keep refrigerants where they belong . . . in the system!

GLOBAL WARMING

Earth is heated by the sun's radiation and, ideally, this heat is radiated back to the atmosphere. **Global warming**, also referred to as the **greenhouse effect**, results from Earth's inability to do this. Pollutants and other gases, including water vapor, carbon dioxide, CFCs, and other chemicals prevent Earth's heat from leaving Earth's atmosphere. These gases, referred to as greenhouse gases, result in the warming of Earth's surface.

OZONE DEPLETION POTENTIAL AND GLOBAL WARMING POTENTIAL

Different substances have different effects on the atmosphere and stratosphere regarding ozone depletion and global warming. The **ozone depletion potential (ODP)** is defined as the ratio of the impact on ozone of a chemical compared to the impact of a similar mass of R-11. Therefore, the ODP of R-11 is 1.0. The ODP is higher for substances that contribute to ozone depletion, and those substances that have little effect on the ozone have low ODPs. For example, R-12 contains two chlorine atoms and has an ODP of 0.93, while R-22, which has one chlorine atom, has an ODP of 0.05. Substances that do not contain chlorine or do not result in the depletion of the ozone layer have an ODP of zero. The

following outlines the various types of refrigerants and their ODPs:

- CFC refrigerants have high ozone depletion potentials
- HCFC refrigerants have lower ozone depletion potentials
- HFC and HC refrigerants have an ozone depletion potential of zero

Similarly, the **global warming potential (GWP)** represents the amount of impact a substance has in contributing to global warming. GWP represents how much a given mass of a chemical contributes to global warming over a given period of time when compared with the same mass of carbon dioxide. The GWP for carbon dioxide is therefore equal to 1.0. Various HCFC and HFC refrigerants have GWPs that range from 93 to 12,100. For example, R-12 has a GWP of 3.0 and R-22 has a GWP of 0.34. R-134a, the popular replacement for R-12, has an ODP of zero and a GWP of 0.28. The higher the GWP of a substance, the more it contributes to global warming. The following outlines the various types of refrigerants and their global warming potentials:

- CFC refrigerants have high global warming potentials
- HCFC refrigerants have lower global warming potentials
- HC refrigerants have low global warming potentials
- HFC refrigerants have very low global warming potentials

REFRIGERANT OILS

Oil in an air conditioning or refrigeration system helps to lubricate the internal moving parts in the system, primarily the mechanical components within the compressor. In addition to lubrication, the oil helps provide a seal between the pistons and the cylinder walls to increase the pumping efficiency of reciprocating compressors. The oil also acts to reduce sound transmission in the system and acts as a heat transfer medium. There are many different types of oil on the market, but those that function best in air conditioning systems are mineral oils. They can be classified as paraffinics, aromatics, and naphthenics, but the best mineral oils for use in our industry are the naphthenics. They have a low wax content and are manufactured primarily from California and Texas crude oil. As the need for new refrigerants emerged over the past years, so has the need for new oils that can be used effectively with these refrigerants. A new breed of synthetic oils has emerged that fall into one of three categories:

- Alkylbenzenes
- Glycols
- Esters

ALKYLBENZENE OILS

Following are properties and guidelines regarding the use and application of alkylbenzene oils:

- Work well with HCFC blended refrigerants
- Will tolerate up to 20% mineral oil when changing over to alkylbenzene oil

GLYCOLS

Following are properties and guidelines regarding the use and application of glycols:

- Polyalkylene glycol (PAG) is popular in automotive air conditioning systems
- Absorb moisture quickly
- May separate from the refrigerant in the condenser
- Not to be used with refrigerants containing chlorine

ESTERS

Following are properties and guidelines regarding the use and application of esters:

- Work well with HFC refrigerants
- Work well with HFC-based blended refrigerants
- Mineral oils must be completely drained from a system before changing to an ester oil
- Wax-free oils
- Absorb moisture quickly

Figure 4-10 provides a list of commonly used refrigerants as well as the appropriate oil to use with each.

from experience...

When choosing an oil, always refer to the literature supplied with the unit (especially the compressor) as well as the refrigerant literature to ensure that the proper oil is selected.

CFC REFRIGERANTS	APPROPRIATE LUBRICANT		
	MINERAL OIL	ALKYLBENZENE	POLYOL ESTER
R-11	+		
R-12	+	+	✓
R-13	+	+	✓
R-113	+	+	✓
R-114	+	+	✓
R-115	+	+	✓
R-500	+	+	✓
R-502	+	+	✓
R-503	+	+	✓

HCFC REFRIGERANTS	APPROPRIATE LUBRICANT		
	MINERAL OIL	ALKYLBENZENE	POLYOL ESTER
R-22	+	+	✓
R-123	+	+	
R-124		+	✓
R-401A	+	+	✓
R-401B		✓	✓
R-401C		✓	✓
R-402A		+	✓
R-402B	+	+	✓
R-403A		✓	✓
R-403B		✓	✓
R-405A		✓	✓
R-406A	+		
R-408A		+	✓
R-409A	+	+	✓

HFC REFRIGERANTS	APPROPRIATE LUBRICANT		
	MINERAL OIL	ALKYLBENZENE	POLYOL ESTER
R-23			+
R-32			+
R-125			+
R-134a			+
R-143a			+
R-152a			+
R-404A			+
R-407A			+
R-407B			+
R-407C			+
R-410A			+
R-410B			+
R-507			+

+ GOOD SUITABILITY   ✓ APPLICATIONS WITH LIMITATIONS

NOTE: ALWAYS CONSULT WITH THE COMPRESSOR MANUFACTURER FOR THE APPROPRIATE LUBRICANT.

FIGURE 4-10 A list of refrigerants with their appropriate oils.

FIGURE 4-11 Color-coded refrigerant cylinders.

Going Green

Always dispose of waste oil according to local codes and laws.

Going Green

Make certain that any discarded oils are disposed of according to local codes and guidelines.

REFRIGERANT HANDLING AND TRANSPORTING

To ensure the safety of those working on air conditioning and refrigeration equipment as well as those around the equipment, there are a number of guidelines that should be adhered to. These guidelines refer primarily to the storage, handling, and transporting of refrigerants.

REFRIGERANT CYLINDERS

When a refrigerant is purchased, it is supplied in a color-coded vessel that identifies the refrigerant in the container. The color codes are standard throughout the industry. For example, R-22 comes in light green vessels, R-134a is supplied in light blue containers,

from experience...

It is always good field practice to verify the contents of a refrigerant cylinder by checking the pressure in the tank. As long as there is a saturated refrigerant in the cylinder, the pressure in the cylinder and the ambient temperature surrounding the cylinder should correspond to the data on the pressure/temperature chart for that refrigerant.

CAUTION

CAUTION: Refrigerant cylinders come with pressure relief safety devices to reduce the chances of tank explosion. **DO NOT TAMPER WITH THEM!**

R-410A comes in rose-colored containers, and R-407C comes in chocolate-colored containers. (Fig. 4-11).

DISPOSABLE REFRIGERANT CYLINDERS

Most often, when working on residential air conditioning systems, refrigerant will be supplied in small, disposable cylinders (Fig. 4-12). When all of the refrigerant has been removed from these containers, they are to be disposed of and not reused to store other gases. Some guidelines regarding the use and handling of disposable refrigerant containers are as follows:

- These cylinders are disposable "one trip" vessels and should not be reused
- Some cylinders allow for liquid removal when the tank is in the vertical position (typically blended refrigerants)
- Some cylinders must be inverted when liquid refrigerant is desired
- Cylinders should be stored in a cool, dry place
- The painted surface of the cylinder should not become scratched
- The tanks should not become dented

FIGURE 4-12 Disposable refrigerant cylinders.

Photo by Eugene Silberstein.

- Cylinders are equipped with liquid and vapor valves so that either liquid or vapor refrigerant can be removed from the tank.
- Cylinders are labeled with vital information including owner's name, DOT number for the tank, serial number, test date (date the cylinder was pressure-tested), and the capacity of the cylinder.
- When moving a cylinder, be sure to use a hand truck that is intended for cylinder transport.

CAUTION

CAUTION: Be sure to secure cylinders to the hand truck with straps when moving refrigerant cylinders.

CAUTION

CAUTION: Reusable cylinders are equipped with safety caps to protect the valves. Make certain these caps are in place before moving the cylinders.

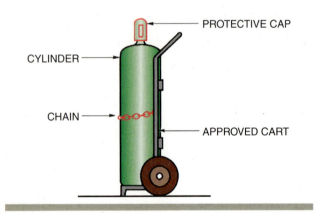

FIGURE 4-13 Reusable refrigerant cylinders.

- Tanks should be secured when being transported
- Never heat a cylinder with a torch to increase the pressure in the cylinder

REUSABLE REFRIGERANT CYLINDERS

Refrigerant also comes supplied in larger, reusable vessels (Fig. 4-13). These containers are also color-coded. Some guidelines regarding the use and handling of reusable refrigerant containers are as follows:

- Cylinders should be stored in a cool, dry place.
- Tanks should be secured when being transported.

RECOVERY CYLINDERS

Cylinders that are used to recover refrigerant from an air conditioning or refrigeration system must be approved by the Department of Transportation. Some guidelines regarding the use and handling of refrigerant recovery containers are as follows:

- Tanks are color-coded grey with yellow tops (Think of a canary sitting on the head of an elephant).
- Tanks should never be filled to more than 80% of the tank's capacity.
- Many tanks are manufactured with sensors that will de-energize recovery equipment should the tank reach the 80% capacity point.
- Evacuated tanks should be tagged, identifying them as being evacuated.
- Tanks containing refrigerant should be labeled with the refrigerant owner's name, amount of refrigerant, type of refrigerant, and date the refrigerant was removed; if possible, include the information (model #, serial #, location of unit) from the unit.
- Information on the tank's tag helps create and maintain the refrigerant log that all companies must keep.
- Cylinders should be stored in a cool, dry place.
- Tanks should be secured when being transported.

REFRIGERANT HANDLING

The following list summarizes guidelines for the safe handling of refrigerant and associated equipment during recovery and transport:

- Always be thoroughly familiar with surroundings.
- Wear **personal protection equipment (PPE)** including safety glasses, gloves and protective clothing.
- Recovered refrigerant may be acidic; BE CAREFUL.
- Do not inhale refrigerant vapors.
- When possible, work in well-ventilated areas.
- Refrigerant containers should never be filled to more than 80% capacity.
- Always secure cylinders before transporting.
- Properly label all refrigerant cylinders.
- Store tanks in a cool, dry place.
- Always maintain equipment and tools including recovery equipment, gauges, hoses, and refrigerant cylinders.
- Dedicate hoses for use with specific refrigerants to reduce cross-contamination.
- Change oil and filters regularly on recovery equipment.

REGULATIONS

For well over 30 years, there has been legislation regarding environmental safety. When concerns were raised regarding the ozone layer, attention was first given to aerosols, leading to the ban of nonessential aerosol use in 1978. However, one of the most aggressive pieces of legislation was the **Montreal Protocol Act** of 1987. In the formation of the Montreal Protocol, 23 nations, including the United States, agreed to reduce the number and quantity of ozone depleting substances.

As a result of the Montreal Protocol Act, production of CFCs has been phased out, starting with the 1987 freeze in production to bring production back to 1986 levels. Gradually, production was reduced, first to 80% in 1993, then to 25% in 1995, until CFC production was halted altogether in 1996.

In addition, HCFC refrigerants are also slated for extinction by the year 2030. The phaseout for HCFCs began in January 2004, when production of HCFCs was reduced to 65% of capacity. Manufacturers have responded to these phaseouts by introducing new refrigerants to replace those that will no longer be available or cost effective. Just as R-134a has become a popular choice for domestic refrigerators and automobiles, R-410A and SUVA AC9000 will likely become the popular choices for air conditioning systems. Equipment

manufacturers already have R-410A equipment on the market and continue to increase production.

A number of regulations have been put into effect as a result of the Montreal Protocol Act and a number of meetings that have taken place since. In addition to the phaseouts mentioned earlier, other regulations include:

- Unlawful to knowingly vent, release, or dispose of CFCs and HCFCs (1992).
- Unlawful to vent alternative refrigerants (1995).
- Technicians must be certified by November 14, 1994, under Section 608 of the Clean Air Act.
- Refrigerant can only be purchased by certified technicians.
- Uncertified individuals can pick up refrigerant for a certified technician provided that prior arrangements were made with the distributor.
- Recovery/recycling equipment must be certified for use.
- Recovered refrigerant cannot change ownership.
- System leak rates have been established for mandatory repair.
- Owners of air conditioning equipment must keep refrigerant records.
- Service companies must keep logs regarding refrigerant purchases, refrigerant usage, recovery equipment maintenance, and refrigerant inventory.

Going Green

Refrigerants cannot be released to the atmosphere! Be sure to recover, recycle, and reclaim whenever possible.

from experience...

This list represents only a sampling of the regulations that are presently in effect. Please refer to the Environmental Protection Agency at http://www.epa.gov for complete descriptions of the laws and requirements that must be followed in the field.

EPA TYPE 608 CERTIFICATION

In order to work on stationary air conditioning systems or handle refrigerants, field personnel must be certified as required by the Environmental Protection Agency's Section 608. Four different certifications can be obtained under section 608 that are classified by the type and amount of refrigerant contained within a particular system. To obtain certification, an individual must pass a core section of a test that deals with refrigerant safety and ozone protection as well as at least one of the other three sections. The four types of certification are:

- Type I—Small Appliances
- Type II—High Pressure and Very High Pressure Appliances
- Type III—Low Pressure Appliances
- Universal—Type I, Type II, and Type III

SECTION 608 TYPE I CERTIFICATION

Type I certification is awarded to individuals who pass the core portion of the examination as well as at least 70% of the questions in the section on small appliances. The Type I questions refer to small appliances, which are defined as hermetically sealed systems that contain less than five pounds of refrigerant. These appliances include refrigerators, freezers, window or wall air conditioners, package air conditioning systems, ice makers, refrigerated vending machines, and water coolers.

SECTION 608 TYPE II CERTIFICATION

Type II certification is awarded to individuals who pass the core portion of the examination as well as at least 70% of the questions in the section on high pressure and very high pressure appliances. High pressure appliances are defined as systems that contain a refrigerant with a boiling temperature between $-58°F$ and $50°F$ at atmospheric pressure. These refrigerants include R-12, R-22, R-500, and R-502. Very high pressure appliances are defined as those with a refrigerant that has a boiling temperature below $-58°F$ at atmospheric pressure. These refrigerants include R-13 and R-503.

SECTION 608 TYPE III CERTIFICATION

Type III certification is awarded to individuals who pass the core portion of the examination as well as at least 70% of the questions in the section on low pressure appliances. Low pressure appliances include systems that contain a refrigerant with a boiling temperature above $50°F$ at atmospheric pressure. These refrigerants include R-11 and R-123.

SECTION 608 UNIVERSAL CERTIFICATION

Universal certification is awarded to individuals who pass the core portion of the examination as well as at least 70% of the questions in each of the sections for Types I, II, and III. Universal certification permits the holder to work on small appliances, high pressure systems, very high pressure systems, and low pressure systems.

CERTIFICATION FOR MOTOR VEHICLE AIR CONDITIONING SYSTEMS

Section 608 of the Montreal Protocol Act covers stationary air conditioning systems and, therefore, does not encompass air conditioning systems that are installed in moving vehicles such as cars. For those technicians who intend to work on and service systems on motor vehicles, a separate certification, known as EPA Section 609 certification, is required.

REFRIGERANT RECOVERY

Refrigerant recovery is the process of removing refrigerant from an air conditioning or refrigeration system and storing it in an external container. The removed refrigerant is not chemically tested, treated, or processed.

If the refrigerant is not contaminated, it can be reused in the same system or another system owned by the same individual. Recovered refrigerant cannot be introduced to a system owned by another individual. In other words, the ownership of recovered refrigerant cannot be transferred. An example of recovery would be having the refrigerant removed from an air conditioning system to have the compressor replaced. After the compressor has been replaced and the system leak-tested and evacuated, the same refrigerant could be reintroduced to the system.

When recovering refrigerant, the following guidelines should be followed:

- The recovery cylinder should be evacuated.
- The recovery cylinder should never be filled to more than 80% capacity.
- A refrigerant scale should be used to monitor the amount of refrigerant that is removed from the system.
- The gauge hoses should be properly purged to prevent air from entering the recovery cylinder.
- The recovery process should be monitored carefully and a system should never be left unattended during the recovery process.

- Refrigerant should be recovered from the system until the required level of vacuum has been achieved (Fig. 4-14)

from experience...

If the refrigerant is to be reused in the system after recovery, it is important that precautions are taken to avoid contaminating the refrigerant. For example, the recovery tank should be properly evacuated to ensure that the refrigerant is not exposed to moisture.

REQUIRED LEVELS OF EVACUATION FOR AIR CONDITIONING, REFRIGERATION, AND RECOVERY/RECYCLING EQUIPMENT (EXCEPT FOR SMALL APPLIANCES, MVACs, AND MVAC-LIKE EQUIPMENT) INCHES OF HG VACUUM		
TYPE OF AIR CONDITIONING OR REFRIGERATION EQUIPMENT	**USING RECOVERY OR RECYCLING EQUIPMENT MANUFACTURED BEFORE NOVEMBER 15, 1993**	**USING RECOVERY OR RECYCLING EQUIPMENT MANUFACTURED ON OR AFTER NOVEMBER 15, 1993**
HCFC-22 EQUIPMENT, OR ISOLATED COMPONENT OF SUCH EQUIPMENT, NORMALLY CONTAINING LESS THAN 200 POUNDS OF REFRIGERANT.	0	0
HCFC-22 EQUIPMENT, OR ISOLATED COMPONENT OF SUCH EQUIPMENT, NORMALLY CONTAINING 200 POUNDS OR MORE OF REFRIGERANT.	4	10
OTHER HIGH-PRESSURE EQUIPMENT, OR ISOLATED COMPONENT OF SUCH EQUIPMENT, NORMALLY CONTAINING LESS THAN 200 POUNDS OF REFRIGERANT.	4	10
OTHER HIGH-PRESSURE EQUIPMENT, OR ISOLATED COMPONENT OF SUCH EQUIPMENT, NORMALLY CONTAINING 200 POUNDS OR MORE OF REFRIGERANT.	4	15
VERY HIGH-PRESSURE EQUIPMENT.	0	0
LOW-PRESSURE EQUIPMENT.	25	29

NOTE: MVAC = MOTOR VEHICLE AIR CONDITIONING

Courtesy of U.S. EPA.

FIGURE 4-14 Required levels of evacuation for air conditioning, refrigeration, and recovery/recycling equipment.

PASSIVE RECOVERY

When refrigerant needs to be recovered from a system that has an operational compressor, passive recovery is often the best method for recovery. The **passive recovery** process uses the system compressor to pump refrigerant from the system into an approved DOT refrigerant cylinder. A self-contained recovery unit is not needed to recover refrigerant using this method. Passive recovery is recommended when the total system refrigerant charge is less than 15 pounds.

To use the passive method of recovery, the gauges, hoses, refrigerant scale, and recovery tank should be connected as shown in Figure 4-15. To recover liquid from the system, the high-side valve on the gauge manifold and the valve on the recovery cylinder are in the open position, while the low-side valve on the manifold is in the closed position. When refrigerant stops flowing into the tank, the high-side valve on the gauge manifold is closed and the low-side valve is opened. This permits vapor refrigerant to flow from the tank back into the system, thereby reducing the pressure in the recovery cylinder. After a period of time, the low-side valve on the manifold should be closed and the high-side valve opened, permitting the recovery of more liquid refrigerant.

If the amount of recovered refrigerant reaches 80% of the tank's capacity, the tank should be closed, removed from the system, and replaced by a new, evacuated tank.

Going Green

In accordance with the guidelines set forth by the EPA, refrigerant hoses must be equipped with low loss fittings (Fig. 4-16). Low loss fittings prevent refrigerant that is trapped in the hoses from being released to the atmosphere when they are disconnected from a system or recovery unit.

from experience...

To speed the recovery process, the recovery cylinder can be placed in an ice bath. This will reduce the pressure in the cylinder, increasing the rate at which the tank is filled.

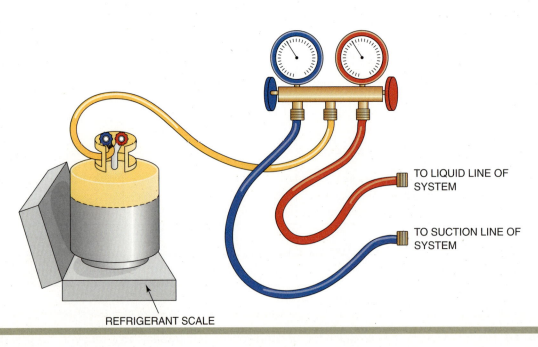

TO LIQUID LINE OF SYSTEM

TO SUCTION LINE OF SYSTEM

REFRIGERANT SCALE

FIGURE 4-15 Setup for passive refrigerant recovery.

FIGURE 4-16 Low loss fitting.

Courtesy of Robinair Division, SPX Corporation.

For step-by-step instructions on removing refrigerant from a system with an operative compressor, see Procedure 4-1 on pages 102–104.

ACTIVE RECOVERY

When the system contains more than 15 pounds of refrigerant or if the system compressor is inoperative, active recovery must be done. **Active recovery** is the process of using a self-contained recovery unit that has been certified for use by the DOT, EPA, and Air Conditioning and Refrigeration Institute (ARI). The recovery unit is equipped with an internal compressor that provides the pressure differential needed to remove the refrigerant from the system as well as the ability to reach the required system vacuum levels.

THE RECOVERY UNIT

Although most certified recovery units on the market operate in a similar manner, it is very important to follow the guidelines and operation instructions that accompany each individual unit. A typical recovery unit is shown in Figure 4-17. Common features on recovery units include inlet and outlet ports, inlet and outlet valves, gauges, and a means to purge the unit of any refrigerant left in it after the recovery process has been completed. Typically, the inlet port (Fig. 4-18) is a ¼ inch male flare fitting that facilitates the connection to the system with a standard gauge hose. The gauges (Fig. 4-19) facilitate taking pressure readings both in the system as well as in the recovery cylinder.

> **CAUTION**
> CAUTION: When operating a recovery unit, the valve at the outlet of the recovery unit as well as the valve on the recovery cylinder must be in the open position to reduce the chances of damage to the recovery unit's compressor.

FIGURE 4-17 Recovery/recycle unit.

Courtesy of Robinair.

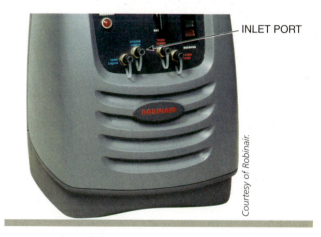

INLET PORT

FIGURE 4-18 Inlet port of a recovery unit.

Courtesy of Robinair.

FIGURE 4-19 Gauges on recovery unit.

Courtesy of Robinair.

HOSES AND RECOVERY CYLINDERS

When recovering refrigerant, it is important to prevent the refrigerant from mixing with a different refrigerant. Mixing refrigerants can result in:

- The contamination of recovered refrigerant
- Reduced system efficiency
- Reduced expected life of equipment
- Higher system operating costs
- Increased costs involved in the disposal of the refrigerant

To reduce the chances of cross-contamination during the recovery process, the tanks and hoses should be labeled and only used for one refrigerant. The hose connected to the inlet of the recovery unit will always be pulled into a vacuum once the recovery process is complete, so the same hose can be used no matter what refrigerant is being recovered.

The hoses at the outlet of the recovery unit, however, are under a positive pressure and should only be used for a given refrigerant. Color-coding or otherwise labeling the hoses will help ensure that cross contamination does not occur. Similarly, recovery cylinders should be permanently marked with the refrigerant they are intended to contain.

from experience...

If a recovery unit is to be used to recover a refrigerant that is different from the refrigerant previously recovered with that machine, the oil in the recovery unit should be changed to reduce cross-contamination. This, of course, does not apply to oil-less recovery units.

USING THE RECOVERY UNIT

When the system has an inoperative compressor or when the system contains more than 15 pounds of refrigerant, using a recovery unit is often the best recovery method. The recovery unit is connected between the unit and the recovery cylinder as in Figure 4-20. To protect the recovery unit and to monitor the state and moisture content of the refrigerant, a filter drier and a sight glass/moisture indicator (Fig. 4-21) should be installed in the line between the system and the recovery unit. This will

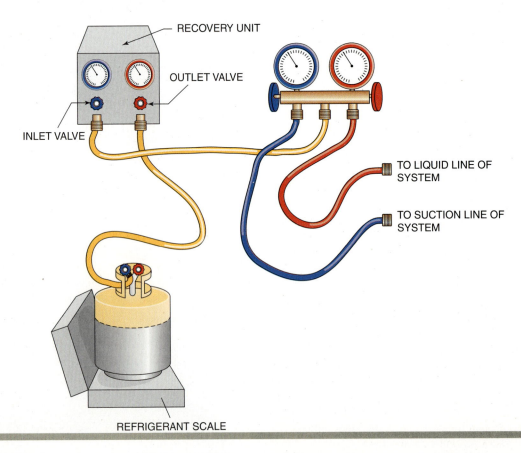

FIGURE 4-20 Setup for active refrigerant recovery.

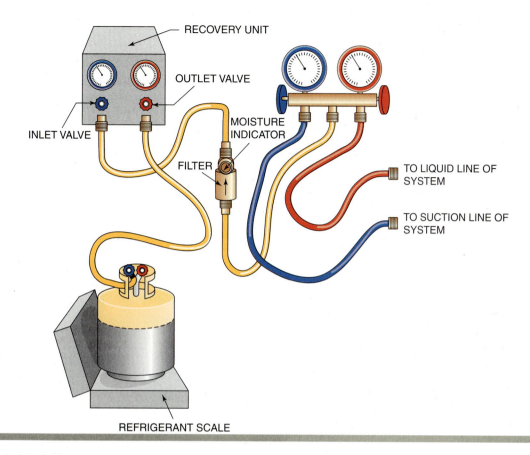

FIGURE 4-21 Active recovery setup with moisture indicator and filter installed.

prevent any particulate matter from entering the recovery unit and will allow the technician to monitor the moisture content and the state of the refrigerant as it passes to the recovery unit.

The recovery process typically begins with the recovery of liquid. Liquid recovery is much quicker than vapor recovery, so it is feasible to remove all of the liquid from the system before switching over to recover vapor. When recovering liquid, the high-side valve on the gauge manifold is in the open position as this hose is connected to the liquid line of the system. When the liquid recovery is complete, the high-side valve is closed and the low-side valve on the gauge manifold is opened to recover the remaining vapor in the system. Some guidelines for using a recovery unit are as follows:

- The recovery unit should contain the same refrigerant that is to be recovered
- Make certain all hoses are purged to remove air from them prior to recovery
- Liquid recovery is faster than vapor recovery
- Warmer ambient temperatures will increase the speed of recovery
- Monitor the sight glass to check the state of refrigerant

- Monitor the moisture indicator to determine the moisture content of the refrigerant
- Follow manufacturer's maintenance guidelines to ensure satisfactory operation
- Change oil and filters often to reduce chances of refrigerant contamination
- Recovery units should be operated only by those trained to do so
- Wear safety glasses and gloves when recovering refrigerant
- Use recovery equipment in well-ventilated areas

For step-by-step instructions on removing refrigerant from a system with a self-contained recovery unit, see Procedure 4-2 section on pages 105–107.

REFRIGERANT RECYCLING

Refrigerant recycling is the process by which refrigerant that has been removed from an air conditioning system is treated to reduce the levels of contamination in the system. Contaminants include acid, particulate matter, and moisture. The treatment of the refrigerant often involves oil separation. The process of recycling refrigerant can be done in the field using equipment designed for refrigerant recycling.

The recycling process can involve multiple passes through the recovery system and is often time consuming. The quality of the refrigerant should, therefore, be evaluated before starting the process to determine if recycling is a viable option. If it is determined that the refrigerant is heavily contaminated, the refrigerant should be taken to be reclaimed and the system should be recharged with new refrigerant.

REFRIGERANT RECLAIMING

Refrigerant reclaiming is the process by which refrigerant that has been removed from a system is restored to new product specifications. These specifications have been set by the ARI and are denoted as the ARI 700 standards. This process takes place only at a reprocessing or manufacturing plant, as the quality of the reprocessed refrigerant must meet certain criteria. As part of the reclaiming process, a chemical analysis of the refrigerant is performed to ensure that the standards have been met. Reclaimed refrigerant can then be resold to the public but it must be labeled as reprocessed. The tax collected on reclaimed refrigerant is less than the tax on newly manufactured refrigerant because the reclaimed refrigerant is not subject to the excise taxes that newly manufactured refrigerant is.

from experience...

If the refrigerant is heavily contaminated or if the refrigerant in the recovery tank is a mixture of refrigerants, the refrigerant may very well need to be destroyed by incineration. Label all recovery tanks with information such as the type of refrigerant, the original refrigerant owner, and the date and the amount of refrigerant removed from the system. This information needs to be entered on the refrigerant log that each service company must keep.

Refrigerant must be reclaimed if it is heavily contaminated, if the quality of the refrigerant cannot be determined, or if recycling is not available. As far as the field technician is concerned, the reclaiming process involves only recovering the refrigerant from the system and taking the refrigerant to a supply house or processing plant, depending on the amount of refrigerant being reclaimed.

SUMMARY

- Refrigerants should be environmentally safe, nontoxic, nonflammable, and chemically stable.
- Halogen refrigerants are created when some hydrogen atoms are replaced with chlorine or fluorine atoms.
- Hydrocarbons (HC) contain only hydrogen and carbon.
- Hydrofluorocarbons (HFC) contain only hydrogen, fluorine, and carbon.
- Chlorofluorocarbons (CFC) contain only chlorine, fluorine, and carbon.
- Hydrochlorofluorocarbons (HCFC) contain only hydrogen, chlorine, fluorine, and carbon.
- Blended refrigerants are mixtures of two or more other refrigerants and can be azeotropic, near-azeotropic, or zeotropic.
- Stratospheric ozone protects the earth from ultraviolet radiation.
- Ozone molecules are destroyed by chlorine.
- Chemical substances are rated by their ODP and their GWP.
- CFCs have high ODPs, while HFCs and HCs have low ODPs.
- Synthetic oils can be alkylbenzenes, glycols, and esters.
- Refrigerant cylinders are color-coded and can be reusable or disposable.
- The Montreal Protocol Act paved the way for a number of regulations regarding refrigerant handling.
- Technicians must be certified according to EPA Section 608 guidelines.
- Refrigerant can be recovered, recycled, or reclaimed.
- Recovery involves the removal and storage of the refrigerant.
- Recycling involves the filtering of refrigerant prior to reintroducing it to the system.
- Reclaiming involves restoring the refrigerant to ARI 700 standards.

GREEN CHECKLIST

☐ A single chlorine atom can destroy up to 100,000 ozone molecules.

☐ Refrigerants cannot be released to the atmosphere!

☐ CFC refrigerants have not been produced since 1995.

☐ CFC refrigerants have very high ozone depletion potentials.

☐ HCFC refrigerants have lower ozone depletion potentials than CFC refrigerants.

☐ HFC refrigerants have no ozone depletion potential as they do not contain chlorine.

☐ Global warming is the result of the Earth's inability to radiate heat back to the atmosphere.

☐ Global warming potential represents a substance's impact on global warming. The higher the GWP, the greater the impact.

☐ Dispose of all oils in accordance with local codes and guidelines.

☐ Low-loss fittings on refrigerant gauge manifold hoses are required by law.

☐ Refrigerant recovery involves the removal and storage of refrigerant.

☐ Refrigerant recycling involves the removal, cleaning, and filtering of system refrigerant.

☐ Refrigerant reclaiming is the process of restoring refrigerant to a "like new" condition.

Recovering Refrigerant from a System with an Operative Compressor

The following procedures should be followed to remove refrigerant from a system with an operative compressor:

- Make certain that the recovery cylinder to be used is evacuated.

A Note the capacity of the recovery cylinder.

B Calculate 80% of the cylinder's capacity.

- Install gauges on the system.

C Connect the center hose from the gauge manifold to the vapor port on the recovery tank.

- Make certain that the air conditioning system is off.

- Open the low- and high-side valves on the gauge manifold.

- Loosen the hose connection on the recovery cylinder for about 1 second. This will purge any air from the gauge hoses.

- Tighten the hose connection of the recovery cylinder.

- Close the high- and low-side valves on the gauge manifold.

- Energize the air conditioning system.

A

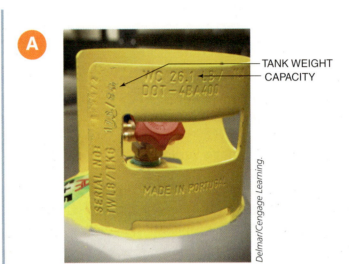

TANK WEIGHT
CAPACITY

Delmar/Cengage Learning.

B

C

Photo by Eugene Silberstein.

- Turn on the refrigerant scale.

- Place the recovery cylinder on the refrigerant scale.

D The setup should resemble that shown in the figure.

- Zero the display on the refrigerant scale.

- Open the high-side valve on the gauge manifold.

- Open the vapor valve on the recovery cylinder. Refrigerant will flow from the system, through the gauge manifold into the recovery cylinder.

- Monitor the system and the display on the refrigerant scale.

- When the reading on the scale's display indicates that no more refrigerant is being introduced to the cylinder, close the high-side valve on the manifold and open the low-side valve on the manifold. This will allow vapor refrigerant to flow back into the system.

- After allowing the pressure in the recovery tank to stabilize, close the low-side valve and open the high-side valve on the manifold. Liquid recovery will resume.

- Repeat the two previous steps until refrigerant recovery is complete.

D

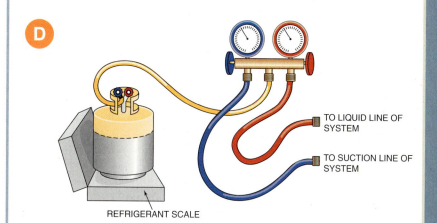

TO LIQUID LINE OF SYSTEM

TO SUCTION LINE OF SYSTEM

REFRIGERANT SCALE

from experience...

Placing the recovery cylinder in an ice bath can help speed the recovery process.

CAUTION

CAUTION: If the display on the scale reaches the 80% capacity point, immediately close the valve on the recovery cylinder, close the high-side valve on the gauge manifold, and replace the cylinder with another evacuated cylinder.

PROCEDURE 4-1 (CONT'D)

Recovering Refrigerant from a System with an Operative Compressor (continued)

- If it is not possible to recover all of the refrigerant using this method, close the valves on the cylinder and gauge manifold, replace the cylinder with another evacuated cylinder, and continue the recovery process.

- When recovery is complete, be sure to close the valves on the gauge manifold and recovery cylinder before disconnecting the tank and gauges from the system.

E Tag the recovery cylinders and label with the date, refrigerant owner, amount of refrigerant in the tank, and the type of refrigerant.

E

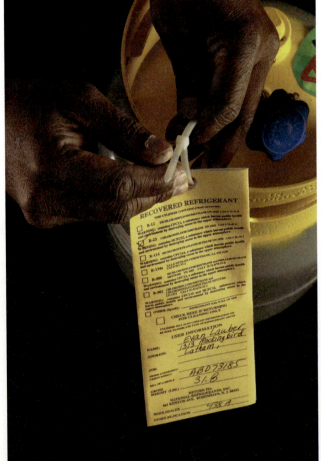

Delmar/Cengage Learning.

PROCEDURE 4-2

Recovering Refrigerant from a System with a Self-Contained Recovery Unit

The following procedures should be followed to remove refrigerant from a system with a self-contained recovery unit:

- Make certain that the recovery cylinder to be used is evacuated.

- Note the capacity of the recovery cylinder.

- Calculate 80% of the cylinder's capacity.

- Install gauges on the system.

- Loosen the hose connections on the blank ports of the manifold to purge any air from the hoses.

- Make certain that both valves on the gauge manifold are closed.

A Connect the center hose from the gauge manifold to the inlet port on the recovery unit.

B Connect a hose from the outlet port of the recovery unit to the vapor port of the recovery cylinder.

- Make certain that both the inlet and outlet valves on the recovery unit are closed.

- Open the high-side valve on the gauge manifold.

Photo by Eugene Silberstein.

Photo by Eugene Silberstein.

Recovering Refrigerant from a System with a Self-Contained Recovery Unit (continued)

- Loosen the hose connection on the inlet valve of the recovery unit to purge any air from the hose.

- Turn on the refrigerant scale.

- Place the recovery cylinder on the refrigerant scale.

C The setup should resemble that in the figure.

- Zero the display on the refrigerant scale.

- Open the inlet and outlet valves on the recovery unit.

- Briefly loosen the hose connection on the recovery tank to purge any air from the line connecting the recovery unit to the recovery cylinder.

- Allow the pressure in the recovery unit to equalize.

- Turn the recovery unit on.

- To recover liquid, open the high-side valve on the gauge manifold and the valve on the recovery cylinder.

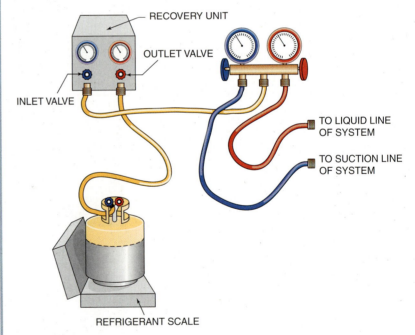

RECOVERY UNIT

OUTLET VALVE

INLET VALVE

TO LIQUID LINE OF SYSTEM

TO SUCTION LINE OF SYSTEM

REFRIGERANT SCALE

from experience...

Placing the recovery cylinder in an ice bath can help speed the recovery process.

- When the liquid refrigerant has been recovered, open the low-side valve on the gauge manifold and continue recovery until the system has reached the required level of vacuum.

- Close the valves on the gauge manifold.

- Follow the recovery unit's instructions to purge refrigerant from the unit into the recovery cylinder.

from experience...

If the gauge manifold being used has a vacuum port, follow the manufacturer's instructions to use the vacuum pump to evacuate the hose that connects the recovery unit to the gauge manifold and the one that connects the recovery unit to the recovery cylinder.

from experience...

Many recovery units are equipped with sensors that will automatically shut the recovery unit off if the recovery cylinder has been filled to 80% of its capacity. If the cylinder is full and needs to be changed, make certain that all valves are closed before removing the cylinder. Make certain that all hoses are properly purged or evacuated before introducing refrigerant to the new cylinder.

REVIEW QUESTIONS

1. **The two molecules on which many refrigerants are based are**
   a. Methane and butane
   b. Methane and ethane
   c. Propane and ethane
   d. Butane and propane

2. **What is the primary reason that hydrocarbons are typically not used as refrigerants in air conditioning systems?**
   a. They are too expensive
   b. They are highly flammable
   c. They are unstable
   d. None of the above are correct

3. **Hydrocarbons contain only**
   a. Hydrogen and chlorine
   b. Chlorine and fluorine
   c. Carbon and hydrogen
   d. Chlorine and carbon

4. **The type of refrigerant that is fully halogenated is**
   a. CFC
   b. HFC
   c. HCFC
   d. HC

5. **What type of refrigerant is R-134a?**
   a. CFC
   b. HFC
   c. HCFC
   d. HC

6. **Refrigerants that experience temperature glides are**
   a. Azeotropic mixtures
   b. Near-azeotropic mixtures
   c. Zeotropic mixtures
   d. Both b and c are correct

7. **A damaged ozone layer can result in**
   a. Skin cancer
   b. Cataracts
   c. Compromised immune systems
   d. All of the above are correct

8. **One chlorine atom has the potential to destroy how many ozone molecules?**
   a. 1,000
   b. 10,000
   c. 100,000
   d. 1,000,000

9. **Ozone is destroyed primarily when**
   a. Hydrocarbons are released to the atmosphere
   b. Ultraviolet rays come in contact with CFCs
   c. Hydrogen atoms come in contact with carbon
   d. All of the above are correct

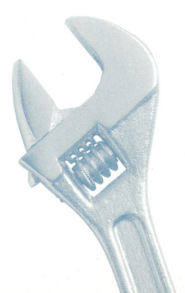

10. **Which of the following refrigerants has the highest ODP?**

a. R-12

b. R-134a

c. R-22

d. Both a and b are correct

11. **Greenhouse gases**

a. Trap the earth's heat in the troposphere

b. Trap the earth's heat in the stratosphere

c. Cause the temperature of the earth to decrease

d. All of the above are correct

12. **Which oil type is recommended for use on systems using HCFC refrigerants?**

a. Glycols

b. Esters

c. Alkylbenzenes

b. All of the above will work equally well in HCFC systems

13. **Which of the following is true regarding "one trip" refrigerant cylinders?**

a. They can be used to store recovered refrigerant

b. They can be used to store pressurized air

c. They can deliver both liquid and vapor refrigerant

d. They are grey in color with yellow tops

14. **It became illegal to knowingly vent CFCs and HCFCs to the atmosphere in**

a. 1987

b. 1992

c. 1994

d. 1995

15. **A technician with Type II certification under EPA Section 608 can work on**

a. All appliances

b. Small appliances

c. High-pressure appliances

d. Low-pressure appliances

16. **A 50-pound recovery cylinder should contain no more than**

a. 30 pounds of refrigerant

b. 40 pounds of refrigerant

c. 50 pounds of refrigerant

d. 60 pounds of refrigerant

17. **When refrigerant must be recovered from a system that has an inoperable compressor**

a. Passive recovery methods must be used

b. Active recovery methods must be used

c. The refrigerant can be released to the atmosphere

d. The refrigerant cannot be recovered from this system

18. While working in the field, it is possible to

a. Recover refrigerant

b. Recover and recycle refrigerant

c. Recover and reclaim refrigerant

d. Recycle and reclaim refrigerant

19. When refrigerant is reclaimed it is restored to

a. ARI 608 standards

b. EPA 608 standards

c. ARI 700 standards

d. EPA 700 standards

20. To prevent cross-contamination of refrigerants

a. Hoses used on recovery equipment should be dedicated to a single refrigerant

b. The oil in the recovery unit should be changed frequently

c. The recovery cylinders should be clearly marked

d. All of the above are correct

KNOW YOUR CODES

It is important that discarded oils be disposed of according to local codes and guidelines. Research oil disposal guidelines in your area and be aware of your responsibility to the environment by following them.

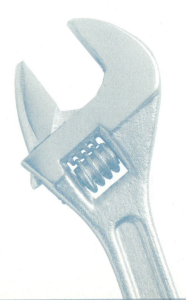

SECTION 2

Tools

Hand Tools

OBJECTIVES   *Upon completion of this chapter, the student should be able to*

- List various hand tools used by air conditioning and heating technicians.

- Explain the proper use of various hand tools.

- Explain the differences among open end, box and combination wrenches.

- List various types of pliers and the appropriate uses for each type.

- List various types of hammers and their uses.

- List and explain tools that are commonly used when performing electrical installations, repairs, or maintenance.

- Explain what must be done to power tools prior to using them.

In addition to gaining knowledge and obtaining a solid understanding of air conditioning, refrigeration, and heating systems, successful technicians in our industry must also have a working knowledge of hand tools. Knowing the various tools that are available is one thing, but knowing how to properly use and care for them is another. In this chapter, we will take a look at various hand tools, describe their purpose, and also discuss how to properly care for them.

It is important for service technicians to understand that tools are to be used only for their intended purpose. Using tools inappropriately can lead to tool damage, personal injury as well as damage to the equipment being serviced. Obtaining a complete set of tools is a major expense that is often absorbed by the individual technician. For this reason, it is important that the technician purchase the appropriate tools, take proper care of them, and use them correctly.

Cheek The side portion of a hammer head

Ethanol Yellow liquid commonly used in levels

Hex heads Six-sided screw or bolt-head shape

Level Term used to indicate a perfectly horizontal line or plane

Mushroomed head Term used to describe the misshapen, damaged head of a chisel or other struck tool

Nut driver Tools that are similar in appearance to screwdrivers but are configured to tighten and loosen screws and bolts with hex heads

Offset Tool configuration that facilitates the reaching of tight or awkward spaces

Peen The rounded back portion of the head of a ball peen hammer

Pigtail splice An electrical wire connection made by twisting two or more wires together

Plumb Term used to indicate a perfectly vertical line or plane

Shank The portion of a screwdriver-like tool that connects the tip of the tool to the handle

Solderless connectors Used to create a nonsoldered, mechanical electrical connection between two wires or between a wire and an electrical component or device

Striking face The portion of a hammer head that comes in contact with the object being struck

Stubby-type Term used to describe tools with very short shanks that are used to get into very tight areas

Torque The twisting force that is exerted by a rotating object. In the case of a screwdriver, the torque is the force put on the screw, causing it to turn

Wire gauge Tool or scale used to determine the gauge or thickness of a wire

SCREWDRIVERS

The screwdriver is one of the tools that is most often misused, so it should come as no surprise that screwdriver breakage is a common occurrence. When used properly, screwdrivers should last for years before replacement is necessary. All too often, though, screwdrivers are used as chisels, scrapers, and pry bars, resulting in chipped or broken tips, loose handles, and even airborne pieces of steel. It cannot be stressed enough that tools, especially screwdrivers, should only be used for the job they are intended to do!

There are many different types of screwdrivers, so it is important to select the proper driver for the screw that is being tightened or loosened. Some common driver tips are shown in Figure 5-1. In addition to the type of tip on the driver, there are other considerations that should be addressed. These include the size of the tip as well as the length of the shaft on the driver. The size of the screwdriver must match the size of the screw. Since screws come in a wide variety of sizes and shapes, the service technician should have a wide selection of screwdrivers on hand. The tip of the screwdriver must fit snugly into the slot (Fig. 5-2). Screwdrivers that have tips that are too large or too small should not be used because tool slippage and personal injury can result.

FASTENER TYPES AND DRIVER TIPS

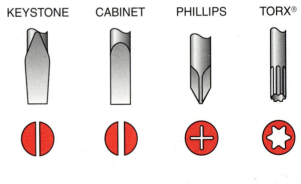

KEYSTONE   CABINET   PHILLIPS   TORX®

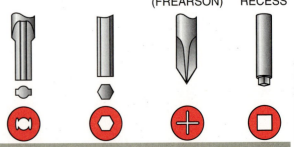

CLUTCH HEAD   HEX HEAD   REED & PRINCE (FREARSON)   SQUARE RECESS

Courtesy Klein Tools.

FIGURE 5-1 Standard screwdriver tip types.

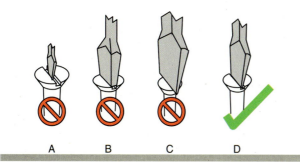

FIGURE 5-2 The screwdriver tip must fit snugly into the slot. (A) This tip is too small. (B) This tip is too thick to fit into the slot. (C) This tip is too wide. (D) This tip fits perfectly into the slot.

The length of the screwdriver's **shank** is also important. The shank is the portion of the tool that connects the head or tip of the screwdriver to its handle. Longer shanks provide more **torque**, or twisting force, so be sure to use a screwdriver with a longer handle if more torque is required. Some screwdrivers are manufactured with square shanks (Fig. 5-3) so a wrench can be applied to it to aid in the removal of stubborn screws. When working in spaces with low clearance, **stubby-type** screwdrivers, which are screwdrivers with very short shanks, are often needed (Fig. 5-4). When turning screws that are located

FIGURE 5-3 Straight slot screwdriver with a square shank.

FIGURE 5-4 Stubby screwdriver.

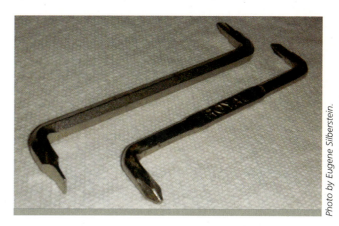

FIGURE 5-5 Offset screwdrivers.

in very tight conditions, **offset** screwdrivers are often a good choice (Fig. 5-5). The term offset refers to the tool configuration that facilitates the reaching of tight or awkward spaces.

CAUTION

CAUTION: Replace screwdrivers that have worn or broken handles, or tips, immediately.

NUT DRIVERS

Nut drivers (Fig. 5-6) are similar in appearance to screwdrivers but are configured to tighten and loosen screws and bolts with **hex heads** (Fig. 5-7). Hex heads refer to the six-sided shape of nuts or bolt heads. Nut drivers are especially useful to the air conditioning technician when service panels on equipment need to be removed or replaced (Fig. 5-8). Most manufacturers color-code the handles of nut drivers for easy identification. In our industry, for example, the most popular sizes of nut drivers are the 1/4″ and the 5/16″, but technicians often refer to them simply as the "red" and the "yellow" (Fig. 5-9).

FIGURE 5-6 Nut drivers.

FIGURE 5-7 Hex head screws.

Photo by Eugene Silberstein.

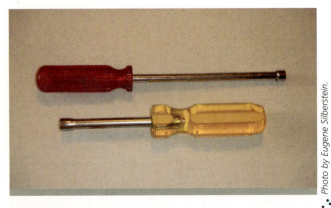

FIGURE 5-8 Technician removing service panel with a nut driver.

Photo by Eugene Silberstein.

FIGURE 5-9 "Red" and "yellow" nut drivers.

Photo by Eugene Silberstein.

As with screwdrivers, nut drivers are available in different lengths for different applications. In addition, some nut drivers are manufactured with hollow shafts to allow the threaded portion of a bolt or screw to extend into the tool itself when it extends past the nut (Fig. 5-10). Some nut drivers are magnetic to hold the fastener in place.

FIGURE 5-10 Hollow handle nut drivers allow the bolt to extend into the tool.

Photo by Eugene Silberstein.

WRENCHES

Wrenches are popular tools and used primarily for turning threaded parts, including screws, nuts, bolts, pipes, and other fastening devices with threads. Since there are many different applications for wrenches, they are available in a multitude of sizes, styles, and handle types geared toward the specific task being performed. Because there is a wrench for every job, there is no need for the technician to struggle with the wrong tool. Investing in a set of wrenches appropriate for the work being performed should be very high on a technician's list of priorities.

from experience...

Wrenches are available in both standard and metric units. Although some metric and standard sizes are similar, be sure to use the proper size wrench for the fastener being worked on.

COMMON WRENCH TYPES

Some common types of wrenches are the open end (Fig. 5-11), the box end (Fig. 5-12), and the combination (Fig. 5-13). Open-end wrenches are often manufactured with a 15 degree offset angle so the user can easily turn a hex head bolt or screw completely by flipping the wrench over (Fig. 5-14). The open-end wrench is not intended to be used for loosening extremely tight fasteners or for the

FIGURE 5-11 Open end wrenches.

FIGURE 5-12 Box end wrenches.

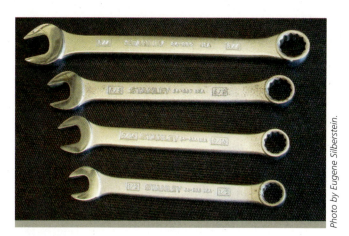

FIGURE 5-13 Combination wrenches.

final tightening of a fastener. Using the open-end wrench for these tasks can result in tool slippage and injury. The box wrench is a general duty wrench but provides more strength than the open end wrench does. The box wrench is less likely to slip during use, as the tool completely surrounds the fastener being tightened or loosened. The box

FIGURE 5-14 The 15 degree offset angle allows for the complete turning of the nut by flipping the wrench over.

wrench can be used for the final tightening of a fastener. Like the open-end wrench, there is often a 15 degree offset on the handle.

CAUTION

CAUTION: Never use a pipe extension to increase the torque or leverage of a wrench.

ADJUSTABLE WRENCH

Another type of wrench found in our industry is the adjustable wrench (Fig. 5-15). The adjustable wrench is very popular for the simple reason that one tool can be used on a number of different fastener sizes. Unfortunately, the adjustable wrench is often misused and, as a result, injuries using this type or wrench are not uncommon. Just as with the open-end wrench, the adjustable wrench is not intended to be used to loosen extremely tight nuts and bolts or to provide final tightening of nuts and bolts.

When using the adjustable wrench it is recommended that the wrench be tightened securely to the nut

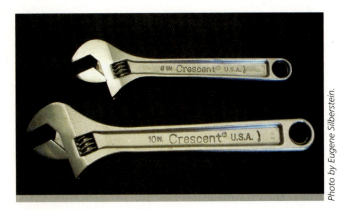

Photo by Eugene Silberstein.

FIGURE 5-15 Adjustable wrenches.

Photo by Eugene Silberstein.

FIGURE 5-16 Adjustable wrenches should be positioned so that the user pulls on the wrench and the force is exerted on the stationary portion of the wrench jaw.

being turned to reduce the possibility of tool slippage. In addition, once secured to the nut being turned, the wrench should be pulled on so that the force is exerted on the stationary portion of the wrench (Fig. 5-16). Putting excessive force on the adjustable portion of the wrench can cause damage to the tool.

CAUTION

CAUTION: Never strike an adjustable wrench with a hammer!

RATCHET-TYPE WRENCHES

The ratchet-type wrench is a popular tool because it allows the user to loosen or tighten a fastener without having to reposition the tool (Fig. 5-17). For this reason, ratchet-type wrenches allow for quick loosening or

Photo by Eugene Silberstein.

FIGURE 5-17 Ratchet-type wrench.

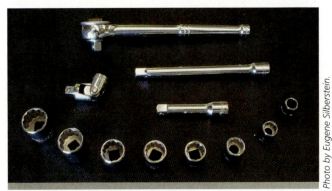

Photo by Eugene Silberstein.

FIGURE 5-18 Socket set.

tightening of the nut or bolt. It should also be noted that the ratchet-type wrench is among the strongest of wrench types and is recommended for the loosening of extremely tight nuts and bolts, and also the final tightening of fasteners.

SOCKET SETS

The socket set is often comprised of a complete set of socket wrenches, a reversible ratcheting handle, at least one extension and a universal joint socket (Fig. 5-18). The socket wrenches themselves are available in many sizes and can be in standard inch or metric measurements (Fig. 5-19). The sockets fit onto the drive mechanism on the ratcheting handle, which can range in size from 1/4″ up to 1″ (Fig. 5-20). General purpose socket sets typically have a 3/8″ drive.

Socket wrench extensions (Fig. 5-21) allow the user to gain access to nuts and bolts that are deep inside equipment (Fig. 5-22). The universal joint socket

FIGURE 5-19 Sockets.

FIGURE 5-20 Socket ratchet wrench.

FIGURE 5-21 Socket wrench extensions.

FIGURE 5-22 Technician using a socket wrench extension to access a nut.

of the three is the hand socket, so it is important that these sockets are not used with power or impact tools. Power and impact socket wrenches provide much more torque and are manufactured to withstand the added stress placed on them.

(Fig. 5-23) is often used in conjunction with an extension to aid in gaining access to hard-to-reach nuts and bolts.

Socket wrenches are manufactured in three common types based on how they are to be used. They are the hand, power, and impact wrenches. The lightest duty

CAUTION

CAUTION: Do not use hand sockets with power or impact wrenches!

FIGURE 5-23 Universal joint.

Photo by Eugene Silberstein.

FIGURE 5-24 Hex keys.

Photo by Eugene Silberstein.

HEX KEYS

Hex keys are also commonly known as allen keys (Fig. 5-24). The hex keys are six-sided tools that are used to turn fasteners that have a hexagonal-shaped recessed opening or slot. Refer back to Figure 5-1 for an image

FIGURE 5-25 Technician using a hex key to remove a blower from the motor shaft.

Photo by Eugene Silberstein.

FIGURE 5-26 T-handle hex keys.

Photo by Eugene Silberstein.

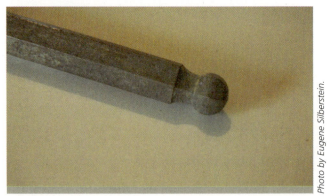

FIGURE 5-27 Ball-tipped hex keys.

Photo by Eugene Silberstein.

of the hex head. Hex keys are commonly used to secure or remove blowers from motor shafts (Fig. 5-25). Hex keys come in a number of different styles, including T-handle (Fig. 5-26), and ball tip (Fig. 5-27). The ball tip

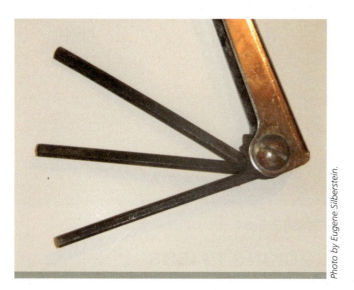

FIGURE 5-28 Fold-out hex key set.

style allows the user to turn a fastener with a hexagonal recessed slot while the tool is at a slight angle from the fastener. Hex keys are also available as a convenient fold-out set, which keeps all keys connected to each other, eliminating the frustration often associated with a missing key (Fig. 5-28).

PLIERS

As with other tools, there are many different types of pliers available and each is intended to perform a specific task. Some pliers are capable of cutting wire, while others are used to access hard-to-reach places. Some are intended for general holding purposes while others are used to remove insulation from conductors. No matter what the task, it is always best to use the correct pliers to perform it. Proper use of any tool will help keep the user safe and also prolong the useful life of the tool. What follows are brief descriptions of a number of different types of pliers.

GENERAL PURPOSE, SLIP-JOINT PLIERS

General purpose pliers are used for many general tasks that involve holding, turning, and bending (Fig. 5-29). These pliers are not intended to perform high torque tasks, final tightening or loosening of extremely tight fasteners. These pliers have a slip-joint which allows the user to adjust the capacity range of the tool.

NEEDLE-NOSE PLIERS

Needle-nose pliers are intended be used in tight places when general use pliers are too awkward or too large to use (Fig. 5-30). This type of pliers is often used for electrical control wiring work and for handling smaller wires and components. Because of the configuration of this pliers type, it is not recommended that the needle-nose pliers be used as a bending tool. Damage to the tool can result. The nose on these pliers can be either straight or curved (Fig. 5-31).

FIGURE 5-29 General purpose, slip joint pliers.

FIGURE 5-30 Needle-nose pliers.

FIGURE 5-31 Curved needle-nose pliers.

SIDE CUTTERS

Side cutting pliers are intended to cut soft wires or other thin metal objects, such as nails and rivets that flush to the surfaces through which they protrude (Fig. 5-32). Side cutters come in different sizes and duty ranges, so be sure that you are not exceeding the tool's limitations when using them.

LINESMAN PLIERS

Unlike general purpose pliers, linesman pliers are intended to perform a number of functions (Fig. 5-33). They can cut and strip wire and the jaws are configured so they can properly twist conductors together to form a pigtail splice. A pigtail splice is an electrical wire connection made by twisting two or more wires together. The handles of linesman pliers should always be properly insulated, as damage to the insulation can result in electric shock.

SLIP JOINT, PUMP PLIERS

Pump pliers have a tongue-and-groove mechanism that, as with the general-purpose pliers, allows the user to adjust the size of the jaws based on the job to be performed (Fig. 5-34). These pliers are often used by plumbers and others in the construction trades as larger pipes and objects are encountered. This tool is very versatile and can be used to grip and hold objects of different shapes and configurations including pipes, large screw-type fittings encountered in the plumbing trades, and nuts or bolts with hex heads. Because of the longer handles, these pliers can provide a fair amount of torque without causing damage to the object being turned.

FIGURE 5-32 Side cutting pliers.

Courtesy of Ideal Industries, Inc.

FIGURE 5-33 Linesman pliers.

Photo by Eugene Silberstein.

FIGURE 5-34 Slip joint, pump pliers.

LOCKING PLIERS

Locking pliers are a very versatile tool that can function as a pair of pliers, a wrench or a clamp (Fig. 5-35). Locking pliers can have either straight or curved jaws, depending on the applications for which they are intended. Some are better for gripping hoses and pipes, while others are better suited for clamping objects such as wood sections being glued together. Hybrid locking pliers are also common, such as the locking needle nose pliers (Fig. 5-36).

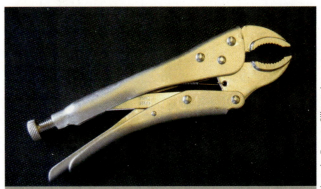

Photo by Eugene Silberstein.

FIGURE 5-35 Locking pliers.

Photo by Eugene Silberstein.

FIGURE 5-36 Locking needle-nose pliers.

WIRE STRIPPERS

Wire strippers come in a wide variety of configurations and should be chosen based on the intended use of the tool (Fig. 5-37). Some wire strippers are designed for simply stripping the insulation from wire. Others are designed to perform tasks such as cutting wire, cutting small screws and bolts, or crimping solderless connectors onto sections of stripped stranded wire. Solderless connectors are used to create a nonsoldered, mechanical electrical connection (Fig. 5-38). Wire strippers can also be automatic (Fig. 5-39). With the automatic wire stripper, all the user has to do is insert the wire into the tool and squeeze the handles together.

FIGURE 5-37 Wire strippers.

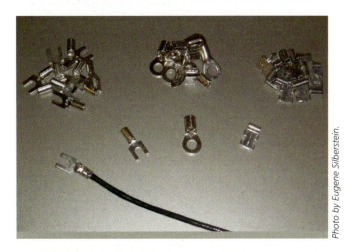

FIGURE 5-38 Solderless connectors.

FIGURE 5-39 Automatic wire stripper.

CAUTION

CAUTION: When stripping wire, it is important to remove only as much insulation as is needed for the task at hand. Removing too much can result in electric shock.

CRIMPING TOOLS

The crimping tool is a multipurpose tool that enables the user to cut wire, strip wire, attach solderless connectors to the stripped ends of stranded wires, and even cut small bolts or screws (Fig. 5-40). Crimping tools often have wire gauges on them so the user can determine the size of wire that is being used. A wire gauge is a tool or scale used to determine the gauge or thickness of a wire.

ELECTRICAL SNAKES

The electrical snake is a roll of flexible, yet somewhat rigid, metal strapping (Fig. 5-41). The snake is approximately 1/8″ wide and is used to help run electrical wires

FIGURE 5-40 Crimping tool.

FIGURE 5-41 Electrical snake.

through walls and ceilings. The end of the snake is bent to facilitate securing the wires to be pulled. It is first run between the points that the wire is to be run. Since it is more rigid than the wire, it is much easier to run the snake than the wire. Once it is in place, the wires to be run are secured to the end of the snake. The snake is then pulled out and, in doing so, the wires are pulled between the desired points.

HAMMERS

Unfortunately, many people think that all hammers are the same. Such is definitely not the case. Hammers are widely used and very often misused and abused. Hammers come in a wide range of sizes, types, and weights, and are configured for very specific tasks. Hammers should be used solely for their intended purposes, as misuse can result in damage to the tool and personal injury. Three common types of hammer are discussed here.

> **CAUTION**
>
> **CAUTION:** Always inspect hammerheads for damage and a secure connection to the tool handle. If any damage is noticed, the tool should not be used.

NAIL HAMMERS

Nail hammers are intended to drive nails (Fig. 5-42). The claw at the back of the hammer is used to remove nails and to rip wood. The claw portion of the hammer, which can be curved or straight, is not intended to be struck against metal surfaces.

> **CAUTION**
>
> **CAUTION:** When striking nails, be sure to use the center of the striking face to reduce the possibility of tool slippage. The striking face is the portion of a hammerhead that comes in contact with the object being struck.

> **CAUTION**
>
> **CAUTION:** Using the nail hammer to strike other hammers or other surfaces can result in tool damage and personal injury.

BALL PEEN HAMMERS

Ball peen hammers have slightly rounded striking faces, which are often beveled (Fig. 5-43). The striking face is used for striking chisels and punches, while the peen portion or the hammer is typically used to shape and/or straighten metals. The peen is the rounded back portion of the head of the hammer. When used to strike chisels, the diameter of the striking face of the hammer should be about 3/8″ larger than the diameter of the chisel.

> **CAUTION**
>
> **CAUTION:** When using a ball peen hammer, only the striking face and peen are to be used for striking objects. Never use the cheek, or side, of the hammer as a striking surface.

FIGURE 5-42 Nail hammer.

Photo by Eugene Silberstein.

FIGURE 5-43 Ball peen hammer.

Photo by Eugene Silberstein.

FIGURE 5-44 Soft face hammer.

Photo by Eugene Silberstein.

SOFT FACE HAMMERS AND MALLETS

Soft face hammers have heads that can be made from a wide variety of materials including plastic, rubber, and wood (Fig. 5-44). These hammers are a popular alternative to those made from steel when the surface being worked on must not become damaged or marred. Wooden mallets are commonly used in conjunction with plastic handled chisels and when hammering other wooden objects. Rubber and plastic mallets are often used in the masonry industry for setting stone. In the HVAC/R industry, rubber mallets are often used to shift sections of metal ductwork during installation to prevent surface damage.

CHISELS

Generally speaking, chisels are cutting tools that are struck with hammers in order to remove materials or, in some cases, cut materials. As with many other types of tools, the proper chisel must be selected for the job being performed. If the incorrect chisel is used, tool damage and personal injury can result. Two types of chisel commonly found in the HVAC/R industry are the cold chisel and the brick chisel.

COLD CHISELS

The cold chisel is made from steel and has a cutting edge on one end and the struck face on the other (Fig. 5-45). The struck face is the end of the chisel that the hammer makes contact with. Cold chisels are used to cut or shape metals that are softer than the chisel itself.

> **CAUTION**
>
> **CAUTION:** Do not use chisels that have become dull or those that have a mushroomed head. The term mushroomed head is used to describe the misshapen, damaged head of a chisel or other struck tool (Fig. 5-46). A chisel should be thrown away if it becomes chipped or is in any way damaged.

FIGURE 5-45 Cold chisel.

Photo by Eugene Silberstein.

FIGURE 5-46 Mushroomed head.

Photo by Eugene Silberstein.

BRICK CHISELS

Brick chisels are made from a single piece of steel which makes up the cutting edge, the handle and the struck face of the tool (Fig. 5-47). These chisels are used to score, cut, and trim bricks. Brick chisels are to be struck with a hand drilling or sledge hammer; never a nail hammer.

FIGURE 5-47 Brick chisel.

Photo by Eugene Silberstein.

As with other chisels, if the struck face is mushroomed, the chisel should be discarded.

CAUTION

CAUTION: Never use a brick chisel on metal surfaces!

FILES

The file is used to remove any rough edges from the exterior surfaces of a cut section of metal or other material (Fig. 5-48). Files can be classified as flat, round, half round, or triangular. Since they come in a wide range of sizes and shapes, always select the right file for the job. For example, a round file can be used to effectively remove burrs from the inside of a cut section of PVC pipe.

HACKSAWS

The hacksaw is often used to cut relatively thin metal objects such as angle iron or metal rods (Fig. 5-49). It can be used to cut piping materials but it is *not* recommended for cutting piping material that is used in a refrigerant-carrying piping circuit. The metal filings and burrs that are created can easily find their way into the piping circuit and cause problems with the components in the refrigeration circuit. The blades used in a hacksaw are secured to the handle by a wing nut that should be tight enough to prevent the blade from moving within the handle. As with any saw or cutting tool, exerting excessive force on the tool can result in damage to the tool as well as personal injury.

UTILITY KNIVES

Utility knives have razor-sharp blades enclosed in a plastic or metal handle (Fig. 5-50). Always handle the utility knife with care, especially if the tool being used does not have a retractable blade. Always cut in a direction away from your body and keep fingers out of the way of the blade. If the tool has a retractable blade, make certain that the blade is completely contained within the tool whenever it is not in use. Also, make certain that the casing of the tools is closed tightly to ensure that the blade is held securely in place.

CAUTION

CAUTION: Sharp blades cut smoothly with less force, so be sure to replace dull blades. Dull blades can bind or get caught, possibly resulting in injury.

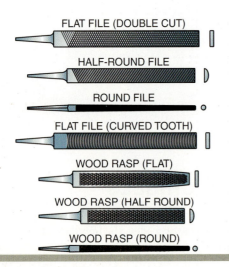

FLAT FILE (DOUBLE CUT)

HALF-ROUND FILE

ROUND FILE

FLAT FILE (CURVED TOOTH)

WOOD RASP (FLAT)

WOOD RASP (HALF ROUND)

WOOD RASP (ROUND)

FIGURE 5-48 Files.

Courtesy of Ideal Industries, Inc.

FIGURE 5-49 Hacksaw.

Courtesy of Ideal Industries, Inc.

FIGURE 5-50 Utility knife.

Courtesy of Ideal Industries, Inc.

FIGURE 5-51 Tape measure.

TAPE MEASURES

Probably one of the most important tools used by the installation crew is the tape measure, which enables the worker to transfer measurements from one place to another (Fig. 5-51). For example, a section of condensate drain piping may need to be cut to a specific length. Once this length is determined, the measurement must be transferred to the pipe to ensure that it is cut to the proper length. The tape measure measures not only feet and inches but fractions of an inch as well. They are commonly found in lengths ranging from 6 feet to 50 feet. Be sure to select a tape measure that will best suit the application for which it is intended.

SQUARES

The square is very versatile in the sense that it acts as a number of separate tools (Fig. 5-52). It can be used as a ruler for short measurements, as a level, and as a guide to mark 45 and 90 degree lines. Squares are available in a number of different configurations depending on the needs of the user. Those who are performing carpentry work, for example, will utilize a carpenter's square (Fig. 5-53).

Photo by Eugene Silberstein.

FIGURE 5-52 Square.

LEVELS

The level is also commonly known as a spirit level or bubble level (Fig. 5-54). This tool is designed to indicate whether a surface, pipe, duct section or other system

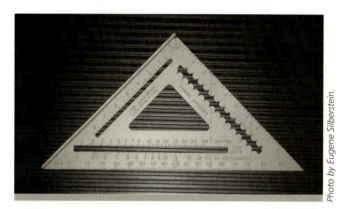

Photo by Eugene Silberstein.

FIGURE 5-53 Carpenter's square.

Photo by Eugene Silberstein.

FIGURE 5-54 Level.

component is **level** or **plumb**. Level simply means at the same height from side to side, while plumb indicates something that is perfectly vertical. This is accomplished by using a slightly curved glass tube which is filled with a yellow liquid, usually **ethanol**. The tube is not completely filled, so a bubble of air remains in the tube. The level is placed in the object to be levelled and then the position of the bulb is observed. When the bubble is in the center of the tube, the object is level. Levels are obtainable in lengths ranging from just a couple of inches to over six feet, depending on what is to be leveled.

DROPLIGHTS AND EXTENSION CORDS

Droplights and extension cords are without a doubt the installation crew's best friends during the installation process (Fig. 5-55). While working in otherwise pitch-black attics and basements, the droplight provides the needed light to help ensure that the job is done right and that the workers remain safe. As with other tools, the ground prong must never be removed from the cord, as

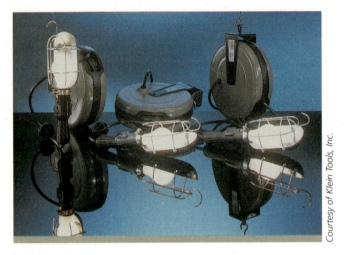

FIGURE 5-55 Droplights and extension cords.

Courtesy of Klein Tools, Inc.

it provides protection from electric shock. When using droplights, try to obtain heavy-duty light bulbs, which will withstand the rough handling that they will be subjected to. Try to position extension cords and droplights away from the path that workers will take to reduce the possibility of tripping.

Going Green

Be sure to use environmentally friendly bulbs in your droplights and other light fixtures on the job.

DRILLS

The electric drill is probably one of the most useful tools used by the air conditioning, heating, and refrigeration technician. From drilling holes to setting screws the drill is a must have for all those working in the field. Drills are available in both corded and cordless models and each has some benefits and drawbacks.

CAUTION

CAUTION: Before using a drill or any other power tool, be sure to inspect it and the cord for any damage. Do not use the tool if any damage is noticed. Never use any power tool if the ground prong on the plug has been removed.

CORDED DRILLS

When working in an area where there is an AC power source, the corded drill is a good choice (Fig. 5-56). Typically, corded drills supply more power than the cordless variety, and there is no need to keep a charged battery on hand. The one drawback of using a corded tool is that mobility ranges are limited to the length of the power or extension cord. Another possible drawback is when working on a construction site where temporary power has not yet been provided for.

CORDLESS DRILLS

Probably one of the most useful tools for the duct installer is the cordless, variable-speed electric drill (Fig. 5-57). When working with ductwork and self-tapping screws, the variable-speed feature enables the user to start the drill at a reduced speed to start the hole. Once the hole has been started, the speed of the drill can be increased. The greatest benefit of using a cordless drill is the increased mobility of the user. There is no need to run extension cords or worry about whether or not there is a power source available. One of the biggest drawbacks, however, is the need to maintain the battery charge. It is a good field practice to keep a spare, charged battery with your tools, just in case. Most manufacturers offer a battery charger that plugs into a vehicle's cigarette lighter.

Courtesy of Milwaukee Electric Tool Corporation.

FIGURE 5-56 Corded drill.

Courtesy of Milwaukee Electric Tool Corporation.

FIGURE 5-57 Cordless drill.

Going Green

When it becomes necessary to dispose of old batteries, be sure to do so in a manner that is in accordance with local codes.

RECIPROCATING SAWS

Reciprocating saws are very useful in the field for cutting wood, metal, piping, or a number of other materials (Fig. 5-58). Reciprocating saws can be fitted with a wide range of blades of different length and style for cutting different materials (Fig. 5-59). Both corded and cordless reciprocating saws are available.

Photo by Eugene Silberstein.

FIGURE 5-58 Reciprocating saw.

Photo by Eugene Silberstein.

FIGURE 5-59 Reciprocating saw blades.

SUMMARY

- Hand tools are to be used only for their intended purpose.
- Inappropriate tool use can lead to tool damage, personal injury as well as damage to the equipment being serviced.
- Select the proper screwdriver for the screw that is being tightened or loosened.
- Longer screwdriver shanks provide more torque than shorter ones.
- Stubby-type and offset screwdrivers are used to get into tight spaces.
- Nut drivers are used to tighten and loosen screws and bolts with hex heads.
- Common wrench types are the open-end, the box-end, the combination wrench, the adjustable wrench, and the ratchet-type wrench.
- The proper size wrench should always to be used so the wrench fits snugly on the fastener being turned.
- The box wrench and the ratchet-type wrench can be used to free tight fasteners and for final tightening of a fastener.
- The open-end wrench and the adjustable wrench *should not* be used to free tight fasteners or for the final tightening of a fastener.
- Hex keys are six sided tools that are used to turn fasteners that have hexagonal-shaped openings.
- Common pliers are the general purpose, needle-nose, side cutters, linesman, pump and locking varieties.
- Wire strippers and crimping tools can perform tasks such as cutting wire, cutting small screws and bolts, or crimping solderless connectors onto sections of stripped stranded wire.
- The electrical snake is used to help run electrical wires through walls and ceilings.
- Common hammer types include the nail hammer, the ball peen hammer, and soft face hammers and mallets.
- Be sure to select the correct hammer for the job being performed.
- Two common types of chisel are the cold chisel and the brick chisel.
- Damaged, chipped or mushroomed chisels should be discarded.
- Files are used to remove any rough edges from the exterior surfaces of a cut section of metal or other material.

- The hacksaw is often used to cut relatively thin metal objects such as angle iron or metal rods.
- The tape measure enables the user to transfer measurements from one place to another.
- The level is designed to indicate whether a surface, pipe, duct section or other system component is level or plumb.

- Electric drills help the technician drill holes and set/remove screws.
- Electric drills can be of the corded or cordless variety.
- All power tools should be inspected for damage prior to use.
- Reciprocating saws can be fitted with a wide range of blades for cutting different materials.

GREEN CHECKLIST

☐ **Be sure to use environmentally friendly light bulbs in droplights and lighting fixtures.**

☐ **Be sure to dispose of used batteries in a manner that will do minimal damage to the environment.**

REVIEW QUESTIONS

1. **If a pry bar is not available, which of the following tools can be safely used instead?**
   a. A screwdriver
   b. A Ratchet-type wrench
   c. A nut driver
   d. None of the above

2. **Which of the following screwdrivers can be used to get into awkward spaces?**
   a. The stubby screwdriver
   b. The offset screwdriver
   c. Neither a nor b is correct
   d. Both a and b are correct

3. **Nut drivers are best used for loosening and tightening**
   a. Straight slot screws
   b. Hex head screws
   c. Phillips head screws
   d. Screws with hex-shaped recessed slots

4. **When loosening extremely tight fasteners, which of the following wrenches is the best choice for the job?**
   a. An open-end wrench
   b. An adjustable wrench
   c. A box wrench
   d. Both a and b are correct

5. **Nut drivers with hollow shafts are especially useful when**
   a. The nut has been over tightened
   b. The bolt extends past the nut
   c. Access to the nut is blocked by a piece of metal
   d. The hex head has been rounded off

6. **Which of the following wrenches provides the most torque?**
   a. The ratchet-type wrench
   b. The open-end wrench
   c. The adjustable wrench
   d. The combination wrench

7. **When is it acceptable field practice to strike an adjustable wrench with a hammer?**
   a. When the wrench is stuck on the nut being turned
   b. When the adjustment screw on the wrench is stuck
   c. When the adjustable wrench is in the completely closed position
   d. It is never acceptable to strike an adjustable wrench with a hammer

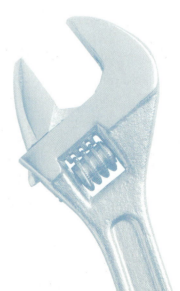

8. **Which of the following best describes a complete socket set?**

   a. Sockets, ratcheting-type wrench, extensions and a ball joint

   b. Sockets, combination type wrench, extensions and a universal joint

   c. Sockets, ratcheting-type wrench, extensions and a universal joint

   d. Sockets, combination type wrench, extensions and a ball joint

9. **All of the following pliers can be used to bend heavy wire** *except*

   a. Locking pliers

   b. Linesman pliers

   c. Needle-nose pliers

   d. General purpose, slip joint pliers

10. **Which of the following type of pliers would most likely be used on a kitchen sink drain?**

    a. General purpose, slip joint pliers

    b. Slip joint pump pliers

    c. Locking pliers

    d. Side cutting pliers

11. **Which of the following type of pliers would most likely be used to cut a small nail flush with a surface?**

    a. General purpose, slip joint pliers

    b. Slip joint pump pliers

    c. Locking pliers

    d. Side cutting pliers

12. **Which of the following hammers would you be most likely to use if you are working with a cold chisel?**

    a. A nail hammer

    b. A ball peen hammer

    c. A soft face hammer

    d. A wooden mallet

13. **Which of the following parts of the ball peen hammer are acceptable for striking?**

    a. Only the striking face

    b. The striking face and the cheek

    c. The cheek and the peen

    d. The striking face and the peen

14. **When referring to a level, which of the following statements is correct?**

    a. The term "plumb" indicates a perfectly vertical plane or line

    b. The term "level" indicates a perfectly horizontal plane or line

    c. Both a and b are correct

    d. Neither a nor b is correct

15. **When working with electric power tools, what should be done before putting the tool into operation?**

    a. Inspect the tool for damage

    b. Inspect the power cord for damage

    c. Inspect the plug for damage

    d. All of the above are correct

KNOW YOUR CODES

Research and discuss with your classmates the safe way to dispose of batteries that are no longer usable. What disposal methods are acceptable in your area? What disposal methods can harm the environment?

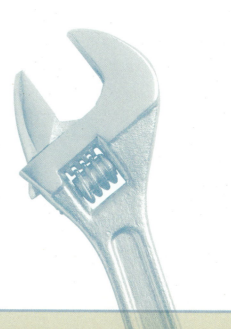

Tubing and Piping Tools

OBJECTIVES *Upon completion of this chapter, the student should be able to*

- List various tubing and piping tools used by air conditioning and heating technicians.

- Explain the function of tubing and piping tools.

- Explain the processes of reaming, flaring, swaging, and bending as they pertain to copper tubing.

- Explain the difference between soft-drawn and hard-drawn piping materials.

- Describe the piping applications that would use pipe fittings and those that would rely on the use of a bending tool.

- Explain the difference between soldering and brazing.

- List the various fuels and gases used for soldering and brazing.

- Explain the difference between an air-acetylene and an oxy-acetylene torch kit.

- List the component parts of an air-acetylene torch kit.

- List the component parts of an oxy-acetylene torch kit.

- Explain the function of a regulator as it applies to an air-acetylene torch kit.

- Compare the temperatures of the flames produced by the various fuels used for soldering and brazing.

- Properly set up the air-acetylene torch kit for use.

- Properly set up the oxy-acetylene torch kit for use.

One aspect of the many tasks that are performed by the HVAC/R technician involves the repairing, installing, and altering of the lines that carry fluids through heating, refrigeration, and air conditioning equipment. These lines include those that carry refrigerant between the system components; those that carry condensate away from the air handler; those that carry water to and from humidifiers, boilers, and other appliances; and those that carry fuel to our heating appliances. In order to keep these fluids contained within the piping arrangements, the individual responsible for assembling these piping configurations must be well versed in the tools used to work with common piping materials. Piping materials commonly used by the HVAC/R technician are soft-drawn copper tubing, hard-drawn copper tubing, steel pipe, and PVC tubing. In this chapter, we will examine many of the tools that the technician uses when working with piping circuits and materials. In Chapter 10, the use of these tools is discussed as they apply to the actual connection of system components.

Acetylene Popular gas used for soldering and brazing

Fittings Preformed piping components which make connections and turns in rigid piping arrangements possible

Flaring Process by which a 45-degree angle is formed on the end of a tubing section

Flaring yoke Portion of the flaring tool used to create the 45-degree angle on the cut portion of the tubing section

Galvanized steel pipe Steel pipe that is coated with zinc to resist rusting

Hard-drawn tubing Term used to describe rigid pipe

Left-hand threads Threaded connection that is tightened by turning the connection to the left

MAPP™ gas Composite gas that burns at temperatures higher than that of propane (methyl acetylene-propadiene)

Process tubes The lines through which hermetically sealed systems are pressurized, leak tested, and charged

Reamer Tool used to remove burrs from piping and tubing

Regulator Torch kit component that regulates the pressure at the outlet of the vessel

Right-hand threads Threaded connection that is tightened by turning the connection to the right

Soft-drawn tubing Refers to piping materials that are typically flexible and easily bent

Soft metal tubing Term used to describe tubing and pipe made from soft metals such as copper and aluminum

Swaging Process of expanding the end of a tubing section so that another section of same size tubing can be inserted into it

TUBING TOOLS

When we refer to tubing, we are discussing piping materials that are classified as soft-drawn. **Soft-drawn tubing** refers to materials that are typically flexible and easily bent. The tools that will be discussed in this section are those that are used on soft-drawn tubing. More information can be found on soft-drawn tubing and its properties in Chapter 10. **Hard-drawn tubing** is the term used to describe rigid pipe. The tools listed in this section are typically not used for hard-drawn pipe except where indicated.

TUBING CUTTERS

The tubing cutter (Fig. 6-1) uses a set of rollers and a round cutting blade. This type of tubing cutter can be used to cut **soft metal tubing** and pipe such as soft- and hard-drawn copper tubing. The piping material being cut is held in place between the rollers and the blade. The position of the blade in the cutter with respect to the tubing is adjusted by the handle on the cutter. As the handle is turned, the blade moves either closer to, or further from, the rollers. Most tubing cutters are equipped with a reamer that removes burrs from the inside of the cut section of tubing. The tubing cutter is the preferred method for cutting refrigeration tubing, as there are no filings produced.

REAMERS

The **reamer** is a tool that is used to remove burrs from a cut section of tubing (Fig. 6-2). The reamer that is often a part of the tubing cutter has the ability to remove burrs from the inside of the cut tubing section. The reamer shown in Figure 6-2 is referred to as an inner-outer reamer and can remove burrs from both the inside and outside edges of the cut tubing section.

Photo by Eugene Silberstein.

FIGURE 6-1 Tubing Cutter.

FIGURE 6-3 Flaring block.

FIGURE 6-4 Components of a flare connection.

FIGURE 6-2 (A) This part of the reamer removes burrs from the inside of the pipe. (B) This part of the reamer removes burrs from the outside of the pipe.

FLARING BLOCKS

The flaring block (Fig. 6-3) is used to hold tubing in place while piping operations, such as **flaring** and **swaging**, are performed. Flaring is the process by which a 45-degree angle is formed on the end of a tubing section. The flared end is then joined to flare fittings or system components equipped with mating flare connections (Fig. 6-4). Swaging is the process of expanding the end of a tubing section so that another section of same size tubing can be inserted into it. The flaring block has holes of different sizes to accommodate tubing of different sizes.

FLARING YOKE

The flaring yoke (Fig. 6-5) is one of the tools used in conjunction with the flaring block. The **flaring yoke** is used to create the 45-degree angle on the end of the cut tubing section. In operation, the flaring yoke is positioned on

FIGURE 6-5 Flaring yoke.

the flaring block over the end of the tubing and tightened in place to create the flare.

SWAGING TOOLS

The swaging tool (Fig. 6-6) is another tool used in conjunction with the flaring block. Swaging tools are used to expand the end of a cut tubing section so that another section of same sized tubing can be inserted into it (Fig. 6-7).

> ### from experience...
>
> When hard-drawn tubing is used, bending is not practical, so straight pipe sections are used in conjunction with fittings. Fittings are piping components that are preformed into angles which make turns in rigid piping arrangements possible.

FIGURE 6-6 Swage punches.

FIGURE 6-7 Copper tubing will fit into the swaged end of the other section.

TUBING BENDERS

In order to minimize the possibility of leaks in a system, it is often desired to bend copper tubing into the desired configuration. Typically, in our industry, only soft-drawn tubing is bent. Two common types of tubing benders are the spring bender (Fig. 6-8) and the lever bender (Fig. 6-9). The spring bender is positioned over the tubing and bent to the desired shape. The spring evenly distributes the pressure on the tubing and prevents it from kinking. The lever bender is a mechanical bending device that enables the user to make precise angles in the tubing.

PINCH-OFF TOOLS

When performing service on hermetically sealed systems, it is often necessary to pinch off **process tubes**. Process tubes are the lines through which sealed systems are pressurized, leak tested, and charged (Fig. 6-10). This enables the technician to once again seal the system, as nonsoldered line taps and service apertures must not remain on the system. The pinch-off tool shown in Figure 6-11 functions as a flaring block as well as pinch-off tool. When tubing is placed in the pinch-off tool and the wing nuts are tightened, the tubing is pinched closed as shown in Figure 6-11(B).

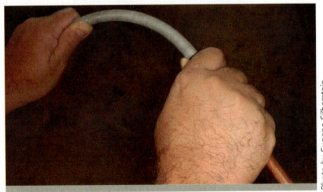

FIGURE 6-8 Tube bending spring.

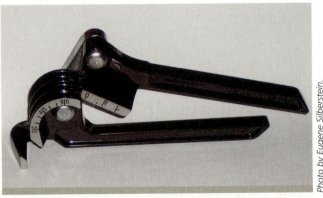

FIGURE 6-9 Lever-type tubing bender.

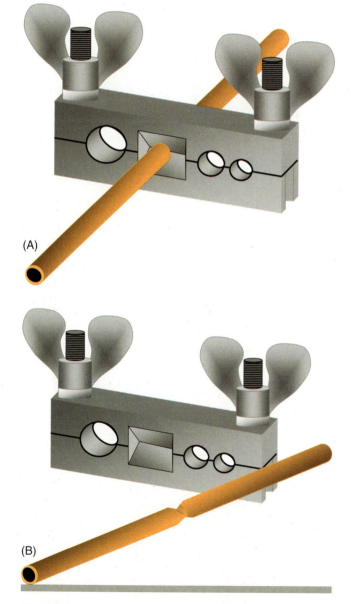

FIGURE 6-10 Process tube.

Photo by Eugene Silberstein.

(A)

(B)

FIGURE 6-11 (A) Pinch-off tool on the tubing.
(B) The pinch-off tool and the final result.

SOLDERING AND BRAZING TOOLS

When sections of tubing or pipe need to be connected, either to each other or to a fitting, the processes of soldering and brazing are often used. Soldering and brazing are two similar, but different processes that involve the introduction of a molten-filler material between the mating surfaces being joined. Soldering involves filler material with melting temperatures below 842°F, while brazing typically involves filler materials with melting temperatures over 842°F. The determination about what process to use depends on a number of factors including the application of the piping arrangement—the expected pressures in the piping arrangement as well as building/construction code requirements. Building and construction codes also determine the piping materials that are to be used for particular applications. The tools and equipment that are covered in this section are those that are typically used to perform the soldering and brazing processes.

FUELS AND GASES

In order to solder or braze, a heat source is needed. We heat the components being joined to the correct temperature and then apply a filler material to the joint in order to make the connection. A number of different substances are used in conjunction with the soldering and brazing processes and include propane, **MAPP™ gas**, nitrogen, oxygen and **acetylene** (Fig. 6-12). MAPP™ gas (methyl acetylene-propadiene) is a composite gas that burns at a temperature higher than that of propane. Acetylene is a very popular gas used for soldering and brazing.

Courtesy of Uniweld Products, Inc.

FIGURE 6-12 Various tanks of pressurized gases used in soldering and brazing processes.

Although acetylene is found naturally in our universe, the acetylene we use is manufactured by the partial combustion of methane. These substances are shipped and stored in vessels of varying sizes to best accommodate the needs of the user. Common sizes for acetylene tanks are the "B" tank and the "MC" tank. The "B" tank measures about 24″ high and 6″ in diameter and has a capacity of 40 ft^3. The "MC" tank measures about 15″ high and 4″ in diameter and has a capacity of about 10 ft^3.

CAUTION

CAUTION: When transporting cylinders, always make certain that protective caps are in place, the tanks are stored vertically and are held securely in place.

Going Green

Make certain that all tanks are closed when not in use and that waste is minimized. Hydrocarbons, although, they do not contribute to ozone depletion, do have substantial global warming potential.

REGULATORS

When dealing with cylinders of pressurized fluids, it is important to regulate the pressure at the outlet of the vessel to help ensure the safety of the user. For this reason, regulators are connected to the tanks (Fig. 6-13). Regulators for acetylene, oxygen, and nitrogen are the commonly used regulators in our industry.

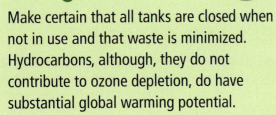

Photo by Eugene Silberstein.

FIGURE 6-13 Oxygen and acetylene regulators.

Photo by Eugene Silberstein.

FIGURE 6-14 Air-acetylene torch kit.

AIR-ACETYLENE TORCH KIT

The air-acetylene torch (Fig. 6-14) is the most popular heat source used by air conditioning, heating, and refrigeration technicians. Typically, the air-acetylene torch provides flames that are in the 5600°F range, making it a great choice for both low- and high-temperature soldering and brazing. The air-acetylene torch kit is comprised of the tank, acetylene regulator, torch hose, torch handle, and torch tip.

For step-by-step instructions on setting up, lighting, and extinguishing the air-acetylene torch, see Procedure 6-1 on pages 151–152.

THE TANK

The acetylene tank has a square-stemmed valve (Fig. 6-15). This stem should only be turned with a wrench that is specifically intended for use on square

from experience...

Do not overtighten the stem on the acetylene tank. You might snap the stem off! This can lead to severe personal injury as well as environmental damage.

FIGURE 6-15 Square stem on the acetylene tank.

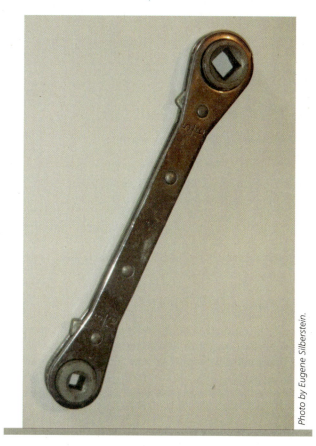

FIGURE 6-16 Refrigeration service wrench.

stems. The refrigeration service wrench is such a tool (Fig. 6-16). Using the improper tool to open and close the valve on the tank can result in the rounding of the stem, making the valve inoperable. Turning the stem clockwise closes the valve, while turning the stem counterclockwise opens it.

ACETYLENE REGULATOR

The acetylene regulator (Fig. 6-17) is connected directly to the acetylene tank. The regulator should be tightened to the tank and the connection should be checked for leaks prior to using the torch. Turning the handle on the regulator counterclockwise lowers the acetylene pressure at the outlet of the regulator, while turning the regulator's handle clockwise increases the pressure. It is at the regulator that any fuel flow adjustments should be made.

CAUTION

CAUTION: When connecting the regulator to the acetylene tank, make certain that the tank is secured to prevent it from slipping or falling over.

FIGURE 6-17 Acetylene regulator.

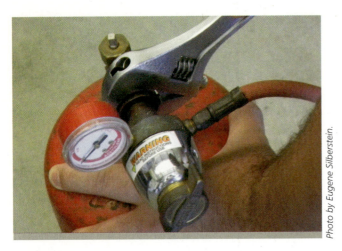

FIGURE 6-18 Acetylene hose connected to the regulator.

TORCH HOSE

The acetylene hose is color-coded red and is connected to the outlet port on the acetylene regulator (Fig. 6-18). The other end of the hose connects to the torch handle. Before using the torch, be sure to inspect the hose for any damage. Replace a damaged hose before using the torch.

TORCH HANDLE

The torch handle (Fig. 6-19) connects the hose to the torch tip and provides a valve, which is used to start and stop acetylene flow to the tip. This valve should not be used to adjust the intensity of the flame. Any fuel flow adjustments should be made at the regulator.

TORCH TIPS

The torch tip (Fig. 6-20) is the torch kit component that determines the amount of acetylene flow, allows air to mix with the acetylene, and produces the flame used for soldering and brazing. The amount of heat required depends on whether the torch is being used for soldering or brazing, and the size of the pipes being joined. Brazing a copper 3″ pipe joint will obviously require more heat than soldering a copper 3/4″ joint. Manufacturers provide charts that indicate the proper size tip to be used for various pipe sizes (Fig. 6-21). Torch tips are also available for special applications. For example, there are tips that allow the user to heat the entire circumference of a pipe or to fit at the same time (Fig. 6-22). There are also torch tips that are used specifically for removing reversing valves from heat pump systems (Fig. 6-23).

MAPP™ GAS TORCH

When soldering or brazing, a MAPP™ torch (Fig. 6-24) is a popular choice because the kit is small, lightweight, and easy to maneuver. The flame produced by MAPP™

FIGURE 6-19 Air-acetylene torch handle.

FIGURE 6-20 Various torch tips.

(methylacetylene-propadiene) is at a temperature of about 5300°F, which is close to the temperature of the air-acetylene flame. Because of the flame temperature, the MAPP™ torch can be used for both soldering and brazing processes. Torch kits like this one are available in different models and styles with different methods for lighting.

ACETYLENE TORCH TIPS								
Tip Size			Gas Flow		Copper Tubing Size Capacity			
Tip No.	in.	mm	@ 14 psi ft³/hr	(0.9 Bar) m³/hr	Soft Solder in.	mm	Silver Solder in.	mm
A–2	³⁄₁₆	4.8	2.0	.17	⅛–½	3–15	⅛–¼	3–10
A–3	¼	6.4	3.6	.31	¼–1	5–25	⅛–½	3–12
A–5	⁵⁄₁₆	7.9	5.7	.48	¾–1½	20–40	¼–¾	10–20
A–8	⅜	9.5	8.3	.71	1–2	25–50	½–1	15–30
A–11	⁷⁄₁₆	11.1	11.0	.94	1½–3	40–75	⅞–1⅝	20–40
A–14	½	12.7	14.5	1.23	2–3½	50–90	1–2	30–50
A–32*	¾	19.0	33.2	2.82	4–6	100–150	1½–4	40–100
MSA–8	⅜	9.5	5.8	.50	¾–3	20–40	¼–¾	10–20

*Use with large tank only.
NOTE: For air conditioning, add 1/8 inch for type L tubing.

FIGURE 6-21 Table showing torch tip sizes and their capacities.

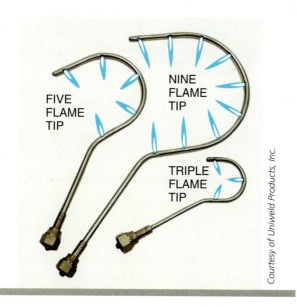

FIGURE 6-22 Torch tip that can be used to heat the entire circumference of a fitting at the same time.

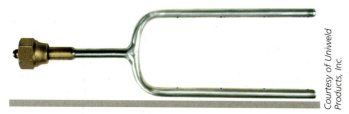

FIGURE 6-23 Torch tip that can be used to remove and replace the four-way reversing valve on a heat pump system.

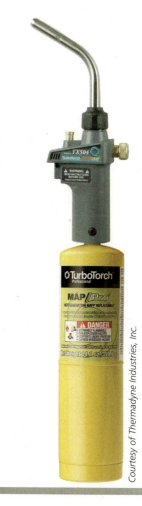

FIGURE 6-24 A MAPP™ gas torch kit.

Going Green

When selecting soldering and brazing supplies and products, be sure to use those that pose the least amount of environmental damage.

PROPANE GAS TORCH

The propane torch (Fig. 6-25) is often used for soldering purposes. The flame that is produced is in the 3600°F range and is therefore not the best choice when brazing. The cost of the propane is much less than that of the MAPP™ gas so it is important to determine the job being performed before making a purchase.

OXY-ACETYLENE TORCH KIT

The oxy-acetylene torch kit (Fig. 6-26) is often preferred by many technicians, especially those who are brazing large diameter pipes. The oxy-acetylene torch, which introduces pure oxygen to the reaction, produces flames in the 6600°F

Courtesy of Uniweld Products, Inc.

FIGURE 6-26 Oxy-acetylene torch kit.

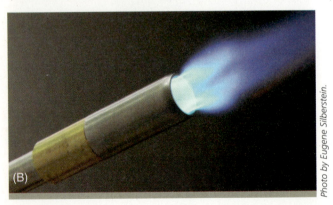

Photo by Eugene Silberstein.

FIGURE 6-25 (A) Propane torch. (B) Flame produced by the propane torch.

to 8000°F range. The oxy-acetylene torch kit is comprised of an acetylene tank, an oxygen tank, an acetylene regulator, an oxygen regulator, a hose set, a torch handle, and a torch tip.

The hose set used for oxy-acetylene torches is made up of two hoses; one red and one green. The green hose is for oxygen and the red hose is for acetylene. To prevent improper tank and torch setup, the hoses are manufactured so that the red and green hose connections cannot be switched. The red hose is manufactured with **left-hand threads** for connection to the acetylene regulator and the green hose is manufactured with **right-hand threads** for connection to the oxygen regulator. Left-hand threads are tightened by turning the fitting to the left while right-hand threads are tightened by turning the fitting to the right.

For step-by-step instructions for setting up, lighting, and extinguishing the oxy-acetylene torch, see Procedure 6-2 on pages 153–155.

OTHER IMPORTANT SOLDERING AND BRAZING TOOLS

In addition to the tools described in the previous pages, there are additional tools that are needed in order to solder and braze. Although some of the following items are not required, they will help make the processes easier to perform.

FIRE EXTINGUISHER

When using a torch to solder or braze, it is extremely important that a fire extinguisher (Fig. 6-27) be nearby. It is also important that the technician is knowledgeable about the use of the fire extinguisher *before* it is actually needed. Be sure to refer back to Chapter 3 for detailed information on fire extinguisher use.

HEAT GUARDS

The heat guard (Fig. 6-28) serves two purposes. It protects the area surrounding the torch flame from fire damage and also helps concentrate torch heat around the pipe. This device clips onto the torch tip.

INSPECTION MIRRORS

Once the soldering or brazing has been completed, the joints must be inspected for leaks prior to system pressurization. The inspection mirror (Fig. 6-29) is used to inspect all sides of a soldered or brazed joint as it might be impossible to do so without the help of a mirror. Some inspection mirrors are manufactured with telescopic

Courtesy of Uniweld Products, Inc.

FIGURE 6-28 Devices such as this heat guard concentrate heat and protect surrounding areas from flame.

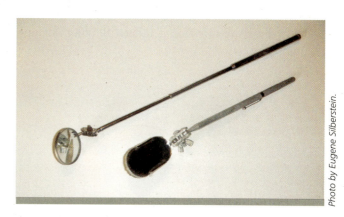

Photo by Eugene Silberstein.

FIGURE 6-29 Inspection mirrors.

handles and magnetic tips to help retrieve small metallic objects that have fallen out of reach.

STRIKER

The striker (Fig. 6-30) is the tool that should always be used to light a torch. The striker creates a spark that ignites the fuel. This tool is the desired method for lighting a torch because it can be operated with one hand and does not contain flammable fluids.

Courtesy of Kidde Safety Products.

FIGURE 6-27 Fire extinguisher.

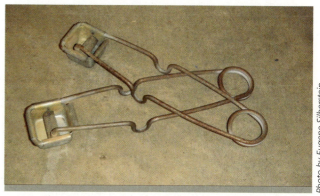

Photo by Eugene Silberstein.

FIGURE 6-30 A striker used to light torches.

FLOW INDICATOR

The flow indicator (Fig. 6-31) is useful when adding nitrogen to the piping arrangement during the brazing process. By adding nitrogen to the piping circuit, oxygen is displaced so no oxidation can occur on the interior surfaces of the pipe. The pressure of the nitrogen being added to the system for this purpose is only 1 psig to 2 psig, and using a traditional gauge to set the nitrogen flow rate, although not critical, is difficult. The flow indicator provides visual confirmation that low-pressure nitrogen is indeed flowing through the piping circuit.

Courtesy of Uniweld Products, Inc.

FIGURE 6-31 The flow indicator provides visual confirmation that nitrogen is flowing through the piping arrangement during the brazing process.

PIPING TOOLS

The tools that will be discussed in this section are those that are used to perform operations on steel piping materials. Steel pipe comes in a number of different types, so the correct pipe should be selected. When manufactured, steel pipe is either seam welded or produced seamless when molten material is drawn through a forming machine. Steel pipe may be painted, left black, or coated with zinc. Steel pipe is coated with zinc to resist rusting and, when coated, is referred to as **galvanized steel pipe**.

PIPE CUTTERS

Similar in appearance to the tubing cutter that cuts copper pipe and tubing, the pipe cutter (Fig. 6-32) cuts steel piping material. These pipe cutters can be equipped with one or more cutting wheels to speed the cutting process. The piping material should be secured in a vise while cutting to ensure that the material is held in place.

PIPE REAMERS

The pipe reamer (Fig. 6-33) is designed to remove burrs from the interior edge of steel pipe. It is inserted into the end of the pipe and rotated to create a smooth edge

Courtesy of Ridge®.

FIGURE 6-32 Pipe cutters.

Photo by Eugene Silberstein.

FIGURE 6-33 The reamer is used to remove burrs from the pipe.

inside the pipe. Burrs on the inside of the pipe can reduce fluid flow through the pipe.

PIPE VISES

The pipe vise (Fig. 6-34) is used to secure piping material during cutting or threading processes. The legs of the pipe stand are spread wide apart to prevent the vise from tipping over. This stand has a chain mechanism to hold the pipe secure. Other pipe vises can use a yoke-type clamp to hold the pipe secure (Fig. 6-35).

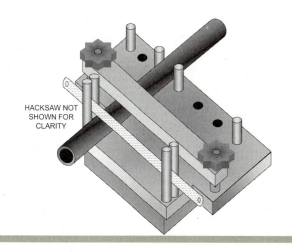

HACKSAW NOT SHOWN FOR CLARITY

FIGURE 6-36 A holding fixture helps keep the saw at right angles to the pipe being cut.

HOLDING FIXTURES

The holding fixture (Fig. 6-36) mounts in a vise and helps ensure a square cut when used with a hacksaw by making certain that the hacksaw remains perpendicular to the pipe during the cutting process.

PIPE THREADERS

Steel pipe sections are joined together to form the desired piping configuration by using fittings. The ends of the fittings are threaded as are the ends of the pipe sections. The fittings and pipe sections are screwed together and tightened to prevent leaks. In the field it is not uncommon for the technician to have to thread a section of pipe for connection into the piping circuit. The tools used to accomplish this task are explained below.

THREADING DIE HEAD

Pipe threading die heads (Fig. 6-37) hold a set of four cutting blades that cut the threads into the end of the steel pipe. These cutters are numbered and must be installed in

Courtesy of Ridge®.

FIGURE 6-34 A pipe stand with a chain vise.

Courtesy of Ridge®.

FIGURE 6-35 A pipe stand with a yoke vise.

Delmar/Cengage Learning.

FIGURE 6-37 Pipe threading dies.

the proper order so that the threads are created properly. Dull blades must be replaced to ensure proper threads and a leak-free installation.

THREADING DIE HANDLES

The pipe threading die handle (Fig. 6-38) holds the threading die head in place. These handles are manufactured with a ratchet to speed the threading process. When the knob is pointing in the clockwise direction, the handle will tighten the dies onto the end of the pipe, creating threads. When the knob is pointing in the counterclockwise direction, the ratcheting handle will remove the threading dies from the end of the pipe.

FIXED DIE-TYPE PIPE THREADER

The fixed die pipe threader (Fig. 6-39) does not use a ratcheting handle. Instead, it has two handles that rotate the threader onto the pipe to create threads. As with any other type of threader, sufficient cutting oil must be applied to the pipe and dies, to ensure proper threading and to prolong the life of the dies.

AUTOMATIC PIPE THREADER

Used on jobs where a large number of pipe threads must be made, the automatic pipe threader (Fig. 6-40) is a great option. This tool automatically lubricates and rotates the pipe section within the dies to create threads. The threader can operate in either the forward or reverse direction to thread and then remove the pipe from the threading dies.

Courtesy of Ridge Tool Company.

FIGURE 6-40 Automatic pipe threader.

CUTTING OIL PUMP

The cutting oil pump (Fig. 6-41) catches oil that falls from the pipe being threaded. This oil passes through a filter screen to separate it from the steel filings and then feeds it to the pipe end during the threading process. Use of an oil pump greatly reduces the amount of oil used to thread pipe as well as the amount of oil that ultimately ends up on the floor.

PIPE WRENCHES

Pipe wrenches (Fig. 6-42) are used to tighten steel pipe and fittings. They have jaws that will grip the pipe when turned towards the open jaw and release the pipe when turned away from the open jaw. Typically, two wrenches are used to create a tight connection between a pipe section and a fitting. One wrench holds the pipe steady, while the other tightens the fitting onto the end of the pipe.

Courtesy of Ridge Tool Company.

FIGURE 6-38 Pipe threading die handle.

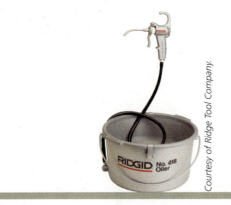

Courtesy of Ridge Tool Company.

FIGURE 6-41 Cutting oil pump.

Courtesy of Ridge® Tool Company.

FIGURE 6-39 Fixed die-type pipe threader.

Courtesy of Ridge Tool Company.

FIGURE 6-42 Pipe wrenches.

ALTERNATIVE PIPING TOOLS

In addition to the tools already mentioned in this chapter, there are others that the technician might find useful. Different applications call for different piping materials and different joining methods.

TUBING SHEAR

PVC and other plastic piping materials are often used for condensate lines and other HVAC/R-related applications. The tubing shear (Fig. 6-43) cuts plastic tubing, small PVC pipe, and other synthetic hoses that are not wire-reinforced. Although hacksaws can be used to cut plastic pipe, the cut ends that result are often not square and are in need of reaming (Fig. 6-44).

PRESS-TYPE PIPING TOOLS

Press-type fittings (Fig. 6-45) are becoming popular for nonrefrigerant carrying applications as well as other low-pressure plumbing applications such as the water lines that are run to evaporative coolers. The crimping tool (Fig. 6-46) is used to secure the fittings onto the piping sections. Various jaw sizes (Fig. 6-47) are available so that the tool can be used on pipes of different sizes. Internal gaskets in the fitting help reduce the possibility of leaks.

Courtesy of Viega.

FIGURE 6-45 Cutaway view of a press-type fitting installed on a pipe.

Photo by Eugene Silberstein.

FIGURE 6-43 Plastic tubing shear.

Photo by Eugene Silberstein.

FIGURE 6-44 When a hacksaw is used to cut PVC piping, many burrs remain so the pipe must be reamed.

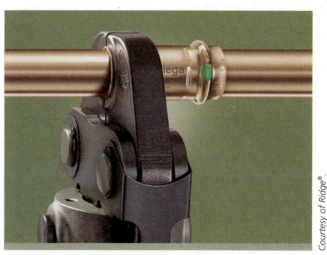

Courtesy of Ridge®.

FIGURE 6-46 Crimping the press-type fitting to join two pipe sections.

Courtesy of Ridge®.

FIGURE 6-47 Various jaw sizes used for installing press-type fittings on pipes of different sizes.

SUMMARY

- Common piping materials used in the HVAC/R industry include, soft-drawn copper tubing, hard-drawn copper tubing, steel black pipe, galvanized steel pipe and PVC.

- Tubing cutters are used to cut copper tubing and rigid pipe without creating filings.

- Reamers are used to remove burrs from the inside and outside edges of piping materials.

- The flaring block is used for both the flaring and swaging processes.

- The flaring yoke creates the 45-degree angle on the end of a section of soft-drawn tubing.

- Flare nuts are used to secure the flared tubing to a fitting or system component.

- Swaging tools are used to expand the end of a cut tubing section so that another section of same size tubing can be inserted into it.

- Tubing benders are used to reduce the number of fittings and soldered or brazed joints in a piping arrangement.

- Fittings are preformed piping components used in conjunction with hard-drawn piping circuits.

- Common types of tubing benders are the spring and lever bender.

- Pinch-off tools are used to seal off process tubes.

- Soldering and brazing often utilize one or more of the following substances: propane, MAPP™ gas, nitrogen, oxygen, and acetylene.

- Soldering is usually done at temperatures below 842°F.

- Brazing is usually done at temperatures above 842°F.

- Nitrogen, acetylene, and oxygen regulators are used to maintain the pressure at the outlet of pressurized tanks.

- The air-acetylene torch kit is comprised of the tank, acetylene regulator, torch hose, torch handle, and torch tip.

- Tanks with square stems should be opened and closed with a refrigeration service wrench.

- Torch tips come in a number of different sizes so the correct one must be used.

- Torch hoses are color-coded red for acetylene and green of oxygen.

- The oxy-acetylene torch kit is comprised of an acetylene tank, an oxygen tank, an acetylene regulator, an oxygen regulator, a hose set, a torch handle, and a torch tip.

- Torches should be lit with a striker, not a cigarette lighter or match.

- Always have a fire extinguisher nearby when using a torch.

- Steel pipe is cut and reamed using pipe cutters and reamers intended for use on steel piping materials.

- Steel pipe sections are threaded, either by hand or by machine, in order to connect them to fittings and other system components.

- Pipe wrenches are used to tighten threaded piping connections.

- Press-type piping connections can often be used on plumbing piping circuits.

GREEN CHECKLIST

☐ **Make certain that all tanks are closed properly to prevent leakage of materials to the atmosphere**

☐ **Do not overtighten stems on acetylene tanks! They can break off leading to personal and environmental damage**

☐ **Always select and use environmentally friendly materials and supplies**

PROCEDURE 6-1

Setting Up the Air-Acetylene Torch

- Secure the acetylene tank in place to prevent it from falling over.

- Using a refrigeration service wrench, quickly open and close the valve on the acetylene tank. This will blow out any particulate matter from the opening of the tank.

A Wipe the tank opening with a clean rag

B Mount the acetylene regulator and hose on the tank.

- Secure the desired torch tip in the torch handle.

- Make certain that the valve on the torch handle is closed.

C Using a refrigeration service wrench, open the valve on the acetylene tank approximately one-half to one full turn.

- Flip the ratchet on the service wrench so that the tank can be closed quickly in the event of an emergency.

- Turn the handle on the regulator clockwise until pressure registers on the regulator's valve.

A

Photo by Eugene Silberstein.

B

Photo by Eugene Silberstein.

C

Photo by Eugene Silberstein.

Setting Up the Air-Acetylene Torch (continued)

D Using a soap bubble solution, check the regulator, hose, and handle connections for leaks.

- Tighten any leaking connections.

- Set the regulator's valve to about the mid-range for igniting the torch.

- Open the valve on the torch handle all the way.

E Using the striker, ignite the fuel. Be sure to keep your hands away from the fuel path and never aim the torch at the tank, other people, or any flammable substances.

F Adjust the flame using the regulator on the tank until the desired flame intensity is obtained. A proper flame will have a bright blue inner flame and a lighter blue outer cone. The hottest portion of the flame is the tip of the inner cone.

- To extinguish the torch flame, simply close the valve on the torch handle.

- For storage, be sure to close the valve on the acetylene tank and bleed the acetylene from the hose by opening the valve on the torch handle.

Photo by Eugene Silberstein.

Photo by Eugene Silberstein.

Photo by Eugene Silberstein.

Setting Up the Oxy-Acetylene Torch

- Make certain that the oxygen and acetylene tanks are secure.

A Connect the acetylene and oxygen regulators to the acetylene and oxygen tanks.

- Make certain that the hoses are properly connected to the two regulators; red hose to the acetylene and green hose to the oxygen.

B Connect the hoses to the torch handle using reverse flow valves.

- Connect the desired tip to the torch handle.

- Make certain that the valves on the torch handle are in the closed position.

Photo by Eugene Silberstein.

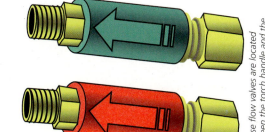

Reverse flow valves are located between the torch handle and the hoses.

CAUTION

CAUTION: Reverse flow valves allow gases to flow through the hoses in only one direction. This prevents the oxygen and acetylene from mixing in the hoses.

C Open the acetylene tank valve one-half turn to introduce acetylene to the regulator.

- The acetylene gauge should register pressure.

Photo by Eugene Silberstein.

Setting Up the Oxy-Acetylene Torch (continued)

  Open the oxygen tank valve one-half turn to introduce oxygen to the regulator.

• The oxygen gauge should register pressure.

Photo by Eugene Silberstein.

  Slightly open the acetylene valve on the torch handle and adjust the acetylene regulator until the pressure gauge reads about 5 psig.

• Close the acetylene valve on the torch handle.

Photo by Eugene Silberstein.

Slightly open the oxygen valve on the torch handle and adjust the oxygen regulator until the pressure gauge reads about 10 psig.

- Close the oxygen valve on the torch handle.

- Open the acetylene valve on the torch handle.

- Using the striker, ignite the fuel. Be sure to keep your hands away from the fuel path and never aim the torch at the tank, other people, or any flammable substances.

 You should observe a large, yellow, smoky flame.

 Slowly open the oxygen valve on the torch handle. The flame will clear up and turn blue. Ideally the flame should be sitting firmly on the torch tip, not blowing away from it.

- To extinguish the flame, start by closing the acetylene valve on the torch handle.

- Close the oxygen valve on the torch handle.

- For storage, be sure to close the valves on both tanks and bleed the pressure from the hoses.

Photo by Eugene Silberstein.

Photo by Eugene Silberstein.

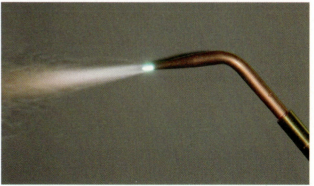

Photo by Eugene Silberstein.

REVIEW QUESTIONS

1. **Which of the following is a property commonly associated with hard-drawn copper tubing?**

   a. It is easily flared

   b. It is used in conjunction with fittings

   c. It is easily swaged

   d. It is easily bent

2. **When cutting copper piping materials for refrigeration applications, it is best to use**

   a. A reciprocating saw

   b. A hacksaw

   c. A tubing cutter

   d. A tubing shear

3. **The purpose of a reamer is to**

   a. Cut hard-drawn copper pipe

   b. Remove burrs from the cut end of a cut pipe section

   c. Create threads on the end of a steel pipe section

   d. Create a 45-degree angle on the end of a copper pipe section

4. **The term "galvanized" is used to describe**

   a. Copper piping materials

   b. The threads on the end of a steel pipe section

   c. Steel pipe that is formed with no seams

   d. Steel pipe that is coated with zinc

5. **In order to create a flare, a technician needs**

   a. A flaring block, a flaring yoke, and a section of soft drawn copper tubing

   b. A flaring block, a hammer, and a section of soft drawn copper tubing

   c. A swaging tool, a hammer, and a section of steel pipe

   d. A flaring block, a swaging tool, and a section of hard drawn copper tubing

6. **The flared end of a section of piping/tubing material is secured to a fitting or system component**

   a. With a flare nut

   b. By soldering

   c. By brazing

   d. With a press-type fitting

7. **When a swaged joint is created to join two sections of pipe, how many soldered or brazed joints must be made?**

   a. 0

   b. 1

   c. 2

   d. 3

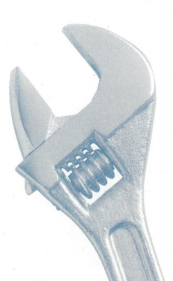

8. **When a section of piping material is bent to create a 90-degree angle, how many soldered or brazed joints must be made at the bent location?**

   a. 0

   b. 1

   c. 2

   d. 3

9. **The piping tool used to seal a process tube is the**

   a. Reamer

   b. Pinch-off tool

   c. Swaging tool

   d. Flaring block

10. **Which of the following is true regarding the temperatures of a propane torch flame, a MAPP™ gas flame, and an acetylene torch flame?**

    a. The acetylene torch flame is the hottest of the three

    b. The MAPP™ gas flame is the coolest of the three

    c. The propane torch flame is the hottest of the three

    d. The acetylene torch flame is the coolest of the three

11. **Which of the following is true regarding the temperatures at which soldering and brazing are done?**

    a. Brazing is done at lower temperatures than soldering

    b. Both soldering and brazing are done at low temperatures

    c. Soldering is done at lower temperatures than brazing

    d. Both soldering and brazing are done at high temperatures

12. **The torch kit component that ensures safe operating pressures at the outlet of a pressurized tank is the**

    a. Torch handle

    b. Refrigeration service wrench

    c. Torch tip

    d. Regulator

13. **Which of the following torches utilizes two pressurized tanks to create the flame?**

    a. MAPP™ gas torch

    b. Propane torch

    c. Oxy-acetylene torch

    d. Air-acetylene torch

14. **Which of the following is true regarding safe torch usage?**

    a. Use only a striker to light the torch

    b. Keep a fire extinguisher nearby

    c. Leak-check the torch kit before lighting

    d. All of the above are correct

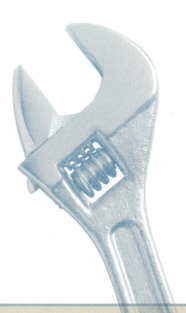

Specialized HVAC/R Tools

OBJECTIVES *Upon completion of this chapter, the student should be able to*

- Describe various service tools used to evaluate the mechanical portions of air conditioning equipment.

- Describe tools and test instruments that are used to evaluate electrical circuits and components.

- List and describe the equipment that is used for the evacuation and charging of air conditioning systems.

- List and describe test instruments that are used to evaluate airflow through a duct system.

- Describe test instruments that are used to set up and evaluate fossil fuel heating appliances.

- Describe tools to fabricate and install air distribution systems.

Proper and efficient air conditioning and heating system operation relies on a number of factors. Air conditioning systems must have the proper refrigerant charge, proper airflow through the condenser and evaporator coils, and a leak free air distribution system. Fossil fuel heating systems must be set up properly to ensure that they operate as intended. Improper fuel-air mixtures on a fossil fuel appliance can lead to improper combustion, reduced operating efficiency, soot buildup, or the introduction of carbon monoxide to the occupied space. It is not possible to obtain and maintain proper system operation without the proper tools and test instruments.

In order to ensure that the air conditioning system is properly charged, accurate pressure and temperature readings must be obtained, not only from the system itself, but also from the areas surrounding the equipment. It is important that the information that is gathered by the technician is accurate. Inaccurate readings can mislead the technician, so it is important that the obtained readings are reliable.

Fossil fuel heating systems must operate with the correct fuel pressures. In addition, once the fuel burns, proper draft helps ensure that the by-products of combustion are quickly removed from the structure. Properly operating and properly used tools are required to ensure that safe conditions are maintained.

Air distribution systems should be leak free. Air leakage, either into or out of the duct system, will lead to reduced system efficiency. By using the proper tools to perform fabrication and assembly processes on duct sections, the leak rate can be greatly reduced.

Alkylbenzene (AB oil) Synthetic oil used with HCFC and CFC refrigerants

Bulldog snip Sheet metal snip with a short nose. Used for cutting multiple layers of sheet metal

Carbon monoxide tester (CO tester) Instrument used to measure the percentage of carbon monoxide in flue gas

CFM Cubic feet per minute

Clamp-on ammeter Test instrument used to measure the amperage in an electric circuit

Combustion analyzer Test instrument used to measure operating conditions on a fossil fuel heating appliance

Continuity Term used to describe a complete path in an electric circuit

Cubic feet per minute (CFM) Term used to describe the volume of airflow

Degassing The process of removing noncondensables from a system

Dehydration The process of removing water vapor from a system

DOT Department of Transportation

GLOSSARY OF TERMS (CONT'D)

DOT-approved cylinders Vessels used to contain recovered refrigerant

Draft gauge Instrument used to measure the pressure over the flame in a heating appliance

Dry-bulb temperature Temperature measured with a standard thermometer

Four-valve manifold Refrigeration gauge manifold that is equipped with four hoses and four valves

Fuel pump pressure gauge Instrument used to measure the pressure at the outlet of the fuel pump on an oil burner

Inches of water column (IWC) Unit of pressure where 27.7 IWC = 1 psig

Infrared thermometer Instrument that measures the infrared energy emitted by an object to determine its temperature

Line set Term used to describe the bundle of refrigerant lines that connects the indoor and outdoor portions of an air conditioning system

Manometer Instrument that measures very low pressures in air conditioning and heating systems

Micron Linear measurement equal to 1/25,400 of an inch

Mineral oil Lubricant used with CFC and HCFC refrigerants

Noncondensables Term used to describe gases that will not condense within the normal operating pressure ranges encountered in an air conditioning system

Polyol ester oil (POE) Lubricant used with HFC refrigerants

Relative humidity Relationship between the amount of moisture in an air sample and the maximum amount of moisture the air sample can hold

Recovery Process of removing refrigerant from a system and storing it in a cylinder without testing, cleaning, or reprocessing it

Smoke tester Instrument used to measure the presence of unburned fuel in flue gases

Two-valve manifold Refrigeration gauge manifold that is equipped with three hoses and two valves

Thermistor An electronic device that changes its resistance as its temperature changes

Vacuum gauge Device used to measure pressures below atmospheric pressure. Used on both air conditioning and heating system

Velocity Term used to describe speed. In the case of airflow, the unit "feet per minute" is commonly used

Volt-ohm-milliammeter (VOM) Electrical test instrument used to measure voltage, resistance, and small currents in electric circuits

Wet-bulb temperature Temperature measured with a thermometer that has a wet sensing bulb

SERVICE TOOLS

In addition to the basic hand tools that are used by most individuals, the HVAC/R technician must also be able to work with specialized tools. Some of these tools are intended to provide the field technician with valuable information, which is then used to evaluate and correct system performance. Knowing how to use these specialized tools is therefore very important if the technician is to effectively troubleshoot equipment. Other tools that are discussed in this section allow the technician to perform various functions such as leak detection, system evacuation, and system charging. Without these tools, performing high quality service would not be possible.

REFRIGERATION GAUGES

Probably one of the most important tools the air conditioning and refrigeration technician uses is the gauge manifold (Fig. 7-1). The manifold is made up of two gauges. One gauge, used to measure low-side system pressure and a vacuum, is color-coded blue. The other gauge, used to measure high pressure, is color-coded red. The gauge manifold is used to determine the saturation temperatures and pressures on the high and low sides of the system. This information, in addition to other readings taken in the system, enables the technician to evaluate system performance. The saturation temperature on the high side of the system is used to help calculate condenser subcooling, while the saturation temperature on the low side of the system is used to help calculate evaporator superheat. Two common configurations of gauge manifolds are the **two-valve manifold** and the **four-valve manifold**. The valves are open when turned counterclockwise and closed when turned clockwise (Fig. 7-2).

FIGURE 7-2 The valves on the manifold are open when turned counterclockwise and closed when turned clockwise.

Going Green

Always make certain that refrigeration gauges are properly calibrated and maintained. Improperly calibrated gauges can lead to improper system charging or evaluating, leading to ineffective and inefficient system operation.

The two-valve manifold has three hoses and two valves (Fig. 7-3). The gauge is divided into two sides, the low side and the high side. The low side is comprised of the low-side valve, the low-side gauge, and the low-side

FIGURE 7-1 Gauge manifold.

FIGURE 7-3 A two-valve manifold.

hose, all of which are color-coded blue. The high side is comprised of the high-side valve, the high-side gauge, and the high-side hose, all of which are color-coded red. The center hose is color-coded yellow and can be connected to

- a vacuum pump for system evacuation,
- a nitrogen tank for leak checking the system,
- a refrigerant tank for system charging, or
- a recovery unit or recovery tank for refrigerant recovery.

The low-side gauge will always be open to the low-side hose and the high-side gauge will always be open to the high-side hose. The position of the blue and the red valves open and close the connection between the center hose port and the low and high sides of the system, respectively. Here are the four possibilities:

- If both the high- and low-side valves are in the closed position, the center hose port is completely isolated from both the high- and low-pressure sides of the system (Fig. 7-4).
- If the low-side valve is open and the high-side valve is closed, the high-side portion of the system is isolated and the low side of the manifold is open to the center hose port connection (Fig. 7-5).
- If the low-side valve is closed and the high-side valve is open, the low side portion of the system is isolated and the high side of the manifold is open to the center hose port connection, and (Fig. 7-6).

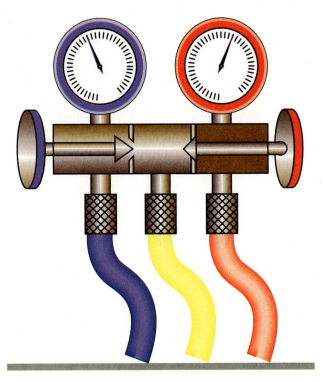

FIGURE 7-5 The high side of the system is isolated when the high-side valve is closed.

- If both the high- and low-side valves are in the open position, the center hose port and both the high- and low-pressure sides of the manifold are all open to each other (Fig. 7-7).

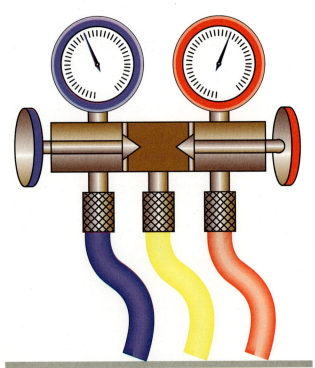

FIGURE 7-4 The center port of the manifold is isolated when both valves are closed.

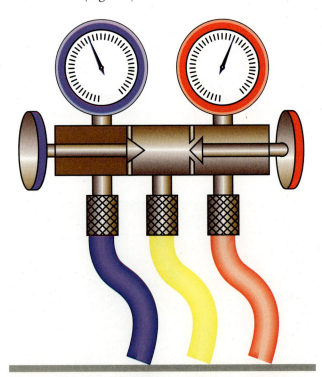

FIGURE 7-6 The low side of the system is isolated when the low-side valve is closed.

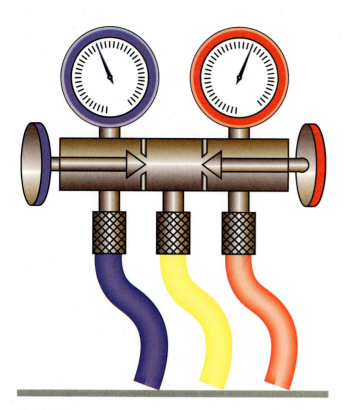

FIGURE 7-7 All ports are open to each other when both valves are open.

Photo by Eugene Silberstein.

FIGURE 7-8 A four-valve manifold.

The four-valve manifold has four hoses and four valves (Fig. 7-8). As with the two-valve manifold, the high and low sides are color-coded red and blue, respectively. The difference between the two configurations is that the four-valve manifold has two center ports. The two center ports are labeled VAC and REF. The center port marked VAC is used for evacuation purposes, while the center port marked REF is used for refrigerant recovery, system leak testing and system charging. The red and blue valves will open and close the connection between the center ports and the high and low sides, respectively. The VAC and REF valves determine which of the two center ports are open to the high and low sides of the manifold.

If the VAC valve is open and the REF valve is closed, the line connecting the manifold to the vacuum pump will be accessed if either the high-side or low-side valves are open (Fig. 7-9). If the REF valve is open and the VAC valve is closed, the line connecting the manifold to

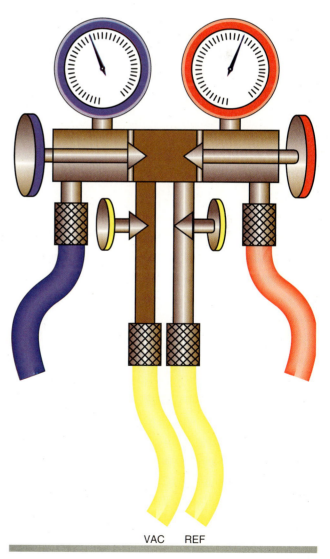

VAC   REF

FIGURE 7-9 Opening the VAC valve accesses the vacuum pump line.

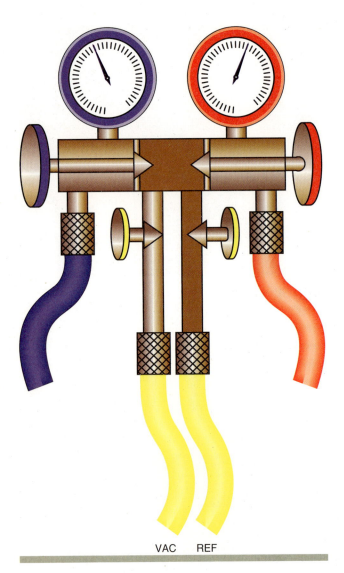

FIGURE 7-10 Opening the REF valve accesses the component connected to that line.

EXAMPLE 1

When the system is being evacuated the VAC hose port will be open and the REF valve will be closed. In addition, both the high-side and low-side valves on the manifold will be open (Fig. 7-11).

FIGURE 7-11 Valve positions for system evacuation.

the refrigerant tank, recovery tank, or nitrogen tank will be accessed if either the high-side or low-side valves are open (Fig. 7-10).

THERMOMETERS

In addition to taking readings of saturation temperatures and pressures in an air conditioning or refrigeration system, taking temperature readings ranks among the most commonly performed tasks in our industry. We need to take temperature readings for a number of reasons and, depending on the application, a number of different types of thermometers are available (Fig. 7-13). Common thermometer types include the stem, thermistor, and infrared. The **stem thermometer** is often carried in the technician's pocket and used to measure the temperature of the occupied space or the temperature of

air in a duct system. The **thermistor thermometer** uses electronic devices that change their resistance as their temperature changes. Thermistor thermometers often have multiple probes. This allows the user to take several different temperature readings at the same time. Some multiprobe thermometers have the capability to calculate and display the difference between two measured temperatures. **Infrared thermometers** determine the temperature of an object by measuring the amount of infrared energy it emits. Be sure to investigate the uses and limitations of each type of thermometer before selecting the instrument to be used. Inaccurate temperature readings can lead to inaccurate conclusions regarding system operation.

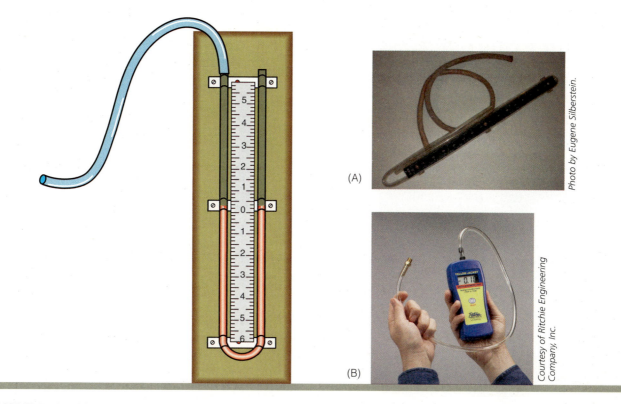

Photo by Eugene Silberstein.

Courtesy of Ritchie Engineering Company, Inc.

(A)

(B)

FIGURE 7-24 Manometers (A) U-tube water manometer. (B) Digital electronic manometer.

measured with the gauge manifold discussed earlier in this chapter. Pressures measured by the manometer are measured in **inches of water column**, or IWC, where 27.7 IWC = 1 psig.

The u-tube water manometer is typically used to measure the pressure of either natural gas or liquefied petroleum, LP, on heating appliances. A typical operating pressure for a residential natural gas heating appliance is about 3.5 IWC, or 0.13 psig. A typical operating pressure for a residential LP heating appliance is about 11 IWC, or 0.40 psig. The digital electronic manometer is often used to measure the pressures in an air distribution, or duct, system. Pressures in a residential duct system are typically less than 0.5 IWC. The digital electronic manometer is a desirable tool, as it provides accurate readings quickly.

MICRON GAUGE

The micron gauge is used to measure the level of vacuum reached during the evacuation process (Fig. 7-25). Micron gauges can be analog, digital, or LED devices. Since the cost of electronic circuitry is constantly decreasing and the accuracy of electronic devices is very high, the most popular micron gauges are the digital and LED versions. It should be noted that the vacuum portion

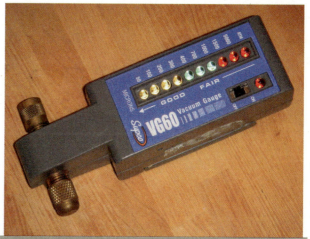

Photo by Eugene Silberstein.

FIGURE 7-25 Micron gauge.

of the compound gauge on the gauge manifold is not an acceptable way to measure the vacuum in an evacuated air conditioning or refrigeration system. It is often required for air conditioning systems to be pulled down to a vacuum of at least 500 **microns**, which cannot be measured on the compound gauge. A micron is a linear measurement that is equal to 1/25,400 of an inch.

from experience...

The micron gauge should not be exposed to system pressure and should be positioned in a line that can be valved off from the system (Fig. 7-26).

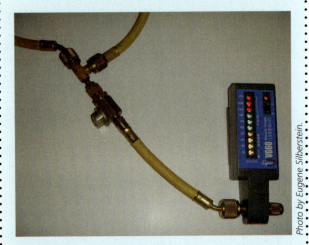

Photo by Eugene Silberstein.

FIGURE 7-26 Micron gauge connected so it can be isolated from the system.

VACUUM PUMP

The vacuum pump is used to evacuate air conditioning and refrigeration systems (Fig. 7-27). The process of evacuation involves the **degassing** and **dehydration** of the system. Degassing involves the removal of air, nitrogen, and other **noncondensables** from the system, while dehydration involves the removal of water vapor from the system. Noncondensables are gases that cannot condense within the pressure ranges typically found in an air conditioning system. Vacuum pumps vary greatly from manufacturer to manufacturer and are rated in CFM. **CFM (cubic feet per minute)** represents the rate at which volumes of vapor can be pumped from the system. Vacuum pumps can also be identified as either one-stage or two-stage pumps. Two-stage vacuum pumps are typically the better choice, especially when it is suspected that moisture is present in the system. Since all vacuum pumps are not created equal, be sure to select the pump that best serves your needs.

RECOVERY UNIT

It is illegal to intentionally release refrigerant from air conditioning or refrigeration systems to the atmosphere. The process of containing refrigerant that is removed

Courtesy of Appion, Inc.

FIGURE 7-27 Vacuum pump.

from systems is called **recovery**. Recovery units (Fig. 7-28) are often used to remove refrigerant from systems and pump the refrigerant into **DOT-approved** (Department of Transportation) **cylinders** (Fig. 7-29). DOT-approved cylinders are color-coded gray with yellow tops and are

Going Green

Never vent or knowingly release refrigerant to the atmosphere. It is illegal to do so and also detrimental to the environment.

Photo by Eugene Silberstein.

FIGURE 7-28 Recovery unit.

FIGURE 7-29 DOT-approved recovery cylinders.

FIGURE 7-30 Line purge kit.

the only vessels that can be used to contain recovered refrigerant. Refrigerant can be recovered from a system in either the vapor or liquid state, depending on how and at what points the recovery unit is connected to the system. Refer to Chapter 4 for more information on refrigerant recovery and other important refrigerant-related environmental issues.

LINE PURGE KITS

When systems are replaced, the existing refrigerant lines are often reused. The **line set**, which is the term used to describe the bundle of field installed lines, is often buried in the walls or otherwise inaccessible. In the past, on noncompressor burnout replacements, residual oil in the lines posed no major problems. This was because the new equipment used oils compatible with the old. Effective January 1, 2010, manufacturers are no longer permitted to manufacture systems that contain R-22.

Since newly manufactured equipment will likely contain HFC single-compound or HFC-blended refrigerants, oil compatibility becomes an issue. Systems that operate with HCFC refrigerants, such as R-22, use **alkylbenzene (AB) oil** or **mineral oil**. Systems that operate with HFC refrigerants and HFC blends, such as R-410A, use **polyol ester (POE) oil**. The POE lubricants are not

compatible with AB or mineral oils. If the line set used was part of a system that contained AB or mineral oil, the lines should be flushed before reusing them. A flush kit can be used to help the technician clean the lines (Fig. 7-30).

LEAK DETECTORS

As environmentally concerned technicians, we owe it to our customers to locate and repair all leaks in air conditioning systems. In order to effectively check for leaks, a number of different leak detection methods are available. Not all methods are appropriate for all applications, so technicians should be prepared to use whichever method is appropriate for the situation at hand. The halide leak detector operates with an open flame and indicates a refrigerant leak when the flame color changes from blue to green (Fig. 7-31). The electronic leak detector uses a pump to pull air samples into the unit and test them for traces of refrigerant (Fig. 7-32). The ultrasonic leak detector uses a microphone to pick up the sounds of escaping refrigerant or nitrogen (Fig. 7-33). The ultraviolet leak detector utilizes a fluorescent refrigerant dye that is added to, and circulated through, the system. The point of the leak is located as it glows when observed under an ultraviolet light (Fig. 7-34). Refer to Chapter 9 for more detailed information including the specific applications for the various leak detection methods.

Courtesy of Uniweld Products, Inc.

FIGURE 7-31 Halide leak detector.

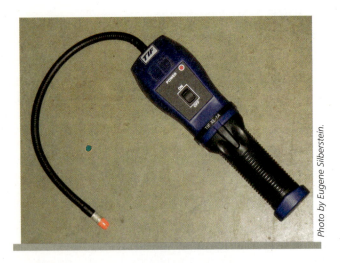

Photo by Eugene Silberstein.

FIGURE 7-32 Electronic leak detector.

Delmar/Cengage Learning.

FIGURE 7-33 Ultrasonic leak detector.

Courtesy of Uniweld Products, Inc.

Courtesy of Uniweld Products, Inc.

FIGURE 7-34 Ultraviolet leak detector.

AIR AND AIRFLOW INSTRUMENTS

In addition to using gauges and thermometers to evaluate an air conditioning system, other instruments help the technician evaluate the properties of the air in the conditioned space and in the air distribution system. Conditions that need to be evaluated include the dry-bulb temperature, the wet-bulb temperature, the relative humidity, the velocity of the air, the volume of the air, and the dew point temperature of the air.

The **dry-bulb temperature** is the temperature measured with a standard thermometer. The **wet-bulb temperature** is obtained with a thermometer that has a wet sensing bulb or element. **Relative humidity** is the relationship that compares the amount of moisture contained in an air sample to the maximum amount of moisture that the sample can hold. **Velocity** is simply the speed of the air, measured in feet per minute, fpm or ft/min. The dew point temperature is the temperature at which moisture begins to condense out of the air. Some instruments used to measure these and other air properties are the sling psychrometer, the velometer, the anemometer, and the flow hood (Fig. 7-35).

HEATING SERVICE TOOLS

The safe and efficient operation of fossil fuel heating appliances requires the use of specialized instruments to test various aspects of system operation. If a fossil fuel-burning appliance fails to operate efficiently, harmful by-products of combustion can be produced, resulting in personal injury or death. If systems are not set up properly, the efficiency of the system can be compromised as well. Not only must the combustion process be evaluated, but the process by which the by-products of combustion are removed from the structure must be checked as well. The **draft gauge** is used to measure very low pressures above the flame in a heating appliance

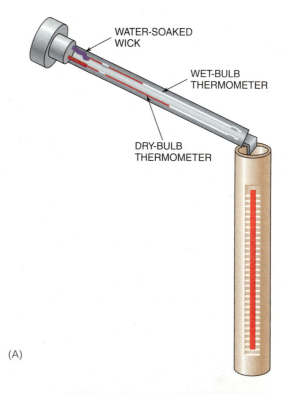

(A) WATER-SOAKED WICK · WET-BULB THERMOMETER · DRY-BULB THERMOMETER

(B) *Photo by Eugene Silberstein.*

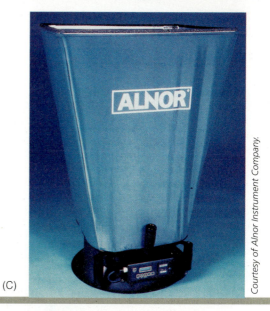

(C) *Courtesy of Alnor Instrument Company.*

FIGURE 7-35 (A) sling psychrometer. (B) Velometer/Anemometer. (C) Flow hood.

to verify the flow of flue gases up a chimney or vent (Fig. 7-36). The **smoke tester** is used to determine the presence of unburned fuel in a sample of flue gases taken from an oil burner (Fig. 7-37). Oxygen and carbon dioxide testers are used to evaluate the combustion process of gas or oil-fired systems (Fig. 7-38). The **carbon monoxide tester** is used to detect the percentage of carbon monoxide in a flue gas sample (Fig. 7-39). The **fuel pump pressure gauge** is used to measure the oil pressure at the outlet of the fuel pump (Fig. 7-40). The **vacuum gauge** is used to measure the pressure/vacuum in the oil line

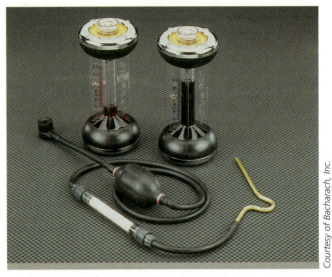

Courtesy of Bacharach, Inc.

FIGURE 7-38 Oxygen and carbon dioxide testers.

Courtesy of Bacharach, Inc.

FIGURE 7-36 Draft gauge.

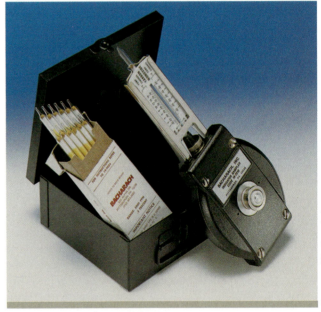

Courtesy of Bacharach, Inc.

FIGURE 7-39 Carbon monoxide tester.

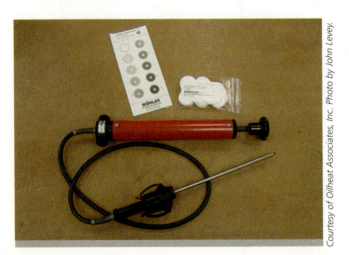

Courtesy of Oilheat Associates, Inc. Photo by John Levey.

FIGURE 7-37 Smoke tester.

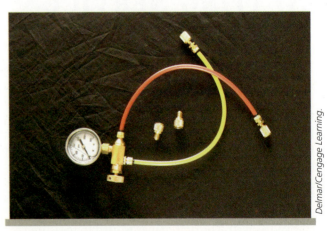

Delmar/Cengage Learning.

FIGURE 7-40 Fuel pump pressure gauge.

FIGURE 7-41 Vacuum gauge.

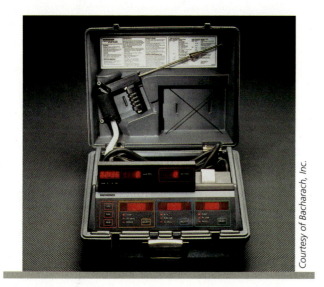

FIGURE 7-42 Combustion analyzer.

feeding the oil burner (Fig. 7-41). The **combustion analyzer** is a vital instrument that performs the functions of multiple individual instruments. The combustion analyzer measures the levels and concentrations of oxygen, carbon monoxide, carbon dioxide, as well as determining the flue gas temperature and system efficiency (Fig. 7-42). The combustible gas detector sounds an audible alarm

FIGURE 7-43 Combustible gas detector.

if combustible gases are sensed (Fig. 7-43). This instrument is often used to help technicians identify cracks in the heat exchanger of a heating appliance.

SHEET METAL TOOLS

A vital part of an air conditioning and heating system is the air distribution, or duct, system. The air distribution system performs two major functions. First, it provides a path for air from the occupied space to flow to the air conditioning equipment. Second, it distributes the conditioned air back to the various areas in the occupied space. The air distribution system should be leak free in order to maintain optimal system performance. Many air distribution systems are, at least in part, made from metal duct sections. Working with metal ductwork is as much art as it is science. Using the correct tools will help ensure proper duct connections and also help keep the technician safe. The items that are listed here are some of the basic tools used by the installation technician.

> **CAUTION**
>
> **CAUTION:** Sheet metal is very sharp! Be sure to wear appropriate hand protection when working with duct sections.

SNIPS

Tin snips are available in a number of styles and each is intended for a specific use (Fig. 7-44). Typically, tin snips are designed to make straight cuts, cuts that curve to the left, and cuts that curve to the right. Tin snips are

Going Green

Heating equipment that is properly adjusted will use less fuel, provide a more comfortable home and have less of an impact on the environment.

often color coded for easy identification. Yellow snips are intended for cutting sheet metal in a straight line or for making slightly curved cuts (Fig. 7-45). Red snips are intended for cutting sheet metal at sharp left angles (Fig. 7-46). Green snips are intended for cutting sheet metal at sharp right angles (Fig. 7-47). The snips can also be configured as either straight or offset (Fig. 7-48).

Another type of tin snip is the **bulldog snip** (Fig. 7-49). This snip can have a shorter nose and is used to cut through multiple layers of metal. Sheet metal shears have longer handles and are available in a number of different sizes for different applications (Fig. 7-50).

SHEET METAL HAMMERS

Sheet metal hammers are easily identified as they have square heads and beveled edges (Fig. 7-51). The peen of the hammer is tapered to the thin flat tip. The handles of the sheet metal hammer are typically wood or covered with vinyl or leather.

Photo by Eugene Silberstein.

FIGURE 7-44 Various tin snips.

Photo by Eugene Silberstein.

FIGURE 7-47 Green snips used to make sharp right angle cuts.

Photo by Eugene Silberstein.

FIGURE 7-45 Yellow snips used to make straight cuts.

Photo by Eugene Silberstein.

FIGURE 7-48 Snips can be conventional (left) or offset (right).

Photo by Eugene Silberstein.

FIGURE 7-46 Red snips used to make sharp left angle cuts.

Photo by Eugene Silberstein.

FIGURE 7-49 Bulldog snips.

FIGURE 7-50 Sheet metal sheers.

FIGURE 7-52 Sheet metal notchers.

FIGURE 7-51 Sheet metal hammers.

FIGURE 7-53 Hand seamers.

NOTCHERS

The notcher is used to remove V-shaped sections from a sheet of metal (Fig. 7-52). This tool is especially useful when laying out and fabricating duct sections.

HAND SEAMERS

Hand seamers are used to make sharp bends in pieces of sheet metal (Fig. 7-53). They are available in various widths for different applications.

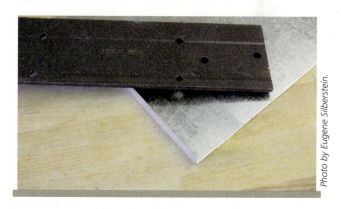

FIGURE 7-54 Folding tools.

FOLDING TOOLS/BENDERS

Folding tools are used to create flanges on the end of a sheet metal section (Fig. 7-54). The tool has predetermined depth gauges to eliminate the need to measure and scribe the line prior to bending. In addition, many folding tools have a sight hole so that the user can confirm that the sheet metal has been inserted all the way into the tool prior to bending (Fig. 7-55).

DUCT STRETCHER

When joining duct sections, it is often necessary to pull the sections together. Sometimes this can be done by brute force, but sometimes additional help is needed. The duct stretcher is an extremely useful tool that enables the technician to use a great deal of leverage to pull the sections together (Fig. 7-56). The two wheels of the tool are positioned in the two "ears" or return flanges on the two duct sections being joined (Fig. 7-57).

Photo by Eugene Silberstein.

FIGURE 7-55 Sight hole on the folding tool.

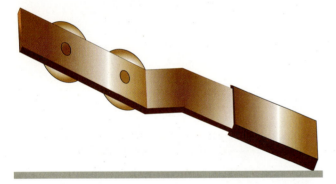

FIGURE 7-56 Duct stretcher.

FIGURE 7-57 The wheels on the duct stretcher are positioned in the "ears" of the duct sections to be joined.

By pushing the handle of the duct stretcher, the two sections are pulled together so they can be joined. Refer to Chapter 9 for more information regarding duct systems and their installation.

CRIMPING TOOL

When constructing duct runs from round pipe sections, the sections must be joined together. This could be a difficult task especially since the pipes are the same size. The crimping tool allows the technician to reduce the diameter of one of the sides of the connection so the sections can fit together (Fig. 7-58).

SCRIBES AND DIVIDERS

When laying out a sheet metal duct section, various tools are used to mark the metal prior to cutting. Some of these tools are scribes and dividers. Scribes are used to scratch straight lines on the metal (Fig. 7-59). Dividers are used to scratch arced lines and circles on the metal (Fig. 7-60). Another tool, the scratch awl can be used to scratch free-form and straight lines on the metal surface (Fig. 7-61).

Photo by Eugene Silberstein.

FIGURE 7-58 Crimping tool.

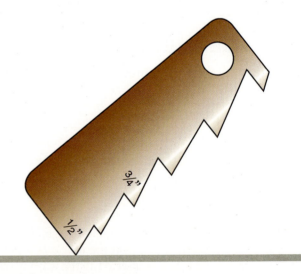

FIGURE 7-59 Scribes.

FIGURE 7-60 Dividers.

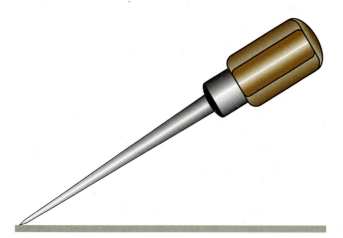

FIGURE 7-61 Scratch awls.

SUMMARY

- The refrigeration gauge manifold is used to read system operating pressures and saturation temperatures.
- The gauge manifold can be of the two valve or four valve variety.
- The center port of the manifold is used for system charging, evacuation, leak testing, or refrigerant recovery.

- Common thermometer types include the stem, thermistor, and infrared.
- The Volt-Ohm-Milliammeter, VOM, is used to measure voltage, resistance, continuity, and very low electric currents.
- The clamp-on ammeter is used to measure current in an electric circuit.
- A refrigeration service wrench should be used to turn valves with square stems.
- Charging scales are used to measure the amount of refrigerant added to or removed from an air conditioning system.
- Manometers are used to measure very low system pressures in the units of inches of water column (IWC).
- The micron gauge is used to measure the vacuum in an evacuated air conditioning or refrigeration system.
- Vacuum pumps are used to evacuate air conditioning and refrigeration systems.
- Evacuation involves degassing and dehydrating a system.
- Recovery units are used to remove refrigerant from an air conditioning system.
- Recovered refrigerant is stored in DOT-approved recovery cylinders.
- Common leak detector types include the halide, electronic, ultrasonic, and ultraviolet.
- The sling psychrometer, velometer, anemometer, and the flow hood are used to measure various air properties.
- Heating system instruments include the smoke tester, the draft gauge, the carbon monoxide (CO) tester, the fuel pump pressure gauge, the vacuum gauge, and the combustion analyzer.
- Various sheet metal tools include snips, hammers, notchers, hand seamers, folding tools, duct stretchers, crimpers, dividers, and scribes.

GREEN CHECKLIST

☐ Refrigeration gauges must be properly calibrated to ensure proper system evaluation and charging.

☐ High quality thermometers should be used to aid in the proper evaluation, start-up and servicing of air conditioning equipment.

☐ Always use a charging scale to determine how much refrigerant is added to or removed from an air conditioning system.

☐ Never knowingly release refrigerant to the atmosphere.

☐ Whenever possible, all refrigerant leaks should be located and repaired.

☐ Properly adjusted heating equipment will use less fuel and will have less of an effect on the environment.

REVIEW QUESTIONS

1. **Which of the following is true regarding the color coding on a gauge manifold?**

   a. The low side is color-coded blue and the high side is color-coded yellow

   b. The low side is color-coded red and the high side is color-coded blue

   c. The low side is color-coded blue and the high side is color-coded red

   d. The low side is color-coded yellow and the high side is color-coded red

2. **The gauge manifold is used to read**

   a. Saturation temperatures and pressures

   b. Superheated temperatures

   c. Only saturation pressures

   d. Subcooled temperatures and pressures

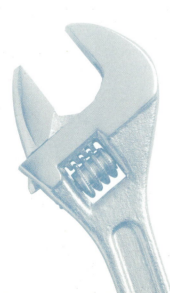

3. **The center hose on the gauge manifold can be connected to a(n)**
   a. Nitrogen tank for leak checking the system
   b. Refrigerant tank for charging the system
   c. Vacuum pump for system evacuation
   d. All of the above are correct

4. **The low-side gauge**
   a. Will be open to the low-side hose when the low-side valve is closed
   b. Will be isolated from the low-side hose when the low-side valve is closed
   c. Will be open to the high-side hose when the low-side valve is closed
   d. Both b and c are correct

5. **What is a main difference between the two-valve and the four-valve gauge manifold?**
   a. The two-valve manifold has two low-side hoses
   b. The four-valve manifold has two center ports
   c. The four-valve manifold has two high-side hoses
   d. The two-valve manifold has four hose connections

6. **Which thermometer type often has the ability to display the difference between two different temperature readings?**
   a. The multi-probe thermistor thermometer
   b. The pocket stem thermometer
   c. The infrared thermometer
   d. Both a and b are correct

7. **Which of the following electrical test instruments can be used to measure voltage?**
   a. Analog VOM
   b. Digital VOM
   c. Clamp-on ammeter
   d. All of the above are correct

8. **When a technician is checking for continuity in an electric circuit, he is**
   a. Determining the amount of resistance in the circuit
   b. Checking to see if there is a complete electric circuit
   c. Checking to see if there is current flowing in the circuit
   d. Determining the amount of voltage being supplied to the circuit

9. **When using a clamp-on ammeter to measure current flow in a circuit**
   a. The meter should be set to the VAC mode
   b. The meter should be clamped around all conductors in the circuit
   c. The meter should be set to the VDC mode
   d. The meter should be clamped around a single conductor

10. **The refrigerant charging scale can be used to**
    a. Determine the amount of refrigerant that has been added to a system
    b. Determine the amount of refrigerant that has been removed from a system
    c. Both a and b are correct
    d. Neither a nor b is correct

11. **The micron gauge is used to measure the**
    a. Positive pressure in an air conditioning system
    b. Vacuum in an air conditioning system in psia
    c. Vacuum in an air conditioning system in inches of mercury
    d. Vacuum in an air conditioning system in units called microns

12. **The process of evacuation involves**
    a. The dehydrating of an air conditioning system
    b. The pressurization of an air conditioning system
    c. The degassing of an air conditioning system
    d. Both a and c are correct

13. **Cylinders used for refrigerant recovery are approved by the**
    a. Air Conditioning, Heating and Refrigeration Institute, AHRI
    b. Environmental Protection Agency, EPA
    c. Department of Transportation, DOT
    d. American National Standards Institute, ANSI

14. **When installing an R-410A split system using the existing refrigerant lines from the old R-22 system, the refrigerant lines should be**
    a. Pressurized with refrigerant for leak testing
    b. Blown out with an air conditioning flush kit
    c. Both a and b are correct
    d. Neither a nor b is correct

15. **Which of the following is measured with a standard thermometer?**
    a. Wet-bulb temperature
    b. Dew point temperature
    c. Dry-bulb thermometer
    d. Both A and B are correct

16. **Which of the following instruments ensures that the by-products of combustion go up the chimney of an oil-fired heating appliance?**
    a. The vacuum gauge
    b. The draft gauge
    c. The fuel pump pressure gauge
    d. The smoke tester

17. **The combustion analyzer performs which of the following functions?**
    a. Determines the flue gas temperature
    b. Determines the system efficiency
    c. Determines the level of carbon dioxide and carbon monoxide
    d. All of the above are correct

18. **Which of the following is true regarding tin snips?**
    a. Red snips are used to cut sharp left angles
    b. Green snips are used to cut sharp right angles
    c. Both A and B are correct
    d. Neither A nor B is correct

19. The tool the sheet metal installer uses to joint two sections of same size round duct is the

a. Duct stretcher

b. Hand seamer

c. Crimping tool

d. Folding tool

20. Scribes and dividers are used to

a. Cut straight lines in sheet metal sections

b. Cut curved lines in sheet metal sections

c. Scratch straight and curved lines in sheet metal sections

d. Scratch freeform lines in sheet metal sections

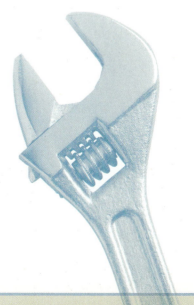

WHAT'S WRONG WITH THIS PICTURE?

Carefully study Figure 7-62 and think about what is wrong. Consider all possibilities.

 **WRONG**

Photo by Eugene Silberstein.

FIGURE 7-62 When using snips to cut sheet metal, use the correct snip for the job you are doing. When making a left-hand cut, be sure to use red-handled tin snips. Green-handled snips are used to make right-hand cuts. Using the wrong snips can result in thin, sharp-pointed protrusions often referred to in the field as "fish hooks". Improperly cutting sheet metal greatly increases the chances of getting injured on the job.

RIGHT

Photo by Eugene Silberstein.

FIGURE 7-63 Correctly using the right snip for the job you are doing will help you stay safe on the job. In addition, your finished work will look much more professional!

Installation

System Location

OBJECTIVES *Upon completion of this chapter, the student should be able to*

- List and explain the requirements for selecting the best location for the air handler of a ducted air conditioning system.

- List the factors that must be considered when selecting the location of a furnace with add-on air conditioning.

- List and explain the requirements for selecting the best location for the air handler of a ductless split air conditioning system.

- List and explain the requirements for selecting the best location for the condensing unit of a split-type air conditioning system.

- Properly select the location for components of air conditioning and heating system components.

The satisfactory operational life of any air conditioning system can be prolonged by a high-quality installation. Systems that are properly installed tend to have fewer malfunctions and, therefore, require fewer service calls. Consequently, a good long-term relationship with a customer starts with the proper installation of a new system, which involves the coordination of all members of the installation crew. A well-trained installation technician has a mastery of general skills, including carpentry, plumbing, and electrical.

The installation crew must also have a working knowledge of the refrigeration system, its components, and the concepts of airflow. Deficiencies in these knowledge areas can result in system component failure as well as a multitude of other problems, including improper airflow to the occupied space, coil freeze-ups, and, more important, customer discomfort.

However, before any of the above tasks can be completed, the equipment location must be carefully selected. Properly selected equipment locations can help minimize waste and increase system efficiency, leading to a greener environment. The selection of the condensing unit location must take into account, among other things, noise transmission, wind factors, ground slope, and proximity to the electrical power. The air handler location is chosen based on the location of the condensing unit, type of air-distribution system, noise levels, serviceability, proximity to the conditioned space, and ease of condensate removal. In this chapter, we will discuss the factors that must be considered when selecting the locations for system components.

Air handler Portion of an air conditioning system responsible for treating air from the occupied space

Building setback The required distance between a property line and structures on the property, or between two structures on a piece of property

Circuit breakers Circuit protection devices that have the ability to be reset

Condensing unit Portion of an air conditioning system made up of the compressor, the condenser coil, and condenser fan motor

Distribution panel Panel that distributes incoming power to protected circuits throughout the structure

Extended-plenum duct system Duct arrangement that facilitates shorter branch duct runs by extending the plenum closer to the branch termination points

Fuses A one-time circuit protection device

Horizontal Configuration of air handlers that causes air to move through the unit from side to side

GLOSSARY OF TERMS (CONT'D)

Natural return Process by which air is returned to the air handler without a physical connection between the return grill and the air handler

Packaged units Systems that have all components in one cabinet

Perimeter duct system Duct system configuration in which the duct system runs around the perimeter of the occupied space

Planning division Local agency that oversees the zoning issues for a particular area

Plenum-type duct system Duct system configuration commonly

used when the air handler is located close to the branch termination points

Reducing-extended-plenum duct system Duct system configuration used to maintain a constant air velocity in the trunk line as air moves through it; not commonly found in residential applications

Return air Air from the conditioned space that is returning to the air handler

Return duct duct that facilitates the return of air to the air handler from the conditioned space

Return grill Decorative grill located in the occupied space that connects to the return duct or provides access for air to return to the air handler

Split-type systems Systems that have the major system components in two different locations

Vertical downflow Air handler configuration where the air enters the unit from the top and is discharged from the bottom

Vertical upflow Air handler configuration where the air enters the unit from the bottom and is discharged from the top

CENTRAL AIR CONDITIONING EQUIPMENT

Central air conditioning systems can be **packaged units** (Fig. 8-1) or **split-type systems** (Fig. 8-2). Packaged air conditioning systems have all components in one cabinet and are also often referred to as self-contained systems. These systems are factory assembled and charged, so the exact amount of refrigerant contained in the system can be found right on the nameplate of the unit. Split systems have the major system components in two

Photo by Eugene Silberstein.

FIGURE 8-1 Packaged air conditioning system for a commercial application. Similar systems are available for residential installations.

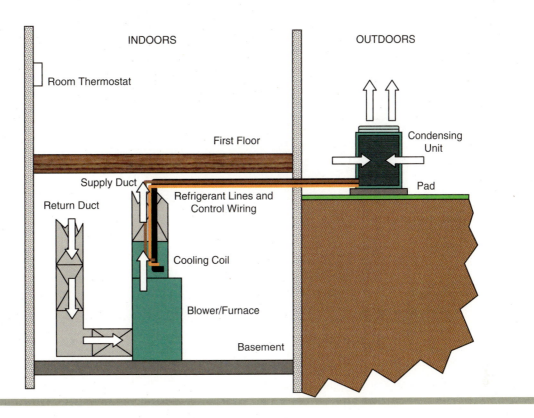

FIGURE 8-2 Split-type air conditioning system.

Courtesy of Carrier Corporation.

FIGURE 8-3 Condensing unit of a split-type air conditioning system.

Photo by Eugene Silberstein.

FIGURE 8-4 Air handler of a split-type air conditioning system.

different locations. Split systems are typically comprised of the **condensing unit** and the **air handler**. The condensing unit is made up of the compressor, the condenser coil and condenser fan motor (Fig. 8-3). The air handler is responsible for treating air from the occupied space. The air handler is made up of the metering device, the evaporator coil and the evaporator fan motor (Fig. 8-4). The refrigeration components in the air handler and condensing unit are connected to each other by field-installed refrigerant lines (Fig. 8-5).

Since the packaged air conditioning system is a one-piece unit, the location of only one piece of equipment must be selected when installing one. When a split system is being installed, the locations for two pieces of equipment, the condensing unit and the air handler, must be determined.

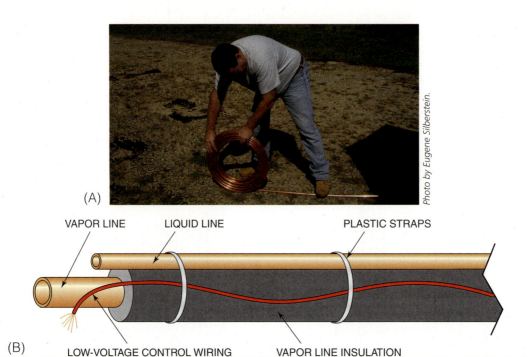

Photo by Eugene Silberstein.

(A)

VAPOR LINE   LIQUID LINE   PLASTIC STRAPS

(B)

LOW-VOLTAGE CONTROL WIRING   VAPOR LINE INSULATION

FIGURE 8-5 (A) Technician unrolling tubing in preparation for insulating, building and installation. (B) Line set bundle.

PACKAGED UNIT LOCATION

For residential applications, packaged air conditioning systems and evaporative coolers are often positioned at ground level—very close to the structure. When selecting the location for the unit, a number of factors must be considered. These include

- noise levels
- ground slope
- local building code requirements
- condenser airflow
- duct system access and connections
- ability to perform service
- appearance
- access to electric power

from experience...

One of the main benefits of using packaged equipment is that the system is factory assembled and charged, so there is no need to run refrigerant lines, leak-check, and evacuate the system.

NOISE LEVELS

Since packaged air conditioning systems often contain, among other things, the compressor and two fan motors, noise is a real concern when selecting the location for the unit. Whenever possible, the unit should be positioned as far as possible from windows, doors, and sleeping areas. In addition, the unit should not be placed close to outdoor living areas, such as decks, patios, and swimming pools. When at all possible, the unit should be positioned so that noise levels do not bother those living nearby. Strategically placed shrubbery can, when needed, help deaden noise transmission to the surrounding outdoor areas.

GROUND SLOPE

As with any other equipment, packaged air conditioning units need to be level. Compressor and fan motor operation can be affected if the unit is not level. If the ground is not level, the pad on which the unit is resting should be (Fig. 8-6). This is not an overly difficult task to accomplish, especially if the pad is constructed by fabricating a wood form and pouring concrete to form the pad. Ideally, the pad should be high enough above grade to prevent water from running into and accumulating in the unit. If the ground is sloped, consider the effects of rainwater runoff when positioning the unit.

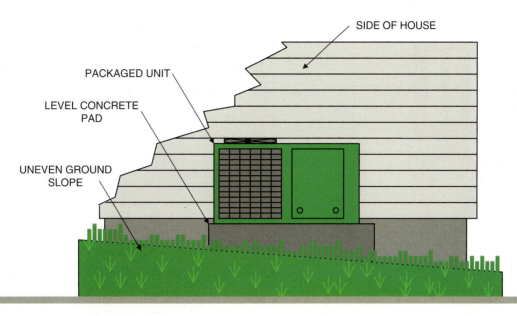

SIDE OF HOUSE

PACKAGED UNIT

LEVEL CONCRETE
PAD

UNEVEN GROUND
SLOPE

FIGURE 8-6 Packaged unit must be level even if the ground is not.

from experience...

Since packaged systems are much heavier than the condensing unit of a split-type system, plastic pads are not often recommended for use with packaged units. Always refer to the manufacturer's literature when preparing and constructing the pad on which the unit will rest.

LOCAL BUILDING CODE REQUIREMENTS

Before the final location for the unit is selected, local building codes should be checked to make certain that the unit will not be too close to the property line of adjoining properties. This cannot be stressed enough as the unit will, in essence, become part of the structure and, moving it later on will result in major expenses. Moving the unit involves moving the electric power supply, the ductwork, the pad, and the unit itself. By investigating the local codes *before* starting work, much grief will be avoided in the long run.

When investigating the building codes for a particular installation, be sure to contact the local **planning** **division** and inquire about the **building setback** requirements for the area. The planning division is the local agency that oversees the zoning issues for a particular area. A building setback is the required distance between a property line and structures on the property, or between two structures on a piece of property. Since building setbacks vary by the geographic area, zoning designation, and/or lot size, contacting the planning division is a must.

CONDENSER AIRFLOW

In order for the air conditioning system to operate properly, there must be ample airflow through both the evaporator and condenser coils. For the most part, the positioning of the packaged unit will not affect the airflow through the evaporator coil, but can affect the airflow through the condenser coil. When selecting the location of the unit, no portion of the condenser coil should be positioned up against the side of the house or any other structure. In addition, shrubs and fences should be located so as not to impede the airflow through the condenser coil. Always refer to the manufacturer's literature for the clearances that should be maintained on all sides of the unit.

Another factor that can affect the airflow through the condenser is vertical clearance. Overhangs and other structural elements such as decks and patios can cause the discharge air from the condenser to recirculate back through the coil (Fig. 8-7). This will result in high head pressure, decreased system efficiency and reduced system performance.

FIGURE 8-7 Air discharged from the condenser must not circulate back to the unit.

DUCT SYSTEM ACCESS AND CONNECTIONS

It is important for the installation crew to keep in mind that the packaged unit will need to have both the supply and return air ducts connected to it. Unlike the split system where the air handler is located near the conditioning space, the packaged unit typically has relatively long duct runs. The location and orientation of the packaged unit must be selected so that the duct connections will not block condenser airflow or access to service panels.

ABILITY TO PERFORM SERVICE

Once installed and put into operation, the system will eventually need to be serviced. Performing service on the packaged unit will be a relatively easy task if the service

from experience...

When installing a system of any kind, the installer should ask himself, "Will I be able to access the system for service once my work is complete?"

access panels are unobstructed. In addition, there should be enough space for the technician to gain access to the system and work comfortably.

APPEARANCE

If at all possible, the unit should be positioned so that it is not visually offensive or distracting. Whenever possible, the system should not be positioned on the side of the house facing the street unless it can be sufficiently concealed. As with the noise considerations, the unit should not be placed close to outdoor living areas, such as decks, patios, and swimming pools.

ACCESS TO ELECTRIC POWER

The location of the unit with respect to the power **distribution panel** will have almost no effect on system operation, but will affect the cost of installing the lines that will supply power to the unit. The power distribution panel is the location where all circuits in the structure originate. It is at the distribution panel where the electric circuit protection devices, such as **fuses** and **circuit breakers**, are located (Fig. 8-8). If the unit is located close to the distribution panel, there will be lower labor and material costs involved when compared to those incurred if the unit is located far from

Delmar/Cengage Learning.

FIGURE 8-8 Power distribution panel.

the panel. When the list of considerations about system location is reviewed, the distance between the unit and the panel is probably the least important of all.

PUTTING IT ALL TOGETHER

When all of the above conditions are taken into account, it is very likely that there will be very few options for unit location. In many cases it will not be possible to find a location if all conditions are to be met. If such is the case, the most important conditions, such as local building code requirements and the ability to perform future service, need to be met first. When selecting a unit location, there are often compromises that must be made and the equipment owner should ideally be a part of the decision-making process. The equipment owner also needs to be made aware of the difference between required conditions and desired conditions.

SPLIT SYSTEMS: CONDENSING UNIT LOCATION

Unlike the packaged system just discussed, the split system is made up of two sections, namely, the condensing unit and the air handler (Fig. 8-9). On heat pump systems, discussed in Chapter 27, the two sections are referred to as the outdoor section and the indoor section. These two sections are connected by two refrigerant lines—one containing vapor and the other containing liquid. Ideally, the field installed refrigerant lines should be as short as possible to maintain system efficiency. For this reason, before the actual job of connecting the components together, the locations of the condensing unit and the air handler must be carefully chosen. The factors that go into the location selecting process for the outdoor unit include

- sound transmission
- wind factors
- location of electric power
- airflow restrictions
- proximity to the indoor unit
- ground slope

SOUND TRANSMISSION

The condensing unit of the air conditioning system contains a compressor and a fan motor, both of which generate noise that could be distracting to those around it. For this reason, the outdoor unit should be positioned far enough away from any bedrooms or other living areas to reduce the inconvenience created by system noise. Ideally, the outdoor unit should be located close to the structure along the side of the house, not the front, in a location that is as far as possible from a window. If applicable, it should also be placed in a location that will cause the least amount of inconvenience for anyone residing nearby.

WIND FACTORS

To prevent prevailing wind from affecting the operation of the system, the condensing unit should be positioned in a well-shielded location. Shrubs provide a very good shield for the unit, preventing wind from blowing through it. The shrubs, however, must be trimmed regularly to prevent them from blocking the unit's coil surfaces. The leaves can also create an airflow problem through the coil itself if they are pulled into the unit. Fences can also be used to block the wind, but the location of an outdoor unit within

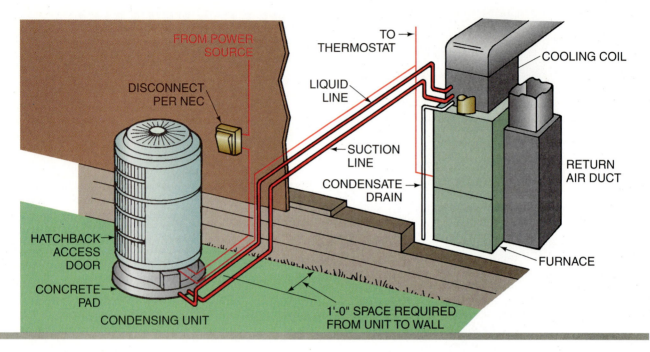

FIGURE 8-9 Indoor and outdoor sections of a split air conditioning system.

a fenced-in area should be carefully considered before the piping process is started. Ample room should be available for a service technician to access the electrical panel, as well as the service panel that provides access to the compressor and other system components. In either case, shrubs and fences provide protection from the wind and help alleviate the eyesore nature of the unit itself.

LOCATION OF ELECTRICAL POWER

The outdoor and indoor units require separate electrical power supplies that are protected by either fuses or circuit breakers. On a new installation, these power supplies must be provided from the existing fuse or circuit-breaker panel, so a new line must be run for each section. Although a relatively minor consideration, the shorter the run, the less expensive the electrical installation will be. Therefore, having the outdoor unit located as close to the electrical service panel as possible is desirable. This line must be installed in accordance with local electrical codes, and a licensed electrician should perform this work.

AIRFLOW RESTRICTIONS

Regardless of where the outdoor unit is located, air must be able to flow freely through the coil. The air flowing through the coil must be able to mix freely with the outside air to prevent the hot discharge air from the outdoor

unit from recirculating back through the outdoor coil. For this reason, the unit should not be located under an overhang or porch. Overhangs trap the discharge air from the outdoor unit, causing the head pressure of the system to rise. Typical clearance minimums are shown in Figure 8-10. The manufacturer's specifications should always be followed when evaluating the clearances for a specific piece of equipment. Under no circumstances should the condensing unit be positioned against a structure; it should be at least 30″ away from any wall.

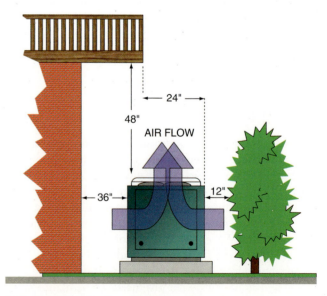

FIGURE 8-10 Condensing unit location and clearances.

PROXIMITY TO THE INDOOR UNIT

Probably one of the most important factors that will ultimately affect the success of the installation is the condensing unit's location with respect to the indoor unit, or air handler. The distance between these two pieces of equipment determines the length of the refrigerant lines that must be installed. The shorter these lines, the better.

Longer refrigerant lines can lead to a decrease in system efficiency. In the cooling mode, the temperature of the refrigerant flowing back to the compressor from the evaporator coil in the air handler will increase, reducing the operating efficiency of the system and increasing the cost of operating the system. To reduce the effects of unwanted heat transfer, the vapor, or suction, line should always be insulated.

For aesthetic purposes, some home owners opt to locate the outdoor unit far away from the structure. The problems with this are threefold. First, the cost of running the electrical power to the unit is higher because of the added distance. Second, the added length of the refrigerant lines can have a negative effect on system performance and operating efficiency. Finally, the refrigerant lines would need to be buried under the ground. This makes servicing the equipment much more difficult, especially in the event of a refrigerant leak.

GROUND SLOPE

Wherever the unit is located, it should be perfectly level. This will prevent unnecessary strain on the condenser fan motor and will help prevent any undesired noise created by excessive vibrations. The unit should never be set directly on the ground. This can cause the unit to sink and the coil to be blocked with leaves or other debris. The unit should be located above ground level and should sit on a slab made of one of a number of materials, including concrete and plastic (Fig. 8-11). Prefabricated pads or slabs can be purchased in a wide range of sizes, depending on the physical size of the unit. The chosen pad should be large enough to completely support the unit with enough room left on all sides to prevent the unit from vibrating off the pad. If need be, a poured concrete slab can be constructed on site by fabricating a wooden form and then pouring the concrete into it. Pouring a concrete slab should be done long before the installation is started to give the concrete ample time to set up.

The outdoor unit should be completely isolated from the structure to prevent the transmission of noise into the structure. For example, placing the unit on a patio or deck that is connected to the structure itself would not be wise. In areas that experience heavy snowfall, the outdoor unit should be elevated well above ground level.

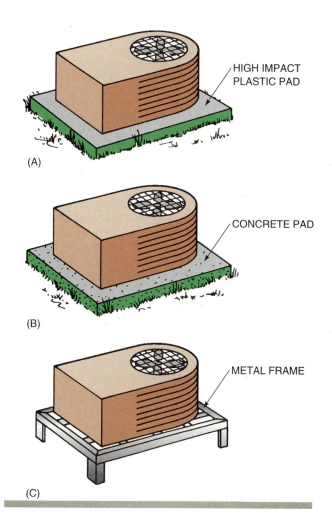

(A)

HIGH IMPACT PLASTIC PAD

(B)

CONCRETE PAD

(C)

METAL FRAME

FIGURE 8-11 Various types of unit pads.

SPLIT SYSTEMS: AIR HANDLER LOCATION

Just as with the outdoor unit, a number of factors must be considered when choosing the location for the indoor unit of the air conditioning system. These factors include

- the type of air-distribution system
- location of the electric power supply
- location of the outdoor unit
- serviceability
- indoor unit configuration
- ease of condensate removal
- noise level
- return air
- location of space to be conditioned

TYPES OF AIR-DISTRIBUTION SYSTEMS

A number of different air-distribution systems can be used on any given system installation. A few of these configurations are shown in Figure 8-12. In the radial or **plenum-type duct system**, (Fig. 8-12(A)) the air handler is located in a position that is central to the supply register locations. On this type of duct configuration, the supply registers are normally located on the interior walls of the structure. The **extended-plenum duct system** (Fig. 8-12(B)) is used primarily when the duct system and the air handler are not located on the same level (e.g., if the indoor unit is located on the first floor of the space and the duct system is located in the attic overhead). A large supply plenum is used to bring the supply air up to the level of the remainder of the duct system. The **reducing-extended-plenum duct system** (Fig. 8-12(C)) is used on systems in which the air handler is located at one end of the structure to be conditioned. In this type of duct system, the size of the trunk line gets progressively smaller as the end of the run is reached in an effort to maintain the velocity of the air moving through the duct. In the **perimeter duct system** (Fig. 8-12(D)) the air handler is typically located below the conditioned space and the ductwork runs around the perimeter of the space. The supply registers, which are floor mounted, are located around the perimeter of the space as well.

Going Green

A simple, compact duct layout will reduce the materials needed to install heating and cooling systems. An energy efficient building envelope keeps the heating and cooling loads lower than a typical home so the duct systems can be smaller. Compact systems use fewer materials, take less time to install, and usually cost less.

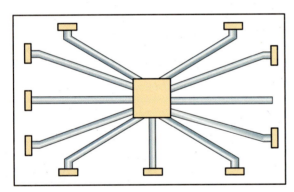

(A) PLENUM OR RADIAL DUCT SYSTEM

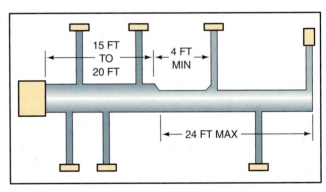

(C) REDUCING-EXTENDED-PLENUM SYSTEM

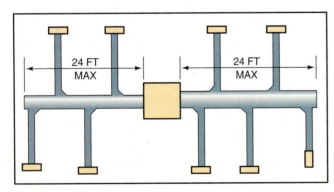

(B) EXTENDED-PLENUM SYSTEM

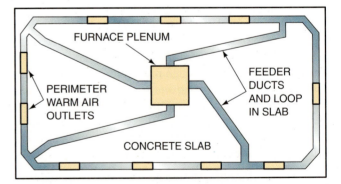

(D) PERIMETER LOOP SYSTEM WITH FEEDER AND LOOP DUCTS IN CONCRETE SLAB

FIGURE 8-12 (A) Plenum system. (B) Extended-plenum system. (C) Reducing-extended-plenum system. (D) Perimeter loop system.

LOCATION OF THE ELECTRICAL POWER SUPPLY

Another factor to consider when choosing the location of the air handler is its location with respect to the electrical panel. As in the case of the condensing unit location, power must be brought to the unit from the circuit breaker or fuse panel. The longer the run, the more costly the installation will be.

Running the power line must be done in accordance with all local electrical codes and should be done by a licensed electrician. The electrical panel should also be inspected to make certain that the power requirements of the system can be satisfied.

LENGTH OF THE REFRIGERANT LINES

Because the refrigerant lines should be as short as possible, the air handler should be located as close to the condensing unit as practical. Excessively long refrigerant lines will have a negative effect on both the operation and the efficiency of the entire system. Long refrigerant lines will increase the system superheat in the cooling mode of operation, which reduces the pumping efficiency of the compressor. The suction line should always be well insulated to reduce this effect. In addition, insulating the suction line reduces the amount of sweating on the line during system operation.

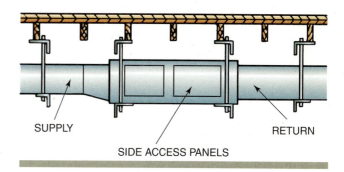

SUPPLY   SIDE ACCESS PANELS   RETURN

FIGURE 8-13 Access panels must be unobstructed.

handler must be located so all the service panels are unobstructed (Fig. 8-13). In addition to this, ample clearance must be available for the technician to gain access to the unit. Although crawl spaces under the conditioned space may be ideal locations for the unit, it should be positioned as close to the access of the crawl space as possible to make servicing the equipment easier.

When located in an attic, the path leading to and under the unit should be clear and unobstructed. If possible, plywood sheets should be laid down on the path to the unit to reduce the possibility of damage to the ceiling and to help ensure the safety of the technician. In addition, a light fixture should be permanently mounted in the attic to help facilitate servicing. When installing, members of the crew should consider the fact that any component within the unit may need to be replaced in the future, and they should therefore plan accordingly.

Going Green

The shorter the refrigerant lines the better! Shorter lines reduce the possibility for refrigerant leaks, reduce the amount of materials used, and help increase the efficiency of the system. Longer refrigerant lines result in additional heat returning to the compressor, making the system work harder. More work equates to more power consumption.

from experience...

Always refer to and obey local building codes when planning a system installation. Some municipalities require that a solid, permanent path be installed to the indoor unit when installed in an attic or crawl space. Some codes require a platform at least 30″ wide at the unit location. Most, if not all, municipalities require that a service disconnect be located on or next to the indoor unit.

SERVICEABILITY

Regardless of where the indoor unit is located, it must be serviced periodically—both during preventive maintenance and during system repair. Ultimately, the air

INDOOR UNIT CONFIGURATION

The location of the indoor unit also relies, to a lesser degree, on the configuration of the air handler. Common configurations of the indoor unit include

- vertical upflow
- vertical downflow
- horizontal

VERTICAL UPFLOW

The **vertical upflow** unit is designed for applications in which the unit is located below the conditioned space—in a basement, for example—or on the same level as the conditioned space—such as in a utility closet (Fig. 8-14). Depending on the manufacturer, the blower can be positioned either upstream or downstream of the evaporator coil.

When installed in a utility closet, the supply plenum often extends straight up into the attic or enclosed ceiling where the duct system is located. When the unit is located in the basement, the duct system is normally located in the basement as well, with the supply registers located in the floor. In two-story dwellings, a system located in the basement often serves the first floor and a unit in the attic serves the second floor.

VERTICAL DOWNFLOW

The **vertical downflow** unit is designed primarily for use on air-distribution systems with floor registers. This type of indoor unit discharges conditioned air to the basement-mounted duct system (Fig. 8-15). The return air enters the unit from the top and is then discharged

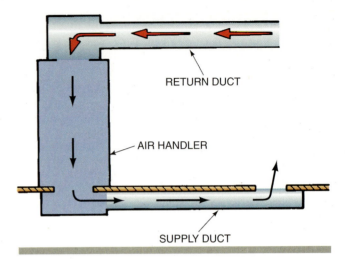

FIGURE 8-15 Vertical downdraft configuration.

downward. One benefit of this type of unit is its serviceability. The unit can be serviced from its location in a closet without having to access the basement or crawl space. This also makes replacing any unit components easier. This configuration is not as common as the vertical upflow and the horizontal configuration.

HORIZONTAL

The **horizontal** configuration is probably the most versatile as far as installation location is concerned. These units can be configured for right-hand or left-hand flow with only minor alterations to the position of the internal drain pan located within the unit. Horizontal units can be mounted in attics, dropped ceilings, utility closets, and basements. When installed in attics, they are normally suspended by cradles that are attached to the rafters of the structure itself (Fig. 8-16). On these installations, the duct system is normally located in the attic as well. When used in a basement, they are also supported by cradles suspended from the floor joists (Fig. 8-13). Using horizontal units in basements is beneficial, especially when the potential for flooding exists. By keeping the unit off the floor, the possibility of water damage is greatly reduced. In addition, by suspending the unit, more floor space is available for storage. These units can also be positioned in utility rooms, once again making more storage space available.

EASE OF CONDENSATE REMOVAL

Regardless of where the air handler is located, it is designed to perform the same function; it moves cooled air through the duct system in the warmer months. In addition to removing heat from the air, the system also dehumidifies the air, thereby creating condensation,

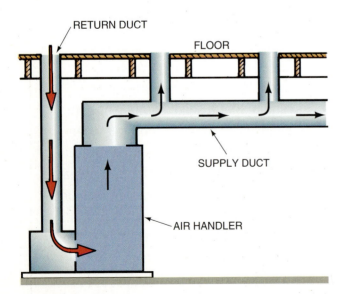

FIGURE 8-14 Vertical upflow configuration.

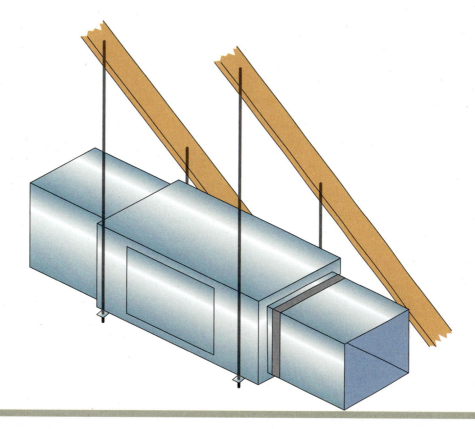

FIGURE 8-16 Air handler is supported from the rafters in the attic.

which is the moisture removed from the air. This condensate must be effectively removed from the structure to prevent water damage. To reduce the possibility of water-related damage, the length of the condensate drain line should be as short as possible.

When the air handler is located in an attic or overhead area, the condensate drain line should be routed to the outside of the structure, if at all possible. If an outside wall is not accessible, the condensate should be directed to a nearby waste line or drain as long as the piping work complies with all local plumbing codes. The same holds true for units installed in utility closets.

When the indoor unit is located in the basement, routing the condensate outdoors is not an easy task, mainly because the basement is below ground level. In this case, a condensate pump (Fig. 8-17) is needed to pump the condensate up to an overhead line, which, in turn, is routed to a waste line or utility sink. If the basement is not finished and a floor drain is available, it can be used to accept the condensate from the system (Fig. 8-18). Piping must be run from the unit to the floor drain to prevent water from accumulating on the floor. In any event, the location of the unit with respect to the ultimate termination point of the condensate should be evaluated carefully before the unit is set in place.

from experience...

Always check the installation literature that is supplied with a unit before piping in the condensate line. Many indoor units require that the condensate line be trapped at the condensate drain pan outlet. Refer to the section of condensate lines for more information regarding condensate removal.

from experience...

Good field practice requires insulating the condensate line for at least the first 10 feet of the run from the indoor unit. This helps prevent sweating on the surface of the drain line, which could result in water-related damage to the structure.

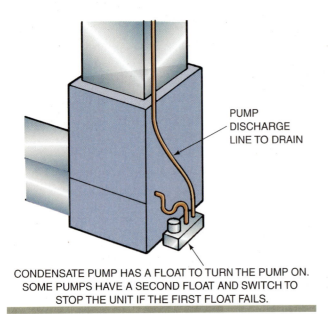

CONDENSATE PUMP HAS A FLOAT TO TURN THE PUMP ON. SOME PUMPS HAVE A SECOND FLOAT AND SWITCH TO STOP THE UNIT IF THE FIRST FLOAT FAILS.

FIGURE 8-17 Typical condensate pump location.

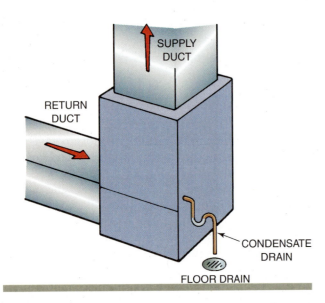

FIGURE 8-18 Condensate piped to nearby floor drain.

NOISE LEVEL

Even though the air handler is typically much quieter than the condensing unit, noise level is another factor that must be considered when choosing the air handler's location. In an attic, the unit should be above a common area, such as a bedroom hallway. If possible, the unit should never be located directly above a bedroom. The cycling of the fan, although not loud, is magnified at night when the house is otherwise very quiet. If the unit must be installed over a bedroom, sufficient insulation should be placed in the attic to reduce the effects

of the unit noise. If the unit is in a basement, it should be positioned so that the inconvenience caused by the noise will be minimized. When the unit is located in a closet, the walls of the closet should be covered with an acoustical material to reduce the amount of noise transmission to the adjacent areas.

RETURN AIR

A method of returning air from the occupied space to the air handler must be provided. The source of the **return air** should be a common location, such as a hallway. When the unit is installed in a closet, the easiest way to ensure return air is to place a **return grill** in the door of the closet. This type of **natural return**, which does not require any physical ductwork, per se, can result in more noise transmission to the occupied space. When the system is in the attic, the return grill is often in the ceiling of the hallway and connected to the indoor unit via a **return duct**, which physically connects the return grill to the unit. The unit should not be directly above the return grill, as the noise levels will be higher. If, however, the unit must be very close to the return grill, good field practice would involve creating a loop, or indirect path, in the return duct. By doing this, the unit noise will be dampened and will have less effect on the occupied space (Fig. 8-19).

from experience...

The return grill should be sized to provide a minimum of 144 unrestricted square inches per ton of capacity. If restrictive or high-efficiency filters are to be used, more unrestricted area is needed.

LOCATION OF THE SPACE TO BE CONDITIONED

Probably one of the most important factors to consider when choosing the location of the air handler is its proximity to the space that is to be serviced. Ideally, the unit should be directly above, below, or beside the space. Excessively long duct runs will have a negative effect on the operation of the system. The interior surfaces of the ductwork offer resistance to airflow, reducing the velocity, or speed, of the air moving through it. In addition, an exchange of heat takes place between the air inside the

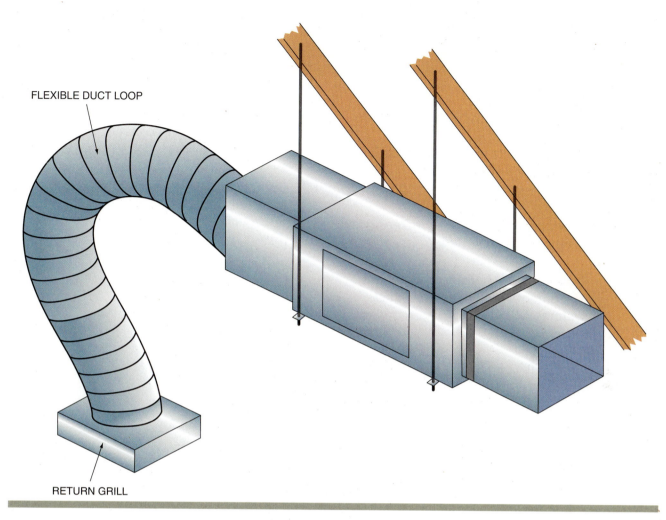

FLEXIBLE DUCT LOOP

RETURN GRILL

FIGURE 8-19 Loop created in the return duct to dampen the noise level.

ductwork and the air surrounding it. Even when insulated, a certain amount of air leakage takes place within the duct system, causing a reduction in the system's operating efficiency. If placement in a nearby location is not possible, the duct system should be constructed of low-resistance sections in an effort to maintain proper airflow rates. For example, flexible or spiral duct offers a great deal of resistance to flow and should be avoided if at all possible. Using an excessive number of transition sections and turning duct sections should be avoided as well, because they add resistance to airflow. Typically, in residential applications, there is no need for multiple transition duct sections.

SUMMARY

- Central air conditioning equipment can be classified as packaged or split-type systems.
- Packaged air conditioning systems have all components in one cabinet.

- Split systems are typically comprised of the condensing unit and the air handler.
- Packaged unit and condenser location considerations include: noise levels, ground slope, local building code requirements, condenser airflow, duct system access and connections, serviceability, appearance, and access to electric power.
- Building setback requirements are among the most important considerations when selecting the unit location.
- Factors affecting the location choice for the indoor unit include the type of duct system, unit configuration, electric power location, noise levels, return air, location of condensing unit, ease of condensate removal, and ease of service.
- Air distribution systems are commonly configured as one of the following: plenum system, extended-plenum system, reducing-extended-plenum system, or perimeter loop system.

GREEN CHECKLIST

☐ Minimize waste whenever possible.

☐ Properly select equipment location to maximize efficiency and minimize environmental impact.

☐ Shorter refrigerant lines increase system efficiency.

☐ Properly sized heating and cooling equipments require smaller duct systems.

☐ Smaller, compact duct systems are more efficient than conventional layouts.

☐ Compact duct systems utilize less materials and cost less to install.

REVIEW QUESTIONS

1. **The main characteristic of the packaged system is that it**
   a. Houses all major system components in one cabinet
   b. Requires the field installation of refrigerant lines
   c. Is made up of a condensing unit and an air handler
   d. Requires field leak checking and evacuation

2. **The system compressor can be found in the**
   a. Packaged unit
   b. Air handler of a split-type air conditioning system
   c. Both a and b are correct
   d. Neither a nor b is correct

3. **The total factory charge of refrigerant in a system is factory imprinted**
   a. On the condensing unit's nameplate of a split system
   b. On the nameplate of a packaged unit
   c. On the air handler's nameplate of a split system
   d. In the manufacturer's literature for the air handler

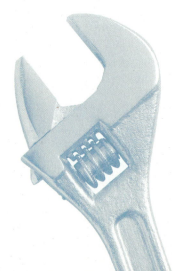

4. **The local planning division is consulted about**

   a. The power requirements of an air conditioning system

   b. The distances between structures and property lines

   c. The distance between property lines

   d. The distance between the condensing unit and the air handler

5. **A low overhang over a condensing unit can result in**

   a. Low-head pressure

   b. Low suction pressure

   c. High-head pressure

   d. Increased system efficiency

6. **Excessively long refrigerant lines can result in**

   a. Increased system superheat

   b. Increased system efficiency

   c. Both a and b are correct

   d. Neither a nor b is correct

7. **Which of the following factors will have the greatest effect on the operating efficiency of a split-type air conditioning system?**

   a. Location of the electrical power supply with respect to the condensing unit

   b. Location of the air handler with respect to the condensing unit

   c. Location of the air handler with respect to the conditioned space

   d. Location of the condensing unit with respect to the conditioned space

8. **Which of the following is the most likely outcome if the condensing unit is not properly leveled?**

   a. The condenser fan motor may fail prematurely

   b. The compressor may fail prematurely

   c. The condenser coil will become blocked or clogged more quickly

   d. The suction pressure will drop

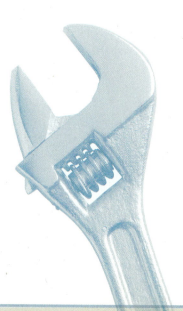

KNOW YOUR CODES

Investigate your local codes regarding the clearances between equipment and property lines. To do this, you will have to contact your local planning division. Be sure to report your findings to your classmates and teacher.

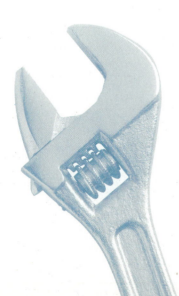

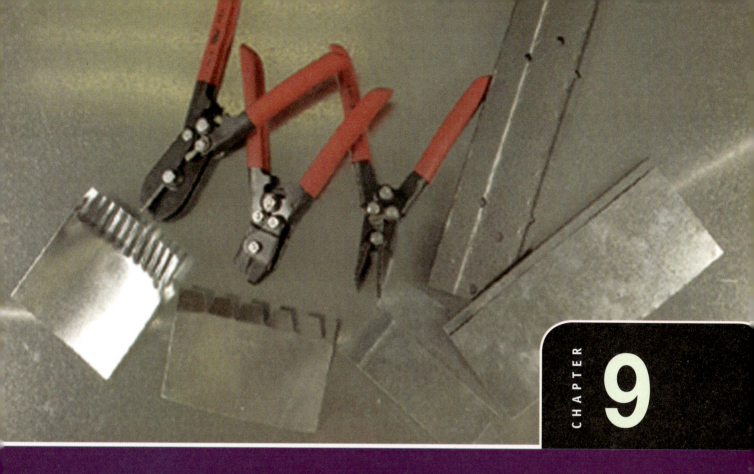

CHAPTER 9

Air-Distribution Systems

OBJECTIVES *Upon completion of this chapter, the student should be able to*

- List and describe four different styles of duct systems.

- List various types of materials used in duct fabrication.

- Describe the materials used to install and support the various types of duct systems.

- Connect, insulate, and support duct sections.

GLOSSARY OF TERMS

Boot Duct section that connects the branch duct to the supply register

Canvas connector A flexible connection between the air handler and main trunk line to lessen vibration and noise transmission in the duct system

Combination duct system A duct system constructed of more than one material

Fiberboard Rigid fiberglass material used to fabricate duct sections

Fiberglass sheeting See fiberboard

Flexible duct Fiberglass duct material that is used to make connections between the main trunk line and the branch termination points

Galvanized sheet metal Sturdy material used to fabricate duct systems

Mastic Sticky putty used to seal air-leaks in metallic duct systems

Offset duct section A duct section that permits a rigid duct to be redirected around an obstacle

Self-tapping screws Sheet metal screws with drill-like tips; eliminates the need to predrill holes in the sheet metal

Slips and drives Materials used to join sections of sheet metal duct

Once the locations for the system components have been selected and the equipment has been set in place, the air distribution system will need to be installed. The air distribution, or duct, system brings air from the occupied space to the equipment and also delivers the conditioned air from the equipment back to the occupied space. The duct system is as important to proper system operation as the refrigeration portion of the system. Systems that are properly installed tend to have fewer malfunctions and, therefore, require fewer service calls. A properly designed and installed air distribution system is a must if the space is to be evenly conditioned. Inadequate airflow to and from areas in the conditioned space can lead to hot spots, cold spots, humidity issues, as well as the overall discomfort of the occupants. Excessively high humidity can result in mold growth, while low humidity can lead to the drying out of home furnishings and skin. In addition, an air distribution system that is not moving the correct volume of air may lead a technician to misdiagnose the system when evaluating its operation. An improper diagnosis leads to an improper repair, which leads to an unhappy customer. In this chapter, we will discuss some common types of air distribution systems and provide insight into the installation of them.

GLOSSARY OF TERMS (CONT'D)

Takeoff Duct section connected to the main trunk line, providing a connection point for the branch duct

Transition duct section Duct section that either reduces, enlarges, or changes the shape of the existing duct section

Trunk line Main run of a duct section coming off the air handler

New Construction and Existing Structure Considerations

Once the indoor unit location has been determined and the unit is actually set in place, the duct system can be installed. This can be a very tricky aspect of the installation and should be performed by qualified individuals. When systems are being installed as part of a new construction project, the problems encountered in ductwork installations are relatively simple to overcome. During new construction, the walls and ceilings are open, giving the installation crew free access to them. Duct sections can be placed in the walls and ceilings before any plasterboard or other covering material is set in place. All duct sections should be properly joined and insulated before they are enclosed, as access to them after construction is complete will be next to impossible. On new construction installations, wall registers are very popular due to the ease of positioning the duct sections between the wall studs and then extending these sections either upward to the attic or down to the basement, depending on the indoor unit location (Fig. 9-1). The duct system layout for the new construction installation is relatively straightforward, because the duct system has been designed along with the structure. For this reason, there are typically fewer bends and twists in the ductwork, making the installation much easier.

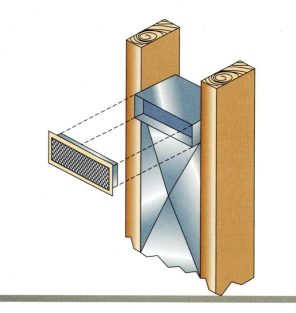

FIGURE 9-1 In new construction, supply ducts are often located in the walls.

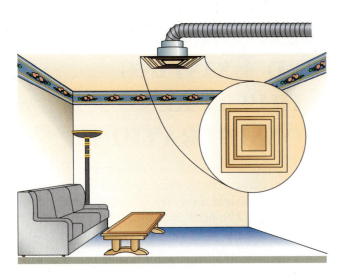

FIGURE 9-2 Ceiling-mounted supply register.

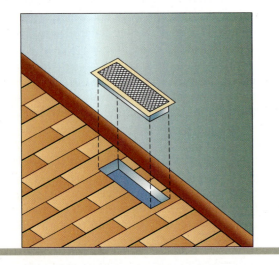

FIGURE 9-3 Floor-mounted supply register.

When an air conditioning system is installed in an existing structure, many more problems must be addressed than in a new construction installation. The duct system must be designed and installed around the limitations that are already present in the structure. The walls are already sealed, and access to them is limited so floor and ceiling-mounted supply registers and grills are the most popular for these installations (Fig. 9-2 and Fig. 9-3).

GREEN DUCT SYSTEMS

Reducing the amount of energy used in a home is one of the most important goals of green building. How energy-efficient a house is depends in large part on the efficiency of the HVAC equipment used, how it is installed, and how well it is maintained. The air distribution system design and installation are also very important to a home's energy efficiency.

Whenever possible, ductwork should be installed within the thermal envelop of the house in order to achieve the best performance. When ductwork is installed in an attic, basement, or crawlspace, it can gain or lose heat through the duct insulation and air-leaks.

Many green homes have HVAC ducts located within specially designed roof trusses, referred to as plenum trusses, or open web floor trusses in two-story homes. Running ducts through plenum-style roof trusses or open-web floor trusses keep them within the conditioned space of the home for the highest efficiency.

Another area to locate ducts is in a dropped ceiling section. In homes with ceiling heights nine feet or greater, sections of the ceiling can be framed lower to create a cavity to run ducts in. A good example is a hallway leading to bedrooms. The hallway ceiling can be lowered creating a route for ductwork. Chases can be designed into soffits along walls or beams to route ducts through to reach other parts of the house. Straight, short, and simple are the keys to efficient duct layout.

DUCT SYSTEM CONFIGURATIONS

As discussed previously, four common duct system configurations are

- plenum
- extended-plenum
- reducing-extended-plenum
- perimeter loop

Regardless of which type is ultimately chosen for installation, its ultimate purpose remains the same: to provide proper airflow to the occupied space. This poses a potential problem since different spaces have different comfort level requirements and, therefore, different airflow requirements. Generally speaking, an air conditioning system delivers 400 ft^3 of air per minute per ton of system capacity, which must be divided among all of the supply register locations. A residential three-ton system, for example, will circulate approximately 1200 ft^3 of air per minute to the occupied space.

PLENUM SYSTEM

The plenum, or radial, air distribution system is probably the easiest of all duct systems to install. This type of system can be installed quickly and efficiently by individuals with a minimum of duct work experience. This system consists of a closed-ended plenum connected to the supply end of the unit. Takeoffs for the individual supply

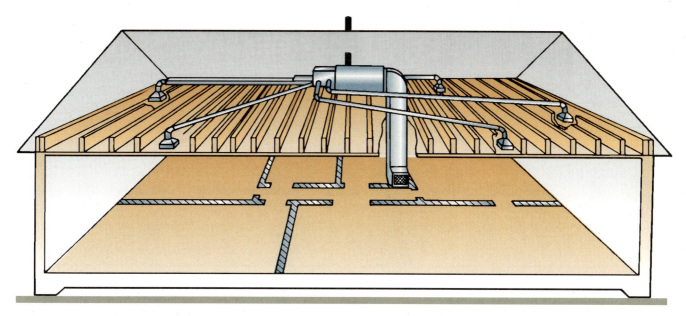

FIGURE 9-4 Plenum duct system.

locations are cut into this plenum (Fig. 9-4). The connections between the plenum and the supply register are often made with insulated flexible ductwork (Fig. 9-5). This type of system has benefits and drawbacks that should be weighed against each other before a final decision is made as to which type of duct system to use. The benefits of a plenum-type system include the following:

- They are very economical from a first-cost standpoint.
- They are extremely easy to install.
- They do not require extensive duct measuring and layout.
- They are a good choice when physical obstacles exist in the area of the unit.

The main disadvantage is that substantial friction occurs within the flexible duct itself. If the individual duct runs are very long, the operational effectiveness of the system will be somewhat compromised. To reduce the frictional losses in the duct system, takeoffs made from rigid sheet metal can be used. Of course, this will greatly increase the cost of the installation. If this turns out to be the case, the extended-plenum system may be a better choice.

EXTENDED-PLENUM SYSTEM

If the individual duct runs are found to be too long, the extended-plenum system is a good alternative to the standard plenum system. It involves the installation of rigid duct sections that are connected to the main plenum box, in effect extending the plenum closer to the individual supply register locations (Fig. 9-6). This extension is commonly referred to as a **trunk line**. It is from this

from experience...

When connecting takeoffs to the plenum, the end of the plenum must be left intact. Cutting a takeoff into the end of the plenum will result in an excessive amount of air being discharged through that opening. Needless to say, the other takeoffs will be starved for air, resulting in uneven conditioning of the occupied space.

plenum extension that the now shorter flexible duct runs are connected. The size of this extended plenum remains the same for the entire length of the run.

Although this type of system uses more prefabricated duct sections, it is still relatively easy to install but more

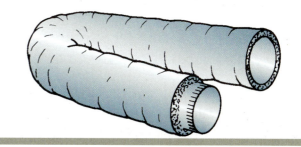

FIGURE 9-5 Flexible spiral duct.

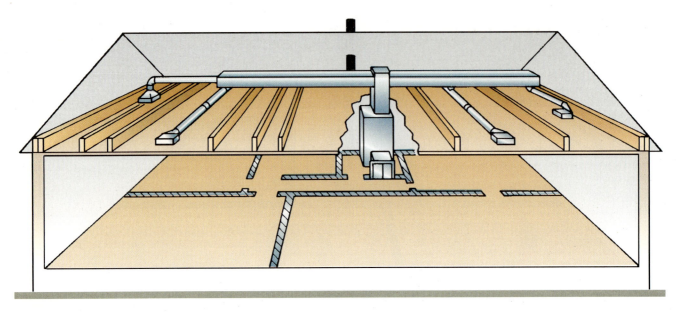

FIGURE 9-6 Extended-plenum duct system.

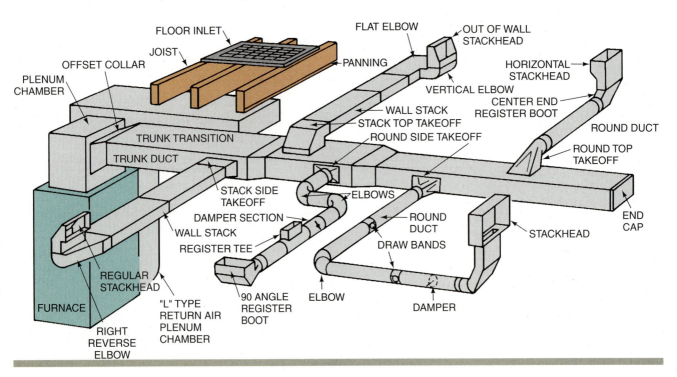

FIGURE 9-7 Reducing-extended-plenum duct system.

from experience...

As with the plenum system, takeoffs should not be located on the end of the plenum on an extended-plenum duct system.

time-consuming than the plenum system described earlier. It is also somewhat more costly than the plenum system, but the frictional losses in the system are greatly reduced.

REDUCING-EXTENDED-PLENUM SYSTEM

Very similar in construction to the extended-plenum system is the reducing-extended-plenum system (Fig. 9-7), which is also designed to bring the supply plenum closer to the supply registers. The main difference is that in the reducing-extended-plenum system, the main trunk reduces in size as the end of the plenum is reached. The main benefit of this system is that the velocity and the pressure of the air flowing through the duct are relatively constant throughout the system. Although the cost of producing smaller duct sections is lower, an added cost is involved because it costs more to fabricate **transition duct sections** than straight sections. Transition pieces are used to connect two sections of different-sized duct to each other (Fig. 9-8). Residential applications very rarely warrant the use of the reducing-extended-plenum system.

> ## *from experience...*
>
> In each of the duct systems discussed, the individual takeoff should be equipped with a damper at the main trunk line by which the volume of air through each branch duct can be controlled. By closing off a branch duct at the register, excessive noise can result.

> ## *from experience...*
>
> In typical residential applications and installations, the relatively small system capacities for the average home rarely warrant the added expense of fabricating transition duct sections. For this reason, the reducing-extended-plenum system is typically not used in new residential construction.

In addition to requiring more transition duct sections, extended-plenum and reducing-extended-plenum systems that are installed in existing structures also require the use of more **offset duct sections** (Fig. 9-9). Offsets are needed when the duct system must be made to fit around and within the existing conditions of the structure. Figure 9-9(A) shows a vertical offset; Figure 9-9(B) shows a horizontal offset.

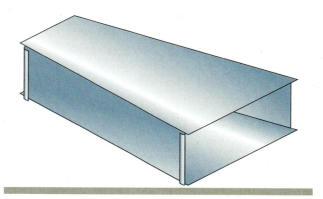

FIGURE 9-8 Transition duct section.

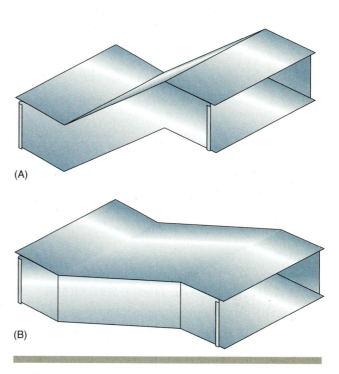

(A)

(B)

FIGURE 9-9 (A) Vertical-offset duct section. (B) Horizontal-offset duct section.

PERIMETER LOOP SYSTEM

In the perimeter duct system, a continuous duct loop runs around the perimeter of the conditioned space (Fig. 9-10). This perimeter loop is fed by a number of feeder ducts that extend from the supply plenum on the indoor unit. Although not commonly found on residential applications, the perimeter loop system is a good choice for commercial structures built on concrete slabs. In this application, the loop is installed before the slab is poured. This system is ideal for conditioning large open areas, as the pressure in the entire loop is the same, delivering a constant flow of air to all the registers. This type of duct system is rarely found on residential systems.

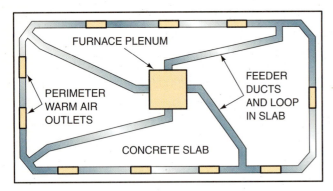

PERIMETER LOOP SYSTEM WITH FEEDER
AND LOOP DUCTS IN CONCRETE SLAB

FIGURE 9-10 Perimeter loop duct system.

DUCT SYSTEM MATERIALS

Ductwork can be made of a number of different materials. The choice of materials is often made by considering the location, the cost, and the intricacy of the installation. Some of the materials found on air-distribution systems are

- galvanized metal
- fiberglass-wrapped, spiral duct (flexible duct)
- fiberglass, fiberboard sheeting
- round sheet metal

GALVANIZED METAL DUCT SYSTEMS

Galvanized sheet metal duct sections are the most costly; they must be prefabricated in a duct shop and then installed on the job. As the name implies, they are fabricated from galvanized metal and are extremely durable. Before the duct system is installed, the space must be carefully measured and each piece must be fabricated exactly according to the plan. If, for any reason, a duct section does not fit properly or if an obstruction is in the way of the duct run, a new piece will have to be made up. This could delay the completion of the installation. The use of galvanized ductwork is the most costly route to take, but the life expectancy of the system is also the longest. It is ideal for applications in which ductwork is to be sealed behind walls and in enclosed ceilings.

When installing a galvanized sheet metal duct system, the most common method for joining the sections is with **slips and drives** (Fig. 9-11). Slips are simply strips of metal formed into an S shape, and drives are strips of metal that are formed into a flattened C shape. The longer edges of the duct slip into the openings in the slip (Fig. 9-12), hence the name. Once the duct sections are

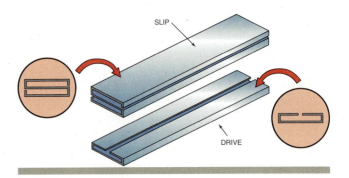

FIGURE 9-11 Slips and drives.

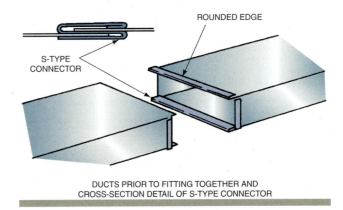

DUCTS PRIOR TO FITTING TOGETHER AND
CROSS-SECTION DETAIL OF S-TYPE CONNECTOR

FIGURE 9-12 Long ends of the duct sections slide into the slip "S" connector.

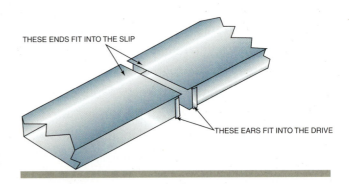

FIGURE 9-13 Shorter ends of the duct sections are folded back to form ears.

pushed together, a drive is used to secure the sections. To use slips and drives, the shorter edges of the duct must be bent over to form ears (Fig. 9-13).

For step-by-step instructions on installing galvanized metal duct systems, see Procedure 9-1 on pages 220–221.

from experience...

Because the metal ductwork can easily transmit noise and vibrations throughout the occupied space, a canvas connector (Fig. 9-14) should be used. This connector, or collar, is made of fireproof canvas or other flexible material and prevents unit vibrations from traveling through the duct sections. The ends are made of metal to facilitate the connection of the collar into the sheet metal duct system. It can be installed right on the discharge of the air handler or on the duct sections that connect to the supply plenum. If installed on the indoor unit itself, proper support for the plenum must be provided. The flexible canvas collar should not be used to support the weight of the duct.

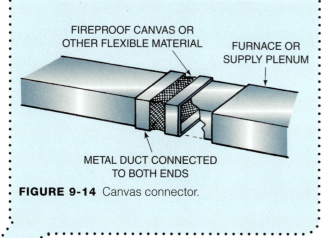

FIREPROOF CANVAS OR OTHER FLEXIBLE MATERIAL

FURNACE OR SUPPLY PLENUM

METAL DUCT CONNECTED TO BOTH ENDS

FIGURE 9-14 Canvas connector.

As part of the installation process, duct systems must be properly mounted and supported. Ducts can be supported from above or below depending on the conditions of the installation and the size of the sections. When they are run in attics or basements, they can be supported from above from either the roof rafters or the floor joists (Fig. 9-15). Strips of heavier gauge metal are often used as the strapping material. If conditions permit, the ductwork can be supported from below, with angle iron legs secured to the side of the section (Fig. 9-16). If multiple duct sections are used, plan to use at least one support for each section in the system.

FLEXIBLE DUCT SYSTEMS

Another material commonly found in duct systems is spiral **flexible duct**, which is constructed of spring-like shaped metal covered with a plastic film. The duct is then sometimes wrapped in fiberglass and covered with another casing of either foil or plastic (Fig. 9-5). Flexible duct is very popular for a number of reasons:

- Ease of installation
- Minimal skill requirements for installation
- No custom fabricated sections are needed for the branch lines
- Shorter installation time
- Low material cost
- No specialty tools required for installation
- Available in a wide range of sizes from 4″ to more than 24″ in diameter

The biggest drawback of this type of duct material is that the friction within the duct is very great. It should, therefore, be used primarily for short runs. Excessively long runs of flexible duct will result in reduced airflow to remote locations.

Flexible duct material is used in conjunction with other materials, such as sheet metal, rigid fiberglass panels, or fiberboard, to make up the complete duct system. Since supply plenums cannot be made from flex duct, these runs are connected to the main trunk line or plenum. Flexible duct is most often used to connect the main trunk line to the termination point of the individual branch ducts. To facilitate the connection of the flexible duct to the trunk line or plenum, a round **takeoff** is used. The other end of the duct run needs to be connected to the termination point of the branch run. This connection is made at the **boot**, which is the duct section connected directly to the decorative supply grill or register.

For step-by-step instructions on installing flexible duct systems, see Procedure 9-2 on pages 222–223.

Whenever possible, minimize the use of flexible ducts. The increased friction reduces airflow and strains the blower. Follow these practices when installing flexible ducts:

- Cut flex duct to fit and fully extend
- Keep flex duct straight
- Use long sweeping bends rather than tight turns
- Support flex duct along the length
- Avoid kinks and sharp bends
- Use bend braces when turning corners

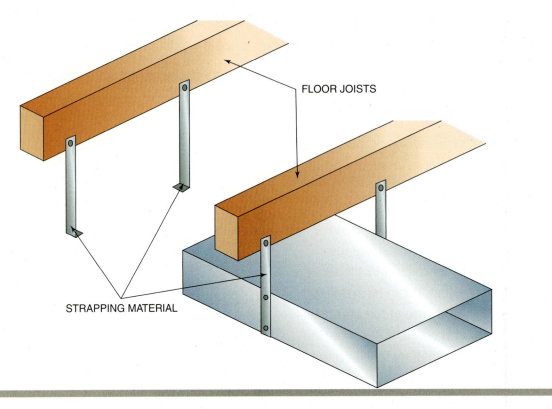

FLOOR JOISTS

STRAPPING MATERIAL

FIGURE 9-15 Sheet metal ducts supported with metal strapping material.

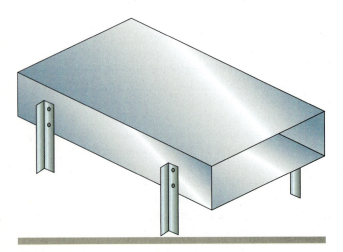

FIGURE 9-16 Angle iron used to form duct supports.

Courtesy of Johns Manville.

FIGURE 9-17 Ducts manufactured from compressed fiberglass with a foil backing.

FIBERGLASS, FIBERBOARD DUCT SYSTEMS

Fiberglass sheeting, also known as ductboard or **fiberboard**, is another alternative for fabricating ductwork. This material is pressed fiberglass with a reinforced foil backing that acts to insulate as well as strengthen the board. Fiberboard is available in rigid sheets (Fig. 9-17). When using rigid sheets, the duct sections can be fabricated in the field. A system made of fiberglass

boards gives members of the installation crew the flexibility to fabricate as they go. Simply put, the sections are measured and fabricated as they are installed, minimizing the chances of making a mistake.

The on-site installation time for this type of system is much greater than for a prefabricated duct system, but the cost of the material is less. Two major benefits of using fiberglass ductwork are that practically no noise is transmitted through the duct system and the ductwork will never sweat. Fiberglass ducts are not as durable as galvanized systems and are, therefore, a good choice when the system is located in a low-traffic area. Ductwork that is to be buried behind walls or in ceilings should not be made of fiberglass. Be sure to check with local building codes before using fiberglass ductwork, as it is not permitted in all municipalities.

For step-by-step instructions on installing fiberglass duct systems, see Procedure 9-3 on pages 224–225.

ROUND SHEET METAL DUCT SYSTEMS

Metal duct systems can also use round sheet metal duct sections for branch runs instead of the commonly used flexible duct. Round duct sections can be purchased in a wide range of sizes from 3″ to well over 20″ in diameter and typically come in 4-foot lengths. Like the flexible duct material, the round duct sections are used to connect the takeoff on the main trunk line to the boot at the supply grill location.

For step-by-step instructions on installing round sheet metal duct systems, see Procedure 9-4 on page 226.

COMBINATION DUCT SYSTEMS

Depending on the application and design of the air-distribution system, the duct system can be constructed of one or more of the materials described in the previous sections. When installing a **combination duct system**, be sure to follow the guidelines for each type of material to ensure the highest quality installation. To recap, the variety of materials, shapes, and configurations of duct materials include the following

Supply plenums and trunk lines

- Square sheet metal
- Rectangular sheet metal
- Fiberboard duct sections

Branch ducts

- Square or rectangular sheet metal
- Round sheet metal
- Rigid round fiberglass sections
- Rigid square or rectangular fiberboard sections
- Flexible duct

LINING AND INSULATING DUCT SYSTEMS

In a sheet metal duct system, there are two major concerns that must be addressed. One is heat transfer between the duct system and the air surrounding it. The other is noise transmission through the duct system.

In the cooling mode, heat from the surrounding air will be transferred to the cooled air inside the duct, resulting in a loss of system effectiveness and efficiency. In addition, metal ductwork will often sweat during the cooling mode of operation, which can result in water damage to the structure. To alleviate these problems, metal air-distribution systems should be insulated. A few options are available to the installation crew for insulating the ductwork. They are

- wrapping the duct with foil-covered fiberglass insulation
- lining the duct with acoustical lining
- wrapping and lining the ductwork

Once the duct system has been installed, the duct can be wrapped with foil-covered fiberglass insulation. This insulation comes in rolls that are typically 4-feet wide. It is secured around the duct with reinforced duct tape that is similar to that used to secure the fiberglass ductwork sections described earlier. The entire duct run must be insulated to prevent any condensation from forming on the duct surface. Also, the duct system must be wrapped tightly to prevent the formation of large air pockets between the duct and the insulation, which can result in excessive condensate accumulation.

To address the noise transmission issue, the duct can be lined with an acoustical material referred to as acoustical liner. Duct lining is installed on the sheet metal before the duct section is fabricated and is held in place with pins, as well as a contact adhesive designed specifically for this purpose. The lining must be secure on the inside surface of the duct and must be able to withstand the effects of the high velocity air moving across it. If the lining comes loose during system operation, the ducts can become blocked, causing system malfunction. The decision to line the duct system must be made when the system is initially designed. The duct size will have to be made larger to accommodate the lining.

Acoustical lining, depending on the application, is approximately 1 inch thick and will reduce the effective cross-sectional area of the duct run. For example, a 10-inch by 10-inch, unlined duct will have a cross-sectional area of 100 square inches, while a 10-inch by 10-inch duct lined with 1-inch thick insulation will have a cross-sectional area of only 64 square inches. Using fiberboard duct sections will have the same result (Fig. 9-18).

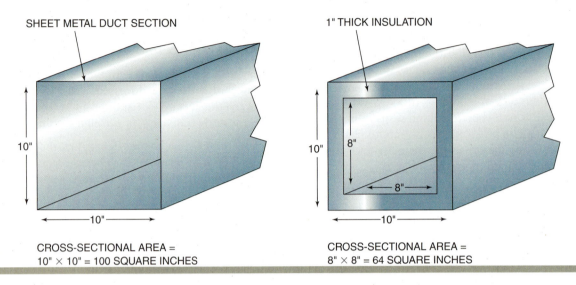

SHEET METAL DUCT SECTION

1" THICK INSULATION

10"

10"

10"

10"

8"

8"

10"

CROSS-SECTIONAL AREA =
10" × 10" = 100 SQUARE INCHES

CROSS-SECTIONAL AREA =
8" × 8" = 64 SQUARE INCHES

FIGURE 9-18 Insulation reduces the cross-sectional area of the duct.

Fiberglass duct sections do not need to be insulated after installation since the duct material itself acts as an insulator. This is another benefit of the fiberglass duct system. Spiral flexible ductwork, on the other hand, can be purchased as either insulated or non-insulated. If the determination is made that the duct run should be insulated, then it should be purchased and installed as such.

SEALING DUCT SYSTEMS

No matter what type of material is used to construct the air-distribution system, the finished product should have a minimum number of air leaks. Air leaks are typically

Delmar/Cengage Learning.

FIGURE 9-19 Technician applying mastic to a joint between two duct sections.

more common in metal duct systems than in fiberglass systems, given the nature of the methods used to fasten the duct sections together. Quite often galvanized sheet metal systems will leak at the corners of the joints, while round metal duct runs can leak at the joints. Thick putty called **mastic** can be applied to the joints and corners of metal duct sections to create an airtight seal. Mastic is desirable because it can be applied relatively easily with a paintbrush and dries to a hard but pliable finish (Fig. 9-19).

Fiberboard duct systems are typically sealed as the sections are connected to each other. Reinforced duct tape is used to secure the joints and, because the fiberglass sections butt up against each other during installation, the possibility of air leakage is very low.

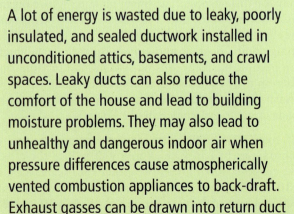

Going Green

A lot of energy is wasted due to leaky, poorly insulated, and sealed ductwork installed in unconditioned attics, basements, and crawl spaces. Leaky ducts can also reduce the comfort of the house and lead to building moisture problems. They may also lead to unhealthy and dangerous indoor air when pressure differences cause atmospherically vented combustion appliances to back-draft. Exhaust gasses can be drawn into return duct air leaks and enter the house.

SUMMARY

- Air-distribution systems are commonly configured as one of the following: plenum system, extended-plenum system, reducing-extended-plenum system, or perimeter loop system.

- Duct systems are often fabricated from one or more of these materials: galvanized metal, fiberglass-wrapped spiral duct (flexible duct), fiberglass or fiberboard sheeting, or round sheet metal.

- Sheet metal duct systems are typically installed with slips, drives, sheet metal screws, and metal strapping material.

- Fiberboard duct sections are joined together with reinforced duct tape and staples.

- Sheet metal duct connections should be sealed with mastic.

- Metal duct systems are wrapped with insulation to prevent heat transfer between the air in the duct and the air surrounding it.

- Metal duct sections can be lined with an acoustical material to reduce noise transmission through the duct system.

TECH TOOL BELT • TECH TOOL BELT

Courtesy of Milwaukee
Electric Tool Corporation.

Courtesy of Milwaukee
Electric Tool Corporation.

Duct Hammer

Hammers are available in a wide variety of styles depending on the tasks they are to perform. For the duct installer, the most commonly used hammer has a square face and a chisel-like back. The hammer facilitates not only the fabrication of duct sections, but also the installation of air-distribution systems.

Delmar/Cengage Learning.

Tin Snips

Tin snips are used to cut sheet metal material and are designed for straight, left, or right cutting. The snips are color-coded yellow, green, and red, respectively. Using the proper set of snips for the task at hand helps reduce sharp edges and burrs on the cut ends of the duct section, thereby reducing the chances of getting cut by the jagged edges.

Variable-Speed Cordless Drill

Probably one of the most useful tools for the duct installer is the cordless, variable-speed electric drill. When working with ductwork and self-tapping screws, the variable-speed feature enables the user to start the drill at a reduced speed to start the hole. Once the hole has been started, the speed of the drill can be increased. The greatest benefit of using a cordless drill is the increased mobility of the user. There is no need to run extension cords or worry about whether or not there is a power source available. One of the biggest drawbacks, however, is the need to maintain the battery charge. It is a good field practice to keep a spare, charged battery with your tools, just in case. Most manufacturers offer a battery charger that plugs into a vehicle's cigarette lighter.

Variable-Speed Drill

When working in an area where there is an AC power source, the variable-speed drill is a good choice. Typically, corded drills supply more power than the cordless variety, and there is no need to keep a charged battery on hand. The one drawback of using a corded tool is that mobility ranges are limited to the length of the power or extension cord.

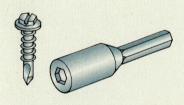

Self-Tapping Screws and Screw Holder

The most commonly used sheet metal screws are **self-tapping screws.** They have drill-bit-type tips that are able to make the hole in the metal and install the screw. These screws typically have a hex head, which reduces the chances of slippage. Special magnetic holders that fit directly into the drill hold the screws in place while they are being positioned. The typical head sizes for these screws are ¼" and 5/16".

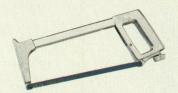

Courtesy of Ideal Industries, Inc.

Hacksaw

The hacksaw is often used to cut relatively thin metal objects such as angle iron or metal rods. It can be used to cut piping materials but it is *not* recommended for cutting piping material that is used in a refrigerant-carrying piping circuit. The metal filings and burrs that are created can easily find their way into the piping circuit and cause problems with the components in the refrigeration circuit. The blades used in a hacksaw are secured to the handle by a wing nut that should be tight enough to prevent the blade from moving within the handle. As with any saw or cutting tool, exerting excessive force on the tool can result in damage to the tool as well as personal injury.

Courtesy of Klein Tools, Inc.

Droplights and Extension Cords

Droplights and extension cords are without a doubt the installation crew's best friends during the installation process. While working in otherwise pitch-black attics and basements, the droplight provides the needed light to help ensure that the job is done right and that the workers remain safe. As with other tools, the ground prong must never be removed from the cord, as it provides protection from electric shock. When using droplights, try to obtain heavy-duty light bulbs, which will withstand the rough handling that they will be subjected to. Try to position extension cords and droplights away from the path that workers will take to reduce the possibility of tripping.

Courtesy of Ideal Industries, Inc.

Tape Measure

Probably one of the most important tools used by the installation crew is the tape measure, which enables the worker to transfer measurements from one place to

another. For example, a section of condensate drain piping may need to be cut to a specific length. Once this length is determined, the measurement must be transferred to the pipe to ensure that it is cut to the proper length. The tape measure measures not only feet and inches but fractions of an inch as well. They are commonly found in lengths ranging from 6 feet to 50 feet. Be sure to select a tape measure that will best suit the application for which it is intended.

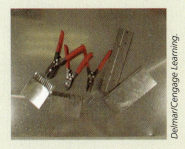

Delmar/Cengage Learning.

Specialty Sheet Metal Benders

Sheet metal benders come in a wide variety of styles and configurations based on the job they are to perform. Some are designed to bend the ears on duct sections; others are designed to make bends in strapping or related duct material.

PROCEDURE 9-1

Installing Galvanized Metal Duct Systems

 A Inspect the two duct sections to be joined and make certain that the short ends of the duct have been properly folded back to form ears.

- Place a slip connector on the longer ends of one of the duct sections.

Delmar/Cengage Learning.

 B Insert the long ends of the second duct section into the slip connectors mounted on the first section.

Delmar/Cengage Learning.

C Place the drive connector over the ears of the shorter ends of both duct sections and tap the drive with a hammer until the drive is centered on the duct. The drive should extend approximately one inch over each edge of the duct.

- Using another drive connector, repeat the previous step on the other edge of the duct.

Delmar/Cengage Learning.

D Hammer the ends of the drives over so that they are flush against the duct.

• Using a variable-speed drill, screw self-tapping screws on the longer ends of the duct making certain that the tabs on both duct sections are penetrated to ensure a stronger connection. To prevent the self-tapping screw from slipping, be sure to operate the drill at low speed until the screw tip has formed an indentation in the sheet metal.

D

Delmar/Cengage Learning.

E Using strapping material, the duct should be attached to ensure that it is securely held in place.

CAUTION

CAUTION: When installing self-tapping, sheet metal screws, remember that small metal filings will become airborne as a result of the drilling action. These filings can be very hot! Be sure to wear proper eye protection.

E

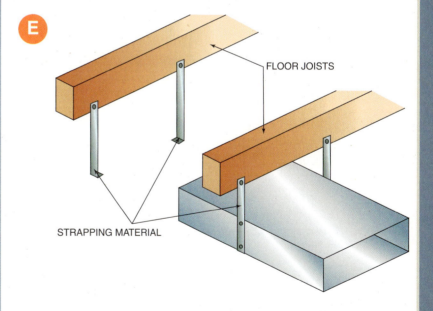

FLOOR JOISTS

STRAPPING MATERIAL

PROCEDURE 9-2

Installing Flexible Duct Systems

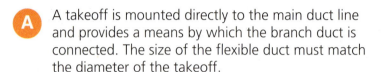
TECH TOOL BELT

Courtesy of Ideal Industries, Inc.

Utility Knife

Utility knives have razor-sharp blades enclosed in a plastic or metal handle. Always handle the utility knife with care, especially if the tool being used does not have a retractable blade. Always cut in a direction away from your body and keep fingers out of the way of the blade.

Photo by Eugene Silberstein.

Nylon Strap Tightener

The nylon strap tightener actually performs two tasks. First, by squeezing the handle on the tool, it tightens the nylon strap around the flexible duct that has been positioned over the takeoff on the duct system. Second, the tightener cuts the excess material from the strap.

A A takeoff is mounted directly to the main duct line and provides a means by which the branch duct is connected. The size of the flexible duct must match the diameter of the takeoff.

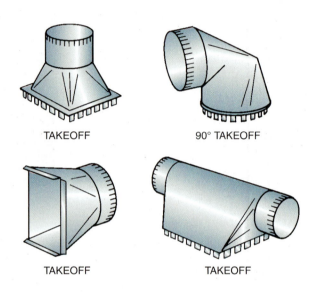

TAKEOFF   90° TAKEOFF

TAKEOFF   TAKEOFF

- One end of the flexible duct is connected to the take off by means of duct tape and nylon straps. The proper procedure for connecting the duct to the takeoff is as follows:

B Inner liner of the duct should first be secured to the takeoff with a nylon strap.

Delmar/Cengage Learning.

- Insulation and vapor barrier should be secured over the inner liner with another strap.

C Wrap the end of the flexible duct with duct tape.

D The other end of the duct run needs to be connected to the boot, which is the duct section connected directly to the decorative supply grill or register. Depending on the size and shape of the supply grill or register, the boot can be any one of a number of styles. As in the case of the takeoff, the boot transitions to a round collar that connects directly to the duct.

• Before making the connection to the boot, the flexible duct must be cut to the correct size. To do this, the duct should be stretched as tightly as possible without putting undue stress on the material.

E When the proper length has been determined, the outer layers of the duct can be cut with a razor or utility knife. The metal coil spring must then be cut as well. A heavy pair of wire cutters will do the trick here. Once the duct has been cut to length, it can be connected to the boot in a manner similar to that used to connect the duct to the takeoff.

C

Delmar/Cengage Learning.

D

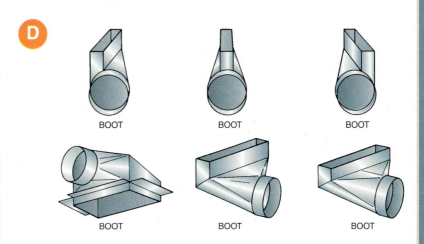

BOOT BOOT BOOT

BOOT BOOT BOOT

E

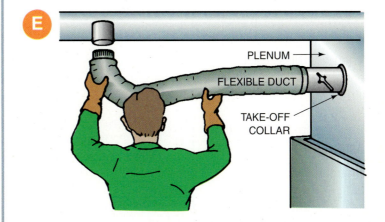

PLENUM

FLEXIBLE DUCT

TAKE-OFF COLLAR

Installing Fiberglass Duct Systems

TECH TOOL BELT

Fiberglass-Cutting Knives

This set of cutting knives is color-coded based on the cuts the knives are designed to make. The knives can cut corners, male edges, female edges, and V-grooves in the fiberboard section.

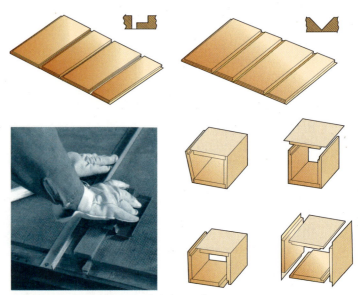

Courtesy of North American Insulation Manufacturers Association.

- The fabrication and installation of fiberboard duct systems requires a great deal of skill. Specialty tools are needed to properly cut the duct material, and accurate measurements are a must. The tools are color-coded and each makes a unique cut in the fiberglass side of the material while leaving the foil backing intact. The board can then be folded to form the duct section.

A When fiberglass duct sections are fabricated, a flap of the foil material is left on the edges of the section so that one section can be stapled to the next.

- Once stapled together using outward clinching staples, reinforced duct tape is used to complete the connection. This tape must be UL181 type,

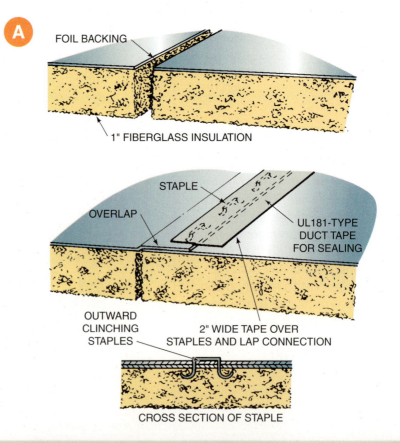

A FOIL BACKING

1" FIBERGLASS INSULATION

STAPLE

OVERLAP

UL181-TYPE DUCT TAPE FOR SEALING

OUTWARD CLINCHING STAPLES

2" WIDE TAPE OVER STAPLES AND LAP CONNECTION

CROSS SECTION OF STAPLE

which is often installed with the aid of a heat iron to activate the adhesive. This tape also has fibers within it, making it very durable. These fibers match the foil backing on the fiberglass boards themselves, creating an airtight seal. One of the major benefits of using fiberglass duct sections is that air leakage is almost completely eliminated.

D As with any other duct system, the ductwork must be properly supported. Since these duct sections are not made of metal, screwing strapping material to them will not work very well. In this case, a cradle for the duct to rest on can be easily fabricated from strapping material.

- This cradle can be secured to the floor joists or rafters, depending on the installation.

- Because this duct system is not rigid, the supporting cradles should be located close together to prevent the duct system from sagging.

D

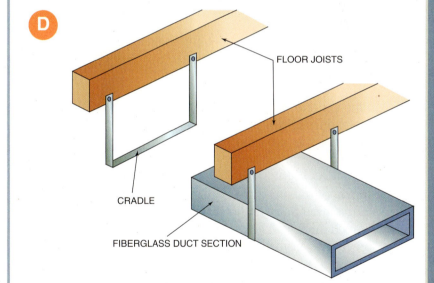

FLOOR JOISTS

CRADLE

FIBERGLASS DUCT SECTION

PROCEDURE 9-4

Installing Round Sheet Metal Duct Systems

A Installing round sheet metal duct sections is relatively straightforward because the sections are very easy to handle, given their short length. The individual sections are slipped together and then secured with sheet metal screws.

• At least three screws should be used on each joint to ensure that the run remains straight. Once again, self-tapping screws and a variable-speed drill help make the installation easier.

• When connecting round metal duct sections together, the male end of the section should always face away from the air supply or toward the termination point on the branch duct. This helps prevent air from leaking from the duct system.

B If bends need to be made in the duct run, round elbows can be used. These elbows are fabricated from four separate sections of metal and can be rotated to create a fitting ranging from 45 degrees to 180 degrees.

• As in the case of joining straight sections, round elbows should be secured with at least three screws to maintain the desired duct configuration.

A

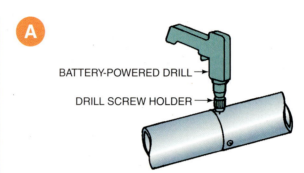

BATTERY-POWERED DRILL →
DRILL SCREW HOLDER →

B

ROUND ELBOW

• As with any other duct system, the round metal duct sections must be properly supported as they are installed. Strapping material can be used to effectively support the duct runs.

C Because this type of duct material is uninsulated, the entire run of the duct system will need to be wrapped with insulation once the sections have been mounted. Refer to the section regarding insulating duct systems for more on this.

REVIEW QUESTIONS

1. **Which of the following duct systems would be used if an attic-mounted air handler was located very close to all of the branch duct termination points?**

   a. Plenum system

   b. Extended-plenum system

   c. Reducing-extended-plenum system

   d. Perimeter loop system

2. **Which of the following duct sections is least likely to be found on an air conditioning installation in a new house?**

   a. Transition duct section

   b. Offset duct section

   c. Round takeoff

   d. Boot

3. **Mastic is a material used to**

   a. Support fiberboard duct sections

   b. Join sheet metal duct sections

   c. Seal fiberboard duct sections

   d. Seal metal duct sections

4. **Which of the following would likely be found on a galvanized sheet metal extended-plenum duct system that has been installed in an existing structure?**

   a. Transition duct sections

   b. Offset duct sections

   c. Excessively long spiral flexible duct takeoffs

   d. Both b and c are correct

5. **The most common method of joining galvanized sheet metal duct sections is**

   a. Fiber-reinforced duct tape

   b. Self-tapping sheet metal screws

   c. Metal strapping material

   d. Slips and drives

6. **When installing a reducing-extended-plenum duct system, which of the following is used to prevent noise transmission and vibration in the duct system.**

   a. Canvas connector

   b. Offset duct section

   c. Mastic

   d. Metal strapping material

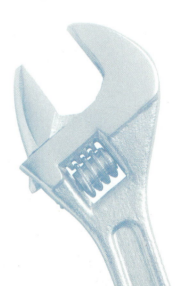

7. One of the biggest drawbacks of using spiral flexible duct is that

a. The spiral duct is more costly than prefabricated sheet metal sections

b. There is a great deal of friction within the spiral flexible duct

c. A highly skilled technician must perform the installation

d. The spiral flexible duct is only available in a limited number of sizes

8. Which of the following is a benefit of fiberboard duct systems?

a. They are more durable than galvanized sheet metal duct systems

b. They do not need to be insulated after installation

c. Sections can be fabricated as they are needed right on the job site

d. Both b and c are correct

KNOW YOUR CODES

Ductwork must meet minimum air tightness requirements set by the building code (IRC section N1103.2.2) but more stringent requirements may be required by green building certification programs. Duct tightness can be tested using a special fan used with calibrated instruments to pressurize the ducts and determine how tight the system is.

Ductwork in unconditioned spaces must also be wrapped with insulation to reduce heat loss and heat gain and prevent condensation. The IRC 2009 requires a minimum of R-8 insulation wrapping ducts within attics and R-6 insulation on all other ducts (R1103.2.1) except those that are within the conditioned living space of the house.

CHAPTER 10

System Connections

- List the piping materials commonly used to install air conditioning systems.

- Explain the difference between soft-drawn and hard-drawn copper piping.

- Explain the purpose of pipe fittings.

- Cut, flare, swage, and bend copper tubing.

- Prepare an air-acetylene torch for use.

- Solder or braze copper piping sections.

- Join sections of plastic piping material.

- Explain the importance of condensate lines, traps, and drain pans.

- Describe the function of a refrigerant trap.

- Install refrigerant lines between the air handler and condensing unit.

- Run the line-voltage wiring from the disconnect switches to the equipment.

- Run the low-voltage wiring between the thermostat, indoor unit, and outdoor unit.

- Make electrical connections on the line and low-voltage circuits.

- Describe the color-coding of the low-voltage wiring between the thermostat, indoor unit, and outdoor unit.

At this point in the installation process, it is assumed that the locations for the air handler and condensing unit have been selected and the equipment has been set in place. Members of the installation crew can then begin to make the necessary system connections. These connections include both electrical connections that control the operation of the equipment and plumbing connections that can carry refrigerant from one part of the system to another or help remove unwanted condensate from the structure.

The importance of the quality of these connections cannot be stressed enough. Improper refrigerant piping can lead to refrigerant leaks and component damage. Improper condensate drain piping can lead to water damage to the structure. Improper electrical wiring can lead to system malfunction or fire. Quality system connections will help ensure not only a safe system but one that will provide years of satisfactory service.

The plumbing connections made in the refrigerant circuit are often made by using a process called brazing. Refrigeration technicians and installation personnel alike must develop good brazing techniques to prevent the system refrigerant from being released to the atmosphere.

ACR tubing Nitrogen-charged piping used in air conditioning applications

Acrylonitrilebutadiene styrene (ABS) pipe Rigid plastic piping material used for drain lines

Auxiliary drain pan Installed under the air handler in the event the primary condensate pan overflows

Brazing Process of joining metallic piping sections using filler materials with melting points in the range of 1500°F

Condensate The moisture removed from the air by the air conditioning system

Condensate drain line Piping arrangement that carries condensate from the structure

Condensate drain trap "U" bend piping arrangement at the outlet of the condensate drain pan that facilitates proper condensate removal

Condensate pump A mechanical pump that carries condensate from the structure

Fittings Piping components that facilitate making bends and turns in a piping system comprised of rigid or hard-drawn piping materials

Flaring Process by which the end of a soft-drawn tubing section is flared to a 45-degree angle

Flux Material applied to piping materials prior to soldering to help reduce oxidation and remove particulate matter from the joint

Hard-drawn pipe See rigid pipe

Polyvinyl chloride (PVC) pipe Rigid plastic material commonly used for drain lines

Reamer Used to remove burrs from the inside and outside of a pipe section

Refrigerant trap Device located at the outlet of the evaporator when the air handler is located below the compressor; used to facilitate oil return to the compressor

Rigid pipe Piping material that is not intended to be bent

Soft-drawn copper tubing Piping material that has the ability to be bent into the desired configuration,

reducing the number of fittings required to connect a system

Soldering Process of joining metallic piping sections using filler materials with melting points in the range of 500°F

Swaging Process of expanding the end of a piping section to accept another section of same-size piping material

PIPING MATERIALS

To complete any air conditioning system installation project, a number of system components need to be connected, by means of piping, to each other. For example, on split-type air conditioning systems the suction and liquid lines need to be run between the condensing unit and the air handler. In addition, while operating in the cooling mode, condensate needs to be removed from the equipment location via piping arrangements that must be field installed. Humidifiers and evaporative coolers need water lines connected to them as well in order for them to perform their functions. Finally, heating systems fueled by oil or gas need to have a means to supply this fuel to the unit. To make certain that the system will provide years of satisfactory service, the correct piping materials and fittings must be chosen, and proper techniques must be used when working with them. Commonly used piping materials include

- Soft-drawn copper tubing
- Rigid copper piping
- Black steel pipe
- Galvanized steel pipe
- PVC plastic pipe
- ABS plastic pipe

Black and galvanized steel piping materials are typically used on heating and hot water systems to carry the fuel to the furnace or boiler. The operations and skills needed to work on steel pipe systems will, therefore, be discussed in the heating unit.

Going Green

When selecting materials for use in system installation, be sure to select the proper materials for the specific application. Also, be aware of alternative, more environmentally friendly materials that achieve the same result with less of an environmental impact.

FITTINGS

No matter what piping material is used in the installation of an air conditioning system, there will undoubtedly be places where the piping arrangement will need to make bends or turns around obstructions or be connected to other system components. A variety of **fittings** are available to make these turns and connections possible. Also, it may be necessary to connect a threaded section of piping material to another threaded section of piping or to a section of pipe that is not threaded, as in the case of copper pipe.

Figure 10-1 shows a number of fittings that are available in a variety of materials, including copper, black steel, galvanized steel, PVC, and ABS. In addition, the fittings are manufactured in a wide range of sizes to accommodate the needs of the tradespeople performing the work.

from experience...

Since black steel and galvanized steel piping arrangements are most commonly used for gas heating equipment, the piping operations for these materials are discussed in the section on installing gas heating equipment.

The most commonly used pipe fittings include

- 90-degree elbows—Used to join two sections of same-size pipe at a 90-degree angle. The diameter of the female ends of the fitting is larger than the pipes to accommodate the pipe sections.

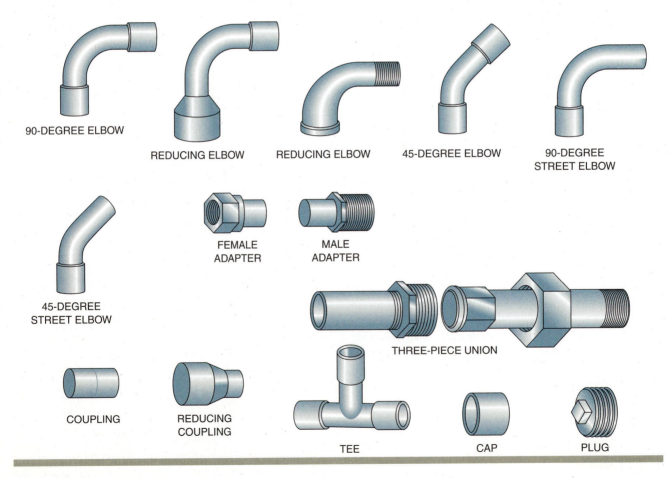

90-DEGREE ELBOW REDUCING ELBOW REDUCING ELBOW 45-DEGREE ELBOW 90-DEGREE STREET ELBOW

45-DEGREE STREET ELBOW FEMALE ADAPTER MALE ADAPTER THREE-PIECE UNION

COUPLING REDUCING COUPLING TEE CAP PLUG

FIGURE 10-1 Commonly used pipe fittings.

- 90-degree reducing elbows—Used to join two sections of different-size pipe at a 90-degree angle.
- 45-degree elbows—Used to join two sections of same-size pipe at a 45-degree angle.
- 90-degree street elbows—Used to make a 90-degree bend in a piping circuit. One end of the fitting is female to accept a section of piping material, while the other end is the same size as the piping material itself.
- 45-degree street elbows—Used to make a 45-degree bend in a piping circuit. One end of the fitting is female to accept a section of piping material, while the other end is the same size as the piping material itself.
- Female adapters—In the HVAC industry, the most common female adapter joins a male threaded pipe section to a section of copper pipe or tubing. The female adapter screws onto the male pipe, providing a means by which a copper pipe can be connected to the circuit.
- Male adapters—In the HVAC industry, the most common male adapter connects a female threaded fitting to a section of copper pipe or tubing. The male adapter screws into the female thread, providing a means by which a copper pipe can be connected to the circuit.
- Unions—Three-piece components used to connect two sections of similar pipe that may need to be disconnected in the future. Unions are also commonly used to repair existing piping circuits.
- Couplings—Similar to unions, couplings are used to connect two sections of same-size pipe. The main difference between the two is that couplings permanently joint the two sections.
- Reducing couplings—Used to connect two sections of pipe that are not the same size.
- Tees—Used to connect three sections of piping material. The three sections may or may not be the same size.
- Caps—Used to terminate a pipe in the circuit. The cap fits over the outside of the fitting.
- Plugs—Used to terminate a pipe in the circuit. The plug is typically made with a male thread and fits inside a fitting with a female thread.

COPPER PIPING

Copper piping material is commonly used in the air conditioning industry to carry refrigerant from one system component to another, and can also be used to carry condensate from the air handler of a system. It is relatively easy to work with and, when installed properly, will provide a leak-free arrangement through which system fluids will flow. Copper pipe is rated by the thickness of the pipe wall.

Some Common Grades of Copper	
K	Thick-walled copper piping used for heavy duty applications
L	Medium-walled copper piping is the most frequently used
M	Thin-walled copper piping is rarely used in the HVAC industry
DWV	Drain, waste, and vent piping is commonly used in the plumbing industry
ACR	Air conditioning and refrigeration piping is commonly used in the HVAC Industry

In the air conditioning industry, **ACR tubing** and piping is the number one choice among service contractors, installation companies, and manufacturers. The primary reason for this is that ACR piping material is dehydrated, sealed, and pressurized with nitrogen to keep the interior of the pipe clean and dry by preventing moisture and particulate matter from entering the piping. Both moisture and dirt are enemies to every air conditioning system.

Copper piping material used in the air conditioning industry is measured by its outside diameter. Simply stated, this means that the opening in the pipe that is available to carry fluids is smaller than the size classification of the pipe. In the plumbing field, however, piping material is measured by its nominal, or inside, diameter. Figure 10-2 shows the difference between ½-inch nominal piping material and ½-inch ACR tubing. Note that the ½-inch nominal pipe used in plumbing application has an outside diameter of ⅝ inch.

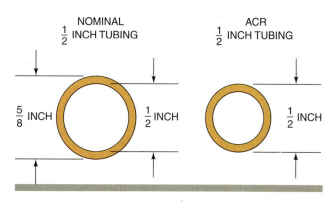

FIGURE 10-2 Copper tubing used for plumbing and heating is sized by its inside diameter. ACR tubing is measured by its outside diameter.

Copper tubing can be obtained in a wide range of sizes from ⅛ inch to more than 4 inches. For residential applications, however, typical pipe sizes will rarely be larger than ⅝ inch. Copper piping material is available as either hard-drawn or soft-drawn. Typically, soft-drawn tubing is used on residential systems, primarily because of the ease of installation and the reduced number of solder joints that are needed to install it.

HARD-DRAWN COPPER PIPING

As the name indicates, **hard-drawn** or **rigid copper piping** is manufactured in 10-foot or 20-foot lengths that are not intended to be bent. This material is commonly used when long, straight piping runs are needed, such as in the installation of a condensate line.

from experience...

When dealing with condensate lines, always follow the local codes regarding the pitch of the line as well as the material used. Most urban areas require that rigid copper pipe be used for condensate lines, but some local municipalities allow the use of PVC pipe, discussed later in the chapter, or similar piping material.

If the piping arrangement must bend up, down, or side to side, pipe fittings must be used to create the desired turns. Pipe fittings come in a wide range of sizes and configurations that enable the installing technician to create the desired piping contour. Refer to the section on pipe fittings for more on the variety of fittings that are available. One drawback to the use of hard-drawn piping is an increased chance that a system leak will develop due to the increased number of joints in the circuit. Whenever a fitting is used, there are at least two joints that must be properly sealed. If the pipe and fitting are not sealed properly, the fluid will leak from the system. Methods by which the joints are sealed are discussed later in this chapter.

SOFT-DRAWN COPPER TUBING

To alleviate the problem of increased leak potential with hard-drawn piping, **soft-drawn copper tubing** (Fig. 10-3) is frequently used in the installation of residential air conditioning systems. Because it can be easily bent into different configurations, the number of fittings required is drastically reduced, thereby reducing the possibility of

FIGURE 10-3 Soft-drawn copper tubing typically comes in 50-foot rolls.

refrigerant leaks. Soft-drawn tubing is especially desirable when lines are to be run through walls or ceilings that will ultimately be sealed. A leaking fitting behind a finished wall can lead to extensive damage and repair costs. Soft-drawn tubing is also desirable because it typically comes in 50-foot rolls that are much easier to transport and carry than are the 10- or 20-foot lengths of rigid pipe.

One thing that must be kept in mind is that soft-drawn tubing must be unrolled from the coil. The best way to do this is to place the loose end on the ground or other flat surface and hold it in place with your hand or foot. Then, carefully unroll the tubing, keeping the roll flat as it is unrolled (Fig. 10-4). Simply pulling the tubing off the coil will result in twisting the tubing, creating waves in the line. Although this will not have a major effect on system operation if the refrigerant line run is vertical, small refrigerant traps will be formed if the run is horizontal. An excessive number of refrigerant traps can affect system operation and hinder the process of compressor oil return.

FIGURE 10-4 Tubing should be unrolled carefully to avoid kinking the pipe.

TECH TOOL BELT • TECH TOOL BELT

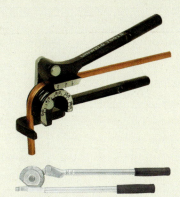

Courtesy of Uniweld Products.

Tubing Cutter

The tubing cutter uses a set of rollers and a cutting blade. The piping material to be cut is held in place between the rollers and the blade. As the handle is tightened, more pressure is place on the pipe and, as the cutter is rotated around the pipe, the cutting process is achieved. Most tubing cutters are also equipped with a reamer that removes burrs from inside the cut tubing. The reamer is inserted into the end of the tubing and rotated to scrape off the burrs.

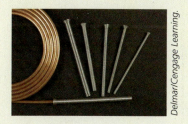

Delmar/Cengage Learning.

Bending Spring

The bending spring is used to bend soft-drawn copper tubing by simply placing the spring over the tubing and bending the spring into the desired shape. The spring will evenly distribute the pressure, preventing the tubing from getting crimped.

Courtesy of Uniweld Products.

Tubing Bender

The lever-type bender facilitates the bending of tubing at precise locations and to the desired angle. Using a tubing bender reduces the number of solder joints that are needed to install a piping system. Tubing benders are available in sizes ranging from ¼ inch to more than ⅞ inch.

Courtesy of Uniweld Products.

Air-Acetylene Torch Kit

The air acetylene torch kit consists of a pressurized tank of acetylene, an acetylene regulator, a hose, a torch handle, and a torch tip. The torch flame is adjusted at the regulator, which meters the flow

of gas to the torch handle and tip. Always make certain that the torch tank is secured to prevent it from falling over. In addition, the torch kit should be carefully leak-tested with a soap bubble solution.

Courtesy of Uniweld Products.

Oxy-Acetylene Torch Kit

The oxy-acetylene torch kit consists of pressurized acetylene and oxygen tanks, acetylene and oxygen regulators, a two-hose set, a torch handle, and a torch tip. The oxy-acetylene torch flame is much hotter than that of the air-acetylene torch and can be used for a wide range of applications. As with any torch kit, make certain that the torch tanks are secured to prevent them from falling over and carefully leak check the setup before use.

ACETYLENE TORCH TIPS									
Tip Size			Gas Flow		Copper Tubing Size Capacity				
Tip No.	in.	mm	at 14 psi ft³/hr	(0.9 Bar) m³/hr	Soft Solder in.	mm	Silver Solder in.	mm	
A–2	³⁄₁₆	4.8	2.0	.17	⅛–½	3–15	⅛–¼	3–10	
A–3	¼	6.4	3.6	.31	¼–1	5–25	⅛–½	3–12	
A–5	⁵⁄₁₆	7.9	5.7	.48	¾–1½	20–40	¼–¾	10–20	
A–8	⅜	9.5	8.3	.71	1–2	25–50	½–1	15–30	
A–11	⁷⁄₁₆	11.1	11.0	.94	1½–3	40–75	⅞–1⅝	20–40	
A–14	½	12.7	14.5	1.23	2–3½	50–90	1–2	30–50	
A–32*	¾	19.0	33.2	2.82	4–6	100–150	1½–4	40–100	
MSA–8	⅜	9.5	5.8	.50	¾–3	20–40	¼–¾	10–20	

*Use with large tank only.
NOTE: For air conditioning, add ⅛-inch for type L tubing.

Courtesy of Thermadyne Industries, Inc.

Torch Tips

Depending on the method being used to join the piping sections, a larger or smaller flame may be required. Larger torch tips will produce the heat required for larger soldering or brazing jobs, while smaller tips will produce smaller amounts of heat. Using the proper size torch tip for the proper job will not only save gas, it will also help to increase the quality of the solder joint.

Photo by Eugene Silberstein.

Refrigeration Service Wrench

The refrigeration service wrench is used to turn valves with square stems, such as those found on acetylene B-tanks. The wrench is a ratchet-type device and can be used on square stems of different sizes. Square stems should NOT be turned with an adjustable wrench as the stem can become stripped.

Courtesy of Uniweld Products.

Striker

The striker is the *only* device that should be used to light a torch of any sort. The striker emits a spark when the handle is squeezed. During the sparking process, a flint is rubbed across a rough metallic surface, creating the spark that ignites the fuel. For safety reasons, matches or cigarette lighters should never be used to ignite the fuel.

Courtesy of The Mill-Rose Company.

Fitting Brushes

These specialty brushes are designed to clean the interior surfaces of pipe fittings prior to soldering. Any dirt that accumulates inside the fitting will reduce the quality of the solder joint. Fitting brushes come in a variety of sizes to accommodate a wide range of fitting sizes. Simply insert the brush into the fitting and turn the brush until the interior is clean.

Courtesy of Uniweld Products.

Flaring Yoke

Used in conjunction with the flaring block, the flaring yoke creates a flare (45-degree angle) on the end of a section of copper tubing. The yoke is secured around the block after the tubing has been placed in the block. The flare is created by tightening the handle of the yoke, which pushes the cone of the yoke into the tubing.

Photo by Eugene Silberstein.

Swaging Tool

Used in conjunction with the flaring block, the swaging tool expands the end of a tubing section so that two sections of same-size tubing can be connected together. After securing the tubing in the flaring block, the swaging tool is inserted into the tubing and then struck firmly with a hammer. Swaging tools are available in sizes ranging from ¼ inch to more than ⅞ inch so be sure to use the right size tool for the tubing being used.

Courtesy of Uniweld Products.

Flaring Block

The flaring block is used to secure a section of copper tubing while being either swaged or flared. The end of the tubing that is being worked on should stick out of the block from the beveled or angled end. The flaring block has multiple holes so it can be used to secure tubing of different sizes.

COPPER PIPING OPERATIONS

To successfully construct and install a copper piping system, a number of operations need to be performed on the piping material to ensure the integrity of the system. Some of the operations that will be discussed in the following sections refer to both hard-drawn and soft-drawn piping, but some refer only to soft-drawn tubing. When such is the case, appropriate indications will be made.

Cutting. Copper tubing and pipe can be cut with either a tubing cutter (see Tech Tool Belt p. 235) or a hacksaw, depending on the application for which the piping material is to be used. If the copper piping will carry refrigerant between the component parts of an air conditioning system, it is recommended that the pipe or tubing be cut with the tubing cutter. The tubing cutter does not produce metal filings, as does the hacksaw. These metal filings can block capillary tubes or find their way to the compressor and cause system problems. If the copper lines are to be used for condensate removal, the hacksaw is a good alternative, but the tubing cutter is always the number one choice because no filings are made. In any event, the filings that are produced by the hacksaw should be removed from the pipe before connecting.

If a hacksaw is used to cut the copper pipe, the cut should be made at a perfect 90-degree angle. To minimize the amount of filings, the hacksaw blade should have as many teeth per inch as possible. The finer the blade, the better the cut will be and the smaller the filings will be. Any burrs left on the pipe need to be removed with a reamer. Refer to the next section.

The tubing cutter is made up of a set of rollers that help hold the pipe straight, a reamer to remove any burrs that may be left on the pipe, a cutting wheel to cut the pipe, and an adjusting handle to tighten the wheel against the surface of the copper.

For step-by-step instructions on using the tubing cutter, see Procedure 10-1 on page 256.

Reaming. The **reamer** (Fig. 10-5), located on the back of the tubing cutter, is used to remove the burrs left behind during the cutting process. When used, the reamer should be pivoted from the back of the cutter so that the blade portion extends past the cutter. The reamer is inserted into the pipe and, while applying pressure, is rotated to remove any burrs from the interior surface of the pipe. Any burrs on the outside surface of the pipe can be removed with the file portion of the reamer.

Swaging Soft-Drawn Copper Tubing. **Swaging** is the process by which two pieces of soft-drawn copper tubing with the same diameter can be joined together by stretching the end of one of the sections (Fig. 10-6). Hard-drawn tubing is not intended to be swaged. Once the end has been swaged, or stretched, the unswaged end will fit into it. The joint can then be soldered or brazed. Soldering and brazing are discussed later in this chapter.

Delmar/Cengage Learning.

FIGURE 10-6 Copper tubing will fit into the swaged end of the other section.

To perform the swage, a flaring block, hammer, and proper size swaging tool are needed.

For step-by-step instructions on swaging soft-drawn copper tubing, see Procedure 10-2 on page 257.

Flaring Soft-Drawn Copper Tubing. The flare connection is used to join a section of soft-drawn copper tubing to a male threaded flare fitting or system component. As the name suggests, **flaring** means that the end of the soft-drawn tubing is flared to a 45-degree angle. This flared portion of tubing will rest against the male portion of the fitting being connected, and will be held in place with a flare nut (Fig. 10-7). To perform the flaring process, a flaring block and a flaring yoke are needed.

For step-by-step instructions on flaring soft-drawn copper tubing, see Procedure 10-3 on page 258.

Soldering. **Soldering**, also called soft soldering, is a process that is frequently used to join a section of piping material to another piping section or a pipe fitting or system component. Materials that are commonly

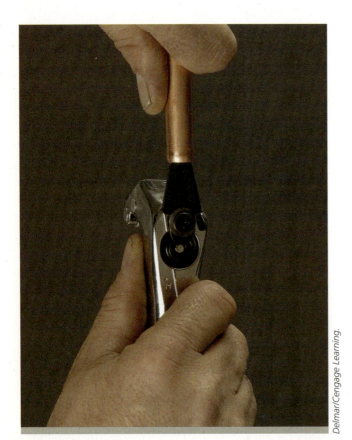

Delmar/Cengage Learning.

FIGURE 10-5 Removing burrs from the inside of the tubing.

Photo by Eugene Silberstein.

FIGURE 10-7 Components of a flare connection.

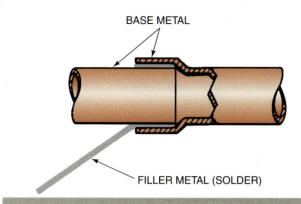

BASE METAL

FILLER METAL (SOLDER)

FIGURE 10-8 The molten filler material flows in the gap between the two pipe sections.

soldered are often made of either copper or brass. During the soldering process, the pipe joint is heated to the desired temperature, usually in the 500°F range, and a filler material is introduced to the space between the pipe sections (Fig. 10-8). The heat of the pipe causes the filler material to melt, filling the gap between the two sections. As the filler material cools, it hardens to seal the joint between the pipes. Different filler materials are used for different applications.

Two Commonly Used Filler Materials	
50/50	50% tin, 50% lead
	Used for lower pressure applications
	Not to be used on potable water supplies
95/5	95% tin, 5% antimony
	Used for higher pressure applications
	Provides greater joint strength

To perform the soldering process, pipe brushes, sand cloth or steel wool, a striker, solder, flux, flux brushes, a refrigeration service wrench, and a heat source such as an air-acetylene torch kit are needed (Fig. 10-9). **Flux**, a material that is applied to the male portion of a joint prior to soldering, reduces the amount of oxidation on the joint and helps remove dirt or other particulate matter from the joint while soldering.

For step-by-step instructions on soldering, see Procedure 10-4 on pages 259–262.

Going Green

Never use lead-bearing solder or filler materials on lines that carry potable water!

Delmar/Cengage Learning.

FIGURE 10-9 Tools and equipment needed to solder: 95/5, 50/50, acetylene torch kit, striker, flux, fitting brushes, steel wool, and a refrigeration service wrench.

Brazing. **Brazing** is another method used to join two sections of piping or to connect a pipe section to a fitting or a system component. Brazing is called silver soldering or hard soldering by many field personnel. It is the method of choice when there will be high pressures inside the piping circuit. Typically, air conditioning system installers use brazing techniques on the refrigeration circuit piping and soft-soldering techniques on other plumbing lines, such as condensate lines, if needed. Some significant similarities and differences exist between soft soldering and brazing.

Some of the similarities include the following:

- A filler material is used to fill the space between the sections being joined.
- A heat source such as an air-acetylene torch is used.
- The heat from the torch is used to heat the pipe surfaces, which in turn heats the filler material causing it to melt.
- The molten filler material will flow toward the source of heat.

Differences between the two methods include the following:

- When brazing, the pipe surfaces do not need to be cleaned and fluxed when joining copper to copper. Joining dissimilar metals does require the use of flux.

from experience...

The flux and filler materials used in the brazing process are different from those used for soft soldering! Do not confuse them!

- When brazing, the pipes need to be heated to a higher temperature, in the range of 1500°F, to melt the filler material.
- Oxidation often occurs on the surface of brazed pipes, resulting in the accumulation of particulate matter on both the interior and exterior surfaces of the pipes.

from experience...

Allowing 1 to 2 psig of nitrogen to flow through the pipe while brazing will reduce the oxidation on the interior surfaces of the pipe. This will help prevent the clogging of metering device and system strainers.

- When brazing, the pipes are typically heated until they begin to glow cherry red. This indicates that the pipes are hot enough to permit the filler material to flow into the space between the piping sections.

To perform the brazing process, a striker, refrigeration service wrench, brazing rods, and a heat source such as an air-acetylene torch kit are needed.

For step-by-step instructions on brazing, see Procedure 10-5 on pages 263–264.

Bending Soft-Drawn Tubing. In the refrigeration industry, one of the biggest problems the service technician faces is dealing with refrigerant leaks. Quite often, leaks occur at joints that have been soldered or brazed during initial system installation. To reduce the possibility of a system leak from a solder joint, the best thing the installation crew can do is reduce the number of solder joints on the job. By bending soft-drawn tubing into the desired configuration, the number of fittings required will be greatly reduced.

The bending spring and the lever-type tubing bender (see Tech Tool Belt p. 235) help accomplish the bending process. Smaller diameter tubing can easily be bent by hand [Fig. 10-10(A)], but larger tubing should be bent by mechanical means. When bending tubing by hand, care should be taken not to kink the tubing [Fig. 10-10(B)]. Kinks in the tubing restrict flow through the pipe and can have adverse effects on system operation.

from experience...

When using a lever-type tubing bender, be sure to read the literature that comes with the tool. Practice is needed to be able to bend tubing sections properly. Inaccurate measurements, calculations, and bends can result in excessive amounts of material waste.

PLASTIC PIPING

Another material commonly found in air conditioning systems is plastic pipe. Plastic piping material is often used for the condensate line that is needed to dispose of the moisture that the system removes. Two types of plastic pipes that are often found in residential applications are ABS and PVC. Plastic piping material is easily cut and joined, making it a popular choice for piping circuits when its use is permitted by local codes.

PVC PIPE

Polyvinyl chloride piping, more commonly known as PVC, is probably the most commonly used material for condensate removal piping systems. PVC is a lightweight material but is supplied as a straight, rigid pipe. As in the case of rigid copper pipe, fittings are needed if the piping run is to make bends. When used to remove condensate, the PVC needs to be connected to the female pipe fitting installed in the air handler's drain pan. To accomplish this, a male adapter similar to the one illustrated in Figure 10-1 is used to change from the threaded fitting to a female PVC fitting. Just as with copper fittings, many different fittings are available for use with PVC pipe.

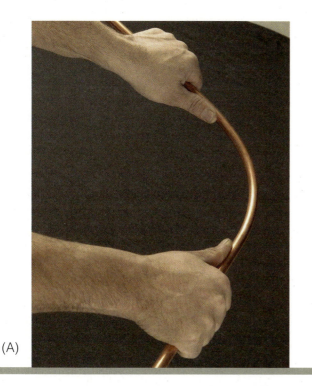

(A) (B)

Delmar/Cengage Learning.

FIGURE 10-10 (A) Small-diameter tubing can be bent safely by hand. (B) Larger tubing can easily kink when bent by hand.

TECH TOOL BELT • TECH TOOL BELT

Courtesy of Ridge Tool Company.

Plastic Tubing Shear

The plastic tubing shear is a very handy tool for cutting PVC pipe. The pipe being cut is placed in the cutter and, by applying pressure to the handle of the tool, the metal blade cuts through the pipe in a single action. Compared to cuts made with a hacksaw, those made with the shear are much smoother and produce fewer burrs and rough edges.

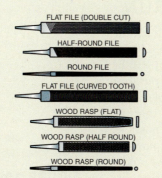

FLAT FILE (DOUBLE CUT)

HALF-ROUND FILE

ROUND FILE

FLAT FILE (CURVED TOOTH)

WOOD RASP (FLAT)

WOOD RASP (HALF ROUND)

WOOD RASP (ROUND)

File

Just as the reamer was used to remove burrs from the interior of a pipe, the file is used to remove any rough edges from the exterior surfaces of a cut section of pipe. Files come in a wide range of sizes and shapes, so always select the right file for the job. For example, a round file can be used to effectively remove burrs from the inside of a cut section of PVC pipe.

ABS PIPE

Acrylonitrilebutadiene styrene pipe, known simply as ABS, is somewhat less durable and not as common as PVC but is used in a number of applications. It is extremely easy to cut and join, as is PVC; and in some local municipalities, it is the desired material for condensate removal.

For step-by-step instructions on joining plastic pipe, see Procedure 10-6 on page 265.

INSTALLING, SUPPORTING, AND INSULATING REFRIGERANT LINES

One of the most important tasks the installation crew can perform to ensure the long-term satisfactory operation of the air conditioning system is the proper installation of the refrigerant lines. Poorly installed refrigerant lines can result in premature compressor failure, as well as a system that does not operate effectively and efficiently. The individuals installing these lines should keep a number of things in mind, including the following:

- The run should be as short as possible
- No excessive piping should be used
- The number of fittings should be kept to a minimum
- When needed, long radius elbows should be selected
- The run should be as straight as possible
- All solder joints should be as perfect as possible
- The suction line should be pitched back toward the compressor
- Refrigerant traps should be installed when necessary
- The suction line should always be insulated
- All structural penetrations must be sealed

LENGTH OF THE PIPING RUN

Although the preceding list seems to be quite long, the items listed summarize a satisfactory piping job. If any one of the items is neglected, system operation will be affected in a negative way. The indoor and outdoor units should be as close to each other as possible. If the location selection is done properly, it follows that the length of the refrigerant lines will be as short as possible. To ensure maximum efficiency, though, the piping run should be as direct as possible, with no excessive solder joints or pipe sections. Remember that the shortest distance between two points is a straight line.

CHOICE OF PIPE FITTINGS

During the piping process, a number of pipe fittings will need to be installed, such as 90-degree elbows, 45-degree elbows, and couplings. When choosing 90-degree elbows, select those with a wide or long radius (Fig. 10-11). Wide-radius elbows provide less resistance to refrigerant flow than those with a tighter radius. This helps maintain constant refrigerant velocity. Each fitting that is added to the refrigerant piping circuit adds to the resistance encountered by the refrigerant; therefore, the number of fittings should be kept to a minimum. Reducing the number of fittings will also reduce the number of solder joints required to connect the air handler and condensing unit. The smaller the number of solder joints, the smaller the chance of a refrigerant leak occurring.

SOLDER AND SOLDER JOINTS

To ensure a high-quality installation, high-quality materials should be used. This includes high-quality solder and brazing materials. This is one of the easiest areas for the installation crew to cut corners, as it is next to impossible for an equipment owner to tell the difference between brazing material that contains 0% silver

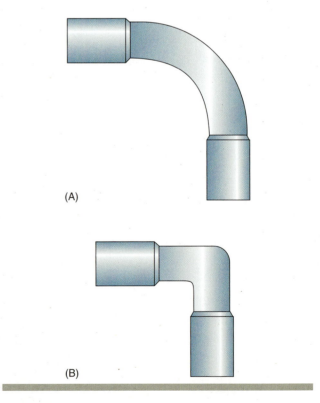

(A)

(B)

FIGURE 10-11 Long (A) and short (B) radius elbows.

and one that contains 15% silver. However, refrigerant lines vibrate to some degree, and a silver-bearing solder will stand up much better to these vibrations than a non-silver-bearing product. All solder joints should be carefully inspected before a new system is put into operation. Refrigerant lines are often sealed behind walls and in ceilings, so access to them will be very limited after the installation is completed. In addition, the suction line is insulated, which would make finding a leak later on that much more difficult. Any joint that appears to be questionable as to its integrity should be resoldered or brazed.

REFRIGERANT TRAPS

Refrigerant traps are designed to aid in the return of oil to the compressor. Approximately 5% of the compressor's oil travels through the refrigerant circuit at any given point in time when the system is operating. The trap allows this oil to accumulate in one location; then, when the line becomes blocked or the trap becomes liquid-locked, the oil is pulled back to the crankcase by the suction of the compressor. Traps, however, are not required on all system installations. Traps are typically installed on systems in which the indoor unit is located below the outdoor unit. One common example of this is when the air handler is located in the basement of a house and the outdoor unit is located in the backyard. In this instance, a refrigerant trap should be located as close to the outlet of the evaporator coil as possible (Fig. 10-12).

The exact measurements of the trap depend on the capacity of the system and the line sizes used. When installing a refrigerant trap, the installation crew should always refer to the manufacturer's installation book for exact trap sizing information.

REFRIGERANT LINE INSULATION

In order to limit the amount of heat entering the system from locations other than the evaporator, it is important that all low-pressure, low-temperature refrigerant lines be insulated. For most split-type central air conditioning systems, the line being referenced is the suction line. An improperly insulated suction line has a negative effect on system operation and efficiency.

At near-design conditions, the suction line temperature at the condensing unit (assuming a 40°F evaporator saturation temperature and 15°F of system superheat) is about 55°F, while the outdoor ambient temperature is in the 95°F range. If the suction line is not properly insulated, the 55°F refrigerant will absorb heat from the 95°F ambient (Fig. 10-13). This will result in an increase in the temperature of the suction gas entering the compressor. Since many compressors are cooled by suction gas, the cooling effect on the compressor will be reduced. In addition, any heat that is added to the system must be rejected by the condenser, so valuable condenser capacity is being used to remove heat that should not, ideally, have been added to the system in the first place.

Special consideration must be given to the insulation of refrigerant lines on ductless split systems. Quite often, the metering device is located in the outdoor unit,

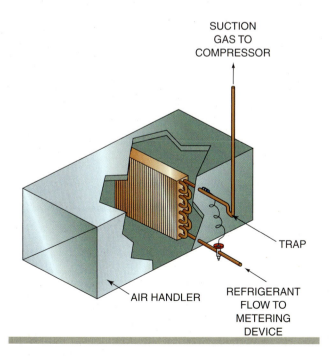

FIGURE 10-12 Refrigerant trap at the outlet of the evaporator.

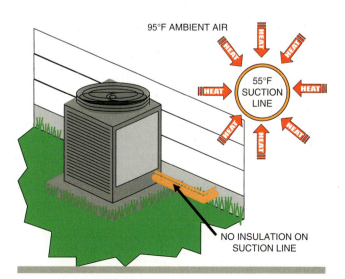

FIGURE 10-13 Cool suction line absorbs heat from the ambient air.

so the line that carries refrigerant from the metering device to the evaporator coil contains a low-pressure, low-temperature saturated refrigerant. Because this is a low-temperature line, it must be insulated. The suction line, which carries refrigerant from the evaporator back to the compressor, must also be insulated. Therefore, it is not uncommon to see insulation on both field installed refrigerant lines on ductless split systems. Figure 10-14(A) shows a conventional split system with the suction line insulated. Figure 10-14(B) shows the configuration of a ductless split system with the metering device located outside. Always refer to the manufacturer's installation literature for proper installation procedures.

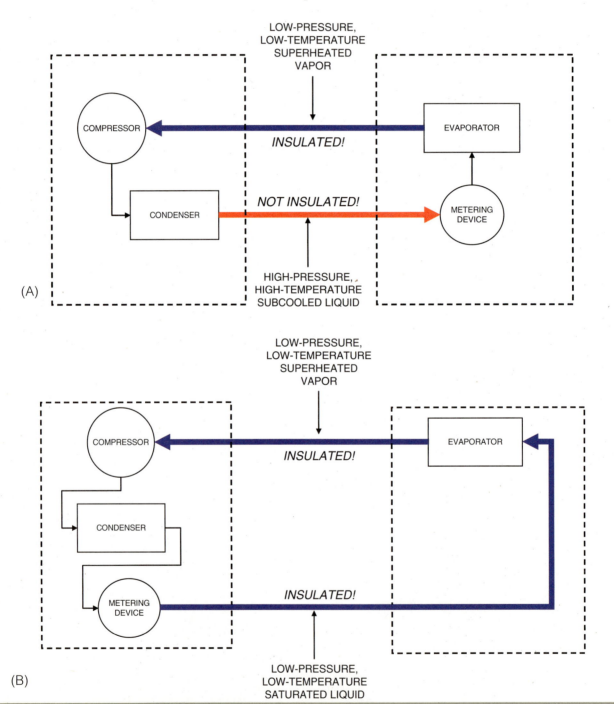

FIGURE 10-14 (A) On conventional split systems, the two field installed lines are the liquid line and the suction line. The liquid line is not insulated, while the suction line is. (B) On ductless split systems, the metering device is often located outside, so both field installed lines contain low-temperature, low-pressure refrigerant and should be insulated.

SEAL ALL STRUCTURAL PENETRATIONS

An air barrier separates the conditioned indoor space from the unconditioned outdoor space. The air barrier may be the drywall covering the inside walls and ceiling, the exterior wall sheathing, the house wrapping materials, the floor sheathing, rigid insulation, framing, and other materials that block air movement. Best practice is to avoid cutting or drilling holes through the air barrier of the house, but that's not always possible. The HVAC installer should attempt to minimize the size and number of penetrations whenever possible. When penetrations are required, care should be taken to properly seal them.

On new construction jobs, the structural insulation is typically installed after the rough ducts, pipes, and equipment are installed. However, during the finish, the HVAC installer has to be careful working around the insulation not to disturb or compact it and reduce its efficiency.

HVAC piping installed outside the thermal envelope of a house in an attic, basement, or crawlspace leads to numerous penetrations through the air barrier. Every hole cut through the drywall or floor sheathing for a supply or return register can leak air. This is one reason why the best practice is to install HVAC ducts, pipes, and equipment within the thermal envelope whenever possible.

Penetrations through the roofing or siding must be sealed against water leaks. Caulking may be part of the sealing process but the primary weather block is flashing. There are specially made flashings for pipes and exhausts of almost every size and type. Refrigerant lines penetrate the wall to reach the condenser and should be routed through a specially made flashing to prevent water leaks. The flashing also protects the lines from damage.

INSTALLING, SUPPORTING, AND INSULATING CONDENSATE DRAIN LINES

While operating in the cooling mode, the air conditioning system will remove moisture from the conditioned air while lowering the air's temperature. The moisture from the air must be safely removed from the structure. This moisture is called **condensate**. To prevent water damage, a **condensate drain line** must be run to carry this condensation away. A number of factors must be considered when laying out the condensate piping for a system, including:

- Drain line size
- Drain line material
- Pitching of the line
- Traps
- Auxiliary drain pans
- Condensate pumps
- Safety float switches

DRAIN LINE SIZE

The drain line of the system must be able to handle the water flow that is generated by the removal of moisture from the air. All air handlers are equipped with internal drain pans and female pipe connections to which the field-installed drain line is to be connected. The drain line that is run should be the same size, or larger than the fitting provided with the system. For example, if the unit is equipped with a ¾-inch female pipe thread connection at the unit, under no circumstances should the size of that line be reduced. The line size is determined by the system manufacturer and should not be altered. In areas of extremely high humidity, however, the line size can be made larger than the system design warrants. Common sense dictates that larger units will require larger condensate lines. For example, a 2½-ton air conditioning system will produce, on average, 1 gallon of condensate per hour, while a 5-ton air conditioning system will produce approximately 2 gallons of condensate per hour. Under no circumstances should a drain line be less than ¾ inch in diameter.

> ## from experience...
>
> To prevent drain line condensation, the primary drain line should be insulated at least 10 feet from the indoor coil's drain pan outlet.

DRAIN LINE MATERIAL

The drain line should be constructed of materials that are in accordance with local plumbing codes. Normally, PVC piping is acceptable for most residential applications.

When the condensate line is to be located within a concealed wall or ceiling, it is recommended, and sometimes required, that copper be used. When installing a PVC condensate drain line, all fittings and pipe ends should be cleaned first with a PVC primer and then joined using PVC cement. Using cement that is not designed for use with PVC pipe can result in a weak joint that may leak in the future. Refer to the section on joining PVC pipe sections for more on this. When installing a copper drain line, all solder joints must be checked for leaks before the system is put into operation. Other lightweight rigid plastic materials should be avoided but can be used if the length of the line is very short.

PITCHING OF THE LINE

Regardless of which material is used, the line should always be pitched toward the termination point of the line. The minimum pitch on the line should be ¼ inch per 1 foot of horizontal run. For example, a 10-foot horizontal drain line should be pitched a minimum of 2½ inches. Drain lines should be adequately supported to ensure that the proper pitch is maintained. To facilitate proper draining of the condensate, the pitch should be made as great as possible. Improperly pitched drain lines can result in water damage to both the system and the structure.

TRAPS

Traps in condensate lines help prevent air from flowing up through the drain line, thereby preventing the condensate from draining properly. The **condensate drain trap** functions in the same manner as the trap under a bathroom sink, which prevents waste fumes from entering the home. A trap, however, is not required on all air handlers. If the refrigerant coil is located downstream of the blower, meaning that air is being pushed through the coil, a trap is not needed. In this case, the pressure at the inlet of the drain line is at a higher pressure than at the outlet and the condensate will drain properly. If, however, the blower is downstream of the coil, a trap must be used. In this case, air is being pulled through the coil and also through the drain line. This causes the pressure at the inlet of the drain line to be lower than the pressure at the outlet, thereby pulling air through the line in the direction opposite to the direction of condensate flow. If the drain line is under a negative pressure and no trap is installed in the line, the condensate will accumulate in the pan and eventually cause overflow, which can lead to water damage. This applies to both horizontal and vertical units. When in doubt, install a trap. The chart in Figure 10-15 provides guidelines for determining whether or not a trap is needed. For example, the system in Figure 10-16 requires a trap on the condensate line because it is an upflow unit with the blower located downstream of the coil.

DRAIN LINE TERMINATION

The condensate that is removed from the system must ultimately be deposited in a location that is not objectionable to the occupant of the occupied space. This location is commonly either outside the structure or down a waste line. If the unit is located in an attic or a closet, the drain line usually leaves the structure through the same building penetration that the refrigerant lines and low-voltage wiring will ultimately pass through. The drain line should be as short as possible to reduce the possibility of a blockage that would cause the line to back up and cause damage. If running the line to the outside of the structure is not possible, the line should be run to a nearby waste line or utility sink. Once again, all local codes regarding the installation of drain lines should be followed.

If the indoor unit is located in a basement, running the condensate line to the outside may not be as easy as in the case of the attic or utility closet installation.

When in a basement, the unit is below ground level and a gravity-type drain is not possible. In this case, two other options are available to the installation crew:

- Use a floor drain
- Pump out the condensate

	Blower above Coil	Blower below Coil
Vertical Upflow	Trap needed	Trap not needed
Vertical Downflow	Trap not needed	Trap needed
	Blower to the Right of the Coil	Blower to the Left of the Coil
Horizontal Left-to-Right Flow	Trap needed	Trap not needed
Horizontal Right-to-Left Flow	Trap not needed	Trap needed

FIGURE 10-15 Condensate trap usage.

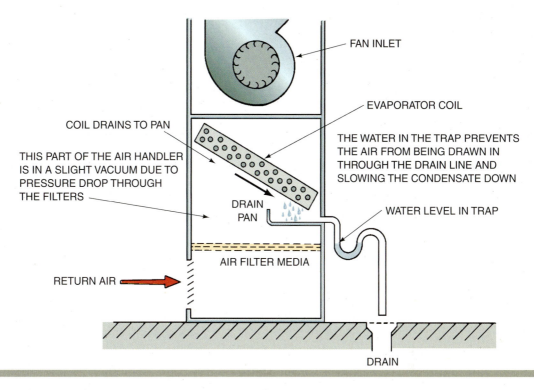

FAN INLET

EVAPORATOR COIL

COIL DRAINS TO PAN

THIS PART OF THE AIR HANDLER IS IN A SLIGHT VACUUM DUE TO PRESSURE DROP THROUGH THE FILTERS

THE WATER IN THE TRAP PREVENTS THE AIR FROM BEING DRAWN IN THROUGH THE DRAIN LINE AND SLOWING THE CONDENSATE DOWN

DRAIN PAN

WATER LEVEL IN TRAP

AIR FILTER MEDIA

RETURN AIR

DRAIN

FIGURE 10-16 This air handler needs a condensate trap.

If the unit is installed in an unfinished basement and a floor drain is nearby, the condensate should be directed toward it (Fig. 10-16). The line should be run from the unit, along the floor of the basement, making certain that the line is secured so the water is directed to the drain. Whether or not a trap is required depends on the configuration of the unit. Before the system is started up, the floor drain should be tested to be certain that it is still operational.

If a floor drain is not available or if the basement is finished, the condensate may need to be pumped from the location. **Condensate pumps** come in a wide range of styles based on voltage, holding capacity, and body style. In operation, the pump is normally in the off position and is controlled by a float switch. Once the water in the pump's reservoir reaches a predetermined point, the pump switches on, removing the water from the pump. When the water level is lowered, it switches off.

The pump is located close to the unit, and the water is pumped either outside or to a utility sink for disposal. The piping run from the unit to the pump is therefore very short. Once again, the need for a trap is determined by the configuration of the indoor unit. Some condensate pumps are equipped with an overflow control. This will prevent water damage from occurring if the pump should fail. This overflow control is a normally closed switch that will open if the water level rises above the maximum level permitted by the pump. If, for example,

the pump is designed to turn on if the water reaches a level of 4 inches, the water level should never be higher than that if the pump is operating correctly. The overflow control will open its contacts if the water reaches, for example, 4½ inches. This overflow control is normally wired in series with the R wire coming off the control transformer. When the overflow control opens its contacts, the system shuts down, preventing any further moisture removal from the conditioned space. The customer will then place a service call indicating that the system is not operating at all. Shutting down the system prevents the water from overflowing the pump and causing damage.

AUXILIARY DRAIN PANS

If a problem arises with a drain line—such as a blockage—the condensate cannot be carried away from the unit. This would lead to an overflow condition and could result in water damage to property located under the unit. Preventive maintenance could help prevent the drain line from getting clogged, but this occurrence cannot be predicted. One way to effectively eliminate the possibility of water damage from overflow is to install an **auxiliary drain pan** under the unit (Fig. 10-17) to catch and remove any overflow condensate.

The pan should be large enough to catch water leaking from anywhere in the unit; in other words, the pan

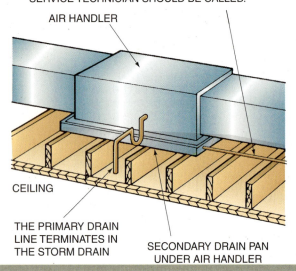

THE SECONDARY DRAIN LINE TERMINATES IN A CONSPICUOUS PLACE. THE OWNER IS WARNED THAT IF WATER IS SEEN AT THIS LOCATION A SERVICE TECHNICIAN SHOULD BE CALLED.

AIR HANDLER

CEILING

THE PRIMARY DRAIN LINE TERMINATES IN THE STORM DRAIN

SECONDARY DRAIN PAN UNDER AIR HANDLER

FIGURE 10-17 Auxiliary drain pan under unit.

SAFETY FLOAT SWITCHES

Although auxiliary drain pans are designed to help prevent water damage, the possibility exists that the auxiliary line can become clogged or blocked as well.

from experience...

The auxiliary drain line must be completely independent of the primary drain line. Never join the auxiliary drain line and the primary drain line into a common line. This will defeat the purpose of the backup line.

must be larger than the air handler. Prefabricated plastic drain pans can be purchased in a wide range of sizes that are suitable for most residential applications. Pans can be made of sheet metal as long as the corners are sealed to ensure that the pan will hold water. The auxiliary pan should be mounted securely under the unit in the event it actually has to hold water. In an ideal situation, this pan should remain perfectly dry at all times.

The auxiliary drain pan and the pan manufactured in the air handler have two main differences. The primary condensate line, the one installed on the unit at the time of manufacture, is piped to the outside of the structure or to a waste line to dispose of the water discretely. The home owner will not be aware that condensate is being removed from the structure. The auxiliary drain pan, on the other hand, is piped to a location that is very conspicuous. This is done for one very good reason. If water is seen coming from that line, there is water in the auxiliary pan and there is a problem with the primary drain on the unit. The equipment owner is instructed upon system installation and startup that, if water is seen coming from the line, a service call must be placed to prevent damage to the structure. This drain line is typically piped to a location over a doorway or window to alert the occupant of a potential problem as quickly as possible. The second difference between the two drains is that the drain line from the auxiliary pan does not need a trap. No pressure difference exists between the two ends of the drain line, so water will flow freely from the pan. Of course, the drain line from the pan must also be pitched downward toward its termination point to ensure proper drainage.

Although this is very unlikely, it is possible. In this instance, when the auxiliary pan fills up, the water will overflow and potentially cause damage to the area below. To prevent this, a safety float switch can be installed in the auxiliary pan. Similar to the safety float switch in the condensate pump, this switch is a normally closed device that will open its contacts if the level of the water in the auxiliary drain pan reaches a predetermined level. The switch is wired in series with the low-voltage control circuit and will shut the system down if its contacts open. Once the system shuts down, the customer will call the service company and report that the system will not operate. Some float switches are designed to maintain a desired water level in a pan or sump. Evaporative coolers and commercial icemakers are examples where a minimum water level is desired.

SYSTEM WIRING

We have discussed a number of topics, including selecting unit location, types of duct systems, piping materials, and piping operations. All of these areas are extremely important, and one cannot have a successful installation if any are lacking. In addition, one of the most important tasks that must be performed during installation is the electrical wiring. No matter how well the system is installed, it cannot operate if the wiring, both power circuits and control circuits, is not connected properly. On a typical residential installation, two types of wiring must be installed. Each will be addressed separately. They are:

- Line-voltage power circuit wiring
- Low-voltage control circuit wiring

LINE-VOLTAGE POWER CIRCUIT WIRING

Typically, supplying the line-voltage power supply to the unit location is the job of a licensed electrician and should not be performed by those not qualified to do so. Electricians are typically responsible for running the power lines from the circuit breaker panel or fuse box to the service disconnect switches (Fig. 10-18) located at both the condensing unit (Fig. 10-19) and the air handler. These disconnect switches can be either fused or unfused. It is the installation crew's responsibility to run

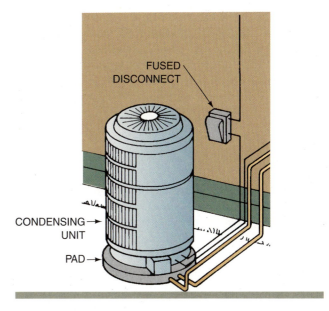

FIGURE 10-19 The fused disconnect is located at the condensing unit location.

the line-voltage wiring from the disconnect switches to the individual pieces of equipment.

CONDENSING UNIT

Since condensing units and evaporative coolers are located outdoors, the electrical connections must be able to withstand a wide range of weather conditions, particularly rain. For this reason, materials used to make this connection must be liquidtight (Fig. 10-20). Depending on the local electrical requirements and codes, the conduit can be either

(A)

(B)

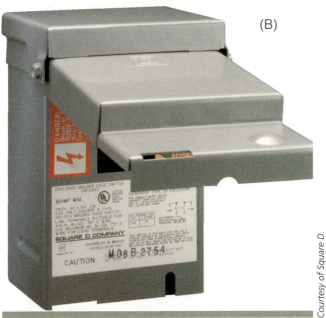

FIGURE 10-18 This service disconnect is equipped with fuses as well as a manual disconnect.

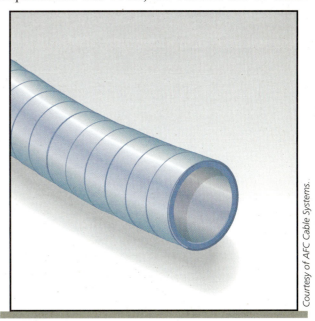

FIGURE 10-20 This cable is liquid-tight and suitable for outdoor use.

liquidtight flexible metallic conduit, LFMC, or liquid-tight flexible nonmetallic conduit, LFNC. Metallic flexible armored (AC) cable, commonly known as BX cable (Fig. 10-21), is not permitted for outdoor use. In addition to the cable itself, the fittings used to make the connections at the disconnect switch and the condensing unit must be liquidtight (Fig. 10-22).

Wire Sizing. The size of the wires that must be run between the condensing unit and the service disconnect switch is determined by the amperage draw of the equipment. The amount of electrical current that the unit draws is provided on the nameplate of the condensing unit (Fig. 10-23). The *National Electrical Code®* (NEC) provides information regarding the conductor size needed to carry specific amounts of current. Since conductors made of different materials, such as copper or aluminum, can be used, consult the *NEC®* to be sure that the wire is sized properly.

CAUTION

CAUTION: To be sure that the installation is safe, follow all the guidelines regarding system installation as set forth in the *National Electrical Code*. Always use the most recent version of the code.

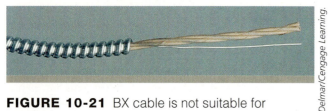

Delmar/Cengage Learning.

FIGURE 10-21 BX cable is not suitable for outdoor use.

Courtesy of AFC Cable Systems.

FIGURE 10-22 Various weather-tight fittings.

Making the Connection. Once the conduit has been run between the condensing unit and the service disconnect switch, the electrical connections can be made. Following are suggestions regarding the connection of these wires:

- Start at the condensing unit and make your final connections at the service disconnect.
- Make certain that the green ground wire is connected to a screw terminal on the chassis of the condensing unit, as well as to the interior of the disconnect (Fig. 10-24).

Photo by Eugene Silberstein.

FIGURE 10-23 Air conditioning unit nameplate.

Delmar/Cengage Learning.

FIGURE 10-24 Ground wire is connected to the chassis of the condensing unit.

TECH TOOL BELT • TECH TOOL BELT

Courtesy of Ideal Industries, Inc.

Linesman Pliers

Unlike regular slip-joint pliers, linesman pliers are intended to perform a number of functions. They can cut and strip wire, and the jaws are configured so they can properly twist conductors together to form a pigtail splice. The handles of the linesman pliers should always be properly insulated. Damage to the insulation can result in electric shock.

Courtesy of Ideal Industries, Inc.

Wire Strippers

Wire strippers come in a wide variety of configurations and should be chosen based on the intended use of the tool. Some wire strippers are designed for simply stripping the insulation from wire. Others are designed to perform tasks such as cutting wire, cutting small screws and bolts, or crimping solderless connectors onto sections of stripped wire. When stripping wire, it is important to remove only as much insulation as is needed for the task at hand. Removing too much can result in electric shock.

Courtesy of Ideal Industries, Inc.

Electrical Snake

The electrical snake is a roll of flexible, yet somewhat rigid, metal strapping. The snake is approximately ⅛-inch wide and is used to help run electrical wires. The end of the snake is bent to facilitate securing the wires to be pulled. It is first run between the points that the wire is to be run. Since it is more rigid than the wire, it is much easier to run the snake than the wire. Once it is in place, the wires to be run are secured to the end of the snake. The snake is then pulled out and, in doing so, the wires are pulled between the desired points.

- Make line voltage connections to the line side-top-of the compressor contactor (Fig. 10-25).
- Make certain all screw terminals are tight.
- Strip only enough insulation from the wire to make a good connection. Removing too much insulation increases the chances of electric shock or short circuit.

CAUTION

CAUTION: Be sure that all disconnects and power sources are de-energized before performing any electrical work on a system.

INDOOR UNIT

Unlike the condensing unit, the air handler is located inside the structure, either in an attic or in the basement. Because of the location of the air handler, the connection between it and the service disconnect does not have to be liquidtight. Metallic flexible BX cable can be used safely for this application.

Wire Sizing. Just as with the condensing unit, the size of the wires that must be run between the air handler and the service disconnect switch is determined by the amperage draw of the equipment. The amount of electrical current that the air handler draws is typically much

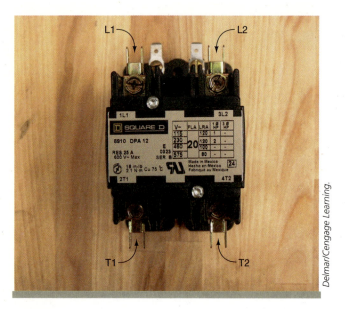

Delmar/Cengage Learning.

FIGURE 10-25 Line-voltage power is brought into the top of the compressor contactor.

less than the amperage draw of the condensing unit. The amperage rating of the equipment is provided on the nameplate of the air handler.

Making the Connection. The method used to make the line voltage connections to the indoor unit is similar to that used to make the connection at the condensing unit, except that liquidtight connectors are not needed. Be sure to follow all safety and code guidelines when making these connections.

LOW-VOLTAGE CONTROL CIRCUIT WIRING

Once the line-voltage power circuits have been installed, the system control circuits can be installed and connected. The control circuits tell the system components what to do and when to do it. Without accurate controls, the system will fail to operate properly. Control wiring cable is made up of a number of very thin, color-coded wires (Fig. 10-26). On a typical residential

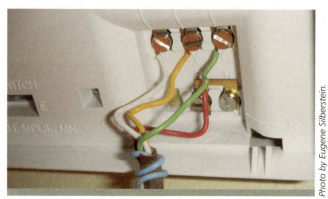

Photo by Eugene Silberstein.

FIGURE 10-26 Color-coded thermostat wire.

air conditioning system installation, there are two sets of low-voltage control wires that must be run to and from various points in the system:

* From the condensing unit to the air handler
* From the thermostat to the air handler

CONDENSING UNIT TO THE AIR HANDLER

The easiest way to run the low-voltage wiring between the air handler and the condensing unit is to secure the low-voltage cable to the liquid and suction lines as they are installed. This bundle (Fig. 10-27) is made up of the insulated vapor line, the liquid line, and the low-voltage cable. Enough excess wire should be available to ensure that the line connecting the indoor and outdoor units is continuous. Joining sections of wire to make this run poses potential service problems in the future. Typically, only two wires are needed from the air handler to the condensing unit.

THERMOSTAT TO THE AIR HANDLER

Running the low-voltage wiring, typically three wires, between the thermostat and the indoor unit is somewhat more involved. Unlike the air handler/condensing unit wiring where the wire is simply getting a free ride from the piping, the thermostat line must be run by itself. This involves either running the wire up within a wall from a basement location or down within a wall from an attic location. In either case, the exact wall location must be determined from either the basement or the attic, and the exact location of the thermostat must be determined. Then, a hole must be drilled directly above for attic installations or directly below for basement installations to gain access to the interior of the wall. Once done, the electrical snake can be used to pull the wire through the wall. Running the wire from the thermostat to the indoor unit is very often a two-person job.

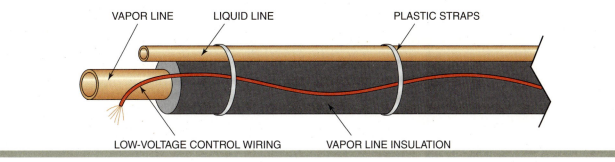

FIGURE 10-27 Tubing bundle consists of the liquid line, suction line, and the low-voltage cable.

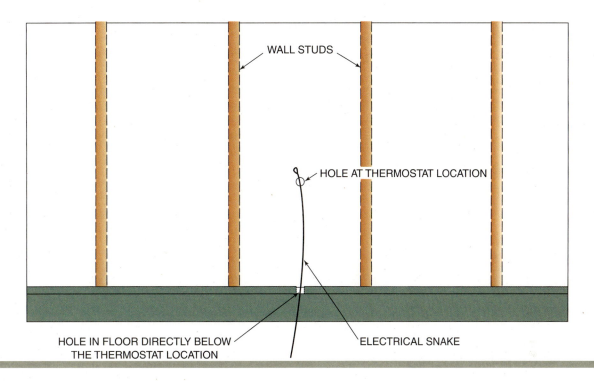

FIGURE 10-28 The hole at the thermostat location must be in the same bay as the hole in the foot of the wall.

When running the line between the indoor unit and the thermostat, the technician must remember that the walls have studs that are typically 16-inches apart from center to center. It is extremely important that the hole made in the wall for the thermostat wire penetration is in the same bay as the hole made in the attic or the basement (Fig. 10-28). By using a flashlight and a small piece of hooked electrical snake material, one member of the installation crew can catch the snake as it is fed up from the basement by another crew-member (Fig. 10-29). Once the electrical snake is brought through the hole in the wall, the low-voltage cable should be attached securely to the snake using electrical tape and pulled down to the basement location. The same principle can be applied to systems that have their indoor units located in the attic.

CONNECTING THE LOW-VOLTAGE CONTROL WIRING

The color coding on the low-voltage control circuit cable makes connecting the low-voltage circuit a breeze. Most manufacturers have embraced a common color-coding system that makes the job of the installation crew that much easier. For a simple residential system, the color codes are as follows:

- Red Wire, denoted "R" on terminal boards, represents the hot power wire from the control transformer
- Yellow Wire, denoted "Y" on terminal boards, represents the cooling circuit
- Green wire, denoted "G" on terminal boards, represents the fan circuit
- White wire, denoted "W" on terminal boards, represents "common" or "heat"

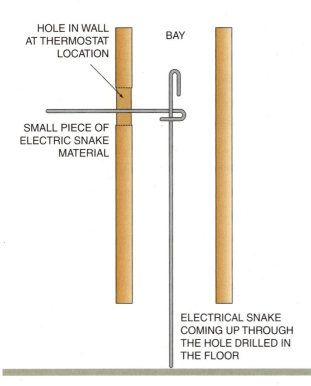

HOLE IN WALL AT THERMOSTAT LOCATION

BAY

SMALL PIECE OF ELECTRIC SNAKE MATERIAL

ELECTRICAL SNAKE COMING UP THROUGH THE HOLE DRILLED IN THE FLOOR

FIGURE 10-29 Fishing the electric snake out of the wall.

> *from experience...*
>
> Not all manufacturers have adopted the same low-voltage color-coding strategy so be sure to check the installation materials supplied with the system to find out the specific wiring codes used by that company.

Condensing Unit. The two wires that are run from the air handler to the condensing unit are connected to the control circuit wires. These wires are often left hanging loose in the control panel of the condensing unit and are tagged as being the low-voltage connections. These wires should be secured using wire nuts and electrical tape.

Thermostat. On a simple system, three wires—red, yellow, and green—should be run between the thermostat and the air handler. On the thermostat, there are screw terminals labeled with the corresponding letters, R, Y and G. The three wires should be connected to these terminals, respectively.

Air Handler. At the air handler, the line from the thermostat and the line from the condensing unit will come together. Once again, following the color coding, the

> *from experience...*
>
> Different thermostats may use different terminal identification codes. Be sure to check the installation materials supplied with the device prior to installation.

wires are connected to the terminal board in the unit. The following are some guidelines for connecting a simple residential air conditioning system:

- There will be two yellow wires: one from the condensing unit and one from the thermostat. Both of these wires get connected to the "Y" terminal on the air handler's control board.
- The green wire from the thermostat will get connected to the "G" terminal on the air handler's control board.
- The red wire from the thermostat will get connected to the "R" terminal on the air handler's control board.
- The white wire from the condensing unit will get connected to the "C" terminal on the air handler's control board.

SUMMARY

- Plastic piping materials are often joined together with special cements.
- Refrigerant lines should be run with as few solder joints as possible to reduce the chance of refrigerant leaks.
- Condensate can be removed from the structure by gravity, a floor drain, or a pump.
- Auxiliary drain pans are intended to protect property in the event the primary drain fails.
- Installation crews are often responsible for connecting the line-voltage power from the disconnect switches to the air handler and the condensing unit.
- Line voltage connections to the condensing unit must be done with liquidtight materials and fittings.
- Low-voltage wiring does not need to be performed by a licensed electrician.
- In a simple low-voltage control circuit, three wires need to be run between the thermostat and the air handler, and two wires are needed between the thermostat and the condensing unit.
- Low-voltage control wires are color-coded to facilitate the wiring process.

GREEN CHECKLIST

☐ Use environmentally friendly materials for system installation.

☐ Keep refrigerant lines as short as possible to minimize waste and maximize system capacity.

☐ Never use lead-bearing solders on lines that carry potable water.

☐ Seal all structural penetrations.

PROCEDURE　10-1

Using the Tubing Cutter

 Mark the pipe with a pencil at the desired cutting point.

• Place the tubing cutter around the pipe and tighten the handle so that the cutting wheel is snug against the pipe surface at the pencil mark. The wheel should be tight enough to hold the cutter in place, but not so tight as to create an indentation in the pipe.

• Slowly rotate the tubing cutter around the pipe, one or two turns, until the pressure of the cutting wheel on the pipe is reduced.

• Tighten the adjusting handle ⅛ to ¼ turn to snug the cutting wheel against the pipe surface again.

• Repeat the tightening and turning of the cutter until the tubing is cut.

 Using the reamer on the back of the cutter, remove the burrs from the interior surface of the pipe.

A

Delmar/Cengage Learning.

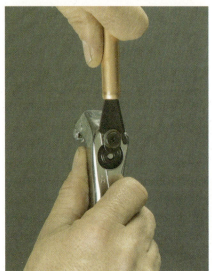

B

Delmar/Cengage Learning.

PROCEDURE 10-2

Swaging Soft-Drawn Copper Tubing

- Locate the proper hole in the flaring block. The proper hole is marked with the measurement that corresponds to the outside diameter of the tubing.

A Insert the section of soft-drawn tubing to be swaged into the proper hole in the flaring block so that the tubing extends a distance equal to the diameter of the tube plus about ⅛ inch. Make certain that the end of the tube to be swaged protrudes from the side of the block that has the beveled or angled edge.

- Tighten the flaring block around the tubing, applying pressure and tightening the wing nuts equally to ensure that the two pieces of the block secure the tubing evenly.

B Insert the swaging tool into the tubing and strike the swaging tool firmly with a hammer.

- Rotate the flaring block ½ turn and strike the swaging tool again.

- Repeat the previous step until the swage has reached a length approximately equal to the diameter of the tubing.

 A

Photo by Eugene Silberstein.

 B

Photo by Eugene Silberstein.

from experience...

A drop or two of refrigeration oil can be placed on the swaging tool during the swaging process to reduce the friction between the swaging tool and the copper tubing. This oil, however, must be completely wiped away before soldering or brazing the connection.

PROCEDURE 10-3

Flaring Soft-Drawn Copper Tubing

- Cut the tubing to the desired length and ream.

 Place the flare nut over the end of the tubing with the threaded portion of the nut facing the end of the tubing.

- Insert the section of soft-drawn tubing to be swaged into the proper hole in the flaring block so that the tubing extends a distance about equal to the diameter of a nickel.

 Make certain that the end of the tube to be swaged protrudes from the side of the block that has the beveled or angled edge.

- Tighten the flaring block around the tubing, applying pressure and tightening the wing nuts equally to ensure that the two pieces of the block secure the tubing evenly.

 Place the yoke around the block and position the cone over the opening of the tubing.

- Turn the screw handle on the yoke until the cone comes in contact with the tubing. Tighten the screw handle an additional ¾ of a turn.

- Loosen the screw handle ¼ of a turn.

Photo by Eugene Silberstein.

Photo by Eugene Silberstein.

Photo by Eugene Silberstein.

- Repeat the tightening process until the cone of the yoke is tight against the flaring block.

D Inspect the flare for cracks or other imperfections. If the flare is cracked, cut the flared end off and re-flare the tubing. Cracked flares will leak!

Photo by Eugene Silberstein.

PROCEDURE **10-4**

Soldering

- Properly cut and ream the sections to be joined. Refer to the cutting and reaming procedure.

- Using sand cloth or steel wool and the correct size pipe brush, clean the male and female portions of the joint being soldered.

A Using a flux brush, apply flux to the male portion of the joint.

- Insert the male portion of the joint into the female end.

- Before connecting the acetylene regulator to the tank, quickly open and close the stem on the tank using the refrigeration service wrench. This will blow any particulate matter from the opening of the tank.

B Mount the acetylene regulator and torch kit to the tank, making sure that the connections are tight.

C Making certain that the valve on the torch handle is closed, open the stem valve on the acetylene tank ½ to 1 turn using the service wrench. Flip the ratchet on the service wrench so the tank can be closed quickly in the event of an emergency.

Photo by Eugene Silberstein.

Photo by Eugene Silberstein.

Photo by Eugene Silberstein.

PROCEDURE 10-4

Soldering (continued)

 Using a soap bubble solution, leak-check the regulator, hose, and torch assembly, making certain that no acetylene is leaking from the kit. Tighten any leaking connections.

 Set the regulator on the tank to the middle range to start. It can always be adjusted later on.

• Open the valve on the torch handle most of the way and ignite the fuel with the striker.

• Adjust the flame using the regulator on the tank until the desired flame intensity is obtained. A proper flame will have a bright blue inner cone and a lighter blue outer cone. The hottest portion of the torch flame is the tip of the inner cone.

Delmar/Cengage Learning.

Delmar/Cengage Learning.

Apply heat to the joint, placing the tip of the inner cone on the surface of the joint. The flux will begin to melt and flow into the joint. Be sure to keep the flame moving to heat the entire joint. After heating the joint for a short period of time, apply the solder to the joint. The solder should be melted by the heat of the copper, not the heat of the torch flame. If the solder does not immediately begin to flow, remove the solder and heat the joint a little more.

Photo by Eugene Silberstein.

CAUTION

- Always wear safety glasses when soldering or brazing because bubbling flux and liquefied solder can splatter and result in personal injury.

- Never point the torch handle toward the tank or another person while igniting or using the torch!

- Never use a match or cigarette lighter to light the torch. Lighting a match requires the use of two hands and also places your hand very close to the torch flame. Cigarette lighters contain a fuel supply that can easily ignite when the torch is lit.

from experience...

The size of the flame and the amount of heat generated by the torch is directly related to the size of the torch tip used. Larger size pipes will require more heat and will therefore require a larger torch tip. Residential applications typically require the use of an A-3 or A-5 tip for soft soldering.

from experience...

Overheating the joint prior to introducing solder will cause the solder to run off the joint instead of sticking to it. If this should occur, the pipes should be re-cleaned and fluxed to ensure a good solder joint.

PROCEDURE 10-4

Soldering (continued)

 When the solder begins to flow, feed enough solder into the joint to completely fill it. You should use no more solder than is needed to fill the joint. Using too much solder will result in buildup on the inside of the pipe. Using too little solder will result in leaks.

- To extinguish the torch, simply close the valve on the torch handle. When the torch is no longer needed, close the stem on the acetylene tank and bleed off any acetylene from the hoses by opening the valve on the torch handle.

- While the joint is still hot, it is good practice to wipe the joint with a rag. This removes excess solder and improves the appearance of the solder joint. Applying a small amount of flux to the joint while the pipe is hot also helps clean the joint

Photo by Eugene Silberstein.

from experience...

For most residential installations, the amount of soft solder used on any given joint should not exceed a length equal to twice the diameter of the pipe.

from experience...

Molten solder will flow toward the source of heat. To help ensure that the solder flows into the joint, apply heat to the base of the female portion of the joint.

PROCEDURE 10-5

Brazing

- Properly cut and ream the sections to be joined. Refer to the cutting and reaming procedure.

- Insert the male end of the joint into the female end.

- Prepare and light the torch. Refer to the beginning steps in the soldering procedure.

- Apply heat to the joint, placing the tip of the inner cone on the surface of the joint. Molten brazing rods will flow toward the source of heat. To help ensure that the material flows into the joint, apply heat to the base of the female portion of the joint.

CAUTION

CAUTION: All Cautions regarding the torch kit apply to brazing as well as soft soldering.

 After heating the joint until the pipes glow cherry red, apply the brazing rod to the joint. If the filler material does not immediately begin to flow, remove the rod and heat the joint a little more.

- When the filler material begins to flow, feed enough into the joint to completely fill it. You should use no more material than needed to

from experience...

The procedures for preparing and lighting the torch kit are the same regardless of the filler material used so please refer to the section on soft soldering for the procedures regarding torch preparation.

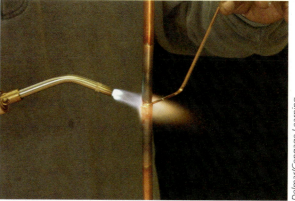

Delmar/Cengage Learning.

PROCEDURE 10-5

Brazing (continued)

fill the joint. Using too much will result in buildup on the inside of the pipe. Using too little will result in leaks. The amount of filler material used on any given joint should not exceed a length equal to twice the diameter of the pipe.

B To extinguish the torch, simply close the valve on the torch handle. When the torch is no longer needed, close the stem on the acetylene tank and bleed off any acetylene from the hoses by opening the valve on the torch handle.

B

Delmar/Cengage Learning.

from experience...

The size of the flame and the amount of heat generated by the torch is directly related to the size of the torch tip used. Larger size pipes will require more heat and will therefore require a larger torch tip. Residential applications typically require the use of an A-5 or A-8 tip for hard soldering or brazing. Recommended torch tip sizes for various applications are shown in the Tech Toolbelt on page 235. Note that brazing requires a larger torch tip than soft soldering for similar size piping material.

PROCEDURE 10-6

Joining Plastic Pipe

- Mark the pipe at the appropriate point with a pencil.

 Cut the pipe using either a hacksaw or tubing shear.

- Remove the burrs from both the inside and the outside of the pipe.

- Apply primer, if required, to both the male and female portions of the joint.

 Apply cement to both the male and female portions of the joint.

 Insert the male end of the fitting into the female end and rotate the pipe ¼ turn.

- Hold the pipe and fitting together for approximately one minute to prevent the pipe from pulling out of the fitting.

A

Photo by Eugene Silberstein.

B

Photo by Eugene Silberstein.

C

Photo by Eugene Silberstein.

from experience...

When working with plastic pipe, always try to dry fit the piping arrangement before cementing. Once a joint is cemented, you don't get a second chance!

REVIEW QUESTIONS

1. **The process of removing burrs from a section of piping material is called**
   a. Reaming
   b. Flaring
   c. Swaging
   d. Brazing

2. **When brazing,**
   a. The heat from the torch is used to melt the filler material
   b. The pipe joint must be heated until it glows cherry red
   c. The filler material used will have a melting temperature of about 500°F
   d. All of the above are correct

3. **When used to classify piping, the term ACR implies that**
   a. The tubing is measured by its inside, nominal, diameter
   b. The pipe or tubing has been capped at both ends and often contains a nitrogen charge
   c. The piping is only intended for use on drain systems
   d. All of the above are correct

4. **Which of the following can result in an overflowing condensate drain pan?**
   a. An improperly mounted air handler
   b. An undersized condensate drain line
   c. A missing trap in the condensate line
   d. All of the above

5. **If an air handler is equipped with a ¾-inch female pipe connection on the condensate drain pan, which of the following pipe sizes could be used effectively to remove the condensate from the structure?**
   a. ½-inch
   b. 1 inch
   c. Either a or b
   d. Neither a or b

6. **When the air handler is located in a basement, which of the following methods could be used to remove the condensate from the structure?**
   a. Use an available floor drain
   b. Use a condensate pump
   c. Pipe the condensate through the foundation of the structure to the outside
   d. Both a and b are possible.

7. **Explain the purpose of an auxiliary drain pan. Explain how it should be sized and how the drain line from the pan should be terminated.**

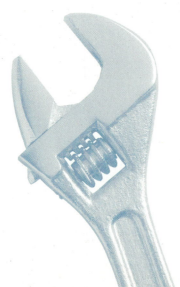

8. **When running refrigerant lines between the air handler and condensing unit, why are long radius elbows preferred over short radius elbows?**

   a. Long-radius elbows are more economical from a first-cost standpoint

   b. Long-radius elbows offer less resistance to refrigerant flow

   c. Long-radius elbows are easier to braze than short-radius elbows

   d. All of the above are correct

9. **On new construction installations, explain why it is extremely important for all solder joints to be inspected and reinspected.**

10. **Explain how an uninsulated suction line can affect system operation.**

11. **Soft-drawn copper tubing is ideal for use behind sealed walls or ceilings because**

    a. The walls of soft-drawn copper tubing are thicker than rigid pipe and less likely to leak

    b. The number of solder joints will be greatly reduced

    c. The resistance to refrigerant flow is much less when soft-drawn copper tubing is used

    d. All of the above are correct

12. **All of the following are true regarding the line-voltage wiring to the condensing unit except**

    a. BX cable is not acceptable for outdoor applications

    b. The size of the wire is determined by the amperage draw of the condensing unit

    c. The amperage draw of the condensing unit is typically greater than the amperage draw of the air handler

    d. The installation crew is responsible for running the line-voltage wiring from the circuit breaker panel to the condensing unit

13. **Which of the following is true regarding the low-voltage control wiring on a standard residential air conditioning system?**

    a. The red wire controls the evaporator fan operation

    b. The yellow wire controls the evaporator fan operation

    c. The green wire controls the evaporator fan operation

    d. The white wire controls the evaporator fan operation

KNOW YOUR CODES

Local electrical codes vary greatly so be sure to investigate the codes in your area to ensure that you are in accordance with them. In addition to the codes regarding wiring specifics, different geographic areas have different rules in place regarding types of work that must be done by licensed electricians. Make certain you know what your local codes are!

Plumbing codes also differ in different regions. Some codes, for example, do not allow the use of plastic piping materials, such as PVC, for use on condensate drains. Make certain you know your codes before installing!

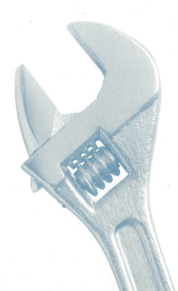

WHAT'S WRONG WITH THIS PICTURE?

Carefully study FIGURE 10-30 and think about what is wrong. Consider all possibilities.

✖ WRONG

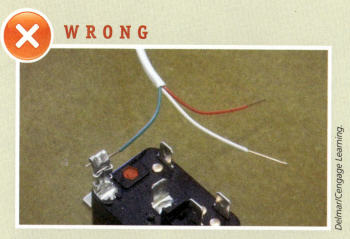

Delmar/Cengage Learning.

FIGURE 10-30 When connecting solid low-voltage or thermostat wire to a solderless connector, crimping the connector onto the thermostat wire is not the acceptable method. These connections are not tight and can easily come loose. Crimp-on solderless connectors should only be used on stranded wire.

✔ RIGHT

Delmar/Cengage Learning.

FIGURE 10-31 When you need to connect solid low-voltage or thermostat wire to a solderless connector, be sure to use a connector that has a screw-type connector. This will ensure a tight electrical connection.

Leak Testing

- Explain the importance of maintaining a leak-free system.

- Describe the process of leak testing a system with a soap bubble solution.

- Describe the process of leak testing a system with a halide leak detector.

- Describe the process of leak testing a system with an electronic leak detector.

- Describe the process of leak testing a system with an ultrasonic leak detector.

- Describe the process of leak testing a system with an ultraviolet leak detector.

- List the benefits and drawbacks of various leak detection methods.

- Explain why leak testing in a vacuum is not an acceptable method for leak checking a system.

Once all the installation work has been completed, the system is almost ready to be put into operation. There are, however, a few more important things that need to be done before system start-up. One of these is leak checking the system to ensure that all soldered, brazed, and flared connections are made properly and are leak free. A leaking air conditioning system will not operate effectively, will consume more electric power, and will result in refrigerant releases to the atmosphere. Therefore, leak checking a system prior to startup is an absolute must, especially when refrigerant lines are to be sealed behind walls or ceilings. Accessing these refrigerant lines and connections will be extremely difficult once the walls and ceilings are finished.

Once the air conditioning system has been properly leak tested, the system can then be started up and charged. This chapter will discuss various methods of leak testing a system. Each of the methods that will be covered is to be used in a particular situation. It is therefore important for the technician to fully understand each type of leak detection method, its intended use, and any limitations that may exist.

Grains Weight measurement where 7,000 grains equals 1 pound

Reactor plate The copper disc in a halide leak detector

Relative humidity The amount of moisture in an air sample compared to the maximum amount of moisture the air sample can hold

Schrader valve A self-sealing valve similar to those found on car or bicycle tires

Schrader-valve stems The portions of Schrader valves that create the seal

Standing-pressure test Leak detection method that involves the pressurization of system and the evaluation of the system pressure

Underfed evaporator An evaporator coil that has insufficient liquid refrigerant being fed into it

IMPORTANCE OF A LEAK-FREE SYSTEM

Air conditioning and refrigeration systems are intended to operate with the correct amount of refrigerant to help ensure efficient operation. If systems operate with a deficiency of refrigerant, either due to improper initial charging or a refrigerant leak, the efficiency of the system will be sacrificed. The results of an undercharged system include

- Inadequate cooling
- Increased system run time
- Increased energy usage
- Decreased equipment life
- Improper compressor cooling
- Increased humidity in the occupied space
- Mold issues

An air conditioning system that operates with a deficiency of refrigerant will have an **underfed evaporator** coil. The term underfed evaporator refers to a coil that has insufficient liquid refrigerant being fed into it. Since the amount of liquid refrigerant entering the coil is reduced, less heat from the occupied space will be absorbed into the system. Because less heat is being transferred from the space to the system, the space temperature will not drop enough to satisfy the setting on the thermostat. Therefore, the system will continue to operate, but not cool the space. Increased run time translates into increased energy usage, higher electric bills, and decreased system life.

It is important to keep in mind that an undercharged system *may* still be able to cool the space. This is because of two main reasons. First, the heat gain of the space is often not accurately calculated so, many times, the equipment capacity is too large for the space. This can be avoided by performing accurate heat gain and heat loss calculations before installing the system. Second, systems do not operate under design conditions most of the time. Since the system was ideally designed to operate under design conditions, there is often excess capacity. So, if the heat gain on the structure is low, a slightly undercharged system might still be able to cool the space.

Another important issue that results from a system undercharge involves inadequate compressor cooling. Many compressors rely on the cool suction gas returning from the evaporator to keep the motor windings cool. When a system operates with an undercharge, the temperature of the returning suction gas is higher. This is because the liquid refrigerant boils faster in the evaporator, increasing the amount of evaporator superheat. Since the temperature of the returning vapor is higher than desired, its cooling effect on the compressor is reduced. This can result in overheated compressor motor windings, damaged winding insulation, and premature compressor failure.

An undercharged system can also have negative effects on the interior environmental space. In earlier discussions on basic system operation, it was mentioned that the evaporator performs two major functions. These functions were the cooling and dehumidifying of the occupied space. If the system is operating with an undercharge of refrigerant, both the cooling and dehumidifying functions of the evaporator coil will be affected. If the coil cannot dehumidify the air in the space, excessive water vapor, or moisture, will be present. This can lead to uncomfortable occupants and, more importantly, an increase in the chances of having mold grow in the space. The desired **relative humidity** in an occupied space is in the 40% to 60% range, with 50% being ideal. A relative humidity over 60% greatly increases the chances of allowing mold to grow in the space. Relative humidity is defined as the amount of moisture in an air sample compared to the maximum amount of moisture the air sample can hold.

EXAMPLE 1

An air sample holds 50 **grains** of moisture, where 7,000 grains is equal to 1 pound. The maximum capacity of the air sample is 100 grains of moisture. The relative humidity is calculated as follows:

Relative Humidity = (Moisture content ÷ Maximum moisture capacity) × 100%

Relative Humidity = (50 grains ÷ 100 grains) × 100%

Relative Humidity = (0.5) × 100%

Relative Humidity = 50%

Going Green

Refrigerant should not be released to the atmosphere. Chlorine, a component part of CFC and HCFC refrigerants causes damage to the ozone layer. HFC refrigerants, even though they do not contain chlorine, can damage our environment as well because they contribute to global warming. Let's keep our systems leak free!

INTRODUCTION TO LEAK-TESTING

Leak testing a system is often one of the most difficult tasks a service technician can face. The leak can, in all actuality, be anywhere along the piping circuit. Quite often, the entire piping arrangement is not accessible, making the process that much more difficult. Leaks can be present deep within the heat transfer coils, such as the evaporator or condenser, or they can be located behind sealed walls and ceilings. The leak detection process is more easily accomplished when the technician is armed with the proper tools. There are a number of different methods available for leak testing an air conditioning or refrigeration system. As mentioned at the beginning of the chapter, leak detection is a task that should be performed whenever a system is newly installed and assembled or when a system has been opened for repair. In addition, if it is determined that refrigerant has leaked from a system, the source of the leak should be located and repaired whenever possible. Which leak detection method or methods that are used in any given situation depends on the conditions that are present. For example, on newly assembled systems, there is no refrigerant present in the system, so the method used to leak test the system must be capable of detecting leaks in the absence of refrigerant. The paragraphs that follow present a number of different methods to leak test a system along with the guidelines, benefits, and drawbacks of each. Some of the most common leak detection methods include

- Audible leak detection
- Soap bubble solutions
- Standing pressure test

TECH TOOL BELT • TECH TOOL BELT

Courtesy of Robinair.

Gauge Manifold

The gauge manifold is probably the most useful tool used by air conditioning service personnel. It is manufactured with two separate gauges: one for the high-pressure side of the system and one for the low-pressure side of the system. The manifold enables the technician to read both saturation temperatures and pressures within the system. The low-side gauge also provides a means by which the technician can read vacuum levels in the system during the evacuation process. For leak testing the system, the gauge manifold is used during the standing pressure test.

Delmar/Cengage Learning.

Nitrogen Tank and Regulator

Nitrogen is typically the gas of choice for pressurizing and leak-checking air conditioning systems prior to system startup. The regulator is used to reduce the pressure leaving the tank to a useful pressure. The pressure in the nitrogen tank can be over 2,500 psi, so simply opening the tank will result in a large amount of gas leaving the tank, possibly causing damage to the system or personal injury. Adjusting the regulator will ensure that only the desired amount of gas is permitted to leave the tank. When working with pressurized vessels, always wear proper eye protection.

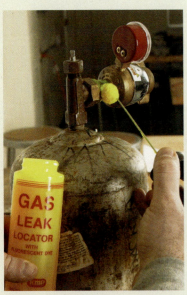

Photo by Eugene Silberstein.

Soap Bubbles

A soap bubble solution is a very effective way to check a pressurized system for leaks.

The solution is simply brushed onto the piping circuit, primarily at solder joints. The formation of bubbles indicates that there is a leak in the system. This method of leak detection is particularly popular since smaller leaks will produce larger bubbles, making them easier to locate.

Photo by Eugene Silberstein.

Electronic Leak Detector

The electronic leak detector is able to sense very small amounts of refrigerant in air samples. Using a small internal pump, air is pulled into the device and, if refrigerant is detected, the audible tone that the detector emits will change. Typically, the electronic leak detector emits a slow beeping sound that increases in frequency depending on the amount of refrigerant detected. The sensing tip on the device should be kept clean at all times and should be moved slowly on

and around soldered joints to increase the effectiveness of the process.

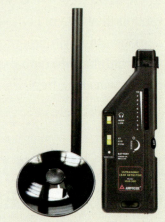

Courtesy of Amprobe.

Ultrasonic Leak Detector

The ultrasonic leak detector reacts to the sound of the escaping gas. It is a good choice when the leak is very small and difficult to locate. One of the benefits of the ultrasonic detector is that it can be used on systems that are pressurized with only nitrogen as well as on systems that have already been put into operation and contain a refrigerant charge.

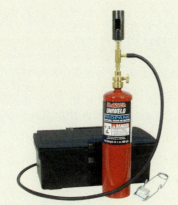

Courtesy of Uniweld Products.

Halide Leak Detectors

The halide leak detector uses an open flame to detect leaks of chlorinated refrigerants. When a leak is located, the flame changes color, as chlorinated refrigerant burns with a green flame. This method of leak detection is primarily used on systems that have already been put into operation since newly installed systems are leak checked before the refrigerant charge is added to the piping circuit. This method of leak detection will not work on a system that has a holding charge of nitrogen in it or on systems that operate with HFC refrigerants.

Photo by Eugene Silberstein.

Ultraviolet Leak Detectors

The ultraviolet leak detection method relies on a dye that is added to the system. The dye is permitted to circulate throughout the system and, in the event that a system leak develops, refrigerant and dye will leak out. The refrigerant will vaporize, but the dye will stain the pipe at the point of the leak. This stain will be visible only when observed under ultraviolet light. Once again, this method of leak detection is primarily used on systems that have already been put into service, as the dye is not added until the initial leak check and system startup have been completed.

- Electronic leak detectors
- Halide leak detectors
- Ultraviolet leak detectors
- Ultrasonic leak detectors

AUDIBLE LEAK DETECTION

In order to leak test a system, vapor must be introduced to the system in order to create a positive pressure within the system. It is the detection of the escaping vapor that enables the technician to identify the location of the leak. The method of leak detection that is initially performed utilizes the sense of hearing. As the system is pressurized for leak testing, larger leaks will make their presence known by loud hissing sounds (Fig. 11-1). The exact location of the leak can be determined by slowly moving your hand over the surface of the pipe where the sounds seem to be coming from. When such leaks are detected, the introduction of pressure to the system should be stopped and

FIGURE 11-1 Technician listening for leaks.

the leak should be repaired. Once all the large, audible leaks have been located and repaired, other leak detection methods can be used to locate smaller, harder to find, leaks.

SOAP BUBBLE SOLUTIONS

A soap bubble solution is a very effective way to check a pressurized system for leaks. The solution is commercially available and often comes in a spray bottle for easy application to the piping connections (Fig. 11-2). Some manufacturers provide an applicator as part of the bottle lid (Fig. 11-3). The solution is simply brushed or sprayed onto the piping circuit, primarily at solder joints. The formation of bubbles indicates that there is a leak in the system. This method of leak detection is particularly popular since smaller leaks will produce larger bubbles, making them easier to locate. The solution is relatively thick, so it coats the pipe and gives the technician enough time to visually inspect the pipes for bubbles. Some bubble solutions contain ultraviolet dye (Fig. 11-4). This makes it even easier to spot leaks if a black light is used. A large glowing bubble is a sure sign that a leak is present (Fig. 11-5). As part of the bubble solution, some manufacturers add antifreeze to prevent freezing when used on low-temperature refrigerant lines.

from experience...

Bubble solutions can be used on newly installed systems that are pressurized with nitrogen or on systems that have already been in service and are already charged with refrigerant. All that is needed is enough pressure to cause bubble to form.

FIGURE 11-2 Bubble solution in a spray bottle.

FIGURE 11-3 Soap bubbles.

FIGURE 11-4 Bubble solution with fluorescent dye.

FIGURE 11-5 Fluorescent dye causes bubbles to glow when observed under black light.

STANDING-PRESSURE TEST

The leak detection method that is widely accepted in the field for performing a leak check on a newly installed, assembled, or repaired system is the **standing-pressure test**. The standing-pressure test involves the pressurization of a system to a safe yet practical level and the evaluation of the test pressure over time to determine if the system is leak free. The steps for performing the standing pressure test include the following:

- Evaluating the system
- Pressurizing the system
- Marking the gauge
- Waiting period
- Evaluating the pressure

EVALUATING THE SYSTEM

Before doing any part of the standing pressure test, the technician must evaluate the system and determine a number of things. The first thing is to find out what the factory test pressure is on the system components. For a split system, these are the factory test pressures of the air handler and the condensing unit (Fig. 11-6). The pressure that is used during the standing pressure test must not exceed the *lowest* factory test pressure. For example, if the air handler has a factory test pressure of 250 psig and the condensing unit has a factory test pressure of 400 psig, under no circumstances should the standing pressure test pressure be brought to a level higher than 250 psig. The second thing that must be done is evaluate the piping of the system and find out if there are any solenoid valves or other refrigerant flow controls that might be closed during the test. It is important that all valves and flow controls be open during the test. Closed valves can prevent the pressurization of the entire system and give the impression of a leak-free system when a leak is actually present.

Photo by Eugene Silberstein.

FIGURE 11-6 The nameplate on the unit provides test pressure information.

PRESSURIZING THE SYSTEM

After determining a safe pressure for the test, the system can be pressurized. Nitrogen is a great choice for accomplishing this. Nitrogen is relatively inexpensive and, after the leak check is complete, it can be safely released into the atmosphere.

> ### CAUTION
> **CAUTION:** Never use pressurized air or oxygen to leak check a system! Some refrigerant oils can react with pressurized air and cause an explosion. Pressurized air contains water vapor, which should not be present in air conditioning systems.

> *from experience...*
> A small amount (a trace) of refrigerant can be added to a system that is pressurized with dry nitrogen to aid in further leak detection if needed. Some other leak detection methods require that refrigerant be present in the system. Those methods are discussed in later sections in this chapter.

> *from experience...*
> Never pressurize a system with 100% refrigerant for leak detection purposes. This is against the laws set forth by the EPA.

Remember that the manufacturer's test pressures should never be exceeded. Exceeding these pressures could result in damage to the equipment. The test pressure on a typical R-22 indoor unit is approximately 150 psig, while the pressure on a typical R-410A system is higher that that. When introducing nitrogen to the system, be sure to use a regulator, as the pressure in the nitrogen tank can be as high as 2,500 psig. The outlet of the regulator should be connected to the center hose of the gauge manifold (Fig. 11-7). The valves on the gauge manifold will be used to pressurize the system. The gauges on the manifold will be used to set and observe the pressure in the system.

FIGURE 11-7 Nitrogen regulator outlet connected to the center hose of the gauge manifold.

Photo by Eugene Silberstein.

from experience...

Before opening the nitrogen tank, make certain that the regulator is connected to the tank and that the regulator handle is turned completely counterclockwise. This will prevent damage to the gauge manifold (Fig. 11-8).

FIGURE 11-8 Before opening the nitrogen tank, the regulator should be closed to prevent damage to the gauge manifold.

Photo by Eugene Silberstein.

MARKING THE GAUGE

After the pressure has been introduced to the system, good field practice requires letting the system stand for a short period of time, about five minutes. This gives the

FIGURE 11-9 Marking the gauge for the standing pressure test.

Delmar/Cengage Learning.

gas time to settle and reach all portions of the system. This helps ensure that the pressure in the system will not change during the test. Once the gas has settled, the face of the gauge should be marked, indicating the exact position of the needle (Fig. 11-9). By marking the gauge, any small movements in the needle's position will be easy to spot. Placing a sticker on the gauge cover and marking the needle position on the sticker with a pen is the easiest way to accomplish this.

WAITING PERIOD

After the system has been pressurized and the gauge has been marked, a period of time should pass before the system pressure is observed and evaluated to see if the needle has moved. The time that should be allowed to elapse depends on the size of the system and the length of the refrigerant lines. Obviously, the longer the system stays under pressure, the better and more reliable the leak check will be. On residential systems, an hour should be more than sufficient for even the smallest leak to cause movement on the gauge. If at all possible, the system should be under pressure overnight to ensure that the system is leak free prior to startup. If a large leak is present, the pressure in the system will begin to drop rapidly (Fig. 11-10). Smaller leaks, however, may require much more time to show up on the gauge. When possible, digital gauges should be used for leak checking a system, as very small changes in the system pressure will be immediately noticeable.

EVALUATING THE PRESSURE

If the pressure in the system drops over time, a leak exists in the system. In this case, the solder and flare joints, if any, should be carefully inspected. If the system was pressurized with 100% nitrogen, a liquid-bubble-type, leak-detecting solution should be used. This solution

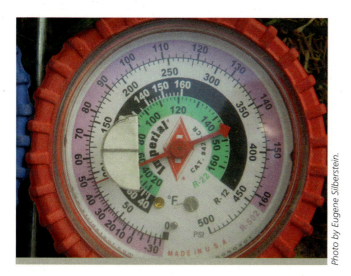

FIGURE 11-10 This system is leaking. Notice the position of the needle with respect to the marking on the gauge cover.

Photo by Eugene Silberstein.

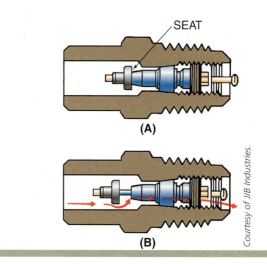

FIGURE 11-12 The Schrader valve is similar to the valve on a car tire. (A) Valve is closed. (B) When the center pin is pushed in, the valve opens.

Courtesy of JB Industries.

FIGURE 11-11 Technician applying bubble solution to system pipe joints and connections.

Photo by Eugene Silberstein.

of nitrogen and a trace gas, the technician can use an electronic leak detector to help locate the leak. Using an electronic device is much faster and neater than using the bubble method. Electronic leak detectors will be discussed later on in this chapter. After the leak is located, the pressure in the system must be released before the leak is repaired. This will prevent the solder from being pushed out of the solder joint as heat is applied. If the leak is found on a flare nut or Schrader valve stem, however, keeping the pressure in the system as the fitting is tightened is a good idea. Once tightened, the area can immediately be rechecked with either the bubble solution or the electronic leak detector. Once the leak has been located and repaired, the system should be repressurized and observed at a later time to ensure that the system is now free from any leaks.

should be applied to all solder joints as well as to all flare connections and **Schrader valve stems** (Fig. 11-11). **Schrader valves** are self-sealing valves similar to those found on car or bicycle tires (Fig. 11-12). The valve stem is the portion of the Schrader valve that creates the seal. Schrader valves and the pins or stems contained in them operate on the same principle as a bicycle or automobile tire valve. The stems can come loose and can develop leaks. The formation of bubbles on the surface of the piping indicates the presence of a leak. Large bubbles indicate a small leak. If the pressure in the system has dropped rapidly and a large bubble is found, chances are that another leak in the system needs to be located as well. If the system has been pressurized with a mixture

from experience...

The technician must remember that the gauge manifold connected to the system may also be a leak source. When checking the system for a leak, the technician should always check the connections on the manifold itself. The system may be leak free, but the manifold may not be.

from experience...

Good field practice entails continuing to look for leaks even after one has been found. The initial leak should be marked while the rest of the system is leak tested. This prevents the need to pressurize a system over and over again as more leaks are located.

ELECTRONIC LEAK DETECTORS

The electronic leak detector is able to sense very small amounts of refrigerant in air samples (Fig. 11-13). The electronic leak detector is probably the most popular leak detection method and has the ability to locate leaks as small as 0.25 ounce per year. Using a small internal pump, air is pulled into the device and, if refrigerant is detected, the audible tone that the detector emits will change. Typically, the electronic leak detector emits a slow beeping sound that increases in frequency depending on the amount of refrigerant detected. The sensing tip on the device should be kept clean at all times and should be moved slowly on and around soldered joints to increase the effectiveness of the process. It is important to properly select the electronic leak detector to be used as different models are designed to sense different refrigerant types. Some detectors have selector switches that allow the user to determine which refrigerant type the detector will sense.

HALIDE LEAK DETECTORS

The halide leak detector is used on systems that contain chlorinated refrigerants (Fig. 11-14). They use a small tank of propane or MAPP gas. MAPP gas is a mixture of liquefied petroleum and methylacetylene-propadiene. The leak detector is attached to the tank. In operation, the valve on the leak detector is opened and the fuel is ignited. When lit, a copper disc called the **reactor plate** in the detector is heated (Fig. 11-15). Air for combustion is provided through a rubber probe hose that connects to the detector and is used to search solder joints for leaks. The probe hose is passed slowly over the piping circuit and air is pulled into the hose. A leak is detected when chlorinated refrigerant is mixed with the air and passes over the heated reactor plate. The flame changes color as the refrigerant burns with a green flame (Fig. 11-16). This method of leak detection is primarily used on systems that have already been put into operation, since newly installed systems are leak checked with an inert gas before refrigerant is added to the piping circuit. This method of leak detection will not work on a system that has a holding charge of nitrogen in it or on a system that contains HFC refrigerants or HFC-blended refrigerants. This is because chlorine must be present for this method to work. The halide leak detector can be used to detect leaks as small as 0.5 ounce per year.

Volume Control

16" Gooseneck Probe

Visual Leak Indicator

Sensing Level Switch

On/Off Balance Control

Contoured Handle

ROBINAIR

Courtesy of Robinair SPX Corporation.

FIGURE 11-13 Electronic leak detector.

Photo by Eugene Silberstein.

FIGURE 11-14 Halide leak detector.

FIGURE 11-15 Heated copper reactor plate.

Photo by Eugene Silberstein.

Courtesy of Uniweld Products.

FIGURE 11-16 The flame on the halide detector turns green when chlorinated refrigerants pass over the heated copper reactor plate.

from experience...

The halide leak detector operates with an open flame and produces small amounts of phosgene gas when refrigerant passes through the device. For these reasons, use of the halide torch in confined spaces or in areas where combustibles are close by is not recommended.

ULTRAVIOLET LEAK DETECTORS

The ultraviolet, UV, leak detection method relies on a dye that is added to the system upon initial startup or at some point thereafter (Fig. 11-17). The dye is permitted to circulate throughout the system and, in the event that a system leak develops, refrigerant and dye will leak out. The refrigerant will vaporize, but the dye will stain the pipe at the point of the leak. This stain will be visible only when observed under ultraviolet light (Fig. 11-18). The ultraviolet lights used in our industry range from large, handheld lamps, to pocket flashlight-sized devices (Fig. 11-19). This method of leak detection is primarily used on systems that have already been put into service, as the dye is not added until the initial leak check and system startup have been completed.

Courtesy of Spectronics Corporation.

FIGURE 11-17 Ultraviolet leak detection equipment.

Photo by Eugene Silberstein.

FIGURE 11-18 Ultraviolet dye stain on pipe at the point of leak.

FIGURE 11-19 Various ultraviolet lights.

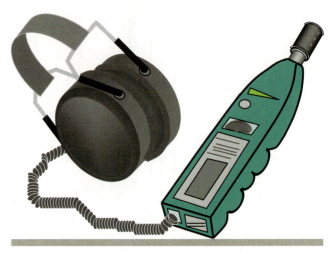

Photo by Eugene Silberstein.

FIGURE 11-20 Ultrasonic leak detector.

from experience...

Use of the ultraviolet leak detection method is ideally intended for use on systems that have expired compressor warrantees. Since air conditioning systems should contain only refrigerant and oil, manufacturers may not honor the compressor warranty if the ultraviolet solution is detected in the system. Be sure to check with the compressor manufacturer before adding the UV solution to the system.

from experience...

When looking for leaks on systems that have already been put into service, it is a good idea to look for traces of oil on the refrigerant piping. Since oil travels with refrigerant as it flows through the system, the oil will leak out with the refrigerant. The refrigerant will vaporize, but the oil will remain behind.

ULTRASONIC LEAK DETECTORS

The ultrasonic leak detector reacts to the sound of the escaping gas (Fig. 11-20). A small microphone and earphones are used to listen to the sound of vapor as it passes through the piping circuit. As the microphone is moved toward the location of the leak, the sound detected will become louder. As the microphone is moved away from the location of the leak, the sound detected will become more faint. It is a good choice when the leak is very small and difficult to locate. One of the benefits of the ultrasonic detector is that it can be used on systems that are pressurized with only nitrogen as well as on systems that have already been put into operation and contain a refrigerant charge.

LEAK-TESTING IN A VACUUM? NOT THE BEST IDEA!

After the leak-testing phase of system installation and startup has been completed, the system needs to be evacuated, or pulled into a vacuum. This process is discussed in the next chapter. Many technicians feel that the processes of leak-testing and system evacuating can be combined into a single step. This is not the case! The idea of leak testing a system in a vacuum is being discussed here so that the reader will gain an understanding of why this is not acceptable field practice.

Some technicians think that leak testing in a vacuum is an acceptable practice and often use the following to support this position:

- By leak checking a system in a vacuum, time will be saved on the installation.
- The system must eventually be pulled into a vacuum before startup anyway.
- If the system holds a vacuum, it must not have a leak.
- Time will be saved by not needing to pressurize the system with nitrogen.
- Time will be saved by not needing to wait to see if system pressure changes.

Although the argument seems to be a valid one, in reality it is not. Granted, the system must be pulled into a vacuum anyway. But that is as far as the argument goes. Even if the system does hold a vacuum, a leak may still be present. Consider the situation in which a small leak exists in a brazed piping connection. If the system is pulled into a vacuum, the flakes that exist around the leak may be pulled into the joint—in effect, sealing it. The gauge now indicates that the system is in a deep vacuum, and the vacuum holds. Once pressure is introduced into the system, the flux is blown out of the joint, reopening the leak. In addition, if a system is leak checked in a vacuum, air can get pulled into the refrigerant piping circuit if a leak is present. The moisture contained in the air will need to be removed during system evacuation. The time required for proper evacuation will be increased because of this added moisture. In addition, if the conditions are correct, moisture that is pulled into the system can freeze, creating ice crystals that can seal the leak until the system is pressurized with refrigerant.

One final reason that a system should not be leak checked in a vacuum is that the pressure differential created by a vacuum pump is very small compared with that created by adding pressurized nitrogen to the system. A leak that may result in substantial refrigerant loss during system operation may not surface during a vacuum-type leak check. Eventually, that leak will have to be repaired. Repairing the leak at this stage of the game, after startup, could involve the following:

- Recovering the system refrigerant
- Opening sealed walls and/or ceilings to access the piping circuit
- Locating and repairing the leak, which would involve pressurizing the system with nitrogen anyway

- Performing a standing-pressure test
- Evacuating the system for a second time
- Going through a second startup procedure
- Supplying refrigerant to replace the amount that was lost due to the leak

By trying to save time on the installation, a great deal of additional time, energy, and money can, in fact, be wasted. In addition, the inconvenience that the customer is being subjected to is also a major factor. The competence level of the entire service company may be questioned if the system begins to leak shortly after installation.

from experience...

Once a system has been leak tested by pressurization, a vacuum test can and should be used. A micron vacuum gauge should be used to determine if the vacuum is holding.

SUMMARY

- Leak checking the system must be done prior to system startup.
- The standing pressure test is the most popular method for leak checking newly assembled and installed systems.
- The standing-pressure test involves determining the lowest factory test pressure, pressurizing the entire system, and evaluating the pressure to determine if a leak is present.
- Common leak detection methods include soap bubble solutions, ultrasonic devices, halide leak detectors, and electronic leak detectors.
- Systems should not be leak tested by pulling the unit into a vacuum.

GREEN CHECKLIST

☐ **Refrigerant should not be released to the atmosphere.**

☐ **CFC and HCFC refrigerants contain chlorine, which damages the ozone layer.**

☐ **HFC refrigerants contribute to global warming.**

☐ **Technicians should make an effort to locate and repair all refrigerant leaks.**

REVIEW QUESTIONS

1. **Leak testing a system prior to initial startup is**
   a. Recommended if the technician does not have field experience
   b. Performed only when flared piping connections are made
   c. Performed only when soldered or brazed piping connections are made
   d. Recommended on all systems

2. **When initially pressurizing a system for leak detection, large leaks are often detected by which of the following methods?**
   a. Ultraviolet leak detectors
   b. Hearing the escaping vapor
   c. Both a and b are correct
   d. Neither a nor b is correct

3. **When performing a standing pressure test, the pressure that is maintained in the system is determined by the**
   a. Low side pressure when the system is operating
   b. High side pressure when the system is operating
   c. The system component with the lowest factory test pressure
   d. The system component with the highest factory test pressure

4. **Which of the following substances is the best to use for system pressurization when performing a standing pressure test?**
   a. Dry nitrogen
   b. Oxygen
   c. Pressurized air
   d. Both a and b are correct

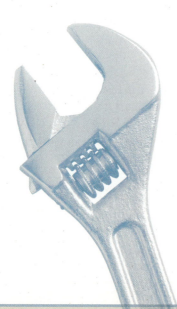

5. **Which of the following is true regarding the use of soap bubble solutions for leak detection purposes?**

   a. Large bubbles indicate a large leak

   b. Small bubbles indicate a small leak

   c. Some soap bubble solutions contain ultraviolet dye

   d. None of the above is correct

6. **Which of the following leak detection methods uses an open flame?**

   a. Electronic leak detector

   b. Halide leak detector

   c. Ultrasonic leak detector

   d. Ultraviolet leak detector

7. **Which of the following is a drawback of the ultraviolet leak detector?**

   a. The dye leaks out of the system when a leak is present

   b. The dye must first circulate in the system for about 24 hours

   c. The presence of an open flame

   d. Both a and c are drawbacks of the ultraviolet leak detector

8. **Which of the following is associated with the ultrasonic leak detector?**

   a. A microphone and earphones

   b. A glowing copper plate

   c. A black light

   d. An open flame

9. **The most popular and most accurate method of leak testing a system that contains refrigerant is the**

   a. Standing pressure test

   b. Electronic leak detector

   c. Halide leak detector

   d. Ultrasonic leak detector

10. **Leak checking the refrigerant piping circuit of a new air conditioning system using the vacuum method is not recommended because**

    a. Air and moisture can be pulled into the system

    b. A small leak may not become evident due to the small pressure differential between the interior of the piping circuit and the surrounding air

    c. Solder flux may be pulled into the leak, temporarily sealing it

    d. All of the above are correct

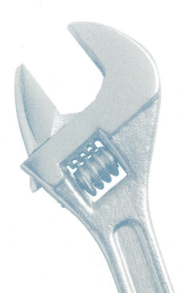

System Evacuation

OBJECTIVES *Upon completion of this chapter, the student should be able to*

- Explain the importance of system evacuation.

- Explain what is meant by the term "deep vacuum."

- Explain the function of the vacuum pump.

- List and describe features commonly found on a vacuum pump.

- Explain the two processes that occur during system evacuation.

- Properly set up an air conditioning system for evacuation.

- Change the oil in a vacuum pump.

- Properly evacuate an air conditioning system.

- Use a micron gauge to measure the level of vacuum in an air conditioning system.

Once the physical installation of the air conditioning equipment has been completed, the final tasks involved in the installation of the system can be performed. This task involves the evacuation, startup, and charging of the system. These tasks are especially important in ensuring that the system and its components remain operational for the expected life of the equipment.

When an air conditioning system is in operation, only refrigerant and oil should be traveling through the piping circuit. During the installation process, however, air and moisture enter the system. These must be removed prior to putting the system in operation. It is during the evacuation process that moisture and undesired gases are removed from the piping circuit. In this chapter, we will discuss the importance of system evacuation, the tools needed to properly evacuate an air conditioning system, and the procedures required to accomplish the task of system evacuation.

Deep vacuum The term used to describe a vacuum that is in the 50 to 250 micron range

Degassing The removal of noncondensable gases from a system during the evacuation process

Dehydrating The removal of water vapor and moisture from a system during the evacuation process

Evacuation The process of removing moisture and noncondensable gases from an air conditioning system prior to startup and charging

Micron A unit of linear measurement equal to 1/25,400 of an inch

Micron gauge The test instrument used to measure the level of vacuum in an air conditioning system

Noncondensable gases Gases that will not change from a vapor to a liquid within the normal pressure ranges commonly found in air conditioning and refrigeration systems

Sludge Organic solids made up of moisture, acid, and refrigerant oil

Two-stage rotary vacuum pump Vacuum pump that is made up of two rotary pumps where the outlet of one pump is the inlet of the second

Vacuum The condition that is present when the pressure in a vessel is below atmospheric pressure, 0 psig

Vacuum pump Piece of equipment used to reduce the pressure in an air conditioning system during the evacuation process

Vacuum pump oil Oil specially designed for use in vacuum pumps

INTRODUCTION TO SYSTEM EVACUATION

Before refrigerant is added to the system during the startup procedure, the system must be properly evacuated. **Evacuation** is the process of removing moisture and noncondensable gases from the system. When moisture is removed from the system, we call that process **dehydrating**. We use the term **degassing** to describe the process of removing noncondensable gases from the system. **Noncondensable gases** are gases that will not change from a vapor to a liquid within the normal pressure ranges commonly found inside air conditioning and refrigeration systems. Evacuation is accomplished with a **vacuum pump**. The vacuum pump is a piece of equipment that is used to reduce the pressure in an air conditioning system during the evacuation process. The vacuum pump pulls air from the closed system, reducing the pressure in the system, creating a **vacuum**. The term vacuum is used to describe the condition that is present when the pressure in a vessel is reduced to a level that is below atmospheric pressure, 0 psig. Pulling a vacuum in the system accomplishes two things. First, the very act of pulling removes unwanted vapors from the system. Second, by lowering the pressure in the system, moisture will vaporize. Having turned into vapor, the vacuum pump can then remove it from the system. If left in the system, moisture and noncondensable gases can lead to system malfunctions.

Noncondensable gases have a negative effect on system performance because, if not removed, they will migrate to the condenser coil. This leads to a number of problems including

- reduced heat transfer capability of the condenser coil (reduced capacity)
- insufficient heat rejection

- increased high side pressure
- increased compression ratio
- increased compressor discharge temperature
- reduced system efficiency
- reduced system effectiveness and reliability

If moisture is not removed from the system prior to startup, system damage can occur. Here are some of the problems that can result:

- Formation of acid in the system
- Freezing moisture

The presence of acid in the system can damage the windings of the compressor and lead to the premature failure of the system. Acid can also eat away at the interior surfaces of the copper refrigerant lines. In addition, moisture can freeze within the system given the low pressures at which the system's evaporator operates; as the pressure of a fluid decreases, so will the temperature. Freezing moisture can lead to a multitude of problems including

- blocked metering devices and strainers
- blocked filter driers
- malfunctioning thermostatic expansion valves
- reduced system effectiveness and efficiency

If acid does form in the system, it is very likely that sludge will also form. **Sludge** is a generic term used to describe organic solids. Sludge is formed when moisture, acid, and refrigerant oil are mixed together. When sludge forms, it often accumulates in the hottest portion of the air conditioning system, which is at the compressor discharge. If sludge does accumulate at the compressor discharge, this port can become blocked. A blocked compressor discharge port can result in major system problems and possible personal injury.

TECH TOOL BELT • TECH TOOL BELT

The following tools are needed to properly evacuate an air conditioning system:

Gauge Manifold

The gauge manifold is used to obtain pressure readings at different points in an air conditioning system. It has two gauges that read the pressure on both the low-pressure side and high-pressure side of the system. The manifold is equipped with valves to isolate either the high-pressure side or low-pressure side from the center port on the manifold. The center port provides the connection for a vacuum pump, refrigerant tank, nitrogen tank, or refrigerant recovery unit. The low-side gauge can read positive pressures as well as pressures in a vacuum.

Photo by Eugene Silberstein.

Gauge Adapter

This tool connects to a Schrader valve and pushes the pin to open the valve during the evacuation process. It can also be used to remove the pin altogether to speed the evacuation process.

Courtesy of Robinair.

Vacuum Pump

The vacuum pump is designed to reduce the pressure in an air conditioning system, causing any moisture in the system to boil off into a vapor. The vapor is then removed from the system by the pump. The vacuum pump connects to the system via the center port on the gauge manifold

and should be sized according to the size of the system. A 4-cfm (cubic feet per minute) pump is sufficient to evacuate a residential central air conditioning system.

Courtesy of Robinair.

Vacuum Gauge

The vacuum gauge is a test instrument that measures the level of vacuum in a system in microns, providing a very accurate reading. The vacuum gauge is connected to the system between the vacuum pump and the system, allowing the gauge to obtain vacuum readings even when the vacuum pump has been disconnected.

from experience...

It is important to note that vacuum pumps will not remove sludge from a system once it has formed. Properly evacuating a system will help prevent sludge from forming.

from experience...

Once a deep vacuum has been reached, the compressor should not be operated. The compressor may overheat, causing damage to the motor windings. Refer to Procedure 12-3 for evacuating the system.

Photo by Eugene Silberstein.

FIGURE 12-1 Vacuum pump.

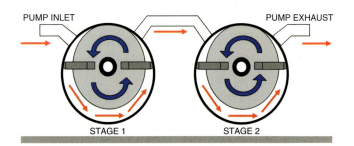

PUMP INLET PUMP EXHAUST

STAGE 1 STAGE 2

FIGURE 12-2 Diagram of a two-stage rotary pump.

THE VACUUM PUMP

The vacuum pump (Fig 12-1) is used to help remove the noncondensable gases and moisture from an air conditioning system prior to adding refrigerant. When connected to the system, the vacuum pump pulls vapor from the system and discharges it into the atmosphere. As the pressure in the system drops, moisture in the system begins to boil off into a vapor which, in turn, is removed from the system by the pump. For the vacuum pump to operate effectively and efficiently, the correct pump must be used and the pump oil must be checked and replaced often. Vacuum pumps are rated by their volumetric pumping capacity in cubic feet per minute, or cfm. For residential air conditioning applications, an oil-sealed, **two-stage rotary vacuum pump** rated at 4 cfm will usually suffice. Some manufacturers of vacuum pumps have relatively lightweight models that have pumping capacities of up to 8 cfm for residential applications. Larger vacuum pumps with capacities over 10 cfm are intended for use on larger commercial systems. Rotary vacuum pumps are typically very efficient and can evacuate smaller systems in a relatively short period of time. The two-stage rotary vacuum pump has two rotary pumps piped so that the discharge of one pump is the inlet of the other (Fig. 12-2).

VACUUM PUMP FEATURES

Although there are many different manufacturers of vacuum pumps, there are a number of features that are commonly found on them. It is important to note that the appearance of these features will differ from manufacturer to manufacturer and that not all vacuum pumps will have all of the features that are described here.

PUMP INLET PORT

The pump inlet is the portion of the vacuum pump that connects to the system by means of gauge hoses (Fig. 12-3). Typically, this inlet port is a ¼-inch male flare connection; by most, manufacturers provide multiple inlet port fitting sizes to allow the user to use hoses of different sizes. Larger hoses reduce the time required to evacuate the system.

PUMP EXHAUST

The pump exhaust is the location where the pump discharges the vapor that is pulled from the system (Fig. 12-4). When the pump is operating, it is not uncommon to see fine oil vapor coming from the port. When

FIGURE 12-3 Inlet port on the vacuum pump. This inlet port can connect to either a ¼″ hose or a ⅜″ hose.

Photo by Eugene Silberstein.

FIGURE 12-5 The isolation valve on allows the user to isolate the vacuum pump from the system.

Photo by Eugene Silberstein.

FIGURE 12-4 The exhaust port on this vacuum pump is incorporated in the pump's handle.

Photo by Eugene Silberstein.

FIGURE 12-6 The oil cap.

Photo by Eugene Silberstein.

working in confined spaces, it might be desirable to vent this exhaust to the outdoors. Some vacuum pumps are equipped with a female hose connection on the exhaust so that the user can connect a garden hose to the exhaust.

> ## CAUTION
>
> **CAUTION: Never block the exhaust of the vacuum pump. If a hose is connected to the exhaust port of the pump, make certain that the hose is not kinked, pinched, or otherwise blocked to ensure proper exhaust flow from the pump.**

ISOLATION VALVE

The isolation valve is a ball-type valve that will isolate the vacuum pump from the system being evacuated as well as from the vacuum gauges and gauge manifolds that might be connected to the system (Fig. 12-5). This valve allows the user to replace the vacuum pump oil during a system evacuation. The valve also provides

confirmation of a positive disconnect between the pump and system when evaluating the vacuum level in an evacuated system.

OIL CAP

When adding oil to the vacuum pump, the oil cap must be removed (Fig. 12-6). Once oil has been added to the pump, the cap must be replaced and hand tightened. Failure to replace the oil cap can result in decreased pump efficiency and increased evacuation time. Refer to the section on vacuum pump oil for more information on how the condition of the vacuum pump oil affects pump operation.

OIL DRAIN PLUG

When removing oil from the vacuum pump, the drain plug on the pump must be removed (Fig. 12-7). Once all of the oil has been drained from the pump, it is important to replace the drain plug on the pump and hand-tighten the connection. Improperly replacing or failing to replace the drain plug will cause obvious problems when oil is reintroduced to the pump!

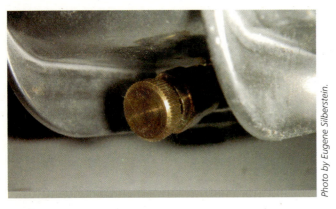

FIGURE 12-7 Drain plug at the bottom of the vacuum pump.

Photo by Eugene Silberstein.

Courtesy of Appion.

FIGURE 12-8 Oil cartridges are found on some vacuum pumps.

OIL CARTRIDGES

Some manufacturers utilize oil cartridges on their vacuum pumps (Fig. 12-8). These cartridges enable the pump user to replace the vacuum pump oil by simply replacing the cartridge.

SIGHT GLASS

The sight glass, located on the side of the pump, allows the pump user to observe the level of oil in the pump (Fig. 12-9). The level of oil in the pump should be such that about half of the sight glass is filled with oil.

GAS BALLAST

As mentioned at the beginning of the chapter, the vacuum pump facilitates the removal of unwanted gases and moisture from the system. In order to remove moisture from the system, it must first be vaporized. This vaporized moisture can condense back into a liquid if it comes

Photo by Eugene Silberstein.

FIGURE 12-9 The sight glass allows the technician to check the level of oil in the vacuum pump.

Photo by Eugene Silberstein.

FIGURE 12-10 The gas ballast is opened slightly when the system being evacuated has been open to the atmosphere for a long period of time or when the system is known to contain moisture.

in contact with a cool surface. The gas ballast is used to help prevent water vapor from condensing inside the vacuum pump (Fig. 12-10). If the system being evacuated has been open to the atmosphere for a long time or if it is known that the system contains moisture, the gas ballast feature can be used. By opening the gas ballast, small amounts of fresh air are introduced to the pumping chamber of the pump. This fresh air helps carry the vaporized moisture from the pump before it can condense back to a liquid.

VACUUM GAUGE

Although not intended to provide accurate vacuum measurements, some vacuum pumps are equipped with a vacuum gauge (Fig. 12-11). These gauges are intended to indicate if the system is capable of being evacuated to the required level. If the vacuum pump is not able to pull the system into the green (vacuum) range, there is likely a leak present in the system. If the vacuum pump is able to pull the system into the green area, more sophisticated and accurate instruments can then be used to measure the vacuum level more precisely.

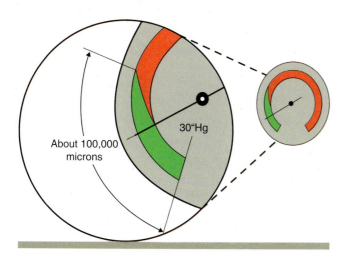

FIGURE 12-11 Some vacuum pumps are equipped with a vacuum gauge.

Photo by Eugene Silberstein.

FIGURE 12-12 Lifting eye on the vacuum pump.

LIFTING EYE

If lifting the vacuum pump is required, it is recommended that a rope or hook be attached to the lifting eye on the pump (Fig. 12-12). This helps ensure that the pump is secure and remains vertical. When lifting, make certain the rope is tight since portable vacuum pumps can weigh as much as thirty pounds.

VACUUM PUMP SAFETY

Technicians should always make every effort to remain safe on the job. Here are a few quick safety tips when working with vacuum pumps:

- Comply with manufacturer's guidelines—Take the time to read the literature that comes with the vacuum pump. When in doubt, READ THE BOOK!
- Inspect power cord—As with all electrical devices and components, inspect the power cord prior to use. If the cord is damaged, do not use the pump.
- Make certain that the pump is properly grounded— It is important to check for proper grounding

prior to use. Quite often, vacuum pumps are used outdoors in a potentially wet environment, increasing the chance of receiving an electric shock.

- Use protective eyewear—Protective eyewear should always be worn when working on air conditioning equipment.
- Keep face clear when starting up the vacuum pump—Keeping your face clear from the pump when starting it will help protect you in the case of a mishap.
- Never block the vacuum pump exhaust port— Blocking the exhaust port can cause the pressure in the pump to reach unsafe levels.
- Use the lifting eye on the pump—When lifting the pump, be sure to secure the pump by the lifting eye.
- Avoid positive system pressure—Vacuum pumps are intended to be used on systems that are at or below atmospheric pressure. Avoid using them on systems that are at pressures over 0 psig.

VACUUM PUMP OIL

When using a vacuum pump, it is very common for the pump to become contaminated with whatever has been removed from the system. If the system contains moisture, acid, or other impurities, they will wind up on the interior surfaces of the pump. These impurities will ultimately find their way into the vacuum pump oil. If the oil becomes saturated with acid or moisture, the pump will not operate effectively and will not be able to pull the system to the desired level of vacuum, which will be discussed in the next section. For this reason, a number of factors must be considered when dealing with vacuum pump oil:

- Only use oil that is rated for use in a vacuum pump.
- Change the vacuum pump oil before each use.
- Keep the oil container closed securely.
- Always maintain the correct oil level in the pump.

from experience...

Some manufacturers recommend draining the oil from the vacuum pump after each use to prevent pump contamination. Be sure to check and follow the manufacturers' recommendations.

Delmar/Cengage Learning.

FIGURE 12-13 Vacuum pump oil.

Vacuum pump oil (Fig. 12-13) is designed with a very low equilibrium vapor pressure, or simply vapor pressure. The vapor pressure of a liquid refers to its evaporation rate. Using oil that has a higher vapor pressure will reduce the pumping capacity of the pump, increasing the amount of time required to properly evacuate the system.

Because the system contaminants will ultimately wind up in the oil, it is important to change the oil frequently. This means, at a minimum, the oil should be changed before each use when installing residential air conditioning systems. If the moisture, acid, and other impurities remain in the pump, the evacuation process will be affected. Follow the pump manufacturer's instructions for draining and replacing vacuum pump oil. Once new oil has been introduced to the pump, the level of the oil should be midway in the sight glass (Fig. 12-14).

For step-by-step instructions on draining and replacing vacuum pump oil, see Procedure 12-1 on pages 299–300.

INSIGHT TO SYSTEM EVACUATION

During the evacuation process, both the high- and low-pressure sides of the system should be connected to the vacuum pump. This will help ensure that the entire system is evacuated to the recommended level. Flow controls such as thermostatic expansion valves and other components have small openings and orifices that will slow the evacuation process if the system is evacuated from only the high- or low-pressure side. Figure 12-14 shows the vacuum pump and gauge connections on a central air conditioning system.

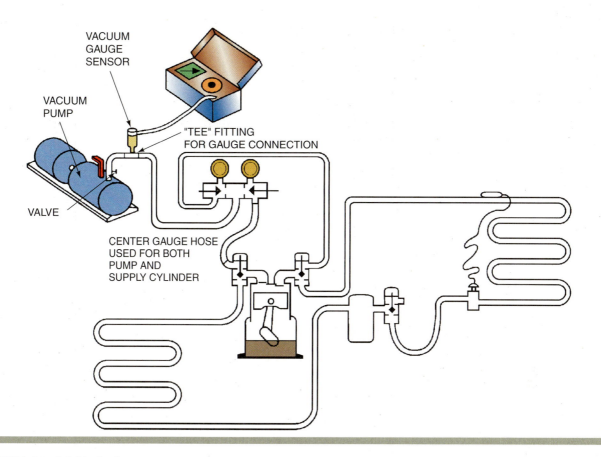

VACUUM
GAUGE
SENSOR

VACUUM
PUMP

"TEE" FITTING
FOR GAUGE CONNECTION

VALVE

CENTER GAUGE HOSE
USED FOR BOTH
PUMP AND
SUPPLY CYLINDER

FIGURE 12-14 Typical vacuum pump and vacuum gauge connections.

To ensure that the system is being evacuated in the most efficient manner possible, the size of the vacuum pump should be adequate for proper evacuation. For most residential applications, a 4-cfm pump is sufficient. Using a pump that is too small will increase the time required to evacuate the system. In addition, an undersized vacuum pump may not be able to remove all the moisture present in the system. The moisture that the vacuum pump helps remove often finds its way into the pump itself. This moisture condenses on the interior surfaces of the body of the pump and dilutes the oil within the pump. If a system containing a large amount of moisture is being evacuated, the pump can eventually displace the oil. This will leave the pump's crankcase filled with water instead of oil, causing damage to the pump. To help avoid this, good field practice requires replacing the vacuum pump oil frequently.

To speed the evacuation process, heat can be added to the system. This will help boil off any liquid moisture into vapor, which is much easier for the pump to remove. If this method is used, the entire system should be heated to prevent moisture from vaporizing in one part of the system and condensing back to a liquid in another. This often happens on split-type air conditioning that has both an indoor component and an outdoor component. The best source of heat in this situation is a heat lamp. If the technician suspects that moisture is present in the compressor crankcase, heat should be applied directly to the crankcase.

Using larger connectors and hoses can also speed up the evacuation process. Hoses with larger internal cross-sectional areas designed especially for vacuum pump use can be purchased at any refrigeration supply house. These hoses have removable pin depressors and are commonly used in conjunction with a gauge adaptor that serves the purpose of depressing the pin on the Schrader valve. The Schrader valves commonly found on system service valves create a restriction even when they are in the open position. Field service valves (Fig. 12-15) can be used to remove the Schrader pins for evacuation and to replace the pins after the proper vacuum has been achieved.

Photo by Eugene Silberstein.

FIGURE 12-15 Core tool used to remove Schrader pins during system evacuation.

from experience...

When setting up to evacuate a system, be sure to check the rubber seals on the ends of the hoses. If the seal is worn, it can create a restriction when the hose is connected to a fitting. Line restrictions increase evacuation time.

from experience...

Removing hose and valve restrictions can reduce evacuation time by up to 30%!

When evacuating a system, the technician should

- make certain that the system is evacuated from both the high-pressure and low-pressure sides
- make certain that the vacuum pump is sized correctly for the system being evacuated
- change the vacuum pump oil frequently, at least before each use
- apply heat to the system to speed the evacuation process
- never operate a compressor while a system is in a deep vacuum
- use the largest valve ports possible to speed the evacuation process
- use the largest diameter hoses available
- use the shortest hoses available
- remove Schrader pins whenever possible
- use copper or corrugated stainless steel hoses as synthetic hoses are permeable (they allow gases to penetrate)
- evaluate and measure the vacuum with a micron gauge (see next section)

MEASURING THE VACUUM

When evacuating an air conditioning system, it is important that an acceptable vacuum level be reached and maintained in order to conclude that the system is leak-free, dry, and ready to be charged and put into operation. Quite often, technicians rely on their gauge manifolds to determine if a sufficient vacuum has been achieved. This is not acceptable field practice. Here's why.

When the system is being evacuated and a vacuum is being pulled on the system, the readings on the gauge manifold will register values below 0 psig. The low-side gauge will indicate a reading in the vacuum range from 0″ Hg to 30″ Hg. (NOTE: A perfect 30″ Hg vacuum is, in reality, unobtainable. So, when reference is made to a 30″ Hg vacuum, we are actually referring to a vacuum of 29.92″ Hg). The high-side needle will indicate a point lower than 0 psig as well, but this value cannot be determined because there are no markings on the high-side gauge in the vacuum region. As the pressure in the system approaches a deep vacuum, close to 30″ Hg, the readings on a compound gauge become both less accurate and less reliable. In addition, gauges are often not properly calibrated, so using the gauge manifold to measure vacuum levels is not acceptable. In addition to the reasons just mentioned, industry and Federal guidelines require that the vacuum level in an evacuated air conditioning system be measured in **microns.**

The micron is the unit of linear measurement equal to 1/25,400 of an inch. It represents a measurement above an absolute pressure of zero, which represents a perfect vacuum of 29.92″ Hg. As the pressure in a system is reduced, the vacuum increases or becomes deeper. A **deep vacuum** is the term used to describe a vacuum range between 50 and 250 microns. At a gauge pressure of 28.9″ Hg, the equivalent reading on a micron gauge is

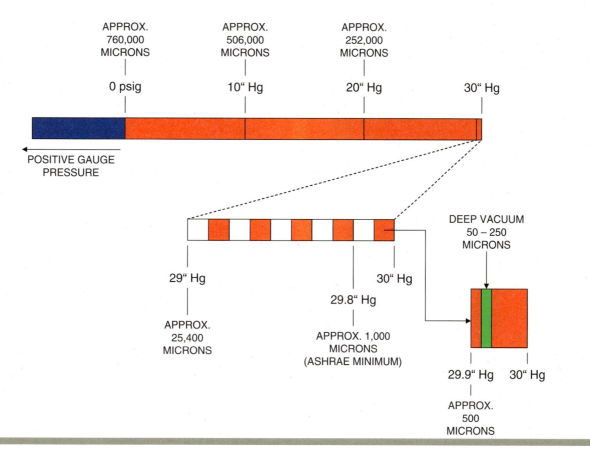

FIGURE 12-16 This diagram shows the relationship between the micron scale and the vacuum scale in inches of Mercury.

about 25,400 microns. To provide a more vivid image of the micron scale, refer to Figure 12-16. In this figure, we can see that the micron equivalent at atmospheric pressure (0 psig, 15 psia) is about 760,000 microns. The deep vacuum range is between 50 and 250 microns, which is the equivalent to about 29.91″ Hg. It is impossible to measure vacuum levels such as these with a compound gauge, so we use a micron gauge.

The **micron gauge** is a test instrument that is designed to measure the vacuum in a system in microns, as opposed to inches of mercury (Fig. 12-17). As the vacuum is increased to 29.91″ Hg, the micron gauge will read approximately 250 microns. In a vacuum of 29.92″ Hg, the micron gauge will read approximately 25 microns. Typically, an acceptable vacuum for most systems is in the 500-micron range, although some equipment manufacturers require that their systems be brought down to a 250-micron vacuum before starting up the system. When evacuating a system, always check with the specific manufacturer's guidelines as well as those set forth by the Environmental Protection Agency (EPA). According to the American Society of Heating, Refrigerating, and Air Conditioning Engineers (ASHRAE), the minimum vacuum level in a system should be 1,000 microns.

Photo by Eugene Silberstein.

FIGURE 12-17 Micron gauge.

from experience...

Measure the vacuum level at the system whenever possible. This will help ensure more accurate readings.

from experience...

Once an acceptable vacuum level has been reached, it is important to isolate the system from the vacuum pump and wait to see if the system maintains the vacuum. A small increase in the vacuum level is normal when the system is initially isolated from the vacuum pump. Large increases indicate the presence of a leak. The larger the leak, the faster the system will return to atmospheric pressure. According to ASHRAE, the vacuum level in a properly evacuated system should not rise to a level over 2,500 microns within a few hours after isolation.

For step-by-step instructions on calibrating the gauge manifold, See Procedure 12-2 on page 301.

For step-by-step instructions on evacuating an air conditioning system, see Procedure 12-3 on pages 302–305.

SUMMARY

- System evacuation is a combination of dehydrating and degassing a system.
- Dehydration involves the removal of moisture from a system.
- Degassing involves the removal of noncondensables from a system.
- Improper system evacuation can lead to a number of system problems including acid and sludge formation as well as reduced system efficiency and reliability.
- System evacuation should be done from both the high- and low-pressure sides of the system.
- Vacuum pump oil should be changed frequently, at least before each use.
- Use the shortest, largest diameter hoses possible to speed the evacuation process.
- Removing Schrader pins and pin depressors will also speed the evacuation process.
- System vacuum levels should be measured with a micron gauge.
- After evacuation, isolate the system from the vacuum pump to make certain the vacuum is maintained.

PROCEDURE 12-1

Draining and Replacing Vacuum Pump Oil

The following procedures are used to drain oil from the vacuum pump:

- Make certain that the pump is warmed up.

A Remove the oil drain cap located at the bottom of one side of the pump and drain the oil into a container.

- Properly dispose of the oil.

- Replace the drain cap.

B Remove the oil fill cap and begin to add oil to the pump.

A

Delmar/Cengage Learning.

B

Delmar/Cengage Learning.

PROCEDURE 12-1

Draining and Replacing Vacuum Pump Oil (Continuted)

C Fill the pump with oil rated for vacuum pump use until the oil level is equal to the middle of the sight glass on the side of the pump.

D Cap the inlet port on the pump.

• Allow the pump to run for approximately one minute and check the oil level in the sight glass with the pump still operating.

• If the oil level is below the oil level mark on the sight glass, slowly add more oil to the pump until the level of the oil is equal to the oil level mark.

E Replace the oil fill cap and tightly close the oil container.

C

Delmar/Cengage Learning.

from experience...

Using oil that is not rated for vacuum pump operation can prevent the pump from pulling the air conditioning system into a deep vacuum.

D

Delmar/Cengage Learning.

E

Delmar/Cengage Learning.

PROCEDURE 12-2

Calibrating the Gauge Manifold

The following procedures are used to calibrate the gauge manifold:

A Remove the high and low side hoses from the blank ports of the gauge manifold.

• Observe the position of the gauge needles.

B If both gauge needles point to 0 psig, the gauges are properly calibrated and no calibration is needed.

• If one or both of the gauges do not point to 0 psig, remove the clear plastic cover from that gauge.

C Using a small straight slot screwdriver, slowly turn the adjusting screw on the face of the gauge until the needle points to 0 psig.

• Turning the screw clockwise will cause the needle to move counterclockwise and vice versa.

• Once the gauges have been calibrated, replace the plastic cover on the gauge.

Photo by Eugene Silberstein.

Delmar/Cengage Learning.

Delmar/Cengage Learning.

Evacuating an Air Conditioning System

The following procedures are used to evacuate an air conditioning system:

- Remove the high and low side hoses from the blank ports on the gauge manifold and make certain that the gauges are calibrated.

- Calibrate gauges if needed.

A Connect the high side hose from the gauge manifold onto the high side service port on the condensing unit of the system.

B Connect the low side hose from the gauge manifold onto the low side service port on the condensing unit of the system.

- Replace vacuum pump oil and make certain that the oil level in the pump is correct.

C Connect the center hose from the gauge manifold to a valve tee connection.

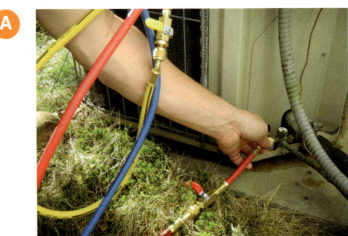

A

Photo by Eugene Silberstein.

B

Photo by Eugene Silberstein.

C

Photo by Eugene Silberstein.

D Connect the micron gauge to the valve port of the tee.

Photo by Eugene Silberstein.

E Connect the remaining side of the tee to the inlet port of the vacuum pump.

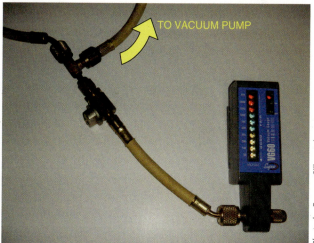

TO VACUUM PUMP

Photo by Eugene Silberstein.

F Make certain that the valve on the tee connection is closed.

Photo by Eugene Silberstein.

Evacuating an Air Conditioning System (Continued)

G Make certain that both valves on the gauge manifold and the valve on the vacuum pump are in the closed position.

- The power to the condensing unit should be disconnected.

- Turn the vacuum pump on.

G

Delmar/Cengage Learning.

H Slowly open the isolation valve on the vacuum pump.

H

Delmar/Cengage Learning.

I Open the high side valve on the gauge manifold.

I

Delmar/Cengage Learning.

Once the needle on the low side gauge falls below 5" Hg, slowly open the low side valve on the gauge manifold.

• Allow the vacuum pump to operate until the needle on the low side gauge indicates a vacuum of lower than 28" Hg.

K Open the valve on the tee connection to the micron gauge

• Turn on the micron gauge

• Allow the vacuum pump to operate until the desired vacuum level is reached

 Once the desired vacuum level has been reached, close the isolation valve on the vacuum pump

• Turn the vacuum pump off

• Monitor the micron gauge

• If the micron level rises slightly and stops, the system is leak-free. (It is normal for the micron reading to rise slightly when the system is initially isolated from the vacuum pump.) If the micron level continues to rise, the system has a leak.

• If the system is leak-free, close the valves on the gauge manifold

• Close the valve on the tee connection

• Turn off the micron gauge

Delmar/Cengage Learning.

Photo by Eugene Silberstein.

Photo by Eugene Silberstein.

REVIEW QUESTIONS

1. **The purpose of evacuating an air conditioning system is to**
   a. Remove moisture from the system
   b. Remove noncondensable gases from the system
   c. Both a and b are correct
   d. Neither a nor b is correct

2. **Moisture left in a system can lead to the formation of _____ in the system**
   a. Acid
   b. Sludge
   c. Both a and b are correct
   d. Neither a nor b is correct

3. **The oil level in the sight glass on the vacuum pump will be _____ when the pump is properly filled with oil.**
   a. Above the top of the sight glass
   b. Below the bottom of the sight glass
   c. In the middle of the sight glass
   d. At the bottom of the sight glass

4. **Which of the following is true regarding the oil in a vacuum pump?**
   a. Any general purpose, lightweight oil can be used successfully in the pump
   b. The oil should be changed frequently
   c. Discarded oil must be disposed off in accordance with local laws
   d. Both b and c are correct

5. **The purpose of the gas ballast on the vacuum pump is to**
   a. Help minimize condensation in the pump
   b. Isolate the vacuum pump from the system
   c. Help maximize condensation in the pump
   d. Increase the amount of moisture present in the oil

6. **According to ASHRAE, the minimum level of vacuum that should be reached when evacuating a system is**
   a. 0 microns
   b. 250 microns
   c. 500 microns
   d. 1,000 microns

7. **The deep vacuum range is typically between**
   a. 0 and 50 microns
   b. 50 and 250 microns
   c. 250 and 500 microns
   d. 500 and 1,000 microns

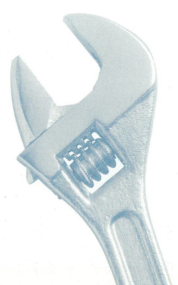

8. **Once the desired vacuum level has been achieved, it is important to**
   a. Close the isolation valve on the vacuum pump
   b. Wait to see if the vacuum level is maintained
   c. Both a and b are correct
   d. Neither a nor b is correct

9. **When evacuating a system, it is good field practice to**
   a. Evacuate from both the high side and low side of the system
   b. Replace the oil in the vacuum pump prior to evacuation
   c. Use copper or corrugated stainless steel hoses whenever possible
   d. All of the above are correct

10. **Which of the following is true regarding the hoses used for system evacuation?**
   a. They should be as long as possible
   b. The diameter of the hoses should be as small as possible
   c. Pin depressors inside the hoses should be removed
   d. All of the above are correct

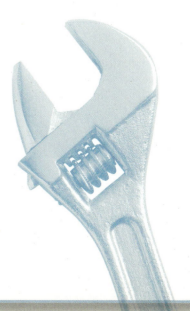

System Startup and Charging

OBJECTIVES

Upon completion of this chapter, the student should be able to

- Perform a prestartup check on an air conditioning system.

- Startup an air conditioning system.

- Measure system superheat and subcooling.

- Determine proper operating pressures in an air conditioning system.

- Add refrigerant to an air conditioning system.

- Remove refrigerant from an air conditioning system.

- Use manufacturers' charging tables to charge an air conditioning system.

- Measure the temperature differential across the evaporator of an air conditioning system.

Once the physical installation of the air conditioning equipment is completed, the final tasks involved in the installation of the system can be performed. These tasks involve the evacuation, startup, and charging of the system. These tasks are especially important in ensuring that the system and its components remain operational for the expected life of the equipment. In the previous chapter, the topic of system evacuation was covered. It is extremely important that the system be properly evacuated before the system is started up and put into operation.

During the startup process, the field technician introduces refrigerant to the system components and the field-installed lines that join them. Once the system is energized, the technician adjusts the refrigerant charge and fine-tunes the installation. An improper refrigerant charge can result in reduced system performance and efficiency. Deficiencies in any of the areas discussed in this chapter can easily result in premature component failure. This chapter will address the tasks that are performed by the technician during the startup process.

Evaporator temperature differential The difference between the temperature of the supply air and the temperature of the return air

Holding charge Initial amount of refrigerant added to the system before it is initially energized.

INTRODUCTION TO SYSTEM START-UP

The installation of an air conditioning system involves the coordination of the work of a number of different individuals. In many cases, the various tasks required for a successful installation are carried out by individuals who specialize in a particular area. For example, the air distribution system might be installed by one team, while the electrical portion of the installation might be completed by another. The start-up technician is responsible for inspecting the entire installation and giving it the final stamp of approval. The sections that follow provide details regarding this all-important final step in the installation process.

from experience...

Some outdoor units are pressurized with a holding charge of nitrogen and contain no refrigerant at all. Always check the manufacturer's literature to determine if the system is charged with refrigerant or nitrogen. Allowing an air conditioning system that contains nitrogen to operate can result in system overheating, greatly reduced efficiency, and component malfunction and failure.

SYSTEM HOLDING CHARGE

After the system has been properly leak-tested and evacuated, a refrigerant **holding charge** can be introduced to the system. A holding charge is the amount of refrigerant that is added to the system before it is initially energized. The holding charge should be released into the system only after the system has maintained the required vacuum level. In many cases, the outdoor unit comes precharged with enough refrigerants for a 25-foot refrigerant line run between the indoor and outdoor units. Since the actual length of the refrigerant line set will never be exactly 25 feet in length, the specific amount of refrigerant in the system will have to be adjusted accordingly.

If the run is longer than 25 feet, additional refrigerant will need to be added. If the run is less than 25 feet, refrigerant will need to be removed.

from experience...

Under no circumstances should the holding charge be added to the system before the system has been evacuated and the vacuum level maintained.

PRE-STARTUP CHECKLIST

Before the system is started up for the first time, the system should be inspected to make certain that all aspects of the installation have been completed. The following is a checklist organized by component that will help ensure that the system is ready to be energized and put into operation.

CONDENSING UNIT

✓ Is the outdoor unit level?

✓ Is the unit mounted on a pad that properly supports the unit?

✓ Is the area around the unit free from debris such as leaves and other potential coil obstructions?

✓ Are all line-voltage connections between the disconnect box and the unit made correctly?

✓ Are all field, line-voltage connections within the unit itself made correctly?

✓ Are all field, low-voltage wiring connections correct?

✓ Have the service valves been positioned to release the holding charge, if any, into the system?

✓ Are all refrigerant lines leading into the structure properly supported and insulated?

✓ Is there ample space around the unit for servicing?

✓ Are all shipping materials (compressor support blocks, etc.) removed from within the unit?

✓ Are the compressor mounting bolts secure?

✓ Is the compressor resting properly on the support springs? (The compressor mountings are sometimes tightened for shipping purposes and should be loosened before startup to allow the compressor to vibrate during operation without transmitting noise through the unit.)

✓ Does the outdoor fan spin freely?

AIR HANDLER

✓ Is the unit correctly mounted and supported?

✓ Is all supporting hardware tight and secure?

✓ Are all access and service panels accessible?

✓ Are all access and service panels in place and secure?

✓ Is there ample space for system servicing?

✓ Are all shipping materials removed from inside the unit?

✓ Do the condensate lines, both primary and auxiliary, run to their proper termination points?

✓ Do the condensate lines drain properly?

✓ Are all line-voltage connections between the disconnect box and the unit made correctly?

✓ Are all field, line-voltage connections within the unit itself made correctly?

✓ Are all field, low-voltage wiring connections correct?

✓ Does the indoor blower spin freely?

DUCT SYSTEM

✓ Are all duct sections properly connected, supported, and insulated?

✓ Are all registers and dampers in the open position?

✓ Are all takeoffs properly connected?

✓ Are canvas connectors, if applicable, properly installed to prevent noise transmission through the duct system?

GENERAL

✓ Are all refrigerant lines properly supported and insulated?

✓ If needed, are refrigerant traps installed and properly located?

✓ Have all electrical lines from the fuse or circuit-breaker panel to the indoor and outdoor units been properly installed?

✓ Has the thermostat been properly secured and leveled?

✓ Are air filters installed?

✓ Has the system been properly leak-checked?

✓ Has the system been properly evacuated?

✓ Is there a holding charge of refrigerant in the system?

✓ Has the crankcase heater been energized prior to system startup?

✓ Is the manufacturer's startup literature for the unit close at hand?

SYSTEM STARTUP

After the installation has been completed and all items on the preceding checklist have been accomplished, it is time to start the system for the first time. Ideally, the system should be started up on a warm day to help simulate the conditions under which it will most likely be operating. In the summer months, this should not pose a problem. In the fall, winter, or spring, however, the problem arises of cooler outdoor ambient temperatures. Starting up a system during times of cooler outdoor temperatures will result in a system that is grossly overcharged once the warmer months arrive. To compensate for the lower ambient temperatures, the startup technician can partially block the outdoor coil, thereby reducing the airflow through the coil. This will cause the head pressure of the system to rise. Ideally, the system should be started up in the cooling mode with the following conditions:

• A condenser saturation temperature of approximately 125°F

• An indoor air temperature of approximately 75°F

A number of factors must be considered when a system is started up for the first time:

• Airflow through the condensing unit

• Airflow through the air handler

- The high-side pressure of the system
- The low-side pressure of the system
- Temperature differential across the evaporator coil
- Evaporator and system superheat
- Condenser subcooling

AIRFLOW THROUGH THE CONDENSING UNIT

On air conditioning systems, the airflow through both the evaporator and condenser coils must be correct. One way to ensure that the airflow is within design range is to take amperage readings of the motors. On smaller systems, the outdoor unit fan is of the direct-drive type, where the fan blade is connected directly to the shaft of the motor. In these cases, the speed at which the fan blade turns is predetermined at the factory by the speed of the motor itself. At any rate, the amperage of the motor should still be checked and compared to the amperage on the nameplate of the motor.

In addition, any obstructions around the condensing unit can result in reduced airflow through the condenser coil. Reduced airflow through the coil will result in an improper refrigerant charge when initially charging the system, higher than normal operating pressures, and decreased cooling. Always make certain that all obstructions are removed from the area immediately around the condensing unit.

AIRFLOW THROUGH THE AIR HANDLER

When checking the airflow through the evaporator coil, the amperage of the blower motor should also be checked. If the blower assembly is direct drive, as in the case of the condenser fan motor, the amperage reading should be similar to the rating on the nameplate.

On larger units, where the blower is driven by a belt and pulley assembly, the driven pulley may need to be adjusted in order to reach the desired amperage draw of the motor.

When a motor's amperage draw is consistent with the nameplate rating, it is an indication that the correct volume of air is passing through the system. Too much airflow through the indoor coil, for example, will result in a higher-than-normal amperage draw. Insufficient airflow will result in a lower-than-normal amperage draw.

When checking the airflow through the indoor coil, make certain that all supply registers are open, the return air duct is unobstructed, and all service panels are secure.

Obstructions on either the supply or return side of the air distribution system will result in reduced airflow across the evaporator coil. Reduced airflow will result in lower operating pressures, reduced cooling, possible coil freeze-ups, and possible slugging of liquid refrigerant back to the compressor.

SYSTEM PRESSURES

Once the proper airflow through both the indoor and outdoor units has been established, the operating pressures of the system can be evaluated. When checking the system's operating pressures, thinking in terms of saturation temperatures rather than pressures is more convenient. Operating pressures vary from refrigerant to refrigerant, but the saturation temperatures are more or less constant from system to system. The condenser saturation temperature should be about 30 degrees higher than the outdoor ambient temperature for a standard-efficiency unit (Fig. 13-1), and 20 to 25 degrees higher than ambient for higher-efficiency models. On cooler days, as mentioned earlier, a portion of the coil can be blocked to increase the pressure to the desired level until the charging process is complete.

The low-side pressure of the system also relates to the temperature at which the refrigerant boils in the evaporator. For an indoor air temperature of approximately 75°F, the evaporator saturation temperature

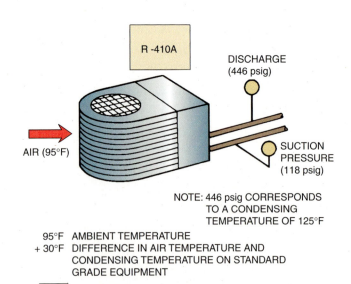

R-410A

DISCHARGE
(446 psig)

AIR (95°F)

SUCTION PRESSURE
(118 psig)

NOTE: 446 psig CORRESPONDS
TO A CONDENSING
TEMPERATURE OF 125°F

95°F   AMBIENT TEMPERATURE
+ 30°F   DIFFERENCE IN AIR TEMPERATURE AND
CONDENSING TEMPERATURE ON STANDARD
GRADE EQUIPMENT

= 125°F CONDENSING TEMPERATURE

FIGURE 13-1 The condenser saturation temperature is approximately 30 degrees higher than the outside ambient temperature.

should be in the range of 38°F to 42°F. A saturation temperature well below that indicates that more refrigerant *may* need to be added to the system, and a saturation temperature well above that indicates that refrigerant *may* need to be removed from the system. Refrigerant, however, should not be added or removed until it has been determined that the charge actually needs to be adjusted. For instance, reduced airflow across the evaporator coil can be the cause for low suction pressure and a dirty condenser coil can cause high operating pressures.

HIGH-SIDE PRESSURE

On a standard efficiency system, the rule of thumb is that the refrigerant should condense at a temperature approximately equal to the outdoor ambient temperature plus about 30 degrees. For example, if the outdoor ambient temperature is 80°F, then the condenser saturation temperature should be roughly 110°F. If the condenser is a high-efficiency model, the condenser saturation temperature may be as low as 20 degrees higher than the outdoor ambient temperature. In this case, if the outdoor ambient temperature is 80°F, then the condenser saturation temperature can be as low as 100°F. On warmer days, the system's operating pressures will be higher.

The desired high-side pressure can be determined by using the pressure temperature relationship for the system refrigerant. For an R-410A air conditioning system, the high-side pressure should be about 365 psig for a standard condenser and approximately 320 psig for a higher-efficiency condenser when the outside ambient temperature is 80°F. If the ambient temperature is 90°F, the condenser saturation temperature will be about 120°F and the expected high-side pressure will be about 446 psig.

Going Green

Improperly charged systems use more energy and cool less effectively. Operating an overcharged system can cause excess moisture accumulation in the home leading to accelerated mold growth rates.

LOW-SIDE PRESSURE

On an air conditioning system that is operating properly, the temperature of the evaporator coil should always be above the freezing point. The design of the evaporator coil typically has the fins of the evaporator very close together to provide a larger heat transfer surface (Fig. 13-2). Should frost begin to accumulate on the surface of the coil, it can easily become blocked with ice. This will greatly reduce the amount of airflow through the coil, resulting in reduced system effectiveness and efficiency.

In an air conditioning system operating with R-401A as its refrigerant, the coil may start to freeze when the pressure in the evaporator reaches approximately 101 psig. This pressure corresponds to a temperature of 32°F on the pressure/temperature chart. Ideally, the temperature of an air conditioning evaporator coil is approximately 40°F, which corresponds to an evaporator pressure of 118 psig. When the outside ambient temperature is high, the operating pressures in the system will also be higher, so the evaporator pressure will be somewhat higher than 118 psig. Conversely, if an air conditioning system is operated when the ambient

from experience...

When starting up an air conditioning system, be sure to use a high-quality thermometer to ensure that the temperature readings being taken are accurate. Since the refrigerant charge is based on the temperature conditions that exist around the system, inaccurate readings can result in an improperly charged system.

Delmar/Cengage Learning.

FIGURE 13-2 Fins on an air conditioning evaporator are spaced close together to maximize heat transfer.

temperatures are low, the operating pressures will also be lower, increasing the chances of a coil freeze-up resulting from the low evaporator coil temperature. When initially starting up an air conditioning system, the evaporator saturation temperature may very well be below the freezing point, but this will change once the charge has been properly adjusted.

TEMPERATURE DIFFERENTIAL ACROSS THE EVAPORATOR COIL

Another factor that will help determine the refrigerant charge in an air conditioning system is the **evaporator temperature differential**. The evaporator temperature differential is the difference between the temperature of the air being supplied to the occupied space and the temperature of the air returning to the air handler. For example, if the occupied space (return air) is at a temperature of 80°F and the air being supplied to the space (supply air) is 61°F, the temperature differential is 80°F–61°F, or 19°F (Fig. 13-3). To ensure accurate readings, these temperature readings should be taken just before and just after the evaporator coil.

At design conditions of 75°F indoor air temperature and 50% relative humidity, the difference between the supply and return air temperatures should be in the range of 17 to 20 degrees. A temperature differential lower than 17 degrees indicates that the system *may* need to have more refrigerant added. A temperature differential of more than 20 degrees indicates that refrigerant *may* need to be removed from the system. Since the evaporator performs the functions of dehumidifying and cooling, a portion of the evaporator capacity is used to remove moisture, a latent process, from the air passing through the evaporator coil. Because of this, we can determine that the relative humidity has an effect on the evaporator's temperature differential.

If the relative humidity of the indoor air is in the 50% range, then the 17 to 20 degree differential is typical. During periods of excessively high humidity, more of the evaporator capacity will be used for dehumidification, so the evaporator temperature differential will be lower. Similarly, during periods of low humidity, evaporator temperature differentials greater than 20 degrees can be expected.

For step-by-step instructions on measuring the temperature differential across the evaporator coil, see Procedure 13-3 on page 326.

EVAPORATOR SUPERHEAT

Earlier in the text, methods for determining both evaporator superheat and condenser subcooling were discussed. These two factors, in addition to those just discussed, will help the technician determine if the refrigerant charge needs to be adjusted. On systems operating with thermostatic expansion valves, evaporator superheat in the range of 8 to 12 degrees is considered to be well within the desired operating range. An evaporator superheat well below 8 degrees is an indication that the system may be overcharged, while an excessively high superheat indicates that the system may be undercharged. Improper superheat can also be caused by other system problems such as reduced evaporator airflow.

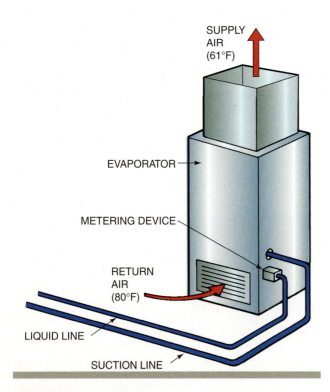

SUPPLY AIR (61°F)

EVAPORATOR

METERING DEVICE

RETURN AIR (80°F)

LIQUID LINE

SUCTION LINE

FIGURE 13-3 The evaporator temperature differential is the difference between the temperature of the supply air and the temperature of the return air. In this example, the evaporator temperature differential is 19 degrees.

from experience...

Certain conditions, such as turning on a system when the heat load is very high, can result in higher than average superheat readings on systems equipped with a metering device other than a thermostatic expansion valve.

For step-by-step instructions on calibrating the gauges and measuring evaporator and system superheat, see Procedure 13-1 on pages 322–323.

CONDENSER SUBCOOLING

To calculate the amount of subcooling that a condenser is operating with, two pieces of information are needed:

- The condenser outlet temperature and
- The condenser saturation temperature

To calculate the condenser subcooling, we subtract the condenser outlet temperature from the condenser saturation temperature. Condenser subcooling should ideally be in the range of 15 to 20 degrees. Always refer to manufacturers' literature and charging tables when starting up a system. The manufacturers' recommended subcooling levels should be used when available instead of general industry guidelines. High subcooling is an indication that the system is overcharged, and low subcooling indicates that the system is undercharged. Subcooling by itself cannot be used to determine if the system is properly charged. For example, low head pressure and low subcooling indicates that the high side of the system has a deficiency of refrigerant, but indicates nothing about the low side of the system. Similarly, high head pressure and high subcooling indicate that there is an excess of refrigerant in the high side of the system, but we know nothing about the low side of the system. To determine if the system charge is correct, other information such as low-side pressure and evaporator superheat are also needed.

Should the condenser coil become blocked or the condenser fan motor become inoperable, the refrigerant in the condenser coil will not be able to condense effectively, causing the head pressure to rise (an indication of an excess of refrigerant) and the subcooling measurement to drop below acceptable levels. Refer to the section on system charging for more on condenser subcooling.

For step-by-step instructions on measuring the subcooling in an air conditioning system, see Procedure 13-2 on pages 324–325.

SYSTEM CHARGING

Ensuring that the system charge is correct is probably the single most important task that the startup technician must perform. The individual who performs the charging process must have a thorough understanding of system operation and the operating conditions that indicate whether a system is properly charged, overcharged, or undercharged.

Most split-type air conditioning systems come from the factory with a holding charge of refrigerant contained within the condensing unit. This holding charge is measured into the unit and is *often* the proper amount of refrigerant for a system with 25 feet of piping between the air handler and condensing unit. If more than 25 feet of piping is used, additional refrigerant will have to be added to the system, and if less than 25 feet is used, refrigerant will need to be removed.

In addition to knowing the amount of refrigerant that comes shipped within the system, knowledge of the system installation, evaporator superheat, condenser subcooling, and temperature differentials across the evaporator coil are also important in ensuring that the system charge is correct upon leaving the job site.

from experience...

Always consult the literature that accompanies the new equipment. Some manufacturers ship condensing units with a holding charge of nitrogen, not refrigerant. In such cases, the nitrogen charge must be released, and the total system refrigerant charge must be added to the system. In addition, some manufacturers add a refrigerant holding charge that corresponds to a piping length that differs from the 25-foot example given.

SYSTEM OVERCHARGE

During the startup process, it may be determined that the system contains too much refrigerant. This may be the result of shorter refrigerant lines than the system's holding charge was intended for or a result of the technician adding too much refrigerant to the system. Symptoms of a system overcharge may include

- high head pressure
- high suction pressure
- high condenser subcooling
- excessively low superheat

- high condenser saturation temperature
- high evaporator saturation temperature
- high temperature differential across the evaporator coil

If the refrigerant charge is higher than intended, most of these conditions will be present. If the system is equipped with a thermostatic expansion valve, the super-heat in the evaporator will remain relatively constant, regardless of whether or not the system is charged properly. If the system is equipped with another type of metering device, the evaporator superheat will be lower than desired if there is an overcharge of refrigerant.

REMOVING EXCESS REFRIGERANT FROM THE SYSTEM

In the event that the system has an overcharge of refrigerant, the excess must be removed in accordance with the laws set forth by the Environmental Protection Agency (EPA). Refrigerant must be removed from the system and stored in a DOT-approved cylinder (Fig. 13-4).

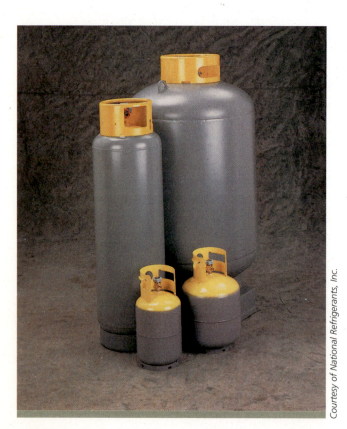

FIGURE 13-4 DOT-approved refrigerant recovery tanks.

Courtesy of National Refrigerants, Inc.

The following guidelines should be kept in mind when removing refrigerant from a system:

- Wear safety glasses when removing refrigerant from a system.
- Remove refrigerant in 4-ounce increments.
- Remove refrigerant from the high side of the system.
- Make certain that the recovery cylinder or tank that is being used to store the refrigerant has been properly evacuated prior to use.
- Give the system ample time to settle after each 4-ounce removal.
- Reevaluate the system temperatures and pressures after each removal.
- Continue to remove refrigerant until the desired temperatures and pressures are reached.

For step-by-step instructions on removing refrigerant from an air conditioning system, see Procedure 13-4 on pages 327–328.

Going Green

Excess system refrigerant must be recovered from the system, never vented to the atmosphere!

SYSTEM UNDERCHARGE

On air conditioning system installations in which the refrigerant lines are longer than the refrigerant holding charge or where the condensing unit is not shipped with a holding charge of refrigerant, additional refrigerant will have to be added to the system during the startup process. Symptoms of a system undercharge may include

- low head pressure
- low suction pressure
- low condenser subcooling
- excessively high superheat
- low condenser saturation temperature
- low evaporator saturation temperature
- low temperature differential across the evaporator coil

TECH TOOL BELT • TECH TOOL BELT

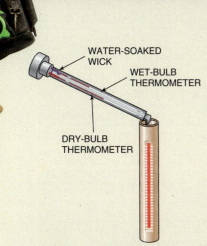

WATER-SOAKED WICK

WET-BULB THERMOMETER

DRY-BULB THERMOMETER

Sling Psychrometer

The sling psychrometer is used to obtain both wet-bulb and dry-bulb temperature readings of the air. The sling psychrometer has two thermometers. One of the thermometers is kept dry, and the bulb of the other thermometer is wrapped with a wet sock or piece of cloth. The dry thermometer measures the temperature of the air; the wet thermometer measures the temperature taking the humidity level into account. When used, the psychrometer is spun in the air for a period of time before the thermometer readings are taken. During periods of low humidity, the water from the sock will evaporate quickly, removing heat from the wet-bulb thermometer. This will cause the wet-bulb reading to drop. During periods of high humidity, the moisture will not evaporate and the wet-bulb temperature reading will be very close to that of the dry-bulb. The wet-bulb temperature is used with manufacturer's charging tables as well as when evaluating system operation and performance.

Photo by Eugene Silberstein.

Digital Thermometer

Digital thermometers are used to take temperature readings at strategic points in an air conditioning system as well as the surrounding air. They provide accurate temperature readings, which help ensure that the charging and troubleshooting processes are performed correctly.

Courtesy of Robinair.

Refrigerant Charging Scale

The refrigerant scale determines the weight of a refrigerant tank or cylinder. It is most useful in determining how much refrigerant is added or removed from an air conditioning system. In addition, the scale is used to determine how much refrigerant a recovery tank contains, helping to ensure that the Department of Transportation (DOT) cylinders are not overfilled.

ADDING REFRIGERANT TO THE SYSTEM

If refrigerant is to be added, it should be done in a slow, deliberate manner. The following guidelines should be followed when adding refrigerant:

- Add refrigerant to the system in vapor form, whenever possible.

- If a blended refrigerant is used, introduce it slowly to the system as a liquid.

- Add refrigerant in 4-ounce increments.

- Allow the system to settle for approximately 5 to 10 minutes after each addition.

- Reevaluate the pressures and temperatures before adding more refrigerant.

- Continue adding refrigerant in 4-ounce increments until the desired pressures and temperatures are reached.

For step-by-step instructions on adding refrigerant to an air conditioning system, see Procedure 13-5 on pages 329–330.

see Procedure 13-5 on pages 329–330.

from experience...

Blended refrigerants must be added to the system in the liquid state. Blended refrigerants are simply two or more refrigerants that are blended together to form another refrigerant. Charging a blended refrigerant in the vapor state will cause the refrigerants to separate, thereby changing its characteristics. Refer to the chapter on refrigerants and refrigerant management for more on blended refrigerants.

Outdoor Temp (°F)	Indoor Coil Entering Air (F) WB													
	50	52	54	56	58	60	62	64	66	68	70	72	74	76
55	9	12	14	17	20	23	26	29	32	35	37	40	42	45
60	7	10	12	15	18	21	24	27	30	33	35	38	40	43
65	—	6	10	13	16	19	21	24	27	30	33	36	38	41
70	—	—	7	10	13	16	19	21	24	27	30	33	36	39
75	—	—	—	6	9	12	15	18	21	24	28	31	34	37
80	—	—	—	—	5	8	12	15	18	21	25	28	31	35
85	—	—	—	—	—	—	8	11	15	19	22	26	30	33
90	—	—	—	—	—	—	5	9	13	16	20	24	27	31
95	—	—	—	—	—	—	—	6	10	14	18	22	25	29
100	—	—	—	—	—	—	—	—	8	12	15	20	23	27
105	—	—	—	—	—	—	—	—	—	5	9	13	17	22	26
110	—	—	—	—	—	—	—	—	—	—	6	11	15	20	25
115	—	—	—	—	—	—	—	—	—	—	—	8	14	18	23

Courtesy of Carrier Corporation.

FIGURE 13-5 Superheat charging table.

CHARGING THE SYSTEM USING MANUFACTURERS' CHARGING TABLES AND CHARTS

Quite often, the equipment manufacturer provides charging tables or charts to aid in the charging process. These tables enable the startup technician to properly charge the system by taking and interpreting temperature and pressure readings at strategic points in *and around* the system. These readings often relate to the system's superheat or subcooling and, in most cases, take the humidity or moisture content of the air into consideration. Systems with fixed-bore metering devices utilize superheat charging tables, while systems with thermostatic expansion valves utilize subcooling tables.

CHARGING WITH MANUFACTURERS' SUPERHEAT TABLES (FIXED-BORE SYSTEMS)

When the manufacturer recommends the superheat method of charging the system, two charts may be supplied with the system (Fig. 13-5 and Fig. 13-6). To use these tables, the system should be permitted to operate for a minimum of 15 minutes before temperature and pressure readings are taken. The required readings for these charts are:

- Outdoor air temperature (dry bulb)
- Indoor air temperature (wet bulb)
- Suction pressure
- Suction line temperature

Superheat Temp (°F)	Suction Pressure at Service Port (psig)								
	61.5	64.2	67.1	70.0	73.0	76.0	79.2	82.4	85.7
0	35	37	39	41	43	45	47	49	51
2	37	39	41	43	45	47	49	51	53
4	39	41	43	45	47	49	51	53	55
6	41	43	45	47	49	51	53	55	57
8	43	45	47	49	51	53	55	57	59
10	45	47	49	51	53	55	57	59	61
12	47	49	51	53	55	57	59	61	63
14	49	51	53	55	57	59	61	63	65
16	51	53	55	57	59	61	63	65	67
18	53	55	57	59	61	63	65	67	69
20	55	57	59	61	63	65	67	69	71
22	57	59	61	63	65	67	69	71	73
24	59	61	63	65	67	69	71	73	75
26	61	63	65	67	69	71	73	75	77
28	63	65	67	69	71	73	75	77	79
30	65	67	69	71	73	75	77	79	81
32	67	69	71	73	75	77	79	81	83
34	69	71	73	75	77	79	81	83	85
36	71	73	75	77	79	81	83	85	87
38	73	75	77	79	81	83	85	87	89
40	75	77	79	81	83	85	87	89	91

Courtesy of Carrier Corporation.

FIGURE 13-6 Suction line temperature table for R-22.

from experience...

Each manufacturer has its own preferred method for ensuring that the proper refrigerant charge is introduced to the system. Be sure to follow the directions that each manufacturer provides.

Using the chart in Figure 13-5, the outdoor air temperature and the indoor air temperature, determine the superheat measurement. This is done by locating the outdoor temperature in the column on the left side of the chart and the indoor temperature across the top of the chart. The point at which these two temperatures cross is the superheat. This temperature is then used in the second table (Fig. 13-6), along with the suction pressure reading. Locate the superheat in the column on the left of the chart and the suction pressure at the top of the chart. The point at which these two measurements cross is the desired suction-line temperature. Compare the temperature on the chart with the actual suction-line temperature of the system. If the actual suction line temperature is higher than the temperature obtained from the chart, refrigerant must be added to the system until the temperatures are the same. If the actual suction line temperature is lower than the temperature obtained from the chart, refrigerant must be removed from the system until the charted temperature is reached.

The chart in Figure 13-6 performs the superheat calculations for systems charged with R-22. If a superheat chart is not available, the calculation can be made manually. For example, if the suction pressure of an R-22 system is 70 psig and the superheat is 10°F, the suction line temperature can be calculated by determining the saturation temperature of R-22 at 70 psig, which is 41°F. At 10 degrees superheat the temperature of the suction line will be 51°F,

from experience...

When removing refrigerant from the system, a DOT-approved recovery cylinder must be used. Refer to the chapter on refrigerant management for more information on the removal of refrigerant from a system.

Superheat Temp (°F)	Suction Pressure at Service Port (psig)								
	105	110	115	118	120	125	130	135	140
0	34	36	38	40	41	43	45	47	49
2	36	38	40	42	43	45	47	49	51
4	38	40	42	44	45	47	49	51	53
6	40	42	44	46	47	49	51	53	55
8	42	44	46	48	49	51	53	55	57
10	44	46	48	50	51	53	55	57	59
12	46	48	50	52	53	55	57	59	61
14	48	50	52	54	55	57	59	61	63
16	50	52	54	56	57	59	61	63	65
18	52	54	56	58	59	61	63	65	67
20	54	56	58	60	61	63	65	67	69
22	56	58	60	62	63	65	67	69	71
24	58	60	62	64	65	67	69	71	73
26	60	62	64	66	67	69	71	73	75
28	62	64	66	68	69	71	73	75	77
30	64	66	68	70	71	73	75	77	79
32	66	68	70	72	73	75	77	79	81
34	68	70	72	74	75	77	79	81	83
36	70	72	74	76	77	79	81	83	85
38	72	74	76	78	79	81	83	85	87
40	74	76	78	80	81	83	85	87	89

Delmar/Cengage Learning.

FIGURE 13-7 Suction line temperature table for R-410A.

which corresponds to the value in the chart. A similar superheat table for R-410A is provided in Figure 13-7.

CHARGING WITH MANUFACTURERS' SUBCOOLING TABLES (TXV SYSTEMS)

When the manufacturer recommends the subcooling method of charging the system, a table similar to that in Figure 13-8 is used. To charge a system in this manner, the high-side pressure of the system and the temperature of the liquid line at the outlet of the condensing unit are needed. The manufacturer will normally specify the desired amount of subcooling for the system. By locating the high-side pressure on the left side of the chart and the desired subcooling across the top of the chart, the proper liquid-line temperature can be found at the point where these two values intersect. If the liquid-line temperature is higher than the value indicated in the chart, refrigerant needs to be added to the system. If the liquid-line temperature is lower than the value in the chart, refrigerant must be removed from the system.

If a subcooling chart is not available, the subcooling can be determined by converting the high-side pressure to the saturation temperature from a pressure/temperature

Pressure (psig) at Service Fitting	Required Subcooling Temperature (°F)					
	0	5	10	15	20	25
134	76	71	66	61	56	51
141	79	74	69	64	59	54
148	82	77	72	67	62	57
156	85	80	75	70	65	60
163	88	83	78	73	68	63
171	91	86	81	76	71	66
179	94	89	84	79	74	69
187	97	92	87	82	77	72
196	100	95	90	85	80	75
205	103	98	93	88	83	78
214	106	101	96	91	86	81
223	109	104	99	94	89	84
233	112	107	102	97	92	87
243	115	110	105	100	95	90
253	118	113	108	103	98	93
264	121	116	111	106	101	96
274	124	119	114	109	104	99
285	127	122	117	112	107	102
297	130	125	120	115	110	105
309	133	128	123	118	113	108
321	136	131	126	121	116	111
331	139	134	129	124	119	114
346	142	137	132	127	122	117
359	145	140	135	130	125	120

Courtesy of Carrier Corporation.

FIGURE 13-8 Subcooling chart for R-22.

Pressure at liquid line service port (psig)	Required Subcooling Temperature (°F)					
	0	5	10	15	20	25
274	90	85	80	75	70	65
295	95	90	85	80	75	70
317	100	95	90	85	80	75
340	105	100	95	90	85	80
365	110	105	100	95	90	85
390	115	110	105	100	95	90
420	120	115	110	105	100	95
440	124	120	115	110	105	100
446	125	124	120	115	110	105
475	130	125	124	120	115	110
507	135	130	125	124	120	115
539	140	135	130	125	124	120
573	145	140	135	130	125	124
609	150	145	140	135	130	125
616	155	150	145	140	135	130

Delmar/Cengage Learning.

FIGURE 13-9 Subcooling chart for R-410A.

chart and then subtracting the condenser outlet temperature from this value. For example, if the high-side pressure of an R-22 system is 223 psig, the condenser saturation temperature is 109°F. If the desired subcooling is 10 degrees, the liquid-line temperature should be 10 degrees less than 109°F, or 99°F. This corresponds to the value in the chart (Fig. 13-8). A similar subcooling table for R-410A is provided in Figure 13-9.

EDUCATING THE CUSTOMER

Once the installation has been completed and the equipment has been put into operation, the customer should be instructed about the key features and operation of the system. It should not be assumed that the customer knows how the system works or how the thermostat is programmed. Here is a list of items to review with the homeowners:

- Keep duct registers clear of furnishings. This will allow the system to maintain the proper airflow rates, keeping the system operating effectively and efficiently.
- How to clean/replace filters. Keeping air filters clean helps increase system efficiency.
- Supply-register adjustments. Instruct occupants not to fully close supply-registers.
- Programmable thermostat settings and operation. Make certain the customer knows how to set the thermostat. Better yet, program it for them and make certain the instruction booklet is left for them.
- System maintenance schedule. Explain the importance of periodic preventive maintenance and service.

A reference manual of the HVAC systems should be compiled so owners can refresh their memories and so the information is carried forward to future owners. The HVAC installer should draft up an instruction manual for custom portions of the HVAC systems in the home. Things to include are

- locations of duct balancing dampers and what each services
- importance of not changing the balancing damper settings
- how to check the condensate drain and pump for proper operation

Save and include manufacturers' printed installation instructions, information, and warrantee forms for the equipment in the reference manual. The documentation will be a reference for the present and future owners.

Teach the owners how their behavior affects energy efficiency and indoor air quality and train them on practices that reduce/improve both, such as

- program the thermostat to lower temperature at night and during the day when no one is home during the winter
- program the thermostat to higher temperature at night and during the day when no one is home during the summer
- leave supply- and return-registers clear of obstructions
- keep doors and windows shut during the heating and cooling seasons
- use bathroom and kitchen exhaust fans properly
- clean or change filters at the specified intervals

SUMMARY

- System evacuation helps remove moisture and noncondensable gases from a system prior to startup.
- Vacuum pump oil should be changed before each use.

- Vacuum levels are most accurately measured in microns.
- Prior to initial startup, all aspects of the installation should be inspected.
- Airflow through the condenser and evaporator coils must be correct before adjusting the refrigerant charge.
- System operating pressures and temperatures should be monitored carefully during startup.
- Superheat and subcooling measurements help determine if the refrigerant charge needs to be adjusted.
- Symptoms of a system overcharge include high operating pressures and temperatures, high subcooling, and a high temperature differential across the evaporator.
- Symptoms of a system undercharge include low operating pressures and temperatures, low subcooling, and a low temperature differential across the evaporator.
- Refrigerant should be added or removed in 4-ounce increments.
- The system must be allowed to settle between refrigerant charge adjustments.

GREEN CHECKLIST

- [ ] **Properly charged air conditioning systems operate more effectively and efficiently.**

- [ ] **Overcharged systems and properly charged systems do not dehumidify.**

- [ ] **Overcharged systems cycle on and off more frequently, using more electrical power.**

- [ ] **If an air conditioning system is overcharged, the excess refrigerant must be recovered; never vented to the atmosphere.**

- [ ] **Use a programmable thermostat to cycle the unit on less frequently when the home is not occupied.**

- [ ] **Keep windows and doors closed when air conditioning equipment is operating.**

- [ ] **Clean or change air filters at regular intervals.**

PROCEDURE 13-1

Measuring the Superheat in an Air Conditioning System

The following procedures are used to measure the superheat in an air conditioning system:

- Turn on the power to the air conditioning system and allow it to operate for approximately 15 minutes.

 Using a high-quality digital thermometer, take the temperature reading of the suction line at the inlet of the condensing unit.

- Remove the low side hose from the blank port on the gauge manifold and make certain that the gauge is calibrated.

- Calibrate gauge if needed. Refer to the procedure for gauge calibration in Chapter 12.

 Connect the low side hose from the gauge manifold onto the low side service port on the condensing unit of the system.

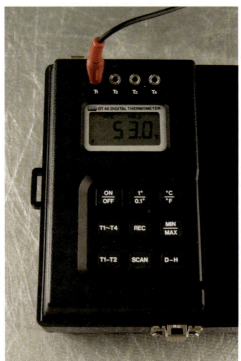

Delmar/Cengage Learning.

Photo by Eugene Silberstein.

C With the system operating, obtain the evaporator saturation temperature from the gauge.

D Subtract the evaporator-saturation temperature from the suction-line temperature to obtain the system superheat.

C

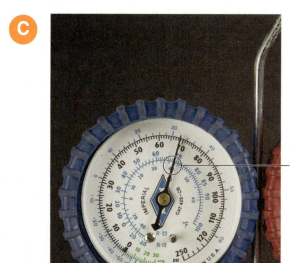

ELEVATOR SATURATION TEMPERATURE

Delmar/Cengage Learning.

D

$$\begin{array}{r} 53°F \\ -\ 40°F \\ \hline 13°F \end{array}$$

PROCEDURE 13-2

Measuring the Subcooling in an Air Conditioning System

The following procedures are used to measure the subcooling in an air conditioning system:

- Turn on the power to the air conditioning system and allow it to operate for approximately 15 minutes.

 Using a high-quality digital thermometer, take the temperature reading of the liquid line at the outlet of the condensing unit.

 Remove the high side hose from the blank port on the gauge manifold and make certain that the gauge is calibrated.

- Calibrate gauge if needed. Refer to the procedure for gauge calibration in Chapter 12.

- Connect the high side hose from the gauge manifold onto the high side service port on the condensing unit of the system.

Delmar/Cengage Learning.

Photo by Eugene Silberstein.

C With the system operating, obtain the condenser-saturation temperature from the gauge.

D Subtract the liquid-line temperature from the condenser-saturation temperature to obtain the system subcooling.

C

Delmar/Cengage Learning.

D

105°F
− 90°F
_____
15°F

PROCEDURE 13-3

Measuring the Temperature Differential Across the Evaporator Coil

The following procedures are used to measure the temperature differential across the evaporator coil:

- Turn on the power to the air conditioning system and allow it to operate.

A Using a high-quality digital thermometer, take the temperature reading of the air in the return duct just before it reaches the evaporator coil.

B Using another thermometer, take the temperature reading of the air in the supply duct after it has passed through the evaporator coil.

C Subtract the temperature of the supply air from the temperature of the return air to obtain the temperature differential across the evaporator coil.

 A

Delmar/Cengage Learning.

 B

Delmar/Cengage Learning.

C

$$79°F$$
$$-60°F$$
$$19°F$$

PROCEDURE 13-4

Removing Refrigerant from an Operating Air Conditioning System

The following procedures are used to remove refrigerant from an operating air conditioning system:

- Turn on the power to the air conditioning system and allow it to operate for approximately 15 minutes.

 Properly evacuate a DOT-approved refrigerant cylinder.

- Connect the center hose of the gauge manifold to the valve on the refrigerant cylinder and place the cylinder on the refrigerant scale.

 Open the high side valve on the gauge manifold. This will push refrigerant from the high side of the system through the center hose.

 Loosen the center hose connection on the refrigerant cylinder for about 2 seconds to purge any air from the center hose.

- Tighten the center hose connection on the refrigerant cylinder.

- Close the high side valve on the gauge manifold.

- Turn the refrigerant scale on.

Courtesy of White Industries.

Delmar/Cengage Learning.

Delmar/Cengage Learning.

PROCEDURE 13-4

Removing Refrigerant from an Operating Air Conditioning System (Continued)

D Push the reset or zero button on the scale and hold it down until the scale reads zero.

- Open the valve on the refrigerant cylinder.

- Slowly open the high side valve on the gauge manifold. Refrigerant will flow from the system into the refrigerant cylinder.

- Monitor the refrigerant scale.

E When the scale indicates that 4 ounces of refrigerant has been removed from the system, close the high side valve on the gauge manifold.

- Allow the system to operate for approximately 5 to 10 minutes and re-evaluate the system charge.

- If the system is still overcharged, remove another 4 ounces of refrigerant from the system.

- Continue removing refrigerant from the system until the operating temperatures and pressures are within the desired range.

D

Delmar/Cengage Learning.

E

Delmar/Cengage Learning.

CAUTION

CAUTION: Never fill a refrigerant cylinder or tank to more than 80% of its capacity. For example, tanks that are designed to hold 50 pounds of refrigerant should never be filled with more than 40 pounds of refrigerant.

PROCEDURE 13-5

Adding Refrigerant to an Air Conditioning System

The following procedures are used to add refrigerant to an air conditioning system:

- Turn on the power to the air conditioning system and allow it to operate for approximately 15 minutes.

A Connect the center hose of the gauge manifold to the valve on the refrigerant tank.

B Place the tank on the refrigerant scale and open the valve on the refrigerant tank.

C Loosen the center hose connection on the gauge manifold for about 2 seconds to purge any air from the center hose.

- Tighten the center hose connection on the gauge manifold.

- Turn the refrigerant scale on.

A

Photo by Eugene Silberstein.

B

Photo by Eugene Silberstein.

C

Photo by Eugene Silberstein.

PROCEDURE 13-5

Adding Refrigerant to an Air Conditioning System (Continued)

D Push the reset or zero button on the scale and hold it down until the scale reads zero.

E Open the low side valve on the gauge manifold. This will introduce refrigerant to the system.

- Monitor the refrigerant scale.

- When the scale indicates that 4 ounces of refrigerant has been introduced to the system, close the low side valve on the gauge manifold.

- Allow the system to operate for approximately 5 to 10 minutes and re-evaluate the system charge.

- If the system is still undercharged, introduce another 4 ounces of refrigerant into the system.

- Continue adding refrigerant to the system until the operating temperatures and pressures are within the desired range.

D

Delmar/Cengage Learning.

E

Photo by Eugene Silberstein.

REVIEW QUESTIONS

1. **If an air conditioning system comes precharged with enough refrigerant for an installation that requires 25-foot refrigerant lines,**

   a. Refrigerant will have to be added to the system if the actual length of the refrigerant lines is 20 feet

   b. Refrigerant will have to be added to the system if the actual length of the refrigerant lines is 35 feet

   c. The refrigerant charge will be correct as long as the actual length of the refrigerant lines is less than 25 feet

   d. The refrigerant charge will be correct as long as the actual length of the refrigerant lines is more than 50 feet

2. **Airflow through the condenser coil can be restricted if**

   a. The filter is dirty

   b. The fan motor on the outdoor unit is not operating

   c. Both a and b are correct

   d. Neither a nor b is correct

3. **Airflow through the evaporator coil can be reduced if**

   a. The filter is dirty

   b. The discharge pressure is higher than desired

   c. The suction pressure is lower than desired

   d. Both a and c are correct

4. **Frost on an evaporator coil will**

   a. Make the system operate more efficiently

   b. Cause the suction pressure to drop

   c. Cause the discharge pressure to rise

   d. All of the above are correct

5. **A high-efficiency system will have a condenser-saturation temperature of 100°F when the outside ambient temperature is approximately**

   a. 80°F

   b. 90°F

   c. 100°F

   d. 120°F

6. **A 25-degree temperature differential across an evaporator may indicate that**

   a. The system is undercharged

   b. The system is overcharged

   c. There is excessive evaporator superheat

   d. Both b and c are correct

7. **As refrigerant is added to an air conditioning system,**

   a. The suction pressure will rise and the subcooling will increase

   b. The discharge pressure will drop and the subcooling will increase

   c. The suction pressure will drop and the subcooling will drop

   d. The discharge pressure will increase and the subcooling will drop

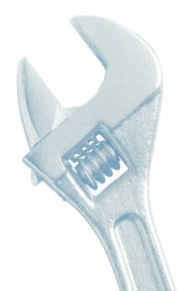

8. **Excessive superheat and a small temperature differential across the evaporator coil are indications that**

   a. The system is overcharged

   b. No noncondensable gases are present in the system

   c. The thermostatic expansion valve is overfeeding the evaporator coil

   d. The system is undercharged

9. **Very low superheat and a large temperature differential across the evaporator coil are indications that**

   a. The system is overcharged

   b. Noncondensable gases are present in the system

   c. The thermostatic expansion valve is overfeeding the evaporator

   d. Both a and c are possible

10. **Upon starting up an R-22 air conditioning system, it is found that refrigerant needs to be added to the system. The best way to introduce new refrigerant to the system is to**

    a. Introduce vapor refrigerant to the high side of the system

    b. Introduce liquid refrigerant to the high side of the system

    c. Introduce liquid refrigerant to the low side of the system

    d. Introduce vapor refrigerant to the low side of the system

11. **When removing refrigerant from an overcharged system**

    a. The excess refrigerant should be removed in 4-ounce increments

    b. The excess refrigerant will be stored in the refrigerant receiver and therefore does not need to be removed from the system

    c. It is quickest to remove vapor from the low side of the system

    d. The system should not be operating to prevent compressor damage from occurring

12. **Once the refrigerant charge has been corrected, the technician should**

    a. Check all modes of operation before leaving the job

    b. Make certain that all debris and rubbish is removed from the site

    c. Take down all pertinent system data including operating temperatures and pressures

    d. All of the above are correct

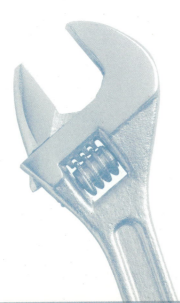

Electricity for HVAC

Electricity Fundamentals

OBJECTIVES

Upon completion of this chapter, the student should be able to

- Describe the structure of the atom.

- Differentiate among electrons, protons, and neutrons.

- Differentiate between conductors and insulators.

- Explain the relationship between magnetism and electricity.

- Explain the electrical quantities of voltage, current, resistance, and power.

- Use Ohm's law to calculate electrical quantities.

- Explain the difference between alternating and direct current.

- Use electric meters to obtain voltage, current, and resistance readings in an electric circuit.

- Identify series and parallel circuits.

- Explain how voltage, current, and resistance behave in series circuits.

- Explain how voltage, current, and resistance behave in parallel circuits.

- Identify electrical components in wiring diagrams.

- Interpret information on schematic and ladder wiring diagrams.

- Determine wire sizes using a wire gauge.

- Explain how fuses, circuit breakers, and GFCIs protect electric circuits.

- Identify single- and three-phase power supplies.

- Determine whether a three-phase power supply was configured as a wye or as a delta.

Nearly 85% of all air conditioning system failures are electrical in nature, so the successful air conditioning technician has not only a solid understanding of refrigeration theory, but electrical theory as well. Electricity is literally the driving force behind system compressors, fans, pumps, heaters, and other components.

Electrical theory begins at the atomic level, and it is here that our discussion will begin. In this chapter, we will cover atomic structure and the atomic properties that enable some substances to become good conductors of electricity, while other substances act as insulators. We will then go on to discuss the electrical characteristics of voltage, current, resistance, and power that help define a particular circuit, as well as Ohm's law, which is the relationship that exists among these characteristics.

Once we have established a general understanding of electric theory, we will examine simple electric circuits and wiring diagrams. Using Ohm's law and circuit theory, we will be able to calculate values of electrical characteristics in both series and parallel circuits.

American wire gauge (AWG) Provides data, including size and characteristics, for various types of electric conductors

Amperes or amps Unit that indicates the amount of current that flows in an electric circuit when 1-volt is applied to a 1-ohm resistance

Analog meter Piece of electrical test equipment that uses a moving needle to indicate the value of electrical characteristics

Atom The smallest quantity of a naturally occurring element

Conductors Materials that have one or two electrons in their outermost shell and facilitate the free flow of electrons

Continuity The term used to indicate that a complete current path exists

Current The amount of electron flow through a conductor

Digital meters Pieces of electrical test equipment that use internal electronic circuits to take voltage, current, and resistance readings; these readings are shown on a liquid crystal display

Electromotive force (EMF) See voltage

GLOSSARY OF TERMS (CONT'D)

Electrons The negatively charged components of an atom

Factory wiring Wiring installed at the factory at the time of system manufacture

Field wiring System wiring that is installed by field technicians or installers

Inductive loads Electric devices that primarily generate a magnetic field

Inductive reactance The additional resistance in an ac circuit that results from the constant building up and collapsing of magnetic fields

Insulators Materials that have a stable atomic structure and do not permit the free flow of electrons from atom to atom

Ladder diagram See line diagram

Legend The portion of a wiring diagram that identifies the abbreviations used in the diagram

Line diagram Wiring diagram that is configured with each circuit on a separate line; used to facilitate effective circuit troubleshooting

Lines of force Fields of force generated by magnets and magnetism

Load Any electric component that uses or consumes electric power

Magnetism The force exerted by a magnetic field; see lines of force

Meters See multimeters

Multimeters Pieces of electrical test equipment that measure a combination of voltage, amperage, and resistance

National Electrical Code® (NEC) A publication that sets the standards for electrical installations

Neutrons The neutrally charged components of an atom

Open circuit The term used when there is no available path for electric current to take

Protons The positively charged components of an atom

Resistance Opposition to current flow

Resistive loads Any electric devices that primarily generate light or heat

Schematic diagram Wiring diagram that provides a representation of every wire in the circuit as well as wire color-coding and electric component location

Short circuit A circuit that offers zero resistance to current flow

Switches Electric devices that open or close electric circuits, starting and stopping current flow

Valence electrons Electrons in the outermost shell of an atom

Voltage The amount of electrical pressure in an electric circuit that causes electron flow

Voltage drop The voltage reading between two points in an electric circuit

Voltmeter Piece of electrical test equipment that measures the voltage or potential difference between two points in an electric circuit

Watt Unit of electrical power that causes 1 amp to flow in a circuit that is powered by 1 volt dc

ELECTRICAL THEORY

For many, electricity is a difficult concept to understand, given the fact that, for the most part, it cannot be seen, smelled, touched, heard, or tasted. However, for components such as compressors, lights, fans, and heaters to operate, electricity must be supplied to them. To keep these items operating properly, the technician must understand electrical theory and have a working knowledge of how electricity behaves under different conditions. The study of electricity begins at the atomic level, as electricity is defined as the flow of electrons, one of the component parts of atoms, through a conducting medium.

ATOMIC THEORY

The **atom** is the basic building block for all matter and is the smallest quantity of any naturally occurring element. If, for example, a quantity of oxygen gas were to be continually reduced by half, eventually there would be a single atom of oxygen. When two atoms of oxygen combine, an oxygen molecule is formed (Fig. 14-1). When three oxygen molecules combine, a molecule of ozone is formed (Fig. 14-2).

At the most basic level, the atom is made up of three component parts: protons, electrons, and neutrons. **Protons** have a positive charge, **electrons** have a negative charge, and **neutrons** have a neutral charge. An atom typically has the same number of electrons as protons and is said to have a neutral charge. Protons and neutrons are located in the center, or nucleus, of the atom, while the

electrons orbit around the nucleus (Fig. 14-3). The atomic model is often expressed in a manner similar to our solar system, where the sun represents the nucleus and the planets that orbit around the sun represent the electrons.

Atoms of different elements have different numbers of electrons, protons, and neutrons. An atom of hydrogen, for example (Fig. 14-4) has one electron and one proton, while an atom of copper (Fig. 14-5) has

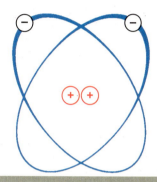

FIGURE 14-3 The electrons orbit around the nucleus.

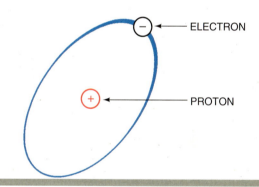

FIGURE 14-4 Hydrogen has only one electron.

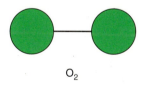

O_2

STANDARD OXYGEN MOLECULE

FIGURE 14-1 Standard oxygen molecule.

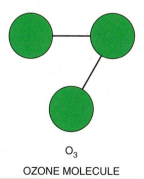

O_3

OZONE MOLECULE

FIGURE 14-2 A molecule of ozone.

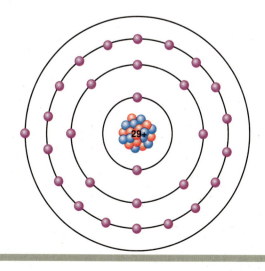

FIGURE 14-5 Copper has 29 electrons.

29 electrons and 29 protons. Both of these atoms are neutral since the number of protons and the number of electrons are the same. If the copper atom were to lose an electron, there would be more protons than electrons and it would have a net positive charge. The copper atom would then want to attract another electron so that it could become neutral again.

ELECTRON ORBITS

Electrons are located in a number of orbits, or shells, that surround the nucleus. Each shell can hold up to a specific number of electrons. The first shell, the one closest to the nucleus, can hold a maximum of two electrons. The next shell can hold a maximum of eight electrons, the third can hold a maximum of eighteen electrons, and so forth (Fig. 14-6). The formula to determine the number of electrons that can occupy any given shell is $2N^2$, where N represents the shell number. For the first shell we would have $2(1)^2 = 2$ electrons. For the second shell we would have $2(2)^2 = 8$ electrons, and so forth.

Ideally, an atom will have full shells, making the atom very stable. Atoms that do not have full shells are less stable and will give up or accept electrons in an effort to obtain full shells. For example, an atom of hydrogen (Fig. 14-4) has one electron in its first orbit. This orbit can hold two electrons, so it will react easily with other atoms in an effort to obtain another electron to fill the first shell. Similarly, the copper atom in Figure 14-5 has only one electron in the fourth orbit and will readily give up that electron to have a full outer orbit.

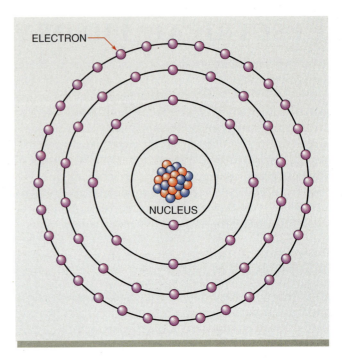

FIGURE 14-6 Electron orbits.

LAW OF CHARGES

The law of charges states that like charges repel each other, while opposite charges attract each other. Using our knowledge of electrons and protons, we can conclude that if two negatively charged electrons were placed close to each other, they would repel each other, just as two positively charged protons would (Fig. 14-7). Conversely, if an electron and a proton were placed together, they would be attracted to each other (Fig. 14-8). We can then see how

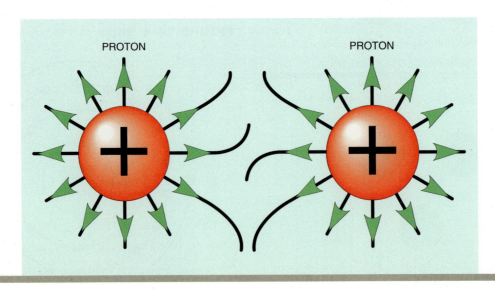

FIGURE 14-7 Like charges repel each other.

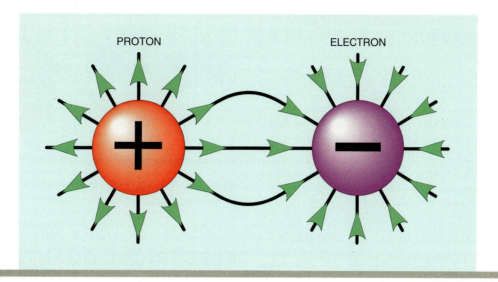

FIGURE 14-8 Opposite charges attract each other.

a negatively charged electron would be attracted to the positively charged copper atom that had lost an electron.

CONDUCTORS

Electrons in the outermost shell of an atom are called **valence electrons**. Atoms with one or two valence electrons, as in the case of copper, are quick to give them up when a small amount of force is exerted on the atom. Materials that have such a property are referred to as conductors. **Conductors**, by definition, have free electrons that can easily travel from one atom to another. If an electron is displaced from an atom, it will travel to another atom and displace another electron. The valence electron in the atom will be displaced and, in turn, displace the valence electron of another atom (Fig. 14-9). This process results in a flow of electrons from atom to atom.

Because electricity is actually the flow of electrons, materials that allow the free flow of electrons are desired when electric current flow is needed. This is why copper wire is typically used to construct electric circuits and motor windings. Gold and silver are also very good conductors but are more expensive than copper, making them less popular choices for use in circuits and electric components.

INSULATORS

Materials whose atoms have full outer shells do not facilitate the free flow of electrons. These materials have very stable atomic structures, do not release electrons easily, and are called **insulators**. Glass, rubber, plastic, and wood are good insulators. Copper wires used in electric circuits

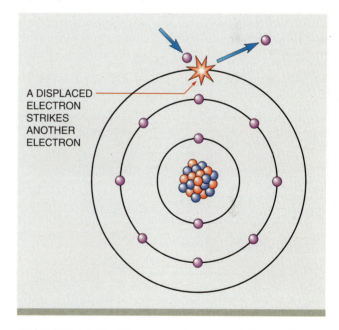

A DISPLACED ELECTRON STRIKES ANOTHER ELECTRON

FIGURE 14-9 A loose electron can strike another electron and knock it out of orbit.

are often encased in rubber or plastic to prevent the wire from coming in contact with other conducting materials. The casing on the wire is called insulation because it insulates the conductor from other conductors.

MAGNETISM

The force needed to start and maintain electron flow through a conductor is often generated by **magnetism**. Although other methods can be used to generate electricity, such as light, heat, friction, and pressure, we will concern ourselves primarily with magnetism, which is the most common.

Just as like charges repel each other and opposite charges attract each other, the poles of magnets also repel or attract one another. Magnets have two poles: a north pole and a south pole. When the north pole of a magnet is placed near the north pole of another magnet, the two forces will repel each other [Fig. 14-10(A)]. However, if the north pole of a magnet is placed near the south pole of another magnet, they will attract one another [Fig. 14-10(B)]. Magnets have fields of force, referred to as **lines of force**, which create the attraction and repulsion between the poles (Fig. 14-11). The strength of a particular magnet is measured by the strength of these lines of force.

When a conductor is passed between two magnets, it cuts through the lines of force that exist between them (Fig. 14-12). It is this cutting of the lines of force that creates current flow in the conductor. Increasing the strength of the magnets will increase the amount of current flow. Forming the conductor into a coil with many turns will also increase the amount of current flow generated when it cuts the lines of force.

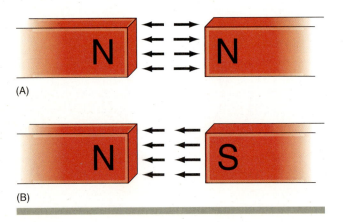

(A)

(B)

FIGURE 14-10 (A) Like poles of a magnet repel each other. (B) Opposite poles of a magnet will repel each other.

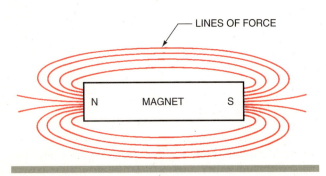

LINES OF FORCE

N   MAGNET   S

FIGURE 14-11 Lines of force around a magnet.

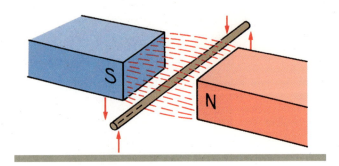

FIGURE 14-12 Moving a conductor through a magnetic field will generate electricity.

ELECTRICAL QUANTITIES

It has already been established that electricity is commonly defined as the flow of electrons through a conductor. To more fully understand this concept, it is important to obtain a working knowledge of other factors that affect the flow of electrons, which ultimately causes motors, pumps, heaters, lights, and other devices to operate.

Consider a bathroom sink. When the faucet is closed, no water flows, but we all know that there is water pressure just before the faucet. With the valve in the closed position, resistance has been created, preventing the water from passing through. Once the faucet is opened, water will flow into the sink as the resistance to flow has been removed. The same concept can easily be applied to electricity.

Consider a standard wall outlet. We know that electric current is not shooting out from the wall, but when we plug a light or other device into the outlet, it is energized. The wall outlet has potential just as the closed faucet has, but no current can flow until there is a path for the current to take. By plugging a light into the outlet and turning the switch on, the resistance is reduced, allowing current to flow and the light to light. This relates directly to our example of opening the faucet, permitting water to flow. Electrically speaking, the potential that exists in a circuit is called **voltage**, the amount of electron flow is called **current**, and, finally, we have **resistance**, which reduces the amount of current flow.

VOLTAGE

Voltage, described as the potential to do work, is also referred to as **electromotive force**, or EMF. It can also be described as the potential difference between two points. For example, a standard electric outlet supplies 110 volts, which is the potential difference between the hot leg and the neutral leg (Fig. 14-13). The more voltage supplied to the circuit, the greater the current flow, assuming all

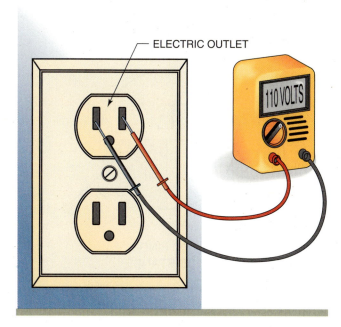

FIGURE 14-13 The potential difference across a residential wall outlet is 110 volts.

other factors are unchanged. Voltage is measured in *volts,* and is denoted by the letter *V* or by the letter *E,* which stands for electromotive force.

CURRENT

The current in an electric circuit is measured in **amperes,** or simply amps (A) and measures the rate of electron flow in a circuit per unit time. Current is denoted by the letter *I,* which stands for the *intensity* of electric current. The current in a circuit is determined by two factors: the voltage supplied to the circuit and the resistance of the circuit. As previously stated, the amount of current flow will increase as the voltage is increased and will decrease as the amount of voltage is decreased. This follows logically if we think of the voltage as the electrical pressure that pushes the electrons through a conductor. More pressure will result in more electron movement, and less pressure will result in less movement.

RESISTANCE

Resistance can be expressed as any material that impedes current flow. As the resistance increases, the amount of current that is permitted to flow will decrease. As the resistance is decreased, the amount of current flow will increase. Electrically speaking, good conductors of electricity have very low resistance, while insulators have very high resistance. Resistance (R) is measured in ohms (Ω). One ohm is defined as the amount of resistance that permits one ampere of current to flow in a circuit when one volt is supplied.

OHM'S LAW

A very predictable relationship exists among voltage, current, and resistance. From the previous three sections, we can conclude the following:

With all other factors remaining unchanged

- the current in a circuit will increase as the voltage is increased
- the current in a circuit will increase as the resistance is decreased
- the current in a circuit will decrease as the voltage is decreased
- the current in a circuit will decrease as the resistance is increased

The relationship that exists among voltage, current, and resistance is called Ohm's law:

$$E = I \times R$$

In this formula *E* represents voltage, *I* represents current and *R* represents resistance. We can see that the voltage in an electric circuit is equal to the current flow in the circuit times the resistance of the circuit. Two other forms of Ohm's law are

$$I = E \div R$$

$$R = E \div I$$

The first formula states that the current in a circuit is equal to the voltage supplied to the circuit divided by the resistance of the circuit. The second states that the resistance of a circuit is equal to the voltage supplied to the circuit divided by the current flow in the circuit. Ohm's law is shown graphically in Figure 14-14, where *E* is at the top and *R* and *I* are on the bottom. This is

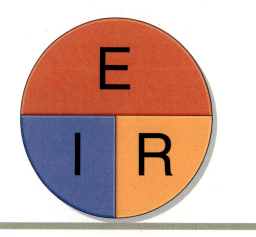

FIGURE 14-14 Graphic representation of Ohm's law.

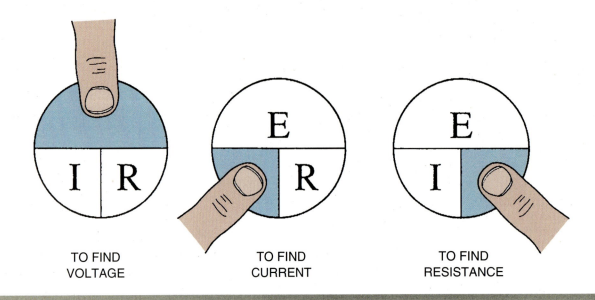

TO FIND
VOLTAGE

TO FIND
CURRENT

TO FIND
RESISTANCE

FIGURE 14-15 To determine the formula for an unknown quantity, cover the letter of the unknown quantity.

useful in recalling the three forms of Ohm's law. If you are trying to find the formula for voltage, all you need to do is cover that letter on the diagram (Fig. 14-15). By covering the E, you can see that the I and the R are next to each other, reminding you to multiply the current and the resistance. Similarly, to find the formula for current, cover the I. You will see that the E is over the R so you must divide the voltage by the resistance.

EXAMPLE 1

From the circuit in Figure 14-16, we can see that there is a current of 3 A flowing through the light that has a resistance of 10Ω. We want to find the voltage that is being supplied to the circuit.

Solution

From the given information, we know the following:

$$I = 3\text{ A}$$

$$R = 10\ \Omega$$

Using the $E = I \times R$ form of Ohm's law we get

$$E = I \times R = 3\text{ A} \times 10\ \Omega = 30\text{ V}$$

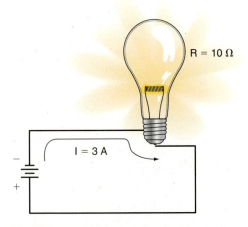

R = 10 Ω

I = 3 A

FIGURE 14-16 Series circuit with a resistance of 10 ohms and a current of 3 amps.

EXAMPLE 2

From the circuit in Figure 14-17, we can see that there is a voltage of 100 V being supplied to the circuit and that the light

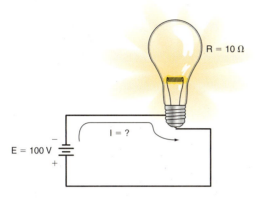

FIGURE 14-17 Series circuit with a voltage of 100 volts and a resistance of 10 ohms.

has a resistance of 10 Ω. We want to find the amount of current that is flowing in the circuit.

Solution

From the given information, we know the following:

$$E = 100 \text{ V}$$

$$R = 10 \text{ Ω}$$

Using the I = E ÷ R form of Ohm's law we get

$$I = E ÷ R = 100 \text{ V} ÷ 10 \text{ Ω} = 10 \text{ A}$$

EXAMPLE 3

From the circuit in Figure 14-18, we can see that a voltage of 100 V is being supplied to the circuit and a current of 5 A is flowing through the light. We want to find the resistance of the circuit.

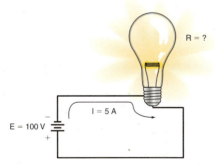

FIGURE 14-18 Series circuit with a voltage of 100 volts and a current of 5 amps.

It is left as an exercise for the reader to confirm that the resistance of the circuit is equal to 20 Ω.

POWER

Power, in general, is defined as the rate at which work is done. Electrically speaking, power (P) is directly related to the voltage, current, and resistance of an electric circuit and is measured in watts (W). One **watt** is the amount of power consumed when a one-volt dc power source causes a current of one amp to flow in a circuit. The relationship between power, voltage, and current can be expressed as

$$P = I \times E$$

Power Formulas and Ohm's law. Since the formula for power is directly related to the voltage and current in an electric circuit, we can incorporate the power formula into Ohm's law to obtain new formulas for power. Since P = I × E and E = I × R, we can substitute to get

$$P = I \times E = I \times (I \times R) = I^2R$$
$$P = I^2R$$

Similarly, using the form of Ohm's law I = E ÷ R, we can substitute to get

$$P = I \times E = (E ÷ R) \times E = E^2 ÷ R$$
$$P = E^2 ÷ R$$

A summary of Ohm's law and the power formulas for dc circuits is shown in Figure 14-19. There are twelve

EXAMPLE 4

How much power is being consumed by a dc circuit that has a supply voltage of 30 volts and a current flow of 2 amps?

Solution

Using the formula $P = I \times E$ and substituting the given values, we get

$$P = I \times E = 2\,A \times 30\,V = 60\,W$$

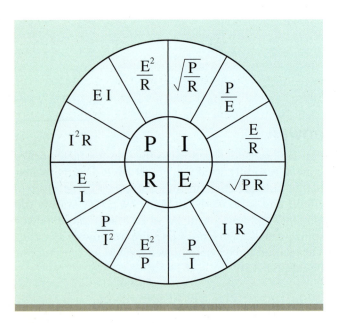

FIGURE 14-19 Chart showing resistance, voltage, current, and power formulas.

formulas altogether: three for power, three for voltage, three for current, and three for resistance. If any two of the four quantities, *E, P, R,* or *I,* are known, the other two can be calculated from the formulas.

DIRECT CURRENT

Direct current, often referred to simply as dc, is the power that is supplied primarily by batteries, a chemical source of electricity. The main characteristic of direct current circuits is that electric current flows in only one direction. Since current flow takes place only when there is a difference in potential between two points, one point must have a surplus of electrons and the other a deficiency.

Two theories exist regarding current flow in dc circuits. The first, referred to as *electron theory,* states that current in dc circuits flows in a direction from negative to positive, or from a point where there are more electrons to a point where there are fewer electrons. An excess of electrons at a given point will result in a charge that is more negative at that point than anywhere else in the circuit. When a conductive path is provided, the excess electrons will flow through the circuit and back to the power source, where the electron deficiency exists (Fig. 14-20). The other theory, referred to as *conventional current flow theory,* states that the direction of current flow is from positive to negative or from a point of higher potential to a point of lower potential (Fig. 14-21). Such is the case with a car battery. The positive terminal of the battery is considered to be *hot,* while the negative terminal is grounded to the chassis of the automobile. The symbol at the left in Figure 14-16 and Figure 14-17 represents the battery, which is the power source for these dc circuits.

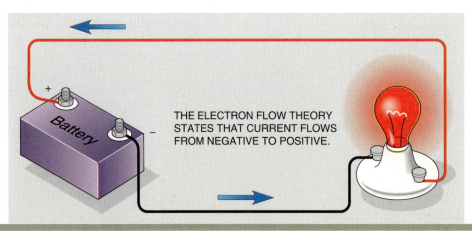

THE ELECTRON FLOW THEORY STATES THAT CURRENT FLOWS FROM NEGATIVE TO POSITIVE.

FIGURE 14-20 Direction of electron flow.

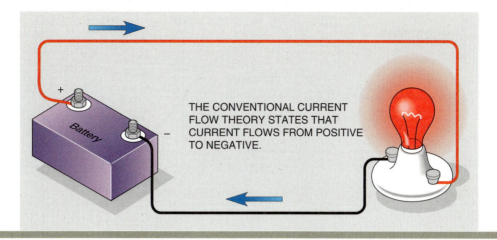

FIGURE 14-21 Conventional direction of current flow.

Since electric current is defined as the flow of electrons through a conductor, we will assume that the first theory, *electron theory*, is correct. However, for calculation purposes, the direction of the current flow in a given circuit does not matter and will not affect our results.

Going Green

Some batteries are intended to be recharged while others are disposable. Use rechargeable batteries whenever possible to reduce waste. When throwing away dead disposable batteries, be sure to dispose of them properly to minimize environmental contamination.

ALTERNATING CURRENT

Alternating current, commonly referred to as ac current, is the power source found in most residences and is much more common in the air conditioning industry than is dc current. Alternating current constantly changes its direction, unlike dc current that flows in only one direction. The changing of direction results from the manner in which the voltage is generated. To generate ac current, a conductor is rotated within a magnetic field. As the conductor is rotated within the field, it is exposed to magnetic fields of varying strengths and polarities. When the conductor is at point "A" in Figure 14-22, the conductor is moving in a direction parallel to the lines of force and the voltage that is generated is equal to zero. The same holds true when the conductor is at point "C." When the conductor is at point "B," it cuts through the maximum number of lines of force and generates the maximum voltage. The same is true at point "D," but the polarity is opposite to that at point "B."

It can then be seen that, as the conductor rotates within the magnetic field, the amount of voltage generated constantly increases and decreases. Figure 14-23 shows a sine wave, which represents the amount of voltage that is generated at each point as the conductor rotates. At 0 degree, the sine wave is indicating that the voltage generated is zero. As the conductor moves, the voltage increases until a maximum has been reached at point "B" or 90 degrees. At point "C" or 180 degrees, the voltage is again zero. As the conductor rotates from point "C" to point "D," the voltage generated again increases to a maximum at 270 degrees, but the polarity, or sign, has been reversed. From point "D" to point "A," the voltage generated again falls to zero. Alternating current will continue to be generated as long as the conductor rotates within the magnetic field.

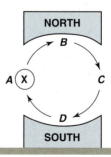

FIGURE 14-22 Conductor rotating within a magnetic field.

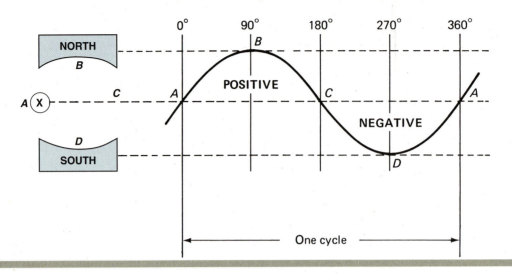

FIGURE 14-23 The voltage generated increases and decreases, following a sine-wave pattern.

Because the voltage generated by an ac power source varies continually, Ohm's law, as originally discussed, does not hold for ac circuits and power sources. This is because there is additional resistance added to the circuit called inductive reactance. **Inductive reactance** is created when the magnetic field constantly builds up and collapses as the alternating current is generated. Since the polarity of the generated ac power is changing constantly, there is often opposition within the circuit, creating an internal resistance of sorts.

from experience...

An additional component, called the power factor, must be factored into the calculations in order for Ohm's law to hold when working with ac circuits. This power factor ranges in value from 0 to 1, depending on the type of circuit and the type of load in the circuit.

MEASURING INSTRUMENTS

Having discussed the electrical characteristics of voltage, current, resistance, and the types of power, ac and dc, that can be generated, we will next discuss the test

instruments that are used in the field to measure these quantities. Some of these test instruments, called **meters**, can measure voltage, some measure amperage, some measure resistance, and most have the ability to measure more than one quantity or all three. Meters that can measure resistance, current, and voltage are called **multimeters**. These pieces of test equipment can be either analog or digital. No matter what the style or configuration of the meter, it is definitely one of the most useful tools in the technician's toolbox, as nearly 85 percent of all air conditioning system failures are electrical in nature.

Most electric meters are able to sense different levels of voltage, current, and resistance by the magnetic field that is generated by current flow. Whenever current flows through a conductor, a magnetic field is generated around the conductor (Fig. 14-24). As the amount of current increases, so does the strength of the magnetic field. This magnetic field, in turn, causes the needle of the meter to move. Consider the example in which a compass needle is placed next to a conductor through which no current is flowing (Fig. 14-25). The compass needle

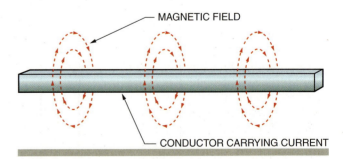

FIGURE 14-24 A magnetic field is generated around a conductor when there is current flowing through it.

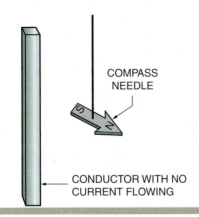

FIGURE 14-25 The compass needle reacts only to the magnetic field of the earth when placed near a conductor that carries no current.

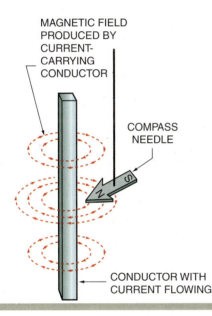

FIGURE 14-26 The compass needle reacts to the magnetic field generated around the conductor when current is flowing.

Delmar/Cengage Learning.

FIGURE 14-27 An analog meter.

Delmar/Cengage Learning.

FIGURE 14-28 Checking voltage with a voltmeter.

will point north as all compass needles do. When current passes through the conductor, though, the magnetic field generated around the conductor causes the compass needle to move (Fig. 14-26).

VOLTMETERS

The voltmeter (Fig. 14-27) is designed to measure the potential difference between two points in an electric circuit. The meter is equipped with two test leads or probes that the technician uses to connect the meter to the circuit at the desired points (Fig. 14-28). Analog voltmeters have internal circuitry that includes resistance with a known value. When a potential difference is sensed between the

test leads, current flows through the internal circuits of the meter, generating a magnetic field (Fig. 14-29A) and causing the needle to move. If the potential is increased, the meter's needle will move more, registering a higher voltage (Fig. 14-29B).

Voltmeters can read both ac and dc voltages. However, the meter must first be set to read the desired voltage range and type. If, for example, one needed to check the voltage at a typical wall receptacle or outlet, the voltmeter would be set to read ac voltage. Since the voltage supplied by the wall outlet is approximately 120 volts, the meter should be set to a scale that is higher than the expected reading of 120 volts. The meter in Figure 14-30 has been set to an ac voltage scale of 200 Vac (alternating current voltage) which is higher than the maximum expected voltage. If, on the other hand, one needed to check the voltage in a dc circuit, the meter would be set to an appropriate Vdc (direct current voltage) scale.

FIGURE 14-29 (A) The voltage reading is low because the generated magnetic field is weak. (B) This voltage reading is higher due to the stronger magnetic field.

FIGURE 14-30 Meter set to the 200-volt scale to read the voltage at a wall outlet.

AMMETERS

The ammeter is an electric meter that measures the amount of current flow in a circuit. The most common type is the clamp-on ammeter (Fig. 14-31). To use it, the user must first set the meter to the appropriate current scale and then clamp the meter around one conductor in the circuit to obtain the current reading. When the ammeter is clamped around the conductor, the meter senses the strength of the magnetic field being generated by the current flow and registers the appropriate reading. Higher current flow through the circuit will result in more needle movement, and lower current flows result in less meter needle movement.

When very low amperages need to be measured, it is very difficult to obtain accurate readings from the display of the meter. To obtain more accurate readings, it

FIGURE 14-31 A clamp-on meter. Clamp the jaws of the meter around only one conductor.

FIGURE 14-32 The 10-wrap method of measuring low amperages.

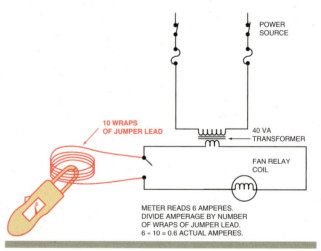

10 WRAPS OF JUMPER LEAD

POWER SOURCE

40 VA TRANSFORMER

FAN RELAY COIL

METER READS 6 AMPERES. DIVIDE AMPERAGE BY NUMBER OF WRAPS OF JUMPER LEAD. 6 ÷ 10 = 0.6 ACTUAL AMPERES.

FIGURE 14-33 The measured amperage is divided by 10 to get the actual current.

is beneficial to create a coil of wire with ten turns from one of the circuit conductors (Fig. 14-32). By clamping the meter around the entire coil, the reading obtained by the meter will be increased by a factor of ten (Fig. 14-33). The meter shown in Figure 14-33 displays a current reading of 6 amps, which is ten times the actual current in the circuit because the conductor was wound around the jaws of the clamp-on meter ten times. The actual current flow in this circuit is 0.6 amps. Most newer ammeters have the ability to read low amperages without having to create a coil.

> ## *from experience...*
>
> Never clamp the ammeter around two or more conductors at the same time when trying to obtain current readings. Doing so will cause the magnetic fields to cancel each other out, leading to inaccurate readings. Remember: Clamp around only one conductor at a time!

OHMMETERS

The ohmmeter determines the amount of resistance that a certain circuit or system component offers to current flow. Although the actual resistance value is sometimes needed, most often the field technician is only concerned with whether or not there is continuity in the circuit. **Continuity** is the term used when there is a path for electric current to take between the two points being tested. If there is continuity between two points and potential is applied across these points, electric current will flow. If there is no continuity, electric current will not flow. When there is no continuity in the circuit or through the component being tested, the resistance is infinite. A circuit that has infinite resistance is said to be an **open circuit**. In a **short circuit**, on the other hand, which exists when there is no resistance to current flow, the current flow will be infinite. This is a dangerous situation and circuits are protected to prevent this from occurring.

CAUTION

CAUTION: The ohmmeter is intended for use on circuits that are not energized. Using the ohmmeter on powered circuits can result in damage to the test instrument. Most meters have an internal fuse that will protect the meter from such a condition.

The ohmmeter is equipped with an internal power supply, a battery, which provides the potential to test the desired circuit or component. As in the case of the

FIGURE 14-34 A bad fuse.

voltmeter, two test leads are used to check the resistance of a circuit or component. The ohmmeter in Figure 14-34 is being used to test a fuse. The needle on the ohmmeter has not moved, indicating that the resistance is infinite and no current can flow through the device. A good fuse will have zero resistance to current flow; therefore, since there is no continuity through the fuse, the fuse is bad. Figure 14-35 shows how a good fuse will register on an ohmmeter.

ANALOG METERS

All of the concepts that have been discussed so far are assuming that the meters are of the analog variety. **Analog meters** are easily identified by the face or dial of the meter as well as a needle that moves to indicate the measured circuit values (Fig. 14-36). When using analog meters, the user must estimate the obtained readings, leading to less accurate measurements. The accuracy of

FIGURE 14-35 This fuse is good.

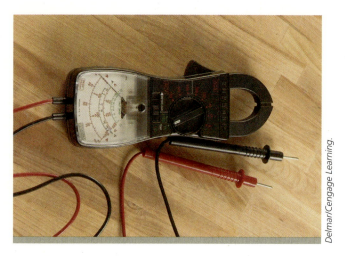

FIGURE 14-36 Dial of an analog meter.

Delmar/Cengage Learning.

FIGURE 14-38 The OL indicates that the meter is out of range or that there is no continuity through the component being checked.

Photo by Eugene Silberstein.

analog meter readings is estimated to be within 2 percent of the actual electrical characteristic, or 98 percent accurate.

DIGITAL METERS

Digital electric meters have become the meter of choice, given their increased accuracy and the fact that the cost of such test equipment has been reduced a great deal as technological advances have been made. The accuracy of digital meters is said to be approximately 99 percent. Digital meters are identified by their liquid crystal displays (Fig. 14-37). They are equipped with an overload feature that protects the meter from internal damage resulting from exposing the meter to excessive voltages or currents. For example, when the meter experiences a voltage that is higher than the setting on the meter, the meter's display will indicate OL, for overload (Fig. 14-38). Digital meters also display the OL when there is no continuity or infinite resistance through a circuit or component.

from experience...

Since different meters are equipped with different features, always refer to the usage manual that is supplied with the meter. Using the meter improperly or for tasks it is not intended to perform will result in damage to the device.

MULTIMETERS

Most meters in field use today are designed to obtain voltage, current, and resistance readings. These meters, called multimeters, can be either analog or digital devices. The multimeter is often referred to simply as a VOM (volt-ohm-milliampere) meter. Since there are so many different types of meters on the market with varying capabilities in measuring voltage, amperage, and current, it is important to do research and locate the meter that will best suit your professional needs.

ELECTRIC CIRCUITS

Electric circuits are the roadmaps that electric current follows to control the operation of system components such as compressors, fans, blowers, heaters, lights, and

FIGURE 14-37 Digital voltmeter.

Photo by Eugene Silberstein.

other electrical devices. There are four component parts of a circuit:

- The power source
- A path for the electric current to take
- The load
- A means of switching or controlling the load

The power source, either ac or dc, supplies the needed voltage or potential to initiate current flow in the circuit. Without a power source, no electric current can flow. For current to flow, there must also be an uninterrupted path for the current to take. If there is no continuity in the circuit, there is not a complete path and, therefore, no current will flow, even if voltage is supplied to the circuit.

The device or component that is ultimately powered by the circuit is called a load. The **load** is defined as any electric component that consumes power. Examples of electric loads are motors, lights, and heaters. Loads can be either resistive loads or inductive loads. **Resistive loads** primarily generate light or heat or a combination of both, while **inductive loads** are characterized by the generation of a magnetic field. An electric motor is an example of an inductive load.

Finally, there must be some means to control the operation of the load. **Switches** are devices that open or close electrical contacts to either stop or start the flow of electric current through the load. Switches do not consume power as in the case of a load and are referred to as pass-through devices. Ideally, electric current should pass through them without affecting the resistance, voltage, or current in the circuit. Switches can be controlled manually or by pressure, temperature, fluid level, air or fluid movement, or a number of other factors.

When these four elements come together, a circuit is formed. Two types of circuits will be discussed: the series circuit and the parallel circuit. A third type, the combination circuit, is made up of both series and parallel circuits.

SERIES CIRCUITS

A series circuit is characterized by the fact that there is only one path for electric current to take. All of the circuits that have been shown thus far in this chapter are series circuits. Three characteristics of series circuits regarding current, resistance, and voltage include

- the current is the same at all points in a series circuit
- the total resistance of a series circuit is the sum of all resistances in the circuit
- the total voltage supplied to the circuit is equal to the sum of all voltage drops in the circuit

Since there is only one path for current to take in a series circuit, the current flow must be constant at every point in the circuit. Consider the example of a highway that has no exits or on ramps. No cars can enter the roadway or leave it, so the number of cars on the road will always be the same, and the rate of traffic flow will always be the same as well. When taking the current reading of a series circuit, the ammeter can be clamped around any conductor in the circuit to obtain the measurement. Consider the series circuit in Figure 14-39. Ammeter #1 reads 5 amps, which is the same as the readings obtained at different points in the circuit by ammeter #2, ammeter #3, and ammeter #4. In general this can be expressed as

$$I_T = I_1 = I_2 = I_3 = I_4$$

where I_T is the total current in the circuit and I_1, I_2, I_3, and I_4 are the current values at various points in the circuit.

In a series circuit, the total resistance that current flow will experience is equal to the sum of all resistances in the circuit. Consider the example, once again, of the highway that has no on ramps or exit ramps. If there are accidents at three different locations on the road, any car on the road will have to pass each of them. Of course there will be a slight traffic jam at each accident site, causing the traffic flow to be reduced. The total amount of time lost by having to pass these accidents will be the amount of time lost at the first accident plus the amount of time lost at the second plus the amount of time lost at the third. When calculating the total resistance of a series circuit, simply add the values of all resistances in the circuit. Consider the circuit in Figure 14-40. In this circuit, there are four resistances with values of 5 Ω, 10 Ω, 10 Ω, and 20 Ω. The total resistance of this circuit

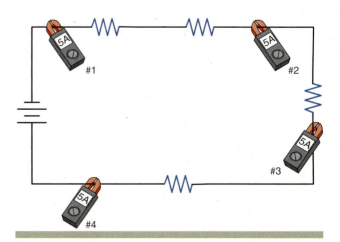

FIGURE 14-39 The current is the same at all points in a series circuit.

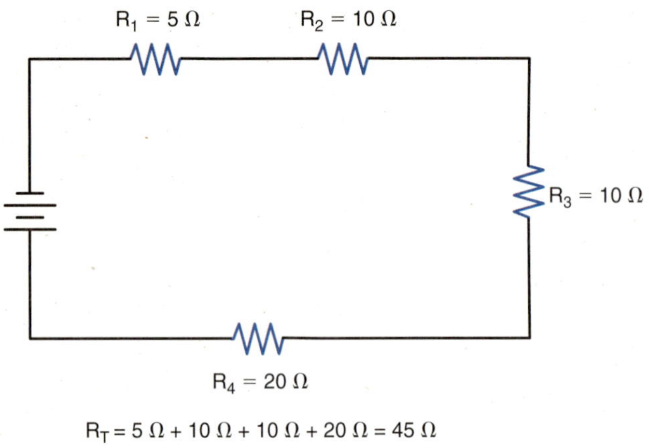

$R_1 = 5\ \Omega$ $R_2 = 10\ \Omega$

$R_3 = 10\ \Omega$

$R_4 = 20\ \Omega$

$R_T = 5\ \Omega + 10\ \Omega + 10\ \Omega + 20\ \Omega = 45\ \Omega$

FIGURE 14-40 Calculating resistance in a series circuit.

will then be 45 Ω, which is the sum of the values of the four resistances. In general, this can be expressed as

$$R_T = R_1 + R_2 + R_3 + R_4 + \ldots + R_N$$

where "N" is the number of resistances in the circuit.

The voltage supplied to a series circuit will be divided among all of the individual resistances in the circuit. If the resistances of all of the loads in the circuit are equal to each other, the amount of voltage supplied to each will be the same. Consider the circuit in Figure 14-41. This circuit has two identical light bulbs. The voltage supplied to this circuit is 120 volts, so the voltage will divide equally between the two bulbs (Fig. 14-42). The measured voltage across each bulb is 60 volts. The circuit in Figure 14-43 has three identical light bulbs connected in series. The voltage supplied to each bulb will be one-third of the 120 volt supply, or 40 volts (Fig. 14-44). The bulbs in Figure 14-44 will be dimmer than the bulbs in

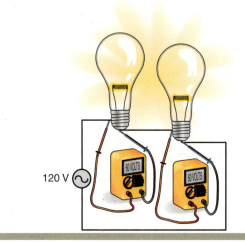

FIGURE 14-42 The voltage supplied to the circuit is divided equally between the two bulbs.

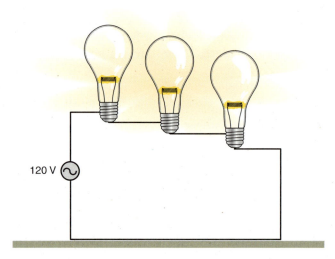

FIGURE 14-43 Series circuit with three identical light bulbs.

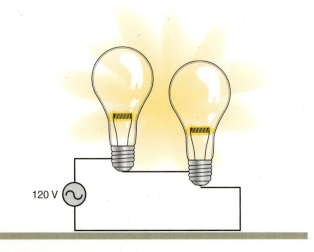

FIGURE 14-41 Series circuit with two identical light bulbs.

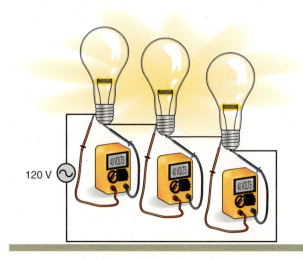

FIGURE 14-44 The voltage supplied to the circuit is divided equally among the three bulbs.

Figure 14-42 because the voltage supplied to each bulb in Figure 14-44 is lower and the current flow in that circuit is lower as well, given the higher resistance. In general, the total voltage in a series circuit can be expressed as

$$E_T = E_1 + E_2 + E_3 + E_4 + \ldots + E_N$$

where E_T is the voltage supplied to the circuit and $E_1 + E_2 + E_3 + E_4$, etc. are the voltage readings across each resistance in the circuit. The term **voltage drop** is often used to refer to the voltage reading between two points in a circuit or across a component in an electric circuit. For example, the *voltage drop* across each bulb in Figure 14-44 is 40 volts.

EXAMPLE 5

Consider the circuit in Figure 14-45. From the information provided, determine the total amount of current that is flowing in the circuit.

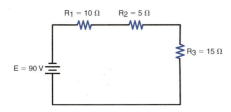

FIGURE 14-45 Series circuit with known resistance and voltage.

Solution

From the circuit, we can determine that the voltage supplied to the circuit is 90 volts. We can also calculate the total resistance of the circuit by adding the individual resistances. The total resistance is 10 Ω + 5 Ω + 15 Ω = 30 Ω. Using the form of Ohm's law

$$I = E \div R$$

we get

$$I = E \div R = \textbf{90 volts} \div \textbf{30 ohms}$$
$$= \textbf{90} \div \textbf{30} = \textbf{3 amps}$$

The total current flow through this circuit is 3 amps.

PARALLEL CIRCUITS

A parallel circuit (Fig. 14-46) is characterized by the fact that there is more than one path for electric current to take. Three characteristics of parallel circuits regarding current, resistance, and voltage include

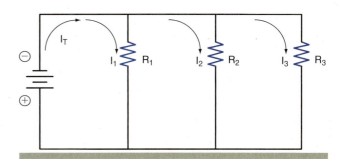

FIGURE 14-46 A parallel circuit.

EXAMPLE 6

Using the same circuit, from Figure 14-45, determine the voltage drop across each of the three resistors.

Solution

From the previous example, we know that there is a current of 3 amps flowing through each of the three resistors. The voltage drop across each resistor can be found using the general form of Ohm's law—E = I × R.

For R_1, we can determine the voltage drop by using $E_1 = I \times R_1 = 3 \times 10$ = 30 volts.

For R_2, we can determine the voltage drop by using $E_2 = I \times R_2 = 3 \times 5$ = 15 volts.

For R_3, we can determine the voltage drop by using $E_3 = I \times R_3 = 3 \times 15$ = 45 volts.

We can check our answers by adding the three voltage drops:

30 volts + 15 volts + 45 volts = 90 volts, which is equal to the voltage supplied to the circuit.

- the total current in a parallel circuit is equal to the sum of the currents in each branch of the circuit
- the total resistance of a parallel circuit is always less than the value of the lowest resistance in the circuit
- the voltage supplied to each branch of a parallel circuit is the same as the voltage supplied to the circuit

CURRENT FLOW IN PARALLEL CIRCUITS

Since there are multiple paths for current to take in a parallel circuit, the amount of current flow through an individual branch circuit may very well be different from the current flow through another branch circuit. In the circuit shown in Figure 14-46, we can see that there are three resistances connected in parallel with the current flow through them denoted as I_1, I_2, and I_3. The total current is shown in the figure as I_T. If we trace the circuit from the negative terminal of the power supply, we have I_T flowing towards the three resistances. When the first branch circuit is reached, an amount of current flow equal to I_1 travels through R_1, while the rest of the current travels on to the other two branches. When the second branch circuit is reached, an amount of current flow equal to I_2 travels through R_2, while the rest of the current travels on to the third branch and

through R_3. Each of the currents, I_1, I_2, and I_3, represent a portion of the total circuit current. In general, it can be stated that

$$I_T = I_1 + I_2 + I_3 + I_4 + \ldots + I_N$$

where "N" is equal to the number of branches in the parallel circuit.

RESISTANCE IN PARALLEL CIRCUITS

The concept of resistance is often difficult to comprehend because, as more branches are added to a parallel circuit, the total resistance of the circuit actually goes down. Consider the following example. There are too many cars backed up on a one-lane road (Fig. 14-47), creating a major traffic jam. At a point down the road, the single lane road becomes a three-lane road. All of a sudden, the traffic jam has disappeared and the cars begin to move at a higher rate of speed (Fig. 14-48). What happened to the traffic jam?

The number of cars is the same, but the number of lanes increased, giving the cars the option of traveling in any one of the three available lanes. The number of cars in each lane will, on average, be reduced to one-third of the total. If there were four lanes, the number of cars traveling in each lane would be one-fourth of the total.

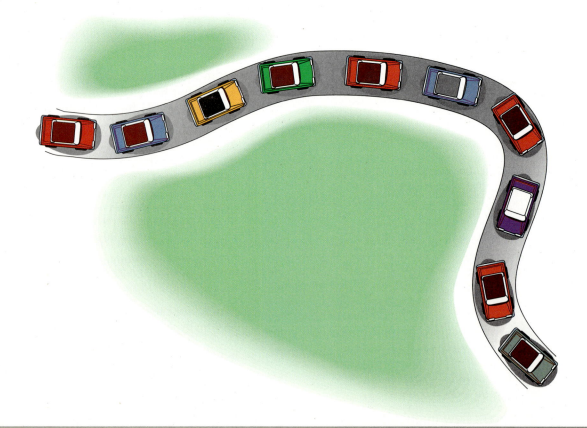

FIGURE 14-47 Cars backed up in a single lane.

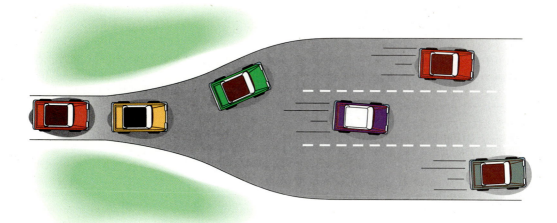

FIGURE 14-48 As the road gets wider, the traffic is reduced.

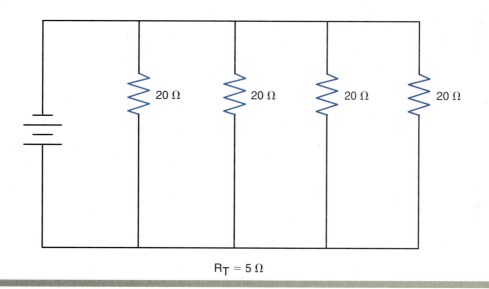

$R_T = 5\ \Omega$

FIGURE 14-49 Four 20-ohm resistors in parallel have a combined resistance of 5 ohms.

This can be applied to parallel circuits in the following manner. In general, for multiple resistances of equal value, the total resistance of the circuit will be equal to the value of one of the resistances divided by the number of paths in the circuit. For example, if a parallel circuit is made up of four paths, each of which has a resistance of 20 Ω, the total resistance will be equal to 5 Ω, which is found by dividing the 20 Ω by 4 (Fig. 14-49). In general, the formula for calculating the resistance of parallel circuits made up of identical branches is

$$R_T = \frac{R}{N}$$

where R is the value of one of the identical resistances and N is the number of paths.

When there are only two branches in a parallel circuit, the total resistance can be obtained by using the product over sum formula, which is

$$R_T = (R_1 \times R_2) / (R_1 + R_2)$$

In general, the total resistance for any parallel circuit can be determined by using the following formula:

$$1/R_T = 1/R_1 + 1/R_2 + 1/R_3 + \ldots + 1/R_N$$

where N is equal to the number of resistances in the circuit.

EXAMPLE 7

Considering the circuit in Figure 14-50, we can calculate the total resistance by

$$R_T = (5 \times 20) / (5 + 20) = 100/25 = 4\ \Omega$$

Note that the total resistance of the circuit is lower than the lowest valued resistance in the circuit.

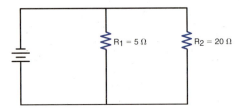

FIGURE 14-50 A 20-ohm resistor in parallel with a 5-ohm resistor.

EXAMPLE 8

Consider the circuit in Figure 14-51. It has three branches with resistances of 3 Ω, 4 Ω, and 6 Ω. Using the formula

$$1/R_T = 1/R_1 + 1/R_2 + 1/R_3$$

We will get

$$1/R_T = 1/3 + 1/4 + 1/6$$

To combine these fractions, we need to find the lowest common denominator

$$1/R_T = 4/12 + 3/12 + 2/12 = 9/12$$
$$1/R_T = 9/12$$
$$R_T = 12/9$$
$$R_T = 1.33\ \Omega$$

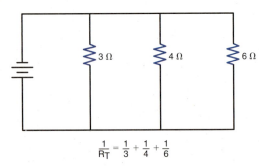

$$\frac{1}{R_T} = \frac{1}{3} + \frac{1}{4} + \frac{1}{6}$$

FIGURE 14-51 Calculating resistance in a parallel circuit.

VOLTAGE IN PARALLEL CIRCUITS

In parallel circuits, the voltage supplied to each branch of the circuit is equal to the total voltage supplied to the circuit. If the circuit has three branches, the voltage in each branch can be shown as

$$E_T = E_1 = E_2 = E_3$$

EXAMPLE 9

Consider the circuit in Figure 14-52. It consists of three resistances with values of 4 Ω, 4 Ω, and 2 Ω and the voltage supplied to the circuit is equal to 20 volts. We want to compute the amount of current flow through each of the branches, as well as the total current in the circuit.

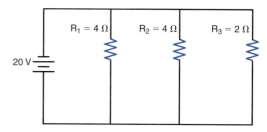

FIGURE 14-52 Parallel circuit with different branch resistances.

Solution

To calculate the current flow through one of the 4 Ω resistances, we use the formula

$$I = E \div R$$

Substituting, we get

$$I = 20 \text{ volts} \div 4\,\Omega = 5 \text{ amps}$$

To calculate the current flow through the 2 Ω resistance, we use the formula

$$I = E \div R$$

Substituting, we get

$$I = 20 \text{ volts} \div 2\,\Omega = 10 \text{ amps}$$

The flow through each of the 4 Ω resistances is 5 amps and the current flow through the 2 Ω resistance is 10 amps (Fig. 14-53).

The total current flow in the circuit is the sum of all the individual branch currents:

$$I_T = I_1 + I_2 + I_3$$

$$I_T = 5 \text{ amps} + 5 \text{ amps} + 10 \text{ amps}$$

$$I_T = 20 \text{ amps}$$

Alternatively, we can calculate the total current by first finding the total resistance of the circuit. We calculate the total resistance as follows:

$$1/R_T = 1/R_1 + 1/R_2 + 1/R_3 = 1/4 + 1/4 + 1/2$$
$$= 1/4 + 1/4 + 2/4$$

$$1/R_T = 1/4 + 1/4 + 2/4 = 4/4$$

$$R_T = 4/4 = 1\,\Omega$$

$$R_T = 1\,\Omega$$

Using the total resistance and total voltage, we can compute the total amperage as

$$I = E \div R = 20 \text{ volts} \div 1\,\Omega = 20 \text{ amps}$$

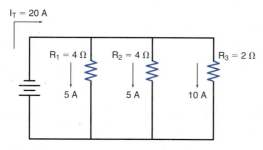

FIGURE 14-53 Calculating branch currents in a parallel circuit.

WIRING DIAGRAMS

The purpose of electric circuits is to direct and control current flow, at the correct time, to the various components in an air conditioning system. A broken wire, open switch, or wiring error can stop the current from flowing, thereby preventing the system, or part of the system, from functioning. A thorough understanding of basic circuit theory is essential to properly evaluate the circuits that, in effect, give the orders that the system components must follow. Electric current will follow the path that is provided, whether or not it is correct. The wiring diagrams that represent electric circuits are like road maps that the current will follow.

Electric circuits are represented by wiring diagrams located in service literature provided with the equipment or on the specific piece of equipment being worked on. These diagrams can take on different appearances depending on their intention and purpose. This portion of the chapter provides a basic understanding of electrical wiring diagrams. Refer to Chapter 17 for more detailed information regarding wiring diagrams. Two commonly used types of wiring diagram are the schematic diagram and the ladder diagram.

SCHEMATIC WIRING DIAGRAMS

The schematic diagram is designed to give the technician the following information:

- Component placement and location within the control panel (on some diagrams)
- A representation of every wire in the circuit
- The color of each wire in the circuit (when necessary)
- A detail of all factory wiring and field wiring

Color-coding the wires in the circuit helps give the field technician a visual advantage. For a technician to try to locate a red wire in a bundle of multicolored wires is much easier than tracing out a black wire in a sea of other black wires. Since every wire is accounted for in the schematic diagram, a technician can find a wiring mistake more easily. The schematic also differentiates between factory and field wiring. Factory wiring is installed in the factory before the unit is shipped out. Window and wall air conditioners have all of their wiring installed at the factory. Other systems come with only some of the wiring factory installed. The most common is the split system, which has the condensing unit in one location and the air handler in another. In these cases, the installation crew must install some of the wiring, referred

to as field wiring. Factory and field wiring are easily distinguished from one another in the wiring diagrams.

The schematic diagram, although a very accurate and complete representation of the actual circuits encountered on the specific system, tends to become confusing and more difficult to navigate as the system becomes more and more complex. Figure 14-54 is an example of a schematic diagram.

This figure shows a portion of a schematic diagram of a split-type air conditioning system, which has the condensing unit in one location and the air handler in another location. Schematic diagrams usually have a legend that describes any abbreviations used in the diagram. The legend for this diagram might look something like this:

COMP	Compressor
C	Compressor—Common terminal
S	Compressor—Start terminal
R	Compressor—Run terminal
CFM	Condenser fan motor
RC	Run capacitor

Using this diagram, you can trace out the circuit and determine how this portion of the system is supposed to operate. Upon initial inspection of the circuit, you can see that the power supply feeds the circuit at Terminal L1 and Terminal L2. You can also see that a normally open set of contacts (Fig. 14-54) is in series with each of the lines entering the unit. This means that, unless both of those sets of contacts are closed, no current can flow through the circuit. The next question you may ask yourself is, "What will make those contacts close?" The answer is that the contacts will close when the holding coil (Fig. 14-54) of the contactor is energized. Refer to the chapter on electric components for more on contactors and their operation.

This holding coil operates on 24 volts and gets its voltage from a location that cannot be seen on this diagram. You do know, however, that when 24 volts is supplied to the holding coil, both sets of contacts will close. Once these contacts close, you can continue to trace out the circuit. Following L1, you can see that after the contacts, the line splits up to feed both the compressor, through Terminal C, and the condenser fan motor. The current flows through each of these devices, and then the two lines leaving them come together again. The current then flows through the other set of contacts, back to the power supply at L2.

This explanation shows that the compressor and the condenser fan motor are designed to operate at the same time, since they are connected in parallel with each other. If one of the components fails, however, the other will

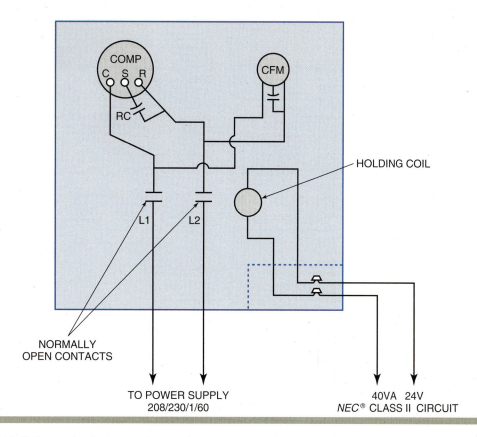

FIGURE 14-54 Schematic diagram of a 208/230-volt condensing unit with low-voltage control wiring.

continue to run because the current will have an alternative path to take. If the holding coil is not energized, neither the compressor nor the condenser fan motor can operate. As systems get more and more complicated, so do the schematic diagrams. For this reason, the ladder diagram proves to be extremely useful.

LADDER DIAGRAMS

A simpler wiring diagram that is often encountered is the **ladder diagram,** also referred to as the **line diagram.** It usually does not provide all of the information found in the schematic diagram, but it is very useful in troubleshooting the system. The ladder diagram depicts each individual circuit separately on its own line, so the diagram tends to look like a ladder. By following each individual circuit from left to right, or right to left, the technician can successfully locate the load that is not operating properly, as well as all switches, contacts, and controls that are in series with the device, or the holding coil that ultimately energizes it. When used effectively together, the schematic and ladder diagrams become a powerful tool that enables the service technician to provide an accurate and quick diagnosis of the system problem.

As an air conditioning system uses more and more components, the interconnecting wiring between the components becomes more complicated as well. The number of wires increases and the schematic diagram becomes more difficult to read. The lines representing the individual wires all seem to come together as one and, to the inexperienced technician, may resemble a plate of spaghetti.

With the ladder diagram, the technician can isolate the desired circuit to determine all of the switches, controls, and wiring connections in that circuit. The ladder diagram in Figure 14-55 is a representation of the same schematic that was shown in Figure 14-54. L1 and L2, representing the power source, are located at the left and right sides of the diagram while the individual circuits make up the *rungs* of the ladder. As more components are added to the system, more rungs are added to the ladder diagram. The basic construction and appearance of the ladder diagram will remain the same.

Tracing the ladder diagram is relatively simple, since you can predetermine the direction of current flow by simply looking at the diagram. This is not as easy with the schematic diagram. In Figure 14-55, start tracing the circuit from L1. Following the line from L1, you first

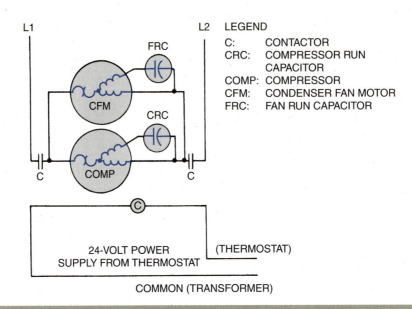

FIGURE 14-55 Ladder diagram of the same circuit in Figure 14-54.

reach the normally open set of contacts. The holding coil located in the control circuit controls these contacts. Once the current flows through this set of contacts, assuming they are closed, it then splits and flows to the condenser fan motor and to the compressor. Once the current leaves the devices and flows through the run capacitors, the lines are connected. The current then flows through the other set of contacts and back to L2. Note that the description of this circuit is the same as the description of the equivalent circuit in Figure 14-54. Most manufacturers provide both a schematic and a ladder diagram on their equipment to ease the job of the servicing technician. In this case, one legend is usually provided that encompasses both diagrams.

WIRE SIZE

When electric circuits are installed, the conductors used must be able to handle the current flow that will be passing through the circuits. Just as a plumber must use pipes that are large enough to handle the quantity of water intended to flow through them, the size of the conductors must be carefully selected to ensure that the circuits do not fail when exposed to the current requirements of the loads being powered. As the current requirements of a circuit increase, so does the size of the conductor. Circuits that are intended to feed power to small motors, for example, would require smaller conductors than circuits that feed power to large, electric heaters. The determination about the size of the conductors to be used for

a given application is made by the *National Electrical Code* (NEC). The NEC® not only governs the sizes and types of conductors that are to be used for given applications, it also sets forth guidelines regarding the manner in which these conductors and associated components are installed. For example, the NEC® provides guidelines for the materials and methods used to install circuits in wet locations, such as in the case of bringing power to a condensing unit that is located outside. Refer to Chapter 18 for more information on the NEC® and how it affects the HVAC/R industry.

The size of the conductor to be used is determined by the amount of current that is to be passed through the circuit. For example, if a typical circuit is to carry no more than 20 amps, the NEC® states that a 14-gauge copper conductor will suffice. If, on the other hand, the circuit will carry as much as 30 amps, the wire size must be increased to a 10-gauge conductor. Charts similar to the one in Figure 14-56 provide useful information about the current-carrying capability of different types of conductors. Standard wire sizes are defined by the **American Wire Gauge (AWG)** which lists the sizes of conductors as well as other data regarding the characteristics of a given conductor.

> **CAUTION**
>
> **CAUTION: Using an undersized conductor can result in the generation of large amounts of heat that can cause fire. Always size conductors carefully!**

Table 310.16 Allowable Ampacities of Insulated Conductors Rated 0 Through 2000 Volts, 60°C Through 90°C (140°F Through 194°F), Not More Than Three Current-Carrying Conductors in Raceway, Cable, or Earth (Directly Buried), Based on Ambient Temperature of 30°C (86°F)

Size AWG or kcmil	Temperature Rating of Conductor (See Table 310.13.)						Size AWG or kcmil
	60°C (140°F)	75°C (167°F)	90°C (194°F)	60°C (140°F)	75°C (167°F)	90°C (194°F)	
	Types TW, UF	Types RHW, THHW, THW, THWN, XHHW, USE, ZW	Types TBS, SA, SIS, FEP, FEPB, MI, RHH, RHW-2, THHN, THHW, THW-2, THWN-2, USE-2, XHH, XHHW, XHHW-2, ZW-2	Types TW, UF	Types RHW, THHW, THW, THWN, USE	Types TBS, SA, SIS, THHN, THHW, THW-2, THWN-2, RHH, RHW-2, USE-2, XHH, XHHW, XHHW-2, ZW-2	
	COPPER			ALUMINUM OR COPPER-CLAD ALUMINUM			
18	—	—	14	—	—	—	—
16	—	—	18	—	—	—	—
14*	20	20	25	—	—	—	—
12*	25	25	30	20	20	25	12*
10*	30	35	40	25	30	35	10*
8	40	50	55	30	40	45	8
6	55	65	75	40	50	60	6
4	70	85	95	55	65	75	4
3	85	100	110	65	75	85	3
2	95	115	130	75	90	100	2
1	110	130	150	85	100	115	1
1/0	125	150	170	100	120	135	1/0
2/0	145	175	195	115	135	150	2/0
3/0	165	200	225	130	155	175	3/0
4/0	195	230	260	150	180	205	4/0
250	215	255	290	170	205	230	250
300	240	285	320	190	230	255	300
350	260	310	350	210	250	280	350
400	280	335	380	225	270	305	400
500	320	380	430	260	310	350	500
600	355	420	475	285	340	385	600
700	385	460	520	310	375	420	700
750	400	475	535	320	385	435	750
800	410	490	555	330	395	450	800
900	435	520	585	355	425	480	900
1000	455	545	615	375	445	500	1000
1250	495	590	665	405	485	545	1250
1500	520	625	705	435	520	585	1500
1750	545	650	735	455	545	615	1750
2000	560	665	750	470	560	630	2000

FIGURE 14-56 Ampacity chart for copper and aluminum conductors.

Reprinted with permission from NFPA® 70, National Electrical Code, Copyright © 2010, National Fire Protection Association®, Quincy, MA. This reprinted material is not the complete and official position of the NFPA® on the referenced subject, which is represented only by the standard in its entirety.

from experience...

As the gauge number of a wire decreases, the size of the conductor increases. A 12-gauge conductor is wider and can therefore carry more current than a 14-gauge wire.

Manufacturers of new air conditioning equipment often provide, as part of the installation instructions, information regarding the proper size conductors to use. If this information is not provided, the installing technician can look at the equipment nameplate to determine the amperage draw of the particular piece of equipment. Once the amperage draw has been obtained, the correct wire size can be determined by checking the table in Figure 14-56.

FIGURE 14-57 Screw-in type fuse.

Courtesy of Bussman, Inc.

CIRCUIT PROTECTION

When excessive current flows in a circuit, a great amount of heat is generated. This heat can cause the conductor's insulation to melt which, in turn, can cause a fire. The current in a circuit can rise above safe levels for a number of reasons. When the resistance of a circuit falls, the amount of current that will flow through the circuit increases. This reduction in circuit resistance can result from improper wiring, a short circuit, damaged wire insulation, or simply placing too many loads in a parallel circuit. How many times has a circuit breaker tripped or a fuse blown in your house when you had too many lights on?

The tripping of the circuit breaker or blowing of the fuse indicated that there was too much current flowing in the circuit. To protect the circuit and reduce the possibility of fire, the circuit protector opened to stop current flow. The most common circuit protectors are fuses, circuit breakers, and ground fault circuit interrupters, GFCIs.

FIGURE 14-58 Cartridge fuse.

Courtesy of Cooper Bussmann, a division of Cooper Industries.

FUSES

Fuses (Fig. 14-57, Fig. 14-58, and Fig. 14-59) are one-time circuit protectors that open their contacts to de-energize a circuit that is operating in an overload condition. Fuses are constructed of a material that has a higher resistance and a lower melting point than the conductors they are to protect and will generate heat faster than the conductors. The amount of heat generated is proportional to the amount of current flow through the device. As the sensed heat rises to unsafe levels, the element in the fuse will melt, stopping current flow. Once the fuse element melts, the fuse must be replaced. The element is designed to melt before damage to the conductors can occur.

FIGURE 14-59 Knife-blade cartridge fuse.

Courtesy of Cooper Bussmann, a division of Cooper Industries.

Fuses are to be sized according to the size and current rating of the conductors that are used in the circuit. For example, if the load and associated wiring are rated for a maximum of 15 amps, a 15-amp fuse should be used. Using a 20-amp fuse for that application will permit 20 amps of current to flow through a conductor that is only capable of carrying 15 amps safely. Excessive heat will be generated, creating an unsafe condition.

Fuses can be either *fast-blow* or *slow-blow* devices. Fast-blow fuses are designed to open whenever an overcurrent condition is sensed, but a slow-blow fuse will tolerate an overcurrent condition for a period of time before the element melts. Slow-blow fuses are typically used to protect devices such as motors that draw large amounts of current when they initially startup. Typical fuse styles include the screw-in type (Fig. 14-57), the cartridge type (Fig. 14-58), and the knife-blade type (Fig. 14-59).

FIGURE 14-60 Single-pole (left) and two-pole (right) circuit breakers.

Delmar/Cengage Learning.

CAUTION

CAUTION: Fuses should never be jumped out or replaced with larger fuses to prevent them from blowing. Excessive heat levels can result, causing fire, severe personal injury, or death.

from experience...

To check a fuse, remove it from the fuse panel and place the test leads of an ohmmeter on each end of the fuse. A good fuse will have continuity or very low resistance, while a defective fuse will have a reading of infinite resistance.

BREAKER IN THE
OFF POSITION

FIGURE 14-61 Circuit breaker in the OFF position.

Delmar/Cengage Learning.

CIRCUIT BREAKERS

Just like fuses, circuit breakers (Fig. 14-60) are intended to protect circuits from an overcurrent condition. The circuit breaker, however, can be reset after it has tripped. Unlike the fuse, the circuit breaker can be used to manually disable a circuit by simply flipping the circuit breaker to the off position (Fig. 14-61). When a circuit breaker trips, the toggle on the breaker will move to a position between the on and off positions (Fig. 14-62) and will, in some cases, display an orange "flag" indicating

that the breaker has tripped. This makes identifying the tripped breaker much easier. Like the fuse, circuit breakers should be sized according to the size and current-carrying capability of the conductors used in the circuit to avoid overheating.

FIGURE 14-62 Tripped circuit breaker.

TRIPPED
CIRCUIT
BREAKER

Delmar/Cengage Learning.

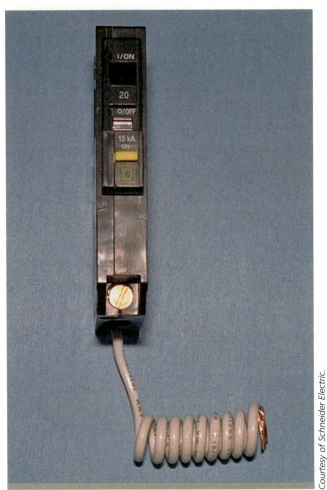

FIGURE 14-63 Ground fault circuit interrupter, GFCI.

Courtesy of Schneider Electric.

from experience...

To reset a circuit breaker, the device must be switched to the off position before switching it to the on position. A circuit breaker *will not* reset if you attempt to simply push the breaker to the on position once it has tripped.

GROUND FAULT CIRCUIT INTERRUPTERS

Ground fault circuit interrupters (GFCIs) were addressed earlier in the chapter on safety. When located in a circuit breaker panel, the GFCI looks very similar to a standard circuit breaker but is equipped with a test button (Fig. 14-63). The GFCI senses small current leaks to ground and will trip when such current leaks are detected. GFCIs are typically used in wet locations such as bathrooms, kitchens, and pool areas and will disable an electric circuit before severe electric shock can occur. The *NEC®* provides guidelines regarding the use of GFCIs as well as the conditions that would require the installation of such a device.

FIGURE 14-64 GFCI in a wet location.

Delmar/Cengage Learning.

GFCIs can also be located at the outlet or receptacle location, enabling the consumer to test the operation of the device without going to the circuit breaker panel (Fig. 14-64).

ELECTRIC SERVICE

To operate electric appliances and components within a structure, electric power must be brought into the structure. This power is referred to as the electric service, which is tied into the fuse or circuit breaker panel. Electric service is rated by the total amperage draw as well as the available voltages. Depending on the structure, electric services can range from less than 100 amps to more than 1,000 amps. Typical residential electric services are in the 100- to 200-amp range. The electric service is brought into the structure at the distribution panel (Fig. 14-65). It is from this panel that the individual branch circuits are run.

120/240-VOLT, SINGLE-PHASE SERVICE

Single-phase alternating current is the most common electric service found in residential applications. The 120/240-volt service permits the occupant to operate

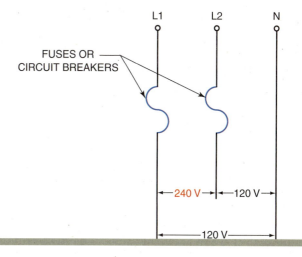

FIGURE 14-66 120/240-volt, single-phase power supply.

devices that require 120 volts as well as those that require 240 volts. The wiring diagram for a typical 120/240-volt, single-phase power supply is shown in Figure 14-66. This service is made up of two *hot* legs, each with a potential of 120 volts, and a neutral that has a potential of 0 volts. If a voltmeter is connected between L1 and Neutral, the meter will read 120 volts. The same reading will be obtained from L2 to Neutral. Measuring the voltage between L1 and L2 will provide a reading of 240 volts.

> ## *from experience...*
>
> The potential at L1 and the potential at L2 are generated out of phase with each other so they do not cancel each other out. If the two sources were in phase with each other, a voltmeter placed across the two would result in a reading of 0 volts.

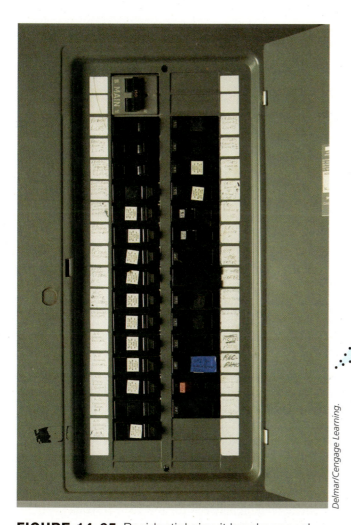

FIGURE 14-65 Residential circuit breaker panel.

Delmar/Cengage Learning.

Each of the two hot power legs is protected by either fuses or circuit breakers that will open to de-energize the circuits should an overload occur. Each individual circuit is protected by a circuit breaker, allowing only the overloaded circuit to be de-energized. Circuit breaker panels also have main breakers that will interrupt power to the entire panel.

When connecting a circuit to the distribution panel, it must be determined if the circuit is to supply 120 volts or 240 volts to the remote location. A 120-volt circuit will require a single pole breaker (Fig. 14-60, left) which will supply one hot leg to the circuit. A 240-volt circuit will require a two-pole breaker (Fig. 14-60, right) which will provide power from both L1 and L2 to the remote circuit.

THREE-PHASE SERVICE

Three-phase alternating current power supplies are commonly found in commercial and industrial structures. These power supplies use three hot legs and a neutral as compared with the single phase electric service just discussed, which was made up of two hot legs and a neutral. All three hot legs must be protected from an overcurrent condition (Fig. 14-67). The only way to obtain 240 volts from the single-phase power supply was to connect a circuit between L1 and L2. When an additional hot leg is added to the electric service, three-phase power is being supplied. Three-phase power supplies are made up of three hot legs, L1, L2, and L3, as well as a ground. Each of these three hot legs is generated 120 degrees out of phase with the next (Fig. 14-68), permitting three separate single phase power supplies to be generated simultaneously. The voltages obtained between the hot legs vary depending on the manner in which the three-phase power is generated. Three-phase power is commonly found as either a delta configuration or a wye configuration.

230-VOLT, THREE-PHASE DELTA SYSTEM

The delta three-phase system is popular for use in commercial and industrial applications that use a large number of three-phase, power-driven pieces of equipment. It is not the best choice when the application calls for a large number of 115-volt circuits, as will be seen shortly.

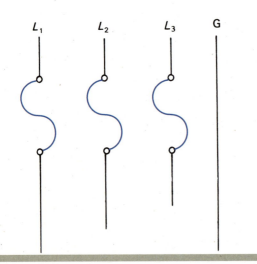

FIGURE 14-67 All three hot legs of a three-phase power supply are fused.

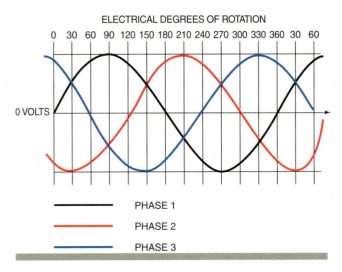

FIGURE 14-68 Sine-wave representation of a three-phase power supply.

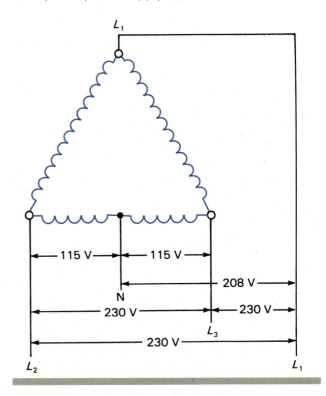

FIGURE 14-69 Delta configuration for three-phase power.

The three-phase delta system is shown in Figure 14-69. The delta configuration is named because the shape of the windings resembles the Greek letter delta. The voltage readings obtained from the delta system are as follows:

Voltage reading between L1 and L2: 230 volts

Voltage reading between L1 and L3: 230 volts

Voltage reading between L2 and L3: 230 volts

Voltage reading between L1 and ground: 208 volts

Voltage reading between L2 and ground: 115 volts

Voltage reading between L3 and ground: 115 volts

Note that the voltage reading between L1 and ground is much higher than the expected value of 115 volts. This is because the windings between the L1 terminal and ground is longer than the other windings and because the windings between these two points are out of phase with each other.

The delta configuration can provide 230-volt, single-phase power (L1 to L2, L2 to L3, or L1 to L3); 230-volt, three-phase power (L1 to L2 to L3); or 115 volts as needed to a remote location. It is very important to note that, with the three-phase delta configuration, connecting a 115-volt load to the hot leg (208 volts to ground) can result in damage to the device being powered.

208-VOLT, THREE-PHASE WYE SYSTEM

The wye three-phase system is popular in applications that require a large number of 115-volt circuits. The three-phase wye system is shown in Figure 14-70. The wye configuration is named because the shape of the windings resembles the letter Y. The voltage readings obtained from the wye system are as follows:

Voltage reading between L1 and L2: 208 volts

Voltage reading between L1 and L3: 208 volts

Voltage reading between L2 and L3: 208 volts

Voltage reading between L1 and ground: 120 volts

Voltage reading between L2 and ground: 120 volts

Voltage reading between L3 and ground: 120 volts

The wye configuration does not have a high leg, so the entire load of the structure can be balanced among the three hot legs. The three-phase wye configuration can deliver 208-volt, single phase power (L1 to L2, L2 to L3, or L1 to L3); 208-volt, three phase power (L1 to L2 to L3); or 120 volts to the remote circuits in the structure.

SUMMARY

- Atoms are made up of protons, neutrons, and electrons.
- Electrons have a negative charge, protons a positive charge, and neutrons a neutral charge.
- Particles with opposite charges will attract each other, but particles with like charges will repel each other.
- Electricity can be generated by light, heat, friction, pressure, and magnetism.
- Electric current is the movement of electrons through a conductor.
- Direct current is supplied primarily by batteries and permits current flow in only one direction.
- Alternating current is characterized by the continual reversal of current flow.
- Voltage is the electrical pressure that pushes current through an electric circuit.
- Current is the rate of flow of electrons through a circuit.
- Resistance is any material that reduces the rate of electron flow in a circuit.
- Ohm's law is the relationship among voltage, resistance, and current.
- Electric meters are used to measure the values of voltage, resistance, and current in electric circuits.
- Electric meters can be either analog or digital devices.
- Electric circuits are made up of power sources, current paths, loads, and switches.
- Electric loads are devices that consume electric power and can be either resistive or inductive.
- Series circuits provide only one path for electric current to take.
- Parallel circuits provide multiple paths for current to take.

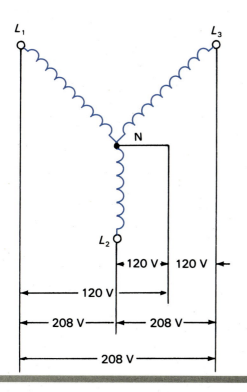

FIGURE 14-70 Wye configuration for three-phase power.

- Wiring diagrams such as the schematic and ladder provide electrical roadmaps for the service technician.
- The *National Electrical Code* (NEC) sets the standards for electrical installations.
- The American Wire Gauge (AWG) provides data on conductors.

- Circuits are protected by fuses, circuit breakers, and GFCIs.
- Single-phase power supplies are generated with two hot legs and a neutral.
- Three-phase power is supplied with three hot legs and a neutral.
- Three-phase power supplies can be configured as either a wye or a delta.

GREEN CHECKLIST

- [ ] **Use rechargeable batteries whenever possible**

- [ ] **Disposable batteries should be discarded properly to minimize environmental contamination**

- [ ] **Properly size electrical conductors**

- [ ] **Undersized wires increase the risk of fire**

- [ ] **Oversized conductors waste natural resources**

REVIEW QUESTIONS

1. **The smallest quantity of any naturally occurring element is an**
   a. Atom
   b. Molecule
   c. Electron
   d. Proton

2. **The part of an atom that has a negative charge is called the**
   a. Nucleus
   b. Neutron
   c. Electron
   d. Proton

3. **Where are electrons found in atoms?**
   a. In the nucleus of the atom
   b. With the protons
   c. In orbit around the nucleus
   d. Electrons are in molecules, not atoms

4. **An atom of a substance with only one valence electron**
   a. Will typically be a good conductor of electricity
   b. Will typically be a good insulator
   c. Will have a full outer orbit or shell
   d. Both a and c are correct

5. **Electric current is the**
   a. Flow of electrons from one atom of an insulator to another
   b. Flow of protons from one atom of an insulator to another
   c. Flow of electrons from one atom of a conductor to another
   d. Flow of protons from one atom of a conductor to another

6. **Cutting magnetic lines of force with a conductor will**
   a. Damage the lines of force
   b. Generate electricity
   c. Strengthen the magnetic field
   d. All of the above are correct

7. **If a dc power supply of 20 volts causes 2 amps of current to flow in a circuit**
   a. The resistance of the circuit is 10 Ω
   b. The power consumed by the circuit is 40 watts
   c. Both a and b are correct
   d. Neither a nor b is correct

8. **If two 10 Ω resistances are connected in parallel, the total resistance will be**
   a. 5 Ω
   b. 10 Ω
   c. 15 Ω
   d. 20 Ω

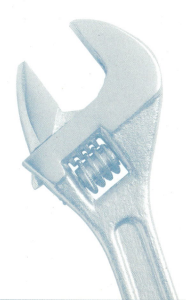

9. **If two 10 Ω resistances are connected in series, the total resistance will be**

    a. 5 Ω

    b. 10 Ω

    c. 15 Ω

    d. 20 Ω

10. **The four components of a circuit are the**

    a. Power source, current path, resistive load, inductive load

    b. Power source, current path, load, switch

    c. Current path, load, switch, light bulb

    d. Load, resistance, current, current path

11. **An example of a resistive load would be a(n)**

    a. Light bulb

    b. Electric heater

    c. Electric motor

    d. Both a and b are correct

12. **Which of the following is true regarding current flow in a parallel circuit?**

    a. The current flow is the same through all branches of the circuit

    b. The paths with higher resistance will have higher current flows

    c. The paths with lower resistance will have higher current flows

    d. As additional branches are added to the circuit, the total current in the circuit will decrease

13. **To read the voltage being supplied to an electrical device such as a light bulb**

    a. Set the meter to read voltage and place the two test leads on both sides of the load

    b. Set the meter to read voltage and clamp the meter around one of the conductors in the circuit

    c. Set the meter to read continuity and place the test leads on both sides of the load

    d. Set the meter to read amperage and clamp the meter around one of the conductors in the circuit

14. **To read the amperage draw of an electrical device**

    a. Set the meter to read voltage and place the two test leads on both sides of the load

    b. Set the meter to read amperage and clamp the meter around both conductors feeding power to the device

    c. Set the meter to read amperage and clamp the meter around one of the conductors feeding power to the device

    d. Set the meter to read amperage and place the test leads on both sides of the load

15. **A defective fuse**

    a. Will have a resistance reading of zero ohms if tested with an ohmmeter

    b. Will have no continuity through the device

    c. Will permit current flow through the device

    d. Must be replaced with another fuse with a higher amperage rating

16. **Circuit breakers used on 240-volt circuits are**
    a. Single-pole devices that close the circuit if an overcurrent condition exists
    b. Two-pole devices that close the circuit if an overcurrent condition exists
    c. Single-pole devices that open the circuit if an overcurrent condition exists
    d. Two-pole devices that open the circuit if an overcurrent condition exists

17. **From the table in Figure 14-57, which of the following wire sizes can carry the most current?**
    a. 10-gauge aluminum wire at 75°C
    b. 10-gauge copper wire at 75°C
    c. 12-gauge copper wire at 90°C
    d. 10-gauge copper wire at 90°C

18. **A single-phase 120/240 volt electric service is made up of**
    a. L1, L2, and neutral
    b. L1 and neutral
    c. L1, L2, and L3
    d. L1, L2, L3, and ground

19. **If a technician is checking a three-phase power supply and finds a voltage reading of 208 volts from L1 to ground**
    a. The three-phase power is configured as a delta
    b. The three-phase power is configured as a wye
    c. She can connect a 120-volt power tool to that circuit without damaging the tool
    d. Both b and c are correct

20. **The legend on a wiring diagram provides all of the following except**
    a. The length of the conductors
    b. The color of the conductors
    c. The placement of the components in the system
    d. The electrical connections

KNOW YOUR CODES

When installing electric circuits, be sure to do so in accordance with your local electrical codes. Your local codes may very well differ from those that are set forth by the *NEC*®. Investigate your local codes regarding the installation of line voltage conductors for the power circuits of an air conditioning system and those governing the installation of wiring for low-voltage control circuits.

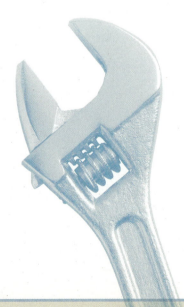

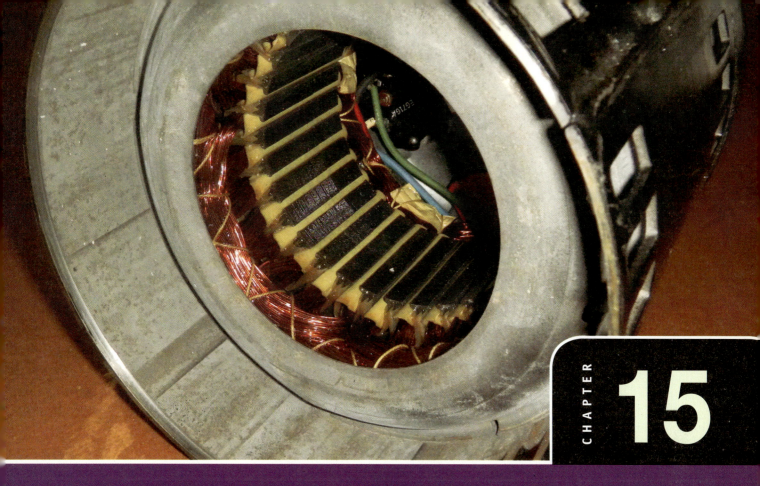

CHAPTER 15

Electric Motors

OBJECTIVES

Upon completion of this chapter, the student should be able to

- Name and identify five types of electric motors commonly found in air-conditioning equipment.

- Explain the operation of a shaded-pole motor.

- Explain the characteristics of the start and run windings in an electric motor.

- Explain the relationship between locked rotor amperage (LRA) and full load amperage (FLA).

- Measure FLA and LRA on single-phase motors.

- Describe the operation of a split-phase or induction-start-induction-run (ISIR) motor.

- Describe the function of start and run capacitors.

- Explain the operation of a current magnetic relay (CMR).

- Troubleshoot a CMR.

- Explain the operation of a potential magnetic relay (PMR) or potential relay.

- Troubleshoot a potential relay.

- Explain the operation of a positive temperature coefficient (PTC) device.

- Describe the operation of a capacitor-start-induction-run (CSIR) motor.

- Describe the operation of a capacitor-start-capacitor-run (CSCR) motor.

- Describe the operation of a permanent-split-capacitor (PSC) motor.

- Describe the electronically commutated, (ECM) motor.

- Connect a single-phase ISIR, CSIR, CSCR, or PSC motor.

- Describe the basic configuration of a three-phase motor.

- Name and describe three commonly used methods for starting three-phase motors.

- Explain the difference between a wye and a delta configuration.

- Connect a three-phase motor.

- Draw electrical schematics for various types of electric motors.

Earlier in the text, when reference was made to electric motors, it was done so in a strictly mechanical manner. Belt tension, pulley alignment, motor lubrication, and other important factors integral to proper motor operation were discussed. However, many electric motors are classified based on a number of factors, including motor winding configuration, design application, starting devices, horsepower, and the number of phases. We will now focus on various types of electric motors as well as the components that help the devices operate as effectively and efficiently as possible, both on startup and after reaching a steady-state condition.

Armature Portion of a relay that moves to open or close sets of electric contacts

Capacitor Energy-storing device that can help increase the starting torque or running efficiency of a motor

Capacitor-start-capacitor-run (CSCR) motor Single-phase motor that starts and operates with the aid of start and run capacitors

Capacitor-start-induction-run (CSIR) motor Single-phase motor that uses only a start capacitor that is removed from the circuit after the motor has started

Centrifugal switch Starting device that opens and closes its contacts based on the speed of the motor

Current magnetic relay (CMR) Starting device that opens and closes its contacts based on the current flow through the run winding of a split-phase motor

Delta configuration Three-phase wiring configuration used primarily for applications that require a large number of three-phase circuits; found primarily in commercial and industrial structures

Drop-out voltage Voltage at which the potential relay returns its contacts to their normally closed position

GLOSSARY OF TERMS (CONT'D)

Heaters Component part of a motor starter that senses the heat generated by current flow, facilitating overload protection for the motor

Horsepower Power rating of motors that is equal to 746 watts

Induction-start-induction-run (ISIR) motor Single-phase motor that starts and operates solely on the imbalance in magnetic field generated by current flow through the windings

Locked-rotor amperage (LRA) Amperage that a conventional split-phase motor draws on initial startup

Motor starter Component used to start three-phase motors and protect them from single-phasing

Part-wind start Three-phase motor designed to have only part of the windings in the circuit upon initial startup

Permanent-split-capacitor (PSC) motor Single-phase motor that has low starting torque but good running efficiency. This motor uses a run capacitor and operates with both the run and start windings in the circuit at all times

Pick-up voltage Voltage at which the potential relay's contacts open

Positive temperature coefficient (PTC) A resistor that reacts to changes in temperature; as the sensed temperature increases, the resistance of the PTC will increase as well

Potential relay Starting device that opens and closes its contacts depending on the induced voltage across the start winding of a motor

Reduced-voltage start Three-phase motor starting technique that adds resistance to the circuit to reduce the amount of voltage supplied to the motor during startup

Rotor Portion of a motor that rotates within the motor shell

Run capacitor Energy storage device that provides additional starting torque and helps improve running efficiency of single-phase motors; it is often oil-filled and is intended to be in the active electric circuit whenever the motor is energized

Run winding Low-resistance winding of a single-phase motor

Running-load amperage (RLA) Amount of current a motor draws after it has been energized and permitted to reach its normal operating speed

Soft start Term used to describe the gradual increase in motor speed on start up

Soft stop Term used to describe the gradual decrease in motor speed on shut down

Split-phase motor Single-phase motor in which the power is split between the start and run windings of the motor

Start capacitor Energy-storing device intended to increase the starting torque of a single-phase motor

Start winding High-resistance winding of a single-phase motor

Starters See motor starters

Starting torque Twisting force used to initiate the motion of an electric motor

Stator Portion of a motor that does not rotate and houses the start and run windings

Switches Component parts of a motor starter that open to protect three-phase motors from single-phasing

Three-phase motor Motor designed to operate when supplied with three-phase power

MOTOR USES

Numerous motors are found on even the simplest of systems, so a good understanding of the various types of motors, as well as their operation, is of vital importance. Motors are commonly used

- in compressors to pump refrigerant through the system
- to move air through an evaporator coil
- to move air through a condenser coil
- to pump water or other liquid through a piping system
- to turn gears in timers and other devices
- to pump water and move air through evaporative coolers

MOTOR POWER AND STARTING TORQUE

In addition to their construction and appearance, motors are also classified by their ability to do usable work. This directly relates to the device that the motor is operating. For example, a motor that is used to turn a large blower must be larger and able to do more work than one used to turn a small fan. The term that describes the motor's power is **horsepower**. It is noted from Ohm's law that power, equal to voltage times current, is measured in watts. The conversion factor between horsepower and wattage is 746. This means that 746 watts = 1 horsepower. The larger the horsepower rating, the more work the motor can do and, typically, the more starting torque that is exerted on initial motor startup. **Starting torque** refers to the force that allows the motor to reach its desired speed effectively and efficiently from the stopped position. Consider this example: If a soccer ball were kicked by an individual who was permitted to swing her leg back as far as she needed, it would travel much farther than if she were not permitted to swing her leg back at all.

Starting torque is often referred to as the extra *push* given to the motor to help it overcome any resistance offered by belts, pulleys, bearings, and internal motor resistance. This allows the motor to start properly. Starting torque is generated in the motor by an imbalance in the magnetic field generated by current flowing through the motor's electric windings. Single-phase motors have two windings; three-phase motors have three. The greater the difference in magnetic field, the greater the starting torque.

MOTOR TYPES

To ensure continued satisfactory motor operation, the type of motor chosen must have been constructed to match the specific conditions under which it is intended to function. Various types of motor construction include the following:

- Open motors
- Enclosed motors
- Drip-proof motors

Open motors (Fig. 15-1) are designed to have their windings cooled by air moving across them and have openings in the motor shell for that purpose. This type of motor would *not* be desirable in a wet location. Enclosed motors (Fig. 15-2) are designed to be used

FIGURE 15-1 An open motor.

Photo by Eugene Silberstein.

FIGURE 15-2 Enclosed motor.

Courtesy of W. W. Grainger, Inc.

in dirty locations. Since air cannot pass over the windings directly, alternative methods must be used to cool them. Drip-proof motors are intended for use in wet locations.

Shaded-pole motors are designed for very light duty and are normally used to turn small fan blades connected directly to the shaft of the motor. They have very low starting torque. Permanent-split-capacitor (PSC) motors have more starting torque than the shaded-pole motor and can, therefore, be used in somewhat larger applications. Capacitor-start-induction-run (CSIR) and capacitor-start-capacitor-run (CSCR) motors are used for even larger applications and employ starting components that allow the motor to start, run, and function efficiently. These starting components include:

- Capacitors
- Current magnetic relays
- Potential relays

Of all the motor types discussed in this text, the three-phase motor has the highest starting torque. Unlike the PSC, the CSIR, and the CSCR motors, which have a start-and-run winding, the three-phase motor has three constantly energized run windings. Three "hot" power lines feed into these motors, and they do not use capacitors or starting relays. Three-phase motors also operate more efficiently than single-phase motors.

SINGLE-PHASE MOTORS

Single-phase motors are powered by either one hot leg or two, depending on the voltage at which the motor is to operate. Single-phase motors are typically constructed with two windings: the start winding and the run winding. The exception to this is the shaded-pole motor. The **start winding** has relatively high resistance and is made up of thin wire formed into a coil with many turns. The **run winding**, on the other hand, has lower resistance and is made of thicker wire formed into a coil with fewer turns than the start winding. The start and run windings are wired in parallel with each other, permitting different amounts of electrical current to flow through each. The varying currents generate magnetic fields of different strengths. The start and run windings are located in the section of the motor referred to as the **stator** (Fig. 15-3).

The stator does not turn and is located directly inside the shell of the motor. The portion of the motor that actually turns is called the **rotor** (Fig. 15-4). The motor's

Photo by Eugene Silberstein.

FIGURE 15-3 Stator made up of start and run windings that do not rotate.

Photo by Eugene Silberstein.

FIGURE 15-4 Squirrel-cage rotor.

shaft is connected to the rotor, which rests inside the stator. The rotor, also called a *squirrel-cage rotor* because of its appearance, is not electrically connected to the stator, but turns as a result of being exposed to magnetic fields produced in the stator.

SHADED-POLE MOTORS

The shaded-pole motor (Fig. 15-5) has the lowest starting torque of all the motor types discussed in this text. It is relatively inexpensive and is used to turn very small fan blades connected directly to the shaft of the motor. The basic construction of this motor is very simple since no start winding is present. The imbalance in magnetic field needed to produce rotation is obtained by *shading* a portion of the run winding with a heavy copper wire or band (Fig. 15-6). When the motor is energized, the strength of the magnetic field is different in the area of the main poles than in the area of the shaded poles,

FIGURE 15-5 Shaded pole motor.

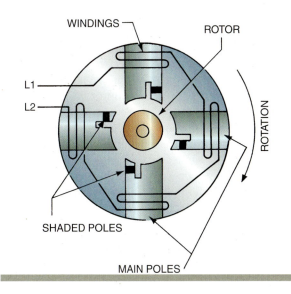

FIGURE 15-7 Layout of a single-speed, shaded pole motor. Note that there is actually one large winding in this type of configuration.

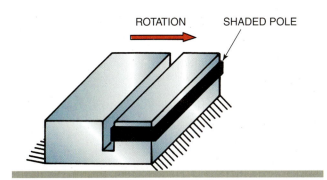

FIGURE 15-6 A portion of the pole is shaded to create an imbalance in magnetic field. The direction of rotation is from the unshaded portion of the pole to the shaded portion of the pole.

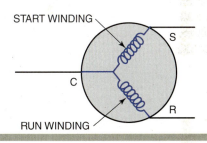

FIGURE 15-8 Schematic diagram of a split-phase motor.

allowing the rotor to turn. The direction of rotation of a shaded-pole motor is determined by the orientation of the main poles and the shaded poles. The rotor will turn in the direction indicated by the *arrows* in Figure 15-7. Shaded-pole motors are used for very small applications and are rated in watts instead of horsepower. They normally range in size from 6 watts to roughly 35 watts. This is the equivalent of 1/125 horsepower to 1/20 horsepower. The direction of rotation is from the unshaded portion of the pole to the shaded portion of the pole.

SPLIT-PHASE MOTORS

Split-phase motors, also referred to as **induction-start-induction-run (ISIR) motors,** have a relatively low starting torque compared with the motors that are discussed later on but more torque than the shaded-pole motor. They range in size from 1/20 horsepower to about 1/3 horsepower. Split-phase motors get their name from the fact that a single power supply is *split* between two individual windings—the run and the start—to produce

the necessary torque to start the motor (Fig. 15-8). The windings and motor wires are labeled as follows:

- COMMON *(C):* The wire or terminal common to both windings
- START *(S):* The wire or terminal unique to the start winding
- RUN *(R):* The wire or terminal unique to the run winding

For step-by-step instructions on how to identify the start and run windings of a split-phase motor using an ohmmeter, see Procedure 15-1 on pages 399–400.

The run winding is energized whenever the motor is energized. This winding has a lower resistance than the start winding, which is only in the circuit long enough to help the motor start. For the motor to start, both the start and run windings must be energized. When both windings are energized, the current flows through each of the windings at a different rate, creating a phase *shift.*

Briefly stated, this phase shift is what creates the imbalance needed to start the motor.

Phase shift is measured in electrical degrees and is often referred to as the phase angle. The larger the phase angle, the more starting torque a motor has. For a point of future reference, a split-phase motor, as just described, has a phase angle of about 30 degrees. If the run and start windings were constructed and configured exactly the same, with the same size wire and the same number of turns, the phase angle would be zero, the magnetic field would have no imbalance, and the motor would not start.

Once the split-phase motor has started, the start winding must be removed from the electric circuit. The start winding is designed to be energized for only a short time and can become damaged if it is not deenergized. One commonly used device to remove the start winding from the circuit is the **centrifugal switch** (Fig. 15-9), which opens and closes its contacts depending on the speed of the motor. The electrical contacts on the switch are connected in series with the start winding and are normally closed. Figure 15-10 shows the wiring configuration. It should be noted that the centrifugal switch is not typically found on air conditioning systems. This is because the centrifugal switch creates sparks when its contacts open and close and locating the device in a refrigerant atmosphere can damage the refrigerant. Centrifugal switches are commonly found on domestic appliances such as washing machines.

When voltage is initially applied to the motor, both the start and run windings are energized and the motor begins to turn. Once the motor has reached a speed equal to about 70 percent of its rated speed, the contacts of the centrifugal switch open, deenergizing the start winding. The run winding, or main winding, is now the only winding energized, and the motor continues to run in this fashion, since the turning rotor now creates the imbalance needed to keep the motor running. When the motor is deenergized, it begins to slow down and the centrifugal switch closes in preparation for the next startup. Other components can be added to the split-phase motor to increase its torque, as well as its range of applications. These components include

- the capacitor
- the current magnetic relay
- the potential relay

Capacitors are devices that store electrical charge and help give a motor the extra torque it needs during startup. Two common types of capacitor, named for their function in the circuit, are the **start capacitor** and the **run capacitor**. The *start capacitor* is intended to be in the electrical circuit only to assist the motor during startup. After the motor is up and running to speed, or at least close to running speed, it is removed from the circuit. *Run capacitors,* on the other hand, are designed to remain in the circuit as long as the motor is energized.

Capacitors and even motor windings can be electrically removed from the circuit as needed using specially designed relays. These relays include the centrifugal switch, the current magnetic relay, and the potential relay. The centrifugal switch, as mentioned earlier, opens and closes its contacts depending on the speed of the motor. The **current magnetic relay (CMR)** opens and closes its

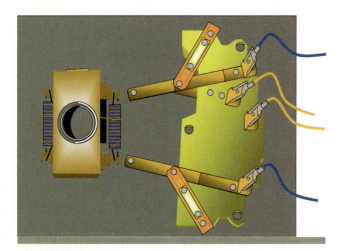

FIGURE 15-9 Centrifugal switch.

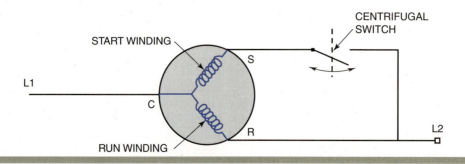

FIGURE 15-10 Split-phase motor wired with a centrifugal switch.

contacts in response to the current flowing through the run winding of the motor. Finally, the **potential relay** opens and closes its contacts depending on the induced voltage measured across the start winding of the motor. The term *induced* refers to the voltage that is generated in the winding as a result of the current flow through the surrounding windings. Another device that is used to remove the start winding from the active circuit is the **positive temperature coefficient**, or PTC device. The PTC is a device whose resistance increases as current flows through it and its temperature increases.

As mentioned earlier, starting torque is the push given to a motor enabling it to startup effectively. Up to this point, torque was obtained by the difference in magnetic fields generated between the run and start windings. Another way this torque can be increased is by using the capacitor (Fig. 15-11). As already noted, the capacitor is a device that stores electrical charge and helps increase

the phase angle in the motor, thereby increasing the starting torque and, in some cases, the running efficiency of the motor. Capacitors are rated in *microfarads,* denoted μ, indicating how much electrical charge they can store. The larger this rating, the more charge a capacitor can store. The capacitor also has a voltage rating, which represents the capacitor's ability to withstand voltage without breaking down.

Capacitors are constructed of two parallel, metal—usually aluminum—plates separated by an insulator called a *dielectric* (Fig. 15-12). The electric symbol for a capacitor (Fig. 15-13) indicates these two plates. The dielectric can be any one of a number of different substances, including air, paper, and aluminum oxide. The ability of the capacitor to store electrical charge is dependent on the size of the plates, the dielectric used, and the thickness of the dielectric.

RUN CAPACITORS

Run capacitors [Fig. 15-11(B)], which are intended to remain in the circuit whenever the motor is energized, are found in the range of 2 to about 60 microfarads. Although the primary function of the capacitor is to increase starting torque, the run capacitor is used mainly to increase the running efficiency of the motor. These capacitors are used on motors that are specially designed to have the start winding remain in the circuit whenever the motor is energized. The run capacitor helps correct and improve the *power factor* of the motor, increasing its efficiency. The power factor is a number, ranging from 0 to 1, that relates the power used, or paid for, to the output power obtained.

Photo by Eugene Silberstein.

Photo by Eugene Silberstein.

FIGURE 15-11 (A) Start capacitor. (B) Run capacitors.

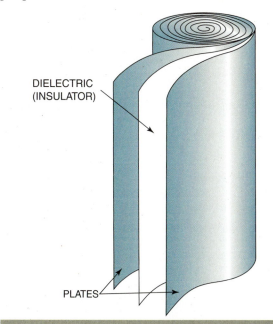

DIELECTRIC (INSULATOR)

PLATES

FIGURE 15-12 An insulator called a dielectric separates the metal plates in the capacitor.

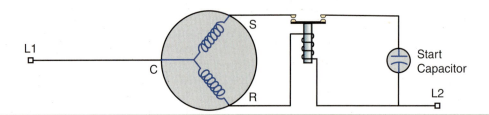

FIGURE 15-13 Note the electrical symbol for the capacitor at the right of the diagram.

The run capacitor, being in the active electrical circuit whenever the motor is energized, has a tendency to get hot. For this reason, it is enclosed in a metal case filled with oil that helps absorb the heat from the plates and transfers it to the metal shell of the device. The heat is then transferred to the air surrounding the device. Therefore, the run capacitor is referred to as a *wet* capacitor. There are some exceptions to this as some manufacturers use newly designed run capacitors that appear much different than the standard run capacitor (Fig. 15-14). Be sure to check the markings on the capacitor to be sure. The distinctions between start and run capacitors are summarized in Figure 15-15.

RUN CAPACITORS AND POWER FACTOR

Electrical circuits can contain a number of different load types. These types include resistive, capacitive, and inductive. When circuits contain only resistive loads, such as electric strip heaters, the current and voltage follow the same sine-wave shaped path (Fig. 15-16). The term used to describe this condition is that the voltage and current are in phase with each other. The term in phase simply states that the current and the voltage rise and fall at the same rate. The power factor of resistive circuits is 1, so that the power that is produced is calculated by multiplying the voltage and the current as shown here:

$$POWER = VOLTAGE \times CURRENT$$

When we energize electric circuits that contain inductive loads, such as electric motors or solenoid coils, the current and voltage are not in phase with each other. In a purely inductive circuit, the current in the circuit lags the voltage by 90 electrical degrees (Fig. 15-17). The term "lag" indicates that the waveform of the circuit current is behind that of the voltage waveform. When this occurs, the power that is produced will be lower than the result obtained by simply multiplying the voltage and current values. The power factor provides the needed correction:

$$POWER = VOLTAGE \times CURRENT \\ \times POWER\ FACTOR$$

Photo by Eugene Silberstein.

FIGURE 15-14 Plastic-cased run capacitor.

	Run Capacitors	**Start Capacitors**
Range	2–60 microfarads	75–600 microfarads
Construction	• Wet type • Oil-filled • Metal casing	• Dry type • Bakelite casing
Function	• In the active circuit whenever the motor is energized • Designed to increase running efficiency of the motor	• Only in the circuit for a few seconds • Designed to increase motor's starting torque

FIGURE 15-15 Summary of start and run capacitors.

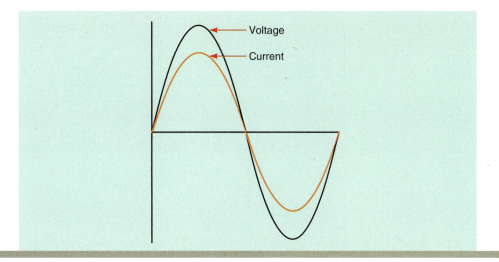

FIGURE 15-16 In a resistive circuit, the voltage and current are in phase with each other.

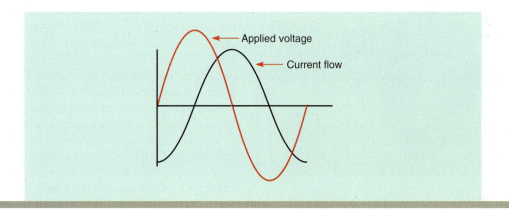

FIGURE 15-17 In an inductive circuit, the current lags the voltage.

For example, if the power factor in an inductive circuit is 0.5, the voltage is 100 volts and the amperage is 5 amps, the calculated power will be:

$$POWER = 100 \text{ V} \times 5 \text{ A} \times 0.5 = 250 \text{ W}$$

If this circuit were purely resistive, then the calculated power would have simply been the voltage multiplied by the current, or 500 Watts.

When we energize electric circuits that contain capacitive loads, which are capacitors, the current and voltage are also out of phase with each other. In a purely capacitive circuit, the current in the circuit leads the voltage by 90 electrical degrees (Fig. 15-18). The term "lead" indicates that the waveform of the circuit current is ahead that of the voltage waveform. When this occurs, the power that is produced will be lower than the result obtained by simply multiplying the voltage and current values.

Since inductive and capacitive loads have opposite effects on the voltage and current waveforms, run capacitors are added to inductive motor circuits to help maximize the power factor and the efficiency of the motor.

START CAPACITORS

Start capacitors are used primarily to add starting torque on motor startup. They are commonly found in ranges from about 75 to 600 microfarads and are easily identified by their *Bakelite* casing [Fig. 15-11(A)], which is a dark, plastic-like substance. Start capacitors are considered *dry* because the plates and the dielectric are placed in the casing with no surrounding fluid. The start capacitor is wired in series with the start winding and is removed from the circuit along with the start winding when the centrifugal switch, or other starting device, opens.

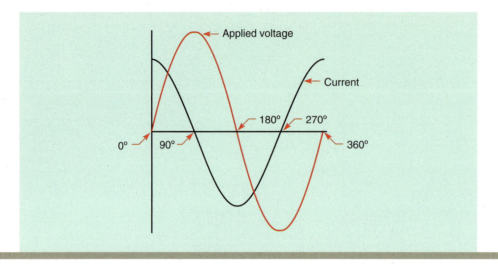

FIGURE 15-18 In a capacitive circuit, the current leads the voltage.

Photo by Eugene Silberstein.

FIGURE 15-19 The current relay is identified by the size of the wire in the holding coil.

CURRENT MAGNETIC RELAYS (CMR)

Another device that can be used in conjunction with split-phase motors is the CMR (Fig. 15-19). This is commonly found on fractional-horsepower motors that require low starting torque. This relay can be used to remove the start winding and the start capacitor from the circuit or, in the cases in which the start winding remains energized during normal operation, just the start capacitor. Unlike the centrifugal switch, which opens and closes its contacts depending on the speed of the motor, the CMR's contacts are opened and closed depending on the current flowing through the run winding of the motor. When a motor is initially energized, the amperage draw is very high, roughly five times the amount of current flowing through the motor when it is up and running up to speed. This amperage is called the **locked-rotor amperage (LRA)**. The amperage that the motor draws after startup is only a fraction of the LRA and is called the **running-load amperage (RLA)**. When the RLA is close to the full-load condition of the motor, it is referred to as the full-load amperage (FLA).

The CMR is made up of two parts: a coil of wire and a normally open set of contacts. The wire in the coil is rather thick, and the number of turns is very low, making the resistance of the coil very low, usually less than 1 ohm. The coil is wired in series with the run winding and, therefore, has the same current flowing through it as the run winding. The coil, with its low resistance, has a small voltage drop across it, making its effect on the voltage across the run winding negligible. The current flowing through the run winding is zero when the motor is de-energized; RLA, when the motor is up and running to speed; or LRA, when the motor is in the process of starting up. When the current drawn by the run winding is equal to the LRA, a strong magnetic field is generated, compared with the field generated when the motor is drawing RLA.

The set of electrical contacts in the relay is held open by either spring pressure or, more commonly, by gravity. This set of contacts is called the bridging contact and is connected to an **armature**, which moves back and forth within the coil (Fig. 15-20). These contacts are wired in

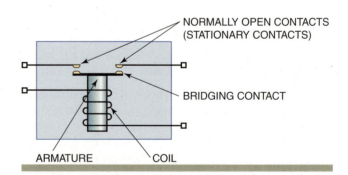

FIGURE 15-20 Coil and contacts in a current relay. The bridging contact is connected to the armature, which moves within the relay's coil.

series with the start winding and, when open, remove this winding from the circuit. The contacts of the relay should be open when the motor is de-energized, as well as when the motor is up and running to speed (Fig. 15-13).

Going Green

Current relays are rated by current. Be sure that the relay being used matches the ratings of the motor. Motors that draw locked rotor amperage for prolonged periods of time use more power and burn out faster.

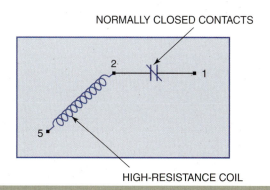

FIGURE 15-21 Potential relays.

In operation, when the motor is initially energized, the run winding is energized but the start winding is not. This will prevent the motor from turning, because no imbalance exists in the magnetic field. The amount of current flowing through the run winding at this point will be LRA. The increased current draw creates a large magnetic field in the coil of the CMR. This force pushes the normally open contacts of the relay closed, completing the electric circuit through the start winding. Having energized the start winding, the necessary imbalance is created, allowing the motor to start. Once started, the current draw of the motor quickly drops from LRA to RLA. This reduced current reduces the strength of the magnetic field holding the contacts closed, and the contacts fall open. The motor then continues to run with only the main winding energized. When the motor is de-energized, the contacts remain open until the next startup.

For step-by-step instructions on how to check a current magnetic relay, see Procedure 15-2 on pages 401–402.

POTENTIAL MAGNETIC RELAYS

Another device that is commonly used in various types of split-phase motors is the potential relay (Fig. 15-21), sometimes referred to as a potential magnetic relay (PMR). The function of the PMR is similar to that of the CMR in that it can de-energize the start capacitor and/or the start winding; but, instead of sensing the current in the run winding, the potential relay controls its contacts by sensing the induced voltage, or potential, across the start winding. The PMR is commonly found on CSCR motors requiring more starting torque than those that use the CMR.

The potential relay consists of a normally closed set of contacts and a coil. The coil is wired in parallel with the start winding and is used to sense the induced voltage across the start winding. *Terminal 2* to *Terminal 5* on the device (Fig. 15-22) identifies the coil. Since the installation of the relay does not reduce the current flow through the start winding, the resistance of the coil is obviously very high, since current takes the path of least resistance. The coil resistance is typically in the range of 15,000 to 50,000 ohms. This concept is similar to that of a voltmeter, where the resistance of the test instrument is very high to allow only a very small amount of current to flow through the meter. The normally closed set of contacts on the PMR is identified by Terminal 1 to Terminal 2 (Fig. 15-22).

The potential relay's coil, in effect, acts as a voltmeter and "measures" the induced voltage across the start winding. The coil then causes the contacts to open and close, depending on this reading. The voltage that is sensed across the start winding can easily be greater than the voltage supplied to the motor. This is because the total voltage that the coil senses is the sum of the actual voltage across the coil and the induced voltage generated by the magnetic field of the run winding.

NORMALLY CLOSED CONTACTS

HIGH-RESISTANCE COIL

FIGURE 15-22 Wiring diagram of a potential relay.

The relay is sensitive to two specific voltages called the **pick-up voltage** and the **drop-out voltage**. The pick-up voltage is the voltage at which the relay's contacts open, and the drop-out voltage is where the relay returns its contacts to their normally closed position.

In operation, when the motor is initially energized, both the start and run windings are energized and the motor starts. As the motor picks up speed, the induced voltage across the start winding increases, since the voltage-generating effect increases. Once this voltage increases to the pick-up voltage, the PMR's contacts open. Generally, this removes only the start capacitor from the circuit. If the motor experiences difficulty running and begins to slow down, the induced voltage will drop and the relay's contacts will close when the voltage falls to the drop-out voltage. This will put the start capacitor back into the circuit and allow the motor to regain its speed. Once again, when the induced voltage rises to the pick-up voltage, the contacts will open. A summary of the centrifugal switch, the CMR, and the potential relay is given in Figure 15-23.

For step-by-step instructions on how to check a potential relay, see Procedure 15-3 on page 403.

POSITIVE TEMPERATURE COEFFICIENT (PTC) DEVICES

The **positive temperature coefficient (PTC)** device is another method that is commonly used to aid in the starting of a split-phase motor (Fig. 15-24). The PTC is a thermistor, or variable resistance device, whose resistance increases as the sensed temperature increases. When cool, the resistance of the device is very low (Fig. 15-25). When cool, both the start and run windings of the motor are in the active electric circuit and full voltage is being supplied to the start winding.

Once the motor has been energized, the PTC quickly heats up and its resistance increases. This greatly reduces the amount of current flow through the PTC device and directs current through the run capacitor (Fig. 15-26).

	Centrifugal Switch	Current Relay	Potential Relay
Controlled by	Speed of the motor	Current flow through the run winding	Induced voltage across the start winding
Coil Characteristics	N/A	• Low resistance • Thick wire • Few turns	• High resistance • Thin wire • Many turns
Coil location	N/A	Wired in series with the run winding	Wired in parallel with the start winding
Normally Open or Normally Closed Contacts	Normally closed	Normally open	Normally closed
Operation	• Contacts open when 70% of total motor speed is reached • Contacts close when motor is de-energized	• Contacts close when locked-rotor amperage (LRA) is drawn • Contacts open when the motor amperage falls to running-load amperage (RLA)	• Contacts open when the included voltage reaches the pick-up voltage • Contacts close when the voltage drops to the drop-out voltage
Application	Used primarily on open motors	• Used primarily on low-torque and low-horsepower motors • Used primarily on CSIR motors	• Used primarily on motors requiring high starting torque • Used in larger-horsepower motors • Used on CSCR motors
Installation	• Not position sensitive • Must be located on the motor	• Must be mounted right-side up • Must be connected directly to the motor	• Normally not position sensitive • Does not need to be located near the motor

FIGURE 15-23 Summary of starting switches and relays.

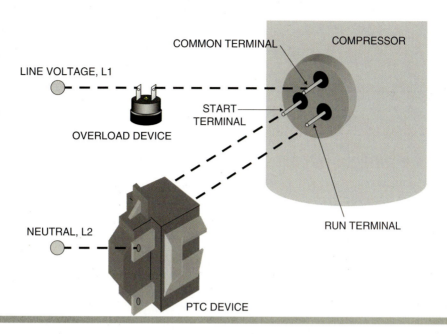

FIGURE 15-24 A positive temperature coefficient (PTC) device.

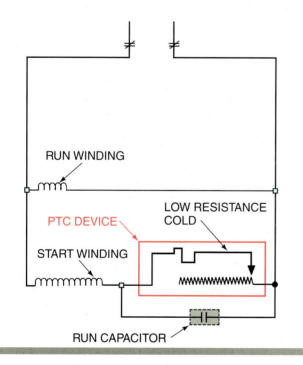

FIGURE 15-25 Cold PTC device in a motor circuit.

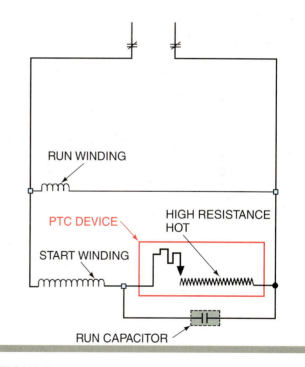

FIGURE 15-26 Hot PTC device in a motor circuit.

Another wiring option is to wire the motor using the PTC as shown in Figures 15-24 and 25, but leaving out the run capacitor. This will allow the motor to operate as a capacitor start induction run motor. When the motor is initially energized, both the start and run windings will be in the active electric circuit. Once the motor is running, the increased resistance of the PTC will, in effect, remove the start winding from the circuit, allowing the motor to operate with only the run winding energized.

CAPACITOR-START-INDUCTION-RUN MOTORS

The **capacitor-start-induction-run (CSIR) motor** is similar in operation to the ISIR motor. The start winding is energized long enough to assist the motor in starting and is then removed from the circuit using either a centrifugal switch or a CMR. Potential relays are not commonly used in conjunction with a CSIR motor,

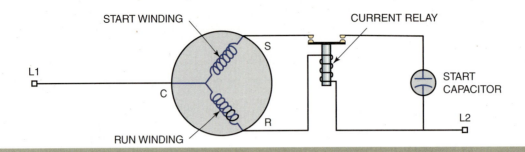

FIGURE 15-27 CSIR motor using a current magnetic relay. When the relay's contacts open, the start winding and the start capacitor are removed from the circuit.

because they are designed for motors requiring more starting torque.

The main difference between an ISIR motor and a CSIR motor is the addition of a start capacitor. The start capacitor, the start winding, and the contacts of the switching device are all wired in series with each other so, when the contacts open, the start winding and the start capacitor are both removed from the circuit. Figure 15-27 shows a wiring diagram of a CSIR motor that is wired with a current magnetic relay.

The CSIR motor's phase angle approaches 90 degrees, compared with the 30-degree phase angle obtained by an ISIR motor without the start capacitor. The CSIR motor is often used for small compressors that must be started under full load, since they provide more starting torque and range in size from about 1/6 horsepower to 3/4 horsepower.

PERMANENT-SPLIT-CAPACITOR MOTORS

Of all the motors that use capacitors, the **permanent-split-capacitor (PSC) motor** has the lowest starting torque. This type of motor has very good running efficiency and is a popular choice for low-torque applications since no

starting relays or switches are needed. The PSC motor uses a small run capacitor that, along with the start winding, is in the active circuit whenever the motor is energized. The run capacitor helps to increase the power factor. A wiring diagram of a single-speed PSC motor is shown in Figure 15-28. The start winding in this type of motor is specially designed to remain in the circuit at all times.

PSC motors are widely used in applications in which the blower is located close to the occupied space, since the low-starting-torque characteristic of the motor causes it to start slowly and quietly.

PSC motors can be one speed or multispeed components. If the motor operates at only one speed, typically it has three wires protruding from the motor's shell labeled Common, Start, and Run. If the motor has multiple speeds, it has a common wire, a start wire, and one wire for each speed. For example, if the motor has three speeds, the wiring diagram will look similar to that in Figure 15-29. As the winding resistance increases, the current draw drops and the strength of the magnetic field drops, thereby reducing the motor speed. It can be seen from the figure that the resistance of the motor is higher when the low speed lead is connected and lower when the higher speed lead is used. The lower resistance allows more current to flow, increasing the motor speed.

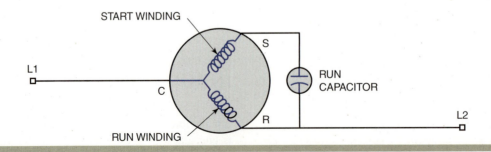

FIGURE 15-28 Schematic of a PSC motor. Note that the start and run windings are energized whenever the motor is running.

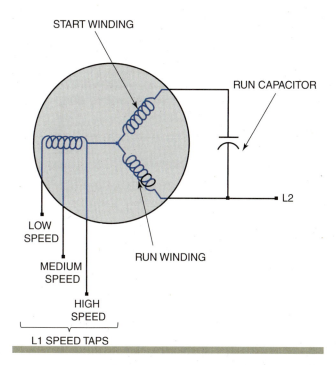

FIGURE 15-29 Schematic of a three-speed PSC motor.

CAPACITOR-START-CAPACITOR-RUN MOTORS

Of all the single-phase motors discussed in this text, the **capacitor-start-capacitor-run (CSCR) motor** has the highest starting torque; hence, the largest phase angle and excellent running efficiency. It ranges from one-half to 10 horsepower. This motor commonly uses a potential relay as a starting device and uses both a large start capacitor and a small run capacitor. The start winding is designed to remain energized all the time. A wiring diagram for a CSCR motor is shown in Figure 15-30. When the potential relay's contacts open, the motor resembles a PSC motor which, as already noted, has excellent running efficiency but very low starting torque. Adding starting torque to the PSC motor creates, in effect, the CSCR motor.

On initial startup, the run and start capacitors are connected in parallel with each other. When wired in this fashion, the effective capacitance obtained is the sum of the values of the individual capacitors. If, for example, the run capacitor is rated at 5 microfarads and the start capacitor is rated at 100 microfarads, then the total capacitance is 105 microfarads on startup but only 5 microfarads once the relay's contacts open.

ELECTRONICALLY COMMUTATED MOTORS (ECM)

Equipment manufacturers, operating under governmental pressure and their own desires to produce more efficient air conditioning systems, are turning to more efficient motors. Energy-efficient systems often incorporate electronically commutated motors, commonly known as ECM motors, to turn the system blowers and fans (Fig. 15-31). ECM motors are very unique. They are powered by single-phase alternating current power, but actually operate as three-phase direct current motors. Through a series of electric circuits, the single-phase alternating current is converted to direct current, which is then altered to produce, in effect, a three-phase power supply to the motor. One of the keys in the success of the ECM motor is the use of permanent magnets to commutate (magnetize) the armature of the motor.

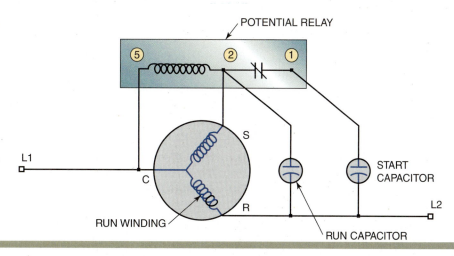

FIGURE 15-30 Wiring diagram of a CSCR motor using a potential relay. The relay removed the start capacitor from the circuit.

FIGURE 15-31 An electronically commutated motor (ECM).

Going Green

The ECM motor is factory calibrated and programmed to best suit the needs of the system it is installed in. A specially designed logarithm incorporates motor speed, torque, airflow requirements, and static pressure for the particular application. This helps provide for efficient operation, improved comfort levels, and improved humidity control.

Presently, the most commonly used motor for evaporator and condenser air moving is the permanent split capacitor (PSC) motor, which was discussed earlier in this chapter. ECM motors are more efficient than the PSC motor and, as a result, can substantially reduce fan-operating costs. The ECM motor is fast becoming the popular motor of choice, both for new equipment as well as for PSC replacement on existing equipment. Some ECM motors are designed to operate at variable speeds, while others are intended for constant CFM applications. Variable-speed motors are beneficial to control airflow through the evaporator coil on an air conditioning system, while single-speed motors can be used as a replacement option for condenser fan motors.

Electronically commutated motors often have **soft start** and **soft stop** features that reduce power consumption, wear and tear on the motor, and noise levels. The soft start feature refers to the starting speed of the motor. For example, a motor might be programmed to start at 50% of its total speed and then, after a predetermined time period, ramp up to its maximum operating speed. This feature reduces the amperage draw at start up, wear and tear on the motor, as well as the noise levels generated by the motor. The Soft stop feature refers to the shut down process of the motor. Instead of going from 100% of total speed to 0%, the motor will gradually step down in speed. Since the motor does not typically operate at full speed all the time, the motor operates with lower surface temperatures than standard motors. This provides an added benefit as less heat will be added to the air passing over the motor and through the air distribution system. In addition to the benefits already mentioned, the soft start and soft stop features also help improve the dehumidification of the occupied space.

THREE-PHASE MOTORS

The motors discussed so far have all been single-phase motors. The difference in resistance between the start and the run windings, as well as the start capacitors, was used to increase the phase angle, creating the effect of more than one phase, thereby generating starting torque. We will now examine a type of motor that uses three phases of power, generating a phase angle of 120 degrees without using capacitors or starting relays. Called the **three-phase motor**, it generates the highest amount of starting torque. It has extremely high running efficiency and is the motor of choice whenever three-phase power is available for use. Although three-phase power and motors are typically not found in the residential setting, there are some rare cases where it is found. So, for the sake of completeness, three-phase power is being discussed here.

Three-phase motors are made up of three run windings, as opposed to the start and run windings found on single-phase motors. Three hot lines identified as L1, L2, and L3 feed into the device to power the motor. For a 240-volt, three-phase power supply (Fig. 15-32), the voltage measured between any two of the hot "legs" is 240 volts. Similarly, a 220-volt, three-phase power supply would provide 220 volts:

- Between L1 and L2
- Between L2 and L3
- Between L1 and L3

Note that each of the three hot legs is responsible for generating a portion of two of the phases. For example, L1 and L2 generate one phase, while L1 and L3 generate

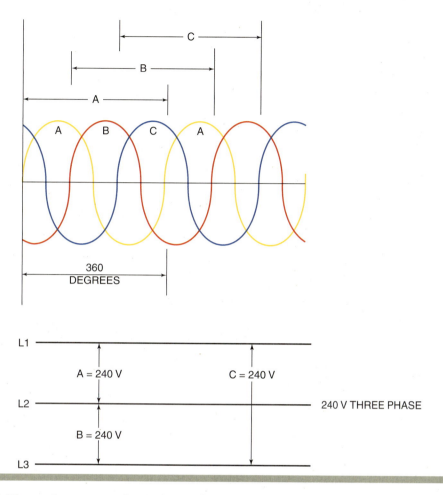

FIGURE 15-32 Three-phase power is made up of three single-phase supplied, each of which is 120 degrees out of phase with the next.

a second phase. The third phase, generated by L2 and L3, does not involve L1 at all. This poses a potential problem.

If one power leg of a single-phase motor opened for whatever reason, the entire motor would shut down because there would be no path for the current to take. In a three-phase motor, this would not be the case. If one leg opened, only two of the three phases would be disabled, leaving the third phase energized.

In this case, the motor may or may not continue to run, depending on the specific application. The motor will, however, overheat due to an increase in the current draw of the motor. The motor will attempt to continue to operate on one phase and will need excessive current to do so. This can cause severe motor damage and, since these motors are typically very expensive, every attempt should be made to prevent this from occurring. This phenomenon is referred to as single phasing and cannot be avoided since something as simple as a blown fuse can cause it. Although a single-phase condition is, at best, difficult to avoid, motor damage is not. By using a device called a motor starter, three-phase motors are protected from this condition.

FIGURE 15-33 Typical three-phase motor starter.

MOTOR STARTERS

Motor starters are, simply stated, contactors with built-in overload protection, (Fig. 15-33). The built-in overload consists of **heaters** and **switches**. The heaters are located in the power circuit, in series with the motor, and the switches are in the control circuit, in series with

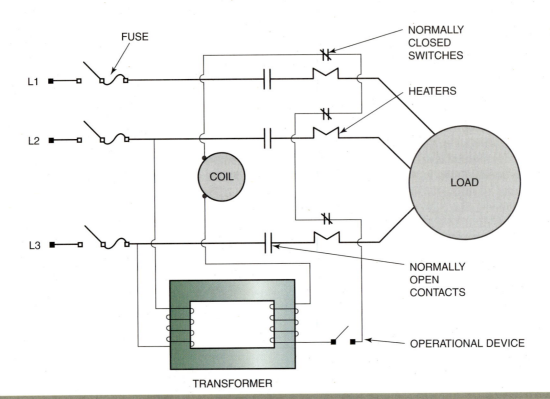

FIGURE 15-34 Wiring diagram for a low-voltage starter.

the holding coil of the starter. If the heaters detect a temperature that is higher than normal, they will cause the switches to open, de-energizing the holding coil of the starter. This, in turn, will open the contacts feeding power to the motor, de-energizing it. The heat that is sensed is a result of the excessive amperage drawn by the motor attempting to operate on one phase, since heat is a by-product of current. A wiring schematic for a starter is shown in Figure 15-34.

The holding coil for the starter cannot be energized unless the operational devices (thermostat or other control) and all three switches in the overload are closed. If, after startup, any one of the fuses blows or one power leg fails to supply voltage, the current draw in that leg will be zero but the current in the other two legs will increase. The heaters in those two lines will sense the excess heat and cause the normally closed switches in those lines to open, de-energizing the holding coil. The starter is reset manually, so the motor will be unable to run until the problem is located and corrected. The starter can then be reset to put the motor back into operation. Once the starter has been reset, current is permitted to flow to the motor windings, and operation resumes. The three-phase motor windings can be configured in one of two ways:

- Wye configuration
- Delta configuration

STATOR WINDINGS: WYE CONFIGURATION

As mentioned earlier, three-phase motors have three run windings, but no mention was made of how these windings were connected to each other. In reality, the windings can be arranged in a number of different configurations, and each one is addressed separately. The first configuration discussed is the wye configuration, which gets its name from the Y shape in the wiring diagram. Figure 15-35 shows a simple, three-phase motor with a wye-configured stator. Each of the three power lines supplying the motor is connected to wires T1, T2, and T3. This type of three-phase motor is designed to operate at a specific voltage and can be easily identified by the fact that only three wires or terminals are available for external wiring connections.

Other types of three-phase motors, called dual-voltage motors, can be used at different supply voltages. For example, three-phase motors that can operate at 220 volts as well as 440 volts are very common. In the case of dual-voltage motors, nine wires protrude from the motor shell instead of the usual three. Each of these wires is numbered, and care must be taken when wiring the motor or serious component damage could occur. The three windings in a dual-voltage motor are each divided into two

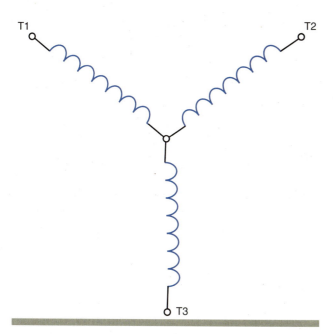

FIGURE 15-35 Three windings connected in a wye configuration.

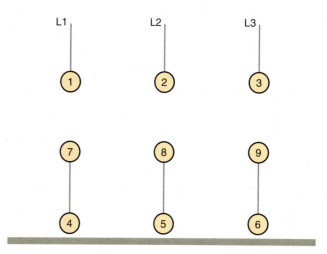

FIGURE 15-37 Wiring diagram for the high-voltage connections on a dual-voltage motor.

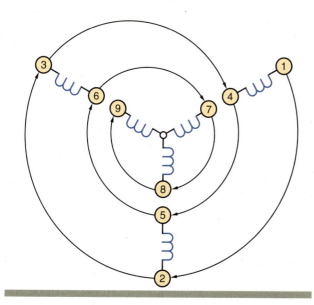

FIGURE 15-36 Methods used to number the wires in a dual-voltage motor.

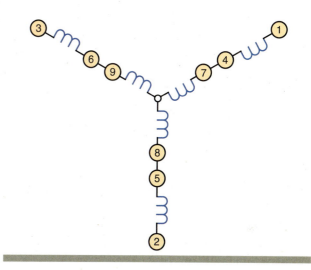

FIGURE 15-38 Windings connected in series.

HIGH-VOLTAGE CONNECTION

When operating at the higher of the two voltages specified on the motor nameplate, the windings are connected in series with each other, forming a large Y. Figure 15-37 shows the wiring diagram as it may appear on the motor nameplate. Figure 15-38 shows how the windings look after the connections are made. In this configuration, the windings resemble those in the simple three-phase motor in Figure 15-35. When operating at the higher voltage, the current draw of the motor is one-half that of the same motor operating at one-half the voltage.

Low-Voltage Connection. When wired to operate at the lower of the two voltages, the windings are connected

smaller windings, making a total of six. Figure 15-36 shows how the windings appear on a wiring diagram and how the wires are numbered. Dual-voltage motors normally have wiring diagrams on the nameplate, providing the field technician with the proper wiring connections for the desired voltage application. The diagrams are normally labeled high voltage and low voltage. In the case of a 220-volt/440-volt motor, high voltage refers to 440 volts and low voltage refers to 220 volts.

in parallel with each other, forming two smaller Y configurations. Figure 15-39 shows the diagram for the low-voltage wiring connections. Figure 15-40 shows how the connections may look. Notice in the figure that *L3* is connected to *Wire 3* and *Wire 9,* putting those two windings in parallel with each other. Similarly, the other windings are also in parallel with each other. Since *Wire 4, Wire 5,* and *Wire 6* are all connected together, another Y connection is formed. The amperage draw of a dual-voltage motor wired to operate at the lower voltage will be twice that of an identical motor wired to operate at the higher voltage. This is because the resistance of the windings is lower, since the windings are connected in parallel.

STATOR WINDINGS: DELTA CONFIGURATION

Another common layout for the windings in three-phase motors is the **delta configuration**. A three-phase motor with the stator wired in the delta configuration will have twelve wires protruding from the motor's shell, instead of the nine wires in the wye configuration. The numbering system used in the delta configuration is similar to that in the wye configuration (Fig. 15-41). Just as in the wye configuration, the high-voltage application places the windings in series, but the low voltage application arranges the windings in parallel. Figure 15-42 shows the high-voltage

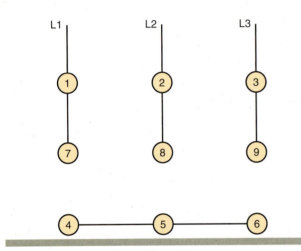

FIGURE 15-39 Wiring diagram for the low-voltage connections on a dual-voltage motor.

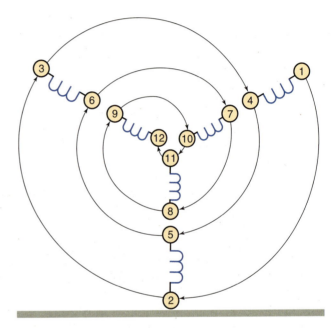

FIGURE 15-41 Numbering system used for a dual-voltage motor with a delta-configured stator.

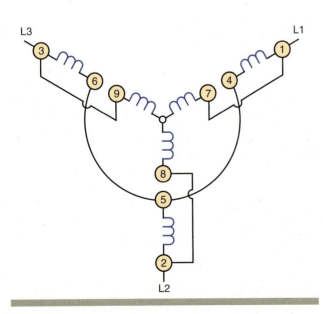

FIGURE 15-40 Windings connected in parallel.

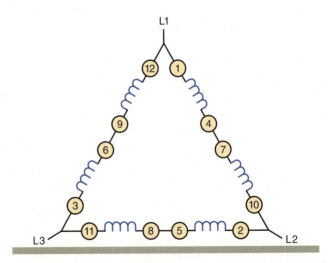

FIGURE 15-42 High-voltage delta configuration.

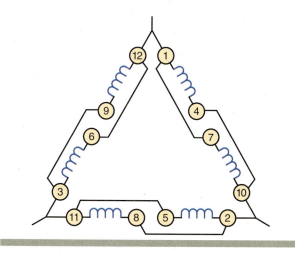

FIGURE 15-43 Low-voltage delta connection.

connections, while Figure 15-43 shows the low-voltage connections.

THREE-PHASE MOTOR STARTING

Since three-phase motors are commonly used for large commercial and industrial applications, the motors themselves tend to be very large. They draw large amounts of current during normal operation and even larger currents during initial startup. For example, a motor that draws 60 amperes during normal operation may draw over 300 amperes during startup. This excessively high startup amperage can cause the contacts on the starters to become pitted and to fail prematurely. For this reason, a number of methods have been implemented to reduce the amount of current that a motor draws on startup.

Smaller, three-phase motors can be effectively started using a method similar to that shown in Figure 15-34. The larger the motor, the more important the effects of the startup amperage become. The following applications are employed primarily on large motors, usually exceeding 25 horsepower. Two of the most common methods used to reduce this amperage are the

- Part-wind start
- Reduced-voltage start

PART-WIND START

To reduce the amperage drawn by the motor on initial startup using a part-wind start, only half of the motor's windings are energized. Figure 15-44 shows a wiring diagram of a part-wind start. The part-wind start is configured as a dual-voltage motor, with a wye-type stator, wired for the low-voltage application. This creates two

small wyes, each of which can be energized at a different time. If the two wyes, which are connected in parallel with each other, are energized at the same time, the resistance will be low and the current will be high.

(Remember from Ohm's law that two resistors connected in parallel with each other will have a combined total resistance that is less than the resistance of each individual resistor.) The LRA of this motor will therefore be very high. By energizing only half of the windings, the resistance is higher and the current is reduced. The LRA drawn by energizing only half of the motor is therefore reduced. Once the motor begins to rotate, the second half of the windings are introduced into the circuit. The sequence of operations is as follows:

1. The operational device (thermostat or other control) closes.
2. As long as the six normally closed switches in the overloads are closed, the holding coil of Starter 1 will be energized.
3. When Coil 1 is energized, the four sets of normally open contacts will close. (Three sets of contacts feed current to the first half of the windings, and one set of contacts completes the circuit through the time-delay relay.)
4. The first half of the windings are energized, as well as the coil on the time-delay relay.
5. The motor starts turning with only one half of the windings energized.
6. After the time-delay period, usually one second, the contacts on the time-delay relay are closed.
7. This completes the circuit through the holding coil of Starter 2.
8. The three normally open sets of contacts on Starter 2 are closed.
9. The second half of the windings are energized.

If any one of the six power leads feeding the motor becomes defective and stops feeding current to the motor, the other heaters will cause the entire motor to shut down. Again, the overloads on the starters are manually reset, so the motor will not operate until the problem is found and corrected.

REDUCED-VOLTAGE START

In the part-wind start, all of the voltage was applied to part of the windings to reduce the current draw on initial startup. The reduced-voltage start accomplishes the same end result but by different means. This type of motor starter supplies a lower voltage to all of the windings until the motor starts, then the voltage is increased to the normal operating voltage. Figure 15-45 shows a schematic of a reduced-voltage start. The underlying

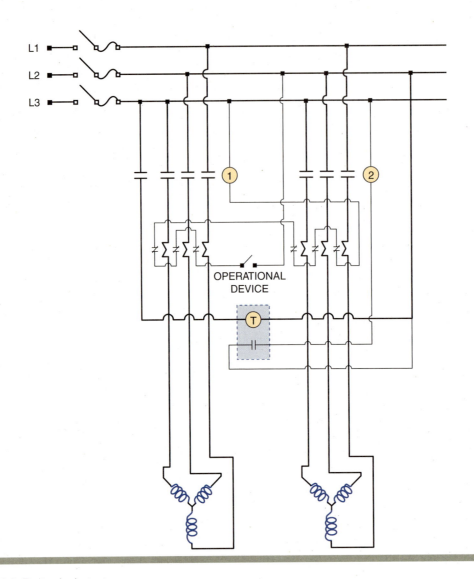

FIGURE 15-44 Part-wind start.

concept of the reduced-voltage start is to place a large resistor in series with each winding of the motor.

Since a voltage drop will occur across the resistors, the voltage that is ultimately applied to the motor will be lower than the supplied line voltage. In addition, since the resistors are added to the circuit, the current in the circuit will be reduced. The motor is started in this fashion. Once the motor begins to turn, the resistors are removed from the circuit. This causes the circuit resistance to drop, thereby increasing the current flowing through—and the voltage supplied to—the motor windings. The sequence of operations of this type of motor starter is as follows

1. The operational device (thermostat or other control) closes.

2. As long as the three normally closed switches in the overloads are closed, the holding coil of Starter 1 will be energized.

3. When Coil 1 is energized, the three sets of normally open contacts will close. These contacts feed current through the three large resistors to the motor windings.

4. Once the three normally open sets of contacts close, the motor starts at a reduced voltage since a substantial voltage drop occurs across the large resistors.

5. The coil of the time-delay relay is also energized when the normally open contacts on the starter close.

6. After the time-delay period has elapsed, usually 2 seconds, the relay's contacts close.

7. The circuit through Holding Coil 2 is now completed.

8. The normally open contacts on Contactor 2 are now closed.

9. Since current takes the path of least resistance, the current now bypasses the large resistors.

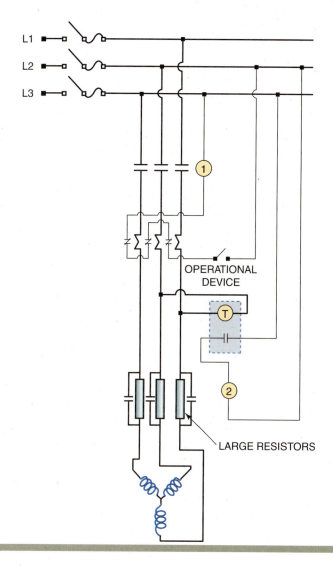

FIGURE 15-45 Reduced-voltage start.

10. The circuit resistance is reduced.
11. The circuit current increases.
12. The voltage applied to the motor is now equal to the supply voltage.

SUMMARY

- Commonly found motors include the shaded-pole motor, the split-phase motor, and the three-phase motor.
- Starting torque is the electric push given to motors to help them start.
- Motors used in lightweight applications typically have low starting torque.

- Motors are made up of a stator, which does not turn, and the rotor, which rotates and is connected directly to the shaft of the motor.
- Split-phase motors have a start winding and a run winding.
- The start winding is a high-resistance coil made up of thin wire with a large number of turns.
- The run winding is a low-resistance coil made up of thick wire with a small number of turns.
- Common types of split-phase motors include the ISIR, the CSIR, the CSCR, and the PSC.
- A split-phase motor can use centrifugal switches, current magnetic relays, or potential relays to remove the start winding and/or start capacitor from the circuit.
- Centrifugal switches operate depending on the speed of the motor.

- Current magnetic relays operate based on the current in the run winding.
- Potential relays operate depending on the induced voltage across the start winding.
- Electronically commutated (ECM) motors are more efficient than standard PSC motors.
- ECM motors operate with lower power consumption and noise levels and less vibration than standard motors.

- Three-phase motors use motor starters, not starting relays or capacitors.
- Motor starters have built-in overload protection.
- Three-phase motors often use part-wind or reduced-voltage starts.

GREEN CHECKLIST

☐ Select the proper motor for the task being performed.

☐ Check to make certain that the starting components being used have ratings that match the motors they are used on.

☐ ECM motors draw less amperage and are more efficient than standard PSC motors.

☐ ECM motors help improve dehumidification in the occupied space.

PROCEDURE 15-1

Identifying the Common, Start, and Run Terminals on a Split-Phase Motor

A Draw a diagram showing the compressor terminals and number them.

• Take resistance readings between terminals 1 and 2, 2 and 3, and 1 and 3.

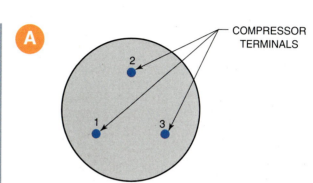

COMPRESSOR TERMINALS

B Record these readings on the diagram.

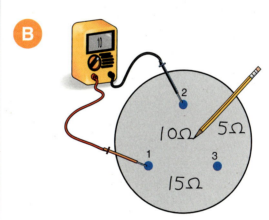

C Locate the highest resistance reading. The terminal opposite the highest reading is the Common terminal.

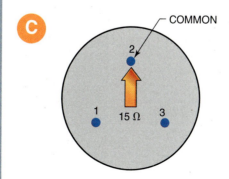

COMMON

PROCEDURE 15-1

Identifying the Common, Start, and Run Terminals on a Split-Phase Motor (continued)

D Locate the lowest resistance reading. The terminal opposite the lowest reading is the Start terminal.

• The remaining terminal is the Run terminal and is opposite the middle resistance reading.

D
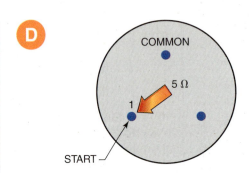

E The internal windings are connected as in the figure. The start winding has a resistance of 10 Ω, while the run winding has a resistance of 5 Ω.

E
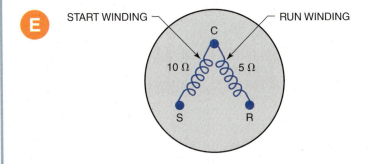

PROCEDURE 15-2

Checking the Coil and Contacts on a CMR

 To check the coil of a CMR, locate the coil terminals on the relay.

- Take a resistance reading of the coil. If the reading is in the 1 Ω range, the coil is good. If the reading is infinite, the coil is open and the relay must be replaced.

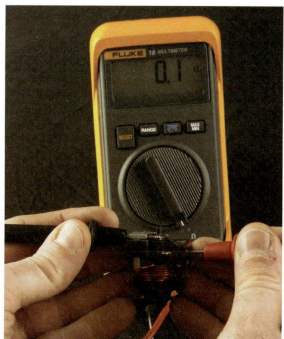

Delmar/Cengage Learning.

from experience...

When checking the resistance of a current magnetic relay coil, use the R × 1 range on the ohmmeter if there is one.

 To check the contacts on the relay, first locate them and then hold the relay in the upright position as it would be mounted on a compressor.

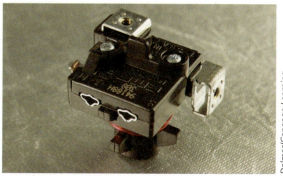

Delmar/Cengage Learning.

PROCEDURE 15-2

Checking the Coil and Contacts on a CMR (continued)

C Take a continuity reading across the contacts. There should be no continuity (infinite resistance) across the contacts.

Delmar/Cengage Learning.

D With the test leads still connected to the contacts of the relay, turn the CMR upside down. There should now be continuity across the contacts. If there is no continuity through the contacts, the relay should be replaced.

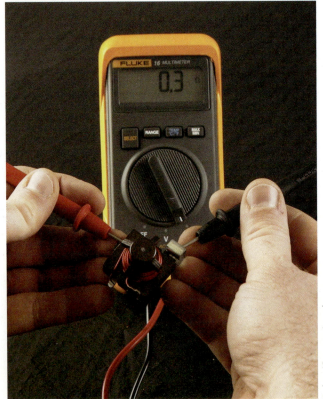

Delmar/Cengage Learning.

CAUTION

CAUTION: Step D refers to CMRs whose contacts are held open by gravity. On relays that are held open by spring tension, there will be no continuity through the contacts if the relay is turned upside down. Gently shaking the relay in your hand will indicate if the contacts are held open by gravity: If you can hear the contacts shifting as the relay is being shaken, the contacts are opened by gravity.

PROCEDURE 15-3

Checking the Coil and Contacts on a PMR

A To check the contacts on the relay, take a resistance reading between terminals 1 and 2 on the relay. A good relay will have a very low resistance reading. A reading of infinity indicates that the contacts are defective.

A

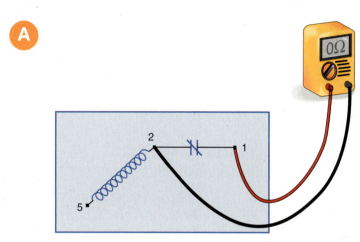

B To check the coil on the relay, first set the meter to the highest resistance scale. Then, take a resistance reading between terminals 2 and 5 on the relay. A good coil will have a very high resistance, typically over 1,500 Ω. A very low resistance reading indicates that the coil is shorted and needs to be replaced.

B

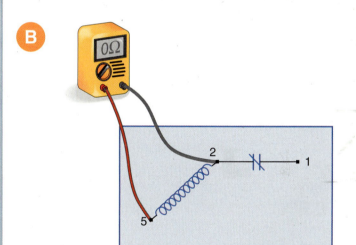

from experience...

Even though it may be determined that the resistance readings of the coil and contacts of the PMR are within acceptable ranges, the relay must be installed in the system to be certain that the contacts of the relay open and close as desired.

REVIEW QUESTIONS

1. **The component part of a motor that is directly connected to the shaft is called the**
   a. Stator
   b. Pulley
   c. Rotor
   d. All of the above are correct

2. **The start and run windings of a motor can be found in the**
   a. Rotor
   b. Current magnetic relay
   c. Potential relay
   d. Stator

3. **The centrifugal switch opens and closes its contacts depending on the**
   a. Amperage flowing through the run winding
   b. Speed of the motor
   c. Induced voltage across the start winding
   d. Induced voltage across the run winding

4. **The coil of a current magnetic relay is wired**
   a. In series with the start winding
   b. In series with the run winding
   c. In parallel with the common winding
   d. In parallel with the run winding

5. **The potential relay**
   a. Has a low-resistance coil
   b. Has a normally closed set of contacts
   c. Is used primarily on larger motors requiring high starting torque
   d. Both b and c are correct

6. **Of the following motors, which has the highest starting torque?**
   a. The CSCR motor
   b. The CSIR motor
   c. The shaded-pole motor
   d. The PSC motor

7. **What is the function of a capacitor?**
   a. Stores electric charge
   b. Provides starting torque for a motor
   c. Helps to correct and maximize the power factor of a motor
   d. All of the above are correct

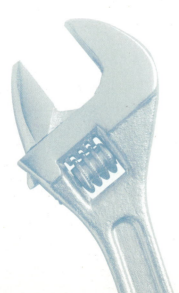

8. **Which of the following motors uses both a start and a run capacitor?**
   a. The three-phase motor
   b. The CSIR motor
   c. The CSCR motor
   d. The ISIR motor

9. **Why are run capacitors considered to be "wet"?**

10. **True or False: Permanent-split-capacitor motors use a CMR to remove the start winding from the circuit.**

11. **A split-phase motor has which of the following windings?**
    a. Common and run
    b. Start and run
    c. Three run windings
    d. Common and start

12. **The power that is supplied to the windings of an ECM motor most closely resembles**
    a. 120 volt, single-phase power
    b. 240 volt, single-phase power
    c. Direct current
    d. Three-phase power

13. **When referring to ECM motors, the term "soft start" refers to the**
    a. Higher amperage on start up
    b. Slower motor speed on start up
    c. Faster motor speed on start up
    d. Both a and b are correct

14. **Three-phase motors operate with a phase angle of**
    a. 0 degrees
    b. 30 degrees
    c. 90 degrees
    d. 120 degrees

15. **Explain the condition called *single phasing*.**

16. **True or False: The purpose of the motor starter is to protect a three-phase motor from single phasing.**

17. **A three-phase, dual-voltage motor with a wye-configured stator**
    a. Will have nine wires that must be field connected
    b. Will have twelve wires that must be field connected
    c. Can only operate at one voltage
    d. Both a and c are correct

18. **The overload portion of a motor starter has the heaters installed in the _____ circuit and the normally closed switches in the _____ circuit.**

19. **True or False: The heaters in the starter are wired in series with the normally closed switches.**

20. A part-wind start uses

  a. Two contactors

  b. Two starters

  c. One starter and one contactor

  d. One starter

21. The part-wind start and the reduced-voltage start are designed to

  a. Reduce the amperage draw of a three-phase motor on initial startup

  b. Reduce the effects of wear and tear on the starting components often caused by excessive heat

  c. Protect expensive pieces of equipment from premature failure

  d. All of the above are correct

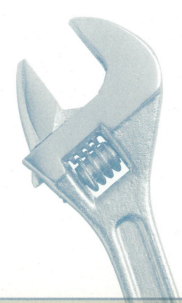

CHAPTER 16

Electric Controls

OBJECTIVES *Upon completion of this chapter, the student should be able to*

- Explain the principle and operation of bimetal strips.
- Explain the importance of limit switches on safe heating system operation.
- Describe the function of a thermostat.
- Differentiate between line-voltage and low-voltage thermostats.
- Properly color code low-voltage control wiring.
- Set, understand, and adjust heat and cool anticipators.
- Accurately measure the amperage in low-voltage control circuits.
- Explain how voltage is obtained for use in low-voltage control circuits.
- Measure voltage at the primary and secondary windings of a control transformer.
- Install and connect a multitap transformer.

- Describe the operation of relays and contactors.
- Explain the difference between a contactor and a motor starter.
- Explain the difference between a cumulative compressor run defrost timer and a continuous run defrost timer.
- Explain uses for normally closed and normally open float switches.
- Describe the construction and operation of flow switches.
- List several types of pressure controls and switches.
- Explain the difference between a manual reset and an automatic reset pressure control.
- Describe the operation of electronic relays, starting relays, and timers.

The underlying concepts of automatic controls are not new. Controls are used to start, stop, or modulate the flow of electric current through a circuit or the flow of fluid through a piping arrangement or duct system. We encounter automatic controls on a daily basis. The float assembly in a toilet bowl tank, for example, stops the flow of water into the tank when the water reaches the desired level. Most electric clothes irons currently produced are equipped with automatic shut-off features that de-energize the appliance if left unattended for a period of time. In the air conditioning and heating industry, automatic controls are used to start and stop system components in response to conditions either within or external to the system. Controls can often be described as switches that open or close their contacts when undesirable conditions exist. Some examples of undesirable conditions are excessive component temperature or pressure, improper fluid levels, excessively warm or cold space temperature, and improper airflow.

In this chapter, we will discuss various components that are designed to enhance system operation, as well as other devices that act as safeties to disable the system in the event unsafe conditions exist. To accomplish these tasks, switches, relays, contactors, and other devices are used to energize and de-energize electric circuits.

Automatic changeover Term used to describe thermostats that switch automatically between the heating and cooling modes of operation

Automatic reset Feature that allows an overload or safety device to reset itself automatically when the dangerous or undesirable condition no longer exists

Bimetal or bimetal overloads See bimetal strips

Bimetal strips Metal strips made up of two dissimilar metals that expand and contract at different rates at different temperatures

Bimetal thermostats Thermostats that open and close electrical contacts as a result of the flexing of a bimetal strip

Cold anticipator Device on a cooling thermostat that causes the cooling contacts to close shortly before the space reaches the cut-in temperature to facilitate even cooling of the space

Continuous run timer Defrost timer wiring that determines defrost cycles by the time the system is operating

Control circuit Circuit that is made up of control devices such as thermostats, relay coils, pressure controls, and safety switches

Control transformer Device that creates the voltage supply for the control circuit

Cumulative compressor run timer Defrost timer wiring that determines defrost cycles by the actual run time of the compressor

Cut-in pressure Pressure at which a pressure switch or control closes its contacts

Cut-out pressure Pressure at which a pressure switch or control opens its contacts

Differential Mathematical difference between the cut-in pressure and the cut-out pressure

Fan switch Device that energizes and de-energizes the blower circuit in response to temperature conditions within the system

Fan-limit switch Control device that incorporates both the fan switch and the limit switch

Heat anticipator Device on a heating thermostat that causes the contacts to open shortly before the space reaches the desired temperature to facilitate even heating of the space

In-line fuse Low amperage fuse in the control circuit that protects the control transformer from excessive current

Limit switch Safety device that de-energizes

heating equipment if unsafe temperature conditions exist

Manual reset Device feature that requires an individual to physically reset the device

Multistage thermostat Thermostat that can control more than one heating or cooling device at any given time

Multitap transformer Control transformer that is wired for use with multiple primary voltages

Negative temperature coefficient (NTC) Thermistor that experiences a decrease in resistance as its temperature increases

Positive temperature coefficient (PTC) Thermistor that experiences an increase in resistance as its temperature increases

Power circuit Circuit that carries power to main system components, such as compressors and fan motors

Primary winding Winding of the transformer that generates the magnetic field that induces the voltage on the secondary side of the transformer

Secondary winding Winding of the transformer that experiences the effects of the magnetic

field generated in the primary winding, producing the voltage used in the control circuit

Sensing bulb thermostat Thermostat that uses a pressurized vessel to push electrical contacts open or closed

Set point Temperature at which a thermostat is set

Solid-state Devices that are made up of electronic circuitry

Step-down transformer Transformer that has a higher primary voltage than secondary voltage

Step-up transformer Transformer that has a higher secondary voltage than primary voltage

Subbase Mounting plate for a thermostat where the electrical connections are made

Temperature differential Mathematical difference between two temperatures. Often referred to as the difference between the temperature of the return air and the temperature of the supply air

Thermistor Resistor that changes its resistance as its temperature changes

Volatile liquid Fluid inside the bulb of a sensing bulb thermostat or control

OVERLOADS

Overloads are designed to protect motors from overload conditions. When a motor is overloaded, the amount of current that the motor draws is greater than the design limit for the device. This excessive current generates greater amounts of heat, resulting in possible damage to the motor's internal windings. The most common type of overload found on residential air conditioning systems is the bimetal type.

The bimetal overload (Fig. 16-1) is wired in series with the motor so the current that flows through the motor also flows through the overload. As the current draw of the motor increases, the heat generated in the overload increases as well. Once the level of heat in the overload reaches a predetermined set point, the overload will open its contacts, de-energizing the motor. **Bimetal overloads** operate on the principle that different metals expand and contract at different rates when exposed to temperature changes. When two dissimilar metals are connected to each other side-by-side and are either heated or cooled, bending will occur (Fig. 16-2). It is this bending of the bimetal strip or disc that opens or closes the circuit that feeds power to the motor. Figure 16-3 shows the wiring diagram of a PSC motor with overload protection. Note that the overload is wired in series with the motor's common terminal so that, if the overload opens, the current flow through the motor will be interrupted.

Overload devices can reset either automatically or manually. The overload in Figure 16-1 resets automatically, but the overload in Figure 16-4 must be reset manually once its contacts have opened.

FIGURE 16-1 Bimetal overload protector.

Photo by Eugene Silberstein.

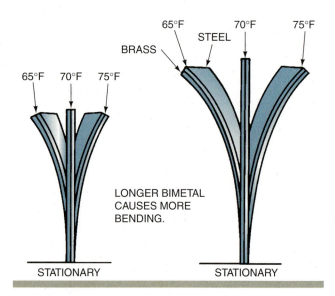

FIGURE 16-2 This bimetal is straight at 70°F. The brass side contracts more than the steel side when cooled, causing a bend to the left. When heated, the brass side expands more than the steel, causing a bend to the right. The longer the bimetal strip, the more the bend.

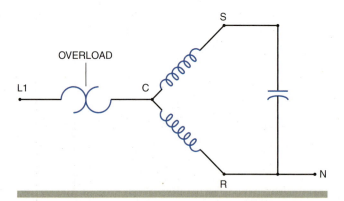

FIGURE 16-3 Bimetal overload in series with the common terminal of the PSC motor.

from experience...

Always be sure that overloads are properly mounted. Overloads used on fractional horsepower motors also sense the temperature of the shell of the motor and will open if the motor gets too hot. Improperly mounted overloads will compromise this protection as the device will remain closed since it would be unable to sense the temperature of the motor.

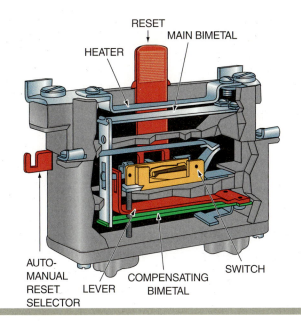

FIGURE 16-4 Manual reset bimetal overload protector.

OVERLOADS ON THREE-PHASE MOTORS

In the previous chapter, three-phase motor starting was discussed. Three-phase motors also rely on overload protection to prevent the motor from operating when unsafe conditions exist. The overloads found on three-phase motor starters are wired in series with the operational control, not the motor itself. Other components, called heaters, are located in series with the motor and sense the heat generated by the current flow through the motor. The heaters, in turn, cause the overloads, referred to as switches on three-phase motors, to open, thereby de-energizing the motor's control circuit. For more on three-phase motor protection, refer to the section on motor starters later in this chapter.

LIMIT SWITCHES

The **limit switch** (Fig. 16-5), found primarily on heating systems, is another bimetal device, used as a safety in the event that too much heat is generated. The limit switch is wired in series with the control circuit of the heat source (Fig. 16-6) and will open its contacts if unsafe temperature levels are sensed. This will de-energize the heat source. Limit switches are normally closed devices and will remain closed during normal system operation. The limit switch will only open when operating conditions are outside the desired range.

FIGURE 16-5 Limit switches.

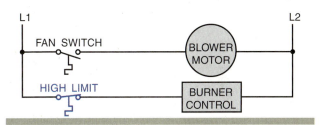

FIGURE 16-6 Limit switch connected in series with the heat source.

> ### CAUTION
> **CAUTION:** Limit switches are safety devices and should never be removed from the active electric circuit. Doing so can result in damage to the equipment, increase the chances of fire, and cause personal injury or death.

FAN SWITCHES

Another bimetal device commonly found on heating systems is the **fan switch**. The fan switch (Fig. 16-7) permits blower or fan operation only when the temperature of the heat source, or heat exchanger, has reached the desired temperature. This prevents the circulation of cool air through the occupied space when heating is desired. At the beginning of the heat cycle, the heat exchanger is cool and the fan switch will be in the open position. Once the heat exchanger reaches the desired temperature, the contacts on the fan switch will close, energizing the blower, and permitting airflow to the occupied space. At the end of the heating cycle, the blower will continue to operate until the heat exchanger has cooled down to the predetermined temperature set point. The fan switch is wired in series with the blower (Fig. 16-8).

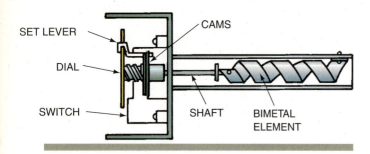

FIGURE 16-7 Fan switch.

FIGURE 16-8 Fan switch controls the operation of the fan motor.

FIGURE 16-9 Fan-limit switch.

FAN-LIMIT SWITCHES

Heating systems that use both fan switches and limit switches are equipped with the **fan-limit switch** (Fig. 16-9). This device incorporates both the fan switch and the limit switch in one component. The wiring diagram of the fan-limit switch is shown in Figure 16-10. The fan switch is often used to control a line voltage circuit, while the limit portion of the control is often used for low-voltage control circuits. When shipped from the factory, a jumper wire is installed, linking the fan and limit switches. This jumper must be removed when the device is used to control circuits with different supply voltages.

THERMOSTATS

Thermostats are temperature-sensing devices that are responsible for maintaining the space temperature at the desired level. They open and close electric circuits at the correct time to energize and de-energize air conditioning and refrigeration system components. Commonly controlled system components are the

- compressor
- condenser fan motor
- evaporator fan motor
- heating circuit

Thermostats can also be used as safety devices that can energize or de-energize system components if temperatures reach unsafe or undesirable levels. For example, if you do not want ice to form on an evaporator coil, you can place a thermostat on the coil that will open its contacts and de-energize the compressor if the coil temperature falls below 32°F. This type of thermostat is a

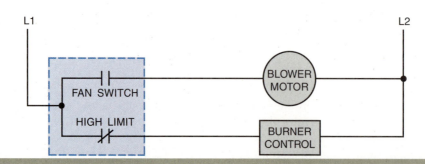

FIGURE 16-10 Schematic for fan-limit switch.

device that opens on temperature drop, which is the same as one that closes on a rise in temperature.

Several methods are employed to convert the temperature that the device senses into mechanical motion that causes the thermostat to open and close its contacts. The various types of thermostat mechanisms are the

- bimetal type
- sensing bulb
- solid-state or electronic

Bimetal thermostats operate on the same principle as the bimetal overload discussed at the beginning of this chapter: Different metals expand and contract at different rates when exposed to temperature changes. When two dissimilar metals are connected to each other side-by-side, bending will occur when the strip is either heated or cooled. The longer the metal strips are, the more exaggerated the bending will be. Unfortunately, longer **bimetal strips** occupy more space, so it is quite common to see bimetal strips bent into various shapes including coils and spirals (Fig. 16-11).

By connecting a mercury-filled glass bulb to the bimetal coil, an electric circuit can be closed or opened depending on the temperature that the device is sensing [Fig. 16-12(A)]. Mercury bulbs consist of a "ball" of mercury, which is an excellent conductor of electricity, inside a sealed glass tube that has had all of the air removed to prevent oxidation. Electric conductors pass through the glass into the chamber, and contacts are opened or closed depending on the position of the mercury within the tube. The bulb in Figure 16-12(B) can complete a circuit either between terminals *C* and *NO* or between *C* and *NC*, but not both. The notations on the contact terminals are defined as follows:

- *C* terminal—Common (This terminal is common to both contacts)
- *NC*—Normally closed
- *NO*—Normally open

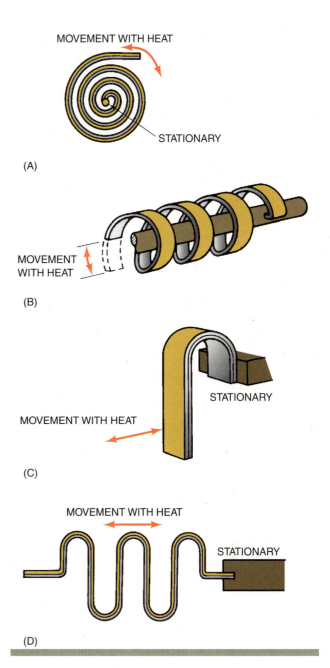

FIGURE 16-11 Adding length to the bimetal. (A) Coil. (B) Wound into a helix. (C) Hairpin. (D) Worm shape.

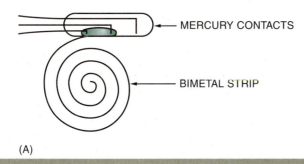

(A)

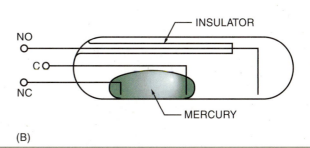

(B)

FIGURE 16-12 (A) Bimetal formed into a spiral to save space. This strip has a mercury bulb attached to it. (B) The mercury shifts within the glass tube depending on the temperature sensed by the bimetal strip.

from experience...

To ensure proper thermostat operation, devices containing mercury bulbs must be installed perfectly level. Improperly mounted thermostats will cause the actual space temperature to differ from the desired set point.

Going Green

The mercury that is contained in many older style thermostats and controls is dangerous to animals, humans, and the environment. High exposures to mercury may cause damage to the gastrointestinal tract, nervous system, and kidneys. Mercury, if not contained and allowed to settle in the ground, can damage the water supply as well. Muscle weakness, mental disturbances, skin rashes, and memory loss are typical symptoms associated with exposure to mercury. Make certain that old thermostats that contain mercury are disposed of in a manner that is environmentally acceptable.

REMOTE SENSING BULB

Photo by Eugene Silberstein.

FIGURE 16-13 Thermostat with a sensing bulb. The temperature switch on the right is a nonadjustable, bimetal-type thermostat.

(A)

(B)

Delmar/Cengage Learning.

FIGURE 16-14 (A) Digital programmable thermostat. (B) The device can be easily programmed with the keypad to allow for multiple temperature settings throughout the day.

The **sensing bulb thermostat** uses a copper or aluminum bulb filled with a **volatile liquid** that follows a pressure/temperature relationship and expands when heated and contracts when cooled. The bulb is connected to a bellows, which can open or close a set of contacts, depending on the amount of pressure exerted on it (Fig. 16-13). The sensing bulb is located within the space for which the air conditioning or refrigeration system is maintaining the temperature. On window air conditioners, for example, the sensing bulb is located in the return air stream.

The **solid-state**, or electronic, thermostat (Fig. 16-14) uses temperature-sensitive resistors called **thermistors**. The resistance of the thermistor changes as the surrounding air temperature changes. If the resistance of the thermistor increases as the temperature it senses increases, the device is said to have a **positive temperature coefficient (PTC)**. If, on the other hand, the resistance of the thermistor decreases as the temperature it senses increases, it is said to have a **negative temperature coefficient (NTC)**. PTCs are commonly used as safety controls and starting components on electric motors and other devices.

The electronic circuitry senses the change in resistance and adjusts the system accordingly. Changes in resistance cause the amount of current flowing in the circuit to change. Lower resistance translates to higher current flow and vice versa. Electronic thermostats are desired by many because they tend to be more reliable and last longer than mercury bulb devices. This is true in part because there are no moving parts to wear out.

LINE-VOLTAGE THERMOSTATS

Line-voltage thermostats typically control 115-Volt or 208/220-Volt circuits, and low-voltage thermostats usually control 24-Volt circuits. The thermostats in Figure 16-13 are typical line-voltage devices. When intended to be located within the occupied space, they are manufactured with tightly secured covers to reduce the risk of electric shock. A licensed electrician may be needed to run electrical lines that connect line-voltage thermostats in accordance with local codes.

Depending on the individual system, the current flowing through line-voltage thermostats can be relatively high when wired in series with the electric load, so there

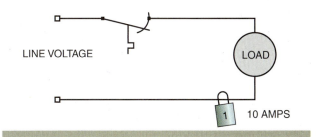

FIGURE 16-15 Ammeter 1 is reading 10 A, which is the current flowing through both the switch and the load.

is a chance of personal injury both to the service technician and to the equipment owner if the device is not properly installed. Recall that a series circuit is one in which the current has one and only one possible path to take, so the current in a series circuit is the same at all points throughout the circuit. For this reason, thermostats wired in series with the electric load (Fig. 16-15) must be able to withstand the current that the load draws.

Thermostats can also be used to energize and de-energize holding coils, which are located in relays and contacts, and are used to control electric contacts in series with the major loads (Fig. 16-16). This greatly reduces the amount of current flowing through the thermostat. In operation, when the thermostat closes its contacts, current flows through the holding coil of the relay or contactor, creating a magnetic field.

This circuit is referred to as a **control circuit**. The current in this control circuit is very low (Ammeter 2), allowing thermostats with lower amperage ratings to be used. Once energized, the holding coil generates a

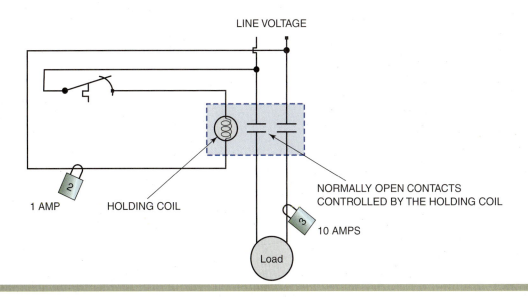

FIGURE 16-16 The holding coil circuit draws much less amperage than the power circuit as can be seen by the amperage readings on ammeters 2 and 3.

magnetic field, pulling the two sets of contacts closed and completing the electric path to the main load. The amperage read by Ammeter 3 represents the amperage drawn by the main load and is much higher than the amperage in the control circuit. The circuit containing the main load is the power circuit and is parallel to the control circuit. Recall that a parallel circuit is one in which the electric current has more than one possible path to take. As can be seen in Figure 16-16, the current can either flow through the circuit containing the thermostat and the holding coil or the circuit containing the main load.

LOW-VOLTAGE THERMOSTATS

Another commonly used type of thermostat is the low-voltage variety (Fig. 16-17). It is used to properly route current through low-voltage control circuits that, in

turn, energize and de-energize the major system loads (Fig. 16-18). Low-voltage wires are colored for easy identification, and most manufacturers in the air conditioning industry have accepted the following color-coding system:

	Low-Voltage	Wire Color-Coding System
R terminal	Red	Hot (power coming from the transformer)
RC terminal	Red	Hot (power feeding the cooling circuit)
RH terminal	Red	Hot (power feeding the heating circuit)
W terminal	White	Heating circuit
Y terminal	Yellow	Cooling circuit
G terminal	Green	Indoor fan circuit

Delmar/Cengage Learning.

FIGURE 16-17 Low-voltage thermostats.

from experience...

The RC and RH terminals are found on isolating subbases that separate the heating and cooling contacts when different control-circuit power supplies are used for each mode of operation. If only one power supply is used, a jumper wire can be placed between the RC and RH terminals.

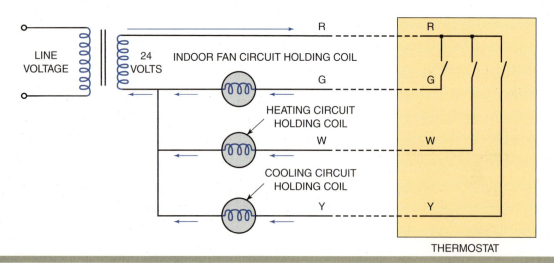

FIGURE 16-18 Simplified low-voltage control circuit. The indoor fan operates when the thermostat closes the contact between R and G. The system will operate in the cooling mode when contacts R and Y are closed and in the heating mode when contacts R and W are closed.

Low-voltage thermostats are connected in a circuit parallel to that of the main loads and, therefore, are not subjected to the same current draws or voltage requirements that the main loads are. These thermostats are often preferred over the line-voltage version because:

- They reduce the risk of severe electric shock because of the low amperage draw
- They tend to be less expensive than line-voltage thermostats
- The associated wiring can be installed without an electrician's license in most geographical areas

Low-voltage thermostats reduce the risk of electric shock because both the supply voltage and the current flow through the circuit are very low. Since the current flow is low, the electric contacts in the device can be manufactured for lighter duty, reducing the overall cost. In addition, the installation of low-voltage thermostats does not generally require the services of a licensed electrician because most cities do not have electrical codes regarding low-voltage wiring. For these reasons, low-voltage thermostats are usually preferred even though they require a transformer to provide the desired voltage.

Low-voltage thermostats are usually mounted on a **subbase** (Fig. 16-19). The subbase is the mounting plate on which the electric connections are made. It is attached directly to the wall or electric junction box, and the wires enter the plate from behind through an access hole provided for this purpose. Subbases are equipped with manual switches that allow the user to switch between the heating and cooling modes of operation and to select either continuous or automatic fan operation (Fig. 16-20).

Thermostats are categorized by the type of equipment on which they are used. They are

- cooling only
- heating only
- heating and cooling combination

Heating-only thermostats are used on independent heating systems, such as baseboard. *Cooling-only* thermostats are used on air conditioning systems that are not equipped with a heating coil. *Heating-and-cooling* thermostats control the heating and cooling cycles of a system equipped with both heating and cooling components. Figure 16-20 shows a heating/cooling subbase.

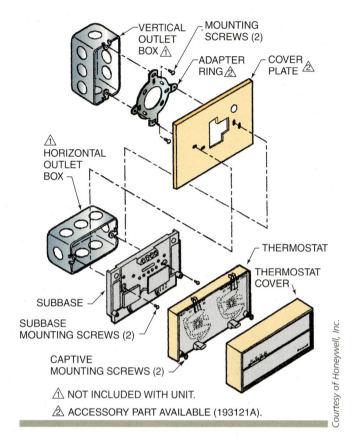

Courtesy of Honeywell, Inc.

FIGURE 16-19 The thermostat subbase mounts to either the wall or an electric outlet box. The electric connections are made on the subbase, and the thermostat is mounted to the subbase.

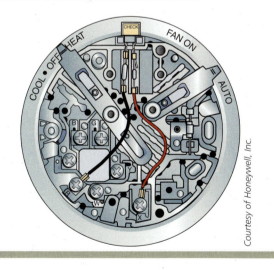

Courtesy of Honeywell, Inc.

FIGURE 16-20 This subbase has manual switches that allow the user to set the HEAT-OFF-COOL switch to the desired mode and to select automatic or constant fan operation.

COOLING THERMOSTATS

Cooling thermostats are designed to close their contacts as the space temperature rises above the desired set point. The cooling thermostat is shown in electrical schematics like those in Figure 16-15. The electrical symbol is drawn to represent the way the device operates. The line segment representing the thermostat "moves" up and down as the temperature it senses rises and falls. The line segment on the cooling thermostat's electric symbol is below the main circuit line. As the temperature rises, the line moves up until, at some predetermined temperature, the thermostat closes the circuit. Similarly, as the temperature falls, the circuit opens. For example, if a room thermostat is set to maintain 74°F (its **set point**), the cooling thermostat closes its contacts if the room temperature rises above 74°F, energizing the cooling portion of the control circuit.

This type of thermostat can be either line-voltage or low-voltage. If a line-voltage thermostat is used, it can be wired directly in series with the compressor (Fig. 16-15). If the compressor is large and draws a large amount of current, the thermostat will most likely energize the holding coil of the cooling circuit. This holding coil will be rated at the supply voltage. If the thermostat is a low-voltage device, closing its contacts will energize the holding coil of the cooling circuit. In this case, the holding coil will be rated at 24 Volts.

If the thermostat closed its contacts at a temperature just above the set point and opened its contacts just below the set point, the system would be turning on and off constantly. To alleviate this situation, most thermostats have a built-in **temperature differential** that provides a temperature swing of about 1 degree above and below the set-point temperature. This means that if the thermostat is set to maintain 74°F, the system will turn on when the space reaches 75°F and turn off when the temperature drops to 73°F.

HEATING THERMOSTATS

Heating thermostats are designed to close their contacts as the space temperature falls below the desired set point. The heating thermostat is shown in electric schematics like those in Figure 16-16. This electrical symbol is drawn to represent the way the device operates. The line segment representing the heating thermostat "moves" up and down as the temperature it senses rises and falls. The line segment on the heating thermostat's electrical symbol is drawn above the main circuit line. As the temperature drops, the line moves down until, at some predetermined temperature, the thermostat closes the circuit. Similarly, as the temperature rises, the circuit opens. For example, if a room thermostat is set to maintain 68°F, the heating thermostat will close its contacts if the room temperature falls below 68°F, energizing the heating system.

Heating thermostats can also be line-voltage or low-voltage devices. They can be wired in series with the heating equipment or used to energize the holding coil of the heating circuit. Like the cooling thermostats,

heating thermostats have a temperature differential built into them to prevent rapid system cycling.

HEATING/COOLING COMBINATION THERMOSTATS

Combination thermostats are used on equipment that has both heating and cooling components. The occupant determines if the equipment is to run in the cooling mode or the heating mode by means of a manual switch. More sophisticated thermostats have an **automatic changeover** that switches the system automatically between the heating and cooling modes. Many thermostats are equipped with fan switches that allow the user to run the indoor fan continuously when switched to the *ON* position, even when the system is off or the room thermostat is satisfied [Fig. 16-21(A) and (B)]. The fan switch can also be set to the *AUTO* position, which is used to cycle the indoor fan on and off with the system compressor [Fig. 16-21(C) and (D)]. Since moving air is somewhat cooler than stagnant air, it is recommended that the fan be in the *ON* position during the warmer summer months and in the *AUTO* position during the cooler winter months.

The thermostat can be programmable or nonprogrammable, digital or analog. Nonprogrammable thermostats are less expensive than programmable models and require more attention from the occupant of the space. Digital thermostats use solid-state circuitry, but analog devices rely on mechanical means to operate. Digital programmable thermostats are generally more expensive than their analog counterparts but can often save the user money in the long run (Fig. 16-22). These thermostats can be programmed to maintain different temperatures at different times of the day. During summer operation, the system can be programmed to maintain a higher temperature when the space is not occupied and to turn on shortly before the occupants return. Newer models of programmable thermostats can be set to maintain up to five different temperatures throughout the day and can even be programmed to turn off on the weekends. This benefits companies that are only open for business Monday through Friday. Since these devices turn the system on and off automatically, they can be mounted with locking covers, which helps reduce tampering by company employees and saves energy dollars.

MULTISTAGE THERMOSTATS

For smaller applications, thermostats with one heating and/or one cooling set of contacts are sufficient to energize the heating and cooling modes, respectively. As the applications get larger and heat loads fluctuate more and more, however, operating only portions of the system at different times may be economically feasible. For example, a 20-ton air conditioning system made up of

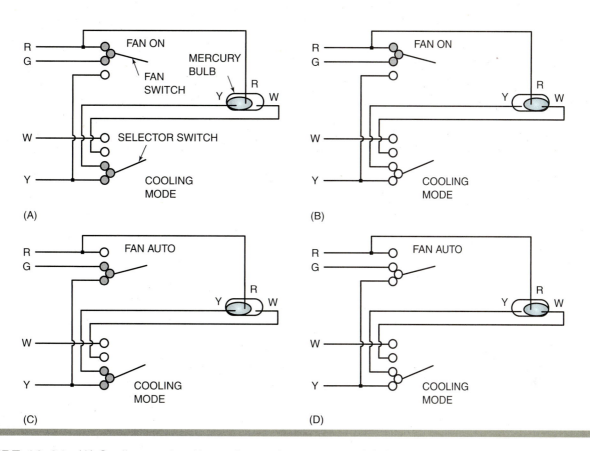

FIGURE 16-21 (A) Cooling mode with continuous fan operation. (B) Continuous fan operation even though the mercury bulb has shifted to open the cooling-circuit contacts. (C) Fan operates when the cooling circuit is energized. (D) Mercury bulb senses that the room has reached the desired temperature and opens the cooling circuit and de-energizes the fan circuit.

FIGURE 16-22 Electronic thermostats.

two, 10-ton circuits may not need to run at full capacity during the early morning hours when the heat load is low. Ten tons may be enough to cool the space while the sun is not at its peak and the building is not crowded with occupants. The **multistage thermostat** allows each circuit to be controlled independently, energizing the second stage only when needed. Multistage thermostats can

have two or more stages of heating and/or two or more stages of cooling. Common configurations of multistage thermostats include the following:

- Two cooling stages, one heating stage
- One cooling stage, two heating stages
- Two cooling stages, two heating stages

A thermostat with two stages of cooling and one stage of heating would have two *Y* terminals, denoted *Y1* and *Y2*, but only one *W* terminal. Similarly, a thermostat with two stages of heating would have two *W* terminals, namely *W1* and *W2*. A multistage thermostat with two heating and two cooling stages would have the following wiring connections:

- *R* Hot (power supplied from the transformer)
- *G* Indoor fan circuit
- *W1* Heating—Stage 1
- *W2* Heating—Stage 2
- *Y1* Cooling—Stage 1
- *Y2* Cooling—Stage 2

The *Y* contacts would open and close at different temperatures, allowing independent operation. The contact for Stage 1 can be set to close at 74°F, and the contact for Stage 2 can be set to close at 76°F. If Stage 1 is capable of handling the heat load, Stage 2 will not be needed and will remain off. If, however, the temperature of the space begins to rise, the Stage 2 contact will close at 76°F, energizing the second half of the system. When the space temperature begins to drop, Stage 2 will de-energize, while Stage 1 continues to operate.

Two-stage heating thermostats are commonly found on heat-pump systems, which are discussed later in the text. The first stage of heating controls the vapor-compression cycle (compressor-operated heat cycle) and the second heating stage controls supplementary heating strips. Stage 1 may be set to close its contacts at 70°F, and Stage 2 at 68°F. If single-stage heating is not sufficient to satisfy the heating requirements of the occupied space, the temperature will continue to drop and Stage 2 will energize the supplementary heating strips. *When the space temperature rises to the desired set point, the second stage will drop out until needed at a later time.*

ANTICIPATORS

In an attempt to minimize the swing in temperature of an occupied space, thermostats are often equipped with **cold anticipators** and/or **heat anticipators**. They allow for a more even conditioning of the space by reducing the extreme temperatures at the beginning and end of the operating cycle.

Cold anticipators are fixed, non-adjustable in most cases, resistances wired in parallel with the cooling contact in the thermostat. When the contacts are open, a small amount of current flows through the anticipator, generating a small amount of heat. This causes the thermostat to close its contacts and energize the cooling cycle shortly before it normally would (Fig. 16-23), preventing the space from getting any warmer.

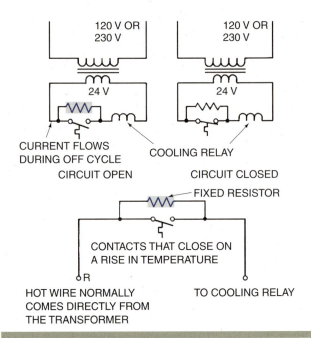

FIGURE 16-23 The cold anticipator is wired in parallel with the cooling contacts of the thermostat. This allows a small amount of current to flow through the device in the off cycle, generating heat. This causes the cooling circuit to become energized early, eliminating the extreme temperatures in the occupied space at the beginning of the cycle.

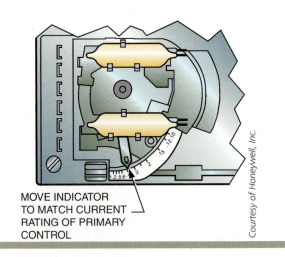

FIGURE 16-24 Adjustable heat anticipator. The indicator must match the current rating of the heating control circuit.

Heat anticipators are usually variable resistances wired in parallel with the heating contacts of the thermostat. Adjustable anticipators must be properly set upon initial installation. The setting on the device must be equal to the current flowing through the heating control circuit (Fig. 16-24). The best way to measure the control-circuit current is to use a coil of wire with

FIGURE 16-25 Using the 10-wrap method to obtain the amperage of the heating control circuit.

10 turns, wrapped around the jaw of a clamp-on ammeter. This wire is then connected across the heating contacts of the thermostat. The current in the circuit will be the reading on the meter divided by 10 (Fig. 16-25). A clamp-on ammeter capable of accurately measuring very low amperages can also be used. The purpose of the heat anticipator is to prevent the occupied space from getting too warm during the heating cycle. This excess heat, or temperature above the desired space temperature, is called *overshoot*. The anticipator will cause the thermostat's contacts to open shortly before it normally would, allowing residual system heat to dissipate. If the anticipator was not there, the thermostat would open at the set point and the residual heat would create a large overshoot. Newer electronic thermostat have a built-in heat anticipator that does not need to be set by the technician upon installation.

MAGNETICALLY OPERATED DEVICES

Magnetically operated devices rely on the magnetic field generated by current flow to perform some desired task. In the chapter on basic electricity, it was established that a magnetic field is generated when current flows through a conductor. This magnetic field, however, is typically not strong enough to perform a task such as opening a mechanical valve. To make use of this magnetic field, it must first be magnified. In the previous chapter, we discussed how to obtain an accurate amperage reading of a circuit carrying a very small amount of current by forming the conductor into a coil of ten loops. The ammeter was then clamped around all ten loops and the resulting measurement was divided by ten to obtain the actual reading. The construction and operation of magnetically controlled

devices is very similar in nature to this scenario. Two commonly used devices that rely on magnetic fields to perform some type of work are the solenoid and the transformer.

SOLENOIDS

The solenoid is formed by creating a coil of a conducting material with many turns. The more turns in the coil, the stronger the magnetic field will be (Fig. 16-26). Consider the example of a simple electromagnet. If a conductor is wrapped around a nail 10 times and the ends of the conductor are connected to a power source, the nail will become magnetized and will be able to pick up small metallic objects. If the wire is wrapped around the nail 20 times, the ability of the magnet to pick up objects will be increased, as the strength of the magnetic field is increased.

Probably the most common use of a solenoid is to open and close a mechanical valve to start or stop the flow of a fluid in a piping arrangement. When used for this purpose, the device is referred to as a solenoid valve (Fig. 16-27), which is simply a valve controlled by a solenoid. The plunger of the valve—the portion of the valve that moves to open or close the device—is positioned within the solenoid and experiences the magnetic field generated when the coil is energized by the control circuit.

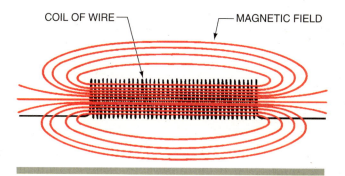

COIL OF WIRE — MAGNETIC FIELD

FIGURE 16-26 There is a stronger magnetic field surrounding wire formed into a coil.

FIGURE 16-27 Solenoid valve.

Solenoid valves can be either normally opened or normally closed. A normally open valve is in the open position when the solenoid coil is de-energized; a normally closed solenoid valve is in the closed position when the solenoid coil is de-energized. When the coil is energized, the position of the valve changes.

CONTROL TRANSFORMERS

Another common device that relies on a magnetic field is the **control transformer** (Fig. 16-28). Most residential air conditioning and heating systems are controlled by low-voltage, typically 24-Volt, circuits. Since a 24-Volt power source is not readily available from a wall outlet, it must be generated within the system. The control transformer, usually referred to simply as a transformer, is made up of two windings: the primary winding and the secondary winding.

The **primary winding** is connected to the line-voltage power source, and the **secondary winding** becomes the power source that provides the voltage for the control circuit. It is important to note that the two windings are not electrically connected to each other. The wiring diagram of a typical control transformer is shown in Figure 16-29.

When power is applied to the primary winding, current flows through the winding, creating a magnetic field around it. The secondary winding, which is located within the magnetic field, is exposed to this magnetic field and voltage is generated in this winding as a result of induction. The amount of voltage generated at the secondary winding is directly related to the voltage applied to the primary winding as well as the number of turns in the primary and secondary windings.

Consider a transformer that has 100 turns of wire in the primary winding and 10 turns of wire in the secondary winding. The ratio between the primary and secondary is therefore 10:1 [Fig. 16-30(A)]. For every 10 Volts applied to the primary, 1 Volt will be generated in the secondary. If the primary voltage is, for example, 100 Volts, the voltage at the secondary will be one-tenth of 100 Volts, or 10 Volts.

STEP-UP AND STEP-DOWN TRANSFORMERS

Just as the strength of the electromagnet in the earlier example was determined by the number of turns in the coil, the voltage produced at the secondary winding is also determined by the number of turns in the coil. The lower the number of turns, the lower the induced voltage. If the primary has 10 times the number of turns as the secondary, the secondary voltage will be one-tenth that of the primary. This transformer is said to have a 10 to 1 ratio, denoted 10:1 [Fig. 16-30(A)]. It is called a **step-down transformer**. The voltage is stepped down

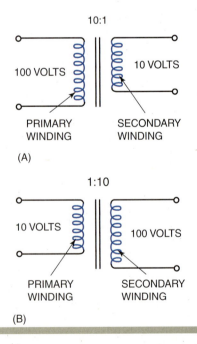

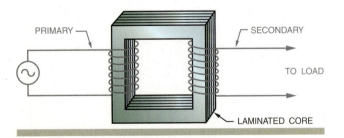

Photo by Eugene Silberstein.

FIGURE 16-28 Transformer.

FIGURE 16-29 Primary and secondary windings of the transformer are wrapped around the core. Note that the windings are not physically connected to each other.

FIGURE 16-30 (A) Step-down transformer. (B) Step-up transformer.

from a higher voltage to a lower voltage. Transformers may also be designed with more turns in the secondary winding than in the primary winding. These transformers are called **step-up transformers** since the voltage is stepped up to a higher potential [Fig. 16-30(B)].

Each transformer is designed to accept a specific primary voltage that must be within design parameters—usually plus or minus 10% of nameplate voltage—to function properly. For example, a transformer with a primary voltage rating of 115 Volts will operate effectively within the range of 103 to 127 Volts. Damage to the device could occur if incorrect voltage were applied to the primary winding. Because of this, being aware of the specific job application is extremely important before selecting a transformer.

If a defective transformer is to be replaced, use the exact replacement whenever possible. This may prove troublesome to the service technician since a number of different transformers would need to be stocked to properly service the customer's equipment. If the exact replacement is not available, the new transformer must match the specifications of the old one.

MULTITAP TRANSFORMERS

The **multitap transformer** is designed with multiple wires, or taps, connected to the primary winding. This enables the device to be used with different supply voltages (Fig. 16-31). Care must be taken to use the correct wires or damage to the device could occur. These primary wires are color-coded and a legend is provided to reduce the possibility of incorrect installation. All unused taps must be isolated from each other and capped to prevent contact with other taps, grounds, or conductors.

> **CAUTION**
>
> **CAUTION:** When installing transformers and other electric components, all safety rules should be observed to avoid personal injury and damage to the components.

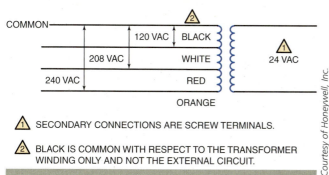

① SECONDARY CONNECTIONS ARE SCREW TERMINALS.

② BLACK IS COMMON WITH RESPECT TO THE TRANSFORMER WINDING ONLY AND NOT THE EXTERNAL CIRCUIT.

Courtesy of Honeywell, Inc.

FIGURE 16-31 Multitap transformer schematic.

The low-voltage circuit connected to the secondary winding must be free from short circuits to prevent winding damage. If a short circuit exists, the circuit resistance becomes very low and causes the current to rise to levels beyond the normal operating range. This can cause the device to overheat and break down. If the secondary winding opens, the device must be replaced. To prevent this, an **in-line fuse** should be installed in the circuit. Then, if the current in the low-voltage circuit is too high, the fuse will blow and the device will not be damaged. This fuse must be sized properly so it does not blow during normal system operation. The cause of the failure should be located and corrected before replacing the fuse.

TRANSFORMER RATING

Transformers are rated in units called *volt-amperes* (VA). This rating represents the power of the transformer or the work that can be done by it. From basic electricity theory and Ohm's law:

$$\text{Power} = \text{Voltage} \times \text{Current}$$

where current is measured in amperes. It directly follows that the VA rating of a transformer is its power rating. This calculation is assuming that the power factor is 1.0, meaning that the load on the circuit is 100% resistive. If the load on the transformer is inductive or capacitive and the power factor is less than 1.0, then the VA rating and the power calculations will not be the same. Refer to the section entitled "Run Capacitors and Power Factor" in Chapter 15 for more information on the power factor.

The VA rating refers to the secondary winding of the device and is used to determine its output capability. For example, a 40 VA transformer supplying 24 Volts can safely operate with a low-voltage-circuit current of 1.67 amps (Fig. 16-32). If too many loads are connected in parallel, the control-circuit resistance will be low, and the current flowing in the circuit will rise. If this current exceeds the rating of the transformer, it can be damaged. If multiple devices need to be connected, a transformer with a larger VA rating must be used. Properly sizing the transformer is important to ensure proper system operation. Transformers used on air conditioning systems are typically rated at 40 VA, while those used on heating-only systems are often rated at 20 VA.

> **CAUTION**
>
> **CAUTION:** Always exercise caution when working on electric circuits. Disconnect power whenever possible to avoid component damage as well as serious personal injury.

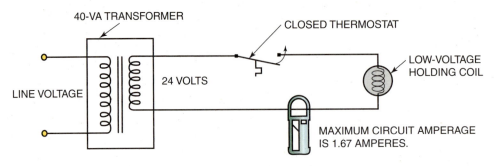

24 VOLTS X 1.67 AMPERES = 40 VA

FIGURE 16-32 Control circuit amperage multiplied by the voltage must not exceed the VA rating of the transformer.

VARIABLE-SPEED MOTOR CONTROL

Quite often, motors are required to operate at more than one speed. As we saw in the previous chapter, changing the size of the motor windings changed the speed of the motor. By introducing a larger portion of the winding to the active electric circuit, we were able to produce a lower motor speed, since the added resistance led to lower current flow through the motor windings.

In a manner similar to the configuration of the multitap transformer, multiple speeds on smaller motors can be achieved. Like the multiple taps on the primary winding of the multitap transformer, multiple taps on the secondary winding can create various voltages at the secondary. This principle is applied here. An autotransformer is a coil of wire connected to the line-voltage power source. By tapping this coil at various points (Fig. 16-33), different voltages can be obtained. By supplying different voltages to the motor, the speed of the motor can be changed.

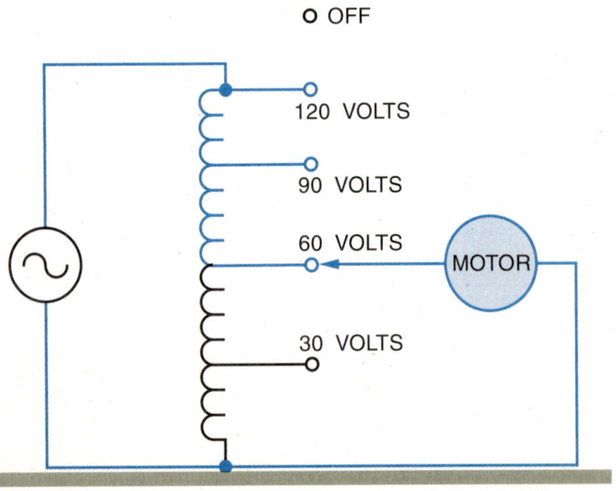

ELECTROMAGNETIC DEVICES

Electromagnetic devices also rely on magnetic fields to perform various tasks, but these tasks typically involve the energizing and de-energizing of electric circuits. These devices are made up of a holding coil, within which is positioned an iron core, and at least one set of electrical contacts. The electrical contacts on the device can be normally open, normally closed, or a combination of both. When voltage is applied to the holding coil of the device, the iron core becomes magnetized and a magnetic field is generated. This causes the electrical contacts to change their position: normally open contacts will close and normally closed contacts will open. The holding coils of the devices are typically rated by the design voltage at which they are to operate (Fig. 16-34).

A set of contacts is often made up of a stationary contact and a movable contact. The movable contact is connected to the armature of the relay, which shifts its

FIGURE 16-33 Autotransformer controls the voltage supplied to the motor.

Delmar/Cengage Learning.

FIGURE 16-34 Voltage rating of a relay.

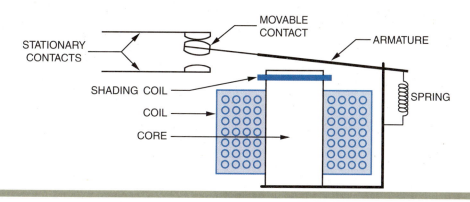

FIGURE 16-35 Simple relay in the de-energized position.

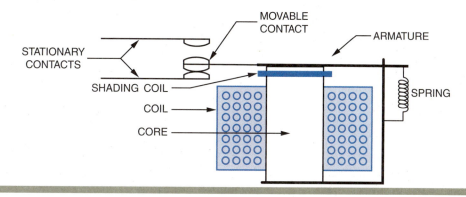

FIGURE 16-36 The coil of this relay is energized and the contacts have changed position.

position as the strength of the magnetic field changes. The armature is usually held in its de-energized position by a spring (Fig. 16-35). The magnetic field that is generated by current flow through the coil overcomes this spring pressure and pulls on the armature (Fig. 16-36).

RELAYS

Electromagnetic devices can be designed for specific purposes, such as the current magnetic relay or the potential relay discussed in the previous chapter, or they can be designed for general purposes. The current relay must be sized properly to ensure that the contacts open and close at the correct amperage, facilitating proper motor operation. The potential relay is rated by the pick-up and dropout voltages at which the contacts open and close. General-purpose relays (Fig. 16-37) can be used for a wide range of applications and are rated by the coil voltage as well as the maximum amperage rating of the contacts. Using a relay that has a higher amperage rating than the intended load will not affect the relay's operation or that of the device being controlled. Relays facilitate the use of low voltage and current to control the operation of devices that run at different voltages and draw larger amounts of electric current.

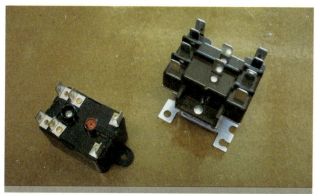

Photo by Eugene Silberstein.

FIGURE 16-37 General purpose relays.

The wiring diagram in Figure 16-38 shows a relay with a holding coil and a normally open set of contacts. The normally open contacts are wired in series with the PSC fan motor. Even though there is power between L1 and neutral, the fan motor does not operate because the contacts on the relay are in the open position. The holding coil is connected in a low-voltage control circuit powered by the 24-Volt secondary winding of the control transformer. There is also a switch or operational device in series with the coil. It is this operational device that determines when the fan motor is to operate.

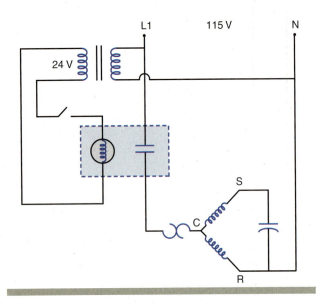

FIGURE 16-38 PSC motor controlled by a relay with one set of normally open contacts.

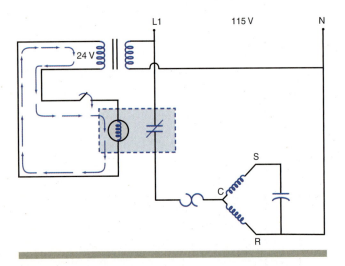

FIGURE 16-39 When the control switch is closed, current flows through the holding coil.

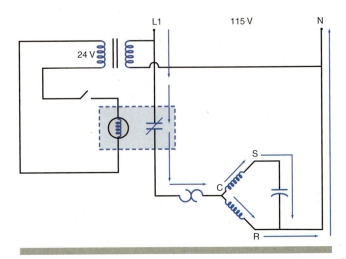

FIGURE 16-40 The holding coil causes the relay contacts to close, energizing the motor.

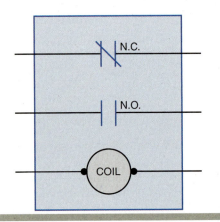

FIGURE 16-41 Relay with one set of normally closed contacts and one set of normally open contacts.

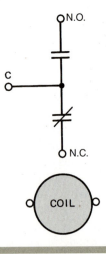

FIGURE 16-42 Schematic of a relay with an SPDT contact arrangement.

When the operational device closes, current will flow in the low-voltage control circuit (Fig. 16-39). This current flow creates a magnetic field around the coil of the relay. This magnetic field, in turn, causes the contacts on the relay to close. Once the contacts close, the circuit through the fan motor is completed, permitting current flow through the fan motor (Fig. 16-40). It will continue to operate until the operational device opens, de-energizing the holding coil on the relay.

A relay with normally closed contacts will open an electric circuit when the holding coil is energized. Most general purpose relays are manufactured with both normally open and normally closed contacts (Fig. 16-41). Some relays are configured as single-pole, double-throw devices (Fig. 16-42). They have a common terminal connected to both—a normally open set of contacts and a normally closed set of contacts. These relays are designed to direct current through one path when the holding coil is de-energized and through another path when the coil is energized. Relays are generally used to control loads that draw less than 10 amps of electric current. When the relay fails, it is thrown away and replaced, given the relatively low cost of the device.

CONTACTORS

Contactors (Fig. 16-43) are basically large relays that control the operation of loads that draw more than 15 amps. The basic operation of the contactor is the same as the relay, but the contacts are of a more substantial construction to handle the higher current requirements. In residential applications, contactors are typically of the two-pole variety, meaning that the holding coil causes two sets of contacts to open and close simultaneously.

Single-pole contactors are used to control devices powered by 120-Volt power sources as they make and break the only hot leg, L1, that feeds power to the device. Three-pole contactors (Fig. 16-44) control three-phase devices and are designed to make and break the L1, L2, and L3 lines that provide power to the device.

One of the main differences between relays and contactors is that the contacts of a relay typically make and break a circuit at one point, while the contactor makes and breaks a circuit at two points. Contactors are said to have double-break contacts (Fig. 16-45). Notice that there are two stationary contacts and a bridging contact. The bridging contact is movable and is connected to the armature of the device. It will electrically connect the two stationary contacts when the holding coil is energized. This concept of the bridging contact is the same as that used in the construction of the current magnetic relay.

Delmar/Cengage Learning.

FIGURE 16-44 (A) Two-pole contactor (B) Three-pole contactor.

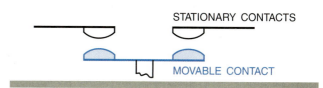

STATIONARY CONTACTS

MOVABLE CONTACT

FIGURE 16-45 A set of double-break contacts.

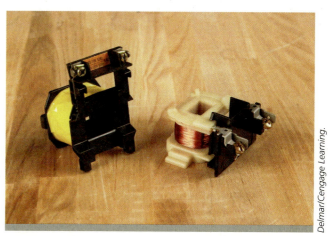

Delmar/Cengage Learning.

FIGURE 16-46 Replaceable contactor coils.

Contactors used to control loads that draw large amounts of current must be equipped with heavier duty contacts and are, therefore, more expensive than those that control the operation of devices that draw lower currents. These larger contactors can often be repaired in the field by replacing the coil (Fig. 16-46) and/or the contacts (Fig. 16-47). Smaller contactors are often discarded because the labor and costs involved in repairing them would be greater than the cost of a new contactor.

Courtesy of Honeywell, Inc.

FIGURE 16-43 Two-pole contactors.

FIGURE 16-47 Contactor contacts can be replaced when they become damaged.

Delmar/Cengage Learning.

MOTOR STARTERS

Motor starters (Fig. 16-48) are found on three-phase power supplies and are used to start and stop the operation of three-phase motors. For the most part, starters operate in a manner very similar to that of the contactor just discussed. The main difference is that the motor starter is equipped with built-in overload protection. Recall that the three-phase power supply is made up of three hot legs. Each of these hot legs must be fused to protect the circuit in the event that an overload condition exists (Fig. 16-49). In the event that one of these fuses blows, current will still be flowing through the motor, as a complete circuit will exist between the other two power lines that are still energized (Fig. 16-50). As a result, the motor will be supplied with single-phase power, as opposed to the three-phase power required for proper operation. This condition was referred to earlier as single phasing.

Single phasing causes the amperage draw of the motor to increase, resulting in the generation of excessive

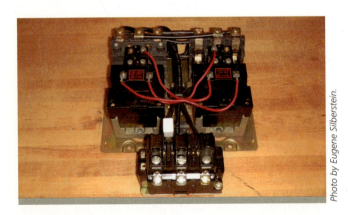

FIGURE 16-48 Typical three-phase motor starter.

Photo by Eugene Silberstein.

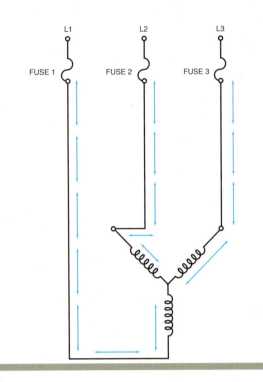

FIGURE 16-49 All three lines of a three-phase power supply need to be protected by fuses.

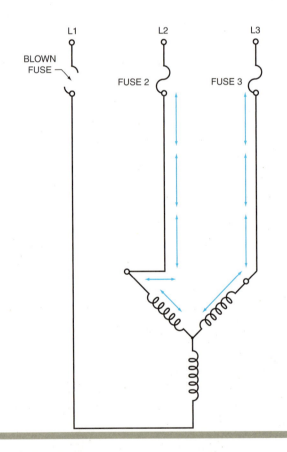

FIGURE 16-50 When one fuse blows, current can still flow through the remaining lines. This is called single phasing.

heat. The overload protection on the motor starter senses this heat and disables the motor by de-energizing the holding coil on the starter. Once the starter de-energizes the motor, it must be manually reset when the reason for circuit failure has been established and corrected.

DEFROST TIMERS

Defrost timers are commonly found in domestic refrigerators that are classified as "frost-free" units. These units periodically shut the compressor down and energize an electric heater to melt any ice that has formed on the evaporator coil. The defrost timer (Fig. 16-51) consists of a synchronous motor that turns at a constant speed, a rotating cam, and a single-pole, double throw (SPDT) switch. A simplified wiring diagram for a typical defrost timer is shown in Figure 16-52 where

- terminal 1 is the common terminal of the single-pole double throw switch
- terminal 2 is the normally open terminal that energizes the defrost heater
- terminal 3 is the normally closed terminal that energizes the compressor
- terminal 4 brings power to the timer motor

During normal operation, the SPDT switch is in the position shown in the figure. After a predetermined period of time, the system goes into defrost. The rotating cam causes the contacts between terminals 1 and 3 to open, de-energizing the compressor. At the same time, contact is made between terminals 1 and 2, energizing the defrost heater. Once the defrost cycle is over, contacts 1 and 2 open and contacts 1 and 3 close, putting the system back into the refrigeration mode.

Consider a sample defrost timer that is manufactured to facilitate a 90-minute refrigeration cycle and

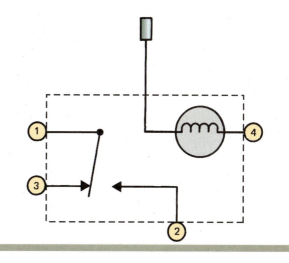

FIGURE 16-52 Schematic of a defrost timer.

a 10-minute defrost cycle. This means that the system will go into defrost for 10 minutes for every 90 minutes of either *system run time* or *compressor run time*, depending on how the system is wired. A **continuous run timer** will have the timer motor energized whenever power is supplied to the refrigerator. The unit will, therefore, go into defrost every 90 minutes, no matter how long the compressor has operated during that period. A **cumulative compressor run timer** will have the timer motor energized only when the compressor is operating. In this case, the defrost cycle will be initiated after the compressor has run for 90 minutes, as the cam on the timer does not turn when the compressor is not operating.

In the field, the technician may encounter a number of different timers, such as those used to defrost heat pump systems or those used to cycle equipment on and off at predetermined times. Although the applications of the devices may be somewhat different, they are, in essence, a series of switches that open and close sets of electrical contacts. The heat pump defrost timer, for example, is often used in conjunction with a thermistor that determines whether there is indeed ice on the coil that needs to be melted. In this case, the timer must be calling for defrost *and* there must be ice on the coil for the defrost cycle to be initiated. If there is no ice on the coil, the system will not go into defrost.

FLOW SWITCHES

The flow switch (Fig. 16-53) is designed primarily to determine if there is air flowing though a duct system. Typically constructed as a normally open set of contacts,

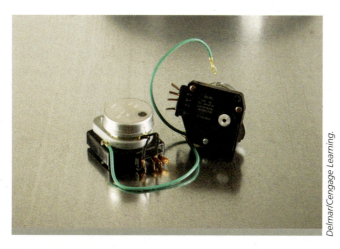

Delmar/Cengage Learning.

FIGURE 16-51 Defrost timer.

Photo by Eugene Silberstein.

FIGURE 16-53 The sail switch detects air movement in a duct system.

the contacts will close once airflow has been established. This is accomplished by a sail that is inserted into the duct. When the blower is in the off position, the sail, which is connected to a spring-loaded switch, will be in a position perpendicular to the direction of airflow. Once the blower is energized and air is flowing in the duct, the sail is pushed by the air. The movement of the sail causes the switch that is connected to the sail to change its position. The sail switch is often used to

- prevent the compressor from operating if the evaporator fan is not operating
- prevent resistive-type duct heaters from operating if the evaporator fan is not operating

FLOAT SWITCHES

Float switches are typically used to energize or de-energize an electric circuit, depending on the water level in a particular vessel or container. Considering the example of the toilet tank described at the beginning of the chapter, we saw that after the toilet is flushed, water will flow into the tank until the desired water level is reached. The device used to start and stop the flow of water into the tank is a float switch. In the air conditioning industry, float switches are typically used to energize or de-energize circuits based on water level.

In earlier chapters, the installation and use of condensate pumps was discussed. Its purpose was to remove the condensate from locations where no drain was available. But what would happen if the condensate pump failed for any reason? Most likely, the condensate would overfill the pump's container and spill out onto the floor. Enter the float switch. Normally closed float switches are often used to prevent this from happening. The float switch is wired in series with the air conditioning

system's control circuit and will de-energize the system compressor if the water level in the condensate pump rises too high.

The float switch is also used to control sump pumps in residential basements. A sump, which is designed to accumulate excess water, will need to be pumped out when the water reaches a predetermined level. In this case, a normally open float switch is used. The pump is normally in the off position; when the water in the sump rises to a certain level, the float switch closes its contacts, energizing the pump's motor. Once the water level has fallen to an acceptable level, the switch opens its contacts, de-energizing the motor.

PRESSURE SWITCHES AND CONTROLS

Pressure controls are switches that open and close electric contacts depending on the pressure within a specific part of the system. They can be used to control the normal operation of a system or as safety elements. These controls help to ensure that the system operates within its designed pressure ranges. Three distinct classifications of pressure controls based on their location in the system and the pressures they sense are the

- high-pressure control
- low-pressure control
- dual-pressure control

HIGH-PRESSURE CONTROLS

High-pressure controls, also known as *high-pressure switches,* are located on the high-pressure side of the system and are designed to open their contacts if the pressure sensed is higher than the desired level (Fig. 16-54). These controls open on pressure rise; they are drawn in schematic diagrams (Fig. 16-55). The schematic symbol is drawn above the main circuit line, indicating that the circuit will open when the system pressure rises.

Consider this example: An air conditioning system is equipped with a mechanical-draft, air-cooled condenser, and the fan motor burns out. What will happen to the system? The compressor, not knowing that the motor is no longer running, will continue to operate, causing the head pressure to rise. At this point, the excess pressure in the system can cause refrigerant lines to burst, possibly causing serious personal injury. The high-pressure control prevents this from happening. The device senses that the system pressure is higher than desired, and it disables

FIGURE 16-54 Nonadjustable pressure switch.

Delmar/Cengage Learning.

FIGURE 16-55 Electrical symbol for a high-pressure switch.

the system. High-pressure controls can be **manually reset** or **automatically reset** (Fig. 16-56). Manual-reset controls prevent the system from operating until the device is reset by hand. If the high-pressure switch trips, the system will stop running, the space temperature will rise, and the customer will place a no-cooling call to the service company. Upon arrival, the technician should notice the tripped control, signaling that a pressure problem exists. The problem should be identified and corrected before resetting the control.

FIGURE 16-56 Pressure switches can be manual-reset or automatic-reset devices.

Photo by Eugene Silberstein.

CAUTION

CAUTION: Some field technicians are under the misconception that they are doing the equipment owner a service by pointing out the manual reset on the pressure control to them. The equipment owner then, in an effort to save money on future service calls, may choose to continually reset the control in the hope that the problem will go away. Remember that the high-pressure control is a safety device and continually resetting it puts the equipment and those around it in a potentially dangerous situation.

High-pressure switches that reset automatically will close their contacts once the pressure drops below a predetermined set point. The pressure at which the contacts open is the **cut-out pressure**, and the pressure at which the contacts close is the **cut-in pressure**. The difference between the cut-out and cut-in pressures is the **differential** (Fig. 16-57). Most high-pressure switches come factory set

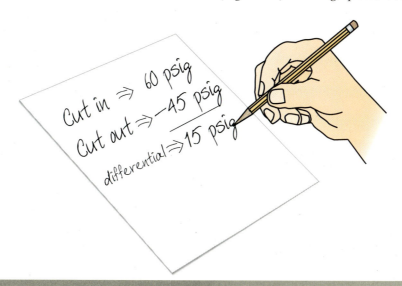

FIGURE 16-57 Differential = Cut in–Cut out.

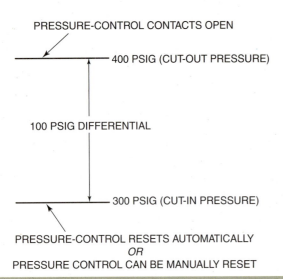

PRESSURE-CONTROL CONTACTS OPEN

400 PSIG (CUT-OUT PRESSURE)

100 PSIG DIFFERENTIAL

300 PSIG (CUT-IN PRESSURE)

PRESSURE-CONTROL RESETS AUTOMATICALLY
OR
PRESSURE CONTROL CAN BE MANUALLY RESET

FIGURE 16-58 High-pressure controls often have a 100-psig differential built into the device.

with a 100 psig differential. For example, if the pressure control is set to cut out at 400 psig, it will automatically reset once the pressure drops below 300 psig. If the pressure control is manually reset, it can only be reset when the pressure drops below 300 psig (Fig. 16-58). The cut-out pressure must be set high enough so that it will not open during normal system operation and low enough to protect the equipment, service technician, and occupants.

Going Green

A frequent need for resetting a high-pressure control indicates a deficiency in system performance. Resetting the device without determining and remedying the cause for the malfunction is doing a disservice to the equipment owner. Improper system operation results in inadequate cooling and increased power consumption.

LOW-PRESSURE CONTROLS

Low-pressure controls are located on the low-pressure side of the system and are designed to open their contacts if the pressure drops below a predetermined point. These devices can be used in many applications, including:

- Low-charge protection
- Temperature control
- Freeze control

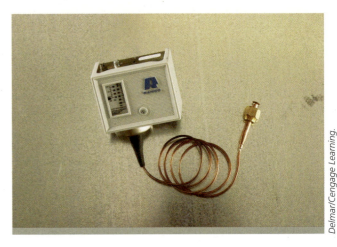

Delmar/Cengage Learning.

FIGURE 16-59 Low-pressure switch.

The low-pressure control, also known as *low-pressure switch*, is shown in Figure 16-59. In circuits, these controls are drawn below the circuit line, indicating that they will open when the pressure they sense drops. When used as a low-charge protector, the cut-out pressure must be well below the normal operating pressure of the system, but above atmospheric pressure so that no atmosphere or moisture is permitted to enter the system. For example, in an R-22 air conditioning system, the low-side pressure is normally in the range of 60 to 70 psig. If the low-pressure, cut-out set point was 30 psig, the system would cut out only if it lost all or part of its refrigerant charge, since the pressure would not normally drop that low. When used for low-charge protection, the device is usually manually reset, which would prevent the compressor from running and reduce the possibility of contaminating the system.

from experience...

Some refrigeration applications use an automatic-reset control so that the system will continue to run, however sporadically, to provide at least some refrigeration to protect the product.

When used as a freeze control, the low-pressure switch is set to cut out at a pressure corresponding to a temperature just below 32°F. For example, if an R-22 air conditioning system normally operates with a 40°F coil, the corresponding suction pressure is 68.5 psig. The coil will not freeze, because its temperature is above 32°F. If the air-flow through the coil were restricted for any reason, the temperature and pressure of the refrigerant inside the evaporator coil would drop, causing the

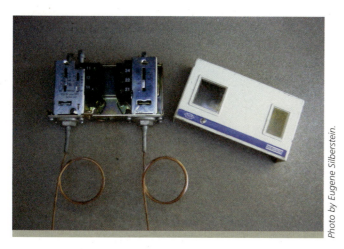

FIGURE 16-60 The dual-pressure switch contains both a high-pressure switch and a low-pressure switch.

Photo by Eugene Silberstein.

temperature of the coil surface to drop as well. The coil would start to frost at a refrigerant pressure of 57 psig. To prevent the coil from freezing, the cut-out set point on the freeze control would therefore be set just below 57 psig. This type of control is normally reset automatically.

DUAL-PRESSURE CONTROLS

Although not commonly found on residential air conditioning systems, the concept of the dual-pressure will be discussed for the sake of completeness. Quite often, refrigeration systems are equipped with adjustable low-pressure controls as well as high-pressure controls (Fig. 16-60). To facilitate the installation and to centralize the controls, dual-pressure controls are often used.

The dual-pressure control is a combination of a low-pressure control and a high-pressure control in a single unit. For the contacts of the dual-pressure control to be closed, both the high and low pressures must be within the specified range. If either the low pressure or the high pressure is out of range, the contacts will be open. It is possible for the low-pressure component of the control to reset automatically, and the high-pressure component to reset manually. This allows the system to operate as though the high- and low-pressure controls are separate devices.

ELECTRONIC CONTROLS

As technology continues to advance, the abilities of electronic circuits cover wider ranges of applications, and the cost of these devices continues to drop. Once

used primarily on commercial and industrial systems, electronic controls and devices, also referred to as solid-state devices, are now commonplace in the residential applications of heating, ventilating, and air conditioning. These devices generate the same results as their non-electronic counterparts, but do so in a manner that is generally more accurate and reliable. Because there are no moving parts in electronic controls, there are fewer component failures. In addition, solid-state controls are resistant to vibrations and are often sealed to protect them from dirt and/or moisture.

RESIDENTIAL AIR CONDITIONING APPLICATIONS

Electronic circuit boards (Fig. 16-61) are becoming more popular on even the most basic of residential air conditioning systems. These devices are even common on window and through-the-wall air conditioners. From measuring the amperage draw of the system compressor to sensing overload conditions to providing a time delay between compressor starts, circuit boards play an integral role in the operation of air conditioning and heating equipment.

Common electronic controls are equipped with features that will disable the system if the voltage supplied to the unit is 10% above or below the voltage at which the system is designed to operate. Also, if the compressor motor is operating under an overload condition, which may be the result of system overcharge, defective condenser fan motor, or non-condensable gases in the system, the sensors in the circuit board will disable the system, preventing any further system damage.

The most common feature of the solid-state circuit board is the time-delay feature. Internal timers prevent the compressor from cycling on and off too quickly. Once

Photo by Eugene Silberstein.

FIGURE 16-61 An electronic circuit board.

the compressor cycles off, the internal timer disables the compressor circuit until a predetermined period of time has elapsed. This is especially useful on systems that are equipped with low-pressure controls that automatically reset. If the system has lost part of its charge, for example, it will operate until the low-side pressure falls to the cut-out pressure setting on the pressure control. The pressure control will then cause the compressor to cycle off and the system pressures will begin to equalize. As the low-side pressure rises, the low-pressure switch will close its contacts, re-energizing the compressor. This can go on indefinitely, causing damage to the compressor. The time-delay feature prevents the compressor from cycling on and off so readily and also allows system pressures to equalize.

ELECTRONIC THERMOSTATS

Electronic thermostats (Fig. 16-14) are a very popular choice for home owners because they allow the equipment owner to maintain different temperatures in the occupied space at different times throughout the day. For example, a home owner can set the thermostat to maintain the following temperatures during the course of a day:

74°F in the morning when the family members wake up

85°F after the family has left the house for the day

72°F in the evening when family members have returned

80°F at night when the family has gone to sleep

The times and temperatures can be programmed into the thermostat to keep the residence at different temperatures throughout the day, thereby saving energy and eliminating the need to manually change the thermostat settings.

Many thermostats have a five-event, five-day, two-day configuration, where the equipment owner can set up to five different times and temperatures for each of the five days of the work week and five different times and temperatures for each of the two weekend days. The thermostat will follow one program for Monday through Friday and the weekend program for Saturday and Sunday.

SOLID-STATE RELAYS

Solid-state relays perform the same tasks as the mechanical relays discussed earlier in this chapter. The method by which the device opens and closes electrical contacts is different though. Solid-state relays (Fig. 16-62) often use

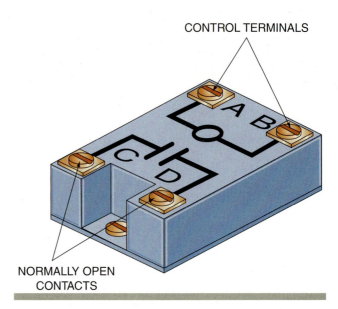

FIGURE 16-62 Solid-state relay.

a light emitting diode (LED) instead of an electromagnetic field to open and close contacts. When an input or control signal is brought to the relay, the LED is lit. The light emitted from the diode is sensed by a power transistor, turning it on, and permitting current flow through the relay.

SOLID-STATE STARTING RELAYS

As you may recall, the current magnetic relay (CMR) was used on single-phase motors to remove the start winding and/or start capacitor from the active electric circuit after initial motor startup. The CMR opened and closed its contacts depending on the amount of current flow through the run winding of the motor. One of the main drawbacks of the CMR is that a large number of relays must be kept on hand as the current ratings of the relay must closely match the current draw of the motor to ensure proper motor operation. The solid-state starting relay, on the other hand, can be used to replace most CMRs. In addition, there are no moving parts on the solid-state starting relay, affording them a longer useful life.

The solid-state starting relay (Fig. 16-63) is actually an electronic component called a thermistor. A thermistor is a resistor that changes its resistance as the temperature it is exposed to changes. The thermistor used in this type of relay is a positive temperature coefficient (PTC) that increases its resistance as the temperature increases. The solid-state starting relay is connected to the motor as shown in Figure 16-64.

FIGURE 16-63 Solid-state starting relays.

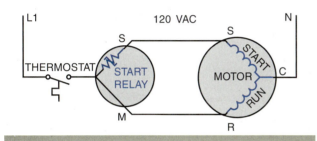

FIGURE 16-64 Solid-state starting relay connection.

Delmar/Cengage Learning.

FIGURE 16-65 Solid-state timer.

6 seconds to 600 seconds, or 10 minutes. The device is typically wired in series with the compressor contactor and, once the thermostat calls for cooling, the delay process is initiated.

When power is initially supplied to the motor, the PTC is cool and the resistance of the device is very low—in the range of 3Ω to 5Ω. Because of this low resistance, both the start and run windings are energized, permitting the motor to start. As the motor begins to operate, the PTC starts to heat up. As the heat level increases, the resistance of the device increases from just a few ohms to several thousand ohms. This increase in resistance drastically reduces the current flow through the start winding, in effect removing it from the circuit. A very small amount of current will still flow through the start winding, keeping the PTC warm enough to keep the resistance high. Once the motor cycles off, the PTC cools down in preparation for the next cycle to begin. It normally takes 2 to 3 minutes for the PTC to cool completely.

SOLID-STATE TIMERS

These devices are commonly used to prevent the air conditioning compressor from short cycling, which is when the compressor cycles on and off too quickly. The solid-state delay timer (Fig. 16-65) can be set to delay the closing of an electric circuit for anywhere from

SUMMARY

- Bimetal overloads protect motors from overload conditions.
- Fan switches are used to start and stop blowers at predetermined system temperatures.
- Limit switches are safety devices that disable heating equipment when unsafe temperature conditions exist.
- Thermostats open and close electrical contacts in response to temperature conditions within the system or occupied space.
- Thermostats can be either line-voltage or low-voltage devices.
- Thermostats can be configured as cooling only, heating only, or combination heating/cooling devices.
- Anticipators minimize the temperature swing in the occupied space by anticipating the heating and cooling requirements of the space.
- Solenoids rely on the strength of a magnetic field to open and close valves.
- Transformers generate a voltage in the secondary winding by induction.

- Transformers can be step-up or step-down devices depending on the number of turns in the primary and secondary windings.
- The VA rating of a transformer determines the work that can be done by the transformer.
- Relays and contactors are made up of holding coils and a number of sets of contacts that can be normally open, normally closed, or a combination of both.
- The holding coils of the relays and contactors are in the control circuit, while the contacts are in the power circuit.
- The voltage of the control circuit is often different from the voltage applied to the power circuit.

- Motor starters are used for three-phase motors and are basically contactors with built-in overload protection.
- Defrost timers can be cumulative compressor run timers or continual run timers.
- Flow switches open and close electrical contacts depending on the amount of air flowing through a duct system.
- Float switches open and close electrical contacts depending on water level.
- Pressure switches open and close electrical contact in response to system pressures.
- Electronic controls perform the same tasks as their mechanical counterparts but have no moving parts and typically operate more accurately and precisely.

GREEN CHECKLIST

- ☐ **Mercury is a dangerous substance and is found in many older thermostats.**

- ☐ **Disposal of old devices containing mercury must be done in an environmentally acceptable manner.**

- ☐ **Mercury exposure can cause muscle weakness, mental disturbances, skin rashes, and memory loss.**

- ☐ **Frequent resetting of a high-pressure control indicates a deficiency in system performance.**

- ☐ **Always determine the cause for system failure before putting the system back into operation.**

- ☐ **Electronic controls are desirable for efficient system operation.**

REVIEW QUESTIONS

1. **If a bimetal overload is used to protect a PSC motor, the overload should be wired**
   a. Between L1 and the motor's common terminal
   b. Between Neutral and the motor's start terminal
   c. Between Neutral and the motor's run terminal
   d. In parallel with the motor's run capacitor

2. **Fan switches, when used on heating equipment**
   a. Open on a rise in temperature
   b. Close on a rise in temperature
   c. Open on a drop in temperature
   d. Both b and c are correct

3. **If the blower on a heating system should fail**
   a. The fan switch will close its contacts and the blower will operate
   b. The heat exchanger will overheat and the limit switch will close its contacts
   c. The heat exchanger will overheat and the limit switch will open its contacts
   d. The heat exchanger will overheat and the fan switch will open its contacts

4. **A cooling thermostat**
   a. Opens its contacts on a drop in temperature
   b. Closes its contacts on a rise in temperature
   c. Both a and b are correct
   d. Neither a nor b is correct

5. **Thermostats that use a mercury-filled bulb**
   a. Rely on thermistors to sense space temperature
   b. Must be mounted perfectly level to ensure proper operation
   c. Both a and b are correct
   d. Neither a nor b is correct

6. **A PTC thermistor**
   a. Increases its resistance as the temperature increases
   b. Increases its resistance as the temperature decreases
   c. Permits more current flow through the device at high temperatures
   d. Decreases its resistance as the temperature increases

7. **In a low-voltage control circuit, the green wire is used for the**
   a. Ground
   b. Heating circuit
   c. Cooling circuit
   d. Indoor fan circuit

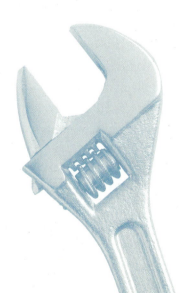

8. **The device used to minimize the temperature swing in an occupied space is the**
   a. Thermostat
   b. Subbase
   c. Anticipator
   d. Thermistor

9. **If 120 Volts is supplied to the primary of a 5:1 step-down transformer, the voltage at the secondary winding will be**
   a. 10 Volts
   b. 24 Volts
   c. 60 Volts
   d. 600 Volts

10. **The VA rating of a transformer represents the**
    a. Work that can be done by the transformer
    b. Current-carrying capability of the transformer
    c. Voltage supplied by the transformer
    d. Size of the conductors used in the transformer

11. **A transformer with a 40-VA rating can safely handle a secondary load of**
    a. 120 Volts at 3 amperes
    b. 24 Volts at 1.67 amperes
    c. 40 Volts at 40 amperes
    d. 240 Volts at 6 amperes

12. **A transformer that can be used at more than one primary voltage is called a**
    a. Multicircuit transformer
    b. Multimeter
    c. Multitap transformer
    d. Bimetal transformer

13. **Relays and contactors**
    a. Have holding coils and sets of normally open contacts
    b. Have holding coils and sets of normally closed contacts
    c. Have holding coils and sets of normally open and normally closed sets of contacts
    d. All of the above are correct

14. **When power is supplied to the holding coil of a relay**
    a. The normally open contacts will close
    b. The normally closed contacts will open
    c. Both a and b are correct
    d. Neither a nor b is correct

15. **Which of the following represents the best application for a contactor?**
    a. A 115-Volt fan motor that draws 3 amps
    b. A 220-Volt fan motor that draws 6 amps
    c. A 220-Volt, single-phase fan motor that draws 20 amps
    d. A 220-Volt, three-phase motor that draws 30 amps

16. **A motor starter can be described as**
    a. A general purpose relay with three poles
    b. A contactor with overload protection built into it
    c. A two-pole contactor with normally closed contacts
    d. A relay with a 115-Volt holding coil

17. **The defrost timer on a residential refrigerator**

   a. De-energizes the compressor
   b. Energizes the defrost heater
   c. Both a and b are correct
   d. Neither a nor b is correct

18. **The differential on a pressure control is**

   a. The difference between the cut-in and cut-out pressures
   b. The cut-out pressure added to the cut-in pressure
   c. Equal to the cut-in pressure of the device
   d. Equal to the cut-out pressure of the device

19. **High pressure controls**

   a. Are used for low-charge protection
   b. Are used as safety devices
   c. Open on a rise in pressure
   d. Both b and c are correct

20. **The solid-state starting relay**

   a. Can be used to replace current magnetic relays
   b. Uses a PTC to reduce the current flow through the start winding
   c. Uses a thermistor that increases its resistance as the temperature increases
   d. All of the above are correct

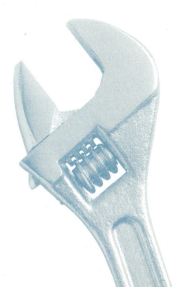

Wiring Diagrams

OBJECTIVES   *Upon completion of this chapter, the reader should be able to*

- Explain the importance of knowing how to read and interpret wiring diagrams.

- Identify symbols commonly used in electric wiring diagrams.

- Sketch the symbols for AC and DC power supplies.

- Sketch the electrical symbols for chassis and Earth grounds.

- Describe the differences in appearance among symbols for pressure and temperature switches.

- Explain how to identify fluid flow and fluid level switches.

- Explain the difference between normally open and normally closed contacts.

- Explain the difference between the symbol for a high-pressure switch and the symbol for a low-pressure switch.

- Identify switching devices based on the number of poles and the number of throws they have.

- Explain the difference between a safety device and an operational device.

- Describe two common applications for a fluid level switch.

- Describe the component parts of relays and contactors.

- Determine the sequence of operations of simple circuits.

In order to be a successful HVAC technician, certain skills are needed. In addition to knowing the basic refrigerant cycle and how it operates, along with important concepts such as superheat and subcooling, the HVAC technician must have a solid understanding of electricity. The three previous chapters in this book discussed electrical theory and how it is applied to electrical devices such as motors and the devices that control them. The successful technician understands not only these important concepts but also the sequence of operations of HVAC equipment. It is not physically possible for a technician to understand the sequence of operations for every piece of equipment manufactured by every single manufacturer. However, if the technician is able to properly read and interpret the wiring diagram for a given piece of equipment, the proper operating sequence can be determined, easing the troubleshooting process.

A wiring diagram can be described as a road map for electric current. The wires represent the roads, while the loads represent the destinations, and the switches represent obstacles along the way. If the technician is able to properly read the road map, evaluating the system will become a relatively easy task. This chapter concentrates on wiring diagram symbols, how to interpret them, and how they are connected to form circuits. In addition, a number of generic circuits will be evaluated to provide the reader with insight on how to properly read and interpret circuits and discover that there is often more than one way to connect an electric circuit to achieve the desired results.

GLOSSARY OF TERMS

Chassis ground Common connection point in a piece of equipment that is not actually driven into the ground

Earth ground The point at which a physical grounding rod or pipe is driven into the Earth

Equipment ground See chassis ground

Normally closed Term used to describe the closed or connected position of a set of electrical contacts when the device is either exposed to atmospheric pressure, de-energized or at room temperature

Normally open Term used to describe the open or disconnected position of a set of electrical contacts when the device is exposed to atmospheric pressure, de-energized, or at room temperature

Operational controls Switches that open and close their contacts during the normal operation of the equipment

Pole Term used to describe the number of sets of contacts that are opened or closed when the position of a switch is changed

Safety device A switch that opens or closes its contacts only when an unsafe condition is present.

Throw Term used to describe the number of active (on) positions on a switch

WIRING DIAGRAMS AS ROAD MAPS

As discussed in Chapter 14, "The purpose of electric circuits is to direct and control current flow, at the correct time, to the various components in an air conditioning system." Wiring diagrams are used to represent electric circuits and serve as a valuable tool for system evaluation and troubleshooting. Just as road maps are valuable tools for the automobile driver, wiring diagrams are useful to the HVAC technician. This statement is only true, however, if the operator of the vehicle knows how to read the road map and the HVAC technician knows how to read the wiring diagram. Consider the simple series circuit in Figure 17-1. In this circuit, there is a generator or power source, a switch, and a load, which in this case is a light bulb. This circuit is relatively easy to identify given that the switch looks like an actual switch, the generator looks like an actual generator, and the light bulb looks like an actual light bulb. In this circuit, we also notice that there are three conductors that connect the circuit components. One of the conductors connects the power source to the switch, one connects the switch to the load, and the third connects the load to the power source. Using this figure as a road map for electric current, we can easily see that current will flow through the circuit along the provided path as long as

- power is being supplied by the power source
- the switch is in the closed position
- the load is operable
- the conductors are connected and not damaged

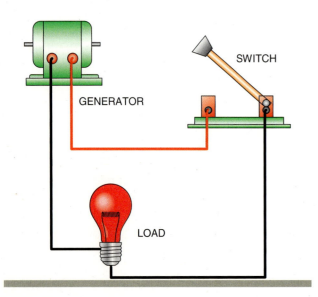

FIGURE 17-1 Pictorial diagram of a simple series circuit.

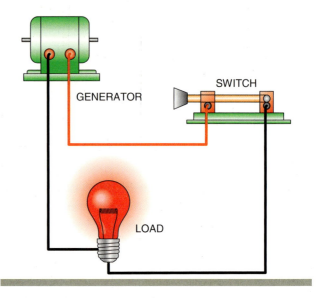

FIGURE 17-2 If the circuit is complete, the load will be energized.

If all of these conditions are met, the circuit is complete and the bulb will illuminate (Fig. 17-2). Unfortunately, not all circuits are as graphically easy to follow as this one. The average wiring diagram for a piece of air conditioning equipment can easily take on the appearance of a plate of overcooked spaghetti to those who are not skilled in the reading of it. Wiring diagrams are made up of symbols that represent various circuit components, instead of images that mimic the actual appearance of them. We will now address some of the most commonly encountered wiring diagram symbols.

ELECTRICAL SYMBOLS

Although some electrical symbols have already been introduced earlier in this text, this section of the chapter provides a more complete compilation of commonly encountered electrical symbols. It should be noted here that electrical symbols for similar components will vary from manufacturer to manufacturer so expect some variations from those that appear in this book. It is important to understand the symbols that are used in each wiring diagram in order to have a solid grasp on what the circuit is intended to accomplish.

POWER SUPPLIES

All electric circuits need a power supply to provide the voltage, or electrical pressure, needed for current flow. Power supplies can be of the alternating current,

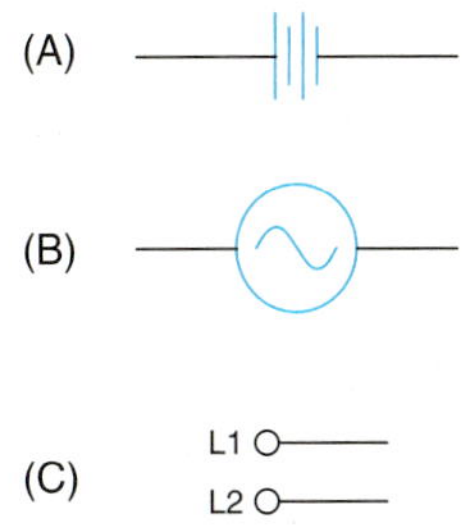

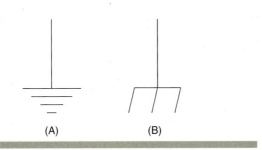

FIGURE 17-3 (A) DC power supply. (B and C) AC power supplies.

AC, or direct current, DC, variety. Direct current circuits are often powered by batteries, and the electrical symbol for a battery-type, DC, power supply is shown in Figure 17-3(A). Alternating current, or AC, circuits have power supplies that are often represented by the symbols shown in Figures 17-3(B) and (C). Alternating power, when generated, creates a sine wave-shaped pattern, which is shown in the Figure 17-3(B) symbol.

GROUND CONNECTIONS

To reduce the risk of receiving an electric shock, equipment and tools should be properly grounded. The term ground indicates an electrical connection between the tool or piece of equipment and the Earth. Ground connections on wiring diagrams are indicated by the symbols shown in Figure 17-4. An Earth ground is indicated by the symbol shown in Figure 17-4(A), while a chassis ground or equipment ground is indicated by the symbol shown in Figure 17-4(B). An **Earth ground** is the point at which a physical grounding rod or pipe is driven into the Earth. A **chassis ground** or **equipment ground** is a common connection point in a piece of equipment that is not actually driven into the ground.

FUSES, CIRCUIT BREAKERS, AND THERMAL SAFETIES

Fuses, circuit breakers, and other thermal safety devices are intended to protect electric circuits from an overcurrent condition (Fig. 17-5). The difference between a fuse and a circuit breaker is that the fuse is a one-time device, meaning that it must be replaced when it blows, while the circuit breaker can be reset when it trips. Other thermal devices may be one-time devices or may reset automatically when they cool down to a safe temperature. Symbols for one-time fuses are shown in Figures 17-5(A) and (B). The symbol for a circuit breaker is shown in Figure 17-5(C). Figure 17-5(D) represents a thermal overload-type device and is commonly found inside motors and similar components. Figure 17-5(E) shows the electrical symbol for an external bimetal overload (Fig. 17-6).

WIRE CONNECTIONS

One of the confusing aspects of reading wiring diagrams is when the lines that represent the conductors cross each other on the diagram. Knowing whether these crossed

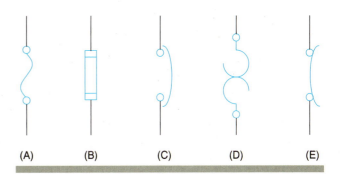

FIGURE 17-5 (A and B) Fuses. (C) Circuit breaker. (D) Internal thermal overload. (E) External bimetal overload.

FIGURE 17-6 External bimetal overload.

FIGURE 17-4 (A) Earth ground. (B) Chassis or equipment ground.

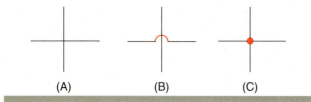

FIGURE 17-7 (A and B) Symbols for wires that are not connected. (C) Symbol for wires that are connected.

lines are connected or not, therefore, is of the utmost importance. There are three common symbols that are used to help clarify this situation. When wires are not connected, they are simply drawn over one another [Fig. 17-7(A)] or drawn as one line "hopping over" the other [Fig. 17-7(B)]. When lines are joined, a dot is used to indicate the connection [Fig. 17-7(C)].

SWITCHES

The goal of any electric circuit is to pass power, at the correct time, to the device being controlled. It is the job of the switch to accomplish this task. Switches are often referred to as power-passing devices, while the load being energized is called the power-consuming device. It is important for the technician to understand the difference between switches and loads in an electric circuit. There are many different types of switches available, and they are grouped by how they open or close their contacts. Common switch types are

- manual
- temperature sensing
- pressure sensing
- flow sensing
- fluid-level sensing

MANUAL SWITCHES

Manual switches are opened and closed when the user physically changes the position of the switch. The light switch in your living room is an example of a manual switch. Manual switches are classified by the number of available positions on the switch as well as the number of contacts that are opened or closed when the position of the switch is changed. The term **pole** is used to describe the number of sets of contacts that are opened or closed when the position of the switch is changed. For example, a single-pole switch opens or closes one set of contacts, while a double-pole switch will open or close two sets of contacts. Figures 17-8(A) and (B) are examples of single-pole switches, while Figures 17-8(C) and (D) are examples of double-pole switches.

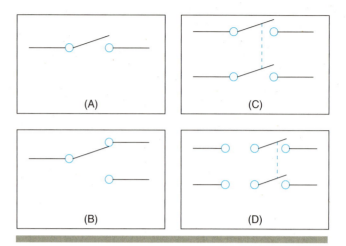

FIGURE 17-8 (A) Single-pole, single-throw switch. (B) Single-pole, double-throw switch. (C) Double-pole, single-throw switch. (D) Double-pole, double-throw switch.

The term **throw** is used to describe the number of active (on) positions on the switch. For example, if the switch is either ON or OFF, the switch is classified as a single-throw switch. If the switch has two separate ON positions, meaning it can energize either one of two separate circuits, the switch is classified as a double-throw switch.

Figures 17-8(A) and (C) are examples of single-throw switches, while Figures 17-8(B) and (D) are examples of double-throw switches. So, manual switches are classified by both the number of poles and the number of throws. The switch in Figure 17-8(A) is a single-pole, single-throw switch; the switch in Figure 17-8(B) is a single-pole, double-throw switch; the switch in Figure 17-8(C) is a double-pole, single-throw switch; and the switch in Figure 17-8(D) is a double-pole, double-throw switch.

TEMPERATURE-SENSING SWITCHES

In the HVAC industry, a number of temperature-controlled switches are used. The term commonly used to describe these devices is thermostat. There are many different applications for thermostats, but they all have one thing in common: they all respond to changes in temperature to open or close electrical contacts. Temperature-controlled switches are easily identified by the square tail on the device (Fig. 17-9).

Commonly encountered temperature-sensing switches are the heating thermostat, the cooling thermostat, and the limit switch. The heating thermostat is the device that is used to control the operation of a heating system. It stands to reason that we want the heating system to operate when the structure is cold and we want

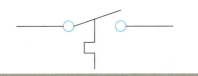

FIGURE 17-9 A temperature-sensing switch.

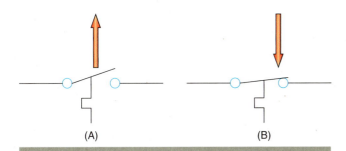

FIGURE 17-10 (A) The heating thermostat is pushed open when the mechanism is pushed up. It opens on a rise in temperature. (B) The heating thermostat is pushed closed when the mechanism is pushed down. It closes on a drop in temperature.

FIGURE 17-11 (A) The cooling thermostat is pushed open when the mechanism is pushed down. It opens on a drop in temperature. (B) The cooling thermostat is pushed closed when the mechanism is pushed up. It closes on a rise in temperature.

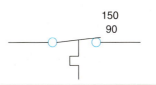

FIGURE 17-12 A limit switch. The symbol often includes information about the opening and closing temperatures of the device.

the heating system to cycle off when the structure is hot. The diagram in Figure 17-9 represents a heating thermostat. The symbols for different thermostats are drawn differently based on their function. Looking at the symbol for the heating thermostat, we can see that the thermostat has two positions, open and closed. In order to "open" the contacts on the symbol that represents the thermostat, the switching mechanism must be pushed up, and to close the thermostat, the switching mechanism must be pushed down (Fig. 17-10). So, this temperature-sensing switch will open on a rise in temperature and close on a temperature drop, which is what the heating thermostat does.

The cooling thermostat is the device that is used to control the operation of an air conditioning or cooling system. Under normal operating conditions, we want the cooling system to operate when the structure is hot and to cycle off when the structure is cooled to the desired temperature. The diagram in Figure 17-11 represents a cooling thermostat. In order to "open" the contacts on the symbol that represents the cooling thermostat, the switching mechanism must be pushed down, and to close the thermostat, the switching mechanism must be pushed up (Fig. 17-11). This temperature-sensing switch will open on a drop in temperature and close on a temperature rise, which allows the cooling system to operate as desired. Both the cooling and heating thermostats are classified as **operational controls**. Operational controls open and close their contacts during the normal operation of the equipment.

The limit switch is a device that is closed under normal operating conditions and will only open when an unsafe condition is present. The limit control is an example of a **safety device**, which is one that opens or closes its contacts only when an unsafe condition is present. The limit switch, similar to the heating thermostat, opens its contacts on a rise in temperature, so it would seem that they can get confused in a wiring diagram. But such is not the case. The limit switch is drawn in wiring diagrams as a normally closed switch and often includes information about the temperature at which the switch contacts open (Fig. 17-12). In this case, the limit switch will be in the closed position until the temperature reaches 150°F. When the temperature rises to the "top" temperature, the switch will shift to its "top" (open) position. Electrical symbols for switches that have set temperatures at which they change position, such as the limit switch, often have their temperature information included in the wiring diagram.

PRESSURE-SENSING SWITCHES

Pressure-sensing switches change their position depending on the amount of pressure sensed, or the difference between two sensed pressures. Common types of pressure switches are the high-pressure switch and the low-pressure switch [Fig. 17-13(A) and (B)]. Notice the difference between the symbols for the high- and low-pressure switches. The mechanism on the high-pressure switch

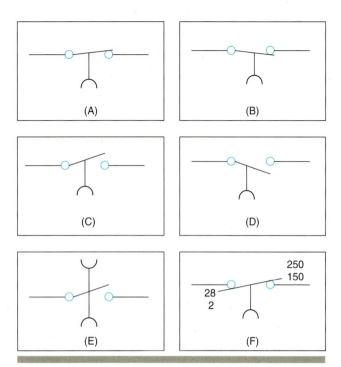

FIGURE 17-13 (A) A high-pressure switch.
(B) A low-pressure switch. (C) Normally open
pressure switch that closed on a drop in pressure.
(D) Normally open pressure switch that closes on
a rise in pressure. (E) A pressure differential switch.
(F) A dual-pressure switch.

from experience...

The terms *normally open* and *normally closed* can be interpreted differently when referring to temperature, electrical, and pressure controls. When referring to temperature controls, the contacts that are closed when the device is at rest or removed from the box are typically the *normally closed* contacts. Those that are open are the *normally open* contacts. When referring to electric controls, such as the contacts on an electric relay or contactor, the contacts that are closed when the coil is de-energized are the *normally closed* contacts. Those that are open are the *normally open* contacts. When referring to pressure controls, the contacts that are closed when the device is exposed to atmospheric pressure are the *normally closed* contacts. Those that are open are the *normally open* contacts.

must be pushed up to open, indicating that the device will open its contacts on a rise in pressure. The mechanism on the low-pressure switch must be pushed down to open, indicating that the device will open its contacts on a drop in pressure. Another commonly encountered pressure switch is the differential pressure switch. This device does not open and close its contacts depending on the sensed pressure, but on the difference between two sensed pressures, Figure 17-13 (E). The electrical symbols for various types of pressure switches are shown in Figure 17-13. Figures 17-13(A) and (B) represent **normally closed** devices, while Figures 17-13(C) and (D) represent **normally open** devices. Normally closed indicates that the contacts are in the closed position when the device is exposed to atmospheric pressure or de-energized. Normally open indicates that the contacts are in the open position when the device is exposed to atmospheric pressure or de-energized.

A special pressure control, the dual-pressure control, is a single device that performs the functions of both the high-pressure switch and the low-pressure switch. Information regarding the opening and closing of the device are often a part of the electrical symbols for

the control. For example, the dual-pressure switch in Figure 17-13(F) has both the low-pressure information and the high-pressure information included. This device will open its contacts if either the low-pressure drops to 2 psig or the high pressure rises to 250 psig. The device will close its contacts if the low-pressure rises to 28 psig or if the high-pressure falls, on automatic reset models, to 150 psig.

FLOW-SENSING SWITCHES

Flow-sensing switches open and close their contacts depending on whether there is fluid flow present. Common flow-sensing devices are the airflow switch and the water flow switch. Flow-sensing switches can be normally open, normally closed, or have both normally open and normally closed sets of contacts (Fig. 17-14). Flow switches are identified by the flag-shaped triangle on the switch.

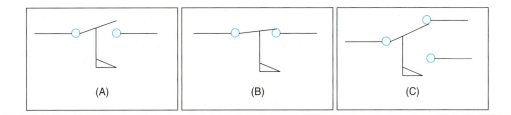

FIGURE 17-14 (A) Normally open flow switch. (B) Normally closed flow switch. (C) Single-pole, double-throw flow switch.

FLUID-LEVEL-SENSING SWITCHES

Fluid-level-sensing switches are also commonly found in the HVAC industry (Fig. 17-15). The symbol in Figure 17-15(A) represents a normally open float switch that will close its contacts on a rise in fluid level. This type of switch is often used in conjunction with a sump pump. If the water in the sump rises, the switch will close and the sump pump will turn on to pump the water away. The symbol in Figure 17-15(B) represents a normally closed float switch that will open its contacts on a rise in fluid level. This type of switch is often used in conjunction with a condensate pump. If the water in the condensate pump rises and the pump fails to operate, the switch contacts will open to de-energize the air conditioning system.

RELAYS AND CONTACTORS

Relays and contactors are devices that open and close sets of electrical contacts when the coil on the device is energized and de-energized. Relays and contactors are made up of coils and contacts, which can be of the normally open or normally closed variety (Fig. 17-16). The electrical symbol for a simple relay includes both the coil and the contacts (Fig. 17-17). Some relays have both normally open and normally closed contacts, and many relays have multiple sets of contacts (Fig. 17-18). Contactors typically have multiple sets of normally open contacts (Fig. 17-19).

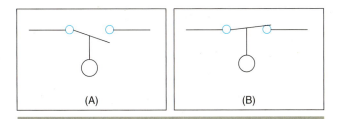

FIGURE 17-15 (A) Normally open fluid level sensing switch that closes on a rise in fluid level. (B) Normally closed fluid level sensing switch that opens on a rise in fluid level.

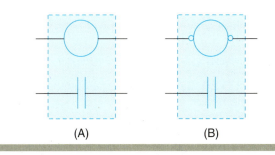

FIGURE 17-16 (A) Set of normally open contacts. (B) Set of normally closed contacts.

FIGURE 17-17 (A and B) Common symbols for relays.

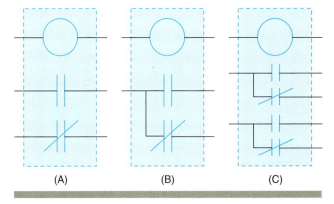

FIGURE 17-18 (A) Relay with isolated normally open and normally closed contacts. (B) Relay with single-pole, double-throw contact configuration. (C) Relay with double-pole, double-throw contact configuration.

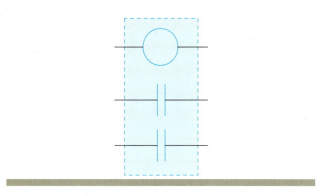

FIGURE 17-19 Symbol for a two-pole contactor.

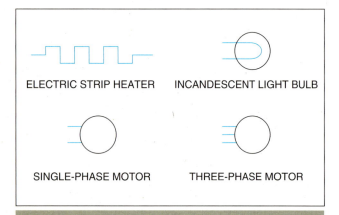

FIGURE 17-20 Symbols for some common loads: Electric strip heater, incandescent light bulb, single-phase motor, and three-phase motor.

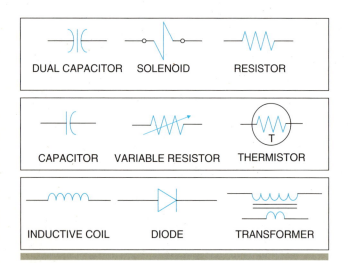

FIGURE 17-21 Symbols for common electrical components.

COMMON SYSTEM LOADS

Common system loads in HVAC equipment include electric heaters and motors. Figure 17-20 shows some common symbols for electric heaters, incandescent light bulbs, as well as single-phase and three-phase motors.

OTHER ELECTRICAL SYMBOLS

There are many electrical symbols that are encountered in our industry, the most common of which have been included in this text. Figure 17-21 provides some additional symbols that are likely to show up in HVAC wiring diagrams.

WIRING DIAGRAMS

In this section, we will examine some wiring diagrams for the sole purpose of developing and mastering the skill of reading them. The basics of wiring diagrams were discussed in Chapter 14 and you are encouraged to revisit that material as needed. What follows are some basic circuits that will be dissected and evaluated. These circuits have been somewhat simplified to provide a straightforward approach to the basics of reading and, just as important, understanding wiring diagrams.

LIGHT BULB RELAY CIRCUIT

This circuit is intended to show the basic operation of a general purpose relay (Fig. 17-22). The relay, which is made up of a coil and a set of contacts, is actually part of both the control circuit and the power circuit. The top portion of the diagram represents the control circuit, while the bottom portion of the diagram represents the power circuit.

The manual switch is used to energize and deenergize the relay coil. When the switch is in the open

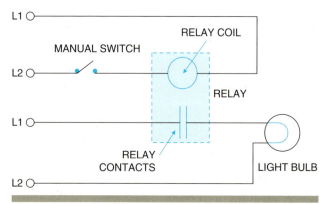

FIGURE 17-22 Simple relay circuit used to control a light bulb.

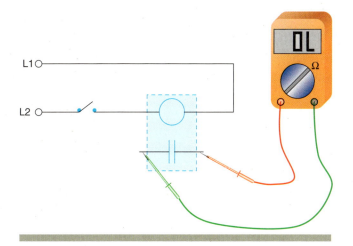

FIGURE 17-23 When the relay coil is not energized, the normally open contacts remain open and there is no continuity between the contacts.

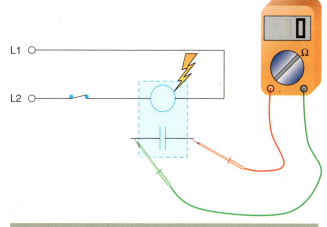

FIGURE 17-24 When the relay coil is energized, the normally open contacts will close and there is continuity between the contacts.

position, the coil is not energized, and the normally open contacts remain open (Fig. 17-23). When the manual switch is closed, power passes to the relay coil, energizing it, causing the relay contacts to close (Fig. 17-24).

The relay contacts now act as the switch that controls the load, which in this case, is an incandescent light bulb. Using a relay for this purpose is not something the field technician is likely to encounter in the field, but this example provides the basic operation and purpose of the device. In most cases, relays are utilized when a low-voltage control circuit is used to control a line voltage

circuit. In our next example, we will use a low-voltage control circuit to control the operation of line-voltage components in an air conditioning system.

BASIC AIR CONDITIONING SYSTEM WIRING DIAGRAM

We will now take a look at the basic control circuit for a straight-cooling air conditioning system (Fig. 17-25). This circuit is actually composed of four separate circuits.

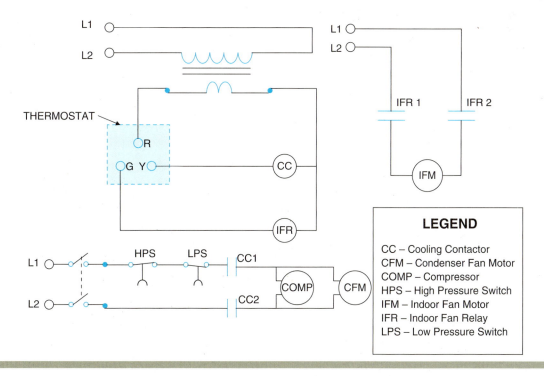

FIGURE 17-25 Wiring diagram of a straight cooling air conditioning system.

One circuit generates the 24 Volts for the low-voltage controls circuit, one circuit is the low-voltage control circuit, one circuit is the line-voltage power circuit for the indoor blower motor, and the last is the line-voltage power circuit for the outdoor condensing unit.

We will first examine the transformer circuit (Fig. 17-26). This simple series circuit is made up of only the power supply and the primary winding of the transformer. The sole purpose of this circuit is to generate the power source for the 24-volt control circuit. By tracing out this path, we can see that current flows from the power source, through the primary winding of the transformer, and then back to the power source. The primary winding of the transformer is the circuit load.

We will next examine the low-voltage control circuit. This circuit comprises the transformer secondary winding, the low-voltage thermostat, the cooling contactor coil, and the indoor fan relay coil (Fig. 17-27). For the cooling mode, three connections are made at the thermostat. These connections are

- R Terminal on the thermostat connected to the 24-volt side of the transformer
- Y Terminal on the thermostat connected to one side of the cooling contactor coil
- G Terminal on the thermostat connected to one side of the indoor fan relay coil

When the thermostat is set for cooling operation, the thermostat passes power from the transformer to the cooling contactor coil, as well as to the indoor fan relay coil.

The low-voltage control circuit is a parallel circuit with the transformer secondary winding supplying power to both the cooling contactor coil circuit and the indoor fan relay circuit.

The line-voltage power circuit for the indoor fan motor is a series circuit as well (Fig. 17-28). When the thermostat calls for indoor fan motor operation, the IFR contacts will close, completing the circuit that powers the indoor fan motor. Current will flow from L1 through contacts IFR2 and on to the indoor fan motor. When current leaves the fan motor, it will flow through contacts IFR1 and back to L2.

The line-voltage power circuit for the outdoor condensing unit is very similar to that of the indoor fan motor with the exception of the addition of some safety device, namely the pressure switches (Fig. 17-29). When the thermostat calls for cooling, the CC1 and CC2 contacts will close, completing the circuit that carries voltage to both the compressor and the condenser fan motor. This, of course, assumes that the power line voltage is supplied at L1 and L2, the low-pressure switch and high-pressure switches are closed, and all conductors are in good shape as are the compressor and condenser fan motor.

It should be noted that the diagrams that were just examined are not the only possible wiring configurations that will allow the system to operate properly.

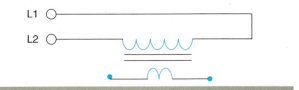

FIGURE 17-26 Transformer primary winding circuit.

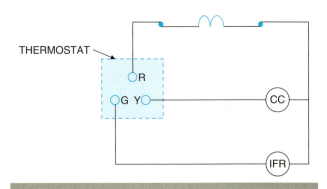

FIGURE 17-27 Low-voltage control circuit for a straight cooling air conditioning system.

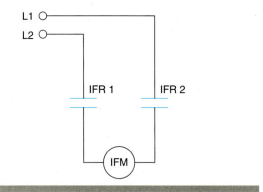

FIGURE 17-28 Line-voltage power circuit wiring diagram for the indoor fan motor.

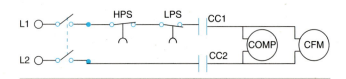

FIGURE 17-29 Line-voltage power circuit wiring diagram for the condensing unit.

The pressure switches, for example, can be located in the low-voltage control circuit instead of the line-voltage power circuit (Fig. 17-30). By placing the pressure switches in the low-voltage control circuit, the pressure switches will not have to carry the current that the compressor and condenser fan motor will draw. This will extend the life of the pressure switches. Placing the pressure switches in the low-voltage control circuit is actually the preferred method of wiring the system.

ELECTRIC HEATING WIRING DIAGRAM

We will next turn our attention to the wiring diagram of a simple electric heater circuit (Fig. 17-31). We are concerned only at this point with the heater itself and how it is energized and de-energized, not the blower motor circuit. As part of this circuit evaluation, we will break the circuit down into three simple series circuits, each of

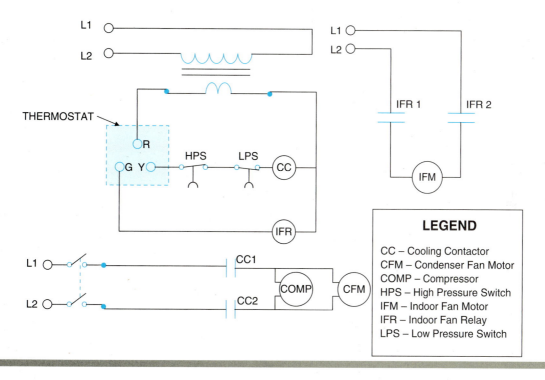

FIGURE 17-30 Wiring diagram of a straight cooling air conditioning system with the pressure switches located in the low-voltage control circuit.

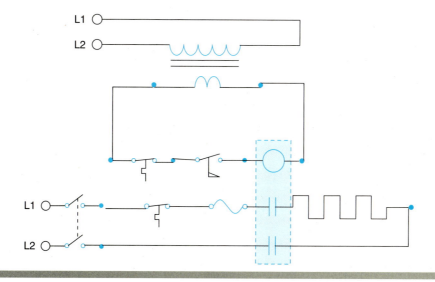

FIGURE 17-31 Wiring diagram for the heater in an electric heating appliance.

POWER SUPPLY

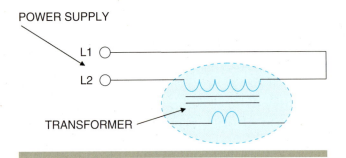

FIGURE 17-32 Transformer primary winding circuit.

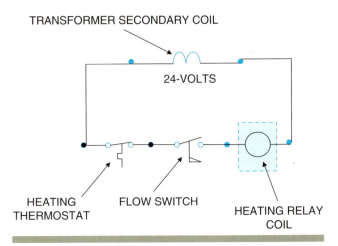

FIGURE 17-33 Low-voltage control for the heater in an electric heating appliance.

which performs its own important part in ensuring the safe, proper operation of the appliance.

We will first examine the transformer circuit (Fig. 17-32). This is the same circuit we examined in the basic air conditioning system and is made up of only the power supply and the primary winding of the transformer. This is the circuit that generates the power source for the 24-volt control circuit. The primary winding of the transformer is the circuit load.

We will next take a closer look at the control circuit (Fig. 17-33). The power supply for this 24-volt control circuit is the secondary winding of the transformer. In addition to the power supply, there are three other circuit components. They are the heating thermostat, the flow switch, and the heating relay coil. The heating thermostat and the flow switch must both be in the closed position in order to energize the heating relay coil. This means that the thermostat must be calling for heating and the blower motor must be operating. If the blower motor is not operating, the flow switch contacts will be open and power cannot pass to the heating relay coil.

If both power-passing devices are closed, the heating relay coil will be energized.

Lastly, we will examine the power circuit (Fig. 17-34). The power circuit is the circuit that actually energizes the electric strip heater. The components of this circuit are the power supply, the disconnect switch, the limit switch, the fusible link, the heating relay contacts, and the electric strip heater. The power supply and the disconnect switch supply power to the circuit. The limit switch and the fusible link are safety devices that are normally closed and will only open if there is an unsafe condition present. The heating relay contacts are the operational switches, as they open and close during the normal operation of the appliance to energize and de-energize the heater as the thermostat requires.

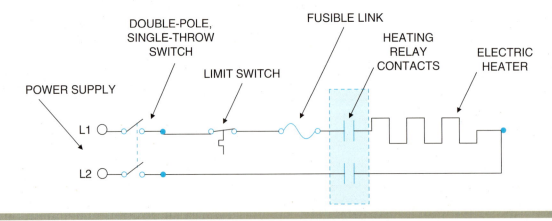

FIGURE 17-34 Line-voltage power circuit wiring diagram for the electric heater.

SUMMARY

- Wiring diagrams are used to represent electric circuits and serve as a valuable tool for system evaluation and troubleshooting.

- Symbols that are used in wiring diagrams can differ from manufacturer to manufacturer.

- Ground connections can be either a chassis ground or an Earth ground.

- Fuses, circuit breakers, and thermal safeties are all represented differently in wiring diagrams.

- Lines that cross each other on a wiring diagram may or may not be connected to each other. Look for dots that indicate junctions or wire connections.

- Switches can be manual devices or can respond to temperature, pressure, fluid flow, or fluid level.

- Switches are classified by the number of poles and the number of throws they have.

- Safety devices are typically normally closed and will open their contacts if an unsafe condition is present.

- Operational devices open and close during the normal operation of the system.

- Cooling thermostats close their contacts on a rise in temperature, while heating thermostats close their contacts on a drop in temperature.

- Flow-level switches are commonly used to energize sump pumps or disable air conditioning equipment in the event of condensate pump failure.

- Relays and contactors are made up of coils and sets of normally open or normally closed contacts.

- Some relays have both normally open and normally closed contacts.

- Contactors have multiple sets of normally open contacts.

- Appliance circuits can often be broken down into simple series circuits for evaluation purposes.

REVIEW QUESTIONS

1. **A wiring diagram is best described as**
   a. The only way that wires can be connected to make an electric circuit operate correctly
   b. A road map for electric current
   c. The sequence of operations for a particular piece of equipment
   d. a and c are correct

2. **Which of the following is required to have an electric circuit operate properly?**
   a. The power supply must supply the correct power to the circuit
   b. All control devices must be in the closed position
   c. The load must be operable
   d. All of the above are correct

3. **Which of the following statements is most likely correct with respect to electrical symbols?**
   a. All equipment manufacturers use exactly the same electrical symbols
   b. Every equipment manufacturer uses completely different electrical symbols

   c. Equipment manufacturers sometimes use electrical symbols that differ from those used by other manufacturers
   d. Electrical symbols are chosen by the National Electric Code® and are all the same, no matter which manufacturer uses them

4. **The symbol for an AC power supply is easily identified by the**
   a. Presence of the letters AC on the symbol
   b. Presence of a sine wave
   c. Absence of a sine wave
   d. Presence of a straight dotted line

5. **A chassis ground**
   a. Is the same as an Earth ground
   b. Is a common equipment connection that is not connected to Earth
   c. Both a and b are correct
   d. Neither a nor b is correct

6. **Circuit breakers and internal thermal overload devices have what in common?**
   a. They are both located inside the motor or electrical device
   b. They both must be replaced when they trip
   c. They both have the same electrical symbol
   d. They both have the capability to be reset

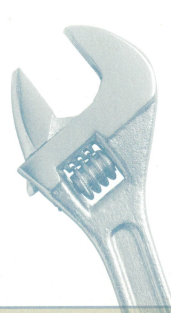

7. **Wire connections in a wiring diagram are easily identified by**

   a. The presence of a dot at the connection location

   b. The absence of a dot at the bypass location

   c. One wire "hopping over" another wire

   d. Both b and c are correct

8. **How are manual devices identified in an electric circuit?**

   a. A square "tail" on the switch symbol

   b. A rounded "tail" on the switch symbol

   c. A triangular "tail" on the switch symbol

   d. No "tail" on the switch symbol

9. **How are temperature-sensing devices identified in an electric circuit?**

   a. A square "tail" on the switch symbol

   b. A rounded "tail" on the switch symbol

   c. A triangular "tail" on the switch symbol

   d. No "tail" on the switch symbol

10. **How are pressure-sensing devices identified in an electric circuit?**

    a. A square "tail" on the switch symbol

    b. A rounded "tail" on the switch symbol

    c. A triangular "tail" on the switch symbol

    d. No "tail" on the switch symbol

11. **A typical wall switch is most likely which type?**

    a. Double-pole, single-throw

    b. Double-pole, double-throw

    c. Single-pole, single-throw

    d. Single-pole, double-throw

12. **A cooling thermostat will**

    a. Open on a rise in temperature

    b. Open on a drop in temperature

    c. Close on a drop in temperature

    d. Both a and c are correct

13. **A heating thermostat is classified as a**

    a. Safety device

    b. Pressure-sensing switch

    c. Normally open switch

    d. Operational device

14. **The low-pressure switch will**

    a. Open on a drop in pressure

    b. Close on a rise in pressure

    c. Open on a rise in pressure

    d. Both a and b are correct

15. **When the coil of a relay is energized**

    a. All normally open contacts will close

    b. All normally closed contacts will open

    c. Both a and b are correct

    d. Neither a nor b is correct

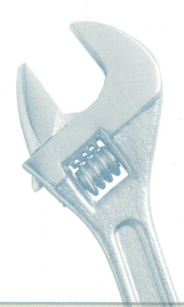

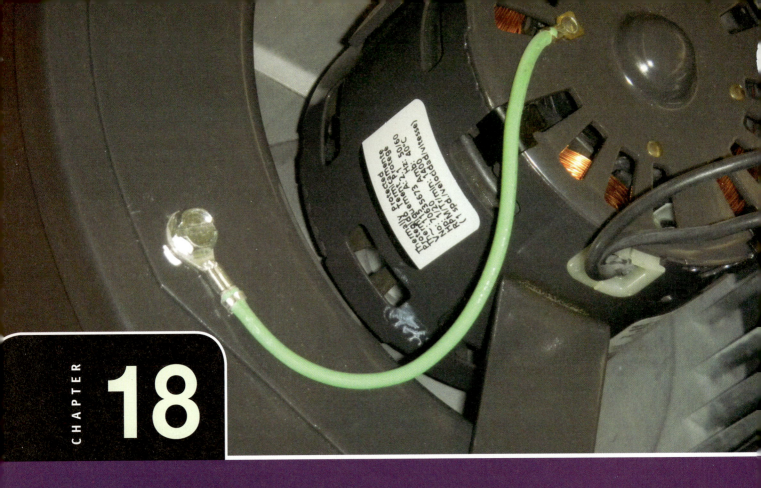

Electric Codes

OBJECTIVES *Upon completion of this chapter, the reader should be able to*

- Explain the mission of the National Fire Protection Association®, NFPA®.

- List some of the NFPA® codes that affect the residential construction industry.

- Explain the purpose of the National Electric Code®, NFPA 70®.

- Explain the difference between mandatory rules and permissive rules as they apply to the National Electric Code.

- Describe some of the sections of the NEC® that directly affect the residential construction industry.

- List and describe some of the important issues that affect the installation of electrical equipment.

- Describe some of the factors that must be considered when installing branch circuits for HVAC accessories and other devices.

- List the applications that require the use of a ground fault circuit interrupter, GFCI.

- Describe some of the factors that must be considered when dealing with circuit overcurrent protection.

- Explain the importance of proper tool and equipment grounding.

- Describe the acceptable wiring methods to use when running nonmetallic conductors through wood beams, joists, or rafters.

- Describe the acceptable wiring methods to use when running nonmetallic conductors through holes or slots in metal members.

- List various types of electrical conductors and their classifications.

- List important considerations that pertain to electrical boxes, their installation, and the number of conductors that they can contain.

- List code-related guidelines that affect general appliance circuits.

- List important factors that affect heating, air conditioning, and refrigeration systems.

- List important factors that apply to motors and motor controllers.

Previous chapters in this text discussed electricity in one way, shape, or form. Early on in the text, electrical safety was discussed, as it is of vital importance to remain safe on the job. It is extremely important to understand how electricity behaves and the possible results if it is not handled correctly. Chapters 14 through 16 provided the reader with general information regarding basic electric characteristics as well as specific information about electric components including motors, switches, controls, and operational and safety devices. Chapter 17 provided information regarding the road maps, namely electric circuits, and wiring diagrams, that electric current follows as it flows from one system component to the next. All of the information that has been presented thus far in the text is extremely important for the HVAC installer and technician. The information that is about to be presented is just as important, as it refers to the guidelines that are to be followed when installing, servicing, and maintaining electrical equipment.

The National Electric Code® (NEC®) provides detailed information regarding many factors that affect electrical installations, maintenance, and electrical system design and layout. The purpose of the NEC® is to provide a set of rules that, if followed properly, will help maximize the safety factor and reduce the possibility of fire caused by improper installation or the use of substandard or unapproved electrical devices. It should be noted that this chapter is not intended to replace the NEC®, but to provide the reader with some of the highlights of the code as they pertain to the HVAC industry.

Accessible When applied to equipment such as disconnect switches, accessible means that these devices are not guarded by locked doors or other means that would prevent easy access to them. For example, you should not need a ladder to gain access to a service disconnect switch

Ampacity The current, in amperes, that a conductor can continuously carry

Appliance branch circuit Circuits that are used to supply power to outlets that will have appliances connected to them

Bare conductor Term used to describe a conductor that has no covering at all.

Branch circuit The wiring that is installed between an overcurrent device and an outlet

Covered conductor Term used to describe a conductor that has a covering *not* recognized by the NEC® as electrical insulation

General purpose branch circuit Circuit that supplies power to two or more outlets for lighting and appliance use

Ground The Earth

Grounded Connected to ground or to a conductive body that is connected to ground

High-voltage Term used to describe voltage levels greater than 600 Volts

Individual branch circuit Circuit that supplies power to only one piece of equipment

Insulated conductor Term used to describe a conductor that has a covering that is recognized by the NEC® as electrical insulation

Knockout plug Commonly used to close unused openings in electrical boxes

Line-voltage Term used to describe voltage levels between 50 Volts and 600 Volts

Low-voltage Term used to describe voltage levels lower than 50 Volts

Mandatory rules Practices that are either allowed or not allowed

Motor controllers Devices that are normally used to start and stop a motor by making or breaking electrical contacts

National Electric Code, (NEC®) Codebook that provides specific safety guidelines for installation and service personnel working in our industry

Permissive rules Practices that are allowed, but not required

Within sight Term used to describe the relative location of two pieces of equipment that implies visibility of one from the other and a distance that is no further than 50 feet

THE NATIONAL FIRE PROTECTION ASSOCIATION, NFPA®

The National Fire Protection Association, NFPA®, is a leading authority on electrical, fire, and building safety. The NFPA® is responsible for the development and publishing of hundreds of codes and standards that are intended to minimize the risk and effects of fire. Virtually, every building, process, service, design, and installation in society today is affected in some way, shape, or form, by documents prepared or approved by the NFPA®. As stated on the NFPA® Web site, "The mission of the international nonprofit NFPA®, established in 1896, is to reduce the worldwide burden of fire and other hazards on the quality of life by providing and advocating consensus codes and standards, research, training, and education" (www.nfpa.org).

The list of codes that the NFPA® oversees is very long, but some of the codes that affect the residential construction industry are

- NFPA® 10   Standard for Portable Fire Extinguishers
- NFPA® 13D Standard for the Installation of Sprinkler Systems in One and Two Family Dwellings and Manufactured Homes
- NFPA® 31   Standard for the Installation of Oil-Burning Equipment
- NFPA® 54   National Fuel Gas Code
- NFPA® 58   Liquefied Petroleum Gas Code
- NFPA® 70   National Electric Code
- NFPA® 86   Standard for Ovens and Furnaces

INTRODUCTION TO THE NATIONAL ELECTRIC CODE (NEC®)

The purpose of the **National Electric Code (NEC®)** (Fig. 18-1) as stated in Section 90.1, is "the practical safeguarding of persons and property from hazards arising from the use of electricity". The NEC® provides specific safety guidelines for installation and service personnel working in our industry. The NEC® is not intended to be used as a design manual, nor is it intended to be used by untrained individuals.

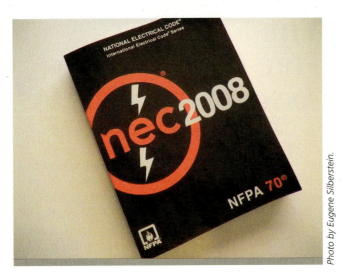

Photo by Eugene Silberstein.

FIGURE 18-1 The National Electric Code, NFPA 70®.

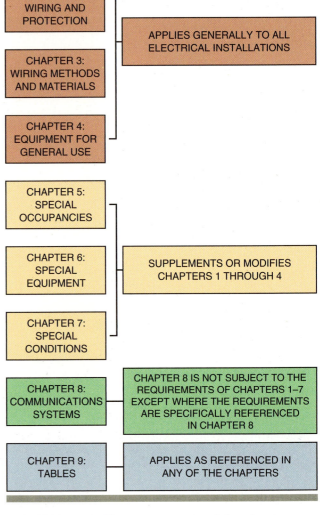

| CHAPTER 1: GENERAL |
| CHAPTER 2: WIRING AND PROTECTION |
| CHAPTER 3: WIRING METHODS AND MATERIALS |
| CHAPTER 4: EQUIPMENT FOR GENERAL USE |

APPLIES GENERALLY TO ALL ELECTRICAL INSTALLATIONS

| CHAPTER 5: SPECIAL OCCUPANCIES |
| CHAPTER 6: SPECIAL EQUIPMENT |
| CHAPTER 7: SPECIAL CONDITIONS |

SUPPLEMENTS OR MODIFIES CHAPTERS 1 THROUGH 4

| CHAPTER 8: COMMUNICATIONS SYSTEMS |

CHAPTER 8 IS NOT SUBJECT TO THE REQUIREMENTS OF CHAPTERS 1–7 EXCEPT WHERE THE REQUIREMENTS ARE SPECIFICALLY REFERENCED IN CHAPTER 8

| CHAPTER 9: TABLES |

APPLIES AS REFERENCED IN ANY OF THE CHAPTERS

FIGURE 18-2 The arrangement of the chapters in the National Electric Code.

from experience...

The National Electric Code is *not* to be used as a design manual and is not intended for use by untrained persons.

The NEC® is composed of nine chapters. The first four chapters apply to general wiring practices and equipment. Chapters 5, 6, and 7 apply to special conditions. Chapter 8 refers to communications systems, while Chapter 9 contains applicable tables (Fig. 18-2).

The language used in the NEC® is very specific and needs to be clearly understood. Article 90.5 in the electric code defines the differences between mandatory rules and permissive rules. **Mandatory rules** identify practices that are either allowed or not allowed. Phrases that identify mandatory rules include *shall* and *shall not*. **Permissive rules** identify practices that are allowed, but not required. Passive rules also apply to alternative acceptable methods or practices. Phrases that identify permissive rules include *shall be permitted* and *shall not be required*.

The sections that follow describe some of the most relevant portions of the code as they apply to the residential portion of our industry. The National Electric Code is made up of over 800 pages and countless codes, guidelines, and rules, so attempting to summarize the entire volume here would not prove beneficial to the reader.

Going Green

Safe electrical installations are green electrical installations. Follow the codes, do the right thing, protect the environment!

ARTICLE 110: REQUIREMENTS FOR ELECTRICAL INSTALLATIONS

Article 110 provides information regarding general requirements for electrical installations. The items that follow represent a sampling of the contents of Article 110 and in no way are intended to represent a complete summary of it.

- Electrical equipment must be inspected before installation. Items to be inspected include mechanical strength and durability of the equipment, electrical insulation, heating effects on equipment, arcing effects, current rating, and any other safety-related items that can affect the users of the equipment. By inspecting the equipment, the installer verifies that the particular component is the proper one for the particular application at hand. In addition, by performing a preinstallation inspection, any physical damage will be noticed.

Going Green

Damaged equipment is inefficient. Be sure to properly inspect all electrical equipment prior to installation.

- Wiring methods used must be suitable and industry accepted. Completed wiring must be free from shorts to ground, faults, and other deficiencies. Care should be taken to ensure that electrical connections are tight, so they do not come loose during system operation.

- Equipment rated for "indoor-use only" is not to be used outdoors. In addition, indoor-use only devices must be protected from weather damage during construction.

- Unused openings on equipment must be closed. The sealed opening must provide the same protection as the wall of the original equipment. A **knockout plug** is commonly used to close unused openings in electrical boxes (Fig. 18-3).

- If openings in nonmetallic boxes are removed and not used, the entire box must be replaced.

Delmar/Cengage Learning.

FIGURE 18-3 Knockout plug used to close openings in electrical boxes.

- Internal parts of electrical equipment such as terminal strips and busbars must not be damaged.

- All electrical equipment must be mounted securely.

- Terminal, splicing, and wire connectors need to be identified for use with the conductor material. Conductors of different materials, such as copper and aluminum, cannot be mixed in a splice connector unless the connector is identified for that purpose. Figure 18-4 shows a number of different types of wire connectors.

Going Green

Be sure to use the proper materials for the job at hand. Using incorrect materials can waste valuable natural resources.

- Conductors are often classified as bare, covered, or insulated. **Bare conductors** are those that have no covering at all. **Insulated conductors** are those conductors that have a covering that is recognized by the NEC® as electrical insulation. **Covered conductors** are those conductors that have a covering that is *not* recognized by the NEC® as electrical insulation.

- All connections made between a conductor and a terminal must be tight, but not too tight so as to damage the conductors.

CRIMP CONNECTORS USED TO SPLICE AND TERMINATE 20 AWG TO 500 KCMILS ALUMINUM-TO-ALUMINUM, ALUMINUM-TO-COPPER, OR COPPER-TO-COPPER CONDUCTORS.

PROPERLY CRIMP
THEN TAPE

CONNECTORS USED TO CONNECT WIRES TOGETHER ON COMBINATIONS OF 18 AWG THROUGH 6 AWG CONDUCTORS. THEY ARE TWIST-ON, SOLDERLESS, AND TAPELESS.

*WIRE-NUT and WING-NUT are registered trademarks of IDEAL INDUSTRIES, INC. Scotchlok is a registered trademark of 3M.

WIRE CONNECTORS VARIOUSLY KNOWN AS
WIRE-NUT, WING-NUT, AND SCOTCHLOK.

CONNECTORS USED TO CONNECT WIRES TOGETHER IN COMBINATIONS OF 16, 14, AND 12 AWG CONDUCTORS. THEY ARE CRIMPED ON WITH A SPECIAL TOOL, THEN COVERED WITH A SNAP-ON INSULATING CAP.

CRIMP-TYPE WIRE CONNECTOR
AND INSULATING CAP

SOLDERLESS CONNECTORS ARE AVAILABLE IN SIZES 14 AWG THROUGH 500 KCMIL CONDUCTORS. THEY ARE USED FOR ONE SOLID OR ONE STRANDED CONDUCTOR ONLY, UNLESS OTHERWISE NOTED ON THE CONNECTOR OR ON ITS SHIPPING CARTON. THE SCREW MAY BE OF THE STANDARD SCREWDRIVER SLOT TYPE, OR IT MAY BE FOR USE WITH AN ALLEN OR SOCKET WRENCH.

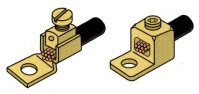

SOLDERLESS CONNECTORS
(LUGS)

COMPRESSION CONNECTORS ARE USED FOR 8 AWG THROUGH 1,000 KCMIL CONDUCTORS. THE WIRE IS INSERTED INTO THE END OF THE CONNECTOR, THEN CRIMPED ON WITH A SPECIAL COMPRESSION TOOL.

COMPRESSION CONNECTOR

SPLIT-BOLT CONNECTORS ARE USED FOR CONNECTING TWO CONDUCTORS TOGETHER, OR FOR TAPPING ONE CONDUCTOR TO ANOTHER. THEY ARE AVAILABLE IN SIZES 10 AWG THROUGH 1,000 KCMIL. THEY ARE USED FOR TWO SOLID AND/OR TWO STRANDED CONDUCTORS ONLY, UNLESS OTHERWISE NOTED ON THE CONNECTOR OR ON ITS SHIPPING CARTON.

SPLIT-BOLT CONNECTOR

FIGURE 18-4 Different types of wire connectors.

- Only one conductor is to be connected under a screw terminal unless it is rated for use with multiple conductors.
- Make certain that the temperature limitations of electrical equipment are not exceeded.

- All circuit disconnects must be clearly and legibly marked.
- Electrical equipment must be **accessible**. Accessible means that these devices are not guarded by locked doors or other means that would prevent easy access

to them. For example, you should not need a ladder to gain access to a service switch.

- Ample lighting must be provided for service equipment located indoors.
- Outdoor electrical equipment must be protected from accidental contact by unauthorized individuals.

ARTICLE 210: BRANCH CIRCUITS

Branch circuits are defined as the wiring that is installed between an overcurrent device and an outlet. Typical branch circuit types include appliance branch circuits, individual branch circuits, and general purpose branch circuits. **Appliance branch circuits** are used to supply power to outlets that will have appliances connected to them. **General purpose branch circuits** supply power to two or more outlets for lighting and appliance use. The **individual branch circuit** supplies power to only one piece of equipment. Individual branch circuits are also often referred to as dedicated lines and are used, for example, to supply power to a window or through-the-wall air conditioning unit. The bulleted list that follows contains items that the HVAC technician is likely to encounter in the field.

- Branch circuits must be sized based on the rating of the overcurrent device.
- Ground fault circuit interrupters, GFCI, are required on all 15- and 20-amp, 125-volt receptacle circuits located in bathrooms, in garages, outdoors, in kitchens near countertops, and near laundry and utility sinks (Fig. 18-5).

Delmar/Cengage Learning.

FIGURE 18-5 A ground fault circuit interrupter.

Circuit Rating (Amperes)	Receptacle Rating (Amperes)
15	Not over 15
20	15 or 20
30	30
40	40 or 50
50	50

FIGURE 18-6 Receptacle ratings for various size circuits. Reprinted with permission from NFPA 70®, *National Electrical Code*®, Copyright © 2007, National Fire Protection Association, Quincy, MA. This reprinted material is not the complete and official position of the NFPA® on the referenced subject, which is represented only by the standard in its entirety.

Circuit Rating (Amperes)	Receptacle Rating (Amperes)	Maximum Load (Amperes)
15 or 20	15	12
20	20	16
30	30	24

FIGURE 18-7 Maximum cord-and-plug connected load to receptacle. Reprinted with permission from NFPA 70®, *National Electrical Code*®, Copyright © 2007, National Fire Protection Association, Quincy, MA. This reprinted material is not the complete and official position of the NFPA® on the referenced subject, which is represented only by the standard in its entirety.

- Outlets and receptacles must have amperage ratings that exceed the amperage of the load being served.
- Receptacles need to be rated at amperages higher than or equal to the amperage rating of the branch circuit.
- When installed, receptacle ratings cannot exceed those shown in Figure 18-6.
- The maximum circuit for cord-and-plug connected equipment cannot exceed 80% of the receptacle amperage rating (Fig. 18-7).
- Circuit loads are not to exceed the branch circuit amperage rating.

ARTICLE 240: OVERCURRENT PROTECTION

Electric circuits need to be protected in the event that too much current flows through them. A byproduct of current is heat and too much heat can cause a fire. The purpose of overcurrent protection is to stop current flow through a circuit if unsafe current levels are detected. As stated by the NEC®, an unsafe current level is one that results in an unsafe temperature in the conductor or

(A)

(B)

Delmar/Cengage Learning.

FIGURE 18-8 (A) Fuses. (B) Circuit breakers.

Fuse or Circuit Breaker Size	Minimum-Size Wire (CU)	Minimum-Size Wire (AU)
15 amp	14 AWG	12 AWG
20 amp	12 AWG	10 AWG
30 amp	10 AWG	8 AWG
40 amp	8 AWG	6 AWG
50 amp	6 AWG	4 AWG

FIGURE 18-9 Minimum copper and aluminum conductor sizes for a given circuit breaker size.

in the conductor insulation. Common overcurrent protectors are fuses and circuit breakers (Fig. 18-8). What follows are some important items to keep in mind when dealing with overcurrent devices.

- Individual circuit breakers or fuses should not be wired in parallel with each other unless they are factory assembled as such and labeled as a unit.

- Thermal overloads and other devices that are not designed to open short circuits or ground faults are not used to protect the conductors against an overcurrent condition.

- Overcurrent protection is to be installed in series with each ungrounded conductor.

- Circuit breakers should open ungrounded conductors manually and automatically.

- Overcurrent protection for ungrounded conductors is to be located at the point where the conductor receives its power.

- Overcurrent devices are to be protected from physical damage.

- Overcurrent devices are selected based on the size of the conductors used in the circuit (Fig. 18-9).

ARTICLE 250: GROUNDING

In terms of electrical safety, one of the most important things the HVAC installer or service technician can do to help ensure his safety, as well as that of those around him, is to ensure that all tools and pieces of equipment are properly **grounded**. The term grounded means that the tool, device, or piece of equipment is connected to **ground**, meaning the Earth, or to a conductive body that is connected to ground. If properly handled by properly trained individuals, electricity is safe. It is important to always respect electricity and its potential, as it can be extremely dangerous in the wrong hands.

The reason for properly grounding tools and equipment is simple. If the insulation on a conductor becomes damaged or an energized conductor comes loose from a terminal connection in a piece of equipment, it is possible to electrify the surfaces that these conductors come in contact with. If the equipment is not properly grounded and someone comes in contact with an electrified surface, severe shock or electrocution could occur. If the equipment is properly grounded, the electric potential will flow through the ground connection to the Earth, instead of through the individual.

Grounding is such an important topic that the NEC® devotes close to 30 pages to the guidelines and codes that govern grounding and its methods. It would not be beneficial to attempt to highlight this portion of the code, given the importance of it. However, some important issues regarding grounding in general are provided here:

- Never cut the ground prong off any plug, extension cord, power cord, or appliance power cable!

- Ground wires in equipment are not intended to carry electric current during the normal operation of the system.

- Alternating current systems need to be grounded.

- Ground wires are typically color-coded green, but can also be bare copper conductor. Bare conductors are conductors that have no covering on them at all.

- Ground wires need to be securely fastened to the equipment to provide the safety factor it is intended to provide.

- Motors that are mounted with rubber or other insulating materials that isolate them from the metallic chassis of the equipment need to have a ground wire connection between the motor and the equipment chassis (Fig. 18-10).

- If a box has more than one ground wire in it, they must be properly spliced together and then connected to the equipment grounding screw with a jumper (Fig. 18-11).

- In metallic boxes, the grounding wire is connected to the equipment with a green grounding screw or other grounding-approved connector.

Photo by Eugene Silberstein.

FIGURE 18-10 Ground wire connecting motor to the blower housing.

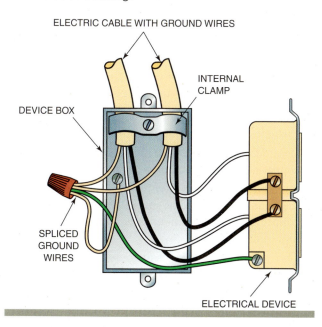

ELECTRIC CABLE WITH GROUND WIRES

INTERNAL CLAMP

DEVICE BOX

SPLICED GROUND WIRES

ELECTRICAL DEVICE

FIGURE 18-11 Ground wires are spliced together and connected to the box by means of a jumper wire.

ARTICLE 300: WIRING METHODS

On occasion, the HVAC installer or technician may need to run conductors from one location to another. When this is required, a number of factors need to be taken into consideration. Some of the important things to keep in mind are

- Conductors need to be protected when there is the possibility of physical damage to them.

- It is permissible to run cables through holes that have been made in beams, rafters, joists, or other wood members. However, the hole that has been made must be no closer than 1.25 inches from the edge of the wood member (Fig. 18-12).

- If the 1.25 inches clearance cannot be maintained, the cable must be protected from damage by a steel plate at least 1/16 of an inch thick (Fig. 18-13).

Going Green

Protecting conductors from physical damage extends the useful life of the materials. This will help save raw materials and reduce waste associated with manufacturing the end product.

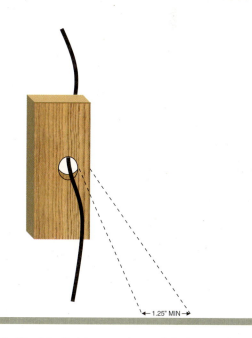

◄— 1.25" MIN —►

FIGURE 18-12 Cables run through holes in wood members such as joists, studs, and beams, must be at least 1.25 inches from the edge of the member.

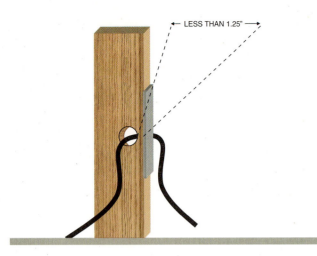

FIGURE 18-13 If the 1.25 inches clearance cannot be maintained, a steel plate at least 1/16-inch thick must be used to protect the cable.

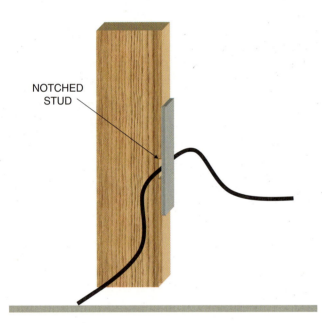

FIGURE 18-14 If cables are positioned in notched wood members, the cable must be protected by a steel plate at least 1/16-inch thick.

- Cables can be positioned in notches in wood beams, joists rafters, studs, and other wood members, as long as they are protected from nails and screws by a steel plate at least 1/16 of an inch thick (Fig. 18-14).
- If nonmetallic cables are run through slots or holes in metal members, approved bushings or grommets must be used to protect the cables from damage (Fig. 18-15). The bushing or grommet must cover all edges of the hole.

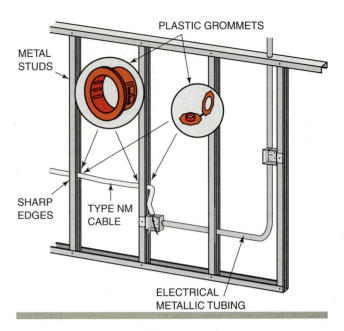

FIGURE 18-15 Grommets or bushing must be used to protect nonmetallic cable that is run through holes or slots in metal framing members.

ARTICLE 310: CONDUCTORS FOR GENERAL WIRING

When wiring air conditioning and heating equipment, proper conductors need to be used. In order to select the proper conductor for a particular application, important factors include insulation, wire type, ampacity rating, and mechanical strength. **Ampacity** is the current, in amperes, that a conductor can continuously safely carry. The following items are of particular importance to the air conditioning heating and air conditioning installer and technician.

- Current-carrying conductors must be insulated.
- Conductors used are typically made of copper, copper-clad aluminum, or aluminum.
- For conductors rated at voltages from 0 to 2,000 the minimum conductor size (AWG) is 14 gauge for copper and 12 for aluminum or copper-clad aluminum (Fig. 18-16).
- Three commonly used terms to describe voltage levels are high voltage, line voltage, and low voltage. The term **high voltage** is used to describe voltage levels greater than 600 Volts. The term **line voltage** is used to describe voltage levels between 50 Volts and 600 Volts. The term **low voltage** is used to describe voltage levels lower than 50 Volts.

Conductor Voltage Rating (Volts)	Minimum Conductor Size (AWG)	
	Copper	Aluminum or Copper-Clad Aluminum
0–2000	14	12
2001–8000	8	8
8001–15,000	2	2
15,001–28,000	1	1
28,001–35,000	1/0	1/0

FIGURE 18-16 Minimum conductor size. Reprinted with permission from NFPA 70®, *National Electrical Code*®, Copyright © 2007, National Fire Protection Association, Quincy, MA. This reprinted material is not the complete and official position of the NFPA® on the referenced subject, which is represented only by the standard in its entirety.

- Some of the common conductor types that are used in dry or damp locations include:

    THHN–Thermoplastic High Heat Resistant Nylon Coated

    THHW–Thermoplastic High Heat and Water Resistant

    THWN–Thermoplastic Heat and Water Resistant Nylon Coated

    THW–Thermoplastic Heat and Water Resistant

- Some of the less commonly encountered conductor types that are used in dry or damp locations include:

    FEP–Fluorinated Ethylene Propylene

    MTW–Machine Tool Wire

    PFA–Perfluoroalkoxy

    RHH–Rubber High Heat Resistant

    RHW–Rubber Heat and Water Resistant

    XHHW–Cross-linked High Heat and Water Resistant

    Z–Ethylene Tetrafluoroethylene (ETFE)-insulated conductor for dry applications

    ZW – ETFE-insulated conductor for wet applications

- When conductors are to be located in wet locations, they should be listed for use in wet locations, properly protected, and, ideally, be one of the following conductor types: MTW, RHW, TW, THW, THHW, THWN, XHHW, or ZW.

- Conductors that will be located in direct sunlight must be rated as sunlight resistant or be covered with a protective material that is rated to be sunlight resistant.

- Conductors should not be exposed to temperatures that exceed the conductor's temperature rating.

ARTICLE 314: OUTLET, DEVICE, PULL, AND JUNCTION BOXES; CONDUIT BODIES; FITTINGS; AND HANDHOLE ENCLOSURES

Article 314 covers a wide range of topics regarding electrical boxes and conduit bodies that are used for electrical junctions, outlets, and electrical devices. Most of the time, the electric circuits needed for the installation of HVAC equipment are installed by licensed electricians. However, there are times that the installation technician will need to run wiring for other system components or accessories. With this in mind, the following information will prove invaluable in ensuring the safe installation of conductors.

- Round boxes are not to be used in conjunction with locknut-type conduits or connectors (Fig. 18-17).

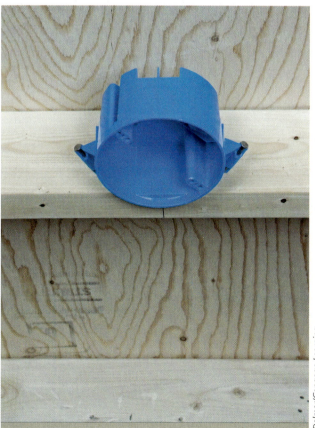

Delmar/Cengage Learning.

FIGURE 18-17 Round electrical box.

- Any boxes used in wet locations must be rated for use in wet locations.
- Boxes should be positioned in a manner that will prevent moisture from entering them.
- Boxes are to be large enough to provide adequate space for all conductors. Figure 18-18 provides the maximum number of conductors that are permitted in various metal boxes.
- Conductors that enter electrical boxes must be properly protected with appropriate bushings to prevent damage to the conductor.
- Any openings in electric boxes through which conductors pass are to be adequately closed.

- When installing boxes with flush-type coves or faceplates in concrete, tile, gypsum, or plaster walls or ceilings, the front end of the box must be within 0.25 inch of the finished surface (Fig. 18-19).

from experience...

If you have a doubt about the wall's construction material, mount the box flush with the finished wall.

Box Trade Size			Minimum Volume		Maximum Number of Conductors* (arranged by AWG size)						
mm	in.		cm³	in.³	18	16	14	12	10	8	6
100 × 32	(4 × 1¼)	round/octagonal	205	12.5	8	7	6	5	5	5	2
100 × 38	(4 × 1½)	round/octagonal	254	15.5	10	8	7	6	6	5	3
100 × 54	(4 × 2⅛)	round/octagonal	353	21.5	14	12	10	9	8	7	4
100 × 32	(4 × 1¼)	square	295	18.0	12	10	9	8	7	6	3
100 × 38	(4 × 1½)	square	344	21.0	14	12	10	9	8	7	4
100 × 54	(4 × 2⅛)	square	497	30.3	20	17	15	13	12	10	6
120 × 32	(4¹¹⁄₁₆ × 1¼)	square	418	25.5	17	14	12	11	10	8	5
120 × 38	(4¹¹⁄₁₆ × 1½)	square	484	29.5	19	16	14	13	11	9	5
120 × 54	(4¹¹⁄₁₆ × 2⅛)	square	689	42.0	28	24	21	18	16	14	8
75 × 50 × 38	(3 × 2 × 1½)	device	123	7.5	5	4	3	3	3	2	1
75 × 50 × 50	(3 × 2 × 2)	device	164	10.0	6	5	5	4	4	3	2
75 × 50 × 57	(3 × 2 × 2¼)	device	172	10.5	7	6	5	4	4	3	2
75 × 50 × 65	(3 × 2 × 2½)	device	205	12.5	8	7	6	5	5	4	2
75 × 50 × 70	(3 × 2 × 2¾)	device	230	14.0	9	8	7	6	5	4	2
75 × 50 × 90	(3 × 2 × 3½)	device	295	18.0	12	10	9	8	7	6	3
100 × 54 × 38	(4 × 2⅛ × 1½)	device	169	10.3	6	5	5	4	4	3	2
100 × 54 × 48	(4 × 2⅛ × 1⅞)	device	213	13.0	8	7	6	5	5	4	2
100 × 54 × 54	(4 × 2⅛ × 2⅛)	device	238	14.5	9	8	7	6	5	4	2
95 × 50 × 65	(3¾ × 2 × 2½)	masonry box/gang	230	14.0	9	8	7	6	5	4	2
95 × 50 × 90	(3¾ × 2 × 3½)	masonry box/gang	344	21.0	14	12	10	9	8	7	4
min. 44.5 depth	FS — single cover/gang (1¾)		221	13.5	9	7	6	6	5	4	2
min. 60.3 depth	FD — single cover/gang (2⅜)		295	18.0	12	10	9	8	7	6	3
min. 44.5 depth	FS — multiple cover/gang (1¾)		295	18.0	12	10	9	8	7	6	3
min. 60.3 depth	FD — multiple cover/gang (2⅜)		395	24.0	16	13	12	10	9	8	4

*Where no volume allowances are required by 314.16(B)(2) through (B)(5).

FIGURE 18-18 Maximum number of conductors permitted in metal boxes. Reprinted with permission from NFPA 70®, *National Electrical Code*®, Copyright © 2007, National Fire Protection Association, Quincy, MA. This reprinted material is not the complete and official position of the NFPA® on the referenced subject, which is represented only by the standard in its entirety.

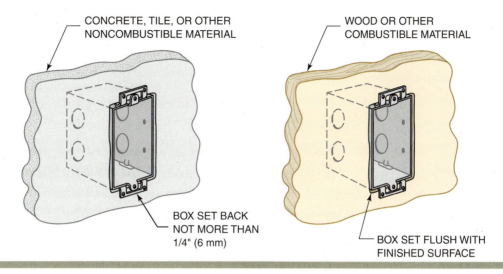

CONCRETE, TILE, OR OTHER NONCOMBUSTIBLE MATERIAL

WOOD OR OTHER COMBUSTIBLE MATERIAL

BOX SET BACK NOT MORE THAN 1/4" (6 mm)

BOX SET FLUSH WITH FINISHED SURFACE

FIGURE 18-19 Electrical boxes installed in concrete, tile, gypsum, or plaster walls must have their front edge no further than 0.25 inch back from the finished edge of the wall. If the wall is wood, the box must be flush with the finished wall.

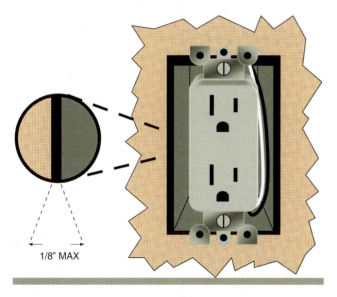

1/8" MAX

FIGURE 18-20 The gap between an electric box and the wall material is not to exceed 1/8 inch.

- When installing boxes with flush-type coves or faceplates in wood (or other combustible material) walls or ceilings, the front end of the box must be flush with the finished surface.

- The maximum allowable gap between a box and the surrounding plaster, drywall, or plasterboard surface is 0.125 inch (one-eighth of an inch) (Fig. 18-20).

- Any electrical enclosure that is to be mounted in a suspended ceiling cannot be greater than 100 in^3 in size.

- When mounting an electrical enclosure in a suspended ceiling, it must be supported and secured to the main structure with either framing members or support wires.

ARTICLE 422: APPLIANCES

This portion of the NEC® covers general guidelines regarding the installation and disconnect methods used for appliances. Information specific to heating and air conditioning equipment are discussed in the three remaining sections in this chapter, Articles 424, 430, and 440. What follows in this section are some useful pieces of information regarding appliances in general.

- Appliances are not to have any exposed live parts. The exception to this are appliances with open-resistance heating elements (Fig. 18-21).

- The rating of a branch circuit must not be less than the rating of the appliance being connected to the circuit.

- The circuit must safely handle the amperage draw of the appliance without overheating.

- If an appliance has a protection device rating marked on it, the branch circuit overcurrent device cannot exceed the appliance rating.

- Electric heating appliances with heating elements rated at more than 48 amperes must have the heating element circuits subdivided into individual circuits, each of which does not exceed 48 amperes. Protection for each of these circuits is not to exceed 60 amperes.

- Electric water heaters and steam boilers are to have heating element circuits that do not exceed 120 amperes. Protection for each of these circuits is not to exceed 150 amperes.

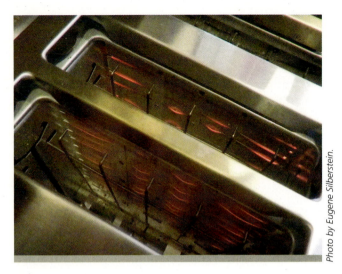

Photo by Eugene Silberstein.

FIGURE 18-21 Heater elements on a toaster.

- Central heating equipment is to be supplied by an individual branch circuit.
- Disconnect switches are to be provided to disconnect an appliance from all ungrounded conductors.
- For appliances under 1/8 horsepower, the branch circuit overcurrent device (circuit breaker) can serve as the disconnect device.
- For appliances rated over 1/8 horsepower, the branch circuit overcurrent device (circuit breaker) can serve as the disconnect device provided it is located **within sight** of the appliance and can be locked in the open (de-energized) position. The term within sight describes the relative location of two pieces of equipment that implies visibility of one from the other and a distance that is no further than 50 feet.

ARTICLE 424: ELECTRIC HEAT

Although some electric heat guidelines are discussed in Article 422 of the NEC®, Article 424 is devoted to electric heat and its unique requirements. Here are some of the highlights of this very important section of the code. The information in this Article of the code includes heating cable, unit heaters, boilers, central systems, and other fixed electric space-heating equipment.

- All fixed electric space-heating equipment must be installed in an approved manner.
- All fixed electric space-heating equipment must be protected from physical damage.

- All fixed electric space-heating equipment that is to be located in damp or wet locations must be rated for this purpose.
- All fixed electric space-heating equipment that is to be located in damp or wet locations must be protected so that water or other liquids cannot enter the appliance.
- All fixed electric space-heating equipment must be installed to provide the proper spacing between the equipment and combustible materials.
- Unit disconnects must simultaneously de-energize the heater, motor controller, and supplementary overcurrent devices.
- For electric heating equipment, in many cases, the branch circuit protector can serve as the unit disconnect if the branch circuit protector (circuit breaker) is within sight or if the branch circuit protector (circuit breaker) can be locked in the open (de-energized) position.
- Thermostatically controlled switching devices on electric heating equipment can function as the required disconnect switch if the device has a clearly marked "OFF" position, opens all ungrounded conductors when in the OFF position, is designed so the system cannot run on automatically from the OFF position, and is properly located as described in the code.
- Resistance-type heating element circuits are not to be rated at more than 48 amperes and protected at more than 60 amperes. Circuits that exceed these limits need to be subdivided and protected accordingly.
- Proper airflow must be provided through electric duct heaters.
- If duct heaters are positioned less than 4 feet from an air conditioner or heat pump system, both the duct heater and the system must be identified as suitable for such use.
- Duct heaters that are used in conjunction with cooling equipment that produces condensate must be identified as being suitable for that application.
- Fan interlock circuits are needed to ensure that the fan circuit is energized whenever a heater circuit is energized.
- Each duct heater is to be equipped with an approved automatic-reset temperature-limiting control.
- Duct heaters and all heating appliances must always be installed according to manufacturer's instructions.
- Adequate clearances must be provided to permit replacement of controls and heating elements and to allow for cleaning and adjusting of the controls and other components requiring such attention.

ARTICLE 430: MOTORS, MOTOR CIRCUITS, AND CONTROLLERS

Motors are integral parts of heating, air conditioning, and ventilation equipment. Properly installing and commissioning motors will help ensure that they provide years of uninterrupted, efficient service. The items in this section of the code provide useful information regarding the safe and proper methods that are to be used when installing, maintaining, and servicing electric motors and their associated components.

- Manufacturers are required to provide certain information on the motor's nameplate. This information includes (but is not limited to) manufacturer's name, motor's full-load current, and voltage ratings, rated frequency and number of phases, motor's full-load speed, rated temperature rise or insulation class, and rated horsepower (1/8 horsepower or more).

- Motor controllers are also required to be marked. **Motor controllers** are devices that are normally used to start and stop a motor by making or breaking electrical contacts. Information commonly found on motor controllers include the manufacturer's name, voltage, horsepower, or amperage rating, and other pertinent information for the intended application.

- Motor controllers and terminals of control circuit devices are to use copper conductors unless specifically identified for use with conductors of other materials.

- Enclosures for motor controllers are not to be used as junction boxes unless designed for that purpose.

- Exposed current-carrying parts of motors are to be protected from dripping or spraying oil, water, or other liquids unless the motor is designed for that particular purpose.

- Motor terminal housings are to be made of metal and be of substantial construction.

- Motors are to be located so that they receive adequate ventilation.

- Motors are to be located in a manner that makes lubrication and servicing possible.

- Conductors that supply voltage to a motor are to have an ampacity of no less than 125% if the motor's full-load current rating.

- For multispeed motors, the conductors on the line side of the motor controller must be sized according to the highest of the full-load current ratings for the motor. On the load side of the motor controller, the conductors are sized according to the current rating of the motor windings that are energized.

- Motor controllers must be able to interrupt the locked-rotor current of the motor.

- Motor controllers must have horsepower ratings at the application voltage not lower than the horsepower rating of the motor.

- Motor controllers do not need to open all of the conductors to the motor.

ARTICLE 440: AIR CONDITIONING AND REFRIGERATING EQUIPMENT

This portion of the NEC® applies to motor-driven air conditioning and refrigeration systems as well as the branch circuits and controllers that are used in conjunction with this equipment. Some of the data presented in this section of the code have been addressed, at least in some form, in other portions of the codebook, but are compiled here to provide the reader with a concise overview of the electrical requirements.

- For hermetic compressors, the rated load current on the compressor nameplate is the value used to determine the ampacity of the circuit disconnect, the size of the branch circuit conductors, the rating of the motor controller, the size of the branch circuit protection, and the motor overload protector to be used.

- For multi-motor equipment using shaded pole or permanent split capacitor motors, the full-load current indicated on the nameplate of the equipment in which the fan or motor is housed is the value used to determine the ampacity of the circuit disconnect, the size of the branch circuit conductors, the rating of the motor controller, the size of the branch circuit protection, and the motor overload protector to be used.

- An air conditioning or refrigeration system is considered to be a single machine and motors are allowed to be located remotely from each other.

- Disconnects that are capable of disconnecting all of the motors and motor controllers in an air conditioning or refrigeration system are to be used.
- Equipment disconnects are to be located within sight and readily accessible from the equipment.
- Equipment disconnects can be located on or in the equipment, but not on a panel of the equipment that must be removed when equipment service is performed.

SUMMARY

- The National Fire Protection Association is responsible for developing codes and standards intended to minimize the risks and effects of fire.
- Some common NFPA® codes include the Standard for Oil-burning Equipment Installation (NFPA 31®), the National Fuel Gas Code (NFPA 54®), the Liquefied Petroleum Gas Code (NFPA 58®), and the Standard for Ovens and Furnaces (NFPA 86®).
- The National Electric Code is also known as NFPA 70®.
- The NEC® is concerned with safeguarding people and property from hazards that arise from the use of electricity.
- Mandatory rules in the NEC® identify practices that are either allowed or not allowed.
- Permissive rules in the NEC® identify practices that are allowed but not required.
- Wiring methods and techniques must be industry accepted.
- Equipment must be rated for its intended purpose.
- Electrical equipment must be mounted securely and accessible.

- Electrical equipment must operate within its rated ampacity and temperature ranges.
- Ground fault circuit interrupters, GFCI, are required on 15- and 20-amp, 125-volt receptacles located in potentially wet locations.
- Overcurrent devices are designed to de-energize circuits in the event unsafe current levels are detected.
- Overcurrent devices are to be installed at the point where the conductor receives its power and should be protected from physical damage.
- Electrical tools and equipment must be properly grounded.
- Ground wires are typically color-coded green but can also be bare conductors.
- Current-carrying conductors must be insulated and protected from physical damage.
- Conductors must be rated for the particular application such as dry, damp, wet, in direct sunlight, and high temperature.
- Electric boxes are to be rated and suitable for their particular application.
- Electric boxes must be large enough to provide adequate space for the conductors.
- Electric circuits must safely handle the amperage draw of the appliance without overheating.
- Central heating equipment is to be powered by an individual branch circuit.
- Adequate clearance must be provided when installing electric duct heaters to allow for servicing and unit repair.
- Equipment disconnects are to be located within sight and readily accessible from the equipment.
- Equipment disconnects can be located on the equipment, but not on any panel that needs to be removed for service.

GREEN CHECKLIST

- [ ] **Safe electrical installations are green electrical installations.**

- [ ] **Follow the codes, protect the environment!**

- [ ] **Damaged equipment is inefficient. Be sure to properly inspect all electrical equipment prior to installation.**

- [ ] **Be sure to use the proper materials for the job at hand.**

- [ ] **Using incorrect materials can waste valuable natural resources.**

- [ ] **Protecting conductors from physical damage extends the useful life of the materials. This will help save raw materials and reduce waste associated with manufacturing the end product.**

REVIEW QUESTIONS

1. **The National Fire Protection Association, NFPA®, was established in**
   a. 1896
   b. 1936
   c. 1986
   d. 2006

2. **The National Electric Code is also known as**
   a. EPA 70
   b. NFPA 70®
   c. DOT 70
   d. NEC 70®

3. **Electrical equipment must be inspected before installation in order to**
   a. Check for any physical damage to the component
   b. Make certain that the device ratings are correct for the particular application
   c. Neither a nor b is correct
   d. Both a and b are correct

4. **If an opening in a nonmetallic box is removed and not used**
   a. A knockout plug can be used to close the opening
   b. The opening does not need to be sealed
   c. The box must be replaced
   d. Both a and c are correct

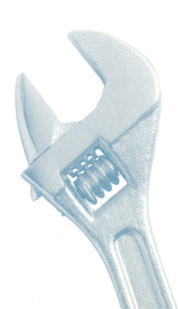

5. **A conductor that has no insulation on it is classified as a**
   a. Bare conductor
   b. Insulated conductor
   c. Uncovered conductor
   d. Covered conductor

6. **A branch circuit that is intended to provide power to a through-the-wall air conditioning unit should be of which type?**
   a. General branch circuit
   b. Individual branch circuit
   c. Appliance branch circuit
   d. Any of the above are acceptable for this application

7. **Ground fault circuit interrupters are required in which of the following areas?**
   a. Living rooms, bathrooms, and near kitchen sinks
   b. Near laundry sinks, pantries, and in bathrooms
   c. Bathrooms, garages, and outdoors
   d. Near utility sinks, in garages, and near fireplaces

8. **Which of the following is true regarding a branch circuit?**
   a. The outlet or receptacle must have an amperage rating that is higher than the amperage rating of the load being powered
   b. The outlet or receptacle must be rated at an amperage that is greater than or equal to the amperage rating of the circuit
   c. Both a and b are correct
   d. Neither a nor b is correct

9. **Removing the ground prong from an electric tool or appliance**
   a. Will make the tool or appliance operate more efficiently
   b. Will remove the protection that the ground circuit provides
   c. Will cause the tool or appliance to stop operating
   d. Will cause the amperage draw of the tool or appliance to rise

10. **If a hole needs to be drilled through a wood joist for the purpose of passing a cable through it, how far from the edge of the joist must the hole be if no metal plates are to be used?**
    a. 0.5 inch
    b. 1.0 inch
    c. 1.25 inches
    d. 2.0 inches

11. **What must be done if an electric cable is run through a notched joist or beam?**
    a. The beam or joist must be reinforced with another piece of wood
    b. The cable must be protected with a metal plate at least 1/16-inch thick
    c. The cable must be cut and sliced at the notched location
    d. Electric cables are never to be run through a notched beam or joist

12. **According to the National Electric Code, a 220-volt power supply is classified as a**
    a. Very high-voltage power supply
    b. High-voltage power supply
    c. Line voltage power supply
    d. Low-voltage power supply

13. **What is the difference between THHN and THHW wire?**

    a. THHN is acceptable for wet applications

    b. THHW is acceptable for wet applications

    c. Both THHN and THHW are acceptable for wet applications

    d. Neither THHN nor THHW are acceptable for wet applications

14. **According to the National Electric Code, which of the following electrical enclosures can be mounted in a suspended ceiling?**

    a. An enclosure that measures 5″ × 5″ × 5″

    b. An enclosure that measures 5″ × 5″ × 3″

    c. An enclosure that measures 6″ × 6″ × 3″

    d. An enclosure that measures 6″ × 5″ × 4″

15. **What is the maximum amperage on a single circuit of an electric heating appliance?**

    a. 24 Amperes

    b. 36 Amperes

    c. 48 Amperes

    d. 60 Amperes

16. **What safety component is required on all fan-assisted electric heating equipment?**

    a. A fan relay

    b. A fan interlock

    c. A thermostat

    d. Both a and c are required

17. **The National Electric Code requires that electric motors**

    a. Be located so they receive adequate ventilation

    b. Be located in a manner that allows for motor lubrication

    c. Be located in a manner that allows for motor service

    d. All of the above are correct

18. **Conductors that feed power to a multispeed motor are sized according to the**

    a. Amperage draw of the motor when operating at the fastest motor speed

    b. Amperage draw of the motor when operating at the slowest motor speed

    c. Sum of the amperages of all motor speeds

    d. Average of the amperages of all motor speeds

KNOW YOUR CODES

The installation of electric equipment and circuits must be performed in the proper manner. As you have experienced in this chapter, there are many guidelines that must be followed to ensure the safety of individuals and to reduce the possibility of fire. This chapter, however, provides only a very small sampling of what the National Electric Code contains.

As an exercise for you and your classmates, using your classroom laboratory or work area, create a two-column list of electrical-related issues that you think are in line with acceptable codes and those that you think are not. Discuss your findings with your instructors and classmates and have them discuss their findings with you. You might have to do some research to get some answers, so be sure to visit the NFPA® Web site at www .nfpa.org.

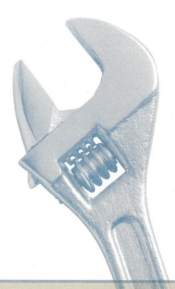

WHAT'S WRONG WITH THIS PICTURE?

Carefully study Figure 18-22 and think about what is wrong. Consider all possibilities.

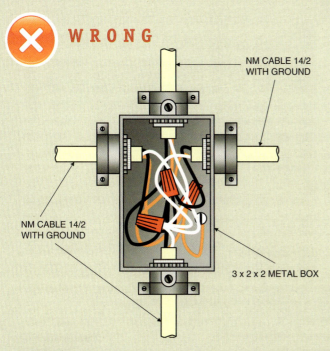

✖ WRONG

NM CABLE 14/2
WITH GROUND

NM CABLE 14/2
WITH GROUND

3 x 2 x 2 METAL BOX

FIGURE 18-22 When selecting an electric box, it is important to properly determine the size of the box that is needed. This box is a 3″ × 2″ × 2″ box and, according to Figure 18-18, it can be filled with no more than five 14-gauge conductors. This box contains 13 conductors: 4 hot (black wires), 4 neutral (white) wires, 4 ground (bare conductor) wires, and 1 ground jumper to connect to the box. This box has been grossly over-filled.

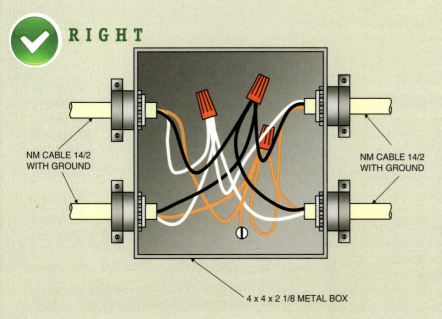

✔ RIGHT

NM CABLE 14/2
WITH GROUND

NM CABLE 14/2
WITH GROUND

4 x 4 x 2 1/8 METAL BOX

FIGURE 18-23 Although the number of wires in this box is the same, the box has been properly sized at 4″ × 4″ × 2 1/8″. This box, according to Figure 18-18, can hold up to fifteen 14-gauge conductors.

Testing, Adjusting, and Balancing

OBJECTIVES *After completing this Chapter, the reader should be able to*

- Define the terms "testing," "adjusting," and "balancing" as they apply to residential structures.

- Explain the importance of system testing and balancing.

- List the three pressures that are found in a duct system.

- Describe the tools used by the TAB technician.

- Explain how to measure duct system pressures.

- Determine air velocity in a duct by using the traverse method.

- Describe the stepwise method of system balancing.

- Balance an air distribution system using the stepwise method.

- Describe the proportional method of system balancing.

- Balance an air distribution system using the proportional method.

The layout and installation of an air conditioning system are important steps in helping to ensure proper system operation. However, there are other tasks that must be performed to ensure that all system components operate as efficiently as possible and that all occupants of the structure are comfortable. System start-up involves the inspection of the installation as well as system charging. Proper system charging is a major task that cannot be overlooked, as it is the refrigerant charge that ultimately determines the efficiency of the equipment.

Another important aspect of the start-up process is the "testing, adjusting and balancing" portion of the installation. Testing, adjusting and balancing is the combined process of determining if the desired amounts of treated air or water are being delivered to the various locations in the structure. Improper fluid flows can lead to uneven heating or cooling of the space. The desired outcome, therefore, of the testing, adjusting and balancing process is to have all areas of the space reach their ideal temperatures at the same time. This way, when the thermostat cycles the heating or cooling equipment off, all occupants will be in their comfort zone. This chapter will address some important issues regarding the testing, adjusting and balancing techniques that are used on the residential side of our industry.

Adjusting The process of varying fluid flow by changing the settings on fluid flow controls such as valves, dampers, and blowers

Balancing The process of matching the actual fluid flow rates to the desired fluid flow rates within an acceptable range

Pete's Plug A self-sealing access port by which system temperature and pressure readings can be obtained

Proportional balancing The balancing process which ensures that all outlets supply the same percentage of design air to the occupied space

Static pressure The pressure that pushes against the sides of the duct sections

Stepwise balancing Method of air balancing that involves stepping down the airflow to the outlets with the greatest amount of excess air

Testing The process of using specialized, properly calibrated instruments to measure characteristics of fluid flow and electricity

Total pressure The sum of static and velocity pressures

Velocity pressure The pressure that results from the speed, or velocity, and the weight of the air

WHAT IS TAB WORK?

TAB work is how the HVAC industry commonly refers to testing, adjusting, and balancing. By definition, according to the National Environmental Balancing Bureau, TAB is the "systematic process applied to HVAC systems to achieve and document air and hydronic flow rates." **Testing** is the process of using specialized, properly calibrated instruments to measure characteristics of fluid flow and electricity. **Adjusting** is the process of varying fluid flow by changing the settings on fluid flow controls such as valves, dampers, and blowers. **Balancing** is the process of matching the actual fluid flow rates to the desired fluid flow rates within an acceptable range.

One of the main key elements of TAB work is the actual documentation of the work performed. For the most part, TAB reporting is limited to the commercial end of the industry. In the residential construction industry, balancing reports and their associated documents are not often required, but this does not negate the importance of the testing, adjusting, and balancing work that should be performed.

The very nature of TAB work applies to all fluid flows, including airflow through a duct system and hot water flow through a hydronic heating system. In the domestic environment, TAB work is typically not performed on the hot water heating system. In order to do so, pressure readings at various points in the hydronic system would need to be obtained. This would be done via a **Pete's Plug**, which is a self-sealing access port by which system temperature and pressure readings can be obtained (Fig. 19-1). In most cases, these access ports are not available on residential hydronic heating systems, so the balancing of these systems is typically not done. The scope of TAB work as discussed in the context of this book will, therefore, be limited to the air distribution system of heating and air conditioning systems.

In practicality, TAB work for residential systems is the process of measuring airflow rates and adjusting the volume control devices and/or blower speeds to maintain the desired airflow in the air distribution system. For the most part, the acceptable air volume is within 10% of the desired airflow rate.

WHY IS TAB WORK NEEDED?

Even a perfectly installed air conditioning system will not operate at peak efficiency if it is not properly balanced. An improperly balanced system will be subject to increased system run times, resulting in increased energy consumption, reduced system life, and increased operating costs. The National Comfort Institute recently reported that approximately 57% of air conditioning systems are operating at less than maximum efficiency (www.nationalcomfortinstitute.com). It is estimated that approximately 50% of residential central air conditioning systems are oversized, about 45% are improperly charged with refrigerant, and that approximately 70% of residential systems have inadequate airflow.

Going Green

Improper system sizing, airflow, and charging result in a significant reduction in capacity that can be as great as 20%. National studies estimate that the results of these system deficiencies can cost consumers as much as 30% extra in energy costs.

The purpose of TAB work is to fine-tune the installation. A properly balanced system provides a number of benefits including increased energy savings and prolonged equipment life.

Going Green

Proper testing, adjusting, and balancing helps ensure that the HVAC system is providing comfort in the most efficient manner.

Going Green

Building green encompasses energy efficiency and making certain that homeowners get what they are paying for. It is the technician's responsibility to do everything in his or her power to ensure that the system is operating as effectively and efficiently as possible.

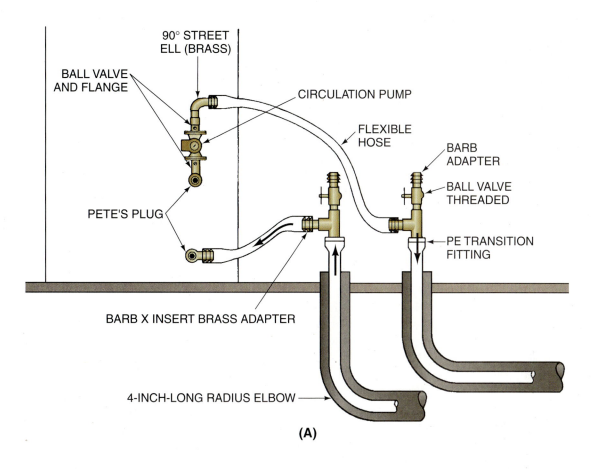

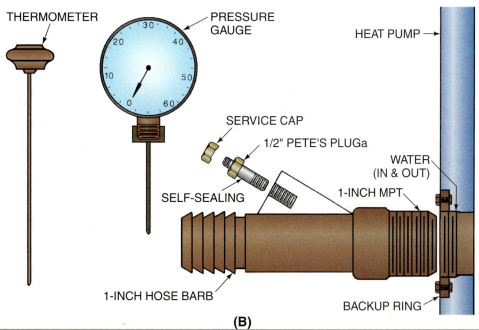

FIGURE 19-1 Pete's plug used to obtain temperature and pressure measurements of fluid flows.

SYSTEM DEFICIENCIES OFTEN IDENTIFIED

Blowing cold air is not enough! In the past, all too often, air conditioning technicians would start-up air conditioning systems by simply grabbing the suction line and feeling the air coming from the supply registers. If the suction line was cold and there was at least a 17-degree temperature difference between the room temperature and the supply air temperature, many technicians concluded that the system was fine. Practices such as these are not acceptable. By performing TAB work, many system deficiencies that would otherwise go unnoticed can be uncovered and remedied.

Before TAB work can begin, the air distribution system must be evaluated to ensure that the proper air volumes are available. There is no sense in balancing a system that does not have adequate airflow to begin with! By initiating the TAB process, a number of system deficiencies that might otherwise go unnoticed can be uncovered and remedied. These system deficiencies include:

- filter bypass
- over-amping motors
- dirty evaporator coils
- pulley alignment issues
- leaking ducts
- improperly designed duct systems

Before the balancing process can begin, the available air volume must be determined. The available air volume should be within 10% of the desired airflow. If the airflow is too high or too low, the cause for this condition must be found and remedied. Excessive airflow can be caused by a missing filter or a filter that allows air to bypass it. Excessive airflow will cause the amperage draw of the motor to rise, possibly to unacceptable or unsafe levels. Inadequate airflow can be caused by duct restrictions, dirty evaporator coils, dirty blower wheels, improper pulley alignment, closed supply registers, blocked return grills, and improperly designed air distribution systems. Leaks in the supply side of the duct system will prevent the required air volumes from reaching the conditioned space even though the blower is moving the correct amount of air.

DUCT SYSTEM PRESSURES

The next section of the Chapter describes some of the tools that are used by the TAB technician. Some of these tools measure the pressures that are present in the duct system. It is here that these duct pressures will be discussed. A duct system is pressurized by three pressures: static pressure, velocity pressure, and total pressure. **Static pressure** is the same as the pressure in a closed vessel and can be thought of as the pressure that pushes against the sides of the duct sections (Fig. 19-2). The **velocity pressure** is the pressure that results from the speed, or velocity, and the weight of the air. **Total pressure** is simply the sum of the static and velocity pressures.

$$\text{TOTAL PRESSURE} = \text{VELOCITY PRESSURE} + \text{STATIC PRESSURE}$$

Consider a duct section with no openings connected to a blower. There will be pressure pushing against the walls of duct section, but no air is moving (Fig. 19-3). If a

FIGURE 19-2 Static pressure is the pressure that pushes against the walls of a duct section.

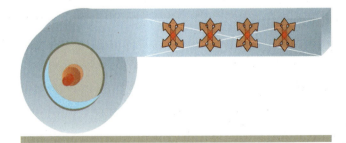

FIGURE 19-3 If air has nowhere to go, there is only static pressure present.

Going Green

Building green means building with long-term efficiency and high-quality operation in mind. Short cuts don't cut it!

hole was made in the wall of the duct, air would flow out. The static pressure in the duct has become velocity pressure as the air is now flowing through the hole (Fig. 19-4). A simple way to visualize the concepts of static and

velocity pressure is to picture a toy balloon. When the balloon is blown up with its opening held closed, there is static pressure, but no velocity pressure (Fig. 19-5). If the balloon is released, the pressure in the balloon rushes out of the balloon, creating velocity pressure (Fig. 19-6).

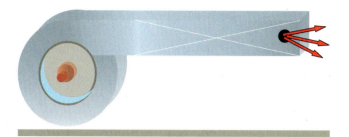

FIGURE 19-4 If an opening is present, the static pressure creates velocity pressure as the air moves through the opening.

FIGURE 19-5 If a balloon is blown up, there is static pressure pushing in all directions.

TAB TOOLS OF THE TRADE

In order to properly perform TAB work, the technician uses a wide variety of hi-tech instruments to measure and adjust the airflow through duct systems. Here are some of the commonly used pieces of TAB equipment.

U-TUBE MANOMETER

The U-tube manometer is a simple device that can measure very small pressures (Fig. 19-7). Pressures in an air distribution system are so small that they are not expressed in pounds per square inch, psig, but in units called inches of water column, IWC. The conversion factor between IWC and psig is 27.7 IWC = 1 psig. The U-tube manometer can be connected to either one or two points in an air distribution system and the manometer will indicate the net pressure by the different levels of liquid in the tube. The manometer in (Fig. 19-8) indicates a pressure of 4 IWC, since that is the distance between the two liquid levels. If one port is open to the atmosphere, the manometer will indicate

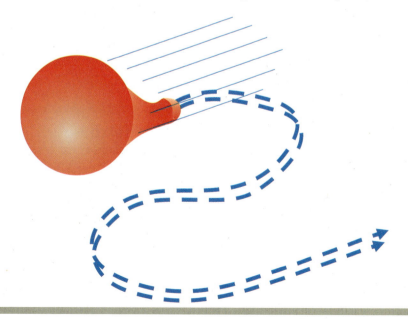

FIGURE 19-6 If the balloon is released, the air rushes out and velocity pressure is created.

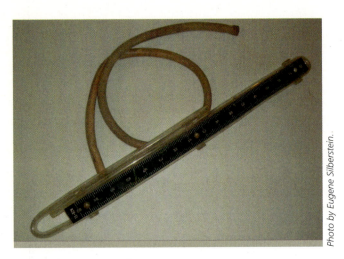

Photo by Eugene Silberstein.

FIGURE 19-7 U-tube manometer.

DUCT PRESSURES

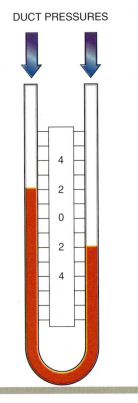

FIGURE 19-8 The manometer displays low pressures by changing the level of liquid in the tube.

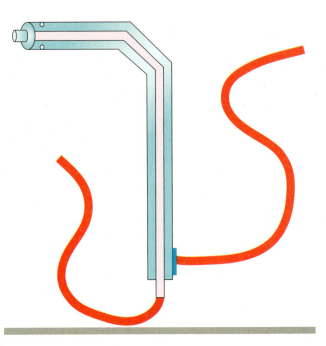

FIGURE 19-9 Pitot tube.

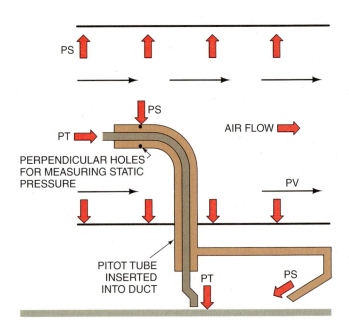

FIGURE 19-10 Holes on the side of the tube are used to measure static pressure.

only the sensed pressure. Since atmospheric pressure is 0 psig, the difference between the sensed pressure and 0 will be the sensed pressure.

PITOT TUBE

The pitot tube is a test instrument that is used to measure the air pressure in a duct system (Fig. 19-9). The pitot tube is constructed of two concentric tubes and has a

90-degree bend in it. The tube is inserted into a hole in the duct and the open end of the pitot tube is directed into the airstream, so air blows into the device. There are small holes in the side of the outer tube. These holes allow for the measuring of the static pressure in the duct system (Fig. 19-10).

The U-tube manometer and the pitot tube can be used to measure the static, velocity, and total pressures in a duct system. To measure the static pressure, only the pressure pushing against the walls of the duct is measured (Fig. 19-11). The velocity pressure measurement requires a differential pressure measurement (Fig. 19-12). The total velocity measurement, which is the sum of the static and the velocity pressures, is obtained by connecting the instruments as shown in (Fig. 19-13).

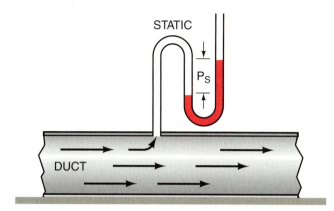

FIGURE 19-11 Measuring static pressure.

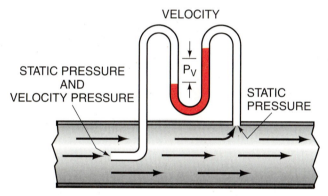

THE STATIC PRESSURE PROBE CANCELS THE STATIC PRESSURES AT THE VELOCITY PRESSURE END AND INDICATES TRUE VELOCITY PRESSURE.

FIGURE 19-12 Measuring velocity pressure.

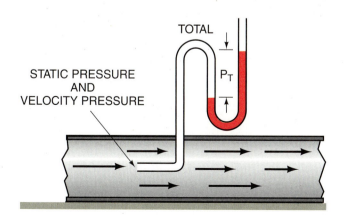

FIGURE 19-13 Measuring total pressure.

from experience...

In order to help ensure accurate velocity readings in a duct section, it is recommended that the location selected for duct testing be at least 7 duct diameters downstream and at least 4 duct diameters upstream of any duct turns, bends, or obstructions. For example, if the duct diameter is 8 inches, the selected point for testing should be at least 56 inches downstream and at least 24 inches upstream of any obstructions (Fig. 19-14). The air will be less turbulent, and more accurate reading will be obtained.

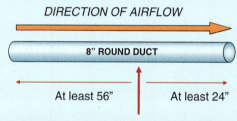

FIGURE 19-14 Recommended distances for obtaining accurate readings in a duct section.

FLOW HOODS

The flow hood, also known as a capture hood or a balometer, is a popular instrument used to measure air volumes at registers and grills (Fig. 19-15). They can be used to measure supply air as well as return air. The flow hood captures the air that exits a grill or diffuser and passes it over a grid that has a number of evenly spaced inlets in it. The air passes through the grid to the "brains" of the instrument to determine a number of characteristics that can include temperature, relative humidity, air velocity, and air volume.

Refer to Procedure 19-1 on pages 491–492 for step-by-step instructions on how to use the flow hood to measure airflow at a supply register location.

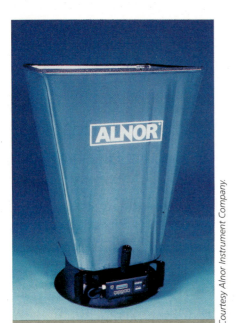

FIGURE 19-15 Flow Hood.

Courtesy Alnor Instrument Company.

from experience...

Flow hoods offer resistance to airflow. This resistance will result in readings that are slightly lower than the actual airflow through the register or grill. Since the readings at all register locations will be off by the same factor, this loss is often ignored.

Refer to Procedure 19-2 on pages 493–494 for step-by-step instructions on how to calculate the correction factor for flow hood readings.

ANEMOMETERS AND VELOMETERS

Anemometers and velometers are used to measure the velocity, or speed, of an airstream in a duct system (Fig. 19-16). Different models have different features. Some units have the capability to measure not only velocity but also temperature and relative humidity (Fig. 19-17).

VOLTMETERS AND AMMETERS

Voltmeters and Ammeters may not seem to play an important role in the job of the TAB technician, but they actually do. Before any balancing can be done, the blower must be delivering the proper amount of air to the duct system. Improper voltage supplied to the blower motor will have an effect on the amount of air delivered. In addition to the voltage, the amperage of the motor must also be checked to make certain that the motor's amperage limits are not exceeded. Excessive airflow in the duct system will cause the motor's amperage to rise, causing the motor to overheat. This can result in premature motor failure.

FIGURE 19-16 Velometer/Anemometer.

Photo by Eugene Silberstein.

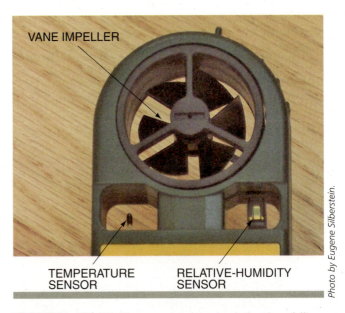

VANE IMPELLER

TEMPERATURE SENSOR

RELATIVE-HUMIDITY SENSOR

FIGURE 19-17 Temperature and relative humidity sensors.

Photo by Eugene Silberstein.

How Do We Balance an Air Distribution System?

The concept of balancing the flow in an air distribution system involves dividing the airflow, in the right proportions, to the desired locations. Before the TAB work can begin, it is important to know what the airflow requirements for the system are. It would be impossible to balance an air distribution system if the desired airflow values are not known. Once the desired values are known, the TAB work can start.

The first step to balancing an air distribution system is to make certain that there is ample airflow being provided by the blower. If the blower is not moving the correct amount of air through the main trunk of the system, obtaining correct air volumes throughout the air distribution system will not be possible. Measuring the air volume supplied by the system blower can be accomplished in a number of ways. The two most popular ways to accomplish this are

- measuring the average velocity in the main trunk line and multiplying the velocity by the cross-sectional area of the supply duct. Determining the average velocity in the duct system is accomplished by using the traverse method.
- measuring the airflow at each supply register location and adding up the individual airflows to obtain the approximate system airflow.

Refer to Procedure 19-3 on page 495 for step-by-step instructions on how to determine the average velocity in an air distribution system using the traverse method.

Refer to Procedures 19-4 and 19-5 on pages 496–497 for step-by-step instructions on how to determine the total airflow in an air distribution system.

Once the airflow has been established, it is important to determine if the actual airflow is acceptable for the particular system. System specifications should be checked to ensure that the airflow is within acceptable ranges. Once the airflow has been established, there are two common methods that can be used to balance the air distribution system. These methods are the stepwise method and the proportional method.

STEPWISE BALANCING

Stepwise balancing involves stepping down the airflow to the outlets with the greatest amount of excess air. By doing this, the air volume to those outlets with a deficiency of air will increase, bringing the system into balance.

Refer to Procedure 19-6 on pages 498–500 for step-by-step instructions on how to balance a system using the stepwise balancing method.

PROPORTIONAL BALANCING

Proportional balancing is a very popular method for ensuring that the airflow rates at all outlets are proportional to each other. The main concern of proportional balancing is to have the airflow at any given outlet at the same percentage of design airflow as any other outlet in the system. For example, if a particular outlet has a design airflow of 100 cfm and an actual airflow of 90 cfm, the outlet is delivering 90% of the design airflow. If the air distribution system is proportionally balanced, another outlet with a design airflow of 200 cfm will have an actual airflow rate of 180 cfm, or 90% of 200 cfm.

Refer to Procedure 19-7 on pages 501–504 for step-by-step instructions on how to balance a system using the proportional balancing method.

HOW DO WE KNOW WHEN THE SYSTEM IS BALANCED?

A system is said to be balanced when two conditions are met. First, the air volume present at each outlet or supply register is within 10% of the desired value. Second, the air pathway with the greatest resistance, which is often the longest, has its volume damper wide open. With the volume damper of this pathway in the full-open position, the blower will be supplying its minimum output to properly meet the needs of the system. By providing minimum output, the energy required to operate the blower is also reduced. As the amount of air moved by a blower increases, the amperage of the motor increases as well. This increases the power consumption of the motor and shortens the useful life of the motor.

SUMMARY

- Testing, adjusting, and balancing is the systematic process of achieving and documenting airflow rates.
- Testing is the process of using specialized instruments to measure system characteristics.
- Adjusting is the process of varying fluid flow by changing the settings on fluid flow controls.
- Balancing is the process of matching the actual fluid flow rates to the desired flow rates.
- TAB work is required for systems to operate at peak efficiency.
- The purpose of TAB work is to fine-tune air distribution systems.
- Improper system sizing, airflow, and charging result in significant reduction in system capacity and efficiency.
- System deficiencies such as air-filter bypass, dirty evaporator coils, pulley alignment issues, leaking ducts and improperly designed air distribution systems are often uncovered as a result of the TAB preliminary work.
- The three duct system pressures are static, velocity, and total pressure.
- Total pressure is the sum of the static pressure and the velocity pressure.
- Commonly used TAB instruments include the U-tube manometer, the pitot tube, and the flow hood.
- Air distribution systems can be balanced by the stepwise or proportional method.

GREEN CHECKLIST

☐ **Proper TAB work helps ensure that the air conditioning or heating system provides comfort in an efficient manner.**

☐ **Improper system sizing, airflow, and charging result in significant reduction in system capacity and efficiency.**

☐ **Building green means building with long-term efficiency in mind.**

☐ **Properly balanced systems are happy systems!**

Using a Flow Hood to Measure Airflow at a Supply Register

- Set up the flow hood by following the manufacturer's instruction literature.

- To minimize errors, be sure to use the capture hood size that matches the size of register being evaluated.

- Set the flow hood to read air volume in the desired units, cfm.

A Position the flow hood up against the register. If the flow hood is larger than the register, try to center the register within the flow hood.

- Obtain and record the airflow reading.

B Rotate the flow hood 90 degrees.

- Position the flow hood up against the register.

- Obtain and record the airflow reading.

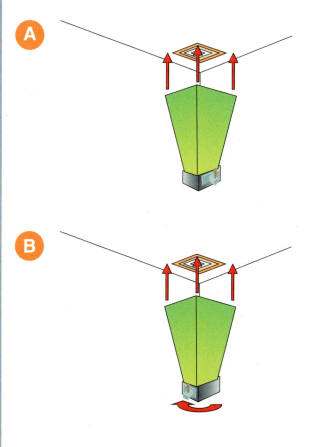

PROCEDURE 19-1

Using a Flow Hood to Measure Airflow at a Supply Register (continued)

C Repeat the previous three steps a total of four times, recording the air volume values.

C

137 cfm
105 cfm
135 cfm
103 cfm
─────
480 cfm

D Average the four readings to obtain the average airflow, in cfm, for that register.

D

480 cfm ÷ 4
120 cfm

from experience...

If the first two readings are the same, there is no need to take the remaining two readings. This means that the flow is not turbulent and the other two readings will most likely be the same as the first two.

PROCEDURE 19-2

Calculating the Flow Hood Correction Factor

A Using a pitot tube, traverse the branch duct and record the velocity values.

A

685 ft/min
500 ft/min
570 ft/min
475 ft/min
486 ft/min
498 ft/min
_____
3,214

B Calculate the average air velocity in the duct.

B

3,214 ÷ 6 =
536 ft/min

C Calculate the cross-sectional area of the duct section, in square feet

C

6" ROUND DUCT
RADIUS = 3 inches
Area = $\pi r^2 = \pi (3^2)$
Area = $9\pi = 9 \times 3.14$
Area = $28.26 \, in^2$

$$\frac{28.26 \, in^2}{144 \, in^2/ft^2} = 0.196 \, ft^2$$

PROCEDURE **19-2**

Calculating the Flow Hood Correction Factor (continued)

D Multiply the average air velocity and the cross-sectional area of the duct to obtain the calculated airflow rate.

D

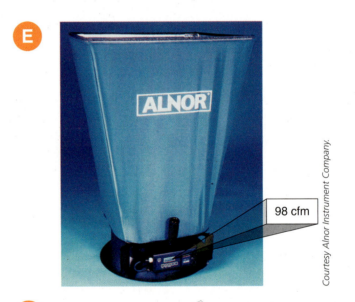

$$0.196 \text{ ft}^2$$
$$\times 536 \text{ ft/min}$$
$$\overline{105 \text{ ft}^3/\text{min}}$$
$$\boxed{105 \text{ cfm}}$$

E Measure the airflow at the supply register with a flow hood.

E

ALNOR

98 cfm

Courtesy Alnor Instrument Company.

F Divide the calculated airflow rate by the measured airflow rate to get the correction factor

- The correction factor can now be used for any supply register on the same trunk line. If another flow hood reading is, for example, 150 cfm, the actual airflow will be 150 cfm × 1.07, or 160 cfm.

F

$$105 \text{ cfm}$$
$$\div 98 \text{ cfm}$$
$$\overline{1.07}$$

PROCEDURE **19-3**

Determining Air Velocity in a Duct Using the Traverse Method

- Select the best location in the duct for taking velocity readings. This location will ideally be 7 duct diameters downstream and 3 duct diameters upstream of any bends, takeoffs, or other airflow restrictions (refer to Fig. 19-14 on page 487).

- The velometer will need to be inserted into the duct to obtain readings, so a series of holes will need to be made in the duct. Follow the guidelines of the manufacturer regarding the size of the hole needed. The hole should only be large enough to accommodate the velometer probe.

A Following a uniform pattern, take a series of air velocity readings in the duct. The goal is to obtain readings across the entire cross-sectional area of the duct. Larger ducts will require more readings, as many as 25, but ducts for residential applications do not require nearly that many.

- Record the velocity measurements.

- Calculate the average velocity in the duct.

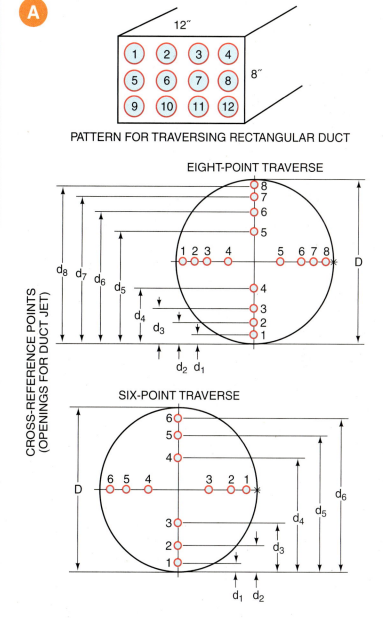

PATTERN FOR TRAVERSING RECTANGULAR DUCT

EIGHT-POINT TRAVERSE

SIX-POINT TRAVERSE

CROSS-REFERENCE POINTS (OPENINGS FOR DUCT JET)

PROCEDURE 19-4

Determining Airflow in the Duct System: Method 1

- By traversing the duct, determine the average air velocity in the duct system.

- Measure the height and width of the supply duct, in inches.

- Multiply the duct height by the duct width to obtain the cross-sectional area of the duct in square inches.

- Divide the cross-sectional area of the duct in square inches by 144 to get the cross-sectional area of the duct in square feet.

- Multiply the cross-sectional area of the duct in square feet by the velocity of the air to determine the total cfm of the system.

- Follow along with the following example:

Air Velocity readings:

650 feet/minute	750 feet/minute	790 feet/minute
745 feet/minute	595 feet/minute	550 feet/minute
645 feet/minute	780 feet/minute	700 feet/minute
800 feet/minute	725 feet/minute	685 feet/minute

Average air velocity = 8415 ÷ 12 = 701 feet/minute
Duct size = 18" × 12" = 216 in^2 ÷ 144 = 1.5 ft^2
Airflow in cfm = Average air velocity × Cross sectional area
Airflow in cfm = 701 feet/min × 1.5 ft^2
Airflow in cfm = 1051 cfm

PROCEDURE 19-5

Determining Airflow in the Duct System: Method 2

This method is to be used as an alternate method in the event Method 1 is not possible or practical.

- Make a list of all supply registers in the occupied space.

- Using the flow hood, take and record airflow measurements at each outlet.

- Add up all of the measurements to obtain the total airflow in the system.

Note: If the duct system is leaking, the sum of the airflow measurements will not be equal to the actual amount of air being moved by the system blower.

Balancing a System with the Stepwise Balancing Method

This procedure will use a numerical example to clarify the process.

A Create a balancing table, listing all outlets, desired airflow, and the acceptable range (plus or minus 10%) for each outlet.

ROOM	DESIGN CFM	ACCEPTABLE RANGE	PASS 1	% CHANGE	PASS 2	% CHANGE	PASS 3	% CHANGE
LR 1	150	135–165						
LR 2	150	135–165						
MBR 1	150	135–165						
MBR 2	150	135–165						
BR 2	150	135–165						
BR 3	150	135–165						
DEN	100	90–110						
BATH 1	100	90–110						
BATH 2	100	90–110						
KIT	250	225–275						
DR	150	135–165						

B Take airflow readings at each outlet and record values in the "PASS 1" column of the table.

ROOM	DESIGN CFM	ACCEPTABLE RANGE	PASS 1	% CHANGE	PASS 2	% CHANGE	PASS 3	% CHANGE
LR 1	150	135–165	190					
LR 2	150	135–165	140					
MBR 1	150	135–165	170					
MBR 2	150	135–165	190					
BR 2	150	135–165	110					
BR 3	150	135–165	190					
DEN	100	90–110	150					
BATH 1	100	90–110	50					
BATH 2	100	90–110	50					
KIT	250	225–275	260					
DR	150	135–165	100					

C Calculate the percentage change for each outlet. This is accomplished by determining the difference between the measured cfm and the design cfm and dividing this difference by the design cfm. For example, for the LR1 outlet, the design cfm is 150 and the measured cfm is 190. This gives us (190–150)/150, or 40/150, which is 0.267, or +26.7%.

ROOM	DESIGN CFM	ACCEPTABLE RANGE	PASS 1	% CHANGE	PASS 2	% CHANGE	PASS 3	% CHANGE
LR 1	150	135–165	190	+26.7%				
LR 2	150	135–165	140	−6.7%				
MBR 1	150	135–165	170	+13.4%				
MBR 2	150	135–165	190	+26.7%				
BR 2	150	135–165	110	−26.7%				
BR 3	150	135–165	190	+26.7%				
DEN	100	90–110	150	+50%				
BATH 1	100	90–110	50	−50%				
BATH 2	100	90–110	50	−50%				
KIT	250	225–275	260	+4%				
DR	150	135–165	100	−33%				

D Starting with the outlets with the highest airflow, throttle down the outlets with the greatest percentage of excess airflow until the airflow is about 10% lower than the desired air volume. Record these values.

ROOM	DESIGN CFM	ACCEPTABLE RANGE	PASS 1	% CHANGE	PASS 2	% CHANGE	PASS 3	% CHANGE
LR 1	150	135–165	190	+26.7%	**140**			
LR 2	150	135–165	140	−6.7%				
MBR 1	150	135–165	170	+13.4%	**135**			
MBR 2	150	135–165	190	+26.7%	**145**			
BR 2	150	135–165	110	−26.7%				
BR 3	150	135–165	190	+26.7%	**140**			
DEN	100	90–110	150	+50%	**90**			
BATH 1	100	90–110	50	−50%				
BATH 2	100	90–110	50	−50%				
KIT	250	225–275	260	+4%				
DR	150	135–165	100	−33%				

E Measure and record the airflow at all outlets. Recalculate the percentage change for each outlet.

ROOM	DESIGN CFM	ACCEPTABLE RANGE	PASS 1	% CHANGE	PASS 2	% CHANGE	PASS 3	% CHANGE
LR 1	150	135–165	190	+26.7%	140	−6.7%		
LR 2	150	135–165	140	−6.7%	160	+6.7%		
MBR 1	150	135–165	170	+13.4%	135	−10%		
MBR 2	150	135–165	190	+26.7%	145	−3%		
BR 2	150	135–165	110	−26.7%	125	−16.7%		
BR 3	150	135–165	190	+26.7%	140	−6.7%		
DEN	100	90–110	150	+50%	90	−10%		
BATH 1	100	90–110	50	−50%	120	+20%		
BATH 2	100	90–110	50	−50%	110	+10%		
KIT	250	225–275	260	+4%	275	+10%		
DR	150	135–165	100	−33%	160	+6.7%		

F Once again, starting with the highest airflow, throttle down the outlets with the greatest percentage of excess airflow until the airflow is about 10% lower than the desired air volume.

ROOM	DESIGN CFM	ACCEPTABLE RANGE	PASS 1	% CHANGE	PASS 2	% CHANGE	PASS 3	% CHANGE
LR 1	150	135–165	190	+26.7%	140	−6.7%		
LR 2	150	135–165	140	−6.7%	160	+6.7%	140	
MBR 1	150	135–165	170	+13.4%	135	−10%		
MBR 2	150	135–165	190	+26.7%	145	−3%		
BR 2	150	135–165	110	−26.7%	125	−16.7%		
BR 3	150	135–165	190	+26.7%	140	−6.7%		
DEN	100	90–110	150	+50%	90	−10%		
BATH 1	100	90–110	50	−50%	120	+20%	90	
BATH 2	100	90–110	50	−50%	110	+10%	95	
KIT	250	225–275	260	+4%	275	+10%	230	
DR	150	135–165	100	−33%	160	+6.7%	140	

PROCEDURE 19-6

Balancing a System with the Stepwise Balancing Method (continued)

 Measure and record the airflow at all outlets.

ROOM	DESIGN CFM	ACCEPTABLE RANGE	PASS 1	% CHANGE	PASS 2	% CHANGE	PASS 3	% CHANGE
LR 1	150	135–165	190	+26.7%	140	−6.7%	145	
LR 2	150	135–165	140	−6.7%	160	+6.7%	140	
MBR 1	150	135–165	170	+13.4%	135	−10%	165	
MBR 2	150	135–165	190	+26.7%	145	−3%	160	
BR 2	150	135–165	110	−26.7%	125	−16.7%	155	
BR 3	150	135–165	190	+26.7%	140	−6.7%	155	
DEN	100	90–110	150	+50%	90	−10%	105	
BATH 1	100	90–110	50	−50%	120	+20%	90	
BATH 2	100	90–110	50	−50%	110	+10%	95	
KIT	250	225–275	260	+4%	275	+10%	240	
DR	150	135–165	100	−33%	160	+6.7%	140	

 Recalculate the percentage change for each outlet.

- At this point, the system should be balanced, with all outlets delivering the desired amount of air, within 10%.

ROOM	DESIGN CFM	ACCEPTABLE RANGE	PASS 1	% CHANGE	PASS 2	% CHANGE	PASS 3	% CHANGE
LR 1	150	135–165	190	+26.7%	140	−6.7%	145	−3.3%
LR 2	150	135–165	140	−6.7%	160	+6.7%	140	−6.7%
MBR 1	150	135–165	170	+13.4%	135	−10%	165	+10%
MBR 2	150	135–165	190	+26.7%	145	−3%	160	+3.3%
BR 2	150	135–165	110	−26.7%	125	−16.7%	155	+3.3%
BR 3	150	135–165	190	+26.7%	140	−6.7%	155	+3.3%
DEN	100	90–110	150	+50%	90	−10%	105	+5%
BATH 1	100	90–110	50	−50%	120	+20%	90	−5%
BATH 2	100	90–110	50	−50%	110	+10%	95	0
KIT	250	225–275	260	+4%	275	+10%	240	−4%
DR	150	135–165	100	−33%	160	+6.7%	140	−6.7%

PROCEDURE **19-7**

Balancing a System with the Proportional Balancing Method

This procedure will use a numerical example to clarify the process.

A Sketch out the air distribution system, labeling the location, and the desired airflow at each outlet.

• Obtain an airflow reading at each outlet and record the values.

B Calculate the percentage of measured airflow to the desired airflow. List them in order from lowest to highest.

• The volume damper for the outlet with the lowest percentage of design airflow, in this case outlet 8, is not adjusted.

• The volume damper for the outlet with the next lowest percentage of design airflow, in this case outlet 4, will be adjusted so that the airflow through it will be about 110 cfm. This is the average of the measured airflows from outlet 8 (100 cfm) and outlet 4 (120 cfm).

A

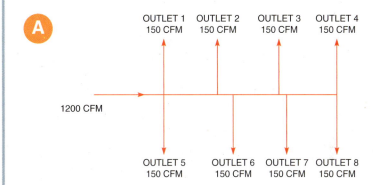

B

OUTLET NUMBER	DESIRED AIRFLOW	MEASURED AIRFLOW	PERCENTAGE OF DESIRED AIRFLOW
8	150 CFM	100	67%
4	150 CFM	120	80%
7	150 CFM	120	80%
3	150 CFM	130	87%
6	150 CFM	140	93%
2	150 CFM	150	100%
5	150 CFM	200	133%
1	150 CFM	240	160%

PROCEDURE 19-7

Balancing a System with the Proportional Balancing Method (continued)

C This should proportionally balance the airflow in outlets 4 and 8.

C

OUTLET NUMBER	DESIRED AIRFLOW	MEASURED AIRFLOW	PERCENTAGE OF DESIRED AIRFLOW
8	150 CFM	110	73%
4	150 CFM	110	73%
7	150 CFM	120	80%
3	150 CFM	130	87%
6	150 CFM	140	93%
2	150 CFM	150	100%
5	150 CFM	200	133%
1	150 CFM	240	160%

D The volume damper for the outlet with the next lowest percentage of design airflow, in this case outlet 7, will be adjusted so that the airflow through it will be about 113 cfm [(110 cfm + 110 cfm + 120 cfm) ÷ 3]. By doing this, the airflow through outlets 4 and 8 will rise slightly, putting outlets 4, 7, and 8 in balance with each other.

D

OUTLET NUMBER	DESIRED AIRFLOW	MEASURED AIRFLOW	PERCENTAGE OF DESIRED AIRFLOW
8	150 CFM	113	75%
4	150 CFM	113	75%
7	150 CFM	113	75%
3	150 CFM	130	87%
6	150 CFM	140	93%
2	150 CFM	150	100%
5	150 CFM	200	133%
1	150 CFM	240	160%

E Repeat the previous step for volume damper 3. The volume damper for outlet 3 will be throttled down to about 118 cfm [(113 cfm + 113 cfm + 113 cfm + 130 cfm) ÷ 4]. This will balance outlets 3, 4, 7, and 8.

E

OUTLET NUMBER	DESIRED AIRFLOW	MEASURED AIRFLOW	PERCENTAGE OF DESIRED AIRFLOW
8	150 CFM	118	79%
4	150 CFM	118	79%
7	150 CFM	118	79%
3	150 CFM	118	79%
6	150 CFM	140	93%
2	150 CFM	150	100%
5	150 CFM	200	133%
1	150 CFM	240	160%

F Repeat the previous step for volume damper 6. The volume damper for outlet 6 will be throttled down to about 122 cfm [(118 cfm + 118 cfm + 118 cfm + 118 cfm + 140 cfm) ÷ 5]. This will balance outlets 3, 4, 6, 7, and 8.

F

OUTLET NUMBER	DESIRED AIRFLOW	MEASURED AIRFLOW	PERCENTAGE OF DESIRED AIRFLOW
8	150 CFM	122	81%
4	150 CFM	122	81%
7	150 CFM	122	81%
3	150 CFM	122	81%
6	150 CFM	122	81%
2	150 CFM	150	100%
5	150 CFM	200	133%
1	150 CFM	240	160%

G Repeat the previous step for volume damper 2. The volume damper for outlet 2 will be throttled down to about 127 cfm [(122 cfm + 122 cfm + 122 cfm + 122 cfm + 122 cfm + 150 cfm) ÷ 6]. This will balance outlets 2, 3, 4, 6, 7, and 8.

G

OUTLET NUMBER	DESIRED AIRFLOW	MEASURED AIRFLOW	PERCENTAGE OF DESIRED AIRFLOW
8	150 CFM	127	85%
4	150 CFM	127	85%
7	150 CFM	127	85%
3	150 CFM	127	85%
6	150 CFM	127	85%
2	150 CFM	127	85%
5	150 CFM	200	133%
1	150 CFM	240	160%

PROCEDURE 19-7

Balancing a System with the Proportional Balancing Method (continued)

H Repeat the previous step for volume damper 5. The volume damper for outlet 5 will be throttled down to about 137 cfm [(127 cfm + 127 cfm + 127 cfm + 127 cfm + 127 cfm + 127 cfm + 200 cfm) ÷ 7].

H

OUTLET NUMBER	DESIRED AIRFLOW	MEASURED AIRFLOW	PERCENTAGE OF DESIRED AIRFLOW
8	150 CFM	137	91%
4	150 CFM	137	91%
7	150 CFM	137	91%
3	150 CFM	137	91%
6	150 CFM	137	91%
2	150 CFM	137	91%
5	150 CFM	137	91%
1	150 CFM	240	160%

I Repeat the previous step for volume damper 1. The volume damper for outlet 1 will be throttled down to about 150 cfm [(137 cfm + 137 cfm + 137 cfm + 137 cfm + 137 cfm + 137 cfm + 137 cfm + 240 cfm) ÷ 8].

• Take air volumes at all outlet locations. The system should be properly balanced within an acceptable range.

I

OUTLET NUMBER	DESIRED AIRFLOW	MEASURED AIRFLOW	PERCENTAGE OF DESIRED AIRFLOW
8	150 CFM	150	100%
4	150 CFM	150	100%
7	150 CFM	150	100%
3	150 CFM	150	100%
6	150 CFM	150	100%
2	150 CFM	150	100%
5	150 CFM	150	100%
1	150 CFM	150	100%

REVIEW QUESTIONS

1. **Balancing is the process of**
   a. Making certain that the airflow at all outlets is the same
   b. Matching actual fluid flow rates to desired fluid flow rates
   c. Both a and b are correct
   d. Either a nor b is correct

2. **Using specialized, properly calibrated instruments to measure characteristics of fluid flow is known as**
   a. Balancing
   b. Testing
   c. Adjusting
   d. Both a and c are correct

3. **When performing TAB work, the acceptable percentage of airflow error is**
   a. 0%
   b. 5%
   c. 10%
   d. 20%

4. **Which of the following can result if a system is not properly balanced?**
   a. Increased system run time
   b. Increased power consumption
   c. Premature system failure
   d. All of the above are correct

5. **Excessive blower motor amperage can be caused by**
   a. Too much airflow
   b. Too little airflow
   c. Dirty evaporator coil
   d. Closed supply registers

6. **Which of the following is true regarding duct pressure?**
   a. Total pressure = Velocity pressure − Static pressure
   b. Velocity pressure = Static pressure − Total pressure
   c. Static pressure = Total pressure − Velocity pressure
   d. Total pressure + Static pressure = Velocity pressure

7. **A pressure of 5 IWC is closest to**
   a. 0.18 psig
   b. 0.36 psig
   c. 0.54 psig
   d. 0.72 psig

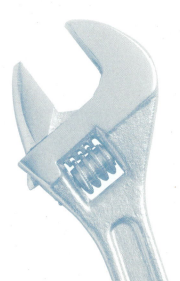

8. **Which of the following best describes the traverse method of establishing the air velocity in a duct system?**

   a. The sum of all the air velocity readings

   b. The highest of all the air velocity readings

   c. The average of all the air velocity readings

   d. The lowest of all the air velocity readings

9. **What is the airflow in a 12″ × 12″ duct section that has an air velocity of 500 feet per minute?**

   a. 144 cfm

   b. 500 cfm

   c. 720 cfm

   d. 72,000 cfm

10. **An air conditioning system has been proportionally balanced. Outlet 1 has a design airflow of 150 cfm and an actual airflow of 125 cfm. What is the actual airflow through outlet 2 if its design airflow is 200 cfm?**

    a. 125 cfm

    b. 167 cfm

    c. 200 cfm

    d. 225 cfm

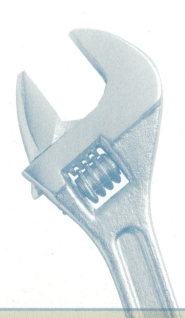

WHAT'S WRONG WITH THIS PICTURE?

Carefully study Figures 19-18(A) and (B) and think about what is wrong. Consider all possibilities.

❌ WRONG

(A)

(B)

Photo by Eugene Silberstein.

FIGURE 19-18 When performing TAB work, it is important to avoid making initial air volume adjustments at the supply grill location. By throttling the airflow at the supply grill location, excessive noise can result.

✅ RIGHT

Photo by Eugene Silberstein.

FIGURE 19-19 When performing TAB work, airflow adjustments should be made at the volume dampers along the main trunk line of the air distribution system (Fig. 19-19). This practice makes for a quieter, more efficient air distribution system.

Indoor Air Quality

OBJECTIVES   *Upon completion of this chapter, the student should be able to*

- Explain the difference between air contaminants and air pollutants.

- Explain how the effects of air pollutants can be reduced.

- Differentiate between permanent and disposable filters.

- Identify various types of air filters.

- Explain how an electrostatic air filter functions.

- Explain the stages of filtration used in an electronic air cleaner.

- Install air filters.

- Install a duct-mounted humidifier.

- Explain how ultraviolet (UV) air treatment helps clean the air.

The two main components of the air we breathe are nitrogen (78%), and oxygen (21%). The remaining 1% is comprised of various other gases. Air also holds a certain amount of moisture that varies as atmospheric conditions change. Maintaining proper oxygen and moisture levels, in addition to the quantity, temperature, and cleanliness of the air, contributes to a more general term called comfort.

Different people have different requirements for comfort. For example, those who reside in Las Vegas have grown accustomed to living in a warm climate with very low humidity, while those who live in New York City have acclimated themselves to a climate with high humidity levels and ever-changing temperatures. The purpose of an air conditioning system is to provide and maintain the desired levels of humidity, temperature, and air cleanliness for the occupants of the conditioned space. This is accomplished through humidification, dehumidification, air filtering, air movement, and cooling.

Air filters are available in a number of different styles and types that provide varying degrees of air filtration. Those who suffer from severe allergies may need more sophisticated filtration devices that remove smaller particles from the air, while those who do not suffer from such ailments may be quite comfortable with standard filtration methods. Hospital operating rooms require a high degree of air filtration as even the smallest particles in the air may lead to contamination.

Humidification equipment is intended to add moisture to the occupied space and is most popular in the winter months when heating equipment is in use. Humidifiers can be freestanding units or can be incorporated into the heating system of the structure. When used in conjunction with a heating system, the humidifier adds moisture to the air to maintain the desired relative humidity in the space.

Atomizing The process of spraying a fine mist of water

Bypass humidifiers Humidifiers that use the difference in pressure between the supply and return ducts to move air across the humidifier media

Contaminants Unwanted particulate matter in air that may or may not lead to health problems

Electronic air cleaners See electronic air filters

Electronic air filters Electrically operated filters that offer multiple stage filtering as well as odor removal

Electrostatic air filters Filters that capture particulates with electric charges opposite to that of the filter media

Element See media

Extended surface filters Filters that have pleated filter media to increase the surface area of the filter

Filter-backed return grill Return grill in the occupied space that supports an air filter

Filters Means by which particulate matter is captured from the air

Filtration Process of removing particulate matter from the air

Flow-through humidifiers Humidifiers that have a constant flow of water through them

Heat recovery ventilators Mechanical ventilation systems that exchange air between the inside and outside of a structure while recovering heat from the discharged air.

HEPA filters (high efficiency particulate arresters) Filters capable of capturing a minimum of 99.7% of all particulate matter at least 0.3 microns in size

High efficiency particulate arresters See HEPA filters

Humidification Process of adding moisture to the air

Humidifiers System components that facilitate the addition of moisture to an air sample

Humidity Amount of water vapor in an air sample

Media Humidifier component that holds water and extends the surface area of the water

Mold Microscopic organism that can cause allergic symptoms similar to those caused by plant pollen

Nanometer Distance equal to 1×10^{-9} meter, or 0.000000001 meter

Pleated filters See extended-surface filters

Pollutants Unwanted particulate matter in air that often leads to health-related problems

Relative humidity Ratio that compares the amount of moisture in an air sample to the maximum amount of moisture the same sample can hold

Saddle valves Two-piece valves that clamp around a domestic water line to provide supply water to the humidifier

Steel filters Permanent, cleanable air filters

Wick Component part of a humidifier that increases the surface area of the water, increasing the efficiency of the humidifier

INDOOR AIR POLLUTION

As building-related technologies in the home construction industry advance, homes are becoming more energy efficient than those constructed only a few decades ago. Newer, tighter-fitting windows and doors and increased insulation all contribute to greater energy efficiency and reduced air seepage into and out of the structure. However, this increased "tightness" of new homes reduces the structure's ability to breathe and increases the levels of indoor pollution and indoor air contamination. It is also important to note that the average individual spends upwards of 90% of his or her time indoors, increasing the negative effects of these air **contaminants** and **pollutants**. The main difference between contaminants and pollutants is that contaminants may or may not cause personal injury, while pollutants do cause health problems. Some common sources for air contamination and pollution include:

pipe insulation	dust mites	cigarette smoke
aerosol sprays	carpets	household cleaners
radon gas	humidity	fireplaces
automobile exhaust	pet dander	paint

To keep the levels of these contaminants and pollutants to a minimum, there are three alternatives:

- Remove the source of the contaminant
- Provide adequate air cleaning
- Provide adequate ventilation

REMOVE THE SOURCE OF THE CONTAMINANT

In some cases, removing the contamination source may be an easy task, but in other cases, the task may be next to impossible. For example, the easiest way to eliminate contamination from cigarette smoke is to stop smoking indoors and not permit house guests to smoke in the home.

If there is asbestos contamination in the home, a contractor qualified to perform asbestos removal should be hired to safely and completely remove it from the structure. Asbestos has been shown to cause various types of cancer, asbestosis, and mesothelioma. A major misconception about asbestos is that it is no longer used in this country and was banned for use in products many years ago. The fact is that asbestos-containing products are still being imported and sold in this country, continuing to endanger people who may come in contact with the substance.

Radon gas is another concern. The presence of this radioactive gas is not easily abated. It is often produced as a result of the breaking down of uranium, which is sometimes found in the earth on which houses are built. Houses should be checked for radon gas and, if it is found within the structure, radon-certified contractors should be brought in to aid in the radon-reduction process.

Going Green

In many houses, there are countless bottles, jars, and containers containing old cleaners, solvents, paints, and other substances that are no longer being used. Keeping these old chemical items in the house is often a fire hazard as well as a contributor to unhealthy air in the space. By removing these items from the structure and disposing of them properly, the air will be cleaner and the occupants will be healthier.

PROVIDE ADEQUATE AIR CLEANING

Since new buildings are tighter than ever before, the air in the residence should be kept as clean as possible. It is recommended that the air be filtered at a rate of six air changes per hour, meaning that the evaporator blower should be able to handle six times the volume of air in the occupied space each hour. To keep the air as clean as possible, it is also recommended that the evaporator fan motor operate continuously to recirculate and filter the air in the space.

PROVIDE ADEQUATE VENTILATION

Since the air within the residence is much dirtier than the air outside, it is further recommended that outside air be allowed to enter the structure and mix with the air in the structure. Keeping windows open while the blower is operating will help bring outside air into the space. Keeping windows wide open during the cooling season will obviously decrease the operating efficiency and cooling ability of the air conditioning system, but keeping one or two windows cracked will supply at least some fresh air to the space. Refer to the section on Heat Recovery Ventilators for more on mechanical ventilation methods.

MOLD

Over the past years, mold has taken its place at the top of the list of potential health-related issues. Homeowners, equipment manufacturers, and service companies have all taken note of the potential dangers mold can present if allowed to get out of hand. Although mold should be, and is, a genuine concern, the media has made the issue appear to be much graver than it actually is. **Mold** is a simple microscopic organism that is found practically everywhere, both indoors and outdoors. When molds are present in large quantities, they can cause allergic symptoms similar to those caused by plant pollen. For mold to grow, there must a food source, which can be wood, dirt, or dust; moisture; and a place to grow.

It is estimated that excessively high humidity levels, over 60%, in a residence can contribute to the accelerated growth of mold. Some major factors that contribute to mold growth are:

- flooding and standing water within the residence
- leaky roofs
- humidifiers
- damp basements
- plumbing leaks

- excessive plant watering
- improperly vented clothes dryers
- improperly vented combustion appliances

Air conditioning duct systems can become contaminated with mold. If the air distribution system has had water damage as a result of a blocked or clogged condensate line, for example, mold can breed within the duct system. Metal duct systems wrapped with fiberglass insulation can be cleaned and disinfected if mold is found in them. If the metal duct system is insulated with internal acoustical lining, the ductwork normally will need to be removed and discarded. Ductwork that is located in difficult locations may have to be abandoned. When in doubt, a contractor trained in mold abatement should be brought in to inspect and evaluate the conditions.

from experience...

Properly maintaining air conditioning equipment can greatly reduce the rate at which mold will grow. Setting humidification equipment to maintain humidity levels in the 30% to 50% range, and making certain that all condensate lines are clear, will help prevent the air conditioning system from becoming a breeding ground for mold.

from experience...

When air conditioning systems are initially designed, it is important to properly size the equipment. Oversized air conditioning systems can actually contribute to mold growth. They will actually recirculate moisture-laden air back to the conditioned space, increasing the moisture levels in the occupied space as well as in the duct system.

AIR FILTRATION

Filtration is the process of removing particulate matter from the air, but can also refer to removing odors as well. The removal of odors from the air will be discussed later in this chapter. The filtering of particulate matter from the air can be accomplished by using various types of filters and filter media. **Filters** capture particulate matter as air flows through them and include the following types:

- Foam filters
- Fiberglass filters
- Extended surface filters
- Steel filters
- HEPA filters

FOAM FILTERS

Foam filters (Fig. 20-1) are often found on window or through-the-wall units. This type of filter is supplied in sheets that must be cut to the proper size with either a razor knife or scissors by the end user. They range in thickness from about ⅛″ to ¼″ and are designed to remove large dust and dirt particles but not smaller airborne particles. Some foam filters, when supplied by the manufacturer, are cut to size and mounted in a plastic frame. These filters can be cleaned periodically, but the foam will eventually break down and tear, at which point the filter should be replaced.

from experience...

Foam filters are intended to be disposable, and it is not recommended that they be cleaned. If, however, no new filter is available and the filter is blocked with dirt, it can be cleaned by placing it, dirt side down, under the faucet and running water through it. Ideally, however, a new filter should always be on hand.

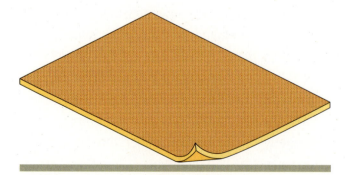

FIGURE 20-1 Foam filter.

from experience...

When installing foam filters on window or wall air conditioners, make certain that the filter media completely covers the surface of the evaporator coil to maximize the filtering benefit. Cutting the filter too small will permit air to bypass the filter, increasing the amount of dirt accumulation on the coil itself as well as the blades of the blower.

Going Green

Be sure to install air filters with the arrow pointing in the direction of airflow. This helps increase the efficiency of the filter by allowing it to capture the larger particles in the less dense portion of the filter.

FIBERGLASS FILTERS

Fiberglass filters are often used for residential applications. Fiberglass filters come mounted in cardboard frames (Fig. 20-2) and may be manufactured with a perforated metal backing (Fig. 20-3) to help the filter keep its shape. These filters are directional and have arrows on the side of the filter that point in the direction of air flow through the filter (Fig. 20-4). The filter gets progressively denser as air passes through it (Fig. 20-5) to catch the larger particles first and then the smaller particles as the air reaches the outlet of the filter.

Framed fiberglass filters come in a wide range of sizes and should be selected to completely fill the rack that holds the filter in place. Using a filter that is too

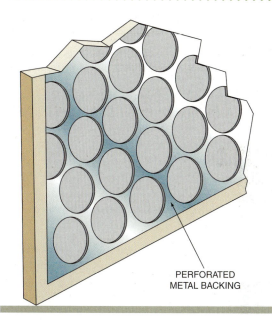

PERFORATED
METAL BACKING

FIGURE 20-3 Perforated metal backing on fiberglass filter.

Courtesy of AAF International.

FIGURE 20-2 Fiberglass filter media in cardboard frame.

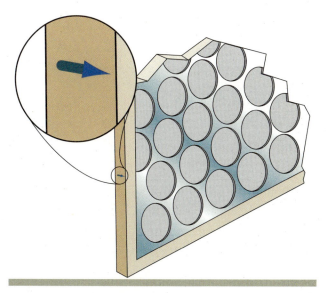

FIGURE 20-4 The arrow on the cardboard frame points in the direction of airflow.

DIRECTION OF AIRFLOW

FIGURE 20-5 Fiberglass filters become denser as air reaches the outlet of the filter.

small will permit air to bypass the filter, increasing the amount of dirt accumulation on the surface of the evaporator coil as well as on the blower blades. This also increases the amount of particulate matter that will be circulated back to the conditioned space as well as the amount of dirt accumulation in the duct system itself. Dirt accumulation in the ductwork can lead to an increase in mold growth since the dirt serves as food for the mold.

Framed fiberglass filters, when used in residential applications, are one inch thick and are designed to slide into a slot in the air handler or into an access panel installed in the duct system (Fig. 20-6). Filters can also be located within the occupied space at the return air grill. When used in this manner, the return grill is referred to as a **filter-backed return grill** (Fig. 20-7). The occupant of the space can easily access the filters in the return grill for both inspection and replacement. Fiberglass filters are not intended for permanent use

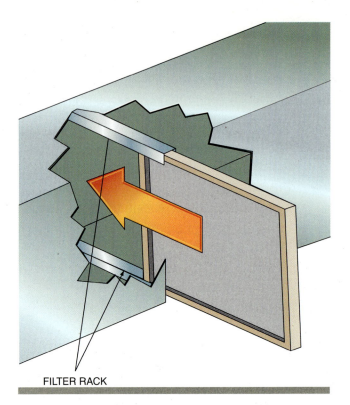

FILTER RACK

FIGURE 20-6 The filter should completely fill the rack to reduce air bypass.

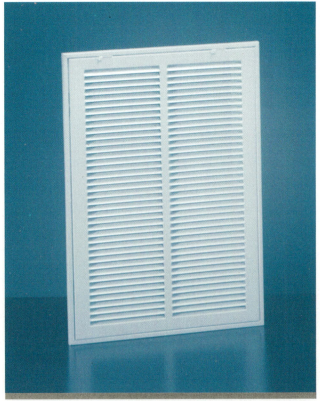

Courtesy of Hart & Cooley, Inc.

FIGURE 20-7 Filter-backed return grill.

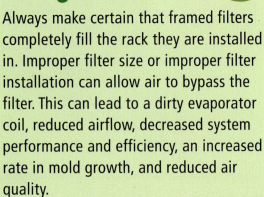

Going Green

Always make certain that framed filters completely fill the rack they are installed in. Improper filter size or improper filter installation can allow air to bypass the filter. This can lead to a dirty evaporator coil, reduced airflow, decreased system performance and efficiency, an increased rate in mold growth, and reduced air quality.

and should be replaced when they become clogged with dirt. Attempting to clean a fiberglass filter will reduce its effectiveness.

EXTENDED SURFACE FILTERS

When additional filtration is needed or desired, **extended surface filters** (Fig. 20-8) can be used. They are made of a cotton material that provides for higher air-filtering efficiency. Also referred to as **pleated filters** because of the configuration of the filter media, these filters are available in one-inch thicknesses so that they can fit into the existing filter rack or panel. These filters are desired in applications where filter access is limited as they do not get clocked as quickly as the non-pleated

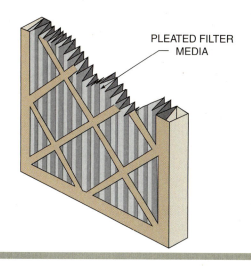

FIGURE 20-8 Extended-surface or pleated filter.

variety. In addition, the air passing through the filter is in contact with the filter medium for a longer period of time and will provide better air filtration than non-pleated filters.

STEEL FILTERS

Unlike fiberglass and pleated filters, **steel filters** (Fig. 20-9) are permanent and can be cleaned repeatedly. Although the initial cost of steel filters is higher than for disposable filters, they can actually result in monetary savings in the long run. Steel filters can be washed in the dishwasher and are often part of a complete air cleaning unit, such as an electronic air cleaner. Although these filters are most commonly found on commercial systems, many homeowners are opting for them, eliminating the need to replenish and maintain a stock of filters.

FIGURE 20-9 Permanent steel filter.

HIGH EFFICIENCY PARTICULATE ARRESTER (HEPA) FILTERS

High efficiency particulate arrester (HEPA) filters are becoming very popular despite their high cost. HEPA filters must be able to capture at least 99.7% of all particulate matter that is at least 0.3 microns in size. A micron is equal to one-millionth of a meter. For comparison, the width of a human hair is about 50 to 100 microns. HEPA filters were originally developed during World War II by the Atomic Energy Commission and were designed to remove and capture radioactive dust particles from the air that could escape and present a health hazard to the researchers working on the atomic bomb projects.

HEPA filters are rated by the percentage of particulate matter they can capture as well as the size of the particles they are intended to capture. To be classified as a "true" HEPA filter, it must be lab tested and labeled with information including:

- the serial number of the filter
- the percentage of particles captured
- the size of the particles

Manufacturers of true HEPA filters must test *all* of their filters and clearly affix the results of this test to the filter itself. Many filters on the market claim to be HEPA filters, but lack the required documentation to be legally classified as such. These filters are often referred to as HEPA-style filters.

Going Green

Although proper air filtration is very important, caution must be used when changing from one filter type to another. Filters with higher filtering capabilities offer more resistance to airflow. Greatly increasing the filtering capability of a filter therefore reduces the amount of airflow through the air distribution system of the equipment. This can lead to frozen evaporator coils in the cooling mode and overheating heat exchangers in the heating mode. In any event, reduced airflow can have negative effects on system performance and efficiency.

ELECTROSTATIC AIR FILTERS

Electrostatic air filters operate on the principle that unlike charges attract one another. The filter media holds an electric charge and attracts particulate matter with an opposite charge. Although electrical charges are often referred to when discussing electric current flow, electrostatic air filters do not need an electric power source to create the necessary charge. They are referred to as self-charging electrostatic filters. The filter media used in electrostatic filters has a natural, permanent charge. One commonly used media in electrostatic filters is polystyrene. When used as a shredded fiber, the polystyrene has about the filtering capacity as a standard, disposable filter but has the advantage of being able to capture smaller, charged particulate matter that a normal filter would permit to pass through.

One of the disadvantages of electrostatic filters is that the efficiency of the filter decreases as the relative humidity of the air passing through the filter increases. This is primarily because the charge held by the filter media as well as the charge of the particulate matter in the air both become weaker as the humidity increases.

ELECTRONIC AIR FILTERS

Electronic air filters, also referred to as electronic air cleaners, typically use a multistage filtration process to remove particulate matter as well as odors from the air as it passes through the cleaner (Fig. 20-10). Electronic

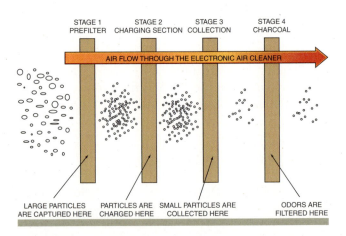

FIGURE 20-10 Four stages commonly used on electronic air cleaners.

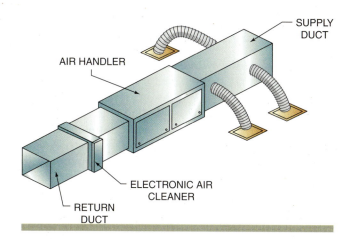

FIGURE 20-11 Electronic air cleaner located in the return duct of the central system.

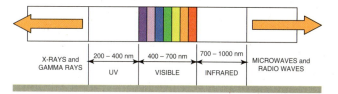

FIGURE 20-12 Diagram showing the wavelength of UV light with respect to that of visible light.

FIGURE 20-13 An Ultraviolet (UV) air treatment unit.

air cleaners are located in the return duct, as are other filters (Fig. 20-11). As air enters the filter, it passes through a prefilter that captures the large particles of dirt and dust that are present in the air. In the second stage of the filtering/cleaning process, the air passes over a series of electrically charged panels. During this stage of the process, particles that were small enough to pass through the prefilter are given an electric charge. The next stage is comprised of a series of collection plates carrying an electric charge that is opposite to the charge transferred to the particles in the previous stage. As the air and dirt pass over the collection plates, the charged dust particles cling to the collection plates, thereby removing them from the air. The final stage is made up of a charcoal filter that removes odors from the air. Once the air has passed through the electronic air cleaner, it is circulated back to the occupied space.

ULTRAVIOLET (UV) AIR CLEANERS

Ultraviolet, UV, light represents a very specific range in the light spectrum. Ultraviolet light has a wavelength that ranges from 200 **nanometers** to 400 nanometers. The nanometer, nm, is a unit that represents a length equal to 0.000000001 meter. To put it in perspective, the visible light spectrum ranges from 400 nm to 700 nm, the infrared spectrum starts at 700 nm and goes up from there, and just below the UV spectrum is the microwave portion of the spectrum (Fig. 20-12). The lower end of the UV spectrum represents light waves that we refer to as UV-C waves, and it is these light waves that help clean the air in forced air heating and cooling systems.

Ultraviolet air treatment systems (Fig. 20-13) are used on forced air conditioning or heating systems and

are most effective when used in conjunction with a high-efficiency air filtration media. The UV-C light has the ability to penetrate the walls of microbes that can have negative effects on indoor air quality. Once penetrated, the UV-C light can damage the DNA of the microbe,

CAUTION

CAUTION: Although it cannot be seen, ultraviolet light can cause temporary or permanent loss of vision. Never look at UV lights when they are illuminated.

Going Green

Ultraviolet lamps contain mercury, which can pose many serious health problems following prolonged exposure. Be sure to dispose of used lamps properly. Do not place them in the trash. Never use a household vacuum to clean up debris from a broken lamp.

killing smaller ones and preventing larger ones from reproducing. The location of the UV light is very important and the manufacturer's guidelines should be followed. These devices are typically positioned either close to the air filter or the evaporator coil, as both of these are prime locations for microbes such as mold to accumulate.

CAUTION

CAUTION: UV lights can cause sever burns. Before removing or handling the lamps in a UV air cleaning system, be sure to disconnect power to the unit and wait at least 15 minutes to minimize the chance of personal injury.

HUMIDITY AND HUMIDIFICATION

As discussed in earlier chapters, the purpose of the evaporator in an air conditioning system is to remove both latent and sensible heat from the air as it passes over the coil. Removing sensible heat reduces the measurable temperature of the air, while the removal of latent heat lowers the moisture content of the air. The amount of water vapor in the air is referred to as **humidity**. It can then be concluded that air conditioning systems lower the humidity of the space that is being conditioned. The term **relative humidity** refers to the amount of moisture an air sample holds, compared with the maximum amount of moisture that the same sample could hold under the same conditions. As the temperature of the air sample increases, the amount of moisture that a given air sample can hold is greatly increased. For example, if a given air sample holds 1 gram of water per cubic foot at 40 degrees and has the ability to hold 2 grams of water under the same conditions, the relative humidity will be ½ or 50%. If the air sample is heated, it may be able to hold 4 grams of water. This will cause the relative humidity to drop to ¼ or 25%.

During the cooler months when heating systems are in operation, the temperature of the conditioned space is increased, thereby increasing the amount of moisture the air can hold, resulting in a lower relative humidity. Lower humidity levels can result in drier skin, brittle hair, cracking of wood furniture, and the like. For this reason, it is beneficial to add moisture to the air to keep the relative humidity levels in the occupied space within an acceptable range, which is typically between 35% and 45% for outside ambient temperatures in the 30-degree range. **Humidification** is therefore the process of adding moisture to an air sample.

HUMIDIFIERS

Humidifiers are the system components that add moisture to the air and are classified based on the means by which they add moisture to the air. The two most common methods are atomizing the moisture and evaporating the moisture. **Atomizing** refers to the process by which the water is sprayed as a fine mist into the air. This process creates very small droplets of liquid that are easily absorbed by the air passing through the humidifier. Evaporative humidifiers operate on the principle that moisture will be readily absorbed by drier air. Quite often humidifiers are equipped with a foam or fabric **wick** that increases the surface area of the water, speeding the evaporation process. The wick is also referred to as the **media** or **element** by some manufacturers and contractors.

FREESTANDING, SELF-CONTAINED HUMIDIFIERS

Commercially available self-contained humidifiers are equipped with a reservoir that must be manually refilled when the water evaporates. Freestanding humidifiers contain small fans that circulate air through the humidifier to speed the evaporation/humidification process. These humidifiers do not have any specific installation procedures as they are simply placed in the desired location and plugged into a nearby outlet or receptacle.

Many self-contained humidifiers operate by atomizing the water in the reservoir against plates or pads that absorb the moisture. As air passes over these pads, the moisture is absorbed by the drier air, increasing the humidity level of the air.

DUCT-MOUNTED HUMIDIFIERS

Although freestanding humidifiers are very popular, most individuals with central air conditioning and heating systems prefer to install humidifiers in the ductwork of their central systems. The most common types of duct-mounted humidifiers are the plenum-mounted humidifier and the bypass humidifier.

PLENUM-MOUNTED HUMIDIFIERS

Plenum-mounted humidifiers (Fig. 20-14) can be mounted in either the supply or return duct of the central system. When installed in the return duct, moisture is added to the air before it passes through the heat

FIGURE 20-14 Plenum-mounted humidifier.

Courtesy of EWC Controls.

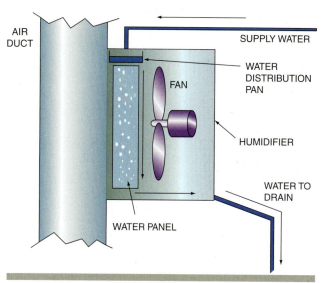

FIGURE 20-16 This humidifier uses a fan to move moisture-laden air into the supply duct. It also has a constant flow of water through the media.

exchanger. When installed in the supply duct, the air is heated first and then moisture is added to it. Although many manufacturers specify that plenum-mounted humidifiers can be installed in either the supply or return ducts, some only recommend installation in the supply duct. One of the main reasons for this is that, as stated earlier, warmer air has the capacity to hold more moisture than cooler air. If the air is heated prior to the introduction of moisture, it will typically absorb more moisture, increasing the rate at which moisture is added to the air. Plenum-mounted humidifiers are often equipped with either a rotating media (Fig. 20-15) or a propeller-type fan (Fig. 20-16) that blows moisture-laden air into the duct.

FIGURE 20-17 Evaporator wheel assembly.

Courtesy of EWC Controls.

Humidifiers with rotating media are equipped with an evaporator wheel (Fig. 20-17) connected to a small motor that rotates whenever the humidifier is energized. The media, which is the humidifier pad (Fig. 20-18), is placed around the wheel like a sleeve. The evaporator wheel and pad are positioned so the media is kept wet by water in a reservoir at the bottom of the humidifier (Fig. 20-19). As the pad rotates, it continuously picks up water from the reservoir, providing moisture to the air in the plenum. The water level is maintained by means of a float valve (Fig. 20-20) that operates much like the float valve found in many residential toilet tanks. One of the drawbacks of this type of humidifier is that, as the water vaporizes, any minerals that are present in the water will remain behind and accumulate in the humidifier.

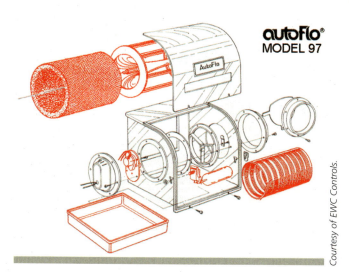

autoFlo®
MODEL 97

FIGURE 20-15 Exploded view of plenum-mounted humidifier and rotating media pad.

Courtesy of EWC Controls.

FIGURE 20-18 Humidifier pad.

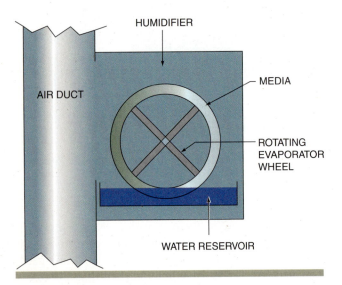

FIGURE 20-19 The media is kept wet by rotating in the water reservoir.

Humidifiers equipped with fans are typically equipped with media through which water flows continuously. These are referred to as **flow-through humidifiers**. Water is supplied to the top of the humidifier where it enters a distribution pan. This pan distributes water over the top of the media and gravity pulls the water through

FIGURE 20-20 Float valve assembly.

it. After passing through the media, the water is directed to the drain on the unit. One of the benefits of this type of humidifier is that the running water helps carry any mineral deposits down the drain (Fig. 20-16). These humidifiers are often equipped with solenoid valves that open when power is supplied to the humidifier and close when the humidifier is not in operation.

BYPASS HUMIDIFIERS

The airflow through **bypass humidifiers** (Fig. 20-21) is achieved by the difference between the pressure in the supply duct and the pressure in the return duct. Bypass humidifiers can be mounted in either the supply or return duct, but a section of piping material must be run between the humidifier and the plenum of the other duct (Fig. 20-22). In this figure, the bypass humidifier is mounted in the supply duct, but a section of piping has

FIGURE 20-21 Bypass humidifier.

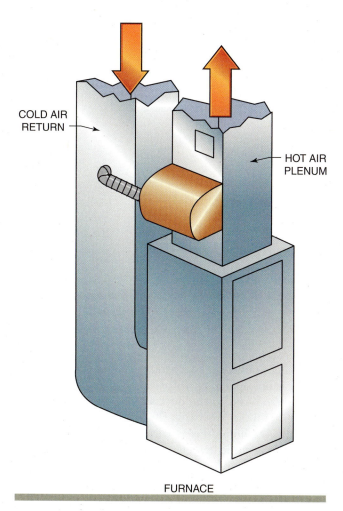

COLD AIR RETURN

HOT AIR PLENUM

FURNACE

FIGURE 20-22 Bypass humidifier installation.

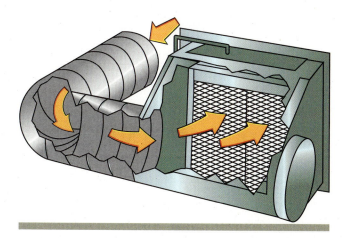

FIGURE 20-23 Airflow through a bypass humidifier.

been run from the return duct to the humidifier. A small amount of air passes from the return duct, through the humidifier, and back to the conditioned space through the supply duct (Fig. 20-23).

INSTALLING AIR QUALITY DEVICES

When installing air quality devices, it is important to follow the instructions supplied by the manufacturer. Each manufacturer has specific guidelines to help ensure that the particular piece of air quality equipment will function as desired. When installing air filters, electrostatic filters, or electronic air cleaners, it is important that they are properly sized so they will fit into the air-distribution system. When installed as part of new construction, the ducts are designed and fabricated to accommodate the air cleaning equipment. For example, wider, extended-surface filters may require special cabinets and tracts to support the filters properly and to reduce the amount of air permitted to bypass the filters. Electronic air cleaners are designed to be installed in the return duct and manufacturers have their own specifications regarding the appropriate distance between the cleaner and the air handler. Sections of ductwork must be fabricated to provide the necessary spacing. Electronic air cleaners also require power supplies, which must be installed according to unit specifications and local electric codes.

Water supplies, drains, and, in most cases, electric power supplies are needed when installing humidifiers. Providing a water supply to the humidifier is usually not a difficult task, as the amount of water required for humidifier operation is relatively small. The water supply is most often tapped off a nearby domestic water line by means of a saddle valve. **Saddle valves** (Fig. 20-24) are two-piece valves that are clamped around a water

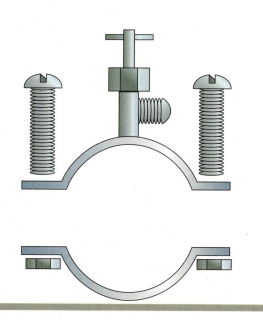

FIGURE 20-24 A saddle valve.

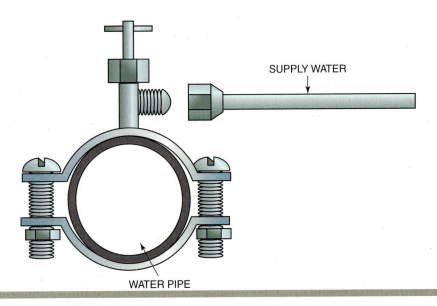

FIGURE 20-25 Water connections on a saddle valve.

line with nuts and bolts. A supply line is then connected between the saddle valve and the humidifier with flare nuts (Fig. 20-25). Typically, this line is no larger than ¼″ O.D. After the valve is secured around the pipe, and the water line has been connected between the valve and the humidifier, the pipe is pierced by tightening the handle on the valve. Opening the valve will start the flow of water.

CAUTION

CAUTION: Never pierce the water line until the line has been installed between the saddle valve and the humidifier.

The drain from the humidifier can be routed to a nearby floor drain, slop sink, or condensate pump. It is important to make note of the type of humidifier being installed and plan the drainage accordingly. Flow-through humidifiers will have a constant flow of water from the drain, but those equipped with a reservoir will have very little water flowing from the drain.

When supplying electricity to humidifiers, be sure that the voltage requirements for the humidifiers are met. Most humidifiers require 115-volt power supplies, but the solenoid valves and humidistats may very well be low-voltage devices. Refer to the product literature during the installation process. When connecting the electric circuits, make certain that the humidifier will be energized only when desired. For example, wiring an

atomizing humidifier to operate even when the system blower is not operating, will result in excessive moisture being introduced to the duct system. This can lead to accelerated mold growth in the air-distribution system. Typically, the humidifier is wired to operate only when the system is in the heating mode and the blower motor is running, although individual requirements may differ.

from experience...

Many new furnaces have humidifier and electronic air cleaner connections on the control board.

FRESH AIR AND VENTILATION

It has been proven that indoor air is more heavily contaminated than the air outdoors. For this reason, infants and the elderly who spend much of their time indoors are more likely to suffer from airborne illnesses and allergies resulting from dirt, dust, and other particulate matter in the air.

In residential applications, however, there is rarely any means by which fresh air is introduced to the occupied space through the air conditioning or heating system. Air leakage through windows and doors is usually sufficient to introduce a fair amount of outside air to the interior of the structure. Exhaust fans in kitchens, bathrooms, and attics are often used to remove excessive heat and odors from the structure. They also reduce the air pressure in the structure and increase the amount of air leakage into the structure through windows, doors, etc.

Whenever possible, windows should be left open to introduce fresh air to the occupied space. If this is not possible or desired, operating the exhaust fans on a regular basis will help introduce outside air to the space.

HEAT RECOVERY VENTILATORS (HRV)

In an effort to increase the efficiency of air conditioning and heating systems, homes have become much tighter than those constructed in the past. Partially because of more energy efficient windows, doors, and insulation, the amount of air leakage, both into and out of the structure, has been dramatically reduced. The benefits of tighter, better insulated homes include lower heating and cooling costs as well as a more comfortable environment due to reduced drafts. It is estimated that newer homes can operate with heating and cooling bills 25% to 50% lower than older homes. The major drawback with a tighter, better sealed structure is that the same air remains in the structure for a longer period of time, increasing the concentration of air contaminants and pollutants such as smoke, carbon dioxide, and other compounds found in cleaning supplies and building materials.

In order to ensure that there is ample fresh air and ventilation, many newer homes are equipped with mechanical ventilation systems that remove stagnant air from the structure and replace it with fresh outside air. Some local building codes even require that these systems be installed on all new residential construction projects. It has been determined by the American Society for Heating, Air Conditioning and Refrigeration Engineers, ASHRAE, that there should be a minimum of 0.35 air changes per hour in the structure to help ensure proper ventilation. To facilitate proper ventilation, heat recovery ventilators are often used. **Heat recovery ventilators**, HRV, are mechanical ventilation systems that perform a number of functions that include:

- removing stagnant air from the structure at a controlled rate
- introducing fresh air to the structure at the same rate at which the air is removed

- recovering the heat from the air being removed and transferring this heat to the entering fresh air in the heating season
- in some cases, recovering the humidity from the air that is removed from the structure and transferring the humidity to the fresh, entering air
- transferring heat from the intake air to the exhaust air while operating in the cooling season

from experience...

Heat recovery ventilators that have the ability to recover humidity as well as heat are referred to as energy recovery ventilators, ERV (Fig. 20-26).

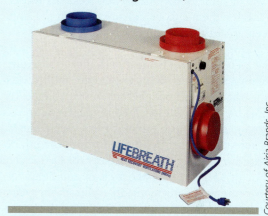

Courtesy of Airia Brands, Inc.

FIGURE 20-26 Energy recovery ventilator.

The heat recovery ventilators are referred to as balanced systems, meaning that the amount of air removed from the structure and the amount brought into the structure are the same. If the system was unbalanced and the amount of air introduced and removed were different, a number of problems could result. Removing more air from the structure can result in a negative pressure in the structure. This can cause improper venting of fossil-fuel system flue gases, resulting in the introduction of carbon monoxide to the occupied space. Introducing too much air into the home will create a positive pressure in the structure. This pressure can push moisture into the walls and ceilings, resulting in the slow but definite deterioration of the building materials.

HRV COMPONENTS

Heat recovery ventilators are made up of a number of components including:

- insulated fresh air ducts
- insulated exhaust ducts
- ductwork within the structure
- blowers to move air through the ventilator
- a heat exchange surface
- air filters on both the intake and exhaust ducts
- operating controls

HRV OPERATION

Consider the HRV system diagram in Figure 20-27. The ventilator has four duct sections connected directly to it. Two of these ducts are for the fresh air intake and two are for the air to be exhausted from the home. In the center of the unit, there is a heat exchange surface that allows the heat from the structure to be transferred to the incoming air. The heat exchanger, therefore, acts to remove heat from the exhausted air and transfer this heat to the incoming air before it is introduced to the occupied space. It is estimated that HRVs can recover approximately 75% of the heat from the exhausted air. Also note that the air is filtered in both the intake and exhaust ducts before the air can enter the heat exchanger. This helps keep the exchanger clean and helps the system operate more efficiently. In addition, there are two separate blowers in the system. One is for the intake section of the system and the other is for the discharge. In the months when the heating system is operating, heat is transferred from the exhausted air to the intake air, while during the cooling season, heat is transferred from the intake air to the air being exhausted.

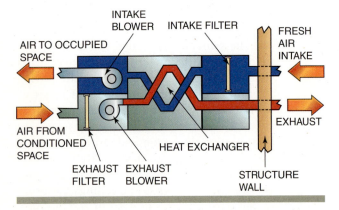

FIGURE 20-27 Heat recovery ventilator configuration.

DUCT SYSTEMS FOR THE HRV

Heat recovery equipment can be a completely separate system or can be incorporated into the forced-air heating system, if the home is heated with one. When installed as part of the forced-air system, the fresh air intake is connected to the return duct system on the furnace. The air is then heated and introduced to the occupied space by means of the furnace blower. If the ventilator is installed as a separate system, which is needed if there is no forced-air heating system in the home, supply and exhaust ducts must be installed.

HRV CONTROLS

Heat recovery ventilators can operate in a number of different modes to meet the needs of the structure. The system often has multiple speeds that allow the user to change the rate at which air is exchanged in the home. For example, when there are a lot of people in the home, a higher ventilation rate may be needed. Many HRVs are also equipped with a circulation mode that facilitates the circulation of air without exhausting any air to the outside. Other controls and features that can be found on HRVs include:

- maintenance lights to alert the home owner that service is needed
- clogged filter indicators
- timers to control the on and off cycles of the unit
- humidistats (found on ERVs)
- automatic changeover controls to switch system between high- and low-speed operation
- pollutant sensors

TROUBLESHOOTING AND MAINTAINING AIR QUALITY DEVICES

Air quality problems are often the result of improperly sized or maintained filtration, air cleaning, or humidification equipment. To keep air quality devices operating properly, preventive maintenance and inspections should be performed. Preventive maintenance should include the following:

✓ Inspect the filters. Partially blocked filters result in reduced airflow and reduced air movement through the occupied space, as well as reduced heating and/or cooling system efficiency.

✓ Clean electrostatic air filters. Always follow manufacturer's recommendations when cleaning electrostatic filter media. Dirty electrostatic filters reduce the filter's ability to capture particulate matter, increasing the amount of dirt and dust in the occupied space.

✓ Clean the charging section, accumulation plates, and charcoal filters of electronic air cleaners. These components must be kept clean to keep them operating in tip-top shape. Some manufacturers recommend cleaning the charging sections in a dishwasher but, as always, follow the individual manufacturer's recommendations.

✓ Inspect the interior of humidifiers. As water evaporates, minerals in the water remain behind. Accumulation of minerals within the humidifier can result in malfunctioning float valves, blocked drain lines, and improper water flow through the humidifier media. If the humidifier leaks, this is the place to start looking for the problem.

✓ Inspect/change humidifier media. The life expectancy of various humidifier media varies depending on a number of factors, including type of humidifier, humidifier run time, and type of media. Evaporator pads are typically made of foam and have a relatively short useful life. At a bare minimum, humidifier media should be replaced each heating season.

✓ Drain water from the humidifier. If the humidifier is intended to operate only during the winter months and the humidifier has a water reservoir, close the saddle valve on the water line and drain the water from the reservoir. Then clean the interior of the humidifier, replace the media, and remove any mineral deposits from within the unit.

SUMMARY

- The amount of air contamination and pollution can be reduced by removing the source of contamination, cleaning the air, or providing adequate ventilation.
- Mold can lead to increased allergic symptoms and its growth rate is affected by the amount of dirt and moisture in the air-distribution system and in the residence.
- Particulate matter is removed from the air by filtering.
- Filters can be made of foam, fiberglass, steel, or cotton material.
- Extended surface area filters and high-efficiency particulate arresters are more efficient than standard air filters.
- HEPA filters capture at least 99.7% of particulate matter at least 0.3 microns in size.
- Electrostatic filters rely on an electrical charge to capture particulate matter.
- Electronic air cleaners are multi-stage cleaners that not only filter the air but remove odors as well.
- Humidifiers add moisture to the air and can be freestanding units or incorporated into the central air conditioning and heating system.
- Humidity is the amount of water an air sample holds; relative humidity is the ratio comparing the actual moisture content of an air sample to the maximum amount of moisture the sample can hold.
- Humidifiers can be manufactured with reservoirs that hold water or with flow-through features that carry mineral deposits from the unit.
- Some humidifiers rely on fans to move air through them and others rely on the pressure difference between the supply and return ducts.

GREEN CHECKLIST

☐ Old chemicals and household products should be removed from the house and properly disposed of to help increase the air quality in the house.

☐ Air conditioning systems must be properly sized to ensure maximum comfort and system efficiency.

☐ Oversized air conditioning systems cause increased humidity levels in the house, which leads to an increase in the rate of mold growth in the structure.

☐ Maintaining relative humidity levels in the 50% range will help reduce the rate of mold growth.

☐ Install air filters correctly, with the arrow pointing in the direction of airflow.

☐ Use the correct size filter to prevent air from bypassing the filter.

☐ Ultraviolet light (UV-C) has the ability to penetrate and alter the DNA of microbes such as mold.

☐ UV lamps contain mercury and should be disposed of carefully and properly.

☐ UV light can cause severe burns. Only handle UV lamps when they are completely cool.

☐ UV lamps can cause damage to the eyes. Never look directly at illuminated UV lamps.

REVIEW QUESTIONS

1. The main characteristic of air contaminants is that
   a. They are much larger in size than pollutants
   b. They may or may not cause health-related problems
   c. They are only gaseous materials
   d. All of the above are correct

2. To reduce the amount of air contaminants and pollutants in the air
   a. The source of the contamination can be removed
   b. The air can be cleaned
   c. Ventilation can be provided
   d. All of the above are correct

3. **The rate of mold growth accelerates when the moisture level is above**
   a. 10%
   b. 30%
   c. 40%
   d. 60%

4. **Which of the following is a permanent-type air filter?**
   a. Fiberglass filter
   b. Foam filter
   c. Steel filter
   d. Extended-surface filter

5. **When installing air filters**
   a. Air should not be permitted to bypass the filter
   b. They can be located in the occupied space
   c. They can be located at the air handler location
   d. All of the above are correct

6. **Extended-surface air filters are typically made of**
   a. Foam
   b. Steel
   c. Cotton material
   d. Fiberglass

7. **HEPA filters must be able to remove at least _____ of all particulate matter at least _____ microns in size.**
   a. 97.7%, 3.0
   b. 99.7%, 3.0
   c. 97.7%, 0.3
   d. 99.7%, 0.3

8. **To be classified as a true HEPA filter**
   a. Eighty percent of the filters must be tested at the time of manufacture
   b. The filter must be labeled with the percent and size of particle captured
   c. The filter's tag must contain the composition of materials used in the manufacture of the filter
   d. All of the above are correct

9. **The efficiency of electrostatic air filters will**
   a. Remain constant as the relative humidity of the air increases
   b. Decrease as the relative humidity of the air decreases
   c. Decrease as the relative humidity of the air increases
   d. Increase as the relative humidity of the air increases

10. **The final stage of an electronic air cleaner is the**
    a. Pre-filter
    b. Collection plate
    c. Charging stage
    d. Charcoal filter

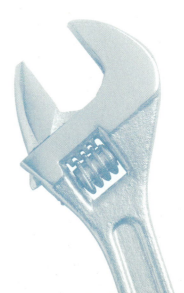

11. **As the temperature of the air in the occupied space rises**
    a. The relative humidity will fall
    b. The amount of water the air can hold will decrease
    c. The amount of humidity will remain the same
    d. Both a and c are correct

12. **If the relative humidity of an air sample is 100%**
    a. The humidity is greater than the amount of moisture the sample can hold
    b. The air can still hold more moisture under present conditions
    c. Additional moisture can be added to the air if the air is heated
    d. Both a and c are correct

13. **The recommended level of relative humidity in a residence when the outside temperature is 30°F is approximately**
    a. 10% to 20%
    b. 25% to 35%
    c. 35% to 45%
    d. 45% to 55%

14. **Atomizing humidifiers**
    a. Use a rotating media pad to facilitate the humidification process
    b. Spray a fine mist of water into the air supplied to the occupied space
    c. Are also referred to as evaporative humidifiers
    d. Are designed to operate when the system blower is not operating

15. **Duct-mounted humidifiers**
    a. Can be mounted in either the supply or return plenum
    b. Can have a reservoir to hold the water in the unit
    c. Can be designed to have a constant water flow through them
    d. All of the above are correct

16. **Flow-through humidifiers are often desirable because**
    a. They accumulate water at the bottom of the unit
    b. They reduce the amount of mineral accumulation in the unit
    c. They use substantially more water than reservoir types of humidifier
    d. All of the above are correct

17. **Bypass humidifiers**
    a. Require piping connections between the supply and return plenums
    b. Require a fan to move air across the humidifier media
    c. Must be mounted in the supply plenum
    d. Are typically easier to install than plenum-mounted units

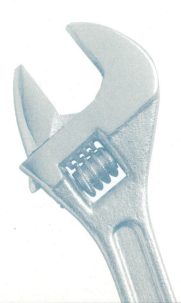

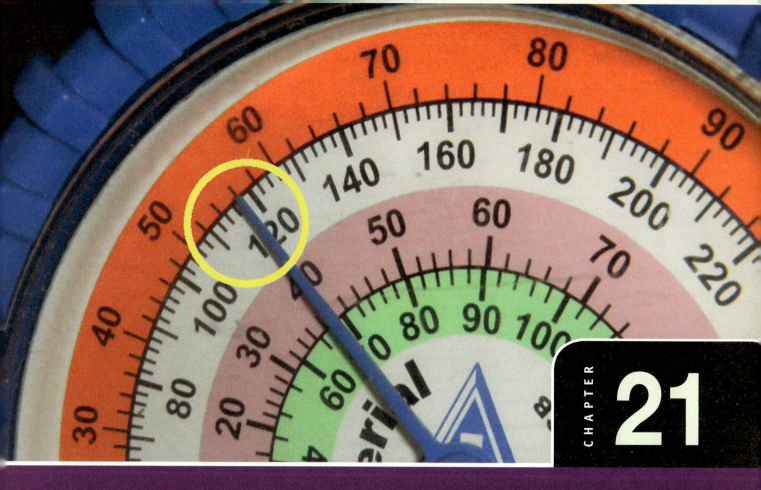

Mechanical Troubleshooting

OBJECTIVES   *Upon completion of this chapter, the student should be able to*

- List several common reasons why an air conditioning/refrigeration system may fail.

- List some reasons why refrigerant escapes from a refrigeration system.

- Identify the effects of a defective condenser fan motor on system operation.

- Identify the effects of a defective evaporator fan motor on system operation.

- Explain how a refrigerant overcharge affects system operation.

- Explain how a refrigerant undercharge affects system operation.

- Explain the importance of a correctly mounted thermostatic expansion valve thermal bulb.

- List the seven steps in performing a successful service call.

- Perform a service call using the seven step method.

- Determine if an air-conditioning system has an overcharge of refrigerant.

- Remove refrigerant from an overcharged air conditioning system.

- Determine if an air conditioning system has a refrigerant undercharge.

- Add refrigerant to an undercharged air conditioning system.

Having obtained a general understanding of the basic vapor-compression refrigeration system, the technician now needs to begin acquiring the skills that help reduce system down time when a system failure exists. Two important issues that separate the excellent service technician from the average technician are the speed and accuracy with which the technician is able to properly evaluate a system and determine the cause for system failure. Although the technician may not be able to immediately repair the problem, an important first step is to identify the problem and decide what path to take to resolve it. Good troubleshooting techniques enable the technician to do this.

An accurate diagnosis allows other company personnel to order needed parts, schedule the repair, get customer approval for the repair, and see it to completion. On the other hand, an inaccurate diagnosis can lead to ordering incorrect parts, increased repair time, lost company revenue, and, more important, a dissatisfied customer. This chapter deals with mechanical problems within the refrigeration circuit that prevent the system from operating properly. Mechanical problems include refrigerant charge related issues, physical damage to a particular component, foreign matter entering the system, improper installation, and airflow problems through either the evaporator or condenser coil.

GLOSSARY OF TERMS

Belt drive The blower arrangement where the blower is connected to the motor by means of belts and pulleys

Direct drive The blower arrangement where the blower is connected directly to the motor shaft

Drive pulley The pulley that is connected directly to the shaft of the motor

Driven pulley The pulley connected to the blower shaft

Endplay The side-to-side motion of a motor shaft that indicates that the motor bearings are damaged

Floodback The condition that exists when liquid refrigerant leaves the evaporator and flows back to the compressor

Flooded See overfed evaporator

Noncondensable gas Any gas that is unable to condense at the pressures and temperatures normally encountered in air conditioning and refrigeration systems

Overfed evaporator An evaporator that is being fed too much refrigerant; these evaporators operate with very low, or zero, superheat

Overload protector Internal thermal device used to protect motors from overheating

Permanently lubricated Term used to describe motors that typically will not require lubrication during their expected useful lives

Preventive maintenance Tasks, such as oiling motors and changing air filters, that are performed on a system to help reduce the chances of future system failure

Sleeve bearings Type of motor bearing often used on residential systems because of their quiet operation

Starved The condition that exists when an evaporator is fed less refrigerant than needed to provide adequate cooling; also known as an underfed evaporator

System charge The total amount of refrigerant that an air conditioning system holds

Temperature differential Term used to describe the difference between two temperatures

Ultraviolet (UV) solution Dye additive that is introduced to air conditioning systems for leak detection purposes.

EVAPORATOR AND CONDENSER FAN MOTOR PROBLEMS

Motor problems can be either electrical or mechanical in nature. Electrical problems are covered later in the text. The most common mechanical problems with motors arise from

- improper airflow
- improper lubrication
- improper pulley alignment
- improper belt tension

Proper **preventive maintenance** on an air conditioning system should help eliminate mechanical motor failure. It is also important that the proper type of motor be used for the specific application. For example, if a motor is to operate in a wet environment, it should be sealed to prevent water from entering it. If an open motor is used instead of a sealed motor, its expected service life will be greatly reduced. A motor that is used within its design and application range will provide years of satisfactory operation; one that is properly maintained will last even longer. Motors on the air handlers of residential air conditioning systems are usually open-type devices, but those used on condensing units and evaporative coolers are sealed to prevent water from entering the shell of the motor.

IMPROPER AIRFLOW

Motors may exhibit symptoms that lead the technician to conclude that the device is defective when, in fact, another system problem is at fault. A common factor that leads to the unnecessary replacement of motors is insufficient airflow. Many motors rely on air passing over them to help keep them cool and within safe operating ranges. If the airflow is restricted, the motor may overheat and cut off on an internal **overload protector**. The overload protector is a device that will open its contacts and de-energize a motor when temperatures exceed the design ratings. The technician may conclude that the motor is not operating properly, causing the reduction in airflow, when the opposite is actually the case. Before concluding that the motor is defective, the technician must first establish that the air path is not restricted. To accomplish this, the following must be checked.

For an *evaporator fan motor,* make certain that

✓ *All air filters are clean.* If the system has no air filters, the return side of the evaporator coil must be visually inspected.

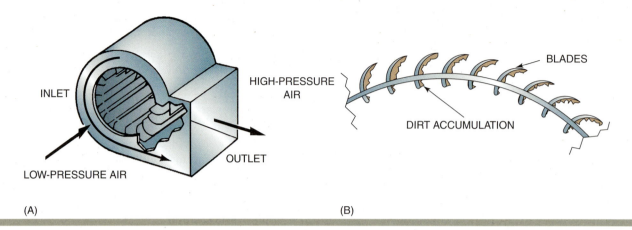

(A)   (B)

FIGURE 21-1 (A) Forward-curve centrifugal blower and housing assembly. (B) Dirt accumulation on the blades of the blower reduces the ability of the blower to move air.

✓ *The evaporator coil is clean.* Dust and condensation accumulating on the coil will form a glue-like substance on the coil surface that restricts airflow. Since the evaporator coil is usually inaccessible from both sides, shining a light through the coil to visually inspect it may be possible. If the light shines through the coil, it should be clean.

✓ *All duct lining is intact.* If not properly installed, acoustical lining has a tendency to come loose and block the air stream. Duct lining can often be found in both the supply and return ducts.

✓ *The return-air grills are not blocked with furniture, boxes, or other obstructions.*

✓ *The supply registers are open and not blocked.* Closed or blocked supply registers will restrict airflow.

✓ *The fan blades are clean and not caked with dirt.* If the blades are caked with dirt (Fig. 21-1) the ability of the blade to scoop and move air is greatly reduced. This usually occurs on the blades of forward-curved centrifugal fans, or squirrel cages.

For a *condenser fan motor,* make certain that

✓ *The condenser coil is not dirty and/or blocked.* If the condensing unit is located outside, leaves, dirt, bushes, and other debris must be cleared from the coil's surface. It is good field practice to mount the condensing unit on a pad or frame (Fig. 21-2) that will lift it above ground level.

✓ *The condensing unit has at least as much clearance as indicated by the manufacturer.* If too little space is provided, the hot discharge air may recirculate back through the coil (Fig. 21-3).

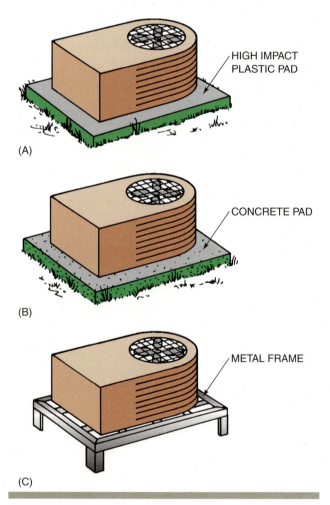

FIGURE 21-2 Condensing units should be mounted level on pads.

✓ *If multiple condensing units are located next to each other, they do not discharge into the other units.* This can cause high-pressure-related problems and inefficient cooling.

✓ *The fan blades are clean and not caked with dirt.* If the blades are caked with dirt, the ability of the blade to scoop and move air is greatly reduced.

✓ *The condensing unit is not located under any low overhangs or other overhead obstructions.* This will cause the condenser's discharge air to recirculate through the coil (Fig. 21-4).

Once the air path has been cleared, the motor should operate properly. If the air path is unobstructed to begin with, other mechanical problems may be the cause of a motor's premature failure.

IMPROPER MOTOR LUBRICATION

Depending on the type and style, motors may require periodic lubrication. Motors installed in equipment that is not easily accessible are usually **permanently lubricated** and do not need periodic oiling. These motors are rare in residential applications. Most motors on residential systems use **sleeve bearings**, which tend to be very quiet in operation, and should be oiled as part of annual or semi-annual preventive maintenance. The type and amount of

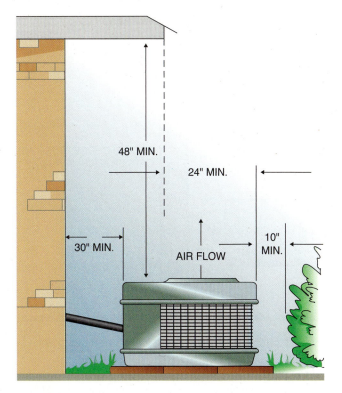

FIGURE 21-3 Sufficient clearance should be provided around the condensing unit to allow proper airflow through the coil.

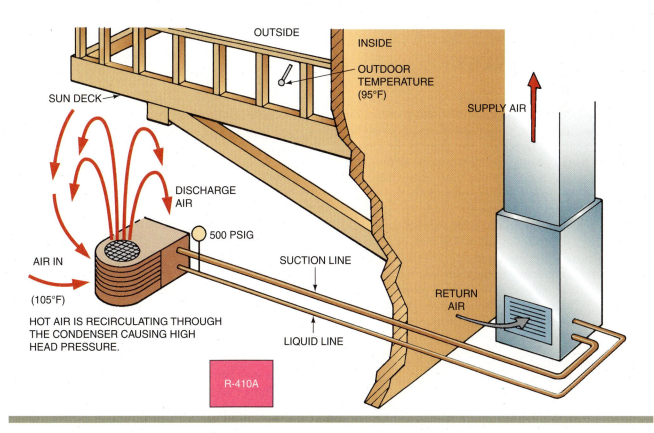

FIGURE 21-4 Low overhangs can cause the discharge air from the condensing unit to recirculate through the coil.

oil needed are usually specified by the manufacturer. If no lubrication data are available, four or five drops of a medium-grade, nondetergent-lubricating oil should be used in each oil port. Motors with sleeve bearings are usually located near the occupied space because of the low noise level and easy access.

Unfortunately, many equipment owners do not call for service until a system problem arises. Improper motor lubrication will become evident if the motor does not turn freely when all pulleys, if there are any, have been removed. Saving the motor may be possible by properly oiling and lubricating the bearings, provided the motor has not operated for an extended period of time without proper lubrication. If the motor has been overheating and bearing damage has occurred, the motor will need to be either repaired or replaced, depending on the size of the motor. Motors on residential systems are usually replaced, since rebuilding smaller motors is not economically feasible.

To determine if excessive bearing damage has occurred, the shaft of the motor should be inspected. The shaft should have some play in and out of the motor but almost no play from side to side. This side-to-side play is referred to as **endplay**. If there is endplay, the bearings of the motor are no longer functional and the motor must be replaced. Oiling or lubricating a motor that has defective bearings will not fix the problem.

IMPROPER PULLEY ALIGNMENT

The majority of residential air conditioning system air handlers use **direct drive** blower assemblies, in which the blower is connected directly to the shaft of the blower motor. On these systems, the blower turns at the same speed as the motor. Some systems, however, are manufactured with a pulley arrangement that allows the blower to turn at a different speed than the motor. This arrangement is called a **belt drive**. Such systems are equipped with a **drive pulley** and the **driven pulley**. The drive pulley is the pulley connected directly to the shaft of the motor, and the driven pulley is connected to the shaft that turns the fan or blower. If these pulleys are not aligned properly, excessive pulley wear and motor bearing damage can occur. Blowers on evaporative coolers are often configured with a belt driven blower assembly.

Excessive pulley wear can be identified when an actual groove has been worn into the interior wall of the pulley. If inspecting the pulley is visually difficult, feeling the groove of the pulley will quickly indicate any imperfections in its surface, which should normally be flat with no waves, gaps, or notches. When pulley wear is observed, the pulley must be replaced. Upon replacement

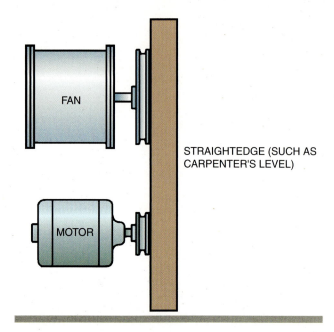

STRAIGHTEDGE (SUCH AS CARPENTER'S LEVEL)

FIGURE 21-5 Pulleys must be properly aligned.

of the pulley, the alignment between the drive and driven pulley must be checked with a straightedge (Fig. 21-5).

Bearing damage, as stated earlier, can be identified by side-to-side play in the motor shaft. Again, after the repair or the replacement of the motor has been performed, the pulley alignment must be carefully checked to reduce the possibility of another failure.

Key indicators that pulleys are not aligned properly include

- uneven belt wear
- excessive belt breakage
- belts coming loose from the pulleys without breaking

Normal preventive maintenance should always include belt inspection. Any imperfections or uneven wear patterns in the belts should alert the technician to a potential problem. Changing the defective belt is only part of what should be done at this point. The cause of the uneven wear must be identified and corrected as well. The term *preventive maintenance* implies maintenance that is geared toward the prevention of future system problems. Because system downtime is always shorter during a preventive maintenance service call than during an emergency repair service call, system downtime is greatly reduced when proper preventive maintenance is performed.

Excessive belt breakage should also alert the technician that an adjustment problem exists. If the broken belt shows uneven wear or imperfections, chances are an alignment problem exists. If the belt shows even wear, the belt may have been adjusted too tightly. Improper belt tension is discussed in the next section. In either case, the cause for the breakage must be identified and corrected to prevent future failure.

On occasion, a technician will arrive on a service call to find that the belt has come loose from the pulleys. This is usually the result of a pulley misalignment and a belt having been installed too loosely. This problem is more common than one may expect, for a very simple reason. When pulleys are misaligned, the belt tends to squeal or make a loud whining sound. To alleviate this sound, some technicians loosen the belt tension to reduce the noise level. This remedies the symptom but not the problem. The belt may now come loose from the pulleys very easily, leading to a "no-cooling" call sometime in the future.

IMPROPER BELT TENSION

Once the pulleys are aligned correctly, the belt tension must be checked. The rule of thumb most field technicians use is that there should be less than one inch of play in the belt when the belt is installed correctly. Incorrect belt tension can lead to a number of system problems including the following:

Belt Tension Too Tight

- Bearing damage
- Excessive belt breakage
- Excessive noise levels
- Motor overheating, premature motor failure

Belt Tension Too Loose

- Belt slippage
- Reduced cooling
- Reduced airflow (possibly leading to low suction pressure and a frosted evaporator coil)
- Belt coming loose from the pulleys
- Excessive pulley wear
- Motor overheating

The service technician can easily determine if the belt tension is too loose by inspecting the grooves of the pulley. If the grooves have been polished to a near-mirror finish, the belt is slipping and is too loose. The belt tension must be increased, and the pulleys should be replaced. The reason for replacing the pulleys is simple. If the next technician arrives at the location and sees the mirror-like surface, the tension may be increased when, in fact, the tension was correct to begin with. In addition, pulleys tend to grip belts better when the groove surfaces are not perfectly smooth. If the problem is caught early enough, a simple adjustment of the belts and pulleys can eliminate a very costly repair and excessive system downtime.

from experience...

When checking a belt, it should be completely removed from the pulleys by loosening the pulley arrangement or motor mounts. Turning the belt inside out will make inspecting the belt easier. Figure 21-6 shows a good belt; Figure 21-7 shows a cracked belt.

Delmar/Cengage Learning.

FIGURE 21-6 A good belt.

Delmar/Cengage Learning.

FIGURE 21-7 This belt is cracked and needs to be replaced.

CAUTION

CAUTION: When working on or around belts and pulleys, always keep the following in mind:

- Never try to stop a motor or blower by hand.
- Never grab a belt that is turning.
- Always keep fingers away from the area between the belt and the pulley.
- Do not wear loose-fitting clothing, especially neckties, when working around moving machinery.

REFRIGERANT CHARGE RELATED PROBLEMS

For an air conditioning system to operate properly, the amount of refrigerant in the system—referred to as the system charge—must be correct. The field technician must be familiar with the system and all of its components to ensure that the amount of refrigerant added to the system is well within the acceptable range. Improper refrigerant charge can lead to many system problems, including the following:

Excessive Refrigerant Charge (System Overcharge)

- Reduced condenser efficiency
- Reduced evaporator efficiency (reduced cooling)
- Reduced system efficiency (increased power consumption)
- Reduced evaporator superheat
- Increased possibility of floodback to compressor (liquid refrigerant getting back to the compressor)
- Higher operating pressures
- Higher compressor operating temperatures

Low Refrigerant Charge (System Undercharge)

- Reduced evaporator efficiency (reduced cooling)
- Increased evaporator superheat
- Low operating pressures
- Reduced system efficiency (increased power consumption)
- Higher compressor operating temperatures

SYSTEM PRESSURE READINGS

The installation of a set of refrigeration gauges on the system will provide the technician with information about what is taking place inside the system. The outer, black-numbered scales on the face of the gauges provide the pressure in pounds per square inch gauge (psig). The inner scales show the corresponding saturation temperatures for various refrigerants.

The data on the gauge faces are exactly the same as the data found in the pressure/temperature chart. This valuable data, if interpreted correctly, will help lead the technician to a rapid and accurate system diagnosis. The gauge manifold enables the technician to

- read the system's high-side pressure (also referred to as head pressure)

- read the system's low-side pressure (also referred to as suction pressure or back pressure)
- read the condenser saturation temperature
- read the evaporator saturation temperature
- determine amount of condenser subcooling
- determine evaporator superheat

If the system is equipped with Schraeder valves, the high- and low-side pressure readings of a system can be obtained by connecting the appropriate gauge hose to any available port in the system. When the valve stem is pushed in, the valve opens. When pressure on the stem is released, the valve closes. On residential systems the high-side service port is most commonly located on the liquid line leaving the condensing unit. The low-side service port is commonly located on the suction line at the condensing unit. Hoses on the gauge manifold are equipped with pin depressors that will push in the pins on the service ports, enabling the technician to obtain immediate system pressure readings.

Pressure readings can be taken whether or not the system is operating. Both situations provide the technician with useful troubleshooting information. Taking pressure readings on an inoperative system will help to:

- determine if the refrigerant charge has been completely lost
- determine if noncondensable gases are in the system

Taking system pressure while the system is operating will enable the technician to:

- determine evaporator and condenser saturation temperatures and pressures
- help evaluate evaporator and condenser effectiveness and efficiency
- help evaluate the pumping effectiveness of the compressor
- determine if the refrigerant charge needs to be adjusted

COMPLETE LOSS OF REFRIGERANT CHARGE

If, upon installation of the manifold, the gauges register 0 psig, the system has lost its charge and is said to be flat. The leak must be located and repaired, before evacuating and recharging, in order to resume normal system operation. Systems develop leaks for a number of reasons. Some of the most common follow:

- Refrigerant piping surfaces that rub together and wear holes in the lines

- System vibration that causes stress fractures in fittings and lines
- Poor-quality soldered or brazed joints
- Refrigerant lines inadvertently getting damaged or cut

A number of different methods can be used to locate the leak, including

- listening for an audible hiss from the leak
- using a liquid leak detector, or bubble solution
- using a halide torch leak detector
- using an electronic leak detector
- using an ultraviolet (UV) light

Courtesy of Uniweld Products.

FIGURE 21-9 The flame on a halide leak detector turns green when a leak is present.

THE AUDIBLE HISS

If the system has a total loss of charge, chances are the leak is very large. Quite often, a fitting has cracked, a solder joint has come loose, a compressor terminal has blown off, or some other system fracture has developed. For this reason, it may be very easy to locate the leak by pressurizing the system with dry nitrogen and listening for the escaping gas.

from experience...

Do not pressurize the system with refrigerant for leak-detection purposes. This will violate laws set forth by the Environmental Protection Agency (EPA).

BUBBLE SOLUTIONS

One very popular method of leak detection is the use of a soap-bubble solution. By pressurizing the system with dry nitrogen and applying the solution to all solder joints and seams in the refrigerant piping circuit, the leak will be identified by the formation of bubbles (Fig. 21-8). This method is very popular mainly because smaller leaks, which are generally harder to locate, tend to produce the largest and most visible bubbles.

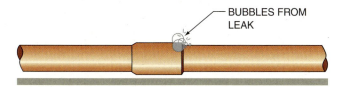

BUBBLES FROM LEAK

FIGURE 21-8 Soap bubble solution causes bubbles to form when a leak is present.

HALIDE TORCH LEAK DETECTORS

Halide torch leak detectors operate on the concept that burning refrigerant will cause a flame to change color. The device usually uses propane, or mapp gas; an open flame; and a hose that is guided over the piping in search of the leak. The flame gets a portion of its needed oxygen through the rubber hose. Refrigerant is pulled through the tubing with the surrounding air, and when a leak is reached, the flame on the device will change color (Fig. 21-9). This type of leak detector is more effective when a larger leak exists. It is a very slow method but still relatively popular, partly because the equipment is relatively inexpensive and the propane refills cost only a few dollars.

Going Green

Halide leak detectors can only be used in conjunction with chlorinated refrigerants.

ELECTRONIC LEAK DETECTORS

Technological advances over the years have made these devices more reliable and accurate. To use the electronic leak detector, the system must contain at least some refrigerant. If the system is operating with a small leak, plenty of pressurized refrigerant is left in the system. If, however, the system is flat, or void of refrigerant, it must first be pressurized with nitrogen, to which a trace of refrigerant has been added. The refrigerant that can be added to the system can be the same refrigerant that the system will normally operate with. For example, if the system being worked on contains R-410A, then a trace of R-410A can be added to the system along with the nitrogen for leak detection purposes. The trace gas must be added since the electronic leak detector will not detect the nitrogen by itself. Electronic leak detectors can detect several different refrigerants, so the literature accompanying the detector should be checked to make certain that the detector being used will indeed detect the refrigerant in the system.

Going Green

Much discussion often arises about trace gas usage and how it is outlined in Section 608 of the Clean Air Act. Many believe that the only refrigerant that can be used in systems as a trace gas for leak detection purposes is R-22. The Clean Air Act does not state this and does not specify which refrigerants can or cannot be used as a trace gas. According to the EPA, the trace gas can be whatever refrigerant the system will contain during normal system operation.

ULTRAVIOLET LEAK DETECTORS

Another popular leak detection method is the UV leak detection system. This system uses a UV light, which causes a refrigerant additive to glow when it is present on the surface of a refrigerant pipe. This additive is introduced into the system either during initial installation or at a later time. If a leak has developed, the technician simply shines the light on the refrigerant piping, which will glow when the leak has been located (Fig. 21-10).

Courtesy of Uniweld Products.

FIGURE 21-10 The pipe glows under ultraviolet light when a leak is present.

This method is very quick and can even locate leaks in the center of evaporator and condenser coils. This method of leak detection tends to be the most expensive from the "first cost standpoint" but tends to pay for itself in a very short time by locating leaks faster. One major drawback with the UV leak-detection method is that its effectiveness is greatly reduced when used outdoors in bright sunlight.

NONCONDENSABLE GAS

Technicians define a **noncondensable gas** as one that cannot condense into a liquid at all, or at least within the normal operating pressure ranges of the equipment they work on. Since these gases cannot condense into a liquid, they simply take up useful space in the condenser, reducing the condenser's effective surface area. This causes the system operating pressures to rise. (On systems with an automatic expansion valve, the low-side pressure will remain the same while the head pressure increases.) Two common non-condensable gases are:

- air
- nitrogen

AIR IN THE AIR CONDITIONING SYSTEM

Air can find its way into an air conditioning system in a number of different ways. The most common are:

- improper system evacuation
- insufficient evacuation time
- low-pressure-side leak

A common mistake made during the system evacuation process is not opening both valves on the gauge manifold. Quite often, the system is evacuated from either the high- or low-pressure side, but not both. To help ensure proper system evacuation, the system should be evacuated from both the high- and low-pressure sides of the system. The technician should therefore check to make certain that the high-side valve and the low-side valve are both in the open position during evacuation.

Another common mistake relates to insufficient evacuation time. The vacuum pump should be permitted to operate for as long as possible to ensure that the system is properly evacuated. During the evacuation process, the field technician can isolate the vacuum pump from the system and micron gauge to determine the status of system evacuation. Upon isolating the vacuum pump from the system, the technician may notice

- the micron gauge reading rises from a deep vacuum reading to ATM or atmospheric pressure
- the micron gauge reading rises from a deep vacuum (in the acceptable micron range) to an unacceptable vacuum level

- the micron gauge reading rises from a deep vacuum (in the acceptable micron range) to another vacuum level still in the acceptable range

In the first instance, there is a system leak. The vacuum pump should be turned off and the system should be pressurized and leak-tested again. In the second instance, the process of system evacuation is still underway. In this case, the vacuum pump should be restarted, and the evacuation process should be continued. In the last instance, the evacuation process is complete. There is no system leak and the system is maintaining the vacuum.

Improper gauge installation and removal can also allow air to enter the air conditioning system. All hoses on the gauge manifold should be properly purged to prevent the introduction of air into the refrigerant lines. Proper purging entails pushing a small amount of refrigerant from the high side of the system or the refrigerant tank through the gauges as they are installed. It is also good field practice to keep the gauge manifold charged with refrigerant when not being used to minimize the chances of foreign matter, including air, entering the hoses. If air is present in the hoses, it will be pushed into the lines as refrigerant is added to the system. In addition, if the system has experienced a complete loss of refrigerant, air can be pulled into the system by the compressor from the low side of the system.

Systems that are not equipped with low-pressure switches are at risk of having atmosphere enter the refrigerant circuit in the event of a low-side pressure leak. The refrigerant will be released from the system until the pressure in the system reaches 0 psig. When the compressor cycles on, the low-side pressure begins to drop, pulling a vacuum. This pulls atmosphere into the system. The low-pressure switch, when used as a low-charge protector, prevents the system from operating when the refrigerant charge has been lost.

NITROGEN IN THE AIR CONDITIONING SYSTEM

Nitrogen is not normally found in a refrigeration system after initial startup. It is, however, found in the system throughout the installation process. Nitrogen is used

- by equipment manufacturers to pressurize various pieces of air conditioning equipment prior to shipment. The release of the nitrogen by the system installer ensures that the equipment arrived at the site, leak-free. Self-contained, or packaged, units are not shipped with nitrogen as they come charged with refrigerant directly from the factory. Carefully read all literature accompanying the piece of equipment since some components come with a holding charge of refrigerant, not nitrogen.

- by field technicians and installation crews to pressure-test refrigerant lines before system evacuation is performed. Pressurizing with dry nitrogen is a popular and preferred method of leak-checking a system. Small leaks can be identified with the aid of liquid solutions that blow bubbles when leaks are present.

In either case, the nitrogen must be completely removed prior to evacuation and the addition of valuable refrigerant to the system.

HOW DO YOU KNOW IF THE SYSTEM CONTAINS NONCONDENSABLE GAS?

If, after installing the gauge manifold on an inoperative system, the pressures are above 0 psig, the system is holding at least some pressure. While off, the high- and low-side pressures should correspond to each other. These pressures should also correspond to a saturation temperature that matches the actual ambient temperature surrounding the unit. If these two temperatures are the same, the system does not contain noncondensable gases.

If the system is operating upon arrival at the service call, turn the unit off and allow it to sit for approximately 30 minutes, longer if at all possible, before comparing the ambient and saturation temperatures. This allows the system pressures to equalize, giving more accurate readings. If the corresponding saturation temperature is higher than the ambient, chances are the system contains noncondensable gas. If this is indeed the case, the refrigerant charge must be properly recovered from the system. The system must then be evacuated and recharged with new refrigerant. **Recovery** involves the removal of the refrigerant from the system, which can then be turned in to an approved facility for reprocessing.

> **CAUTION**
>
> **CAUTION:** Only technicians possessing proper certification granted by the Environmental Protection Agency (EPA) can handle and work on systems containing refrigerant.

DETERMINING THE EVAPORATOR SATURATION PRESSURE AND TEMPERATURE

The evaporator saturation pressure is read directly from the face of the low-side, or blue, gauge. This gauge is typically located on the left-hand side of the manifold. The outer, black dial provides the technician with the suction, or back, pressure, depending on the position of the needle on the gauge. The low-side gauge in Figure 21-11 indicates a pressure of 118 psig on the low side of the

FIGURE 21-11 This gauge reads 118 psig.

Photo by Eugene Silberstein.

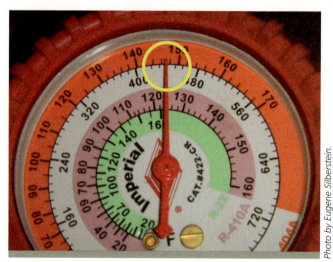

FIGURE 21-13 This gauge reads 440 psig.

Photo by Eugene Silberstein.

FIGURE 21-12 For R-410A, a pressure of 118 psig corresponds to a saturation temperature of 40°F.

Photo by Eugene Silberstein.

system. The evaporator saturation temperature is also read directly from the face of the gauge. If an air conditioning system is operating with R-410A as its refrigerant and the low-side pressure is 118 psig, the evaporator saturation temperature will be 40°F (Fig. 21-12). In this case, the refrigerant is boiling in the evaporator at 40°F. The data contained on the face of the gauge are the same as those provided on the pressure/temperature chart.

DETERMINING THE CONDENSER SATURATION PRESSURE AND TEMPERATURE

The condenser saturation pressure and temperature are obtained from the face of the high-side, or red, gauge, in a manner similar to that of the evaporator saturation pressure and temperature. This gauge is located

on the right-hand side of the manifold, and the system head pressure is indicated on the outer, black scale. The high-side gauge in Figure 21-13 indicates a pressure of 440 psig on the high side of the system.

The condenser saturation temperature is also read directly from the face of the gauge. If an air conditioning system is operating with R-410A as its refrigerant and the high-side pressure is 440 psig, the condenser saturation temperature will be 124°F. In this case, the refrigerant is condensing at 124°F. Just as on the low side of the system, the information on the gauge is the same as that provided on the pressure/temperature chart.

EVALUATING EVAPORATOR EFFECTIVENESS AND EFFICIENCY

For air conditioning applications, the ideal, or design, evaporator saturation temperature is roughly 40°F. This temperature is desired because it is cool enough to provide proper air conditioning and is above the freezing point, which eliminates the need for a defrost cycle. Under normal system operation, frost should never accumulate on the coil's surface.

As with the rest of the refrigeration system, it is impossible to visually look into the evaporator coil and determine at which point the liquid refrigerant completely vaporizes. For this reason, we rely on the evaporator superheat to give us this information. The normal range of evaporator superheat is from 8 to 12 degrees for systems with TXVs and somewhat higher for capillary tube systems. Two temperature readings are needed to determine the evaporator superheat. They are

1. the evaporator outlet temperature
2. the evaporator saturation temperature, measured near the thermal bulb

As stated earlier in the text, the evaporator superheat is determined by subtracting the evaporator saturation temperature from the temperature of the suction line at the outlet of the evaporator. Under many system conditions and configurations, taking a suction pressure reading at the outlet of the evaporator may not be possible because a service port may not be located there. If the suction service valve is not located too far from the evaporator, the evaporator saturation temperature can be read from it, but the superheat calculation may be off by 1 or 2 degrees.

Excessive Superheat. If the evaporator superheat is excessive, more than 15 degrees, the evaporator is not operating at its maximum potential. It is said to be a **starved,** or **underfed evaporator.** Common reasons for high superheat are

- refrigerant undercharge
- improper superheat spring setting on the thermostatic expansion valve
- blocked or clogged strainer on the thermostatic expansion valve
- blocked or clogged capillary tube
- blocked or clogged strainer at the capillary-tube inlet
- indoor fan motor speed too high

Low Superheat. If the evaporator superheat is too low, less than 5 degrees, the evaporator is operating very effectively, but there is an increased possibility of liquid refrigerant flowing back to the compressor. This is called floodback. The evaporator is called an **overfed evaporator.** Common reasons for low superheat are

- refrigerant overcharge
- overfeeding metering device
- improper superheat spring setting on the thermostatic expansion valve
- improperly mounted thermostatic expansion valve thermal bulb
- restricted airflow across the evaporator (blocked coil or inoperative fan motor)
- dirty air filters
- reduced water flow through the evaporator (chiller application)

EVALUATING CONDENSER EFFECTIVENESS AND EFFICIENCY

As with the evaporator evaluation, we must rely on some external method to determine how effectively the condenser is operating. By calculating the subcooling in the condenser, we are able to determine how effectively system heat is being rejected. The greater the degree of subcooling, the more efficient the condenser and vice versa.

As with the evaporator, two measurements are needed to calculate subcooling:

1. The condenser saturation temperature
2. The condenser outlet temperature

To briefly recap, condenser subcooling is calculated by subtracting the condenser outlet temperature from the condenser saturation temperature. The normal range for condenser subcooling is from 15 to 20 degrees. Higher-efficiency condensers can operate with somewhat higher subcooling.

Low Subcooling. If the amount of condenser subcooling is low, the condenser is not rejecting its heat effectively. A number of factors can cause the condenser subcooling to drop. A few of them are

- system undercharge
- dirty or blocked condenser coil
- sefective condenser fan motor
- service panels removed from condensing unit
- reduced water flow through the condenser (water-cooled applications)
- defective water-regulating valve (water-cooled applications)

Excessive Subcooling. As mentioned earlier, a condenser's effectiveness is measured by the amount of subcooling; however, too much subcooling could be an indication of a system deficiency. Too much subcooling can actually reduce the effectiveness of the condenser because the heat-transfer surface has been reduced. High subcooling could be a result of

- system overcharge
- system operating in low ambient conditions
- refrigerant-flow restriction
- noncondensable gas in the system
- overfeeding water-regulating valve (water-cooled applications)

DETERMINING IF THE REFRIGERANT CHARGE NEEDS TO BE ADJUSTED

Upon initial inspection of the system, pressure and temperature readings will play an important role in the proper evaluation of an air conditioning system. Superheat and subcooling calculations, along with other information provided in the text, enable the field technician to determine if the refrigerant charge is in need of adjustment.

Note that multiple readings are often necessary to properly evaluate the system. Reaching intelligent conclusions about a system is difficult by taking, for example, a low-side pressure reading and nothing else.

EVALUATING THE METERING DEVICE

The metering device is responsible for controlling the flow of refrigerant to the evaporator. If this device does not feed properly, the evaporator can become either **flooded** or starved. A flooded evaporator is one that has saturated refrigerant at its outlet. This is not desirable because system damage could result if the liquid refrigerant should reach the compressor. A metering device that overfeeds could cause the following:

- Liquid floodback to the compressor
- Low compression ratio
- High suction pressure
- Low head pressure
- Reduced cooling capacity

A metering device that underfeeds the evaporator can cause the following:

- Excessive superheat
- Reduced cooling capacity
- Low suction pressure
- High head pressure
- Increased compression ratio
- Ice formation on the evaporator coil

EVALUATING THE CAPILLARY TUBE

Since the capillary tube is a fixed-bore metering device, it has no moving parts to go bad, no adjustments that need to be made, and no general maintenance to be performed. The number-one enemies of the capillary tube, though, are moisture and dirt.

If moisture is present in the system, it commonly freezes as it enters the capillary tube, because there is a pressure drop in the tube. Ice crystals will cause a restriction in the tube, affecting system operation.

If the system is turned off, the ice will melt and, when energized again, the system will operate until the ice forms again. The moisture must be removed from the system by recovering the refrigerant charge, by evacuating and recharging the system, or by replacing the filter driers.

Dirt or debris in the capillary tube will also cause a restriction in the line. Wax, found in system contaminants, has a tendency to solidify in the capillary tube, so strainers are needed at the inlet of the device

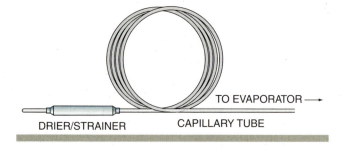

FIGURE 21-14 Capillary tube metering device. The strainer at the inlet of the capillary tube prevents particulate matter from entering the capillary tube.

(Fig. 21-14). System problems that can result from a restricted capillary tube include those listed for underfeeding the evaporator. Capillary-tube systems do not experience problems with overfeeding evaporators unless the capillary tube was replaced with a shorter tube or one with a larger bore.

Going Green

Since the minimum system SEER rating requirements have increased from 10.0 to 13.0, manufacturers have, for the most part, steered away from the capillary tube metering device. Thermostatic expansion valve has fast become the metering device of choice. The thermostatic expansion valve allows for more efficient system operation.

EVALUATING THE AUTOMATIC EXPANSION VALVE

The automatic expansion valve is designed to maintain a constant evaporator pressure and is not generally serviced in the field. If the pressure the device is maintaining is different from the desired pressure, turning the adjustment screw (Fig. 21-15) can change the valve setting. The screw is turned clockwise to increase the evaporator pressure and counterclockwise to reduce the pressure. Adjustments should be made slowly to ensure an accurate and precise setting and to reduce the possibility of

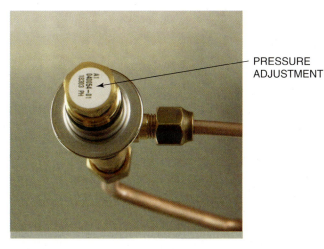

FIGURE 21-15 The adjusting screw on the automatic expansion valve.

Delmar/Cengage Learning.

compressor damage. If the device fails to maintain the proper pressure after adjustment and the system charge is correct, the valve may need to be replaced.

EVALUATING THE THERMOSTATIC EXPANSION VALVE

Unlike the capillary tube and the automatic expansion valve, the thermostatic expansion valve can be serviced in the field, depending on the size and configuration of the valve. The thermostatic expansion valve is designed to maintain constant evaporator superheat and, therefore,

must be able to effectively measure the evaporator outlet temperature. The following should be checked when evaluating this control:

✓ Make certain the thermal bulb is mounted securely to the suction line at the outlet of the evaporator, according to manufacturer's instructions (Fig. 21-16).

✓ Make certain the thermal bulb is wrapped with insulation tape to ensure accurate readings.

✓ Make certain the strainer at the inlet of the valve is clean and free from debris.

✓ If the valve fails to maintain superheat, the thermal bulb should be checked for possible loss of charge. Gently warming the bulb should cause the suction pressure to rise, since the bulb pressure pushes the valve open. Heat the bulb with your hand, or place the bulb briefly in a glass of warm water. If the evaporator pressure does not change, the power head has lost its charge and must be replaced. If the valve does not have interchangeable parts, the entire valve must be replaced.

THERMOSTATIC EXPANSION VALVES THAT OVERFEED THE EVAPORATOR

Thermostatic expansion valves that overfeed the evaporator tend to operate with a superheat that is lower than desired. To alleviate this problem, the superheat spring adjustment should be turned clockwise to increase the spring pressure. This will increase the amount of evaporator superheat. Before adjusting the valve, make certain

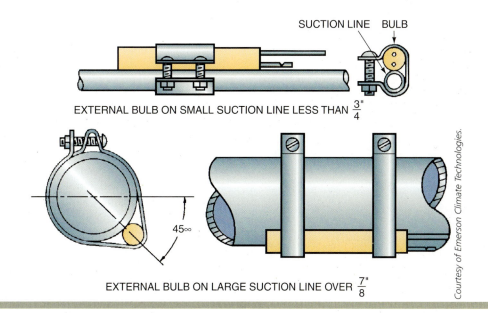

FIGURE 21-16 Proper positions for thermal bulb of the TXV.

Courtesy of Emerson Climate Technologies.

that the four items listed for evaluating this control were checked first.

Superheat adjustments on thermostatic expansion valves should be made with care. They should be made slowly, and ample time must be allowed for the system to stabilize itself after each adjustment. Improper valve adjustment can result in major component—including compressor—damage. One full turn of the superheat spring adjusting screw generally changes the superheat approximately ½ to 1 degree.

THERMOSTATIC EXPANSION VALVES THAT UNDERFEED THE EVAPORATOR

Underfeeding of the evaporator generally causes a reduction in cooling efficiency. An evaporator that is underfed has a higher than desired superheat, which can be caused by

- a system undercharge
- a clogged inlet strainer at the thermostatic expansion valve
- moisture in the system
- improper superheat spring adjustment

A refrigerant undercharge will result in high superheat, low suction pressure, and low discharge pressure. Adjusting the charge should alleviate the problem. Obviously, the reason for the loss of refrigerant must be located and corrected before proceeding.

A low refrigerant charge also reduces the pressure drop across the thermostatic expansion valve and affects its performance.

A clogged inlet strainer (Fig. 21-17) reduces the amount of refrigerant that flows through the thermostatic

FIGURE 21-17 Strainer at the inlet of the thermostatic expansion valve.

Photo by Eugene Silberstein.

expansion valve. The evaporator will be starved and the superheat will be high; but, unlike the low-charge scenario, the head pressure will be high.

Moisture in the system can freeze in the valve, causing the valve to freeze closed. This will create the same symptoms as a clogged strainer, but the evaporator pressure will rise if the valve body is gently warmed, allowing the ice to melt. The filter driers should be replaced to absorb the moisture in the system.

If all else fails, the superheat spring may need to be adjusted. As stated earlier, care must be taken when making adjustments to the valve. Adjusting the superheat spring counterclockwise will reduce the spring pressure, which reduces the superheat and allows more refrigerant to flow to the evaporator.

TROUBLESHOOTING STEPS

The troubleshooting process can be broken down into seven simple steps. These seven steps should be performed in order, and each should be completed before moving on to the next. If followed correctly, even the newest field technician will be able to experience success in the field very quickly. The steps are

1. verify the complaint
2. gather information
3. perform a visual inspection
4. isolate the problem
5. correct the problem
6. test the system operation
7. complete the service call

VERIFY THE COMPLAINT

This step may seem obvious, but it is essential. The technician must realize that the customer is not the service expert. If that were the case, the service company would not have been called in the first place. What the customer sees or thinks may or may not be what is actually taking place within the system. Therefore, the service technician must verify the information that the customer is providing. For example, the customer may tell the technician, "The unit doesn't blow cold air." If this is accepted as the truth, the technician may make the following assumptions:

- The evaporator fan is operating.
- The condensing unit is not running.
- The system is calling for cooling.

The technician may reach the conclusion that the condensing unit is not working properly. This may or may not be correct. The following may, in reality, be the case:

- The thermostat is set to the cooling mode.
- The fan switch is in the ON position.
- The room thermostat is satisfied, and the condensing unit is de-energized.
- Since the fan is operating all the time, even when the compressor is not operating, the air coming from the registers will be warm.

It can be seen from this example that the system is operating correctly, but the customer is under the impression that the air coming from the registers must *always* be cold. He felt the air and, since it was warm, came to the conclusion that something must be wrong with the system. A service call was then placed to the HVAC contractor.

GATHER INFORMATION

Once the service technician has verified the complaint, other information that may help get to the root of the problem quickly must be obtained from the customer. Pertinent information may be obtained by asking questions similar to the following:

- "When did you first notice this problem with the system?"
- "Was anybody working on or around your equipment at the time you first noticed the problem?"
- "When was the last time this unit was serviced?"
- "Prior to this problem, had the system been working reliably?"
- "Have there been any recent power failures?"

Depending on the customer complaint, additional questions may shed more light on the situation at hand. For example, if the customer says that the system just isn't cooling enough, asking when the problem was first noticed may help solve the problem, depending on the answer received. If the customer responds with something like, "It stopped cooling well toward the end of last summer, but since the weather started getting cool, I figured I would wait until this year to have it checked out. Now it's not cooling much at all." This would lead the astute technician to believe that there is most likely a refrigerant leak in the system because the system cooled, but not well, last summer and cools even less now.

Other workers who may have been around the equipment may have had something to do with the improper system operation. For example, if the customer was having the house painted, supply registers may have been covered with plastic and taped closed to prevent paint

drips from damaging the grills. The condensing unit may have been covered with a drop cloth for the same reason. Or perhaps the customer's alarm company was at the house checking out the system and the technician, while crawling around the attic, inadvertently pulled on a low-voltage control wire, causing damage to it.

Knowing when the system was last serviced can also provide useful information. Motor lubrication, filter changes, belt adjustment, and other tasks performed during a preventive-maintenance service call help keep the system in tip-top working order. Neglecting these items can cause system problems that would normally be avoided with proper maintenance.

Asking the homeowner if the system was working properly prior to this problem may or may not provide useful information. It is not possible for a layperson to know whether or not the system was working correctly. The answer to this question will give you the answer to the question, "Was the house comfortable prior to the system failure?" If the answer to the question reflects that the system performed "perfectly" prior to this, a mechanical failure of some type or a total loss of refrigerant can *usually* be suspected. As with any other probable cause, this would need to be confirmed through a thorough system check.

Neighborhood power failures can also result in a system that fails to operate as expected. Programmable thermostats may lose their program, and other digital devices may need to be reset. Knowing if the power was interrupted for any length of time may help reduce service time.

Obviously, these questions are only a small sampling of the possible questions that will provide the technician with the needed extra information to properly evaluate the system. It is up to the technician to decide what questions would best help him to help the customer.

PERFORM A VISUAL INSPECTION

Once the technician has spoken to the customer and has gathered information regarding the system, the technician must gather additional information independently.

This information is obtained by performing a visual inspection of the equipment and includes

- ✓ Checking the thermostat settings
- ✓ Checking the air being discharged from the supply grills
- ✓ Inspecting air filters at the return grill
- ✓ Inspecting the condensing unit

By checking the thermostat settings, the technician can determine if the system is actually calling for the desired mode of operation. If the customer is not getting any cooling, for example, and the thermostat is set to the heating mode, it is quite obvious that there will be no cooling. If the thermostat is set to provide cooling and the fan is not operating, a low-voltage problem may be suspected. The fan blowing warm air may lead the technician to believe the problem lies within the condensing unit. Dirty or blocked filters may also give the technician useful information. If the filters are completely blocked, this means that not only is there an airflow problem with the system, but there may also be a motor-lubrication or belt-tension problem, since the system has obviously not been serviced for some time. Inspecting the condensing unit will tell the technician if the condensing unit is operating when the thermostat is set to the cooling mode. If the thermostat is set to the cooling mode and the condensing unit is not operating, an electrical problem should be suspected. This would include the opening of a pressure switch caused by a loss of refrigerant charge.

ISOLATE THE PROBLEM

Once the visual inspection has been completed, the technician can concentrate on the particular problem. If, for example, on a no-cooling call, it was determined that the indoor fan motor was operating when the thermostat's fan switch was turned to the FAN ON position but turned off when the fan switch was switched to the AUTO setting, the transformer would not need to be checked, since it has already been established that low voltage is being provided by the device. If, on the other hand, the fan failed to operate at all, this would be the first thing that should be checked. If a problem exists with a condensing unit and the condenser fan motor is operating but the compressor is not, checking whether or not voltage is being supplied to the condensing unit is not necessary. The technician already knows that voltage is being supplied because the condenser fan motor is operating. The technician also knows that the cooling contactor is operating properly. The problem, therefore, lies within the compressor or its associated wiring.

CORRECT THE PROBLEM

After the specific problem has been identified, the repair should be performed in a manner that observes all safety guidelines and government laws and regulations. Safety guidelines include the following:

✓ Power to the circuits should be disconnected when repairing or replacing defective components.

✓ Safety glasses and work boots should always be worn.

✓ Loose-fitting clothing and jewelry should not be worn.

✓ Long hair should be tied back.

Government laws and regulations as outlined by the Clean Air Act and the Montreal Protocol Act should be followed, and all service technicians working on air conditioning and refrigeration equipment must be certified by the Environmental Protection Agency and possess EPA Section 608 certification.

The type of equipment worked on determines the type of certification required. For example, low-pressure systems require Type III certification under Section 608, while technicians who work on automotive air conditioning systems are required to possess Section 609 certification.

In correcting the problem, the service technician must always be aware of and respect the customer's property. The technician should work quickly, efficiently, and neatly and should consciously strive to provide the best possible repair in the shortest possible time, thereby providing the best possible value for the customer.

TEST THE SYSTEM OPERATION

Once the repair has been completed, the system must be tested. Good field practice includes checking all modes of operation and allowing the system to operate under supervision for at least 15 minutes. Waiting until the unit cycles automatically to ensure proper operation is a good idea. System pressures and temperatures should be checked, including evaporator superheat, condenser subcooling, and evaporator temperature differential. The amperage draw of the compressor and motors should also be checked after the system repair to ensure that they are within design parameters. While the system is being monitored, final cleanup and paperwork preparation can be done in order to make good use of this time. This helps ensure that, upon completion of the job, the technician will be ready to proceed to the next call.

COMPLETE THE SERVICE CALL

No service call is complete until the paperwork is filled out with all required information and the customer signs off on the job. Descriptions of the work performed should be detailed and written legibly. The technician should not simply write "Replaced Compressor" on the work ticket but should list everything that was done including refrigerant recovery, system evacuation, and so on. A complete list of all materials and parts used must be provided. Remember, this is the only written record of the work performed. The billing department will rely on this work ticket to prepare a detailed bill for the customer. Time spent on the job and materials used are vital pieces of information that *cannot* be left out.

The technician must also explain to the customer exactly what was done to the system. After all, the service technician is the initial—and often the only—contact between the customer and the service company. Making the company look good by acting in a professional and courteous manner is the technician's duty. Before leaving the job site, the technician should perform an inspection of the work area and remove any garbage and debris from the area. Dirt should be cleaned up. The customer cannot see the work performed inside a sealed unit, but fingerprints on or around a thermostat indicate that the technician is a sloppy worker, no matter how well the job was performed. This final inspection also helps reduce the possibility of leaving tools and materials behind. Precious time will be lost if a tool is left behind that is needed on the next job.

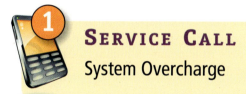

SERVICE CALL
System Overcharge

CUSTOMER COMPLAINT

A customer places a call to her air conditioning service company and tells the service manager that her split-type, air conditioning system is not cooling her house properly. She explains that the air coming from the supply registers is warm, and that her thermostat is displaying a space temperature of 88°F.

Two hours later, Pedro, the service technician, arrives at the house. The home owner explains the situation to him. She also adds that at the end of last year's cooling season, the system developed a refrigerant leak. The system was repaired but, shortly after the repair was made, the weather turned cool and the system was not used any more that year. She adds that she has used the unit two or three times this season during relatively warm weather, but this is the first time it has been used on a very hot day. The outside temperature is 95°F.

TECHNICIAN EVALUATION

After listening to the customer's complaint, Pedro makes his way over to the system thermostat to make certain that the thermostat is set to the cooling mode and is set

to maintain a temperature below the present room temperature. The room temperature is 88°F and the thermostat is set at 74°F. He then goes over to a ceiling supply register and feels the air coming from it. There is a great amount of air coming from the register but, just as the customer stated, the air is cool, but not cold. From this, he is able to establish the following:

- The evaporator fan motor is operating
- There is ample airflow through the evaporator
- The low-voltage control circuit is energized (If there were a problem with the control circuit, the indoor fan motor would not be operating)
- The problem likely lies with either the refrigerant charge or the airflow through the condenser coil

Pedro then asks the homeowner to show him the location of the condensing unit. The customer leads Pedro outside to the side of the house. He immediately notices that the condenser fan motor is operating and can hear that the compressor is operating as well. Concluding that the problem is not electrical in nature, Pedro decides to check the operating pressures of the system. The system contains R-410A and the outdoor temperature is 95°F. Pedro proceeds to install his gauges on the system. He obtains a high-side pressure reading of 500 psig and a low-side pressure reading of 170 psig.

Checking his pressure/temperature chart, he finds that the condenser saturation temperature of R-410A at 500 psig is 134°F. Knowing that the outside ambient temperature is 95°F and that the condenser saturation temperature should be about 25°F to 30°F higher than ambient, Pedro concludes that the high-side pressure was higher than it should be. High head pressure can be a result of:

- System overcharge
- Noncondensable gases in the system
- Defective condenser fan motor
- Dirty or blocked condenser coil

Pedro now needs to determine which of these possible causes is to blame for the system failure. He wants to proceed cautiously, since high system pressure can result in personal injury as well as system component damage. His actions are as follows.

First, he turns off the power to the unit, removes his pocket flashlight and visually inspects the condenser coil. He is able to see clear through the coil and notices no dirt deposits between the coil's fins. He also notices that no leaves or other debris is blocking the coil, and he concludes that the condenser coil is not dirty or blocked. The reason for the failure has now been narrowed down to the following:

- Noncondensable gases in the system
- Defective condenser fan motor
- Refrigerant overcharge

Since the system was operating when he arrived, he is able to determine that the condenser fan motor is not the problem and that the problem is with either the refrigerant charge itself or the presence of noncondensable gases in the system.

To address the issue of noncondensable gases, the system must be turned off so the pressures in the system have a chance to equalize. He had turned the system off to inspect the condenser coil, so he leaves the system off and waits a few more minutes. He then takes a reading of the outside ambient temperature, which in this case is 95°F, and compares it to the condenser saturation temperature on the high-side gauge of his manifold, which is 95°F. The gauge reading is 295 psig, which corresponds to the saturation temperature of 95°F. Since this temperature corresponds to the actual ambient, Pedro rules out the possibility that noncondensable gases are in the system. Pedro has now eliminated all possibilities except the system overcharge and concludes that the system has been overcharged. To confirm this, he starts up the system, allows it to run, and measures the subcooling and superheat in the system. The subcooling measurement is 30 degrees and the superheat measurement is 5 degrees. The high subcooling and low superheat also indicate that the system is overcharged.

Pedro places a call to his service manager to discuss his findings. The customer's file is consulted, and it is found that the system was repaired on a cool day at the end of last year's cooling season. The technician that performed the original work had overcharged systems several times before, primarily due to a lack of proper training, and has since found employment with another service company.

Using a properly evacuated, DOT-approved recovery tank, Pedro begins to remove refrigerant from the high side of the system and store it in the tank. Once he has removed enough refrigerant, he restarts the system and adjusts the charge. Since the outside ambient temperature is 95°F, he knows that the condenser saturation temperature for this system should be approximately 125°F. He takes temperature readings of the return and supply air within the occupied space and finds the **temperature differential** (TD) to be 19 degrees. (Return air = 88°F; Supply air = 69°F; 88 − 69 = 19 degrees). Pedro knows that the normally acceptable temperature differential across the evaporator is between 17 and 20 degrees if the relative humidity in the conditioned space is about 50%.

Pedro monitors the system for about 15 minutes while he completes his paperwork for the job. He calls his manager and inquires about the service charges for this call and is told that, since an improper repair last season caused the failure, the customer has no financial obligation. Pedro puts the customer on the phone with his manager so that the manager can explain the situation to the customer. While he waits, Pedro opens the return-air grill and inspects the air filters.

from experience...

Systems with higher SEER ratings will generally operate with lower condenser saturation temperatures, resulting in a lower temperature differential across the condenser coil. If systems with higher SEER ratings are charged to a 30-degree differential, they will most likely be overcharged. A 20-degree differential is more the order when dealing with more efficient systems.

SERVICE CALL DISCUSSION

This service call provides a perfect example of how to follow the seven-step procedure for successfully completing a service call. Pedro first listened to the customer complaint and then verified what he heard. By doing this, he was quickly able to eliminate a number of possible reasons for the system failure.

Using basic refrigeration theory, common sense, and the process of elimination, he was able to establish that the system was overcharged. Pedro performed his job in a professional manner and used proper techniques to remove the excess refrigerant from the system. If he had been unable to remove enough refrigerant from the system by simply bleeding it from the high side of the

from experience...

Many acceptable methods exist for determining proper refrigerant charge, including weighing in the charge or using the manufacturer's charging chart. Weighing in the proper refrigerant charge is most reliable and effective when the system is of the package type with no field-installed refrigerant piping. Using the manufacturer's charging table will help guarantee that the proper refrigerant charge is obtained.

system, he would have had to use a recovery unit. Refer to the Refrigerant Management chapter for more on the refrigerant recovery process.

Another important aspect of this service call is the handling of the customer. Although everything seemed very calm, the situation could have become difficult. If Pedro, for example, had told the customer that one of his coworkers had overcharged the system, the customer might not have been very quick to accept the credibility of the company and its technicians in the future. Instead of placing himself in this situation, he correctly decided to put the customer on the phone with his service manager, who has much more experience in dealing with customers in situations like this and had access to the customer's entire service history file.

This sample service call illustrates that the service technician needs to be somewhat of a public relations person as well as technically knowledgeable. Remember that the service technician actually represents the company and must always display a professional, courteous, and intelligent attitude toward the work *and* the customers. Is there anything that you, as the technician, would have done differently on this service call?

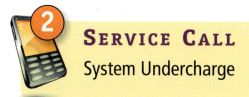

SERVICE CALL
System Undercharge

CUSTOMER COMPLAINT

A residential customer calls his air conditioning contractor and tells the dispatcher that, while out in his garden that day, he noticed that the condensing unit ran continuously. The unit did not shut down at all the entire time he was outside. He tells the dispatcher that the system is keeping the house at the desired temperature, but he is concerned that the system did not cycle off. He is also concerned that the system may not be able to keep the house cool as the weather gets warmer and is concerned about a possible increase in power consumption that can lead to higher utility bills.

SERVICE TECHNICIAN EVALUATION

Linda is given the service call by her dispatcher. She goes over to check out the system as soon as she finishes writing up the paperwork and explaining the charges to the customer at her previous service call.

She arrives at the customer's house and finds the home owner in the garden. She approaches him and

introduces herself, and he explains the problem to her. He tells her that, prior to placing the service call, he thought that the air filters might be dirty so he replaced them. The system still has not cycled off. Since the system seems to be working and Linda's shoes are already somewhat muddy from the damp grass, she decides to check the outdoor condensing unit first to reduce the possibility of tracking dirt into the customer's house. Linda goes to her truck to retrieve her tools and returns to the side of the house to begin her evaluation of the system.

Linda can tell immediately from a visual inspection that both the compressor and the condenser fan motor are operating. She decides to install her gauge manifold on the system so that she can monitor the system pressures. After following the correct gauge installation procedure, she is able to determine that the head pressure is 180 psig and the low-side pressure is 45 psig. Using a good-quality thermometer, Linda takes a reading of the outside temperature and determines that, at the present 80°F temperature, the head pressure for this R-22 system should be approximately 225 psig. She also recalls that the back pressure on an R-22 air-conditioning system ideally should be in the range of 70 psig. Since both the low-side and the high-side pressures are lower than she expected, she is able to determine that the system

- has restricted airflow through the evaporator coil or
- is short of refrigerant

Reduced airflow through the evaporator coil can be the result of a number of situations, including

- dirty air filters
- blocked return-air grill or ductwork
- defective evaporator fan motor
- broken or loose belt
- blocked or closed-off supply registers or ductwork
- frozen evaporator coil
- loose or missing air-handler access panels
- dirty or blocked evaporator coil

Since the customer just replaced the air filters prior to Linda's arrival at the job, she is able to eliminate that possibility relatively quickly. She now needs to enter the house and inspect the air handler for any of the aforementioned conditions. She carefully cleans her shoes before entering. Making her way to the air handler's location in the basement, she is able to hear the evaporator fan running and the movement of air through the ductwork. She turns off the power switch to the unit and removes the service panel at the blower. These panels are provided to give the servicing technician access to the various system components. Noticing that the belt is intact and its tension is correct, these possibilities are eliminated as well. Confirming that all service panels are properly installed and that the return plenum, the section of ductwork connected directly to the return portion of the furnace or air handler, is clear, she is able to eliminate that possibility as well. She then removes the service-access panel at the evaporator coil and inspects it and the supply plenum. The supply plenum is the section of supply ductwork connected directly to the furnace or air handler. The coil and plenum are clean. She notices, however, that a small portion of the evaporator coil is frosted. The rest of the coil is warm. After replacing the service panels, she makes her way up to the living area of the house and inspects the supply registers and finds them to be open. Upon turning the power switch to the furnace back on, she can feel air being discharged from the registers.

Having eliminated the causes for a reduced airflow through the evaporator coil and having noticed that a portion of the evaporator coil is frosted, Linda concludes that the system is short of refrigerant.

Before adding refrigerant to the system, she uses her electronic leak detector to look for a refrigerant leak. She turns the detector on and hears the slow beeping of the device. She begins moving the probe slowly around the soldered copper fittings and across the suction and liquid lines that carry refrigerant to and from the air handler. After checking the condensing unit and finding no leaks, she needs to go to the air handler to continue leak-checking. After wiping her feet thoroughly, she enters the house and continues checking for leaks. She does not find any.

At this point if a leak does exist, it is very small and will be difficult to find. Linda speaks with the home owner and explains that the system is short of refrigerant and that her attempt to locate the leak has failed. Linda tells the home owner that she will add refrigerant as needed but will also introduce an **ultraviolet (UV) solution** (Fig. 21-18) to the system. She explains that this solution will circulate through the system with the refrigerant and, if it escapes from the system, will stain the piping at the point of the leak. This stain will glow when seen under UV light. This will make the leak very easy to find if, indeed, one exists. The customer agrees, and Linda prepares to add the solution and charge the system.

After determining the total system charge, Linda adds the correct amount of UV solution to the system. She makes certain that all of the ice has been removed from the evaporator coil and then begins to add refrigerant to the system. To ensure that no liquid refrigerant enters the compressor, she introduces vapor refrigerant into the suction service valve of the unit.

Both the head pressure and the suction pressure begin to rise. Linda brings the head pressure up to approximately 220 psig and the suction pressure to about 68 psig. She tapes a thermometer probe to the suction line at

FIGURE 21-18 Ultraviolet solution being added to the air conditioning system.

from experience...

When working with blended refrigerants, the refrigerant must leave the tank as a liquid in order to maintain the proper proportions of the various refrigerant components of the blend. Therefore, when charging or adding refrigerant to a system, care must be taken to ensure that the compressor is not flooded with liquid.

the inlet of the condensing unit and reads a temperature of 55°F. Since the evaporator saturation temperature for the system is 40°F (pressure-temperature relationship), she calculates that the system superheat is $55 - 40 = 15$ degrees, which is acceptable. She enters the house, again cleaning her feet, and checks the air handler. She has a temperature differential across the coil of 20 degrees, which is acceptable.

After monitoring the system for approximately 15 minutes, she removes her gauge manifold from the system, packs up her tools, writes up her work order, gets her paperwork signed, and proceeds to her next service call.

SERVICE CALL DISCUSSION

While performing this service call, Linda was very concerned and respectful of the customer's property. She was conscious of the fact that she would most likely track dirt into the house if she went inside first. She tried to minimize the number of times she needed to enter the residence. She also made a point of keeping the home owner informed as to what was being done and what must be done in the future.

> ### from experience...
>
> Some customers may not be interested in the play-by-play explanation of the repair but, instead, may say something like, "Just let me know when it's fixed." The technician must evaluate just how much involvement the customer desires, and adjust accordingly.

As far as the technical aspect of this service call is concerned, Linda was extremely conscious of the fact that refrigerant does not simply disappear. If the system is short of refrigerant, it *must* have leaked out. Government regulations suggest that system leaks be found and repaired. Although she could not locate and repair the leak, she took the proper steps by adding the UV solution. This way, if the system loses refrigerant again, the leak will be found relatively easily. This shows compliance with government policies as well as respect for the customer's wallet.

Linda could have easily spent hours and hours, which are billable, looking for a leak that may not even exist. What if, when the unit was last serviced, a service cap was loose, allowing a portion of the refrigerant charge to escape from the system? Linda spent a reasonable amount of time looking for the leak but did not overdo it.

When adding refrigerant to the system, she used outside ambient temperature, head pressure, back pressure, evaporator temperature differential, and system superheat to help evaluate the system charge. Using only one of these may speed the process of system charging but can easily lead to an improper amount of refrigerant being added. Remember that ideal system and design conditions are just that—ideal—and do not necessarily apply to all systems and installations.

SERVICE CALL
Dirty Air Filter

CUSTOMER COMPLAINT

A residential customer calls his service company and informs the service manager that his house is very warm and that there is no air coming from the supply registers. He tells the manager that the outdoor unit is operating and there is a lot of ice on one of the copper lines coming from the unit. Later that afternoon, William arrives at the customer's home.

TECHNICIAN EVALUATION

When William arrives at the customer's home, he asks the home owner to explain what is happening with the system. The owner explains to William that the system is not cooling the home and that no air is coming from the supply registers. He also mentions the ice on the refrigerant line.

William goes over to one of the supply registers and confirms that there is no air being discharged into the room. The thermostat is set to the cooling mode and the fan switch is set to the on position, indicating that the evaporator fan motor should be operating. He then goes out to the condensing unit and notices the ice formation on the suction line. He gets his tools from his truck and installs his gauges on the R-410A system. The high-side pressure is 275 psig and the low-side pressure is 83 psig. The outside ambient temperature is 80°F and the air conditioning system is a standard-efficiency model. From this, William is able to determine that the condenser saturation temperature should be in the range of 110°F. The condenser saturation temperature is actually 90°F. The 83 psig low-side pressure indicates an evaporator saturation temperature of 23°F.

From this information, William determines that both the low-side pressure and the high-side pressure are lower than they should be. Low operating pressures can result from a number of conditions including

- system undercharge
- blocked return grill
- dirty air filter
- blocked or dirty evaporator coil
- defective evaporator fan motor
- closed or blocked supply registers

- loose acoustical lining or insulation in the supply duct
- loose acoustical lining or insulation in the return duct

Before adjusting the refrigerant charge, William decides to eliminate the other possibilities, which all relate to the airflow through the system. He first goes to the air handler located in the basement. He can hear the evaporator fan motor operating. He turns off the power switch to the unit and removes the service panel on the air handler. He inspects the evaporator coil and notices that the coil is completely frozen. Replacing the cover on the unit, William goes to the system thermostat and turns the unit to the off position but leaves the fan switch in the on position. In this position, the compressor will not operate while the blower is operating. This will help move air over the coil, allowing the ice to melt. While waiting, William decides to check the air filter.

Going back into the occupied space, William uses a step ladder to access the filter grill in the living room. Upon opening the filter rack, he observes that the filter is completely blocked with dirt and that no air can possibly pass through it. William asks the customer if he has any new filters in the home. The homeowner goes to the garage and hands William a new filter, which he places in the rack. He closes the filter rack and then goes back to the air handler. The majority of the ice on the evaporator coil has melted, so William is able to inspect the evaporator coil. The coil is clean. After all of the ice has melted, he replaces the service panel, returns to the living space, and rechecks the air coming from the supply registers. This time, there is ample air being discharged from the grills. From this, William is able to establish that

- the air filter is dirty
- the evaporator fan motor was operating
- the acoustical lining or insulation was intact
- the supply registers were open
- the evaporator coil was not blocked

William suspects that the only problem with the system was a dirty air filter. He turns the thermostat to the cooling position, goes back to the supply register, and feels cool air coming from it. He then goes outside to the condensing unit and checks the operating pressures of the system. The suction pressure is 118 psig and the head pressure is 365 psig. Both pressures are acceptable.

William writes up his work ticket outlining what he did on the service call. Before presenting the work order to the home owner for his signature, William informs him about the importance of keeping the air filters clean. He shows the home owner how to replace the filter and

tells him to check the filter frequently to ensure that it remains clean. The owner thanks William and signs the work order.

SERVICE CALL DISCUSSION

In this service call, William checked the air flow through the evaporator coil before adjusting the refrigerant charge. When he initially checked the operating pressures, he found that both the high and low pressures in the system were lower than desired. This is a tell-tale sign that the system is short of refrigerant. Had William added refrigerant to the system, the system would have been grossly overcharged. Recalling that there was no air coming from the supply registers indicated that there was likely an airflow problem. He addressed that issue first before attempting to adjust the refrigerant charge. If, on the other hand, he had added refrigerant to the system when it was determined that the pressures were lower than desired, he would have had to remove the excess refrigerant from the system. This would have resulted in wasted time. At the end of the service call, William explained the importance of keeping the filters clean to the home owner. A little extra customer service goes a long way in helping to keep the customer satisfied.

SUMMARY

- Motor problems can result from improper airflow, lubrication, pulley alignment, or belt tension.
- A refrigerant overcharge or undercharge can cause reduced system efficiency and cooling capacity.
- Gauge manifolds must be installed and removed properly to reduce the possibility of allowing atmosphere to enter the system.
- Refrigerant leaks can be caused by rubbing or vibrating surfaces or poor-quality soldering or brazing.
- Refrigerant leaks can be located by audible noise, bubble solutions, electronic leak detectors, halide torches, or UV light.
- Noncondensable gases, such as air and nitrogen, increase operating pressures, reduce system performance, and reduce cooling capacity.
- A vacuum gauge should be used during system evacuation to ensure that proper vacuum levels are reached.
- Excessive evaporator superheat reduces cooling capacity.

GREEN CHECKLIST

☐ Halide leak detectors will only work with chlorinated refrigerants.

☐ R-410A and R-407C are HFC refrigerant blends and do not contain chlorine.

☐ When pressurizing a system with nitrogen for electronic leak detection purposes, a trace of the normal system refrigerant can be added.

☐ Newer, high-efficiency air conditioning systems operate with lower condenser saturation temperatures and pressures than older, standard efficiency systems.

☐ Systems with higher SEER ratings will typically use thermostatic expansion valves instead of capillary tubes.

☐ When charging systems with HFC blends, such as R-407C and R-410A, the refrigerant must leave the tank as a liquid and be carefully throttled into the low side of the system.

REVIEW QUESTIONS

1. **Which of the following protects a motor from overheating?**
   a. Crankcase heater
   b. Internal overload protector
   c. Low-pressure control
   d. Low ambient control

2. **Reduced airflow through an evaporator could be caused by**
   a. Dirty air filters
   b. Dirty evaporator coil
   c. Dirty condenser coil
   d. Both a and b are correct

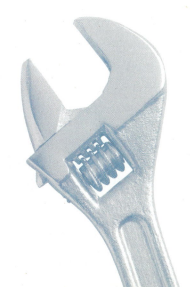

3. **Permanently lubricated motors are equipped with**
   a. Oil ports
   b. Grease fittings
   c. Relief plugs
   d. None of the above is correct

4. **Endplay in a motor shaft is an indication that**
   a. The motor needs to be lubricated
   b. The bearings have been damaged
   c. The grease fitting has come loose
   d. All of the above are correct

5. **The drive pulley is connected to the**
   a. Motor bearings
   b. Motor shaft
   c. Blower shaft
   d. Blower wheel

6. **Belt tension that is too tight can result in:**
   a. Belt slippage
   b. Reduced airflow
   c. Excessive noise
   d. "Polishing" of the pulley groove

7. **A belt that is too loose can lead to**
   a. Belt slippage
   b. Bearing damage
   c. Reduced airflow
   d. Both a and c are correct

8. **True or False: A dirty return air filter can cause reduced airflow in the supply duct.**

9. **Which of the following can cause reduced airflow through an outdoor condensing unit's coil?**
   a. Blocked or closed supply registers
   b. Excessive amounts of fallen leaves or shrubs
   c. Dirty evaporator coil
   d. Both a and b are correct

10. **True or False: The relief plug must never be removed when adding grease to a motor equipped with grease fittings.**

11. **Excessive belt breakage is an indication that**
    a. The belt tension is too tight
    b. The belt tension is too loose
    c. The belt may be too small
    d. Both a and c are possible

12. **Which of the following information can be obtained directly from the dial of a low-side pressure gauge?**
    a. Evaporator saturation temperature
    b. Evaporator superheat
    c. Evaporator outlet temperature
    d. All of the above are correct

13. **A low refrigerant charge will cause**
    a. The discharge pressure to rise
    b. The suction pressure to rise
    c. The evaporator superheat to increase
    d. Excessive cooling capacity

14. Liquid floodback occurs when

a. The evaporator produces a large quantity of condensation

b. Liquid refrigerant travels back to the compressor

c. Liquid refrigerant floods the floor

d. Liquid refrigerant travels from the compressor to the evaporator

15. Refrigerant leaks can be detected by which of the following methods?

a. UV light

b. Electronic leak detectors

c. Halide torch

d. All of the above are correct

16. True or False: Large leaks can be detected by pressurizing the system with refrigerant and listening for the leak.

17. True or False: A noncondensable gas is one that cannot condense into a liquid.

18. If the thermal bulb of a thermostatic expansion valve comes loose from the suction line

a. The evaporator superheat will go up

b. The evaporator superheat will go down

c. The evaporator superheat will remain the same

d. The system refrigerant will leak out

19. One turn of the superheat spring adjust will generally change the superheat how many degrees?

a. 0.5 to 1.0 degrees

b. 1.0 to 2.0 degrees

c. 2.0 to 2.5 degrees

d. 2.5 to 3.5 degrees

20. Reduced water flow through a water-cooled condenser will cause

a. The head pressure to rise

b. The condenser saturation temperature to rise

c. The condenser saturation pressure to drop

d. Both a and b are correct

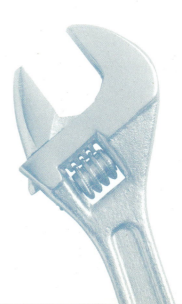

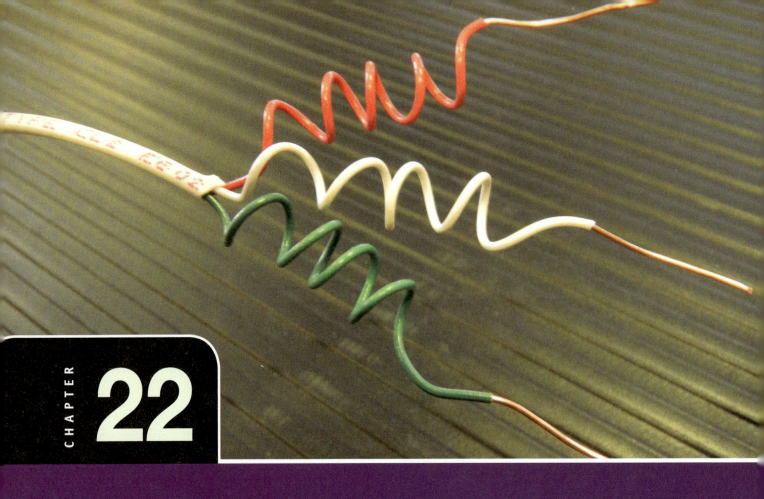

22

Electrical Troubleshooting

OBJECTIVES *Upon completion of this chapter, the student should be able to*

- Check a contactor or relay holding coil to determine if it is defective.

- Check a thermostat to determine if it is functioning properly.

- Check a transformer to determine if the device is operating correctly.

- Explain how to establish the cause of a blown control-circuit fuse.

- Determine the proper operation of pressure and other safety devices.

- Determine the reason for a blown power circuit fuse or tripped circuit breaker.

- Check a motor for grounds, open windings, or internal short circuits.

- Field-test a capacitor.

- List the seven steps in performing a successful service call.

- Perform a successful service call.

Knowing what controls and loads are in any given electric circuit is the first step in effective troubleshooting. Knowing the system's sequence of operations is the next step. Another important step, the application of basic circuit theory, is being able to examine a given circuit electrically and determine if the individual components in the circuit are performing as they should. To accomplish this, the technician must know the characteristics of these components. For example, what resistance reading is too high or too low when checking the coil of a contactor? Or, how does a field technician electrically check a run capacitor to determine whether or not it is defective? These questions and many others may need to be answered by the servicing technician in order to reach a correct diagnosis of the system problem. The intention of system troubleshooting is to evaluate the system and its components prior to making a physical repair.

Electrical troubleshooting is often performed with the aid of a voltmeter, an ammeter, and an ohmmeter. These three pieces of test equipment are just as important as a doctor's stethoscope or a fisherman's hook. Electrical troubleshooting cannot be performed without these instruments. By examining a system's schematic or ladder diagram, the servicing technician can determine which switches and controls are in series with each individual component. If a component, such as a motor, is not operating correctly, the technician will be able to determine *why* this is the case. Is the circuit open? Is a winding in the motor defective? Is power being supplied to the circuit?

Frontseating in the case of the liquid line service valve, the process of closing off the liquid line, leaving the gauge port open to the condensing unit side of the valve

Open motor winding condition in which an internal motor winding experiences a break that prevents current from flowing through the winding

Pitting condition in which the surfaces of relay or contactor contacts become worn

Pumping down process of storing the system refrigerant in the condenser by frontseating the liquid valve at the outlet of the condenser

Short circuit to ground condition in which an internal motor conducting surface comes in contact with the shell of the component

Shorted motor winding condition in which damage to a motor winding exists, creating a reduction in the resistance of the winding

In the previous chapter, we discussed how various system components and controls operate in air conditioning systems. In this chapter we will discuss methods and procedures for evaluating and troubleshooting these system components to ensure that they are operating as intended. It is also our intent to determine the cause for component failure, as there are often underlying conditions and circumstances that cause the failure. Simply replacing the defective component may not remove the actual cause of the problem.

CONTROL-CIRCUIT PROBLEMS

As shown earlier in the text, most electrical schematics are divided into two circuits: the control circuit and the power circuit. This section examines components commonly found in the control circuit and discusses how to properly evaluate them. Like any circuit, the control circuit consists of the following:

- Power supply
- Switches and/or controls
- One or more loads
- Path for the current to take

In low-voltage control circuits, the *power supply* is the transformer. Reading a voltage, usually 24 volts, at the transformer's secondary winding verifies that the control circuit is being powered. *Switches and controls* determine where the current in the circuit flows. The *loads* in control circuits are normally the holding coils of contactors and relays. These holding coils, in turn, control the operation of sets of contacts in the power circuit. The *path* is the interconnecting wiring among the power supply, the switches, and the loads. In this text, the focus is on low-voltage control circuits, although the control circuit can just as easily be 115, 208, 220, or 240 volts.

Low-voltage control circuits are easily distinguishable from the power circuit. The wire used in these circuits is much thinner, since the voltage is low and, more importantly, the current flowing in the circuit is low, normally in the range of 1.7 amperes for a 40-VA transformer. This comes from the power rating of the transformer, VA, which is equal to the voltage times the current. For the 24-volt control circuit, the amperage would be 40-VA/24-v, which is 1.7 amperes. The wire used in low-voltage control circuits is usually brightly color-coded and comes encased in either white or brown

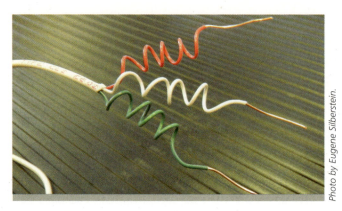

FIGURE 22-1 Thermostat wire.

vinyl (Fig. 22-1). This makes running the wire from one point to another much easier, while keeping all of the individual conductors together. Most localities do not require technicians to have an electrician's license to work on or install low-voltage wiring.

HOLDING COILS

Just as a motor has a voltage rating, the holding coil of a contactor or relay also has a voltage rating. It is important that the voltage supplied to the holding coil matches this rating. If the voltage and the rating do not match, the device will not work properly and, in many cases, can become damaged beyond repair. The relay, for example, is a throwaway device since it is not economically feasible to repair if it becomes damaged. If the coil burns out, the entire relay must be replaced. The contactor, on the other hand, may, in some instances, be repaired if the coil becomes defective. Smaller contactors are usually treated like the relay and discarded. Larger contactors, though, tend to be very expensive. The field technician can quickly replace the holding coil or the contacts of a contactor (Fig. 22-2) provided the proper replacement coil is available. But how can a technician determine if the holding coil is defective or not?

FIGURE 22-2 Replaceable parts of contactors.

The holding coil can best be evaluated when the power is disconnected and the wires leading to the coil are removed. Removing the wires ensures that the readings taken are accurate and reliable by eliminating the possibility of reading through a parallel circuit. When evaluating the coil, the technician can come to one of three possible conclusions:

- The coil is good
- The coil is shorted
- The coil is open

To properly check the holding coil, an ohmmeter should be used. The meter should be set to the lowest scale that will register actual resistance. Using the continuity scale, for example, will only rule out whether the coil is open but will not distinguish between a good coil and a coil that is shorted.

The resistance reading of the coil will determine whether or not the coil is good. If the resistance reading across the coil is 0 ohms, the coil is shorted and should be replaced. Again, if the coil is in a relay or a small contactor, the entire device should be discarded. If the resistance reading of the coil is infinity, double-check the coil by turning the dial on the meter to the next lowest resistance range in case the resistance of the coil is higher than the top limit of the lowest range. If the reading is still infinity, discard the coil. If the ohmmeter registers a resistance value, chances are that the coil is good. Exact resistance values cannot be provided here, since they vary depending on the manufacturer and the voltage rating of the coil. The manufacturer of a specific relay or coil can be consulted if the actual resistance value is needed.

THERMOSTATS

Many inexperienced technicians find that the thermostat is a very confusing system component—both in understanding its operation and troubleshooting it. Part of this confusion arises from the fact that many different types and configurations of thermostats are produced by a number of different companies. The easiest way to understand the thermostat is to realize that it acts as a series of switches that open and close in response to some external and internal conditions. Most of these conditions are manually determined by the end user of the equipment. For example, the owner decides whether the system is going to operate in the cooling or heating mode and whether the indoor fan is going to operate continuously or just when the compressor is operating.

The system thermostat is normally comprised of two separate components: the thermostat proper and the subbase. Most troubleshooting of the thermostat takes place at the subbase, which is where the electrical connections to the device are made. A simple, low-voltage control circuit operates with the following wires and terminals:

Red	R terminal	Hot wire from transformer
Yellow	Y terminal	Cooling circuit
White	W terminal	Heating circuit
Green	G terminal	Indoor fan circuit

Most low-voltage thermostats adhere to the preceding color-coding and terminal designation, which makes the evaluation of the circuits much easier. When the thermostat is removed from the subbase, these terminals can be seen (Fig. 22-3). They are labeled, screw terminals to which the color-coded, low-voltage wires are connected. With the thermostat removed, the majority of the system can be evaluated. It is often helpful to realize that the R terminal feeds power to the thermostat and the G, W, and Y terminals carry power from the thermostat to the fan, heating, and cooling circuits, respectively.

A small, insulated jumper wire connected to the R terminal (Fig. 22-4) is all that is needed to determine if the thermostat and subbase are operating correctly. The first thing that is normally checked on a thermostat is the indoor fan operation. This is done for several reasons. First, the indoor fan can usually be heard when it is energized and de-energized, because the thermostat is often located near the return-air grill—hence, near the fan itself. Another reason for checking the fan first is to verify the presence of low-voltage power being supplied to the thermostat.

Remember that the indoor fan motor cannot operate if there is no low voltage. The following represents a typical, but by no means the only, procedure to evaluate the operation of the indoor fan motor:

- With the thermostat removed, turn the fan switch to the ON position.
- The indoor fan should begin to operate.

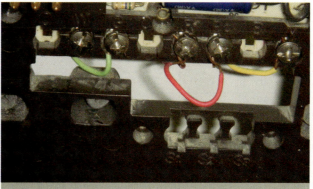

Photo by Eugene Silberstein.

FIGURE 22-3 Electrical connections are made on the subbase.

NOTE: THE FAN SELECTOR SWITCH IS NORMALLY A
PART OF THE SUBBASE.

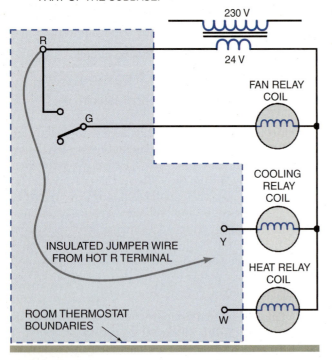

FIGURE 22-4 Jumper wire can be used to bypass
the thermostat to test system component operation.

- If the fan starts, the transformer is producing the
  desired low voltage, and the R terminal on the thermostat has a potential of 24 volts.

- By turning the fan switch to the AUTO position, the
  fan should turn off and there should be a 24-volt
  reading between the R and G terminals on the subbase (Fig. 22-5).

- If the fan does not start, place a jumper from the
  R terminal to the G terminal. If the fan starts, the
  subbase is defective and needs to be replaced. If the
  fan does not start, there is most likely a problem
  with the system transformer or the associated
  wiring. In most cases, this indicates that the
  problem does *not* lie within the thermostat or
  subbase.

Evaluating the cooling portion of the thermostat can be
accomplished as follows:

- If the indoor fan operated when it was checked
  before, place a jumper from R to Y to energize the
  cooling circuit (Fig. 22-6).

- If the cooling circuit is energized (compressor runs)
  when the jumper is in place, but it is not energized
  when the thermostat is replaced and set to the
  cooling mode, the thermostat is defective.

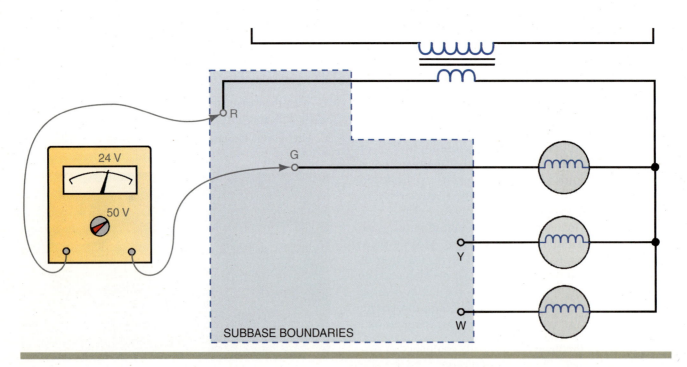

FIGURE 22-5 A multimeter can be used to test the control circuit of a system at the subbase with the
thermostat removed.

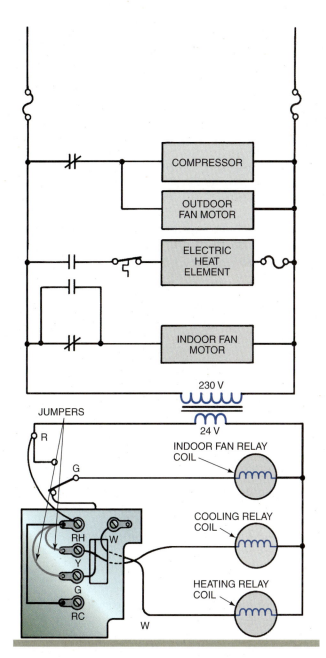

Going Green

Many older style thermostats contain mercury. If it is determined that a mercury thermostat is defective, it must be disposed of in an environmentally acceptable manner.

FIGURE 22-6 Jumper placed between the R and Y terminals should energize the cooling circuits. Jumper placed between the R and G terminals should energize the indoor fan circuit.

Evaluating the heating portion of the thermostat can be accomplished as follows:

- If the indoor fan operated when it was checked before, place a jumper from R to W to energize the heating circuit.
- If the heating circuit is energized when the jumper is in place, but it is not energized when the thermostat is replaced and set to the heating mode, the thermostat is defective.

TRANSFORMERS

Transformers are among the easiest components of the control circuit to evaluate. They have no moving parts and nothing to oil, tighten, adjust, or calibrate (Fig. 22-7). When voltage is applied to the primary winding, an induced voltage is produced at the secondary. The transformer can be easily damaged if care is not taken when the device is being installed or when the system is being serviced. If the transformer is a multi-tap-type transformer, the following guidelines should be observed:

- Make certain that the device is wired according to the wiring diagram on the device.
- Make certain that the proper wire leads are used for the voltage supplied to the device.
- Make certain that all unused wire leads are capped individually and taped to prevent the possibility of a short circuit.

When troubleshooting the transformer, both the primary and the secondary windings can be evaluated. The transformer can be checked while installed in the system with voltage applied to it and when removed entirely from the system. The most obvious way to check the transformer is to apply the correct voltage to the primary

Photo by Eugene Silberstein.

FIGURE 22-7 Transformer.

winding and take a voltage reading at the secondary. If a 24-volt reading is obtained at the secondary, then the transformer is functioning correctly. If there is no secondary voltage, further investigation is necessary and the device should be disconnected from the system and checked as follows:

- Using a multimeter set to read resistance, take a reading of the primary winding. If the resistance is infinity, the winding has burned. Applying the incorrect voltage to the primary winding of the device often causes this. If there is a measurable resistance across the primary winding, it is okay.

- Using a multimeter set to read resistance, take a reading across the secondary winding. If the reading is infinity, the winding is open and the transformer needs to be replaced. An open secondary winding is often the result of a short in the control circuit. Before replacing the transformer, the short circuit must be located and repaired. A measurable resistance across the secondary winding is an indication that it is okay.

CONTROL FUSES

Good field practice includes installing a small fuse, in the range of 3 amperes, in series with the secondary winding of the transformer. This in-line fuse will blow if a short develops in the control circuit, thereby acting as a safety and protecting the transformer from future damage. When the circuit is operating correctly, this in-line fuse will simply act as a closed switch.

Control fuses, also called in-line fuses, are designed, like power-circuit fuses, to protect the components in the circuit. The primary device being protected in the control circuit is the transformer. Control fuses are usually encased in a plastic holder that provides the means by which they are connected in the control circuit. As with any other type of fuse, the reason the fuse blew must be discovered and corrected before replacing it. The most common cause of a blown control-circuit fuse is a shorted holding coil, either on a contactor or on a relay. Three ways that a control circuit fuse can be checked are

- visually
- with an ohmmeter
- with a voltmeter

Since many in-line fuses are made of glass, the element can be clearly seen. If the element is intact, the fuse is good. Even though this is sufficient in most cases, checking the fuse by other means may be necessary, especially when the fuse is not made of glass.

Delmar/Cengage Learning.

FIGURE 22-8 Checking the fuse with an analog ohmmeter.

Checking the fuse with an ohmmeter is a very simple matter. The fuse must be removed from the circuit, the ohmmeter should be set to read continuity, and a test lead should be placed on each end of the fuse. If there is continuity through the fuse, it is good. If there is no continuity, the fuse needs to be replaced (Fig. 22-8).

Checking fuses with a voltmeter is also very effective. The fuse is left in the circuit and the power to the circuit is left on. Since a fuse acts as a switch, there should be 0 volts across a good fuse (closed switch) and 24 volts across a blown fuse (open switch).

PRESSURE CONTROLS AND SAFETY DEVICES

Pressure switches and other pressure-actuated devices can act as safety controls as well as operational controls (Fig. 22-9). Safety controls are designed to disable the

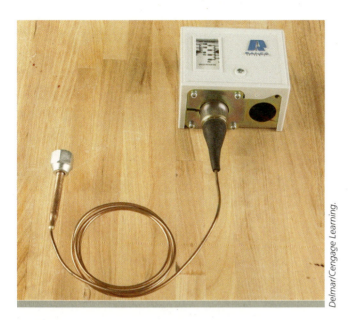

Delmar/Cengage Learning.

FIGURE 22-9 High-pressure switch.

Delmar/Cengage Learning.

FIGURE 22-10 Frontseating the liquid line service valve.

system if unsafe conditions arise. Like any other switch, these devices can be checked with a voltmeter to determine if they are in the open or closed position. A closed pressure switch will have a potential difference across it of 0 volts. However, evaluating these controls electrically is not enough to ensure proper operation of the device.

The mechanical action of the contacts and the bellows in the control must also be checked to ensure that the contacts open and close at the correct time. To do this, a gauge manifold should be installed on the system. By placing a voltmeter across the control, the exact pressure at which the contacts of the control open and close can be determined by observing the display on the meter. For low-pressure controls, the evaluation process would be conducted as follows:

- Install a gauge manifold on the system with the power off.
- Connect a voltmeter across the low-pressure control.
- Make note of the pressure at which the control is set to cut out.
- Turn the system on.
- The voltmeter should be reading 0 volts since the contacts of the control are closed.
- Frontseat the liquid-line service valve on the liquid line at the outlet of the condensing unit (Fig. 22-10). **Frontseating** is the process of turning the stem on a service valve all the way clockwise, closing off the line port on the valve. The compressor will continue to operate and pull refrigerant from the evaporator and liquid line, but the refrigerant in the system will not be permitted to leave the condenser, causing the low-side pressure to drop. This process is called **pumping down**.
- Observe the needle on the low-pressure gauge as the pressure drops.
- When the display on the voltmeter reads 24 volts, the contacts of the low-pressure control have opened. If the low-pressure control is in series with the compressor contactor's holding coil, it should be obvious that the contacts open at the same instant that the compressor cycles off.
- Compare the pressure at which the contacts opened to the pressure that was set on the control.
- Adjust as needed.
- Continue to make adjustments until the contacts open and close at the desired pressures.

For high-pressure controls, the evaluation process is more difficult. Since high-pressure controls are primarily

Going Green

Whenever refrigeration gauges are to be installed on a system, be sure to take all precautions to minimize the loss of refrigerant to the atmosphere. Intentionally releasing refrigerant to the atmosphere is a violation of the laws set forth in the Clean Air Act.

used to disable the system if excessively high pressure exists, simulating that pressure is difficult. These controls can, however, be set with at least some degree of accuracy and certainty using the following guidelines.

IF THE HIGH-PRESSURE CONTROL IS ALREADY INSTALLED ON THE SYSTEM

- Set the control to cut out at a pressure about 20 psig higher than the normal operating head pressure of the system.
- Turn the system on.
- The contacts on the high-pressure control should be closed, and the system should operate normally.
- Slowly adjust the pressure control setting to a lower pressure until the system shuts down.
- Compare the pressure at which the system shut down to the pressure setting on the control.
- If these pressures are the same, the setting of the control can be raised to the desired pressure.
- Since the control is properly calibrated, the cutout pressure should be very close to the reading on the control's face.

IF THE HIGH-PRESSURE CONTROL IS NOT INSTALLED ON THE SYSTEM

- Connect the transmission line from the pressure control to the high-side port of the gauge manifold (Fig. 22-11).
- Connect the center hose of the gauge manifold to the regulator on a nitrogen tank (Fig. 22-12).
- Open the nitrogen tank and set the regulator to a pressure just higher than the desired cutout pressure of the high-pressure control.

FIGURE 22-11 Connecting a high pressure switch to the gauge manifold for calibrating and testing purposes.

Delmar/Cengage Learning.

Delmar/Cengage Learning.

FIGURE 22-12 Connecting the gauge manifold to a nitrogen tank and regulator.

- Slowly open the high-side valve on the gauge manifold to allow the nitrogen to enter the high-pressure control.
- An audible click will be heard when the contacts of the control open. An ohmmeter can also be connected across the device. While the contacts are closed, there should be continuity. When the contacts open, there should be no continuity.
- Compare this pressure to the desired cutout pressure.
- Adjust the control as needed.

POWER-CIRCUIT PROBLEMS

This section examines components commonly found in the power circuit and discusses how to properly evaluate them. The power circuit is generally easier to evaluate and troubleshoot than the control circuit. This is because the loads, whether compressors or fan motors or pumps, are connected directly to the line-voltage power supply. The devices that cause these loads to be energized and de-energized are the sets of contacts that are controlled by the holding coils of the control circuit. Generally speaking, all power circuits are configured basically the same for similar types of equipment.

CONTACTOR AND RELAY CONTACTS

When a set of electrical contacts closes, current flows to the load that is ultimately being controlled. These contacts are made up of smooth, metallic surfaces that, when closed, offer very little resistance to current flow [Fig. 22-13(A)]. Because of this, the voltage reading across a set of closed contacts should be 0 volts. Since the terminals on the line side of a contactor are labeled L1 and L2 and the terminals on the load side of a contactor are T1 and T2 (Fig. 22-14), the potential difference between Terminal L1 and Terminal T1 would be 0 volts.

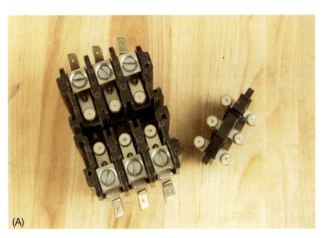

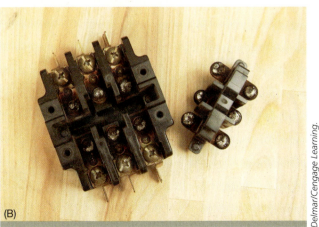

FIGURE 22-13 (A) Good set of contactor contacts. (B) Pitted set of contacts.

Delmar/Cengage Learning.

FIGURE 22-14 The line side of the contactor is labeled L1 and L2, while the load side of the contactor is labeled T1 and T2.

Similarly, the potential difference between Terminal L2 and Terminal T2 would also be 0 volts. If a resistance reading was taken between Terminal L1 and Terminal T2 when the contacts are closed, it would be 0 ohms, or very close to it.

> **CAUTION**
>
> **CAUTION:** Taking the resistance reading in a circuit should only be done when the circuit is de-energized.

As relays and contactors age, the contacts undergo a process called **pitting**, in which the smooth surfaces of the contacts become rough. Pitting [Fig. 22-13(B)] results from the electrical arcing that takes place whenever electrical contacts open and close. This arcing generates high temperatures, which slowly cause the contact surfaces to deteriorate. As the pitting gets worse, the operation of the main load being controlled is affected. The resistance across a set of pitted contacts is measurable. If the resistance is greater than 2 or 3 ohms (Fig. 22-15), the contacts should be replaced. If these contacts are enclosed in a relay, the relay should be discarded and replaced. Resistance across a set of contacts will result in a potential difference greater than 0 volts across the contacts. This can cause problems with a motor or other load, because these contacts are in series with the load and a voltage drop across the contacts will reduce the voltage being supplied to the load. Consider the following simple circuit.

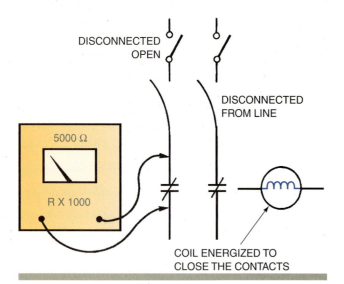

FIGURE 22-15 Checking the contacts of a contactor with an ohmmeter shows resistance across the contacts as a result of pitting.

A 120-volt power supply is feeding a 30-ohm load (Fig. 22-16). The load is controlled by a normally open set of contacts. When closed, the contacts offer no resistance to current flow. The current in the circuit is calculated, using Ohm's law, as:

$$I = 120/30 = 4 \text{ amperes}$$

The load has 4 amperes flowing through it and is supplied with the full 120 volts as provided. Now consider the following: The contacts that control the load are very old and pitted (Fig. 22-17). The resistance across them is 10 ohms (very exaggerated). The total circuit resistance is now:

$$30 + 10 = 40 \text{ ohms}$$

The circuit current is now given by:

$$I = 120/40 = 3 \text{ amperes}$$

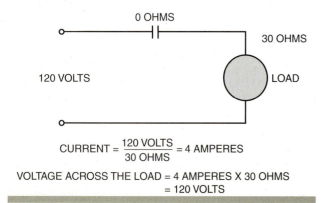

0 OHMS

30 OHMS

120 VOLTS

LOAD

$$\text{CURRENT} = \frac{120 \text{ VOLTS}}{30 \text{ OHMS}} = 4 \text{ AMPERES}$$

VOLTAGE ACROSS THE LOAD = 4 AMPERES X 30 OHMS = 120 VOLTS

FIGURE 22-16 Good contacts add very little or no resistance to the circuit.

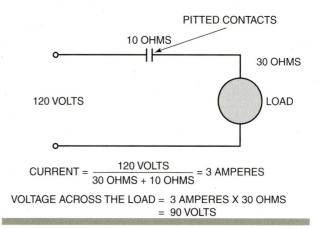

PITTED CONTACTS

10 OHMS

30 OHMS

120 VOLTS

LOAD

$$\text{CURRENT} = \frac{120 \text{ VOLTS}}{30 \text{ OHMS} + 10 \text{ OHMS}} = 3 \text{ AMPERES}$$

VOLTAGE ACROSS THE LOAD = 3 AMPERES X 30 OHMS = 90 VOLTS

FIGURE 22-17 Pitted contacts add resistance to the circuit, reducing the voltage supplied to the load.

The voltage that is now supplied to the load can be calculated, again using Ohm's law, by:

$$V = 30 \text{ ohms} \times 3 \text{ amperes} = 90 \text{ volts}$$

If the load is rated at 120 volts, it will safely operate at 120 volts plus or minus 10 percent. 120 volts minus 10 percent is 108 volts. This load is operating well out of its designed range and will experience operational problems. In this case, the contacts should be replaced.

CIRCUIT BREAKERS AND FUSES

Just as control-circuit fuses protect the components in the control circuit, circuit breakers and fuses are used to protect the components in the power circuit (Fig. 22-18 and Fig. 22-19). Since the amperage draw in the power circuit is generally much greater than the current draw of the control circuit, the power-circuit fuses and circuit breakers are much larger than those in the control circuit. A blown fuse can be identified relatively easily, as

FIGURE 22-18 Fuses.

FIGURE 22-19 Circuit breakers.

Delmar/Cengage Learning.

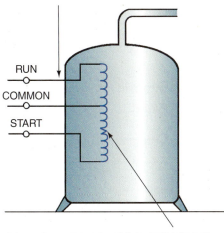

THE MOTOR TERMINALS ARE INSULATED WHERE THEY PASS THROUGH THE COMPRESSOR HOUSING

RUN

COMMON

START

SOME OF THE WINDINGS ARE TOUCHING EACH OTHER AND REDUCING THE RESISTANCE IN THE START WINDING

FIGURE 22-20 Compressor with a shorted winding.

described earlier. A good fuse will allow current to flow freely through it, while a voltage reading taken across it will register 0 volts. Again, a good fuse acts as a closed switch. A reading of line voltage across the fuse indicates that the fuse is bad and needs to be replaced.

Just as with any other component, the reason for the failure needs to be determined and corrected before the device is replaced. Fuses blow and circuit breakers trip because of excessive current flowing in the circuit. Excessive current can be a result of a number of situations, including

- a shorted device (load) such as a fan motor or a pump
- a short in the associated component wiring
- a short circuit to ground

A load or device that is shorted will offer little or no resistance to current flow (Fig. 22-20). This increases the current draw of the device and will cause the power-circuit fuse to blow or the circuit breaker to trip. In this case, the device will need to be either repaired or replaced, depending on the economics of the situation.

A short circuit in the associated wiring is often less costly to the customer but can be very difficult to isolate and correct. A short circuit in the wiring can be the result of

- improper field wiring by a technician
- improper installation of system components or accessories
- worn wire insulation (conductors making contact with other conductors)

In any case, each circuit should be evaluated with an ohmmeter to determine where the short circuit exists. Each circuit should have at least some degree of measurable resistance. If not, chances are that a dead short

Going Green

An occasional tripped circuit breaker or blown fuse can often be the result of a temporary increase in system load, power brown out or other one-time occurrence. When this happens, many call it a nuisance trip. However, when a circuit breaker trips frequently, or when a fuse blows regularly, a significant system problem is likely present. Allowing systems to operate under these conditions will have a negative effect on system operating effectiveness and efficiency. Inefficient systems provide inadequate cooling and cost more money to operate.

must be isolated and repaired. As equipment ages, the wire insulation usually becomes worn or damaged. When this damage occurs, short circuits are likely to surface.

A **short circuit to ground (short to ground)** is a special type of short circuit. This occurs when a current-carrying conductor makes contact with a ground or zero-potential metallic surface (Fig. 22-21). This is quite

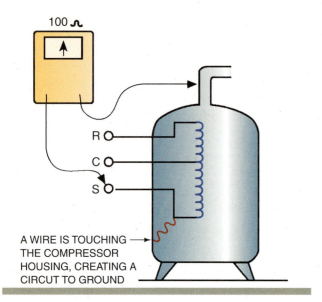

A WIRE IS TOUCHING THE COMPRESSOR HOUSING, CREATING A CIRCUT TO GROUND

FIGURE 22-21 Compressor with a short to ground.

common in our industry, because a number of such surfaces make a perfect ground. They include:

- The compressor shell
- The refrigerant-carrying copper tubing and piping
- The evaporator and condenser coils
- The casing of the condensing unit and air handler

To help prevent possible shorts to ground, all wiring should be inspected regularly and all electrical connections should be tight and sealed properly. A short to ground can be identified by a measurable resistance between a current-carrying conductor and a grounded surface. When checking for a short to ground, the power should be disconnected and the ohmmeter should be set to a high range. If the meter registers resistance between a conductor and ground, the range can always be lowered to get a more accurate reading. If the range selected is too low, a high resistance may not register on the display, leading the technician to conclude that there is no short circuit when in fact there is. Short circuits to ground are easily remedied when they are in the associated wiring but not when they exist within the motor, pump, or compressor. Shorted wires can be either replaced or insulated. Shorted components often need to be replaced. The next section deals with some specifics on the evaluation of fan motors and compressor motors.

FAN MOTORS AND COMPRESSOR MOTORS

Fan motors and compressors are among the most expensive components to replace in an air conditioning or refrigeration system. Therefore, these components must

be properly evaluated. An improper determination can lead to artificially high repair costs to the customer. Motors can fail for a number of reasons that can be broken down into two categories: mechanical and electrical. The mechanical problems were addressed earlier in the text. The electrical problems often encountered are, just as with other devices made up of wire coils, the open winding and the shorted winding. These often result in costly repairs; and, in either case, the device often needs to be replaced or rebuilt.

OPEN MOTOR WINDINGS

The windings of an electrical motor can best be checked with an ohmmeter. The power supply should be off, and the wires supplying power to the motor should be disconnected. All capacitors, if there are any, should be properly discharged and removed from the circuit. Before checking the actual windings, though, it may be suspected that a motor has an open winding when voltage is supplied to the device and it does not operate.

> ### CAUTION
>
> **CAUTION:** Working on electrical circuits that are energized can result in electric shock and personal injury. Capacitors store electric charge and should also be discharged prior to working on circuits.

Single-phase, split-phase motors, as we already know, have both a start and a run winding. The resistance of the start winding can be determined by taking a reading between the common wire and the start wire of the motor. These terminals are easily identified on a compressor motor, but the wiring diagram on the body of a fan motor may need to be checked to identify the proper wires. The resistance of the run winding can be determined in a similar manner by taking a reading between the common and the run terminals of the motor. The resistance of the start winding should be greater than the resistance of the run winding. If either or both of the readings is infinity, there is an **open motor winding** (Fig. 22-22) and the device needs to be replaced or repaired.

Simple, three-phase motors have three windings instead of two. The terminals on these three-phase motors are labeled T1, T2, and T3. The resistances of these windings should all be the same. If any resistance reading between the terminals is infinity, the motor has an open winding and must be replaced or repaired.

Open windings can be caused by a number of factors, including motor age, overheating, improper voltage supply, and improper application. If the technician finds that a motor does, indeed, have an open winding, the

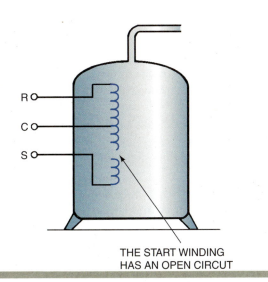

FIGURE 22-22 Compressor with an open start winding.

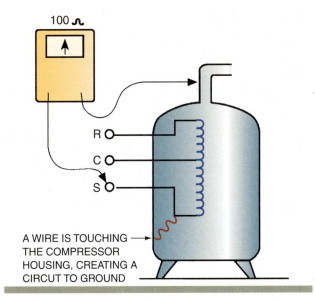

FIGURE 22-23 Grounded winding.

associated wiring, power supply, and intended application need to be checked as well to make certain that the new motor does not fail prematurely.

SHORTED MOTOR WINDINGS

When the resistances of the windings are lower than normal, there is a good possibility that the windings are shorted. For example, if the resistance of the start winding is 8 ohms and the resistance of the run winding is 12 ohms, there is most likely a partially **shorted motor winding**. The short is in the start winding. Part of the coil has been removed from the circuit by a short circuit. A resistance reading of 0 ohms across a winding indicates a complete short. This often becomes obvious when power is initially fed to the device. The circuit breaker will trip, or the fuse will blow, indicating the presence of a short circuit.

GROUNDED MOTOR WINDINGS

Another problem that occurs within motors is the grounded winding (Fig. 22-23). This is often the result of internal motor overheating. The insulation on the winding breaks down and comes in contact with the shell of the device or with other metallic surfaces or conductors, causing a short circuit. In essence, this is the same as a shorted winding; but, in this case, the connection is made to the shell of the device. Again, using an ohmmeter can identify a short to ground. The meter should be set to read high resistance, and readings should be taken between the conductors and the shell of the device. Any measurable resistance reading is an indication that the motor is grounded and needs to be replaced or repaired.

Going Green

If a motor has a shorted winding, be sure to find out why. A shorted motor winding is often the result of improper airflow across the motor, causing it to overheat. Simply replacing the defective motor will not solve the *real* problem, which might be a dirty air filter or closed supply registers. By replacing the motor without resolving the problem, the system will likely fail again, resulting in wasted resources, inefficient system operation, and an unhappy customer.

CAPACITORS

Often technicians new to the field decide that a motor is defective when, in reality, the capacitor is at fault (Fig. 22-24). A defective capacitor can prevent a motor from operating or it can lead to more serious motor problems resulting from overheating. A successful technician must be able to field-test a capacitor to determine if the device is functioning properly. Several instruments are available to field-test capacitors, but they can also be quickly and effectively tested and evaluated using an ohmmeter. An appropriate field check for a capacitor is as follows:

- Make certain that the capacitor is properly discharged.

Delmar/Cengage Learning.

FIGURE 22-24 Start and run capacitors.

- Remove wiring from the capacitor, including bleed resistors, noting the location of the wires to ensure proper reinstallation of the device.
- Set the resistance range to the R × 1,000 or R × 100 range for start capacitors or the R × 1,000 or R × 10,000 range for run capacitors.
- Touch the meter's leads to the capacitor terminals.
- If the capacitor is good, the display should fall quickly toward 0 ohms and then slowly go back toward infinity.
- If the display reads 0 ohms and stays there, the capacitor is shorted and needs to be replaced.
- If the display falls toward 0 ohms and then registers a measurable resistance, the capacitor has a partial short and should be replaced as well.
- If the capacitor is to be checked again, the test leads should be reversed.
- There should be infinite resistance between the capacitor terminals and the casing of the device. If there is continuity, the device should be replaced.

This process is outlined in Figure 22-25.

DISCONNECT BLEED RESISTOR AFTER BLEEDING A 20,000-OHM, 5-W RESISTOR FOR TEST IF DESIRED. WHEN BLEED RESISTOR IS LEFT IN THE CIRCUIT, THE METER NEEDLE WILL RISE FAST AND FALL BACK TO THE BLEED RESISTOR VALVE.

FOR BOTH RUN AND START CAPACITORS

1. FIRST SHORT THE CAPACITOR FROM POLE TO POLE USING A 20,000-OHM 5-WATT RESISTOR WITH INSULATED PLIERS.

20,000 OHMS

2. USING THE R X 100 OR R X 1,000 SCALE, TOUCH THE METER'S LEADS TO THE CAPACITOR'S TERMINALS. METER NEEDLE SHOULD RISE FAST AND FALL BACK SLOWLY. IT WILL EVENTUALLY FALL BACK TO INFINITY IF THE CAPACITOR IS GOOD (PROVIDED THERE IS NO BLEED RESISTOR).

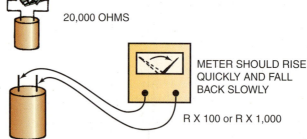

METER SHOULD RISE QUICKLY AND FALL BACK SLOWLY

R X 100 or R X 1,000

START CAPACITOR

3. YOU CAN REVERSE THE LEADS FOR A REPEAT TEST OR SHORT THE CAPACITOR TERMINALS AGAIN. IF YOU REVERSE THE LEADS, THE METER NEEDLE MAY RISE EXCESSIVELY HIGH AS THERE IS STILL A SMALL CHARGE LEFT IN THE CAPACITOR.

4. FOR RUN CAPACITORS THAT ARE IN A METAL CAN: WHEN ONE LEAD IS PLACED ON THE CAN AND THE OTHER LEAD ON A TERMINAL, INFINITY SHOULD BE INDICATED ON THE METER USING THE R X 10,000 OR R X 1,000 SCALE.

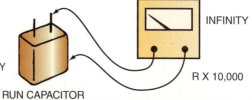

INFINITY

R X 10,000

RUN CAPACITOR

FIGURE 22-25 Testing a capacitor.

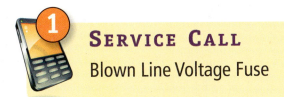

SERVICE CALL
Blown Line Voltage Fuse

CUSTOMER COMPLAINT

A customer calls with a no-cooling complaint. It is the middle of the summer, and the customer tells the service manager that the system has worked fine all season. He had a preventive maintenance service call at the beginning of the spring, at which time the refrigerant charge, air filters, and overall system operation were checked.

SERVICE TECHNICIAN EVALUATION

John arrives at the customer's home and asks the homeowner about the problem he is experiencing with his air conditioning system. After listening attentively, John goes over to the room thermostat and checks the setting. It is set to the cooling mode, and the thermostat is set to maintain a temperature that is lower than the present space temperature (Fig. 22-26). The top arrow on the face of the thermostat, which represents the desired space temperature, is set at 70°F, while the bottom arrow shows the actual space temperature. The fan switch is in the ON position. He then goes over to a wall supply grill and feels the air coming from it. The air is warm, but he can tell that the fan is operating. From the information he was given and his visual inspection, John has already made some conclusions:

- The transformer is functioning
- The low-voltage control circuit is being energized
- The holding coil on the indoor fan relay is energized
- The indoor fan is operating
- The problem most likely is in the condensing unit or the wiring supplying power to it

John leaves the thermostat set to the cooling mode and sets the desired temperature to the lowest setting. He then goes outside to the condensing unit and finds that it is not operating. After putting on his safety glasses, he turns the switch on the fused disconnect off at the condensing unit and opens the cover. He then carefully turns the disconnect switch back on. Using his voltmeter set to the proper range, he checks for voltage at the load side of the fuses in the box and reads 0 volts (Fig. 22-27). Since this system is rated at 220 volts, John has established that power is not being supplied to the condensing unit.

He knows that each hot leg of a 220-volt power supply will give a reading of approximately 115 volts to

SYSTEM SET TO COOL   DESIRED SPACE TEMPERATURE 70F   FAN IN ON POSITION

ACTUAL SPACE TEMPERATURE 76F

Photo by Eugene Silberstein.

FIGURE 22-26 Low-voltage thermostat calling for cooling.

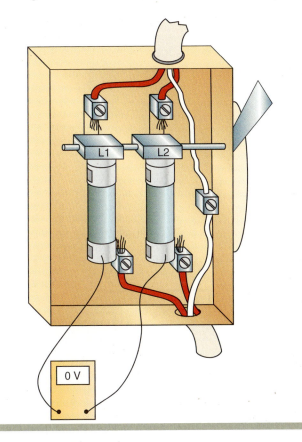

0 V

FIGURE 22-27 Power is not being supplied to the condensing unit.

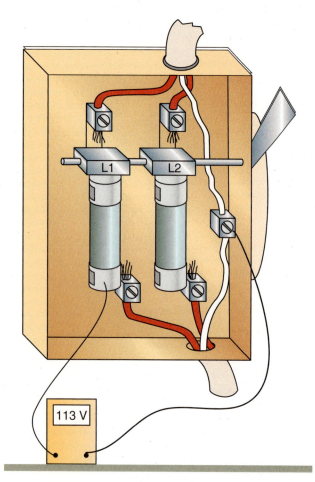

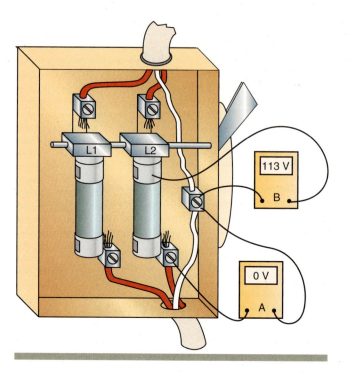

FIGURE 22-28 The fuse on the L1 side is good since power is able to pass through it.

FIGURE 22-29 No power is passing through the fuse on the L2 side. This fuse is bad.

ground. He takes one of the leads from his voltmeter and holds it to the ground terminal in the box. He touches the other test lead to the load side of the fuse on L1 and reads 113 volts (Fig. 22-28). He then does the same for L2. This time, he reads 0 volts (Meter A in Fig. 22-29). He now takes his test lead and touches the line side of the same fuse and reads 113 volts (Meter B in Fig. 22-29). In a schematic diagram (Fig. 22-30), John's findings can be easily visualized.

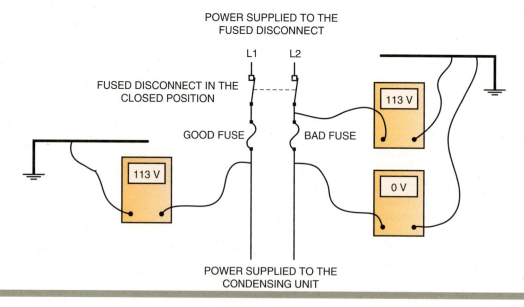

FIGURE 22-30 Schematic showing the bad fuse.

John has now established that the fuse on L2 is bad. He turns the disconnect switch off and removes the defective fuse. To be sure, he takes his multimeter and sets it to read the lowest resistance range and checks across the fuse. There is no continuity through the fuse; so, again, it is proven that the fuse is defective (Fig. 22-31).

John must now ask himself the all-important question, Why did the fuse blow? He knows it blew for a

reason, and he *must* determine the cause *before* replacing it. If he simply installs a new fuse, it will undoubtedly blow again. He now begins his systematic search for the source of the problem.

With the disconnect switch in the OFF position, John uses his multimeter, set to read resistance, to check for continuity between the power lines feeding the condensing unit and ground. When reading from the load side of the fuse on L2 to ground, he reads infinity (Meter A in Fig. 22-32). When reading from the load side of the fuse on L1, his meter reads 0 ohms (Meter B in Fig. 22-32) meaning there is a short to ground. He knows that the short is within the condensing unit, so he evaluates each of the following possible locations:

- The compressor
- The wiring leading to the compressor
- The condenser fan motor
- The wiring leading to the condenser fan motor
- The wiring from the disconnect box to the line side of the cooling contactor

By systematically checking each of the preceding scenarios, John should be able to find the problem efficiently and accurately. He first locates any capacitors in the condensing unit and properly discharges them to avoid getting an electrical shock.

FIGURE 22-31 Checking the fuse with an ohmmeter. The OL display indicates that there is no continuity through the fuse.

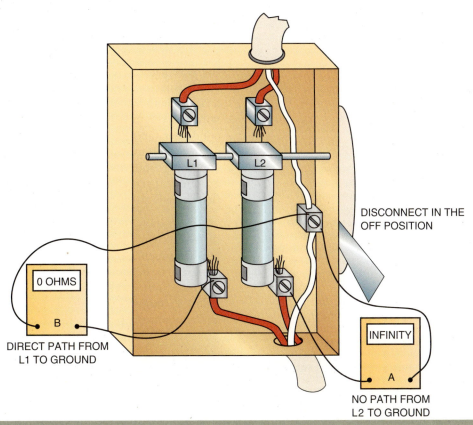

FIGURE 22-32 The load side of the disconnect switch has a short to ground on the L1 side.

CAUTION

CAUTION: Even though the disconnect switch is in the OFF position, it is still possible to receive an electrical shock from the system. Remember that capacitors are energy-storing devices that can discharge even when the system is off. Always discharge them properly before servicing equipment.

John locates the wires that supply power to the compressor. He labels the wires so he knows where they came from and disconnects them from the load side of the cooling contactor (Fig. 22-33). He then checks again for continuity at the disconnect box. He still reads 0 ohms from L1 to ground, so he can eliminate the possibility that the compressor or the associated compressor wiring is at fault. He then locates the wires from the condenser fan motor, labels them, and disconnects them from the load side of the contactor as well (Fig. 22-34). He still reads 0 ohms from L1 to ground and is now able to establish that the problem lies in the wiring from the disconnect box to the line side of the contactor.

John disconnects the two wires from the line side of the contactor and checks each for continuity to ground.

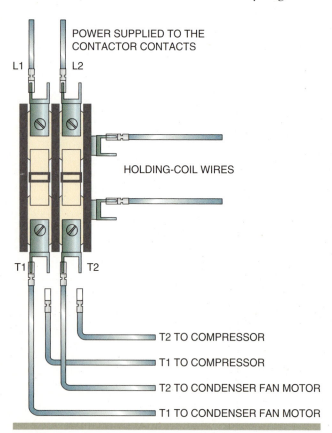

POWER SUPPLIED TO THE CONTACTOR CONTACTS

L1 L2

HOLDING-COIL WIRES

T1 T2

T2 TO COMPRESSOR

T1 TO COMPRESSOR

T2 TO CONDENSER FAN MOTOR

T1 TO CONDENSER FAN MOTOR

FIGURE 22-33 Removing the compressor wires from the load side of the contactor.

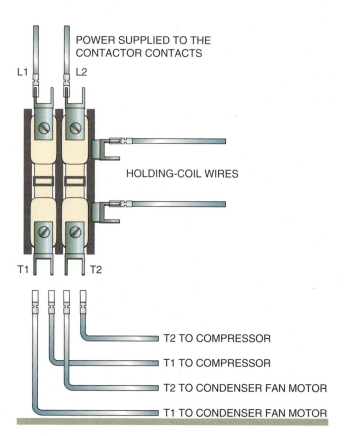

POWER SUPPLIED TO THE CONTACTOR CONTACTS

L1 L2

HOLDING-COIL WIRES

T1 T2

T2 TO COMPRESSOR

T1 TO COMPRESSOR

T2 TO CONDENSER FAN MOTOR

T1 TO CONDENSER FAN MOTOR

FIGURE 22-34 Removing the condenser fan motor wires from the load side of the contactor.

He finds that the wire from L1 is making contact with ground (Fig. 22-35). Upon closer inspection, he sees that the wire has rubbed against a sharp metal edge in the condensing unit. He goes to his truck and brings back a new length of wire and a new fuse and replaces the wire and the fuse. He reconnects all the wires from the condenser fan motor and compressor. As a final check, after all connections are made, John again checks for continuity from the lines feeding the condensing unit to ground and gets a reading of infinite resistance, meaning that there is no electrical connection between the points.

John then closes the cover on the disconnect and turns the switch on. The condensing unit begins to operate. He touches the suction line entering the unit and finds it to be very cool. He feels the liquid line leaving the unit and finds it to be warm. He monitors the system for about 15 minutes while he completes his service ticket and explains the repair to the customer.

SERVICE CALL DISCUSSION

In this service call, John listened to the customer complaint and, by simply feeling the air coming from the supply registers, was able to establish that the problem was most likely in the condensing unit or the associated wiring

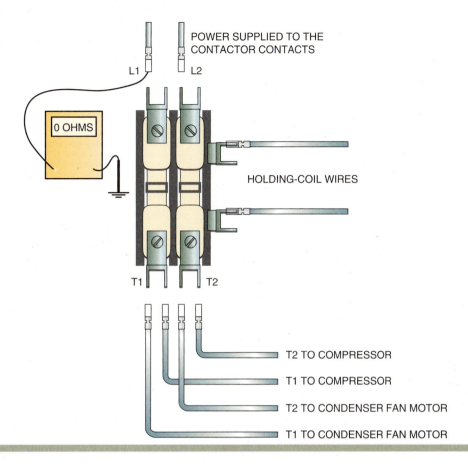

POWER SUPPLIED TO THE
CONTACTOR CONTACTS

L1   L2

0 OHMS

HOLDING-COIL WIRES

T1   T2

T2 TO COMPRESSOR

T1 TO COMPRESSOR

T2 TO CONDENSER FAN MOTOR

T1 TO CONDENSER FAN MOTOR

FIGURE 22-35 A short to ground exists within the conductor feeding the contactor.

leading to it. He was able to establish this within the first 5 minutes of the service call. By applying basic circuit theory, he quickly determined that the fuse was blown.

John did not, however, simply replace the fuse. He first determined the cause. Also, notice how the service technician used two different methods to establish that the fuse was defective:

- He used a voltmeter to read the voltage across the fuse while it was in the circuit, just like reading voltage across a switch. A good fuse acts as a closed switch, and a defective fuse acts as an open switch.

- He also used an ohmmeter once the fuse was removed from the circuit. He knew that if no continuity existed through the fuse, the fuse was bad. Similarly, if continuity existed through the fuse, the fuse was good.

After he determined that there was a short to ground, he proceeded in a logical manner to isolate the location of the short. He did not simply check around haphazardly, hoping that the problem would surface on its own. By using his test instruments correctly and applying common sense, he was quickly able to isolate the trouble and, ultimately, resolve it.

SERVICE CALL
Defective Transformer

CUSTOMER COMPLAINT

A home owner places a service call to the air conditioning service company. She tells the dispatcher that the unit is not cooling and there is no air coming from the supply registers. She explains to the dispatcher that she turned the circuit breaker for the system off and then back on, but the system still did not come on. This unit is a package-type system (Fig. 22-36) and is located behind the house. Package units are self-contained and do not require any field piping during installation. They do, however, require the installation of supply and return ducts to the outdoor location of the unit. The dispatcher informs her that William, the company's lead man, is presently working on a construction job nearby and that he will pull him off the job and send him over right away.

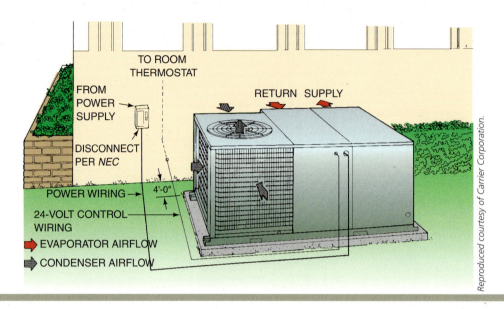

FROM POWER SUPPLY

TO ROOM THERMOSTAT

DISCONNECT PER *NEC*

RETURN SUPPLY

POWER WIRING

4'-0"

24-VOLT CONTROL WIRING

EVAPORATOR AIRFLOW

CONDENSER AIRFLOW

Reproduced courtesy of Carrier Corporation.

FIGURE 22-36 A typical packaged unit.

SERVICE TECHNICIAN EVALUATION

William arrives at the residence within 20 minutes. He immediately asks for a description of the problem. After listening to the details, he goes to the thermostat and sees that it is set to operate in the cooling mode and is set to maintain a temperature below the present temperature of the store. William then goes out back to the unit. Since the unit appears to be off, his first task is to check for line voltage. He removes a multimeter from his toolbox and puts his safety glasses on. He flips the service disconnect switch to the OFF position and opens the cover. He checks for voltage and finds 223 volts being supplied to the disconnect (between points A and B in Fig. 22-37). He closes the disconnect box and turns the switch back on. William's next step is to open the service panel on the unit. Once the cover is removed, he is able to check the voltage on the line side of the contactor. Again, he reads 223 volts (Meter A in Fig. 22-38). He then checks the voltage on the load side of the contactor and reads 0 volts, which indicates to him that the contactor's holding coil is not pulling the contacts closed (Meter B in Fig. 22-38). He then checks the voltage being supplied to the holding coil and finds that at the coil it is 0 volts (Meter C in Fig. 22-38).

William now needs to establish that a low-voltage control circuit exists. He checks the voltage at the primary winding of the transformer and reads 223 volts. He then checks the voltage at the secondary winding and reads 0 volts. Since 223 volts are being applied to the primary winding and no voltage is present at the secondary, William correctly concludes that the transformer is defective. He needs to know, though, which

winding—the primary or secondary—is bad. He then shuts off the power to the unit, discharges the system capacitors, and removes the transformer from the unit.

With a multimeter set to read ohms, William finds no continuity through the secondary winding, indicating that this winding has burned. This indicates that there was, and probably still is, a short in the low-voltage control circuit. Before replacing the transformer, he takes a resistance reading of the control circuit and finds a resistance of 0 ohms, which indicates a direct short.

William notices that there are only two holding coils in the control circuit, so he decides to check each one separately. By isolating the holding coil in the evaporator fan motor control circuit (Meter A in Fig. 22-39), William establishes that the coil has a resistance of 8 ohms. He then replaces the wires on the coil. Doing the same for the cooling contactor's holding coil, he obtains a resistance reading of 0 ohms (Meter B in Fig. 22-39). This coil is shorted, and the contactor needs to be replaced.

William goes to his truck and gets a replacement contactor, a transformer, and an in-line fuse for the low-voltage control circuit. He then tags all of the wires on the old contactor and removes it from the unit. He mounts the new contactor, after double checking the amperage rating of the contacts and the voltage rating of the holding coil, and reconnects the wires. He then removes the old transformer and installs the new one, being careful to cap any unused wires on the new transformer. To prevent the transformer from blowing in the future, William installs a 3-ampere fuse in the control circuit. If a short circuit develops in the future, the fuse will blow instead of the transformer. This will save the customer money on parts.

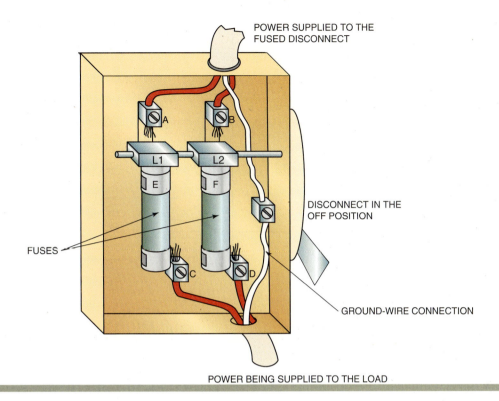

POWER SUPPLIED TO THE
FUSED DISCONNECT

A

B

L1

L2

E

F

DISCONNECT IN THE
OFF POSITION

C

D

FUSES

GROUND-WIRE CONNECTION

POWER BEING SUPPLIED TO THE LOAD

FIGURE 22-37 Even in the OFF position, there should still be a voltage reading between points A and B.

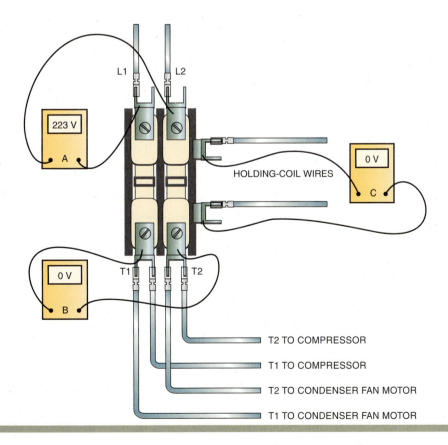

L1   L2

223 V

A

0 V

C

HOLDING-COIL WIRES

0 V

B

T1   T2

T2 TO COMPRESSOR

T1 TO COMPRESSOR

T2 TO CONDENSER FAN MOTOR

T1 TO CONDENSER FAN MOTOR

FIGURE 22-38 The contactor contacts are open. No voltage is being supplied to the holding coil.

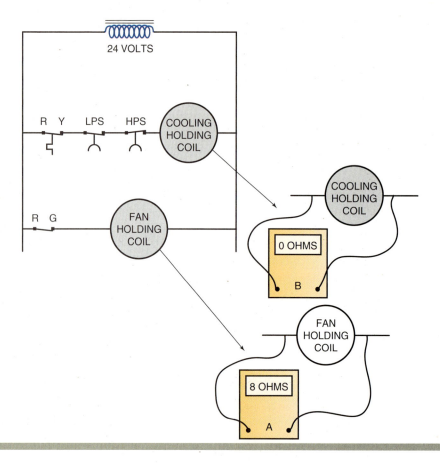

FIGURE 22-39 When removed from the circuit, the resistance of the individual hold coils can be checked. The fan relay is okay, but the cooling circuit holding coil is shorted and needs to be replaced.

Before energizing the circuit, William checks the resistance of the control circuit to verify that the short circuit has been removed and the problem solved. After double-checking his wiring connections, he proceeds to energize the system.

The compressor, the condenser fan, and the evaporator fan all come on and the system begins to cool. William allows the system to operate while he checks the air filters and takes a temperature differential reading across the evaporator. After determining that the filters are clean and the temperature differential across the evaporator is 18 degrees, he completes his work order, making certain that he includes all parts used on the service call. He completes his cleanup, packs his tools, explains the repair to the home owner, gets his work order signed, and returns to the construction site.

SERVICE CALL DISCUSSION

In this service call, the system appeared to be in the OFF position because there was no low-voltage control circuit. After verifying the complaint, William first checked for line voltage at the unit because the most obvious solution to the problem would be the absence of supply voltage. Once he determined that line voltage was being supplied to the unit, the next logical step was to check for low voltage; but William first checked to make certain that voltage was present on the line side of the cooling contactor. This is not incorrect and is meant to show that every technician has his or her own different way of troubleshooting and evaluating a system. No two technicians do exactly the same thing at the same time, and they all must do whatever works for them personally.

Once William was able to establish that the transformer was defective, he determined why the transformer went bad. He realized that components do not fail without cause. Since the circuit had only two low-voltage holding coils, William was able to determine which coil was shorted out. If neither coil had been defective, he would have had to check the interconnecting low-voltage wiring to locate the short circuit. The procedure William used to replace the contactor should serve as a model to the new field technician. Cooling contactors, in general, have at least eight wires connected to them. When replacing a contactor, good field practice includes labeling each wire so that it can be replaced in the proper position.

Many technicians even draw simple wiring diagrams showing the contactor terminals and the wires that should be connected to each. *Do not* rely on memory! Even a small, unexpected distraction is enough to cause a technician to forget exactly which wire goes where. If this should occur, it will obviously take longer to complete the repair.

3 SERVICE CALL
Defective Contactor Holding Coil

CUSTOMER COMPLAINT

A residential customer has just returned from work and has found that her home is warm. She has a digital programmable thermostat that normally turns the air conditioning system on about an hour before she arrives home. She looks at the thermostat and sees that it is set to the cooling mode. She then feels the supply air and finds it to be warm. She immediately places a call to her service company, even though it is late at night, because she is throwing a party the next day.

SERVICE TECHNICIAN EVALUATION

Anthony arrives at the customer's home and listens attentively as the home owner vents her frustration and explains the situation with the system. To confirm her statements, Anthony goes to the thermostat and checks the settings. He then feels the supply air, which is, indeed, warm. He is able to make the following assertions:

- The transformer is providing low voltage
- The evaporator fan is operating
- The system is calling for cooling

Anthony then decides to check the condensing unit, which is located in the back yard. Upon visual inspection, he can tell that the condensing unit is not operating. Making certain his safety glasses are on, he opens the service panel of the unit. With his voltmeter, he checks for voltage at the line side of the cooling contactor and finds 218 volts (Meter A in Fig. 22-40). This system has a 220-volt rating, so this is fine. Upon checking the voltage on the load side of the cooling contactor, Anthony read 0 volts (Meter B in Fig. 22-40). From this, he can establish

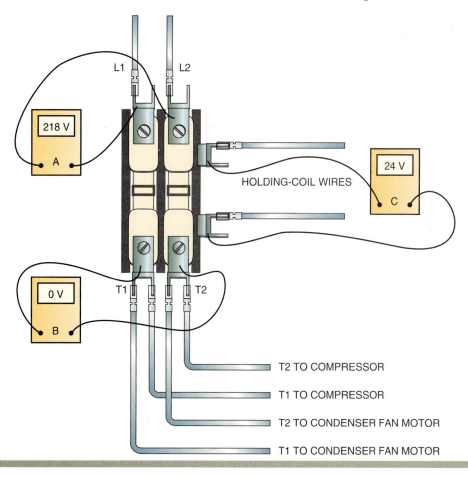

FIGURE 22-40 The contactor contacts are open, and voltage is being supplied to the holding coil. A defective holding coil is immediately suspected.

that the contactor contacts are not closed. Changing the range on his voltmeter, he checks the voltage being supplied to the contactor coil and reads 24 volts (Meter C in Fig. 22-40).

For all intents and purposes, the contactor contacts should be closed. The coil is receiving voltage, but the contacts are not closed. Anthony immediately suspects a defective contactor coil. After turning the service disconnect switch off and discharging the capacitors, Anthony removes the wires from the contactor coil and, with his multimeter set to read resistance, takes a reading of the contactor coil (Fig. 22-41). The meter reads infinity, indicating that no continuity is present across the coil. The contactor needs to be replaced.

Anthony labels all the wires on the old contactor and removes it from the unit. He checks the amperage rating and the coil voltage rating on the new contactor before installing the device. He mounts the new contactor securely and makes all necessary electrical connections. He then turns the service disconnect switch on, and the system begins to operate.

While monitoring the system, Anthony closes all service panels, packs his tools, completes his paperwork, explains the repair to the customer, and gets signed off the job.

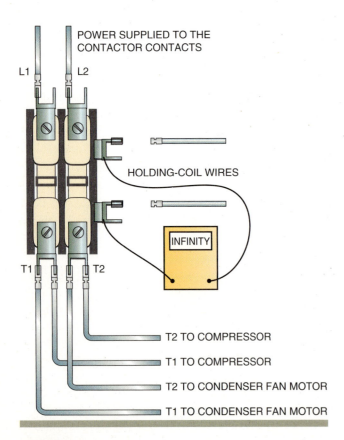

FIGURE 22-41 There is a break in the holding coil. It must be replaced.

SERVICE CALL DISCUSSION

In this service call, Anthony was able to confirm the customer complaint and verify that the transformer was good and that the thermostat was calling for cooling. He knew that, based on the settings on the thermostat, the unit should be operating in the cooling mode. He quickly went out to the condensing unit and saw that it was not operating. Anthony was able to isolate the problem by checking for and verifying the presence of both line and low voltage at the condensing unit. He then checked for line voltage at the line side of the cooling contactor and for low voltage at the contactor coil. Based on this information, the contactor contacts should have been closed but they obviously were not. By checking the coil, he was able to conclude that the coil was defective.

Note that Anthony made it a point to check both the amperage rating and the coil voltage of the new contactor before installing the device. This is vital to ensure that the new component will function properly in the system. If the incorrect part is inadvertently installed, future failure will most likely result.

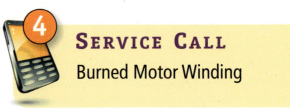

SERVICE CALL
Burned Motor Winding

CUSTOMER COMPLAINT

A home owner who is concerned about her excessively high electric bill places a service call. Although the air conditioning season has just started, she has noticed that her electric charges are higher than they normally have been for peak summer operation. She requests that a service technician come to her house to check out the system.

SERVICE TECHNICIAN EVALUATION

Later that afternoon, Diane arrives at the home. While discussing the system problem, she asks the home owner whether or not the house is able to maintain a comfortable temperature. She is told that when the temperature outside is cool the house is comfortable but a little humid. As the temperature outside begins to rise, the house becomes a little uncomfortable but not unbearable.

Diane goes over to the thermostat and makes certain that it is set properly for cooling. She then feels the supply air, and finds that it is not very cool but not excessively warm either. From this, she is only able to conclude that the indoor fan motor is operating. She needs to inspect the condensing unit.

As she walks around to the side of the house, she can hear a compressor running, but it quickly shuts down. After two minutes or so, the compressor turns back on but the condenser fan motor does not. After running for about a minute, the compressor shuts down again. Diane immediately turns off the service disconnect switch and is now able to evaluate the situation. The facts are

- the customer is concerned about high electric bills
- the system is not properly maintaining the house temperature
- the humidity in the house is higher than normal
- the evaporator fan is operating
- the compressor is cycling on and off at two-minute intervals
- the condenser fan motor is not operating

Diane is able to piece together the puzzle. She knows that the evaporator performs two functions. It cools, and it removes humidity from the air. Since it is warm and humid inside the house, Diane concludes that the evaporator is not functioning effectively. She then notices that the condenser fan is not operating and the compressor is cycling on and off. This explains the high electric bills.

With the service disconnect switch in the OFF position, Diane opens the service panel on the condensing unit. Because the compressor is operating, she concludes the following:

- Line voltage is being supplied to the condensing unit
- Low voltage is being supplied to the cooling contactor's holding coil, and the contacts are closed
- The problem lies within the condenser fan motor or the wiring leading to it

Diane discharges the capacitors and disconnects the condenser fan motor wires from the load side of the contactor. She also disconnects the wires from the motor's run capacitor, making certain to label the wires for future identification. With her multimeter set to read resistance, Diane checks the three condenser fan motor wires for continuity, two at a time. She obtains the following readings:

Common to Run   8 ohms

Run to Start   Infinity

Common to Start   Infinity

Drawing a quick sketch (Fig. 22-42), she notices that the start winding on the condenser fan motor is burned.

Since Diane does not have a replacement motor with her, she needs to order a new part. From her

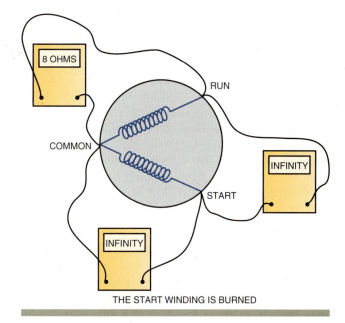

THE START WINDING IS BURNED

FIGURE 22-42 This motor has an open starting winding. Note that there is a reading of infinite resistance across that winding.

experience, though, she knows that removing the condenser fan blade from the motor shaft without damaging it—especially on older units—is difficult, so she decides to try to remove the blade now. This way she will know if a new blade is also needed. (If she waits until the new motor is installed, she risks having to return to the home again to replace the blade. This would ultimately result in customer dissatisfaction, and her boss would not be very happy either.) Fortunately, she is able to remove the blade without damaging it. She then sets the blade gently inside the condensing unit, takes all data tag information from the motor, and writes it on her work ticket.

To ensure that the new motor will be installed correctly, Diane replaces all of the wires in their proper places and leaves the old motor mounted. She leaves the service disconnect switch in the OFF position.

Since the repair cannot be completed today, Diane informs the customer what additional work needs to be done. She also informs the customer that the system fan can be used in the meantime to circulate air through the house. Diane calls her service manager with the news and gives him the motor information.

Diane informs the customer that her service manager will call her back as soon as he knows how much the motor will cost to install and exactly when the job can be done. She cleans up, packs her tools, completes her paperwork, gets signed off the job, and proceeds to her next service call.

SERVICE CALL DISCUSSION

By seeing the compressor operating, although intermittently, Diane was able to quickly conclude that the problem was with either the condenser fan motor or its associated wiring. By checking the continuity in the motor windings, she was able to determine that the condenser fan motor was defective and in need of replacement. This service call was rather straightforward with the exception of her foresight to remove the fan blade from the motor shaft before leaving the job. Consider the following possible situation if she had not removed the blade.

Diane returns in a day or two with a brand new motor. She then begins to remove the old motor from the unit and finds that the *hub*, or portion of the blade that the motor shaft fits into, is rusted. While attempting to free the blade from the shaft, it is damaged. It now needs to be replaced along with the motor. Diane places a phone call to her service manager and explains the situation. Obviously, he is not at all pleased with the situation. Why?

- First, he must tell the customer that the repair cannot be completed as promised.
- Second, the customer must take another day off from work to meet the service technician.
- Third, the service manager must tell the customer that the repair will now cost more than the original estimate.
- Fourth, valuable time has been lost, costing the company money.

By taking the time to remove the blade before the estimate was given to the customer, Diane eliminated the possibility of this awkward situation arising. She was able to tell her service manager exactly what part or parts she needed as well as approximately how long the repair would take.

5 SERVICE CALL
Grounded Compressor

CUSTOMER COMPLAINT

A newlywed couple has just purchased their first home, a handyman's special. Upon turning the air conditioning system on, they find that the air coming from the supply registers is warm. John, the new groom, having some

technical knowledge, goes out back to the condensing unit. He opens the disconnect box and finds a 2-pole, 40-ampere circuit breaker that is tripped. He resets the breaker, and it immediately trips again. Realizing that he is getting in a little over his head, he calls a local air conditioning contractor for help.

SERVICE TECHNICIAN EVALUATION

Upon arriving at the home, David, the technician, is immediately told about the system. Since the circuit breaker at the condensing unit tripped immediately, he concludes that there must be a short circuit somewhere in the outdoor unit. With the disconnect switch in the OFF position and all capacitors discharged, he checks for continuity between the two power lines on the load side of the disconnect switch and ground. David reads 0 ohms from L2 to ground, indicating a direct short to ground within the condensing unit. The short must be in one of the following locations:

- The wiring from the disconnect box to the line side of the cooling contactor
- The compressor
- The wiring from the load side of the contactor to the compressor
- The condenser fan motor
- The wiring from the load side of the contactor to the condenser fan motor

To begin, David removes the wires from the line side of the contactor and again checks for continuity from the load side of the disconnect switch to ground. This time he reads infinity. This indicates that a short circuit is not present in the wires feeding the condensing unit and that the short lies within the condensing unit itself. He then checks for continuity between the load side of the contactor and the ground and reads 0 ohms (Fig. 22-43). David then removes the compressor wires from the load side of the contactor and checks for continuity again. This time he reads infinity (Fig. 22-44). Since he has removed the compressor and its associated wiring from the circuit, he has narrowed the search down to the compressor and its wiring components.

After replacing the wires leading to the compressor onto the correct terminals of the contactor, David removes the wires from the compressor itself, thereby isolating the compressor from the rest of the circuit. He then checks the compressor using a digital VOM set to read resistance. He scrapes a small section of the copper tubing on the compressor and holds one of the meter's test leads on this cleaned surface. This ensures that the meter will read properly. If the copper tubing has paint

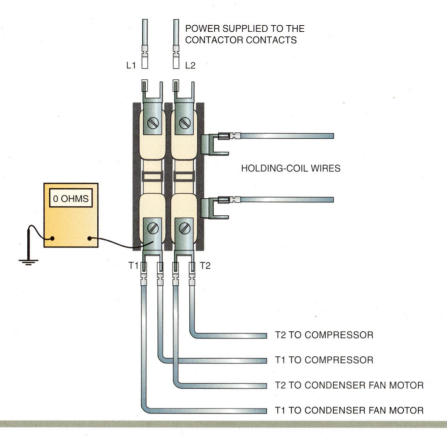

FIGURE 22-43 A short to ground exists on the load side of the contactor.

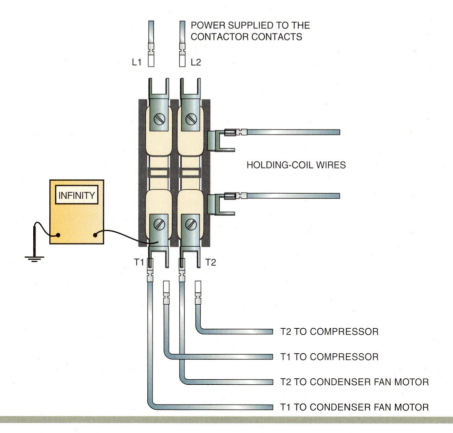

FIGURE 22-44 By removing the compressor from the circuit, the short is no longer present. The problem lies within the compressor or its associated wiring.

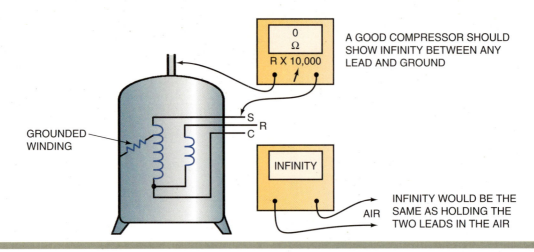

A GOOD COMPRESSOR SHOULD
SHOW INFINITY BETWEEN ANY
LEAD AND GROUND

INFINITY WOULD BE THE
SAME AS HOLDING THE
TWO LEADS IN THE AIR

FIGURE 22-45 There should be no continuity between the compressor terminals and the shell of the compressor or its interconnecting piping.

or some other substance on it, the meter may not register continuity even if a short to ground exists. The other test lead is touched to each compressor terminal, one at a time, to check for continuity. David reads 0 ohms when checking from the start terminal to the copper tubing of the compressor (Fig. 22-45). He concludes that the compressor is grounded and needs to be replaced.

Before placing a call to his service manager, David retrieves all the data tag information from the system and the compressor itself. He also visually inspects the area for any situation or condition that might make the replacement of the compressor more difficult. He then calls his office and reports his findings. The service manager is able to properly price up the job and approach the customer with an estimate for the repair.

SERVICE CALL DISCUSSION

In this service call, David was able to verify exactly what the customer was telling him by witnessing the tripping of the circuit breaker himself. The tripping of the breaker indicated either a short circuit within the wiring circuits or a short to ground. By using his VOM to check the circuit, he was able to determine that there was indeed a ground somewhere in the circuit.

When trying to isolate a problem in a system, knowing where the problem is *not* can be just as helpful as knowing where the problem *is*. By consistently and continually narrowing down the possibilities, the fault can be isolated in a concise and logical manner. By disconnecting the compressor wires from the load side of the

contactor, David established a dividing line with which to work. If, after disconnecting the compressor wires, David still found continuity from the contactor to ground, he would have established that the ground did *not* exist in the compressor or its associated wiring. He would have then checked the condenser fan motor and its circuitry. This is a very systematic method of locating the system fault.

In any event, once the problem is isolated, the repair must still be performed. Replacing a compressor involves a number of different operations that, when performed in a logical manner, can reduce the overall repair time. For example, unless the system has a major refrigerant leak, the refrigerant must be properly recovered.

Since the process of recovering refrigerant is generally time-consuming, good field practice includes beginning the recovery process first. While the refrigerant is being recovered, the old compressor can be disconnected electrically, its mounting hardware can be removed, and other preparation work can be done, as long as the recovery process is continually monitored. Once the recovery is complete, the old compressor will most likely be ready for removal from the system.

A similar situation arises when the new compressor is installed. The evacuation process, when air and moisture are removed from the system prior to system charging, is also time-consuming. For this reason, good field practice involves getting the new compressor in place and piped in as quickly as possible so that the evacuation process can begin. While the system is being evacuated, the compressor mounting hardware can be installed, the electrical connections can be made, and so forth. This way, once the system is properly evacuated, the charging process can begin.

SUMMARY

- Electrical troubleshooting is accomplished with ammeters, voltmeters, and ohmmeters.
- Contactor and relay holding coils can be checked by reading the coil's resistance.
- The voltage rating of the holding coil must match the supplied voltage.
- A defective holding coil can be either shorted or open.
- Thermostats can be easily evaluated by placing a jumper across their contacts.
- In-line fuses are often installed in the control circuit to protect the transformer.
- Pressure controls must be checked to ensure that they open and close their contacts at the correct pressures.
- The resistance across a set of closed contacts should be 0 ohms.
- The potential difference across a closed set of contacts should be 0 volts.
- When contacts become pitted, both the resistance and the potential difference across them increase.
- Fuses blow and circuit breakers trip when a short circuit or ground is present.
- Short circuits can result from improper field wiring, improper component installation, or worn or damaged wire insulation.
- Defective motor windings can be open, shorted, or grounded.
- Capacitors can be field-tested with an ohmmeter.
- Service technicians should always be conscious of and respectful of the customer's property.
- Successful troubleshooting relies on the implementation of a logical and systematic approach to locating the system problem.

GREEN CHECKLIST

☐ **Many older thermostats contain mercury so be sure to dispose of them properly.**

☐ **Install refrigeration gauges with care to reduce the loss of refrigerant to the atmosphere.**

☐ **Frequent circuit breaker tripping is an indication of a system deficiency. Be sure to investigate to find the system problem before simply replacing the fuse.**

☐ **Shorted motor windings are often the result of improper airflow across them. Be sure to identify the cause of the problem to help keep the system operating as efficiently as possible.**

REVIEW QUESTIONS

1. **If the technician reads low voltage at the holding coil of a 24-volt cooling contactor and the normally open contacts are not closed, which of the following is most likely?**

   a. The transformer is defective

   b. The holding coil is open

   c. The thermostat is not calling for cooling

   d. The compressor is grounded

2. **A technician notices that the normally open contacts of a cooling contactor are not closed. After taking a voltage reading at the coil, the technician determines that no voltage is being supplied to the holding coil. Which of the following is a possible cause for this situation?**

   a. The transformer is defective

   b. The thermostat is not calling for cooling

   c. The holding coil is open

   d. Both a and b are correct

3. **A technician takes a resistance reading of a contactor's holding coil and reads 0 ohms. A resistance reading of infinity is also measured between the coil and ground. These two readings indicate that:**

   a. The coil is good

   b. The coil is shorted

   c. The coil is open

   d. The coil is grounded

4. **A reading of 24 volts between the R terminal and the G terminal of a low-voltage thermostat is an indication that**

   a. The evaporator fan motor is not operating

   b. The transformer is defective

   c. The compressor is operating

   d. The thermostat's indoor fan motor contacts are closed

5. **If the _____ is operating and the _____ is not, there is most likely a problem with the evaporator fan motor circuit.**

   a. Evaporator fan motor, condenser fan motor

   b. Evaporator fan motor, compressor

   c. Compressor, indoor fan motor

   d. Compressor, condenser fan motor

6. **A reading of 0 volts between the R and W terminals on a low-voltage thermostat is an indication that**

   a. The system is operating in the cooling mode

   b. The system is operating in the heating mode

   c. The control transformer is defective

   d. Both b and c are correct

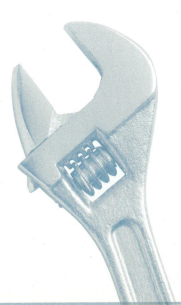

7. **A blown control fuse can be caused by**

   a. A shorted holding coil

   b. An open holding coil

   c. Both a and b are correct

   d. Neither a nor b is correct

8. **Which of the following is a possible indication of a defective control fuse?**

   a. The compressor does not run, but the condenser fan motor does

   b. The evaporator fan runs, but the compressor does not

   c. The evaporator fan does not operate

   d. The compressor and the condenser fan are drawing excessive amperage

9. **A line-voltage reading across a power-circuit fuse**

   a. Is impossible

   b. Indicates that the fuse is defective

   c. Indicates that the fuse is good

   d. Indicates that the compressor should be operating

10. **The potential difference between Terminal L1 and Terminal T1 on a contactor that has its holding coil energized should be**

    a. 0 volts

    b. 24 volts

    c. 115 volts

    d. 220 volts

11. **Which of the following can cause a line-voltage, power-circuit fuse to blow?**

    a. A short circuit in a holding coil

    b. An open winding in a fan motor

    c. A grounded compressor

    d. All of the above are correct

12. **A defective control transformer will**

    a. Make the evaporator fan motor operate continuously

    b. Make the system appear to be in the OFF mode

    c. Cause the compressor to overheat

    d. Cause the evaporator coil to freeze up

13. **If 24 volts are applied to a contactor's 24-volt holding coil**

    a. The normally closed contacts should open

    b. The normally open contacts should close

    c. The coil will burn out

    d. Both a and b are correct

14. **If the evaporator fan motor is operating, it can be assumed that**

a. The transformer is not defective

b. The compressor is operating

c. The control circuit is being powered

d. Both a and c are correct

15. **While testing a split-phase motor with an ohmmeter, the technician reads 0 ohms between the common and the run terminals. The technician can conclude that the**

a. Run winding is open

b. Run winding is shorted

c. Motor will operate normally

d. Motor is grounded

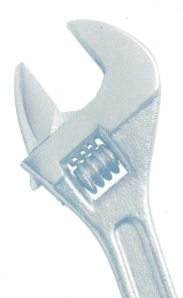

Heating Systems

23

Electric Heat

OBJECTIVES *Upon completion of this chapter, the student should be able to*

- Explain the principles of electric heat.

- Determine proper electric requirements for electric furnaces.

- Explain the function of electric heat elements in air handlers and duct systems.

- Install an electric furnace according to manufacturer's specifications.

- Install electric duct heaters.

- Follow all local electric codes and safety guidelines.

- Install control and power circuits for use on electric heaters and furnaces.

- Troubleshoot electrical and mechanical problems with electric heating systems.

If an air conditioning system should fail to operate for any reason, it is possible to open a window to get a breeze. Even though opening the window is not the same as turning on the central air conditioning system, it is well known that moving air is cooler than stagnant air. Before the advent of air conditioning systems, fans and open windows were the means by which we cooled ourselves. To escape the heat of summer, we would find a shady place under a tree or get a cool glass of lemonade. Many people fan themselves to help evaporate the perspiration that formed on their foreheads.

Heating, on the other hand, has been a requirement for our very survival. Without heating in some form or another, we would suffer the effects of our environment. In the dead of winter, we cannot find a shade tree or open a window to get a warm breeze. From building a fire in a cave to building a fire in the fireplace to turning on our central heating system, we have relied on heating to provide us with warmth for thousands of years. Although the methods for providing heat have changed over the years, the underlying concept is the same: to increase the temperature of the occupied space.

There are many methods for providing heat for comfort. We will begin our discussion of heating equipment with the electric furnace. The theory of electric furnaces builds on many of the concepts we have already discussed; namely motors, controls, relays, and resistive loads. As part of this chapter, we include not only the theory and operation of electric furnaces, but three sample service calls that reflect the most common failures associated with electric furnaces.

Fusible link one-time control device that will melt to open its contacts if the temperature in a furnace reaches an unsafe level; used primarily as a backup should the limit switch fail to open

Heat exchanger in heating, the component that keeps the air from the occupied space separate from gases and by-products of combustion resulting from burning fossil fuels

Interlock any condition or set of conditions that must be met before an unrelated or separate electric circuit can be energized

Kilowatt power measurement equal to 1,000 watts

Limit switch typically used in furnaces, the control device that opens its contacts if an unsafe or undesirable high temperature is sensed

Nichrome wire made up of nickel and chromium that is typically used in the fabrication of resistive heating elements

Sequencers control devices used to energize and de-energize electric heating elements at regular time intervals

THEORY OF ELECTRIC HEAT

In earlier chapters in this book, we referred to various types of electric loads. One type is the inductive load that typically converts electrical energy into electromagnetic energy, which then provides motion, as in an electric motor or a solenoid valve. As a result of creating this motion, some heat energy is also generated. Resistive loads, on the other hand, primarily generate light and/or heat, as in the case of a light bulb. These loads also generate small magnetic fields as a result of current flow through them. Electric heating equipment uses resistive-type heating elements that convert electrical energy to heat energy.

Resistive heating elements (Fig. 23-1) are high-resistance components that do not conduct electricity well. The high resistance of these elements results in the generation of heat that can then be transferred to the air in the occupied space. **Nichrome** wire is typically used to construct these heating elements.

Electric, resistive-type heating is by far the most efficient of all types of heating systems. Nearly all of the power supplied to the heating system is transferred to the heaters which, in turn, is converted to heat. Electricity, however, is a relatively expensive method of heating when compared with other methods, such as fossil fuel heating systems. For this reason, electric heat is often used as a supplementary heat source on heating equipment installed where the cost of electricity is high.

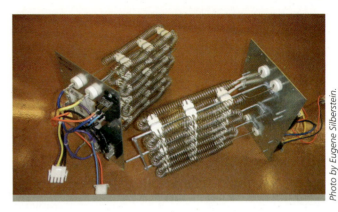

Photo by Eugene Silberstein.

FIGURE 23-1 Electric heating element.

Going Green

The terms *inefficient* and *expensive* should not be confused with each other, especially when discussing electric heat. In many areas, electricity is very expensive to purchase, but this does not mean that electricity is not efficient. The fact is that electric heating is one of the most efficient methods. There are no losses involved in electric heating systems, and 1 watt of electrical power converts to 3.413 Btu of heat energy. When discussing the cost of heating using various fuel and power sources, the cost per Btu will provide an accurate comparison.

THE ELECTRIC FURNACE

One of the benefits of the electric furnace is that no heat exchanger is needed, as is the case of gas and oil heating systems, which will be discussed in later chapters. **Heat exchangers** keep the air from the occupied space separate from the air and by-products of combustion resulting from burning fossil fuels. Since there is no burning fuel, the air from the occupied space can simply pass over the heating elements, absorb heat from them, and then return to the area being heated. The amount of heat produced by the furnace is determined by the size of the heaters installed. Heating elements are rated in **kilowatts**, kw, which is the power unit equal to 1,000 watts.

As discussed earlier in the text, one Btu of heat energy is required to raise the temperature of 1 pound of water 1 degree Fahrenheit. The Btu was also used in air conditioning applications where 12,000 Btu/hr represented 1 ton of refrigeration. Heating systems are rated in Btus also, with the system being sized according to the anticipated load on the structure. The heating elements used in the furnace are selected on the structure load and the following formula:

$$1 \text{ Watt} = 3.413 \text{ Btu}$$

Using this formula, we can conclude that a 1,000 watt (1 kw) heating element will produce 3,413 Btu of heat energy. If a house has a calculated heat loss of 50,000 Btu, the heating element capacity can be calculated as follows:

Electric Heat Requirement in watts =
(calculated heat loss in Btus)(1 watt/3.413 Btu)

Electric Heat Requirement in watts =
(50,000 Btu) (1 watt/3.413 btu)

Electric Heat Requirement = 14,650 watts = 14.65 kw

The electric furnace in this example will require heating elements totaling 15 kw. This furnace will probably be equipped with three 5 kw heaters.

Multiple heating elements are often used instead of one larger element to allow each individual element to become energized after a brief time delay. Energizing the elements one at a time eliminates the immediate energizing of a large load and allows the electric load on the furnace to increase over a period of time. We will now address the wiring, controls, and components used in the electric furnace to facilitate the desired system operation.

ELECTRIC FURNACE WIRING

Electric furnaces are typically very simple as far as the components that comprise the furnace are concerned. In addition to the controls, the main system components are the indoor fan blower and the heating elements themselves. Proper furnace wiring helps ensure the proper continued operation of the electric furnace. We will now examine the components that make up the electric circuits that facilitate electric furnace operation.

INTERLOCKS

Probably the single most important feature on electric furnaces is the **interlock** that exists between the blower motor and the electric heating elements. An interlock is any condition or set of conditions that must be met before an unrelated or separate electric circuit can be energized. In the case of the gas furnace, air movement through the duct system must be established before power can be supplied to the heating elements. This is accomplished by either a sail switch or a pressure switch. The sail portion of the sail switch is located in the air-distribution system and moves in response to air movement, just as the sail on a sailboat moves. This movement closes sets of electric contacts to confirm that there is airflow in the duct. The pressure switch (Fig. 23-2) senses the pressure difference created by air movement through the duct. In either case, once movement of air through the duct has been confirmed and a call for heat exists, the heating elements can be energized. A simplified interlock circuit between the movement of air in the duct and the heating elements is shown in Figure 23-3. For the heating element to be energized, the two sets of normally open contacts, C1, must be closed. These contacts will be in the closed position only when the coil, COIL 1, is

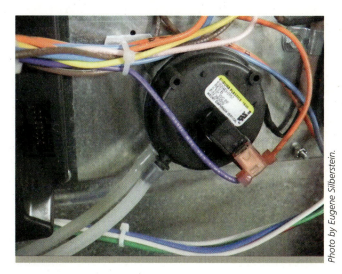

FIGURE 23-2 Pressure switch.

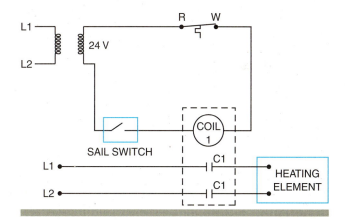

FIGURE 23-3 Sail switch wired in series with the heating coil.

energized. The following conditions must be satisfied for COIL 1 to be energized:

- There must be 24 volts at the transformer secondary
- The thermostat must be calling for heating (contacts R and W are closed)
- The sail switch must be in the closed position

From this diagram we can see that the heating element cannot be energized unless the contacts of the sail switch are in the closed position. Simply stated, the resistive heating elements cannot be energized unless there is airflow across them.

> **CAUTION**
>
> **CAUTION:** Sail switches and other devices intended to ensure the safe operation of HVAC equipment should *never* be jumped out or removed from the circuit. Severe personal injury or property damage could result.

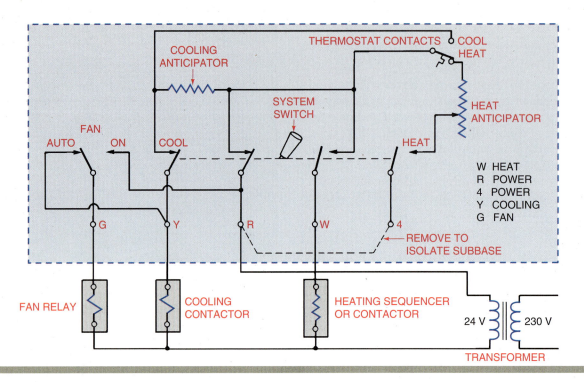

FIGURE 23-4 Wiring diagram of a low-voltage thermostat.

THERMOSTATS

The thermostat is the device that is ultimately responsible for maintaining the occupied space at the desired temperature. Thermostats that control forced air, heating-only equipment are designed to control indoor fan operation as well as the heat source. In the case of the electric furnace, the heat source is the bank of electric heat elements. The wiring diagram of a low-voltage thermostat that controls an electric furnace, as well as the cooling circuit of a central air conditioning system, is shown in Figure 23-4. Refer to earlier sections in this text for more information regarding the operation of room thermostats.

MULTIPLE-STAGE ELECTRIC HEATING

As mentioned in the opening section of this chapter, it is not desirable to energize the entire bank of electric heating elements at once. This would overload the electric service by placing the entire heating load in operation at the same time. If an electric furnace is equipped with three 5 kw heating elements, it is often beneficial to have them energized at regular time intervals. The sequencer is the control device that accomplishes this task. **Sequencers** (Fig. 23-5) are designed to open and close electrical contacts which, in turn energize and de-energize the heating

FIGURE 23-5 Sequencer.

Photo by Eugene Silberstein.

elements in the electric furnace. Sequencers can be configured to control a single heating element or multiple heating elements, depending on the application.

The sequencer is a heat activated, bi-metal device that causes electrical contacts to close when the heating control circuit is energized. The bimetal strip, wired in series with the heating terminal, has current flowing through it when the thermostat calls for heat

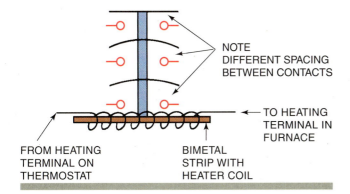

FIGURE 23-6 Sequencer is controlled by a bimetal strip in the heating control circuit.

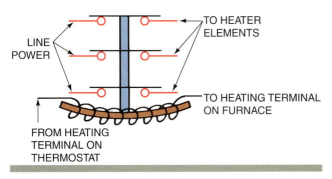

FIGURE 23-7 This three-stage sequencer has energized all three heating elements.

(Fig. 23-6). The current flow through the strip causes it to warp, resulting in the closing of the contacts in series with the heating elements (Fig. 23-7). Multistage sequencers cause the contacts to close at predetermined times, preventing multiple heaters from being energized at the same time.

ELECTRIC HEATING SAFETY DEVICES

Since the main source of heat in electric furnaces is the resistive strip heater, there must be safety devices in place to reduce the chance of starting a fire. Although the sail or pressure switch permits heater operation only when there is airflow through the duct system, it is possible for these devices to malfunction. In the event of a malfunction, additional protection must be inherent in the system. One such device is the **limit switch**, an automatic reset device that opens its contacts if the temperature exceeds the desired operating temperature in the furnace. The contacts on the device close when the temperature drops to an acceptable level. Another control device in the electric strip heater circuit is the **fusible link**, which is designed to melt should the temperature exceed safe levels. The temperature at which the fusible link will melt is higher than the temperature at which the limit switch will open, providing back-up protection. If the sail switch and the limit switch fail to de-energize the heater circuits, the fusible link will melt. The fusible link is a one-time device and must be replaced once the safety melts. The wiring of a single electric heating element is shown in Figure 23-8. Notice that sequencer contacts 3 and 4 are in series with the heater circuit. Additional heater elements would be controlled by contacts 5/6 and 7/8 should they be required. *Each individual heater circuit must be protected by its own fusible link and limit switch.* Such is the case when three heating elements are used (Fig. 23-9).

> ## CAUTION
>
> **CAUTION:** Should you notice that the fusible link has melted, you must check the other safeties in the circuit as fusible link failure is indicative of the failure of at least one of the other safeties.

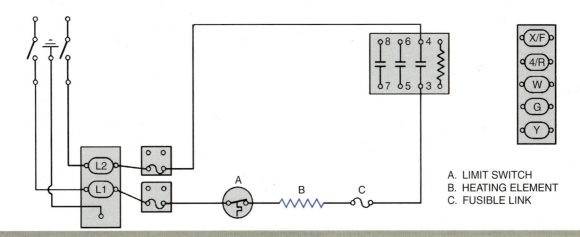

FIGURE 23-8 Single heating element sequencer circuit. Notice that contacts 3 and 4 are in series with the heating element.

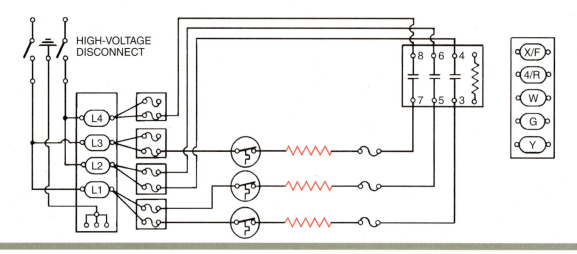

FIGURE 23-9 Three heating element sequencer circuit.

FAN OPERATION

Electric furnaces are often equipped with a two-speed fan motor that is intended to operate at high speed during the cooling mode of operation and at low speed during the heating mode of operation. The speed of the fan motor is controlled by a relay, the coil of which will be energized when the system is operating in the cooling mode (Fig. 23-10). When the coil is energized, the normally closed contacts open, the normally open contacts close, and the fan operates at high speed (Fig. 23-11). When the coil is de-energized, the blower operates at low speed (Fig. 23-12). When operating in the cooling mode, the blower can cycle on and off with the thermostat and no special operational controls are needed. When operating in the heating mode, however, the blower operation must be somewhat modified.

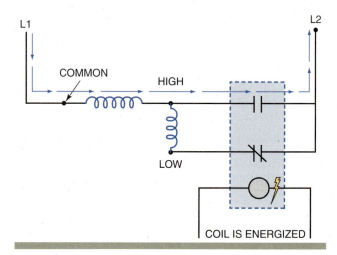

FIGURE 23-11 When the coil is energized the fan operates at high speed.

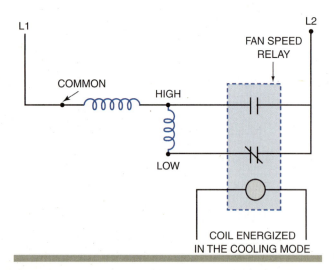

FIGURE 23-10 The fan speed relay permits high fan operation in the cooling mode and low-speed fan operation in the heating mode.

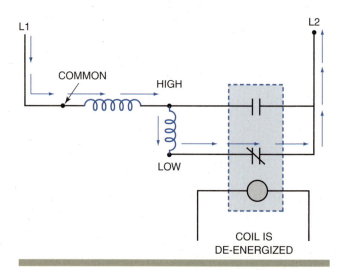

FIGURE 23-12 The fan operates at low speed when the relay coil is de-energized.

To prevent excessive heat from building up in the electric furnace, the indoor blower must operate before the heaters are energized as well as after the heaters are de-energized to dissipate the heat in the furnace. When we discussed the interlock setup earlier, we established that there must first be airflow in the duct system for the heaters to be energized. At the end of the heating cycle, however, the blower would cycle off if it was set to operate automatically. This would cause residual heat to build up in the furnace, possibly melting the fusible links and causing heat-related damage to the equipment. In an effort to alleviate this possibility, the blower is intended to operate until the furnace has cooled down to a predetermined temperature.

The fan is initially energized by the fan control coil, which then starts blower operation by closing the contacts in series with the blower (Fig. 23-13). This fan control is often a time-delay relay or part of an electronic time-delay circuit that will cause the fan contacts to close after approximately 45 seconds. A temperature-controlled device is wired in parallel with these electrical contacts (Fig. 23-14). When initially energized, the fan control contacts close and the temperature control is in the open position. The blower begins to operate. As the temperature in the furnace increases, the contacts on the temperature control close. At the end of the heating cycle, the blower continues to operate because the temperature control contacts are in the closed position as a result of the high furnace temperature. Since the heaters are no longer energized, the blower will dissipate the heat in the furnace and the furnace temperature will drop. Once it has dropped to a predetermined level, the temperature control's contacts will open and the blower will be de-energized. A complete ladder diagram showing the controls of an electric furnace is shown in Figure 23-15.

INSTALLATION OF THE ELECTRIC FURNACE

An electric furnace is installed much like the air handler of a split-type air conditioning system. The following represents a list of items that should be addressed when installing the furnace. This material was discussed in the installation portion of this text, so refer back to those sections for more information.

When selecting the location for the furnace, the following should be considered:

✓ Type of air-distribution system

✓ Location of the power supply

✓ Serviceability

✓ Furnace configuration

✓ Noise level

✓ Return air

✓ Location of conditioned space

The air-distribution system can be constructed of any of the following materials or a combination thereof:

✓ Galvanized metal

✓ Flexible duct

✓ Fiberglass, fiberboard sheeting

✓ Round sheet metal

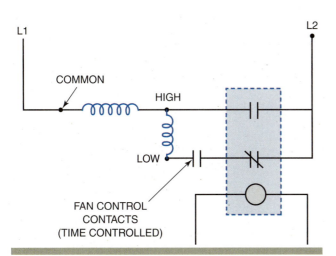

FIGURE 23-13 Time controlled fan contacts energize the motor after a predetermined time period.

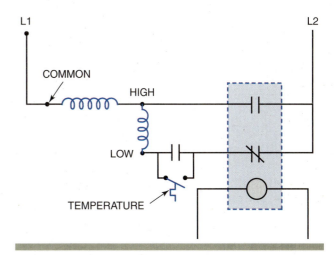

FIGURE 23-14 Temperature-controlled switch in parallel with the fan contacts de-energize the fan motor after heat has been dissipated from the furnace.

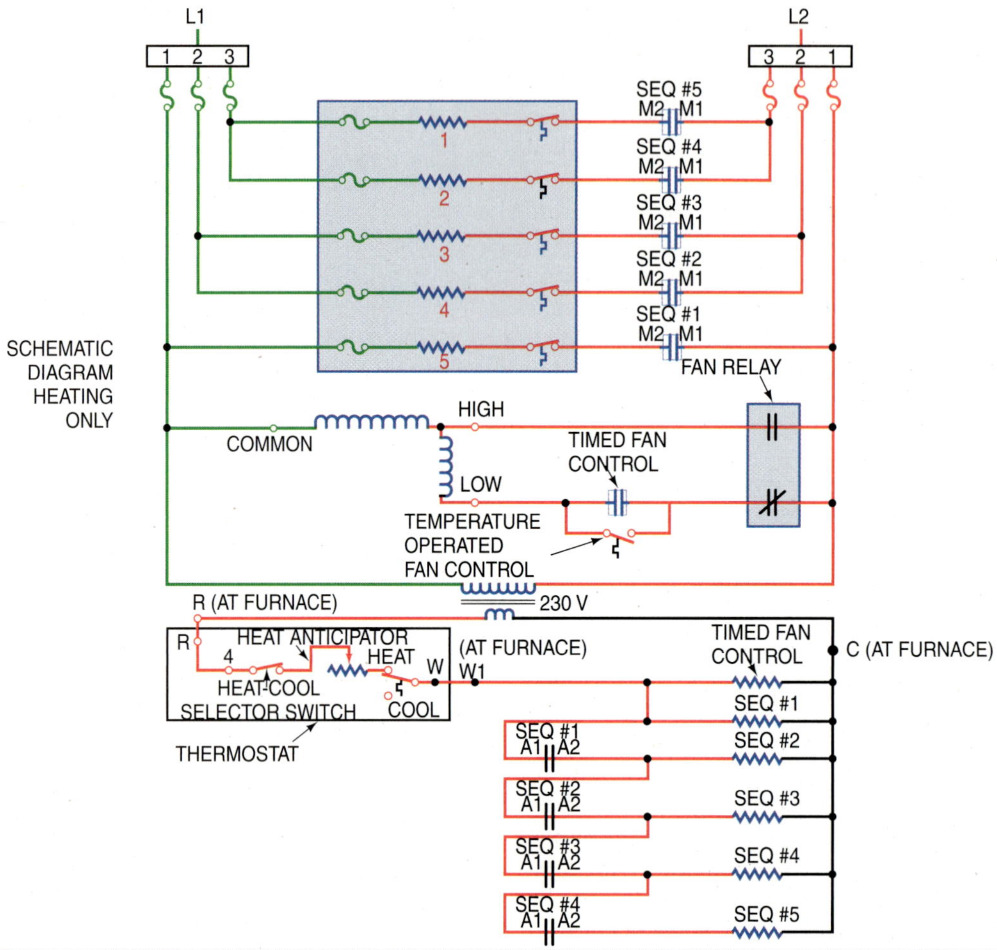

FIGURE 23-15 Line diagram of a typical electric furnace.

When supplying power to the furnace, keep the following in mind:

✓ Be sure that the power requirements of the furnace are satisfied. The heating elements draw much more amperage than the air handler of an air conditioning system.

✓ When running the low-voltage wiring for the thermostat, run extra conductors between the furnace and the thermostat in the event the home owner opts to add cooling equipment at a later date.

ELECTRIC DUCT HEATERS

Electric duct heaters, which are comprised of heating elements and safety controls (Fig. 23-16) can be installed to provide an additional heat source for fossil fuel or

Courtesy of Thermolec.

FIGURE 23-16 Duct heater.

heat pump forced air systems. If the fossil fuel system becomes inoperative as a result of a failed system component or lack of fuel, the electric strip heater can be used as a backup heat source. In a heat pump system, which will be covered later in this text, electric heating elements are often used as a supplementary heat source in case the mechanical heating system cannot satisfy the requirements of the occupied space.

Duct heaters are available with capacities ranging from 0.5 kw to well over 1,000 kw. These heaters often come with all the controls and safeties, such as pressure switches and sequencers, built into the cabinet, so it is relatively simple to install these units. When selecting a duct heater, it is important to choose one that matches the size of the duct in which the heater is to be installed. For example, using a duct heater with a cross-sectional measurement of 10″ × 10″ in a duct with a cross-sectional measurement of 14″ × 10″ will result in a large amount of bypass air, which will reduce the heating efficiency of the unit.

INSTALLING DUCT HEATERS

Installation of electric duct heaters involves two major steps: Mounting the unit and wiring the unit. When mounting the heater, a hole must be made in the duct section in which the unit is to be mounted (Fig. 23-17). The hole in the duct should be the same size as the heating element section of the heater. Once the hole has been made in the duct, the heater should be inserted into the duct and secured in place with self-tapping sheet metal screws. Many manufacturers recommend at least 24″ of straight duct sections on either side of the heater to facilitate even airflow through the heater.

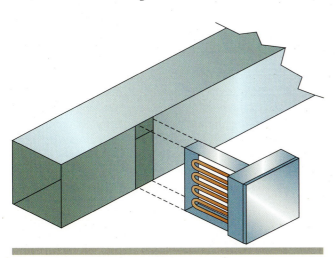

FIGURE 23-17 Hole in duct must accommodate the duct heater.

The electrical connections are made on the terminal block in the unit panel. Electrical connections include the line-voltage power as well as the control wiring, which is often low voltage. It is important to note the power requirements for the heater and to follow all local codes when supplying power to the unit. Although duct heaters are often equipped with overload and fuse protection, the line feeding power to the unit must be protected as well. A disconnect switch should also be installed close to the heater in case the heater needs to be serviced. Always follow the manufacturer's installation instructions provided with the heater.

TROUBLESHOOTING ELECTRIC HEATING SYSTEMS

Since the electric furnace is made up primarily of the heating elements and the blower, system malfunctions can typically be narrowed down to either airflow problems or electrical problems. Airflow problems can result in the failure of electric components, especially since the safeties in the electric furnace are intended to de-energize the heaters and their associated circuits if unsafe conditions exist.

AIRFLOW PROBLEMS

Airflow problems in an electric furnace are the same as those that can arise in any other forced-air system, whether heating or cooling. Airflow across the heating elements, through the return duct, and to the supply registers can be restricted by a number of factors, including the following:

- Defective blower motors
- Defective motor starting components
- Improper motor lubrication
- Broken belts
- Improper belt tension
- Dirty air filters
- Closed supply registers
- Blocked supply or return grills
- Loose duct insulation

DEFECTIVE BLOWER MOTOR

If the blower motor becomes defective and fails to operate, the most likely customer complaint will be that the system is not working. The customer will report no fan operation and that the system is unable to maintain the desired temperature in the occupied space. Since the electric furnace is equipped with interlock devices that prevent the heating elements from becoming energized when no air is moving through the duct system, the likelihood of the homeowner sensing a burning smell is low.

By checking and confirming that the proper voltage is being supplied to the motor, it can be confirmed that the problem lies within the motor itself or with its associated components. If the motor hums but does not turn, there may be a problem with the starting components. On residential systems, the component most commonly found on blowers is the run capacitor, because many of these motors are of the permanent split capacitor (PSC) variety. The capacitor and its associated wiring should be checked to ensure that all is okay. If the motor fails to operate and does not hum or make an attempt to start, resistance readings of the motor windings should be taken. An open winding is an indication that the motor needs to be replaced. However, if the motor has internal thermal overload protection, it should be given a chance to cool down before deciding to replace it.

Another possibility is a lubrication problem. If the motor is a direct-drive motor, it should spin freely. If there is a sufficient amount of friction and the motor does not turn freely, the motor bearings may be dry. In this case, the motor should be lubricated. If the motor is not properly lubricated, it may very well hum and draw locked rotor amperage when it is energized and attempts to start.

BROKEN BELTS

If the blower on the electric furnace is driven by belts and pulleys, a broken belt will result in no airflow through the duct system. This will have the same effect as mentioned in the previous section. The heaters will not be energized because there is no air movement. Belts should always be replaced with the correct width and length to ensure proper operation of the blower. Quite often belts break because of age, improper belt tension, and improper pulley alignment.

from experience...

When installing or replacing belts on the blower, be sure to take amperage readings of the motor. Improper belt tension, either too tight or too loose, will result in improper current flow through the motor.

During periodic routine preventive maintenance, the belts should be checked for cracks and the pulleys should be checked for proper alignment. Improper pulley alignment can cause a belt to break if it is too tight, or skip from the pulleys if the tension is too loose. Loose belt tension often results in polishing the interior pulley surfaces. Loose belt tension also results in reduced airflow through the duct system and across the heating elements in the furnace. This causes a higher temperature differential between the supply and return air temperatures.

Going Green

Proper belt tension helps ensure proper and efficient system operation. Tight belts result in an increase in motor amperage, excessive motor heat, premature motor failure, and premature belt failure. Loose belts result in belt slippage, system overheating, and premature belt and pulley failure. Therefore, improper belt tension increases waste, increases power consumption, and reduces the life expectancy of the equipment.

DIRTY AIR FILTERS

One of the most common causes of electric furnace failure is a dirty air filter. Dirt accumulation on the filter results in reduced airflow through the duct system and across the heating elements. This can result in the opening of the limit switch or, if the limit switch fails, the melting of the fusible link. If the air restriction is severe, the sail or pressure switch may not close its contacts, resulting in not only reduced airflow, but no heat whatsoever.

Reduced airflow through the system will permit the air passing over the heating elements to remain in contact with the heating elements for a longer period of time, increasing the temperature of the air coming off the elements. This will result in a higher temperature differential across the furnace, which is a telltale sign that there is an airflow problem. Typical temperature differentials across the furnace should be in the range of 50°F. Excessive temperatures in the furnace can cause the limit switch, which senses the furnace temperature, to open.

The most important thing that a homeowner can do to ensure the continued satisfactory operation of his HVAC system is to periodically check and replace the air filters.

CLOSED SUPPLY REGISTERS

Quite often, when a room gets too hot, the homeowner will close, either completely or partially, the supply register in that room. Although this action reduces the airflow to that room, it increases the amount of air flowing to the other locations. Eventually, if enough supply registers or grills are closed, the amount of airflow through the system will be reduced. As in the case of the dirty air filter, closed supply registers will result in higher than normal temperature differentials across the furnace. This can cause the limit switch to open.

BLOCKED SUPPLY OR RETURN GRILLS

Blocked supply or return grills have exactly the same effect as dirty air filters or closed supply registers. The airflow becomes restricted and can result in high temperature differentials and open limit switches.

LOOSE DUCT LINING

Although harder to determine, loose duct lining will result in all of the previous situations regarding temperature in the furnace, inadequate airflow, and opening of safeties in the furnace. This problem is less common than those just mentioned and should be checked after all of the preceding items have been checked. Checking the duct lining involves making a hole in the duct through which the lining can be visually inspected. If the lining is intact, the hole must be patched with a panel made from the duct material. If indeed the duct lining has become loose, the air-distribution system must be disassembled and the problem area fixed.

from experience...

If you are the technician who is installing the air-distribution system, be sure to check the integrity of the lining *before* installing it. Doing so will reduce the chance of having the lining come loose after the system has been put into operation.

ELECTRICAL PROBLEMS

The electrical circuits in an electric furnace are relatively straightforward because not many components are being controlled. Electric components and controls typically found on an electric furnace are the following:

- Fan motor and associated starting components
- Fan relay

- Sail or pressure switches
- Thermostat
- Limit switches
- Fusible links
- Heating elements

FAN MOTOR AND ASSOCIATED STARTING COMPONENTS

When power is supplied to the windings of the motor, it should operate. To determine if the problem lies within the motor itself or with the power supply, it is best to check the supply voltage right at the motor. If no voltage is being supplied to the motor, the problem likely lies with the power circuit feeding the motor. If voltage is supplied to the motor and the motor fails to operate, the problem is with either the motor or the starting components. If the motor hums but does not start, the likely culprit is the starting device, although a burned start winding is a definite possibility as well.

Checking the motor with an ohmmeter, with the power disconnected of course, will enable the technician to determine if there is an open or shorted winding in the motor. Also, taking resistance readings from the windings to ground will help determine if the motor windings are grounded. Refer to the motor section in this text for more on motor evaluation and troubleshooting.

FAN RELAY

The fan relay is designed to energize the fan motor once voltage has been applied to the coil of the relay. Fan relay contacts are normally open, and there should be infinite resistance across the contacts when the motor is not energized. A reading of zero ohms indicates that the contacts are closed and that the relay should be discarded and replaced. This is, of course, assuming that the contacts on the relay are normally open. Relays are equipped with either normally open or normally closed contacts, so be sure to determine the type of relay in use before discarding the old one.

In addition to the contacts, there should be a measurable resistance across the coil of the relay. A reading of infinite resistance across the coil indicates that the coil is open, and the relay should be replaced.

By applying the correct voltage to the coil of the relay, the normally open contacts should close. With all wires from the contacts removed, taking a resistance reading across the contacts with the coil energized will indicate if the relay is operating properly. A reading of zero ohms indicates that the contacts are closed. A reading of infinite resistance indicates that the contacts are open. If the voltage is removed from the coil, the contacts on the relay should return to their normal position and the resistance reading should change accordingly.

SAIL OR PRESSURE SWITCHES

The sail or pressure switch is intended to prevent the heating elements from becoming energized if the fan motor is not moving air through the duct system. Both of these devices are in the normally open position and will close when airflow is established. A resistance reading of infinity through the control is desired when there is no airflow in the air-distribution system. A continuity reading through the control when no airflow is present indicates that the control is bad and needs to be replaced.

> ### *from experience...*
>
> As with any other device, when checking continuity or resistance, it is good field practice to remove all wires connected to the device to prevent reading the resistance of a parallel circuit.

THERMOSTAT

The thermostat, as discussed earlier in the text, is a device that simply routes power to components as needed for proper system operation. The thermostat control connections are typically made at the furnace on a terminal board (Fig. 23-18). The low-voltage control wiring color code for a typical heating system is as follows:

R	Red	Power (24 volts)
W	White	Heat
G	Green	Fan

If the thermostat is calling for heating, the voltage readings taken at the terminal board should be as follows:

R to W	0 volts
R to G	0 volts

These readings indicate that both the heating and fan circuits are closed within the thermostat. If the thermostat is set to call for heat and either of the above readings is 24 volts, a problem arises with either the thermostat or the wiring between the thermostat and the furnace. The easiest way to determine the cause for the problem is to remove the thermostat from the subbase and place a jumper wire between the R terminal and the terminal that is giving the reading. For example, if the thermostat is set for heating and there is a reading of 24 volts between terminal R and W, the jumper wire should be placed between these two terminals on the

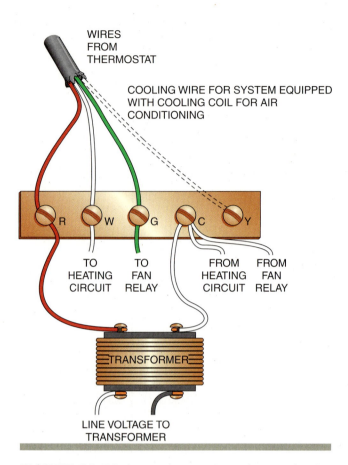

FIGURE 23-18 Low-voltage connections at the furnace.

subbase. If there is still a reading of 24 volts between these two terminals on the terminal board, there is a break in a wire between the thermostat and the furnace. Refer to the section on thermostats for more on troubleshooting these devices.

LIMIT SWITCHES AND FUSIBLE LINKS

The limit switches and fusible links on the electric furnace are safety devices and will only open their contacts if an unsafe condition exists. To check these components, it is best to disconnect the power to the unit and remove the wires connected to them. By removing the wires, the chance of reading continuity through a parallel circuit is eliminated. The resistance reading of each of these devices should be very close to zero ohms. High or infinite resistance readings are indications that the component is bad. In the case of the limit switch, the device should be given time to cool down, at which time the contacts should close. If they don't, the switch should be replaced. If there is no continuity through a fusible link, it must be replaced, as these devices melt and will never reset. Of course, if a melted fusible link is encountered, the sail or pressure switch and the limit switch should be checked carefully because they are likely defective as well.

HEATING ELEMENTS

The heating elements on electric furnaces are resistive-type loads and have no moving parts. Just like a light bulb, if power is supplied to the heater and it fails to generate heat, it needs to be replaced. The heating elements can be checked with an ohmmeter after the power to the elements has been disconnected. A reading of infinite resistance indicates that the heating element is defective and must be replaced.

from experience...

When checking heating elements, be sure to take a resistance reading of only the element. Remember that there are limit switches and fusible links in series with the elements. If the limit switch or fusible link is bad, there will be no continuity through the circuit when a resistance reading is taken. In addition, replacing the fusible link is much cheaper and easier than replacing the entire element.

SERVICE CALL
Defective Sequencer

CUSTOMER COMPLAINT

A homeowner calls her service company and informs the service manager that her electric furnace is not operating properly. She says that the air coming from the supply registers is cold and that she has turned the system off to stop the cold air from coming into the occupied space.

TECHNICIAN EVALUATION

Later in the afternoon, the technician, Anthony, arrives at the residence. The homeowner explains the situation to him, and he goes to the thermostat to check the setting of the thermostat. As the customer has informed him, the unit is in the off position and the fan is set for automatic fan operation. Anthony sets the thermostat to operate in the heating mode and sets the fan to the "on" position. He hears the indoor fan blower come on, goes to one of the supply

registers, and feels that the air is indeed cold. He then goes down to the basement where the electric furnace is located.

Upon reaching the furnace, Anthony decides to check the electric heaters. He has already established that line voltage is being supplied to the furnace since the blower is operating. He opens the service panel and clamps his ammeter around one conductor feeding each of the five 4 kw heating elements. His readings indicate that no amperage is flowing in any of the four circuits. Anthony is then able to establish the following:

- Line voltage is being supplied to the furnace (the blower is operating)
- The control transformer is functioning properly (the blower is operating)
- No amperage is flowing through any of the heater circuits
- The problem most likely lies within the heating element sequencers

Since line voltage is being supplied to the furnace, and the heater circuits are not energized, Anthony decides to check the sequencers.

Using his voltmeter, he checks the voltage being supplied to the coil of the first sequencer and finds that 24 volts is being supplied to it. Since the coil of the sequencer is being powered, the contacts on that relay should be closed, completing the circuit to the first heating element on the furnace, but no current is flowing through that element.

With his voltmeter, he takes a voltage reading across the contacts on the sequencer. He reads line voltage, which indicates that the contacts are in the open position. To confirm his suspicion that the sequencer is defective, Anthony disconnects the power to the unit and removes the wires on either side of the sequencer contacts. He tapes the wires to prevent them from coming in contact with other metallic surfaces and to reduce the chances of receiving an electric shock. He then restores power to the unit, once again checks the voltage at the sequencer coil, and obtains a reading of 24 volts. He now sets his meter to read resistance and checks for continuity across the contacts on the sequencer. Anthony reads infinite resistance, indicating that the contacts are indeed in the open position.

He goes to his truck and retrieves a new sequencer. Returning to the furnace, he disconnects the power to the unit and replaces the sequencer. When he restores power to the unit, the blower begins to operate. He places his ammeter around a wire feeding the first stage heating element and reads amperage. After waiting a minute or so, he checks the amperage of the other element circuits and determines that these stages are now energized as well. He decides to measure the temperature difference across the furnace and finds that the temperature of the return air is 65°F and the air coming off the furnace is 114°F.

Subtracting the return air temperature from the supply air temperature, he concludes that the temperature differential across the furnace is 49°F, which is acceptable. To confirm that the system is operating correctly, he goes to the room thermostat and turns the system off to make certain that the heating elements will de-energize when heating is no longer needed. Amperage readings of the element circuits confirm that the heaters are indeed de-energized. Turning the system back to the heating mode, he begins to write up his work ticket, being sure to include the part number of the new sequencer to ensure proper billing.

SERVICE CALL DISCUSSION

While performing this service call, Anthony was able to quickly establish that line voltage as well as low voltage was present in the unit, because the indoor blower was operating. Since the indoor blower was controlled by a low-voltage control circuit, the control transformer was also functioning as desired. By setting the thermostat to the heating mode of operation and setting the fan to the "on" position, he was able to troubleshoot the system from the furnace location, without making repeated trips between the furnace and the thermostat.

With the thermostat in the heating position, Anthony knew that the system should be operating and the heating elements should be energized. The absence of current flow through the heaters indicated that the contacts were in the open position. By using a systematic approach, he was able to narrow down the cause of system failure very quickly. Once he identified the problem and rectified the situation, he checked system operation to ensure that there were no other problems with the heating system.

SERVICE CALL
Defective Blower Motor

CUSTOMER COMPLAINT

A customer calls her service company and complains that her electric heating system is not operating at all. When she turns the thermostat to the heating mode, nothing happens. She cannot hear the fan come on and there is no heat coming from the supply registers.

TECHNICIAN EVALUATION

Later in the day, Dawn arrives at the customer's home. She notices that the house is very cold and, upon checking the thermostat, finds that the room temperature is

58°F. She checks the setting of the thermostat, making sure it is set to the heating mode, and then goes to the furnace in the attic to check the system further.

Upon reaching the unit, Dawn cannot hear the blower operating. She decides to check if there is power at the unit. She turns the disconnect switch at the unit off, removes the service panel, and locates the power lines coming into the unit. She then turns the disconnect back on and, using her voltmeter, takes a reading across the lines. She reads 230 volts, indicating that power is being supplied to the unit. Knowing that, she decides to check whether the control transformer is operating. She takes a voltage reading across the secondary winding of the transformer and reads 24 volts, which indicates that the transformer is good.

Continuing to check the low-voltage control circuit, Dawn takes voltage readings from the "R" terminal to the "C" terminal and from the "R" terminal to the "W" terminal. She reads 24 volts from "R" to "C" and 0 volts from "R" to "W" (Fig. 23-19). These readings indicate

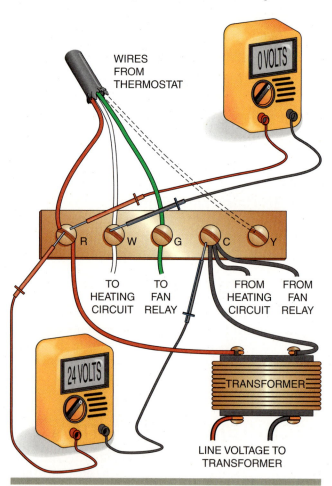

FIGURE 23-19 A 24-volt reading between terminals "R" and "C" indicate that the transformer is putting out 24 volts at the secondary winding. A 0-volt reading between terminals "R" and "W" indicate that the heating contacts on the thermostat are in the closed position.

that there is power at the transformer (R to C), which she already knew, and that the thermostat's contacts are in the closed position (R to W) and calling for heat.

She then turns her attention to the fan relay. She checks the voltage at the coil of the relay and reads 24 volts, indicating that voltage is being applied to the coil of the relay (Fig. 23-20). She also checks the voltage across the relay contacts feeding power to the fan motor. This reading is 0 volts. To confirm that voltage is being supplied to the motor, Dawn takes a voltage reading across the wires supplying power to the motor. A reading of 230 volts confirms that power is being supplied.

Since the motor is being powered but is not operating, she decides to check the motor itself. After disconnecting power to the unit and discharging the capacitor, Dawn removes the motor wires from the terminal block in the

unit. Taking resistance readings across the three wires in sets of two, she obtains readings of 12 ohms, infinity, and infinity. The infinity readings indicate that one of the motor windings is open. The motor is not equipped with an internal thermal overload, so the motor has to be replaced. She obtains the information from the data tag on the motor and goes to the local supply house to get a new motor.

Upon returning with the part, she makes certain that the power is off and proceeds to remove the old motor and install the new one. Once the new motor is in place, she powers the system. The blower comes on immediately and, after a minute or so, she begins to smell the heating elements. She takes amperage readings of the element circuits and finds them operating as they should. Dawn packs her tools, writes up her work ticket, and leaves for her next job.

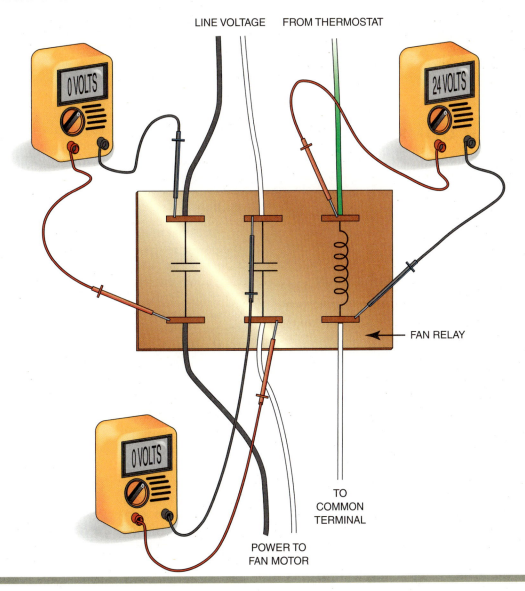

FIGURE 23-20 Twenty-four volts is being supplied to the coil of this relay and there are 0 volts across each set of contacts, indicating that the relay is energized and the fan motor should be operating.

SERVICE CALL DISCUSSION

In this service call, Dawn was very detailed in her evaluation of the system. Since the system was not operating at all, she first assumed that there might be a problem with the power being supplied to the system. She checked the control circuits, which were all in good shape. Concluding that the control circuits were all functioning, she was then able to determine that the motor was defective.

Had she first checked for power to the blower motor, however, she could have quickly determined that line voltage was being supplied to the unit as well as the motor, and that the low-voltage control circuit was intact. Nonetheless, she was able to correctly conclude that the motor was in need of replacing.

SERVICE CALL
Blocked Air Filter

CUSTOMER COMPLAINT

A heating customer places a service call to his HVAC service company stating that there is periodic heat coming from the supply registers. Quite often, however, the air coming from the registers is cold and the house is very cold as well.

TECHNICIAN EVALUATION

A short time later, John arrives at the customer's home. He listens as the customer explains the situation and then goes to the thermostat to check the setting. The thermostat is set to the heating mode and the blower is operating, so he goes to the electric furnace in the basement to check out the system. He removes the cover from the furnace and performs a visual inspection of the unit. He can hear the blower operating. He then checks the heater circuits with his ammeter and reads amperage in each of the three heater circuits. At this point, John has determined that the blower is operating and the heater circuits are indeed energized. He takes readings of the supply and return air to determine the temperature differential across the furnace. The return air temperature is 60 degrees and the supply air temperature is 140 degrees. The temperature differential across the furnace is 80 degrees, which is much too high. The average temperature differential should be in the 50-degree range.

Since the heaters are energized and the blower is operating, he then recalls what the customer told him

when he arrived at the job. There is periodic heat, but most of the time, cold air is coming from the supply registers. This tells John that the heaters must be turning off and he makes a mental list of the components that could turn the heaters off. He comes up with the following:

- Thermostat
- Sail switch
- Fusible link
- Limit switch

Since the fan is operating and the thermostat is set to the heating mode of operation, the thermostat is likely not the cause of the problem. The sail switch is in the closed position when air is flowing through the duct, so as long as the blower is operating, the contacts *should* be closed. John concludes that the sail switch may very well be defective. He quickly rules out the fusible link because, if it had melted, there would be no heat at all. Since the customer mentioned that there is periodic heat, the fusible link is good. Finally, he considers the limit switch. The limit switch opens and closes its contacts based on the temperature sensed in the furnace. If the temperature in the furnace rises above the desired level, the limit switch opens. Once the furnace cools down, the limit switch closes its contacts. This seemed like a reasonable explanation for intermittent heat. John has now narrowed his options down to the sail switch and the limit switch.

Deciding to check the sail switch first, he turns off the power to the furnace and removes the sail switch from the duct. Using his ohmmeter, he checks the sail switch for continuity, finding the contacts open and no continuity through the control. He then moves the sail of the device and reads continuity through the contacts. He moves the sail back and forth and finds he still has continuity through the device, thereby concluding that the sail switch is good. He replaces the sail switch in the duct and reconnects the wires to it. He then restores power to the furnace and turns his attention to the limit switch.

Once again, he places his ammeter around one of the conductors of the first heater circuit and reads amperage. He leaves the ammeter on the conductor and in a short time the amperage reading drops to zero. He quickly gets his voltmeter and takes a reading across the limit switch. There is a voltage drop across the limit switch, indicating that the device is in the open position. He now knows that the limit switch is de-energizing the heater circuits. Soon the limit switch closes and the heaters are once again energized.

Thinking back on the readings he took, he remembers that the temperature differential across the furnace was much too high, which indicated that the air passing over the heaters was moving too slow, allowing the air to pick up more heat from the elements. He then decides that there might be an airflow problem, so he goes upstairs to the return grill, which houses the air filter, and opens the cover. Finding the filter extremely dirty, he replaces it and goes back to the furnace, which is operating. He allows the system to operate for a few minutes and then takes air temperature readings again. The return air temperature is 60°F as before, but the supply air temperature was 111°F, giving him a temperature differential of 51°F. The problem was reduced airflow through the furnace resulting from the dirty filter.

While writing up his work order, John continues to monitor the system and, upon completion, presents the bill to the customer. He also explains the importance of keeping the air filter clean and shows the home owner how to check and replace the filter.

SERVICE CALL DISCUSSION

In this service call, John ultimately found the problem, but he did not proceed in the best possible manner. Once he found that the temperature differential was higher than desired, he should have immediately suspected an airflow problem. The simplest cause of reduced airflow through the system is a dirty air filter. Although intermittent sail switch failure could have been the problem, the chances were slim. In addition, removing the sail switch from the duct to test it requires much more time than checking the air filter. He should have evaluated the information he had obtained in a more precise manner. In addition, had the limit switch not opened while he was there or as quickly as it did, he would have spent much more time searching for the cause of the problem. In reality, had he checked the air filter before checking the sail and limit switches, he would have saved even more time on the service call.

SUMMARY

- Heating elements are high resistance coils made of nichrome wire.
- Electric furnaces require no heat exchanger, and the air passes directly over the heating elements.
- Heating elements are rated in kilowatts (kw) where each kw produces 3,413 Btu.
- Electric furnaces and heaters often use multiple heating elements, as opposed to one large heater.
- Electric furnaces are equipped with interlocks to prevent the heaters from being energized when there is no air movement through the duct system.
- Commonly used interlocking devices are the sail switch and the pressure switch.
- Sequencers are used to stagger the energizing of the heating elements when multiple stages are used.
- Limit switches and fusible links are safety devices used to de-energize the heaters if unsafe temperatures are reached in the furnace.
- The fan in the furnace must be energized before the heaters can be energized.
- The fan must operate at the end of the heating cycle to dissipate the heat in the furnace.
- The fan in the electric furnace is energized by a timed fan control and de-energized by a temperature-sensing switch.
- Electric duct heaters typically come equipped with all safeties and controls.
- Airflow problems can result from defective motors or associated controls and drive mechanisms; blocked return ducts, including dirty air filters; or blockages in the supply duct.
- Electrical problems can result from defective motors, thermostats, controls, safety devices, or heating elements.

GREEN CHECKLIST

☐ Inefficiency and expensive are two terms that are not necessarily interchangeable.

☐ Electric heat may, in many cases, be expensive, but it is 100% efficient.

☐ Systems that are not operating efficiently are definitely more costly to operate in the long run.

☐ Improper belt tension on furnace blowers can result in increased power consumption and reduced equipment life expectancy.

☐ Dirty air filters have a negative effect on system performance and efficiency.

REVIEW QUESTIONS

1. **Nichrome wire typically**
   a. Has a high resistance
   b. Permits a great deal of current flow through it
   c. Both a and b are correct
   d. Neither a nor b is correct

2. **A 5 kw electric strip heater will generate approximately**
   a. 3.413 Btu
   b. 3,413 Btu
   c. 17,065 Btu
   d. 34,130 Btu

3. **Which of the following components is typically used as an interlock in an electric furnace?**
   a. Sequencer
   b. Limit switch
   c. Sail switch
   d. Fusible link

4. **What system component is used to prevent all strip heaters from energizing at the same time?**
   a. Sequencer
   b. Limit switch
   c. Fusible link
   d. Fan control switch

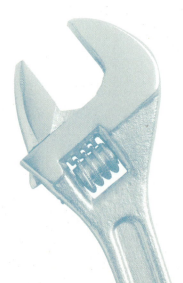

5. **Which of the following best represents the fan operation on an electric furnace?**

   a. The fan comes on after the heating elements are energized and remains on until after the furnace has cooled

   b. The fan comes on after the heating elements are energized and shuts down at the same time as the strip heaters

   c. The fan must be in operation before the heaters are energized and remains energized until the furnace has cooled

   d. The fan must be in operation before the heaters are energized and shuts down at the same time as the heaters

6. **During electric furnace operation, the indoor fan motor is energized by the _____ and de-energized by the _____.**

   a. Fan control contacts, fan control contacts

   b. Temperature-controlled switch, temperature-controlled switch

   c. Fan control contacts, temperature-controlled switch

   d. Temperature-controlled switch, fan control contacts

7. **If there is no airflow in the air-distribution system and the thermostat is calling for heat**

   a. The indoor fan motor is not operating and the heaters are energized

   b. The indoor fan motor is operating and the heaters are energized

   c. The indoor fan motor is not operating and the heaters are not energized

   d. The indoor fan motor is operating and the heaters are not energized

8. **If the limit switch opens and the fusible link remains intact**

   a. The fusible link needs to be replaced

   b. The limit switch is operating under normal conditions

   c. Too much heat has been generated in the furnace

   d. There is likely too much airflow through the furnace

9. **Which of the following is a likely cause of the limit switch opening?**

   a. Dirty air filter

   b. Defective fusible link

   c. Excessive air flow

   d. All of the above are correct

10. **If the fan switch on the thermostat is turned to the "on" position**

   a. Cold air can be circulated to the occupied space when the thermostat is satisfied

   b. Cold air can be circulated to the occupied space if the limit switch opens

   c. Cold air can be circulated to the occupied space if the fusible link melts

   d. All of the above are correct

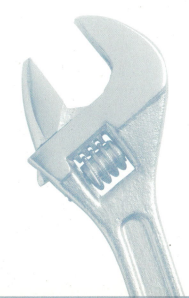

Gas Heat

OBJECTIVES

Upon completion of this chapter, the student should be able to

- Explain the theory of gas heating.

- Explain the operation of the gas furnace.

- List several types of ignition systems used on gas furnaces.

- Explain the benefits and restrictions of each ignition type.

- Install a gas furnace.

- Install piping used to carry gas to the furnace.

- Install vent piping to carry by-products of combustion away from the furnace.

- Start up the gas furnace.

- Test for proper combustion in the gas furnace.

- Troubleshoot problems with the gas furnace.

In the previous chapter, we began our discussion of heating by introducing electric heating elements and the electric furnace. Electric heating equipment is nearly 100% efficient because all of the electric power used in the elements is converted to heat energy. We will now begin our discussion of fossil fuel heating systems with the gas furnace.

Unlike the electric furnace, the gas furnace must burn fuel to generate heat. A certain amount of heat is lost as a result of this burning process, so the efficiency of fuel burning heating appliances is lower than that of electric heating systems. The efficiency with which the fuel is burned will determine the efficiency of the heating system.

In this chapter, we will discuss not only the burning of the fuel, but methods used to ignite the fuel, control the operation of the system, and maintain the safe and continued operation of the system.

Annual Fuel Utilization Efficiency, AFUE measures the amount of fuel used to provide heat to the structure compared to the amount of fuel that is supplied to the appliance

Atmospheric burners burners in which combustion air is supplied at atmospheric pressure

Cold junction in a thermocouple, the loose ends of the dissimilar wires

Combustion reaction that results when fuel, oxygen, and heat are combined in the proper proportions

Complete combustion achieved when only nontoxic and nonpoisonous products of combustion, such as carbon dioxide, water vapor, and nitrogen, are released during the combustion process

Condensing furnaces high-efficiency furnaces that remove large amounts of heat from the flue gases, resulting in the condensing of the flue gases

Cylindrical heat exchanger heat exchanger with only one combustion chamber

Dilution air air that is mixed with the flue gases as they enter the flue pipe to dilute the concentration of products of combustion

Direct spark ignition (DSI) ignition method that does not use a pilot; an electric spark is used to ignite main gas directly

Electronic manometer test equipment that reads gas pressure in inches of water column

Flue gases substances that are created during the combustion process

Flue pipe piping arrangement used to vent products of combustion from the structure

Gas Appliance Manufacturers Association (GAMA) organization that provides venting tables for properly sizing and installing vents for gas appliances

Gas valve furnace component that modulates (starts, stops, or adjusts) the flow of gas in response to system and safety requirements

Hot junction in a thermocouple, the point at which the dissimilar wires are connected

Hot-surface igniter silicon carbide or silicon nitride strip used to ignite fuel in gas

heating appliances; see hot-surface ignition

Hot-surface ignition ignition method by which electric current is passed through a high-resistance material to ignite gas as it passes through the gas valve; see hot-surface igniter

Inches of water column (in. W.C.) units used to measure pressures that are too low to read in psi; pressure readings are typically obtained with a manometer in which one psig is equal to 27.7 in. W.C.

Incomplete combustion combustion that results in the formation and release of carbon monoxide as well as other toxic and poisonous substances

Individual section heat exchanger heat exchange configuration with common openings at the top for common venting and at the bottom for common fuel ignition

Intermittent pilot ignition method that lights the pilot only during a call for heat

Liquefied petroleum liquefied propane, liquefied butane, or a mixture of the two

Manufactured gas gas mixture made from coal, oil, and hydrocarbon gases

Millivolt quantity of voltage equal to 1/1000 of a volt

Natural gas mixture of hydrocarbon gases consisting mostly of methane, with smaller amounts of ethane, propane, and butane

Orifice opening in the spud that delivers the fuel to the burners

Pilostat coil of wire that generates a magnetic field to hold the gas valve open

Pilot in a furnace, a small flame used to light the main gas at the burners; it can be standing or intermittent

Primary air air that is mixed with the fuel prior to combustion

sealed-combustion appliances appliances that draw in outside air for combustion and then exhaust the flue gasses directly back out of the structure

specific gravity the ratio of the density of a particular material compared to the density of another

Spud pipe fitting on the gas manifold that delivers the fuel to the burners

Stack thermometer thermometer that can read the high temperatures commonly found in the stack or flue pipe coming off the furnace

Standing pilot pilot light that remains lit all the time

Thermocouple device formed by joining the ends of two dissimilar metallic wires; a small voltage is generated when the joint is heated

Thermopile multiple thermocouples connected

in series to generate higher voltages

Water manometer piece of test equipment that relies on the level of water in a plastic or glass tube to read gas pressure in inches of water column

COMBUSTION

For a fossil fuel furnace to operate, the fuel being used must be ignited and permitted to burn until the desired space temperature has been reached. This process is called **combustion**. During the combustion process (Fig. 24-1) oxygen is combined with the fuel so the stored energy in the fuel can be released. To release the heat energy in the fuel, the fuel-oxygen mixture must first be ignited by an external source, such as a flame or electric spark. After the combustion process has begun, the source of ignition is typically no longer needed, as the burning fuel provides continued combustion. Once the combustion process has terminated, however, an ignition source will be needed to begin the combustion process again. If fuel, oxygen, or ignition is missing, combustion cannot begin. For example, in a gas furnace, if no gas is being supplied

to the furnace, there will be no combustion. Similarly, if there is no means to ignite the fuel oxygen mix, combustion cannot occur.

COMBUSTION EFFICIENCY

The efficiency of the combustion that takes place in the furnace is determined by comparing the amount of useful heat produced with the total heat produced, including that which ends up in the chimney. If a furnace operates at

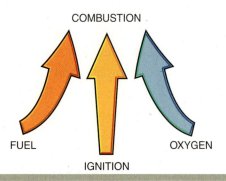

FIGURE 24-1 Fuel, ignition, and oxygen are required for combustion.

Going Green

The **Annual Fuel Utilization Efficiency, AFUE,** rating of fossil-fuel heating appliances is commonly used to evaluate the efficiency of the appliance. This rating represents the percentage of the energy in the fuel that actually becomes heat for the structure. For example, an appliance with a 90% AFUE rating converts 90% of the energy in the fuel to heat energy that is used to heat the home, while the remaining 10% is lost up the chimney.

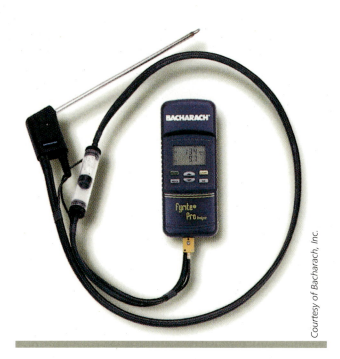

FIGURE 24-2 Piece of test equipment used to measure combustion efficiency.

Courtesy of Bacharach, Inc.

85% efficiency, approximately 15% of the heat produced by the burning fuel is lost up the chimney. The combustion efficiency of a furnace can be measured by field personnel using a tool such as a combustion analyzer (Fig. 24-2).

COMPLETE COMBUSTION

Ideally, when combustion occurs, all of the fuel burns, and the resulting byproducts are harmless to the atmosphere and the occupants of the space being heated. **Complete combustion** is achieved when carbon, which is released during the combustion process, bonds with the oxygen to form carbon dioxide (CO_2). Carbon dioxide is a nontoxic substance that can be safely vented to the atmosphere. In addition to carbon dioxide, water vapor is formed, which can also be safely released to the atmosphere. The byproducts of complete combustion are carbon dioxide, water vapor, nitrogen, and air, none of which are toxic or poisonous.

INCOMPLETE COMBUSTION

If there is an insufficient oxygen supply during combustion, the formation of carbon dioxide is affected, and carbon monoxide (CO) is formed instead. Carbon monoxide is toxic and poisonous. In addition, carbon, hydrogen, and oxygen bond together to form aldehyde, which is a toxic and poisonous substance. Unlike carbon dioxide, which is colorless and odorless, aldehyde has a very unpleasant

odor. When carbon monoxide and aldehyde are produced during the combustion process, **incomplete combustion** is taking place. Incomplete combustion also results in excessive carbon accumulation in the heating system and in the vent that carries the byproducts of combustion away from the furnace and occupied space. All heating systems should be configured and maintained to ensure that complete combustion is taking place as the fuel burns.

TESTING COMBUSTION

As part of an initial system startup and seasonal system checkup, checking the combustion in the furnace and the temperatures and draft within the furnace are as important as any other task the field technician performs. Improper combustion or furnace draft can result in dangerous levels of carbon monoxide and other toxic substances in the occupied space, leading to severe personal injury or death. When testing the combustion of a system, it is good field practice to start the system up and allow it to operate for 5 to 10 minutes prior to testing. As always, when using testing instruments, be sure to follow the manufacturers' guidelines to ensure proper and accurate testing. The sections that follow describe various tests that can be performed to evaluate combustion. Some manufacturers offer kits (Fig. 24-3) that can perform most, if not all, of the tests required for complete combustion evaluation.

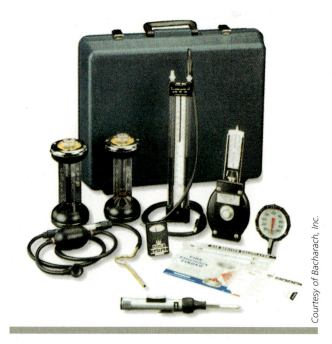

Courtesy of Bacharach, Inc.

FIGURE 24-3 Combustion test kit for use on gas appliances.

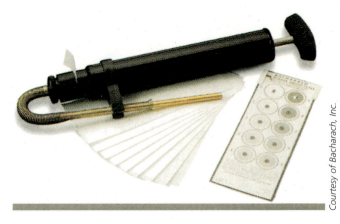

FIGURE 24-5 Smoke test kit.

THE DRAFT GAUGE

The draft gauge (Fig. 24-4) measures the pressure of flue gases in a furnace. The pressure over the fire in the furnace should be negative since the flue gases are moving away from the appliance. The draft over the fire is typically in the range of -0.01 to -0.02 **inches of water column (in. W.C.)**. One psig of pressure is equal to 27.7 in. W.C.

THE SMOKE TESTER

During the combustion process, excessive smoke should not be produced. The smoke tester (Fig. 24-5) is used to collect a sample of smoke, which is then compared to a graduated smoke sample scale (Fig. 24-6). To ensure that the sample size is accurate, a pump is used to move a precise volume of smoke from the flue, roughly 2,000 cubic centimeters (Fig. 24-7), through the filter paper, which is then compared with the smoke scale. The closest match between the smoke sample and the smoke scale indicates the amount of smoke present. A match at smoke spot #0 indicates that no smoke is present; a match at smoke spot #9 indicates that the most extreme smoke condition is present.

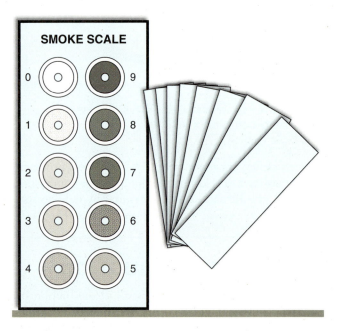

FIGURE 24-6 Graduated smoke scale and smoke test kit.

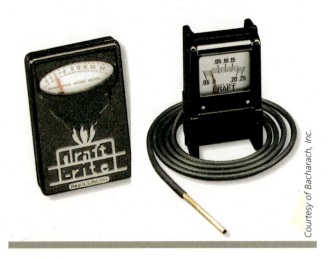

FIGURE 24-4 Draft gauge.

FIGURE 24-7 Technician performing a smoke test.

Courtesy of Bacharach, Inc.

FIGURE 24-8 Carbon dioxide tester.

CARBON DIOXIDE TESTER

Electronic carbon dioxide testers (Fig. 24-8) take air samples and evaluate the concentration of carbon dioxide in the sample. Nonelectronic carbon dioxide testers use chemical compounds that readily absorb carbon dioxide. To obtain the carbon dioxide reading, a known volume of flue gas is introduced to the instrument. As the carbon dioxide is absorbed into the chemical compound, the volume of the remaining flue gas is reduced. The amount of carbon dioxide in the flue gas can be read from the gauge on the instrument. Similar analyzers are used to determine levels of carbon monoxide.

CARBON MONOXIDE DETECTOR

The electronic carbon monoxide detector (Fig. 24-9) determines the level of carbon monoxide in the air by drawing an air sample through electronic sensors.

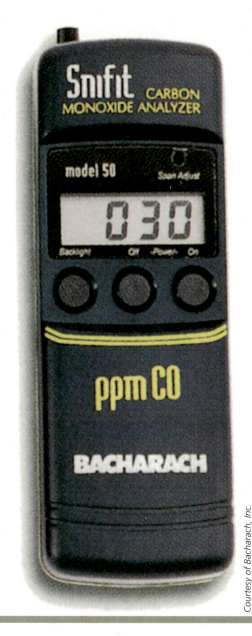

Courtesy of Bacharach, Inc.

FIGURE 24-9 Carbon monoxide tester.

Nonelectronic carbon monoxide detectors draw an air sample through a chemical reagent in a glass tube. A stain in the tube will result if carbon monoxide is present. The length of the stain determines the concentration of carbon monoxide.

STACK THERMOMETER

The **stack thermometer** (Fig. 24-10) can read the high temperatures commonly found in the stack or flue pipe coming off the furnace. The stack temperature reading should be obtained with the furnace operating. The final temperature reading has been reached when it rises no more than 3 degrees per minute.

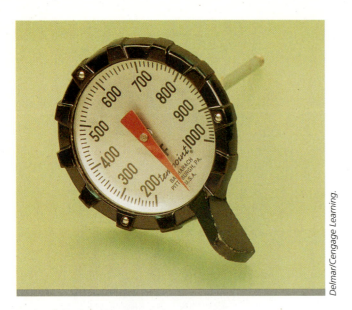

FIGURE 24-10 Stack thermometer.

Delmar/Cengage Learning.

FUELS

Before beginning our discussion of the gas furnace, we will address the fuels commonly found in furnaces and other types of heating equipment. Various fuels are used depending on the availability and accessibility of the fuel, and the geographic location of a specific end user. Individuals who live in a rural area will likely not have natural gas lines running to their homes and will undoubtedly use either liquefied petroleum or propane as their fuel source.

NATURAL GAS

Natural gas is a mixture of 60% to 95% methane and 5% to 40% other hydrocarbons such as ethane, propane, and butane, depending on the geographic location of the gas supply. Butane is not often found in natural gas because, at temperatures below 30°F, the pressure of butane is below atmospheric pressure. Natural gas is lighter than air with a **specific gravity** of 0.6. Specific gravity is the term used to describe the ratio of the densities of two substances. For example, the specific gravity of air is 1.0 and the specific gravity of natural gas is 0.6. This means that natural gas is lighter than air. When 1 ft³ of natural gas is burned, approximately 1,050 Btus of heat energy are released. The actual Btu rating for 1 ft³ of natural gas varies, depending on the geographic location in the country, and can range from 945 Btu/ft³ in Kansas City, Missouri, to 1,121 Btu/ft³ in Abilene, Texas. For general calculations, however, the 1050 Btu/ft³ can be used effectively to calculate the input rating of a

furnace. For example, a natural gas furnace that uses 75 ft³ of fuel per hour will have an input capacity rating of approximately 1,050 Btu/ft³ × 75 ft³ = 78,750 Btu/hr.

Natural gas has no color or odor and can displace oxygen if allowed to accumulate in a confined space. In addition to displacing the oxygen in the space, natural gas can also explode, so care should be taken to ensure that no leaks are left behind when installing or servicing gas heating equipment. To reduce the chance of having natural gas leaks go undetected, odorants are added to natural gas to provide it with a garlic-like smell, which will alert those around the equipment that there is indeed a gas leak.

MANUFACTURED GAS

Manufactured gas is produced from a combination of coal, oil, and a mixture of hydrocarbon gases. The Btu rating for manufactured gas is approximately one-half that of natural gas—500–600 Btu/ft³.

LIQUEFIED PETROLEUM

Liquefied petroleum can be liquefied propane, liquefied butane, or a mixture of propane and butane. Liquefied propane is most commonly used as a space-heating fuel because it boils at −40°F at atmospheric pressure and therefore easily boils off into a vapor for easy use as a heating fuel. Butane, on the other hand, vaporizes at about 30°F at atmospheric pressure and will, therefore, remain in the liquid state at temperatures below 30 degrees. Compared to natural gas, liquefied petroleum provides more Btus per cubic foot. Liquefied propane gives off approximately 2,500 Btu/ft³, while liquefied butane gives off approximately 3,200 Btu/ft³. It should be noted that both liquid propane and liquid butane are heavier than air and will settle in low-lying areas, increasing the chance of explosion in the event of a spark. The specific gravity of propane is about 1.5 and the specific gravity of butane is about 2.0. Since liquefied petroleum can be liquefied propane, liquefied butane or a combination of both, the actual specific gravity of LP will vary, but the value is often close to 1.52.

THE GAS FURNACE

The gas furnace (Fig. 24-11) provides heat primarily for forced-air heating systems. Quite often, the furnace is used in conjunction with a central air conditioning system to facilitate the heating and cooling of the occupied space through one set of ductwork using the same blower. Gas, the fuel source, is burned to convert the fuel to heat energy at a rate of about 1,050 Btu/ft³.

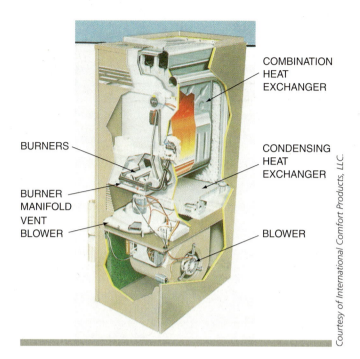

FIGURE 24-11 Condensing gas furnace.

COMBINATION HEAT EXCHANGER

CONDENSING HEAT EXCHANGER

BURNERS

BURNER MANIFOLD VENT BLOWER

BLOWER

Courtesy of International Comfort Products, LLC.

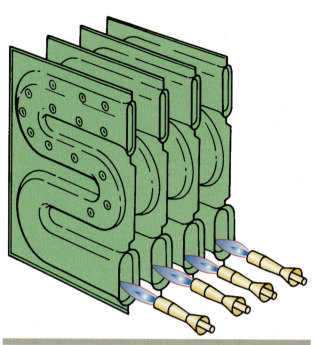

FIGURE 24-12 A four-section heat exchanger.

As the fuel burns, byproducts of combustion such as carbon monoxide and other toxic substances are formed. These byproducts must be kept separate from the air in the occupied space to prevent the poisoning of its occupants. This is accomplished by using a heat exchanger (Fig. 24-12) that keeps the conditioned air separate from the products of combustion. Air from the conditioned space travels over the outside of the heat exchanger and absorbs heat from the surface of the heat exchanger before returning to the occupied space (Fig. 24-13). The products of combustion are vented outside the structure via vent or flue pipe.

Just as air handlers for air conditioning applications are available in a number of different configurations depending on the direction of airflow through them, furnaces are available in a number of styles as well. They include the following:

- Horizontal
- Vertical upflow or highboy
- Lowboy
- Vertical downflow or counterflow
- Multiposition or multipoise

The horizontal furnace (Fig. 24-14) is typically located in either an attic or a basement. In either case, it is secured from above, either from the floor joists in the basement or from the rafters in the attic. Horizontal furnaces are desirable in locations where there is not sufficient clearance above. In operation, the air from the conditioned space is filtered and then enters the furnace

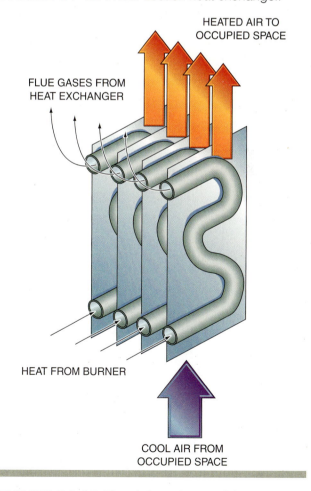

HEATED AIR TO OCCUPIED SPACE

FLUE GASES FROM HEAT EXCHANGER

HEAT FROM BURNER

COOL AIR FROM OCCUPIED SPACE

FIGURE 24-13 The air from the conditioned space and the products of combustion are separated from each other by the heat exchanger.

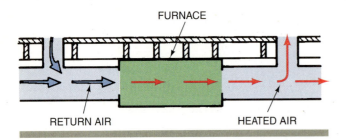

FIGURE 24-14 Airflow through a horizontal gas furnace.

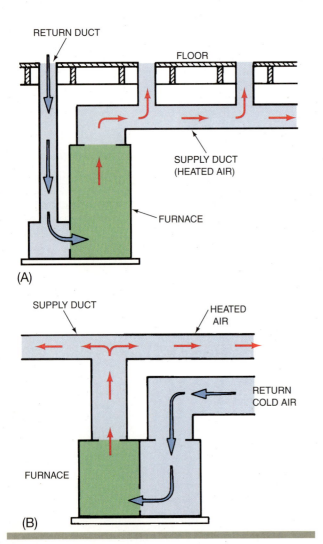

FIGURE 24-15 (A) Airflow through an upflow gas furnace. (B) Lowboy furnace.

from either the right- or left-hand side. The air is heated as it passes over the heat exchanger, and the heated air is sent back to the conditioned space by means of a blower incorporated into the furnace.

The vertical upflow furnace [Fig. 24-15(A)] can be used in a basement or in a utility closet on the same floor as the conditioned space. In this configuration, return

air from the occupied space enters the furnace from the bottom and is discharged from the top of the unit.

The lowboy configuration [Fig. 24-15(B)] is similar to the upflow design but is much shorter. It is typically used in basement locations where there is little headroom. Instead of being located in line with the heat exchanger, the blower assembly of the lowboy is positioned behind the heat exchanger to reduce the height of the unit.

The vertical downflow furnace (Fig. 24-16) introduces return air to the top of the furnace. The heated air is distributed back to the conditioned space from the bottom of the furnace. This configuration is desirable on system installations that have limited basement space and floor-mounted supply registers and grills.

The multiposition or multipoise configuration furnace gives the installation crew the option of configuring the furnace in any number of ways to provide the best possible installation. In this type of furnace, the heat exchangers, vents, and associated components can be repositioned to provide horizontal, upflow, or downflow applications.

FURNACE COMPONENTS

Although this chapter deals with gas furnaces, the basic components of a furnace are the same, regardless of the fuel used to generate the heat. These components include the following:

- Heat exchanger
- Means by which fuel is regulated and delivered to the furnace

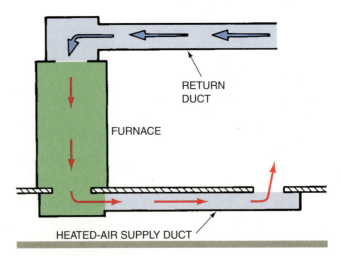

FIGURE 24-16 Airflow through a downflow or counterflow gas furnace.

- Means by which the fuel is ignited
- Means by which heat is transferred to the air from the occupied space
- Means by which combustion is monitored
- Means by which the products of combustion are removed
- Fan motor and blower
- Fan and limit switches

HEAT EXCHANGERS

The heat exchanger in a gas furnace is responsible not only for providing a heat transfer surface from which heat is transferred to the air from the conditioned space, but for keeping the products of combustion separate from the air in the conditioned space as well. The combustion process takes place in the heat exchanger, which is available in a number of different styles, depending on the design of the furnace. Heat exchanger types include:

- Individual section
- Cylindrical

The **individual section heat exchanger** is manufactured with common openings at both the top and bottom to provide for common ignition of the fuel and common venting of the products of combustion. This provides increased surface area for a greater rate of heat transfer between the heat exchanger itself and the air passing over it. This type of heat exchanger can be configured as a series of tubes (Fig. 24-17) or as a stamped-plate type (Fig. 24-12 and Fig. 24-13) that provides a serpentine channel in which the combustion process takes place. The effectiveness and efficiency of the heat exchanger is directly related to the surface area of the exchanger, so manufacturers strive to design heat exchangers with as large a surface area as possible. The **cylindrical heat exchanger**, on the other hand, has only one chamber in which combustion takes place.

FIGURE 24-17 Heat exchanger made up of a series of tubes.

FUEL DELIVERY TO THE FURNACE

One of the most important devices on the gas furnace is the gas valve. The **gas valve** functions to:

- Provide a means to manually shut off the supply of gas to the furnace
- Act as a pressure-regulating valve that will permit the proper gas pressure at the furnace, regardless of the gas pressure before the valve
- Act as a safety cutoff for the unit
- Automatically open and close in response to system requirements

Two common types of gas valves that will be discussed here are the combination gas valve (CGV) and the two-stage gas valve. The CGV (Fig. 24-18) has the ability to control the flow of gas to both the pilot and the main gas line, which feeds gas to the burners. In addition, the CGV has a manual control and a built-in regulator that helps ensure that the pressure in the gas line to the furnace is within the desired range. The two-stage gas valve allows the system to operate at two different capacities (Fig. 24-19). The two-stage gas valve is found on two-stage gas furnaces and has two solenoid valves inside it. When the system calls for first stage heat, the first

FIGURE 24-18 Automatic combination gas valve.

Courtesy of Refrigeration & Air Conditioning Technology, 6 Edition. Photo by John Tomczyk.

FIGURE 24-19 A two-stage valve on a two-stage furnace.

Photo by Eugene Silberstein.

FIGURE 24-20 Manometer.

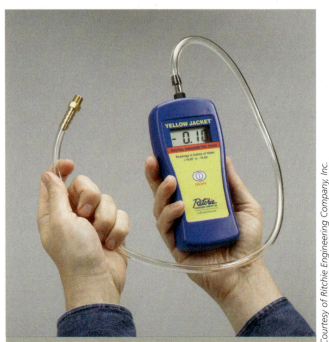

Courtesy of Ritchie Engineering Company, Inc.

FIGURE 24-21 An electronic manometer.

solenoid is energized and the system operates at about 50% of total capacity. When the system calls for full capacity, the second solenoid is energized, allowing full gas flow through the valve.

The desired gas pressure at the furnace varies depending on the fuel used. For natural gas, the desired pressure is approximately 3.5 in. W.C. The desired pressure for systems operating with manufactured gas is 2.5 in. W.C., while the pressure for liquefied petroleum systems is about 11 in. W.C. This measurement can be made with a **water manometer** (Fig. 24-20) or an **electronic manometer** (Fig. 24-21). The water manometer is constructed of clear tubing, bent to form a U-shape, mounted on a graduated inch scale. One side of the manometer is connected to the gas supply, while the other end is open to the atmosphere. The manometer is filled halfway with water. As the gas pressure pushes down on one side of the manometer, the water level on that side drops, while the water level on the other side rises (Fig. 24-22). The difference between the two measurements provides the pressure of the gas in in. W.C. (Fig. 24-23). The electronic manometer is connected directly to the outlet of the gas valve and automatically displays the gas pressure with respect to atmospheric pressure without having to perform any calculations. If the pressure of the gas being

Going Green

The actual gas pressures that will be required for a given area will vary depending on geographic area and altitude. Be sure to check with your local gas supplier for information regarding gas pressures in your area.

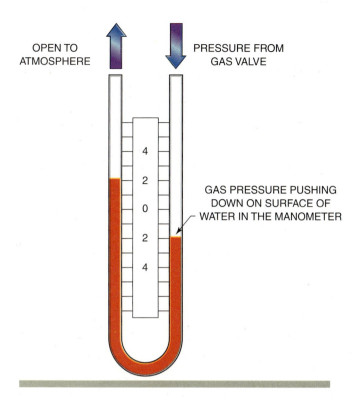

FIGURE 24-22 Gas pressure pushes down on one side of the manometer, while the other side is open to the atmosphere.

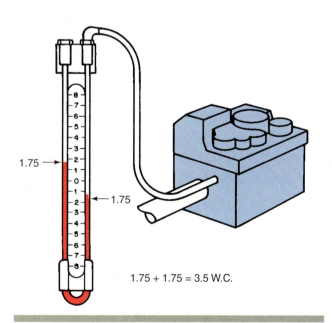

FIGURE 24-23 Calculating the total pressure in inches of water column.

supplied to the furnace is either too high or too low, a spring adjustment on the gas valve can be tightened or loosened to increase or decrease the gas pressure supplied to the furnace.

Refer to Procedure 24-1 on page 646 for step-by-step instructions for how Procedure 24-1 on page 646 to properly set the gas pressure on liquefied petroleum or gas furnaces.

THE GAS MANIFOLD

Once the gas passes through the gas valve, it travels through a number of other components before entering the burners on the furnace. Connected to the outlet of the gas valve is a manifold that feeds fuel to each of the heat exchanger sections. Figure 24-24 shows a heat exchanger with four sections. It can be seen that the manifold is connected to the piping circuit after the gas valve. The manifold is responsible for feeding the same amount of gas to each section of the heat exchanger.

At the end of each branch of the manifold is a fitting called a **spud** (Fig. 24-25). The spud is a threaded fitting that screws onto the manifold at the bottom of each section of the heat exchanger. A hole, called an **orifice**, is drilled into the spud, which permits a specific amount of fuel to flow from each branch of the manifold to the burners. The size of the orifice is determined by the fuel used, the pressure in the manifold, and the amount of gas

Photo by Eugene Silberstein.

FIGURE 24-24 This manifold is feeding a three-section heat exchanger.

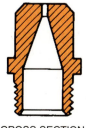

CROSS SECTION

FIGURE 24-25 The spud with a drilled orifice.

required at each burner. Orifices that are used on natural gas systems are sized larger than those used on liquefied petroleum (LP) systems, given the higher pressures of the LP system. The manifold pressure is affected by a number of factors including the altitude at which the system is installed. As mentioned earlier, the pressure of natural gas should be 3.5 in. W.C. and the pressures for manufactured gas and liquefied petroleum are 2.5 and 11 in. W.C., respectively. The size of the orifice is also affected by the altitude of the appliance. For example, an orifice might be sized to provide 975 Btu/ft^3 at sea level, but if the appliance is located at an altitude of 5,000 feet, the same orifice will likely only deliver about 850 Btu/ft^3. Therefore, in order to achieve the required heat output at higher altitudes, a larger orifice will be needed. Equipment manufacturers supply informative tables regarding the proper sizing and selection of orifices. Once the gas leaves the orifice, it is ignited at the burners.

Most burners used in residential applications are **atmospheric burners**, because the air supplied to the furnace is at atmospheric pressure. Burners are configured differently. Some are designed with slotted ports [Fig. 24-26(A)], that produce many individual flames, and others, like the ribbon burner [Fig. 24-26(B)], produce a single ribbon of flame in the furnace. Single-port burners, the upshot and inshot [Fig. 24-26(C)], are manufactured with only one port. Figure 24-27 shows a typical gas manifold.

PRIMARY AND SECONDARY AIR

Most manufacturers require that a certain amount of air be mixed with the gas prior to ignition. This **primary air** (Fig. 24-28) is mixed with the fuel as soon as it leaves the manifold from the orifice in the spud. The amount of primary air can be adjusted. Too little primary air will result in a yellow flame at the burner [Fig. 24-29(A)] and too much primary air will cause the flame at the burner to lift off the burner surface [Fig. 24-29(B)]. The correct amount of primary air will result in orange-tipped, blue flames that rest on the surface of the burner [Fig. 24-29(C)].

Additional air is introduced after the fuel has ignited. This is called secondary air (Fig. 24-30). If insufficient secondary air is being supplied to the burners, carbon monoxide will form.

FUEL IGNITION

In most cases, the fuel supplied to the furnace is ignited by a **pilot**, which is a smaller flame that uses a small quantity of gas as its fuel source. For the main gas valve

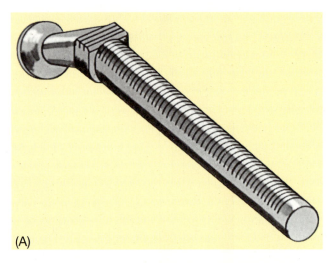
(A)

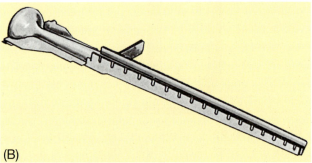
(B)

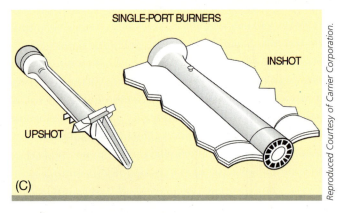
SINGLE-PORT BURNERS

INSHOT

UPSHOT

(C)

Reproduced Courtesy of Carrier Corporation.

FIGURE 24-26 (A) Slotted ports on a cast iron burner. (B) Ribbon-type burner. (C) Single-port burners.

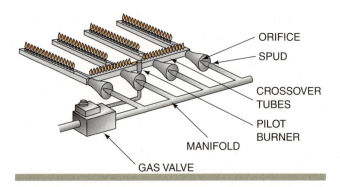
ORIFICE

SPUD

CROSSOVER TUBES

PILOT BURNER

MANIFOLD

GAS VALVE

FIGURE 24-27 Gas valve feeds gas to the burners through the manifold.

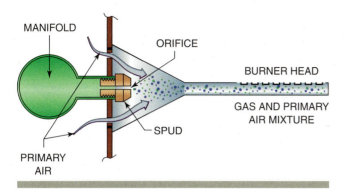

FIGURE 24-28 Primary air mixes with the gas prior to ignition.

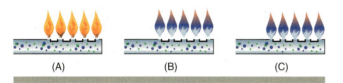

FIGURE 24-29 (A) Yellow flame indicates insufficient primary air. (B) Too much primary air will cause the flames to lift off the burners. (C) Proper flame.

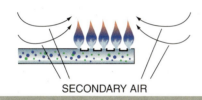

FIGURE 24-30 Secondary air is supplied after ignition.

to open, it must first be established that the pilot is lit. Methods for establishing the pilot will be discussed shortly. On gas heating systems without pilots, ignition systems ignite the gas as it passes through the valve. Common methods for igniting the fuel in gas heating systems include:

- Standing pilot
- Intermittent pilots
- Direct spark ignition (DSI)
- Hot surface ignition

STANDING PILOT

In heating systems with a **standing pilot**, the pilot light is lit all the time, regardless of whether or not the system is operating. Standing pilots can also be found on older stoves, ovens, and water heaters. If the pilot light goes out, it must be relit by hand. Safety devices in the system

prevent the main gas valve from opening when the pilot is out. If this were not the case, the main gas valve could open and introduce gas to the furnace that would not ignite, which could have deadly consequences. A number of methods can be used to establish that the pilot is lit. They include:

- Thermocouples
- Thermopiles
- Bimetallic strips
- Liquid-filled remote bulbs
- Flame rectification

Going Green

Although the HVAC technician will likely see many existing systems with a standing pilot light, manufacturers of new equipment are turning to spark-type and glow-plug methods for igniting the fuel on heating appliances. The spark-type and glow-plug features allow for more efficient appliances.

Thermocouples. Safety protection is incorporated into the gas valve assembly that will cause the gas valve to close if the pilot flame goes out for any reason or fails to be lit in the first place. The pilot flame is sometimes sensed by a **thermocouple** (Fig. 24-31), which is made up of two dissimilar metallic wires welded together at one end. Figure 24-31(A) shows that this thermocouple is made up of a chrome-iron alloy wire and a copper-nickel alloy wire. The end where the two wires are joined together is the **hot junction**, and the loose ends of the thermocouple make up the **cold junction**. When the hot junction is heated, electrons flow in opposite directions through each of the two metals [Fig. 24-31(B)]. The thermocouple is located in the path of the pilot flame (Fig. 24-32). When the thermocouple is heated, a small electric voltage—about 30 millivolts—is generated. One **millivolt** is equivalent to 1/1000 of a volt. If the pilot light goes out or fails to light, the voltage across the cold junction will be zero volts and the gas valve will not open.

This voltage is then sensed by the **pilostat**, or power unit, which is a coil of wire located in the gas valve assembly that holds the gas valve open by means of an electromagnetic field (Fig. 24-33). The pilostat operates on the magnetic field generated by the voltage produced

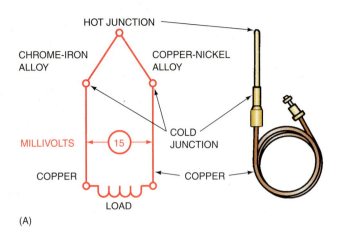

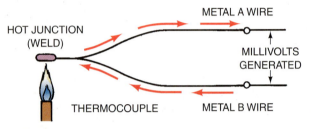

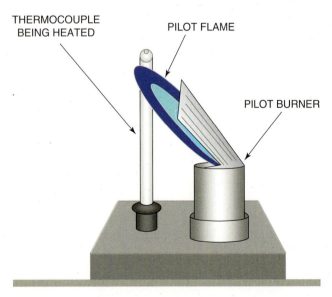

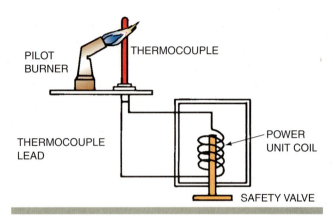

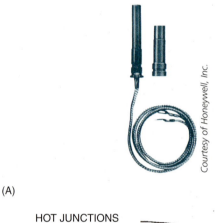

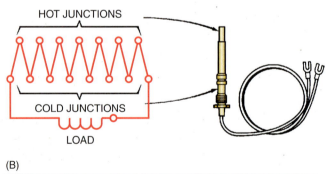

Courtesy of Invensys Controls.

Courtesy of Honeywell, Inc.

FIGURE 24-31 (A) Thermocouple made up of dissimilar wires. A small voltage is generated across the cold junction when the hot junction is heated. (B) The current flows toward the hot junction through one wire and away from the hot junction through the other.

FIGURE 24-32 Electric current is generated when the thermocouple is heated. If the thermocouple does not sense any heat, the gas valve cannot open.

FIGURE 24-33 Millivolt signal from the thermocouple creates a magnetic field in the pilostat or power unit that holds the valve open.

FIGURE 24-34 (A) Thermopile. (B) Thermocouple made up of a number of thermocouples connected in series.

by the thermocouple. The magnetic field that holds the gas valve open results, therefore, from the thermocouple's response to the temperature of the pilot flame.

Depending on the operating voltages of the pilostats and the gas valves, the voltage generated in a thermocouple may not be sufficient to hold the power unit open. Depending on the design of the system, a **thermopile** [Fig. 24-34(A)] may be used instead. A thermopile is

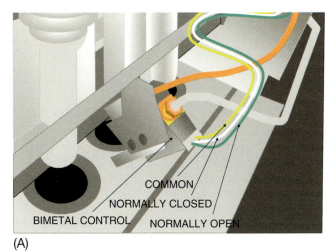

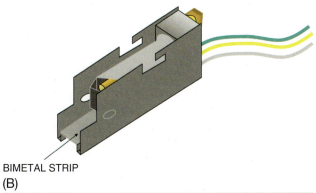

FIGURE 24-35 (A) Bimetal control in a gas furnace. (B) The location of the bimetal strip in the bimetal control.

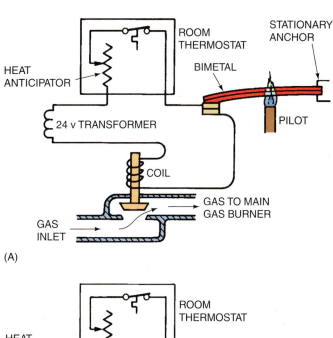

FIGURE 24-36 (A) When the pilot is lit, the bimetal will warp and the electrical contacts will close. This will complete the coil circuit, opening the gas valve. (B) When the pilot is no longer lit, the bimetal strip will straighten and the coil circuit will open, causing the gas valve to close.

made up of a number of thermocouples wired in series with each other [Fig. 24-34(B)]. The voltage generated by a thermopile may be about 800 millivolts, but the voltage generated by a thermocouple may be only 30 millivolts.

Bimetallic Strips. Like other bimetallic controls and devices discussed earlier, the bimetal strip used in conjunction with gas valves (Fig. 24-35) opens and closes electrical contacts in response to heat sensed. When the pilot light is lit, the bimetal strip warps, closing a set of electrical contacts and permitting voltage to be supplied to the gas valve (Fig. 24-36A). The figure shows that the magnetic field generated in the coil lifts the plunger in the gas valve. If the pilot light goes out, the bimetal strip will cool and the electrical contacts will open, causing the gas valve to close (Fig. 24-36B).

Liquid-Filled Remote Bulbs. The bimetallic strips just discussed open and close electrical contacts, depending on the temperature sensed at the pilot location. Remote bulbs (Fig. 24-37) are filled with a volatile fluid that expands and contracts in response to the temperature

sensed by the pilot. The bulb is located close to the pilot and, as the bulb is heated by the pilot, the pressure inside the bulb increases, closing electrical contacts in series with the gas valve (Fig. 24-38). If the pilot flame goes out, the pressure in the bulb will drop, causing the electrical contacts in series with the gas valve to open, thereby de-energizing the gas valve.

Flame Rectification. The concept of flame rectification involves converting alternating current (AC) to direct current (DC). The pilot flame, which results from burning gas, contains combustion gases with charged particulate matter. When the pilot flame is present between two electrodes, electrical current is able to flow

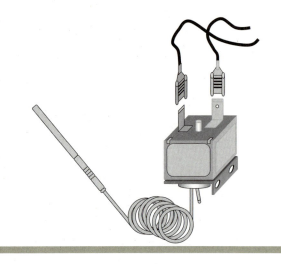

FIGURE 24-37 Liquid-filled remote bulb.

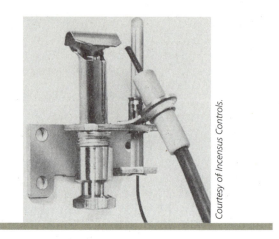

Courtesy of Incensus Controls.

FIGURE 24-39 Spark-to-pilot ignition system.

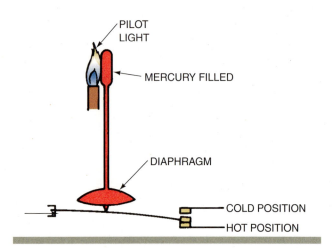

PILOT
LIGHT

MERCURY FILLED

DIAPHRAGM

COLD POSITION
HOT POSITION

FIGURE 24-38 When heated, the diaphragm expands, causing electric contacts to change position.

Courtesy of Refrigeration & Air Conditioning Technology, 6 Edition. Photo by John Tomczyk.

FIGURE 24-40 Intermittent pilot control with 100% shutoff.

freely between them. Current flow between the two electrodes completes a rectification circuit in the control board, generating a DC voltage that powers the main gas valve.

INTERMITTENT PILOT

As the name **intermittent pilot** systems implies, the pilot is only lit when the system is calling for heat. During the off-cycle the pilot is off. On a call for heat, the pilot valve opens and an electric spark ignites the pilot (Fig. 24-39). The spark, which is about 10,000 volts, is generated approximately twice per second. In the path of the pilot flame is a thermocouple or similar flame-proving device that allows the main gas valve to open once the pilot flame has been established. If the pilot does not light

within a predetermined period of time, the system *may* lock itself out. Such systems are referred to as 100% shutoff and must be reset manually. A control module for a 100% shutoff system is shown in Figure 24-40. This system configuration is required for fuels that are heavier than air, since they will settle in low-lying areas. One example of such a fuel is liquefied petroleum. On natural gas systems, however, the pilot valve will open and the sparking will begin at regular time intervals until the pilot flame is established. Since natural gas is lighter than air, the un-ignited fuel that passes through the pilot valve will rise up through the flue pipe and be removed or vented from the structure. Natural gas systems do not require 100% shutoff. A control module for use on natural gas is shown in Figure 24-41. Note that this control is clearly marked that it is for use only with natural gas.

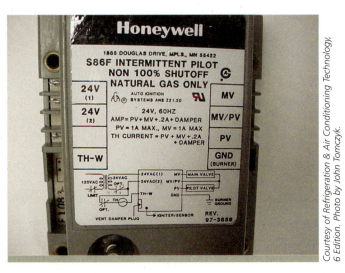

FIGURE 24-41 Intermittent pilot control with non-100% shutoff for use on natural gas only.

Courtesy of Refrigeration & Air Conditioning Technology, 6 Edition. Photo by John Tomczyk.

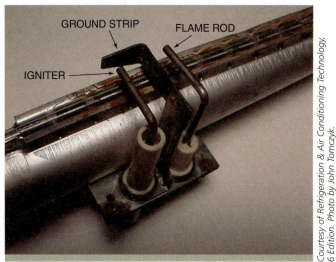

FIGURE 24-42 Direct spark ignition.

Courtesy of Refrigeration & Air Conditioning Technology, 6 Edition. Photo by John Tomczyk.

from experience...

Always refer to and follow local codes regarding the type of fuel ignition system that can be used on any given heating system. Some local municipalities require that 100% shutoff systems be used in conjunction with fuels that are heavier than air, while others do not.

DIRECT-SPARK IGNITION

In the intermittent pilot system just discussed, the spark used to light the pilot, which, in turn, signaled the opening of the main gas valve and ignited the fuel from the main gas valve. In the **direct spark ignition (DSI)** no pilot valve or pilot light is used. Instead, the spark ignites main gas. The igniter and sensor are located near the burner (Fig. 24-42). Since the main gas valve opens on a call for heat, the furnace must fire quickly and a signal must be sent to the ignition module (Fig. 24-43) verifying that the furnace has fired. If furnace firing cannot be verified within a few seconds, no more than ten, the gas valve closes and the system goes into a lockout mode, similar to the 100% shutoff associated with fuels that are heavier than air. This will prevent large volumes of un-ignited gas from being introduced to the furnace.

In operation, the space thermostat calls for heat, and voltage is supplied to the direct ignition module, which supplied power to the gas valve and the spark igniter.

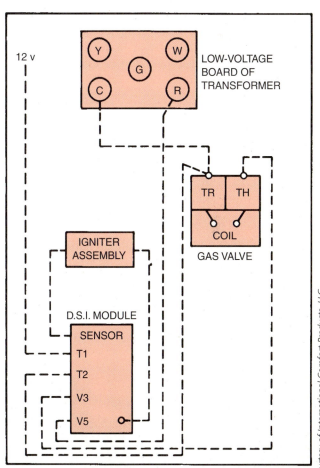

FIGURE 24-43 Components in a DSI system.

Courtesy of International Comfort Products, LLC.

Once the gas rack has ignited, a sensor rod sends a low amperage signal to the ignition module, verifying that the fuel has ignited. Once furnace firing has been established, the sparking sequence will be terminated.

HOT-SURFACE IGNITION

Instead of using an open flame to ignite the gas being supplied to the furnace via the pilot valve or main gas valve, the method of **hot-surface ignition** uses a material such as silicon carbide or silicon nitride that offers a great deal of resistance to electric current (Fig. 24-44). Some newer systems use silicon nitride, as it is less fragile than igniters made of other materials. When subjected to supply voltage, typically 120 volts, it will reach

a temperature that is higher than the ignition temperature of the gas being supplied to the furnace. It is the hot surface of this material that leads to the term **hot-surface igniter**, as the fuel is ignited when it comes in contact with the hot surface of the silicon-based material. Once the hot surface igniter heats up (Fig. 24-45) and the flow of fuel is initiated, immediate ignition should occur (Fig. 24-46). Some furnaces use a hot surface igniter to light the pilot instead of main gas (Fig. 24-47).

CAUTION

CAUTION: Touching the surface of a silicon carbide hot surface igniter will shorten the life of the igniter, since grease and oil from the hand will create "hot spots" on the igniter.

Photo by Eugene Silberstein.

FIGURE 24-45 Energized hot surface igniter.

from experience...

Hot surface igniters are typically very brittle, so be sure to work around them carefully. Newer igniters made from silicon nitride are typically more durable than those made from silicon carbide. In addition, hot surface igniters fabricated in round shapes such as the double helix are more durable than flat igniters.

Photo by Eugene Silberstein.

FIGURE 24-44 Hot surface igniter.

Courtesy of Refrigeration & Air Conditioning Technology, 6 Edition. Photo by John Tomczyk.

FIGURE 24-46 Ignition in the furnace.

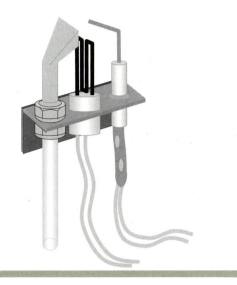

FIGURE 24-47 Hot surface igniter used to light the pilot instead of main gas.

VENTING AND REMOVING PRODUCTS OF COMBUSTION

As discussed in the section on combustion, when fuel burns, compounds such as carbon monoxide are formed, which must be removed from the structure. Improper venting of these compounds can result in severe personal injury or death. These products of combustion are **flue gases**, and the piping that directs them from the structure is **flue pipe** (Fig. 24-48 and Fig. 24-49). The difference in temperature between the flue gas and the air surrounding the flue pipe helps facilitate the venting process. The greater the temperature difference, the faster the rate of venting. Flue gas is hot and, as it travels through the piping, readily gives up sensible heat. As it cools, it can condense into a liquid, which can ultimately damage the vent piping, chimney, and the heat exchanger in the furnace itself. For this reason, the flue gas should be vented

Delmar/Cengage Learning.

FIGURE 24-49 Double-wall, insulated pipe.

to the outside as quickly as possible. However, quickly venting the flue gases to the outside means that a large amount of heat is literally being sent up the chimney. It costs money to burn fuel and the heat contained in the flue gases contains valuable heat that is ultimately rendered useless as it leaves the home.

> ### *from experience...*
>
> When venting appliances, be sure to refer to the proper venting tables. The **Gas Appliance Manufacturers Association (GAMA)** is an organization that provides installation information as well as tables to calculate and properly size venting systems.

Delmar/Cengage Learning.

FIGURE 24-48 Single-wall flue pipe.

To extract as much heat as possible from the flue gases, additional heat exchangers can be used that allow the flue gases to recirculate before they are removed from the structure. The main problem with this configuration is that the flue gases are now able to condense, since they have given up a great deal of sensible heat. For this reason, plastic is used to vent the flue gases on these furnaces because it will not corrode as a result of coming in contact with flue gases. In addition, fans are used to speed the flue gases as they rise from the furnace. These furnaces are higher efficiency models and will be discussed shortly.

Most furnaces of standard efficiency, about 80%, are manufactured with draft hoods that facilitate the natural venting of the flue gases. Natural draft means that the warmer flue gases are lighter than cooler air and will rise out of the structure by natural convection currents. The draft hood allows some of the room air, called **dilution air**, to blend with the flue gases (Fig. 24-50) before entering the flue pipe and chimney. The heated flue gases rise up the chimney creating a draft, which in turn, pulls more room air into the flue pipe, aiding in the venting process and reducing the concentration of the products of combustion.

VENTING CATEGORIES

When referring to the venting of flue gases, gas appliances are separated into four categories. These categories describe the average temperature of the flue gases as well as the static vent pressure.

Category I Appliances. Category I appliances operate with a negative static vent pressure and typically have flue gas temperatures above 275°F, which is roughly

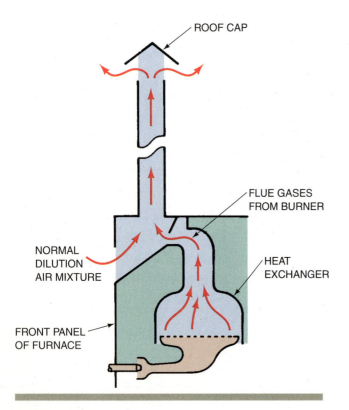

FIGURE 24-50 Dilution air mixed with products of combustion prior to entering the flue pipe and chimney.

150 degrees higher than the condensing temperature of the flue gases.

Category II Appliances. Category II appliances operate with a negative static vent pressure and typically have flue gas temperatures below 275°F. These appliances operate with natural drafts.

Category III Appliances. Category III appliances operate with a positive static vent pressure and flue gas temperatures above 275°F. These appliances rely on the heat of the flue gases to aid in the venting process and often reuse the heat in the flue, resulting in lower flue gas temperatures and condensing of the flue gases. Aluminum piping is often used for venting, given the possibility of condensation.

Category IV Appliances. Category IV appliances have a positive vent pressure and operate with flue gas temperatures well below 275°F—very often below the dew point temperature of the flue gases. Condensing furnaces (see the next section) are classified as category IV appliances and use plastic PVC or CPVC materials for venting, mainly because PVC will not corrode.

FLUE GAS TEMPERATURES

In a typical gas furnace, the average temperature of the flue gas leaving the furnace is about 400°F. As the flue gases leave the furnace, they quickly cool as they come in contact with the surrounding air. If the flue gases are permitted to cool enough, they will begin to condense when the temperature of the gas cools down to approximately 125°F. Chimney size plays a big role in the condensing, or lack thereof, of the flue gases. If the cross-sectional area of the chimney is too large for the application, the velocity of the flue gases will decrease, permitting the flue gases to condense. In this case, a lining should be installed in the chimney, especially if it is a masonry chimney. The mortar in these chimneys will be greatly affected by the corrosive nature of the condensed gases.

Furnaces with higher efficiencies than those described in the previous paragraph operate with lower flue gas temperatures. These furnaces are able to use more of the heat from the flue gases before sending them up the chimney. The flue gas temperature can therefore be reduced to the 170°F to 190°F range. When using such furnaces, it is important that the temperature of the flue gases be kept above the 125°F condensing temperature.

Furnaces that operate at more than 90% efficiency reduce the temperature of the flue gases to about 115°F to 120°F. Since the temperature of these flue gases is less than 125°F, the gases begin to condense. These furnaces are referred to as **condensing furnaces**.

COMBUSTION EXHAUST SAFETY

The exhaust from any heating appliance that burns a fuel contains a combination of water vapor, carbon dioxide, nitrogen oxides, carbon monoxide, and other trace gasses. All of the gasses can have a bad effect on the building and occupants, so they must not be allowed to leak inside the house. Carbon monoxide is colorless, odorless, and deadly. Even in low quantities, it can cause symptoms similar to a bad cold or flu (headache, nausea, vomiting, and fatigue). Nitrogen oxides cause eye, nose, throat, and lung irritation, coughing, nausea, and sometimes, respiratory swelling. Water vapor raises the humidity level inside the house that leads to other indoor air quality problems like mold and biological allergens, and it can condense and cause decay inside building cavities.

In order to reduce the chances of indoor air contamination, **sealed-combustion appliances** can be used. Sealed-combustion appliances draw in outside air for combustion and exhaust the flue gasses directly back out. There is no interaction between the combustion chamber and indoor air so there is little chance that gasses can enter the house. When the building budget does not allow for a sealed combustion model, an induced draft, power-vented model should be installed. Power-vented combustion appliances utilize blowers to efficiently remove the products of combustion from the structure. Refer to the venting sections in Chapter 25 for more information about fossil fuel appliance venting.

VENTING MATERIALS

The most commonly used material for venting non-condensing or conventional furnaces and other fuel-burning appliances is type B vent (Fig. 24-51). Type B venting material is constructed of an outer shell, which can be either galvanized steel or aluminum, and an inner aluminum pipe. The air space between the two layers of a B vent section acts as an insulator that helps keep the flue gases warm and helps reduce the surface temperature of the vent itself. Type B vent materials are typically used on category I appliances and are never used on category III and IV appliances.

Condensing furnaces—category IV appliances—are typically vented with polyvinyl chloride (PVC) or

chlorinated polyvinyl chloride (CPVC) piping material, which does not corrode when subjected to condensing flue gases.

New products on the market help installing vent piping a much easier task. As just described, typical flue pipe material must be cut to size and then crimped on one end in order to join two sections together. These sections must then be fastened and secured using sheet metal screws, clamps, or other approved method. When venting a category II, III, or IV appliance, venting material such as that shown in Figure 24-52 must be used. These piping sections are constructed

SECTION OF TYPE B GAS VENT

FIGURE 24-51 Type B gas vent.

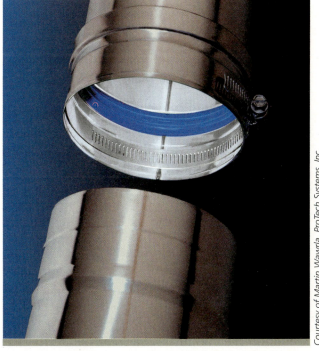

Courtesy of Martin Wawrla, ProTech Systems, Inc.

FIGURE 24-52 Single-wall venting material.

of stainless steel and have mechanical locking bands and gaskets. The locking bands are part of the pipe itself, so there is no more fussing or scrambling to find screws or clamps. Once connected, an air and water-tight seal is created.

To join piping sections, the two pieces are pushed together until the stop bead is reached. Once together, the mechanical locking band is then tightened. This product features adjustable vent lengths at the wall termination point so there is no need to cut the pipe sections. A double wall variety of the product just described (Fig. 24-53), is also available when construction, building, or other local code requires the use of double wall venting material for a particular appliance or application.

THE DRAFT DIVERTER

Furnaces with natural draft venting can experience pilot light outages if high winds push down the chimney and blow across the gas rack and pilot assembly. To remedy this potential problem, manufacturers install draft diverters (Fig. 24-54) in the furnace. These diverters are designed to re-route any downdraft air out of the furnace from the point where the dilution air is introduced to the furnace.

REPLACEMENT AIR

The volume of air that is used during the combustion process in the furnace and the air that goes up the chimney along with the flue gases must be reintroduced to the space surrounding the furnace to avoid creating a partial vacuum. If adequate air cannot be supplied to

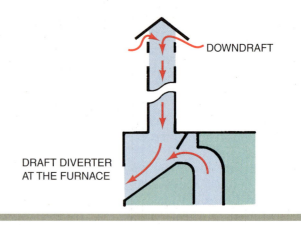

FIGURE 24-54 Draft diverter prevents a downdraft from extinguishing the pilot or burner flames.

the furnace's surrounding area, the pressure at the discharge of the chimney will be greater than the pressure in the structure. This will cause the flue gases to be pulled back into the space around the furnace and could result in severe personal injury or death as a result of inhaling the products of combustion.

To aid in the correct venting of the flue gases from the structure, sufficient quantities of air must be supplied to the area in which the furnace is located. Some of the air is required for combustion and some serves as dilution air that flows up the flue pipe along with the products of combustion. Combining the combustion and dilution air, approximately 30 ft^3 of air is required for each cubic foot of natural gas that is burned. Since each cubic foot of natural gas produces 1,050 Btu/h when burned, a 105,000 Btu furnace would require 3,000 ft^3 of air:

$$\frac{105{,}000 \text{ Btu / h} \times (30 \text{ ft}^3 \text{ of air / ft}^3 \text{ of natural gas})}{1{,}050 \text{ Btu / h / ft of natural gas}}$$

$$= 3{,}000 \text{ ft}^3 \text{ of air}$$

> ### CAUTION
>
> **CAUTION:** Improper air supply at the furnace location can result in the introduction of deadly products of combustion into the occupied space. Always follow manufacturer's recommendations and local codes regarding required air quantities and venting procedures. When in doubt, refer to the National Fuel Gas codes.

Sizing the Flue Pipe. Different local codes allow for the use of different materials for the flue piping that connects the furnace to the chimney, so it is important that the chosen material has been approved for the specific

Courtesy of Martin Wawrla, ProTech Systems, Inc.

FIGURE 24-53 Double-wall venting material.

installation. The size of the flue pipe coming directly off the furnace should be at least the same size as the vent opening on the furnace itself. Sizing the vent too small can result in some of the flue gases remaining in the area around the furnace. Sizing the vent too large can result in the slowing of the flue gases, giving the gases more time to condense and possibly causing rapid deterioration of the flue pipe and chimney.

> ## from experience...
> When venting multiple appliances, always refer to venting tables for the correct flue pipe sizes.

The Chimney. Quite often, masonry house chimneys are sized larger than needed. This can result in the deterioration of the mortar used to hold the bricks together, caused primarily by condensing flue gases from the furnace. To reduce the corrosive effects on the interior surfaces of the chimney, it is recommended that chimneys be properly lined.

The chimney liner can be made of either metal or some type of fireclay tile that is resistant to the corrosive substances in the flue gases. Properly sizing the chimney or liner will also keep the velocity of the flue gases high as they leave the chimney, thereby reducing the corrosive effects.

When properly installed, chimney liners do not come in direct contact with the interior walls of the main chimney. This creates an insulating barrier between the liner and the chimney similar to insulated, multi-pane windows. A properly installed chimney liner helps keep the flue gases hot, allowing them to vent faster and preventing them from condensing in the chimney.

> ## from experience...
> It requires more heat to warm an oversized chimney liner than one that is sized properly. Oversized liners will, therefore, be cooler than desired, permitting flue gases to condense.

FAN MOTOR AND BLOWER

On gas furnaces, the fan motor operates to distribute the heated air through the duct system to the occupied space. To prevent premature deterioration of the heat exchanger and to maximize comfort, the blower is typically wired to operate only after the heat exchanger has reached the desired temperature, and to remain on at the end of the heating cycle to allow the heat exchanger to cool.

Delaying the starting of the blower motor at the beginning of the cycle eliminates the introduction of cool air into the occupied space while the heat exchanger reaches its operating temperature. This is accomplished with a fan switch that will be discussed in the next section.

At the end of the heating cycle, damage to the heat exchanger could result if the blower is de-energized immediately. Excessive heat would accumulate in the furnace, possibly causing the heat exchanger to crack. Permitting the blower to operate for a period of time after the burners have extinguished lets the furnace dissipate heat slowly, allowing the heat exchanger to cool.

FAN SWITCHES

The purpose of the fan switch (Fig. 24-55) is to cycle the blower on and off at the desired times, as mentioned in the previous section. This device energizes and de-energizes the blower depending on the temperature in the heat exchanger. The probe on the control is located close to the heat exchanger and senses the temperature of the air immediately surrounding it (Fig. 24-56).

Delmar/Cengage Learning.

FIGURE 24-55 Fan switch.

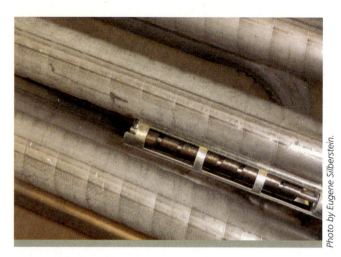

FIGURE 24-56 Bimetal element located close to the heat exchanger.

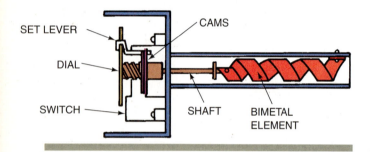

FIGURE 24-57 Cutaway view of fan switch.

FIGURE 24-58 Electronic module that controls time-delayed blower operation.

FIGURE 24-59 Limit switches.

The lever on the fan switch is manually set to the desired temperature for blower operation to start (Fig. 24-57). When the furnace initially lights, the bimetal element senses the increase in temperature within the heat exchanger and causes the cam on the control to rotate. When the cam reaches the set-point on the control, electric contacts close, energizing the blower. At the end of the heating cycle, the temperature of the heat exchanger decreases, causing the cam to rotate in the opposite direction. When the heat exchanger has cooled, the contacts in the control will open, de-energizing the blower. Some furnaces are equipped with electronic fan controls that cycle the blower on and off at fixed time intervals (Fig. 24-58).

LIMIT SWITCHES

The limit switch (Fig. 24-59) is primarily a safety device and should always be in the closed position unless a problem arises within the furnace. If the temperature within the furnace reaches a temperature that is higher than the maximum desired temperature, the contacts on the limit switch will open, de-energizing the gas valve. Excessive temperature in the furnace can result from a number of conditions including a defective or dirty blower, dirty air filter, blocked return duct, blocked supply duct, or improper gas pressure at the furnace. Quite often the fan switch and the limit switch are combined into a single component called the fan limit (Fig. 24-60).

GAS FURNACE INSTALLATION

The installation of the gas furnace must be done in accordance with the manufacturer's guidelines as well as all local codes. Setting the furnace is similar to setting an air

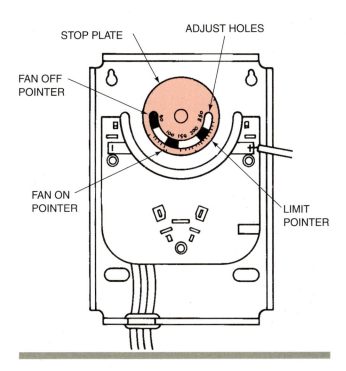

FIGURE 24-60 Fan limit control houses both the fan switch and limit control.

handler on split-type air conditioning systems. The two main factors that should determine the location of the furnace are the gas supply and, more importantly, the location of the chimney or termination point for the flue gases. The closer the furnace is to the chimney, the better. The four major areas of concern for the installation of a gas furnace are:

• Electrical wiring
• Air-distribution system
• Gas piping
• Venting

The electrical wiring and the air-distribution systems on gas furnaces are very similar to those on straight cooling air conditioning systems. For the details on electrical connection and duct systems, please refer to the appropriate sections in this text.

INSTALLING THE GAS PIPING

As with any piping arrangement, the piping for a gas furnace should be as direct as possible, using as few fittings as possible. Short runs of gas pipe with few fittings and connections will greatly reduce the chance of gas leaks within the structure. When installing the gas

piping for a furnace, the following guidelines should be followed:

• Always read the installation information provided with the furnace

• Understand and adhere to all local codes

• Make certain that pipe dope or similar joint compound is used on all fittings

• Make sure there is a manual shutoff valve on the gas line feeding the furnace (Fig. 24-61); refer to local codes for information regarding the maximum distance between the valve and the furnace

• Make sure a drip leg or drip trap is installed on the supply line that will catch particulate matter, condensation, dirt, and other solid debris before it is able to enter the gas valve and furnace (Fig. 24-61)

• Make certain there is a union between the manual shutoff valve and the gas valve (Fig. 24-61) to facilitate the installation of the furnace and make it easier to service and to replace the gas valve in the event of failure in the future

• Pitch gas piping down toward the furnace to allow condensate to run into the drip leg; once again, check local codes, but a pitch of ¼ inch per 15 feet of horizontal run is usually acceptable (Fig. 24-62)

• Check gas piping for leaks and have it inspected by the local gas utility before putting the system into operation

Refer to Procedures 24-2, 24-3, and 24-4 on pages 647–650 for step-by-step instructions for cutting, reaming, threading, and joining steel pipe and fittings.

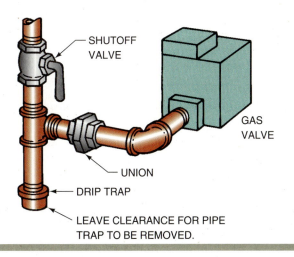

FIGURE 24-61 Gas piping installations should include a manual shutoff valve, drip trap, and union.

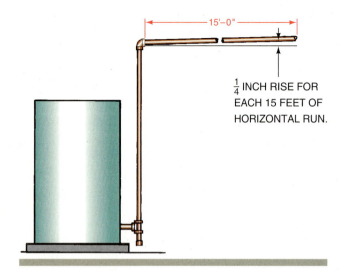

FIGURE 24-62 Gas piping to furnace should be pitched downward approximately ¼ inch per 15 feet of horizontal run.

INSTALLING THE VENT PIPING ON CONVENTIONAL FURNACES

When installing the vent piping between the furnace and the chimney, it is important that the material selected is approved for use on that particular installation based on local codes and the manufacturer's guidelines. In addition, the sizing of the flue piping must be correct. The following is a list of guidelines that should be observed when installing vent piping on a furnace using a natural draft:

- The vent piping should be at least the same diameter as the vent fitting on the furnace
- Horizontal flue pipe runs should be as short as possible
- The minimum slope of flue pipe should be ¼ inch per 1 foot of horizontal run (Fig. 24-63)
- As few fittings as possible should be used
- The flue pipe should be flush with the inside surface of the chimney
- The chimney should be completely clear and free of obstructions

from experience...

Always follow manufacturers' guidelines and local codes when installing vent piping on any furnace.

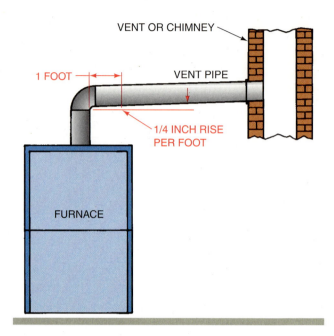

FIGURE 24-63 Vent pipe should be pitched upward from furnace a minimum of ¼ inch per 1 foot of horizontal run.

- Screws can sometimes be used to join B vent sections, but the inner lining should never be penetrated
- Flue piping should be supported every 4 to 5 feet
- Gas appliances using B vent materials typically need to be terminated vertically

INSTALLING THE VENT PIPING ON CONDENSING FURNACES

Just as with conventional furnaces, the material used must be approved for that particular installation based on local codes and manufacturers' guidelines. Typically, PVC or CPVC piping material is recommended because these materials are noncorrosive. The following is a list of guidelines that should be observed when installing vent piping on a condensing furnace:

- The vent piping should be at least the same diameter as the vent fitting on the furnace
- There should be a minimum of six inches of vertical run off the top of the furnace before any bends are made
- Horizontal flue pipe runs should be as short as possible
- The minimum slope of flue pipe should be ¼ inch per 1 foot of horizontal run
- As few fittings as possible should be used

- All plastic pipe joints should be cleaned with purple primer prior to cementing to prove that the pipes were primed prior to cementing; many codes require purple primer as opposed to clear primer for this reason
- The venting should be supported at least every 3 feet to prevent sagging
- Approved vent caps should be used at the structure penetration
- Vent piping passing through unheated space must be insulated

from experience...

Always follow manufacturers' guidelines and local codes when installing vent piping on any furnace.

TROUBLESHOOTING GAS HEAT PROBLEMS

Since each furnace manufacturer incorporates different controls and wiring methods into each furnace, what follows are general guidelines for troubleshooting the gas furnace.

FURNACE FAILS TO OPERATE AT ALL

If the furnace does not operate at all, do the following:

✓ Check the thermostat setting. Make certain that the thermostat is calling for heat.

✓ Check for line voltage at the service disconnect on the furnace. If there is no voltage at the disconnect, check the circuit breaker or fuse panel for a tripped or blown breaker or fuse.

✓ Check for line voltage to the primary of the control transformer. A reading of line voltage indicates that the transformer is being powered.

✓ Check for low voltage across the secondary winding of the control transfer. A reading of

24 volts indicates that there is control voltage to the unit. A reading of 0 volts indicates that the transformer is not being powered, or there is a problem with the transformer windings. Disconnecting the power to the unit and taking continuity readings (after disconnecting the wires from the transformer) through the primary and secondary windings of the transformer will reveal a defective winding. Infinite resistance readings through either winding indicates that the transformer is defective.

from experience...

In the event a control transformer has a defective winding, always find the cause for failure before replacing the component.

✓ If the transformer is good and there is 24 volts at the secondary, a jumper placed across the low-voltage heating terminals (typically R and W) on the furnace terminal board will indicate whether the problem lies within the thermostat and associated wiring or within the furnace. If the jumper wire causes the furnace to come on, the problem is with either the thermostat or the wiring leading to it. Jumping the heating wires at the thermostat will indicate if the problem lies within the thermostat. If the jumper wire at the furnace caused the furnace to come on, but the jumper at the thermostat failed to energize the furnace, the problem lies within the wiring between the furnace and the thermostat.

INOPERATIVE BLOWER

The following is a list of things to check if the gas rack on the furnace lights and the blower fails to operate:

✓ Check for voltage across the motor windings. A reading of line voltage indicates that the problem lies within the motor itself. A reading of 0 volts indicates that the problem most likely lies with the motor control.

✓ Check the voltage across the fan switch. A reading of 0 volts indicates that the control is closed and, if there is line voltage in the furnace, the blower should be operating. If the voltage reading is line voltage, the fan switch is open and should be checked again once the heat exchanger has had a chance to reach the desired temperature. If the fan switch fails to close once the heat exchanger has reached temperature, check and adjust the setting on the fan switch. If the setting is correct and the blower fails to operate, replace the control.

THE GAS VALVE

To make certain the gas valve is being powered, check the following:

✓ Be sure the thermostat is set to the heating mode and that the thermostat is set to a temperature higher than the temperature of the occupied space.

✓ Be sure line voltage is being supplied to the furnace. If there is no voltage at the disconnect, check the circuit breaker or fuse panel for a tripped or blown breaker or fuse.

✓ Check for low voltage at the secondary of the transformer. A reading of 24 volts indicates that there is control voltage to the unit. A reading of 0 volts indicates that the transformer is not being powered or there is a problem with the transformer windings. Disconnecting the power to the unit and taking continuity readings (after disconnecting the wires from the transformer) through the primary and secondary windings of the transformer will reveal a defective winding. Infinite resistance readings through either winding indicates that the transformer is defective.

✓ Check for voltage at the gas valve. A reading of 24 volts at the gas valve indicates that the valve should be open. A reading of 0 volts may be caused by an improperly set or defective thermostat, an open safety or limit switch, a break in the wiring between the furnace and thermostat, or a loose or broken low-voltage conductor within the furnace.

IGNITION PROBLEMS

Just as a variety of methods are used to ignite the fuel, many controls are on the market that will accomplish each task. Therefore, it would be impossible to address ignition problems in this text. However, there are some guidelines that apply to most ignition systems:

✓ On furnaces with thermocouples, a millivolt meter can be used to confirm that the thermocouple is generating the desired voltage. This can be accomplished by disconnecting the thermocouple

wiring from the gas valve and connecting the millivolt meter leads to the thermocouple. Lighting the pilot will heat the thermocouple and generate the voltage needed to open the gas valve. If the thermocouple fails to generate voltage, check the location of the thermocouple in the flame and, if that fails, replace the thermocouple.

✓ A testing adapter can also be used to test the thermocouple under load conditions. The thermocouple is connected to the adapter, which in turn, is connected to the gas valve. A millivolt reading can be taken from the adapter.

✓ On hot surface igniters, make sure the thermostat is calling for heat and that there are 24 volts at the secondary of the transformer. There should be a voltage reading at the igniter. If there is a reading across the igniter and it fails to heat up, replace the igniter. Be careful when handling the new igniter, especially those made from silicon carbide, because the oils from your hands will result in hot spots on the igniter and will shorten the life of the element. If no voltage is being supplied to the hot surface igniter, check the thermostat, transformer, interconnecting wiring, and controls and switches that are in series with the element.

✓ On electronic ignition modules, it is not practical for the field technician to evaluate the component parts of the board. It is only possible for the technician to determine if there is power being supplied to the module and if the remote devices connected to them are functioning. If power is supplied to the control module, the remote devices are operational, and the furnace fails to operate, the module should be replaced.

✓ Always refer to the manufacturers' installation/service literature to evaluate ignition and service issues relating to the specific furnace. Most manufacturers provide troubleshooting procedures for their particular control modules.

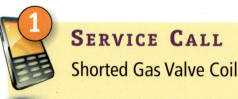

1 **SERVICE CALL**
Shorted Gas Valve Coil

CUSTOMER COMPLAINT

A residential customer calls her service company and informs the service manager that her gas furnace is not operating and that the indoor fan motor does not operate either.

TECHNICIAN EVALUATION

A short time later, Michael arrives at the customer's house. He listens to the customer as she tells him about the problem she is having with her system. He goes to the thermostat, sets the fan switch to the on position, and does not hear the fan come on. He sets the thermostat to the heating position, sets the fan to the automatic setting, and goes to the basement where the furnace is located.

Michael first checks the power to the unit. He disconnects the power to the unit and removes the service panel. He locates the power strip on the unit and restores power to the unit. Using his voltmeter, he finds that 220 volts are being supplied to the unit. He then decides to check the control transformer and reads 220 volts at the primary of the transformer and zero volts at the secondary. Michael quickly concludes that the control transformer is defective, so he goes to his truck to get a new transformer. He disconnects the power to the unit, removes the defective transformer, and installs the new component. He goes to the thermostat and turns the system off, and then restores power to the unit.

At the thermostat, Michael turns the fan switch to the on position and can hear the fan come on. He then turns the thermostat to the heating mode of operation and sets it to a temperature above the present space temperature. After about a minute, the indoor fan motor stops operating. Michael goes back down to the basement and checks the transformer again. There are 220 volts at the primary winding of the transformer and zero volts at the secondary winding. He disconnects the power to the

TECH TOOL BELT • TECH TOOL BELT

Courtesy of Ridge Tool Company.

Pipe Vise/Tripod

The pipe vise is used to secure piping material during cutting or threading processes. The legs of the pipe stand are spread wide apart to prevent the vise from tipping over. This stand has a chain mechanism to hold the pipe secure. Other pipe vises can use a yoke-type clamp to hold the pipe secure.

Courtesy of Ridge Tool Company.

Pipe Cutter

Similar in appearance to the tubing cutter that cuts copper pipe and tubing, the pipe cutter cuts steel piping material. These pipe cutters can be equipped with one or more cutting wheels to speed the cutting process. The piping material should be secured in a vise while cutting to ensure that the material is held secure.

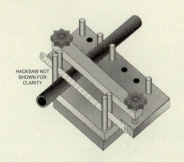

HACKSAW NOT SHOWN FOR CLARITY

Holding Fixture

The holding fixture mounts in a vise and helps ensure a square cut when used with a hacksaw by making certain that the hacksaw remains perpendicular to the pipe during the cutting process.

TECH TOOL BELT • TECH TOOL BELT

Courtesy of Ridge Tool Company.

Pipe Reamer

The pipe reamer is designed to remove burrs from the interior edge of steel pipe. It is inserted into the end of the pipe and rotated to create a smooth edge inside the pipe. Burrs on the inside of the pipe can reduce fluid flow through the pipe.

Pipe Threading Dies

Pipe threading die heads hold a set of four cutting blades that cut the threads into the end of steel pipe. These cutters are numbered and must be installed in the proper order so that the threads are created properly. Dull blades must be replaced to ensure proper threads and a leak-free installation.

Courtesy of Ridge Tool Company.

Pipe Threading Die Handle

The pipe threading die handle holds the threading dies in place. These handles are manufactured with a ratchet to speed the threading process. When the knob is pointing in the clockwise direction, the handle will tighten the dies onto the end of the pipe, creating threads. When the knob is pointing in the counterclockwise direction, the ratcheting handle will remove the threading dies from the end of the pipe.

Courtesy of Ridge Tool Company.

Automatic Pipe Threader

Used on jobs where a large number of pipe threads must be made, the automatic pipe threader automatically lubricates and rotates the pipe section within the dies to create threads. The threader can operate in either the forward or reverse direction to thread and then remove the pipe from the threading dies.

Courtesy of Ridge Tool Company.

Cutting Oil Pump

The cutting oil pump catches oil that falls from the pipe being threaded. This oil passes through a filter screen to separate it from the steel filings and then feeds it to the pipe end during the threading process. Use of an oil pump greatly reduces the amount of oil used to thread pipe as well as the amount of oil that ultimately ends up on the floor.

Courtesy of Ridge Tool Company.

Fixed Die-Type Pipe Threader

The fixed die pipe threader does not use a ratcheting handle. Instead, it has two handles that rotate the threader onto the pipe to create threads. As with any other type of threader, sufficient cutting oil must be applied to the pipe and dies to ensure proper threading and to prolong the life of the dies.

Courtesy of Ridge Tool Company.

Pipe Wrenches

Pipe wrenches are used to tighten steel pipe and fittings. They have jaws that will grip the pipe when turned towards the open jaw and release the pipe when turned away from the open jaw. Typically, two wrenches are used to create a tight connection between a pipe section and a fitting. One wrench holds the pipe steady, while the other tightens the fitting onto the end of the pipe.

---

unit and removes the wires from the secondary winding of the transformer. Using his ohmmeter, he checks the secondary winding and finds infinite resistance, meaning the winding is open. The brand-new control transformer is bad as well.

Before making another trip to his truck for a new transformer, Michael decides to check the control circuit to see why the second transformer went bad so quickly. From what he has already observed, he concludes the following:

- The indoor fan motor operated fine by itself
- When the system was switched to the heating mode, the transformer blew

Michael decides to check the control circuit for the heating cycle. He disconnects the power to the unit and sets the thermostat to the heating mode, making sure that the thermostat is set to a temperature higher than that of the occupied space. With the wires disconnected from the secondary of the transformer, he takes a resistance reading of the low-voltage circuit. The reading is 3.5 ohms. To get a point of reference, Michael goes to his truck and gets a new transformer, and a new gas valve as well. Removing the gas valve from the box, he takes a resistance reading of the coil on the valve and obtains a reading of 43 ohms. He takes a reading of the identical gas valve on the furnace and gets a reading of 3 ohms. The coil on the gas valve is shorted.

With the unit power still disconnected, Michael closes the manual gas valve in the main gas line feeding the furnace. He then removes the defective gas valve and installs the new valve, making sure to leak-check the gas piping prior to starting up the furnace. He replaces the control transformer, restores power to the furnace, and tests the operation of the unit.

SERVICE CALL DISCUSSION

Ultimately, Michael did locate the problem and make the necessary repair. However, this was not accomplished without a casualty: the first control transformer. Michael should have realized at the beginning that transformers do not go bad for no reason. Had he investigated further before replacing the transformer, he would likely have found that the coil on the gas valve was shorted. Even if he had not noticed the short in the coil, by installing a low-amperage, in-line fuse in the control circuit in series with the secondary winding, the fuse would have blown instead of the transformer. In-line fuses cost much less than transformers!

At any rate, Michael's initial response was to alleviate the immediate problem, which was the defective transformer. He failed to investigate and locate the cause of the transformer failure until an additional component had been damaged.

SERVICE CALL
Defective Fan Switch

CUSTOMER COMPLAINT

A residential customer calls her service company and complains that her gas furnace is not operating and that the fan is not operating. She explains to the person at the company that, when she went downstairs to the furnace, she could see and hear the gas rack ignite, but it turned off after about two minutes.

TECHNICIAN EVALUATION

Later that afternoon, John arrives at the customer's house and the customer explains the situation to him. He goes to the thermostat to make certain that the thermostat is set to the heating mode and to a temperature higher than the temperature in the space. He then goes to the basement to check the furnace. Just as the home owner told him, he sees the gas rack is lit, but the blower is not operating. A very short time later, the gas rack goes out. From this, John is able to conclude that the gas rack is operating, but there seems to be a problem with the blower or its associated wiring and/or controls.

Using his voltmeter, John checks the voltage across the fan switch and finds that the reading is equal to line voltage. This indicates that the fan switch is in the open position. He then checks the setting and notices that the setting is about 140 degrees. Since the fan is not coming on when the gas rack is lit, he decides to check the switch itself by lowering the cut-in temperature on the control to the lowest setting. When the gas rack lights a short time later, John notices that the blower still does not come on by the time the gas valve closes and the rack shuts down.

Thinking that the fan switch is not operating as it should, John takes a jumper wire from his toolbox and, after disconnecting the power to the furnace, places it across the wires on the fan switch. He restores power to the furnace and the blower begins to operate. John turns the power off and retrieves a new fan switch from the truck.

After replacing the fan switch and setting it properly, he restores power to the furnace. He can see the gas rack light and, after a short time, the blower cycles on. To make certain that the blower will turn off at the end of the cycle, John goes to the room thermostat and changes the set-point of the thermostat to cycle the heat

off. He goes back to the basement and sees that the gas rack is no longer lit and the blower is still running. After about 60 seconds, the blower cycles off. John turns the thermostat back on, checks the air filters on the furnace, and writes up his work order, making certain to include the new fan switch.

SERVICE CALL DISCUSSION

By observing the gas rack lighting, John knew immediately that there was power to the furnace and that the thermostat was calling for heat. Once the gas rack had lit, the next order of business was for the blower to come on. The very fact that the blower did not cycle on led him to the one control that operates the blower.

SUMMARY

- Combustion occurs when proper amounts of fuel, oxygen, and heat are combined.
- Complete combustion results in the production of nontoxic and nonpoisonous substances.
- Incomplete combustion results in the production of toxic and poisonous substances such as carbon monoxide.
- Combustion efficiency can be determined in the field using specialized pieces of test equipment.
- Natural gas is a mixture of methane, propane, ethane, and butane and produces approximately 1,050 Btu/ft^3 when burned.
- Manufactured gas is a mixture of coal, oil, and hydrocarbon gases and produces approximately 600 Btu/ft^3 when burned.
- Liquefied petroleum is liquefied propane, liquefied butane, or a combination of both, and that produces approximately 3,000 Btu/ft^3 when burned.
- Heat transfer between the burning fuel and the occupied space takes place in the heat exchanger.
- Gas pressure at the furnace is measured with a manometer.
- Typical gas pressures are 3.5 in. W.C. for natural gas, 2.5 in. W.C. for manufactured gas, and 11 in. W.C. for liquefied petroleum.
- The gas valve controls the flow of gas into the manifold and burners, where ignition takes place.
- Primary air is added to the fuel before ignition, while secondary air is added once the burners are lit.
- Fuel ignition can be accomplished by a standing pilot, intermittent pilot, direct spark, or hot surface ignition.

- The pilot flame can be established by a number of methods, including thermocouples, bimetallic strips, liquid-filled bulbs, and flame rectification.
- The products of combustion must be removed from the structure by means of vent piping.
- Flue gases, which can be highly corrosive, must remain warm to prevent them from condensing.
- Venting materials should be selected based on the category and type of furnace being used.

- Sufficient air must be supplied to the furnace to ensure proper combustion and venting.
- The fan switch permits blower operation once the heat exchanger heat reaches the desired temperature at the beginning of the heat cycle. It also keeps the blower operating at the end of the cycle to allow the heat exchanger to cool.
- The limit switch de-energizes the gas valve if the temperature in the furnace reaches unsafe levels.

GREEN CHECKLIST

- [ ] Gas-fired appliances are rated by their Annual Fuel Utilization Efficiency, AFUE.

- [ ] Every attempt should be made to maintain proper combustion.

- [ ] Incomplete combustion affects system efficiency and produces harmful byproducts such as carbon monoxide.

- [ ] Combustion must be tested using properly calibrated instruments.

- [ ] Gas-fired heating appliances cannot be set up for proper operation by using the senses of sight and sound alone.

- [ ] Liquefied petroleum is heavier than air, and precaution must be taken to prevent it from accumulating in low-lying areas.

- [ ] Altitude affects the required gas pressure and the sizes of system orifices. Improper pressures or components can lead to inefficient system operation.

- [ ] Two-stage gas valves are used to provide different capacities as the heating requirements of the structure change.

- [ ] Manufacturers are turning toward direct spark and hot surface ignition in order to increase the efficiency of their appliances.

Procedure for Setting the Proper Gas Pressure on an LP or Gas-Fired Furnace

- Check the furnace nameplate for required gas pressure. Typically this will be 3.5 in. W.C. for natural gas and 11 in. W.C. for liquefied petroleum.

- Turn the main gas valve to the OFF position.

- On the outlet side of the gas valve, locate the plug for the outlet gas pressure.

- Remove the plug.

- Install the male fitting from the manometer.

- Turn the main gas valve to the ON position.

- Start the furnace.

- Obtain the gas pressure reading from the manometer.

- If the gas pressure reading and the required pressures are not the same, locate the gas pressure adjusting knob on the gas valve.

- If the actual pressure is higher than the required pressure, turn the gas pressure adjusting knob counter-clockwise.

- If the actual pressure is lower than the required pressure, turn the gas pressure adjusting knob clockwise.

- In either case, make adjustments slowly and continue to monitor the manometer readings.

- Once the pressure reading matches the required pressure for the furnace, turn the gas valve to the OFF position.

- Remove the manometer from the pressure tap.

- Replace the plug on the gas valve.

PROCEDURE 24-2

Cutting and Reaming Steel Pipe

- Secure the pipe section to be cut in a pipe vise.

- Mark the pipe to the desired length with a pencil.

- Position the pipe cutter around the pipe and tighten the handle on the cutter until the blade is in contact with the pipe surface and the cutter is secure around the pipe.

- Rotate the cutter around the pipe one or two turns until the pipe cutter feels somewhat loose around the pipe.

- Tighten the cutter around the handle by turning the handle clockwise approximately ¼ turn.

- Repeat the previous two steps until the pipe has been cut completely.

A With the pipe still secure in the vise, insert the end of the reamer into the end of the pipe.

- While applying moderate pressure to the reamer, rotate the reamer in the pipe to remove any burrs that have formed on the inside edge of the pipe.

- The pipe is now ready to be threaded.

 A

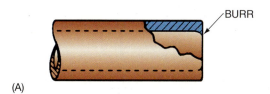

BURR

(A)

(B)

Photo by Eugene Silberstein.

PROCEDURE 24-3

Manually Threading Steel Pipe

- Make certain that the pipe has been cut square and properly reamed.

- Wipe the end of the pipe with a rag to remove any particulate matter from the end of the pipe.

- Apply a small amount of cutting oil to the end of the pipe.

- Install the proper die into the stock of the pipe threader.

- Be sure that the ratchet on the pipe threader is pointing in the clockwise direction.

- Place the pipe threader over the end of the steel pipe with the cutting blade end of the threader pointing away from the end of the pipe.

A While applying moderate pressure on the end of the stock and die, begin to turn the threader onto the pipe.

- Continue to thread the pipe, applying oil liberally to the end of the pipe, until the end of the pipe reaches the end of the cutting dies.

- Once the pipe has been threaded, turn the ratchet knob on the handle of the threader so it points in the counter-clockwise direction.

- Ratchet the threader off the end of the pipe.

A

Delmar/Cengage Learning.

B Inspect the pipe for proper threads. There should be seven perfect threads on the end of the pipe as well as two or three imperfect threads.

- If the threads are not correct, cut the threaded portion off the pipe and rethread.

B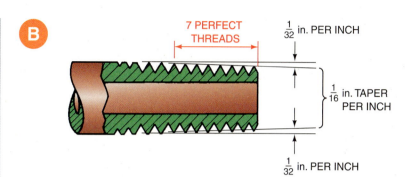

PROCEDURE 24-4

Installing Steel Fittings on Threaded Steel Pipe

- Wipe the pipe threads with a rag to remove any steel filings from the pipe.

A Apply pipe dope to the male threaded portion of the pipe, leaving the two threads closest to the end of the pipe dope free.

- Begin to thread the appropriate fitting on the pipe in a clockwise direction.

- Tighten the fitting on the pipe hand tight.

- Place one pipe wrench on the pipe so that the wrench will grip the pipe when turned in the counter-clockwise direction.

B Place a second pipe wrench on the fitting so that the wrench will grip the fitting when turned in the clockwise direction.

- Using the first wrench to hold the pipe steady, turn the second wrench to tighten the fitting onto the pipe.

- Continue tightening the fitting onto the pipe until the fitting is tight.

A

USE MODERATE AMOUNT OF DOPE

LEAVE TWO END THREADS BARE

B

Delmar/Cengage Learning.

REVIEW QUESTIONS

1. **The three components required for combustion are**
   a. Fuel, gas, and heat
   b. Oxygen, gas, and nitrogen
   c. Fuel, oxygen, and heat
   d. Spark, flame, and fuel

2. **During complete combustion, all of the following are produced except**
   a. Carbon dioxide
   b. Carbon monoxide
   c. Water vapor
   d. Nitrogen

3. **A lack of oxygen during combustion will result in the formation of _____, which is a toxic and poisonous substance.**
   a. Hydrogen hydroxide vapor
   b. Carbon dioxide
   c. Carbon monoxide
   d. Nitrogen

4. **A pressure reading of 10 in. W.C. is approximately equal to**
   a. 277 psig
   b. 27.7 psig
   c. 3 psig
   d. 0.36 psig

5. **Which of the following hydrocarbons makes up the largest percentage of natural gas?**
   a. Methane
   b. Ethane
   c. Butane
   d. Propane

6. **Which of the following is true regarding natural gas and manufactured gas?**
   a. They are both heavier than air
   b. They both provide approximately 1,050 Btu of heat per cubic foot
   c. They both contain hydrocarbon gases
   d. All of the above are correct

7. **Which of the following gives off the most heat per cubic foot?**
   a. Liquefied propane
   b. Manufactured gas
   c. Liquefied butane
   d. Natural gas

8. **In a gas furnace, combustion takes place in the**
   a. Heat exchanger
   b. Gas valve
   c. Blower compartment
   d. Thermocouple

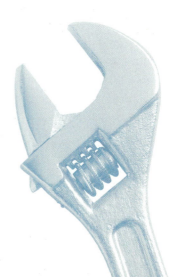

9. **Atmospheric burners are so named because**

   a. They vent the products of combustion to the atmosphere

   b. The air supplied to them is at atmospheric pressure

   c. The pressure in the flue pipe will read 0 in. W.C. on the draft gauge

   d. None of the above is correct

10. **A reduction in primary air will result in**

    a. The flame lifting off the burner

    b. A blue flame with orange tips

    c. A yellow flame

    d. Both a and b are correct

11. **Thermocouples and thermopiles**

    a. Generate a low voltage when heated

    b. Send a "signal" to the gas valve when the pilot is lit

    c. Will close the gas valve if the pilot goes out

    d. All of the above are correct

12. **Which of the following is true regarding thermopiles?**

    a. They generate less voltage than thermocouples

    b. They are made up of multiple thermocouples connected in parallel with each other

    c. Both a and b are correct

    d. Neither a nor b is correct

13. **The ignition method that uses an electric spark to ignite the pilot is called the**

    a. Standing pilot

    b. Intermittent pilot

    c. Direct spark

    d. Hot surface ignition

14. **Insufficient replacement air can result in which of the following?**

    a. Introduction of flue gases to the occupied space

    b. Flames lifting off the burner surface

    c. Rapid venting of flue gases

    d. All of the above are correct

15. **Condensing furnaces are typically vented with**

    a. Type B venting material

    b. PVC plastic piping material

    c. CPVC piping material

    d. Both b and c are correct

KNOW YOUR CODES

Before installing a gas-fired heating appliance, it is very important to be aware of the local codes that are in effect in your area. Factors that are affected include not only the installation of the appliance itself but also the running of the fuel lines and the venting of the flue gasses. The National Fire Protection Association® publishes Code NFPA® 54, which outlines accepted practices regarding the installation of gas-fired appliances. These codes may differ from those in your area.

Research your local codes regarding the installation and venting of gas-fired appliances in your area and report your findings to your colleagues.

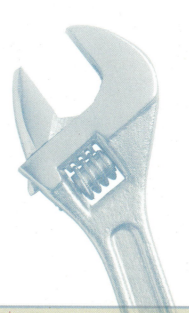

Oil Heat

OBJECTIVES *Upon completion of this chapter, the student should be able to*

- Explain the theory of oil heating.

- Explain the operation of the oil furnace.

- List the component parts of an oil burner.

- Explain the combustion process in an oil furnace.

- Describe the operation of the primary control.

- Explain the difference between a cad cell and a stack relay primary control.

- Explain the function of the nozzle.

- Install an oil furnace.

- Install piping used to carry oil to the furnace.

- Explain the operation of the oil filter and de-aerator.

- List and describe three methods commonly used to remove flue gases from the structure.

- Install vent piping to carry byproducts of combustion away from the furnace.

- Startup an oil furnace.

- Test for proper combustion in an oil furnace.

- Troubleshoot problems with an oil furnace.

The two previous chapters dealt with furnaces that relied on either electricity or burning gas for the heat source. Although it is available in many markets, many home builders fail to consider oil-fired heat and hot water systems when designing and building new homes. Some are surprised to find out that oil-fired heating has been with us since the late 1800s and that more than 13,000,000 homes and businesses throughout America rely on oil for their heating needs. Unlike older systems, modern oil-fired heating systems are clean, efficient, environmentally friendly, and incredibly safe. Although these systems are typically more expensive than other furnace types from a first-cost standpoint, the economics of installing and operating oil heating systems has made it a viable option.

American Society for Testing and Materials (ASTM) agency that established minimum specifications for fuel oils

Atomization process of breaking the fuel oil up into a fine mist, which facilitates combustion

Barometric damper see draft regulator

Bilateral switch used in cad cell primary controls, the electronic control switching device that works in conjunction with the triac to control burner operation

Cad cell cadmium sulfide device that changes its resistance as the amount of light it senses changes

Cloud point temperature at which wax crystals form in fuel oil, typically 20 degrees higher than the pour point

Cold contacts contacts in the stack relay that are in the closed position when the flue pipe is cold

Color pure, fresh heating oil is normally champagne colored but is dyed red to differentiate it from taxable on-road diesel fuel; fuel tends to darken when exposed to light, so color is not a good indication of fuel quality.

De-aerator oil system accessory that facilitates the removal of air from the oil before it passes into the fuel pump

Direct vent venting method that uses the blower on the burner to pull outside air in for combustion and to push flue gases from the structure

Draft regulator automatic damper that opens and closes in response to the amount of draft in the flue pipe; high draft will cause the draft regulator to open, and low draft will cause the draft regulator to close

Electrode assembly portion of the oil burner comprised of the electrodes, electrode clamp, and ceramic insulators

Electrode clamp strap that holds the electrodes in place

Electrodes component part of the oil burner that generates the spark needed for ignition

Fire point see ignition point

Flame retention burner type of oil burner that concentrates the airflow pattern to create a better fuel/air mixture at the time of combustion

Flash point temperature at which fuel oil will momentarily flash and immediately extinguish

Fossil fuel fuel made up of the remains (fossils) of prehistoric plants and animals

Fuel pump component part of an oil burner that is responsible for feeding pressurized oil through the nozzle

Gravity indication of the heat value of fuel; ASTM standards indicate fuel oil's density to be 7.1 pounds per gallon and 139,400 Btus per gallon

High-pressure atomizing oil burner oil burner that delivers atomized fuel oil through the nozzle at pressures above 100 psig

Hollow cone spray pattern where the oil is concentrated on the outer edge of the spray

Hot contacts contacts in the stack relay that are in the closed position when the flue pipe is hot

Ignition point lowest temperature at which the fuel will rapidly burn in the presence of air; the standard for fuel oil is 637°F

Ignition transformer step-up transformer that provides a secondary voltage of approximately 10,000 volts to vaporize and ignite the fuel

Interrupted ignition ignition method that initiates spark only at the beginning of the cycle

Motor-off-delay the process of keeping the burner blower energized for a period of time at the end of the heating cycle; also referred to as post-purge

Nozzle component part of an oil burner that atomizes, meters, and sprays the oil into the combustion area

Nozzle adapter pipe fitting that connects the nozzle to the nozzle line

Nozzle assembly unit comprised of the nozzle line, nozzle adapter, nozzle, and electrodes

Nozzle line section of tubing that connects the outlet of the pump to the nozzle adapter

Nozzle tube see nozzle line

Number 2 fuel oil most common type of fuel oil used in residential applications

One-pipe system fuel oil piping arrangement that uses one supply pipe connected between the oil tank and the oil burner

Pre-purge the process of energizing the burner blower for a period of time before allowing fuel to enter the combustion chamber; also referred to as valve-on-delay

Post-purge the process of keeping the burner blower energized for a period of time at the end of the heating cycle; also referred to as motor-off-delay

Pour point minimum temperature at which the fuel will flow; the standard is 20°F

Power venting venting method that uses a powerful blower to pull flue gases from the furnace

Semi-solid cone spray pattern where most of the oil is concentrated on the outer edge of the spray, while a lesser concentration is found within the spray area

Solid cone spray pattern where the oil is evenly dispersed within the entire spray pattern

Stack relay primary control that controls the operation of the burner by sensing the temperature of the flue gases

Sulfur chemical element found in fuel oil; the maximum amount of sulfur in fuel oil is limited to 0.5%

Triac used in cad cell primary controls; the electronic control switching device that works in conjunction with the bilateral switch to control burner operation

Two-pipe system fuel oil piping arrangement that uses two pipes connected between the oil tank and the oil burner; one serves as the supply, and the other serves as the return for any unburned fuel

Valve-on-delay the process of energizing the burner blower for a period of time before allowing fuel to enter the combustion chamber; also referred to as pre-purge

Vent alarm device that whistles as an oil tank is being filled, but stops when the tank is nearly full

Viscosity the ability of a fluid to flow—higher viscosity results in a lower ability to flow; the viscosity of fuel oil increases as its temperature decreases

Water finding paste substance that changes color when exposed to water; used to determine if water is present in an oil tank

THE OIL FURNACE

The oil furnace (Fig. 25-1) is designed to deliver heated air to the occupied space in order to compensate for the heat lost by the space. Just like the gas furnace, burning fuel is the source of this heat. The oil furnace is comprised of and relies on the following major components:

- Motor and blower assembly
- Filters
- Casing
- Air-distribution system
- Heat exchanger
- Flue piping
- Oil burner
- Oil tank
- Oil filter

To avoid redundancies in this text, we recommend that the reader refer to previous chapters for information regarding the motor and blower assembly, air filters, casing and furnace configurations, and the air-distribution system. They are exactly the same as those used in the electric and gas-fired furnaces covered previously. We will therefore begin our discussion of oil-fired furnaces with the fuel oil and move on to the other system components on which the safe and satisfactory operation of the furnace rely.

Courtesy of Williamson-Thermoflo.

FIGURE 25-1 Oil-fired furnace.

FUEL OIL

Fuel oil as we know it is a **fossil fuel**, as is the natural gas we discussed in the previous chapter. The term *fossil fuel* references the fact that the oil is a product of prehistoric plants and animals. Fuel oil used for residential applications, classified as **number 2 fuel oil**, is made primarily from hydrogen (15%) and carbon (85%), which classifies it as a hydrocarbon like other fuels such as propane, methane, ethane, and butane. Recall from earlier chapters that propane, methane, and other hydrocarbons were classified as good refrigerants because of their good heat transfer capabilities but were not desirable for practical use because of their high flammability. Fuel oil is much more refined now than in previous decades, which makes oil heating equipment cleaner and more efficient than ever before. The **American Society for Testing and Materials (ASTM)** publishes standards for a number of industries including the oil heat industry. It has established minimum standards for the production and refinement of oil used for heating purposes. Although the field technician has no control over the quality of the oil being supplied to the structure, the following list provides information regarding the characteristics of fuel oil:

- The maximum amount of **sulfur**, a naturally occurring element, that can be present in fuel oil is 0.5%.

- The **flash point**, which is the temperature at which fuel oil will momentarily flash and immediately extinguish, must be a minimum of 100°F. Typically, the flash point of fuel oil ranges from 130°F to 214°F.

- The **ignition point**, or **fire point**, which is the lowest temperature at which the fuel will rapidly burn in the presence of air, is a minimum of 637°F.

- The **pour point**, which is the lowest temperature at which the fuel will flow, is about 20°F, but can be as low as −55°F.

- The **cloud point**, which is typically 20°F higher than the pour point, is the temperature at which wax crystals form in the oil. Wax can clog flow devices but heating the oil above the cloud point temperature will cause the wax crystals to go back into solution.

- The warmer the fuel oil, the better the system will operate. At lower temperatures, the oil's **viscosity** increases, making it thicker and less fluid. This can result in inefficient combustion, delayed ignition, and reduced system efficiency.

- The amount of water in the oil should be no more than 0.1%.

- The **gravity** of the fuel oil relates to the density of the oil, which is approximately 7.1 pounds/gallon.

- The Btu rating for #2 fuel oil is approximately 19,560 Btu/pound and about 139,400 Btu/gallon.

- The normal **color** of fuel oil resembles that of champagne. Fuel oil is, however, dyed red to distinguish it from diesel fuel, which is taxed differently.

COMBUSTION

The combustion process is the same regardless of the fuel being burned. For combustion to take place, there must be a supply of oxygen, heat, and fuel. If any of these three components is missing, combustion cannot occur. The fuel we are referring to in this chapter is oil. As oil passes into the combustion chamber, it is broken down into small droplets and mixed with air. This process is called **atomization**. This air/fuel mixture is heated by means of an electric spark that causes the oil portion of the mixture to vaporize and then ignite. Fuel oil will not burn in the liquid state, so the oil must vaporize prior to ignition. The vaporization and ignition of the fuel takes place in the burner, which is discussed later in this chapter.

As mentioned previously, fuel oil is a hydrocarbon made up of carbon and hydrogen. During the combustion process, the oxygen reacts with the fuel oil to produce heat, carbon dioxide, and water vapor (Fig. 25-2). This is very similar to the products of combustion created when natural or manufactured gas is burned. When combustion is not complete, carbon monoxide, soot (free carbon), and other undesirable substances form, in addition to the heat, carbon dioxide, and water vapor (Fig. 25-3). The amount of air that is supplied for

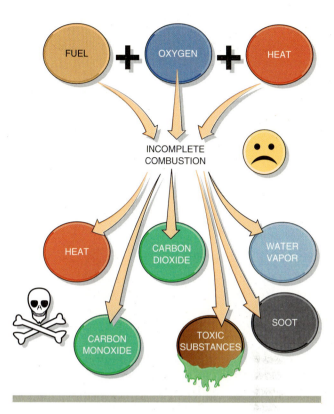

FIGURE 25-3 Incomplete combustion.

combustion greatly affects the combustion process. Too much air can cause lower system efficiency, but insufficient air results in the formation of carbon monoxide. In addition, the better the air and fuel are mixed together, the better the combustion will be. This mixing also takes place in the oil burner.

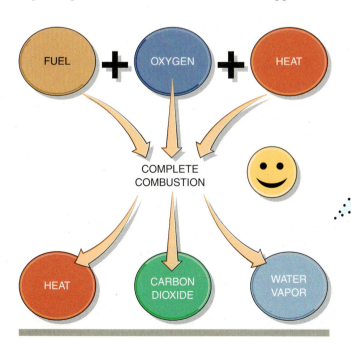

FIGURE 25-2 Complete combustion.

> ## *from experience...*
> When installing and putting a new oil burner into service, follow all manufacturers' guidelines regarding the combustion air supply and properly test the combustion in the system.

TESTING COMBUSTION

Since improper combustion can cause carbon monoxide and other poisonous substances to form, the single most important thing a technician can do is test the flue gases and the combustion process. Failure to check these aspects of system operation can result in inefficient system operation and, more importantly, carbon monoxide poisoning or death. Complete test kits (Fig. 25-4) are available for oil-fired systems.

These test kits typically contain the following instruments:

- The draft gauge
- The smoke tester
- Carbon dioxide tester

- Carbon monoxide detector
- Stack thermometer

This test equipment was discussed in the previous chapter, so it is recommended that the reader refer back to those sections for more information regarding its purpose and use.

Going Green

Combustion testing is not an option. Combustion testing is something that all technicians must do! It is impossible to set up or adjust the firing efficiency of a fossil-fuel heating appliance by sight or sound.

COMBUSTION CHAMBERS AND HEAT EXCHANGERS

The burning of the fuel/air mixture takes place in the combustion chamber, which is located at the bottom of the heat exchanger (Fig. 25-5). Notice that the furnace blower directs air from the conditioned space through the combustion chamber and up through the heat exchanger. The heated air is then redirected back to the occupied

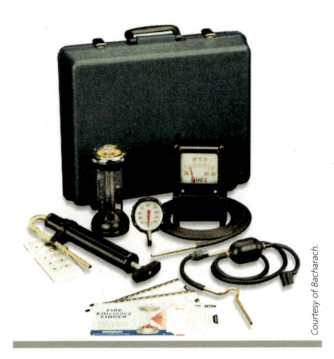

Courtesy of Bacharach.

FIGURE 25-4 Combustion test kit.

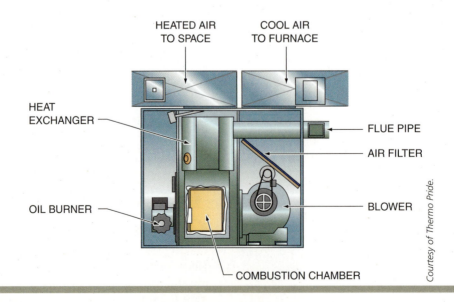

Courtesy of Thermo Pride.

FIGURE 25-5 Combustion chamber and heat exchanger location.

space. In reality, the combustion chamber is an extension of the heat exchanger because heat is being transferred to the air moving through the chamber. Higher-efficiency furnaces have a secondary heat exchanger (Fig. 25-6) to maximize the amount of heat that is extracted from the flue gases before they are vented from the structure. Figure 25-7 shows a cutaway of an oil-fired water heater. Notice that the combustion chamber is located at the bottom and the heat exchanger is located in the center of the water tank. The heat from the combustion chamber aids in the heat transfer process as the bottom of the tank is directly over the combustion chamber. As the flue gases rise through the heater, heat is transferred to the water.

OIL DELIVERY TO THE UNIT

For the oil furnace to operate properly, there must be an ample supply of fuel oil as well as a reliable method of supplying the oil to the appliance. Before the oil actually reaches the appliance, it must be delivered to the residence location, filtered, pumped to the appliance and, in many cases, de-aerated. These processes are accomplished by:

- Oil storage tanks
- Oil lines
- Oil filters
- Oil de-aerators

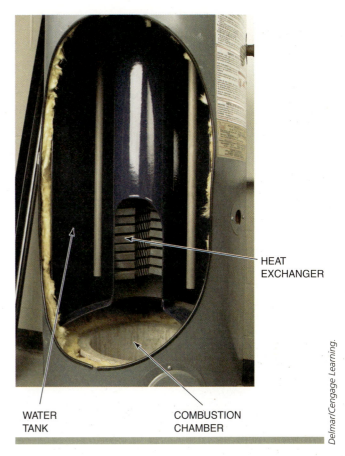

HEAT EXCHANGER

WATER TANK

COMBUSTION CHAMBER

Delmar/Cengage Learning.

FIGURE 25-7 Heat exchanger, combustion chamber, and water tank on a water heater.

OIL STORAGE TANKS

Oil tanks (Fig. 25-8) offer protection from temporary supply disruptions by enabling the consumer to store a quantity of fuel on site for future consumption. While most oil tanks are found in basements, they can also be located in other areas, such as crawl spaces or garages. They can also be installed outside, either above the ground or underground. Regardless of where the tank is installed, it must comply with manufacturers' instructions and all

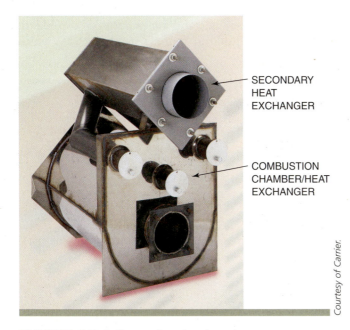

SECONDARY HEAT EXCHANGER

COMBUSTION CHAMBER/HEAT EXCHANGER

Courtesy of Carrier.

FIGURE 25-6 Secondary heat exchanger.

Going Green

Follow all safety precautions when installing oil tanks. The number one concern of tank installers is to keep oil from seeping into the ground. By following proper guidelines, the possibility of a mishap is greatly reduced.

FIGURE 25-8 Oil tank.

local ordinances and regulations. Most local authorities require that a UL listed tank be used and that the requirements of either the International Code Council (ICC) or the National Fire Protection Association® (NFPA) be followed. Clearances must be observed to property lines, electric meters, combustion sources, windows, doors, air intakes, exhausts, and other considerations. Tank piping connections should all be tight and connected with an approved pipe joint compound. The following is a list of things to keep in mind:

✓ Residential oil tanks located indoors should be on the lowest floor of the building.

✓ Oil tanks should be placed on concrete pads or floors.

✓ There should be no fittings on the buried portion of fuel lines.

✓ Buried fuel lines should be placed in a protective sleeve.

✓ Above-ground outdoor oil tanks with a capacity of 275 gallons or less must be at least five feet from a property line; larger tanks must be further from the line.

from experience...

Be sure to adhere to all local codes regarding the oil storage tank installation. These codes designate proper clearances and distances between oil fill or vent pipes and windows as well as the type of piping material needed for fill and vent lines.

CAUTION

CAUTION: Teflon tape should NOT be used on any oil tank, line, or burner fittings as it can damage fuel pumps and void their warranties.

Fill and vent pipes should be made of wrought iron, steel, or brass pipe connected with malleable fittings. PVC pipe and cast iron fittings should never be used! Fill and vent pipes should pitch toward the tank opening to prevent "traps" and should be at least 1¼″ for most residential applications. Tanks holding more than 660 gallons require larger size pipes, although it is very rare to see such large tanks used for residential applications. The fill pipe should terminate with a nipple at a 45- or 90-degree angle and a weatherproof cap identifying it as a fuel oil fill. **Vent alarms** (Fig. 25-9), which

FIGURE 25-9 Vent alarm.

are devices that whistle when the oil tank is nearly full, should be installed in all tanks and the vent piping should then be attached to the alarm. The vent cap should be weatherproof and have a screen no finer than #4 mesh. Although the fill and vent lines can be the same size, it is good field practice to use a 2-inch fill pipe and a 1¼-inch vent pipe to ensure satisfactory operation of the vent alarm.

> ## *from experience...*
> Most local codes require that the fill line for the oil tank pitch toward the tank at a slope of no less than 2 inches per 5 feet of horizontal run.

OIL LINES

Oil lines should be sized according to pump manufacturers' recommendations. For most residential applications, ½″ OD tubing is recommended. Sleeved tubing (Fig. 25-10) is also recommended and secondary containment should be used for any underground lines. Connections should only be made with flare fittings (Fig. 25-10)—never compression fittings—and there should not be any fittings underground. Depending on the installation, one pipe supplying oil to the furnace from the oil tank is adequate. This piping configuration is called a **one-pipe system** since there is only one pipe connecting the oil tank to the appliance. Other installations will need a second pipe between the tank and the appliance to carry any unburned oil back to the tank. This is called a **two-pipe system**.

If the fuel tank is above the burner, the fuel is supplied to the burner by gravity and the pump requires no suction to pull oil from the tank. In this situation, a one-pipe system normally suffices. If the fuel tank is located below the burner, a two-pipe system may be preferable where one pipe brings oil from the tank to the burner and the other pipe returns any unused oil to the tank.

> ## *from experience...*
> Internal bypass plugs must be installed on most pumps connected to two-pipe systems. These plugs *must* be removed when single pipe systems are used. Over the past few years, oil de-aeration devices, covered shortly, that allow one-pipe units to function as two-pipe systems have become popular.

OIL FILTERS

Oil burner nozzles are manufactured to very close tolerances and contain passages that are narrower than a human hair. The installation of an oil filter (Fig. 25-11) helps to keep these passages from clogging or fouling. A wide variety of filters are on the market, so it is important to select one that has the appropriate flow rate for the appliance you're installing.

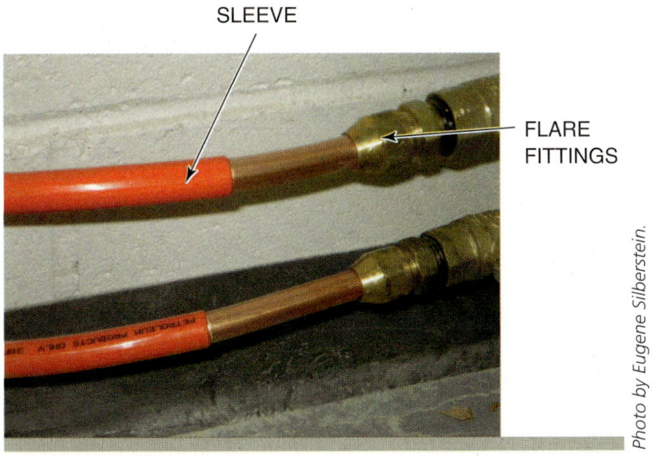

SLEEVE

FLARE FITTINGS

Photo by Eugene Silberstein.

FIGURE 25-10 Sleeve over oil lines and flare connections.

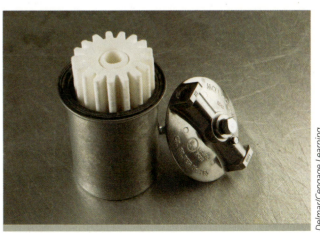

Delmar/Cengage Learning.

FIGURE 25-11 Oil filter.

OIL DE-AERATORS

The oil **de-aerator** (Fig. 25-12) allows a one-pipe system to function as a two-pipe system because two pipes are connected between the de-aerator and the fuel pump (Fig. 25-13). The de-aerator helps separate air from the oil as it passes through the device. In operation, the fuel travels from the oil tank through the oil filter into the de-aerator,

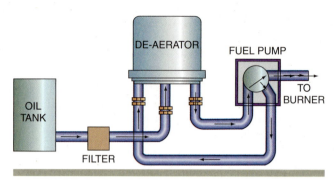

FIGURE 25-14 Piping diagram of the oil de-aerator.

where air that may be present in the oil is separated from the fuel. The fuel then passes on to the fuel pump. Only the fuel that will be burned passes through the pump into the nozzle, while the remaining fuel is recirculated back to the de-aerator instead of the oil tank (Fig. 25-14).

THE OIL BURNER

Just like gas furnaces, the oil furnace requires a means to deliver the fuel to the combustion chamber as well as enough air to support combustion and a means by which the fuel is ignited. In addition, safety devices are employed to ensure the safe and proper operation of the system. The most common type of oil burner found on residential heating systems is the **high-pressure atomizing oil burner** (Fig. 25-15). It uses a fuel pump to deliver

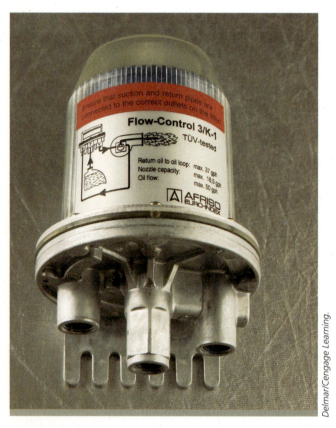

FIGURE 25-12 Oil de-aerator.

Delmar/Cengage Learning.

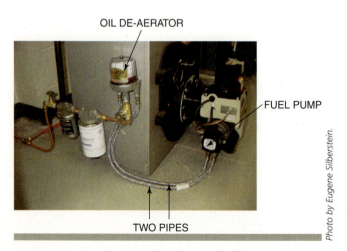

FIGURE 25-13 Two lines connected between the de-aerator and the fuel pump.

Photo by Eugene Silberstein.

FIGURE 25-15 High pressure oil burner.

Courtesy of R.W. Beckett Corp.

high-pressure oil, typically over 100 psig, into the combustion chamber where it is ignited. Many newer oil burners are of the flame retention design. The **flame retention burner** is a radically improved type of burner first introduced in the 1960s and continually improved since then. Compared with conventional burners, flame retention burners offer many improvements and modifications including the following:

✓ Faster motor speed (3450 vs 1725 RPM)

✓ Higher static pressure

✓ Improved air/oil mixing

✓ Lower emissions

✓ Higher efficiency

✓ Cleaner combustion

✓ Lower system maintenance requirements

Flame retention burners produce a 300 to 500°F hotter flame than the conventional burners that were popular through the early 1960s and they produce much less soot and smoke than older, less-efficient burners. The hotter operating flames combined with the radically improved air/oil mixing make them more efficient and desirable to use. Common components of the oil burner are the following:

- Motor
- Fan
- Fuel pump
- Igniter/transformer
- Primary controls
- Nozzles

MOTOR

The motor in the oil burner (Fig. 25-16) is responsible for turning a fan and a fuel pump, which provide ample air and fuel to the combustion chamber, respectively. The motor's speed and frame size are dictated by the manufacturer for each model of burner. Most flame retention burners operate at 3450 RPM and most conventional burners operate at 1725 RPM. The flame retention burner (Fig. 25-17) creates a swirling air pattern that allows the fuel and air to mix completely, resulting in more efficient fuel burning. The increase in efficiency is typically about 10%. This is accomplished in part by motors that turn twice as fast as those found on conventional burners. Conventional burners operate at slower motor speeds, which produce more erratic flame patterns that do not burn fuel as efficiently as the flame retention burners.

Courtesy of R.W. Beckett Corp.

FIGURE 25-16 Oil burner motor.

Delmar/Cengage Learning.

FIGURE 25-17 Flame retention burner.

FAN

The fan or blower delivers the air necessary to support combustion. It is connected directly to the shaft of the motor and, therefore, turns at the same speed as the motor. The fan is a forward curved, centrifugal blower, similar to that used to move air between the furnace and the occupied space, only much smaller in size. The amount of air that is able to reach the fan is controlled by a manually adjusted air shutter or air band.

FUEL PUMP

Just like the fan, the **fuel pump** (Fig. 25-18) is connected to the shaft of the motor by means of a flexible coupler (Fig. 25-19) and turns at the same speed as the motor. The pump serves three major functions in the oil burner. First, it is responsible for moving the oil from the oil storage tank to the burner, where it is ignited. Second, the pump acts to filter, or in some cases, pulverize, any particulate matter that may have found its way into the oil. Finally, the fuel pump regulates the pressure of the oil that is permitted to enter the nozzle. An end-view diagram of a typical fuel oil pump is shown in Figure 25-20. The configuration and connections that exist between the motor, fan, and pump are shown in Figure 25-21.

Delmar/Cengage Learning.

FIGURE 25-19 Flexible couplers.

SINGLE- AND TWO-STAGE FUEL PUMPS

Fuel pumps with one set of gears to pull oil from the tank and pump it through the nozzle are referred to as single-stage. These pumps can be hooked up as one pipe or as two pipe as long as the vacuum necessary to operate does not exceed 12″. Two-stage pumps have two sets of gears, one to pull oil from the tank and the other to push oil through the nozzle. Two-stage pumps operate at a vacuum as high as 17″.

PUMP PRESSURES

Oil pumps develop vacuum when pulling oil from the tank. Vacuum can be measured in inches of mercury. The rule of thumb is that a pump requires 1 inch of vacuum to lift oil 1 foot vertically, 1 inch to pull oil 10 feet horizontally, and 1 inch for filters, valves, and other accessories. Thus, a pump that is 5 feet above the tank and 30 feet away from the tank with an oil filter installed, will create 9 inches of vacuum. A pump that is below the tank will normally require no vacuum. The vacuum can be read using a vacuum gauge (Fig. 25-22) connected to the inlet side of the pump. Conversely, the outlet pressure of the pump can be determined by using a pressure gauge (Fig. 25-23) connected to the outlet of the fuel pump. Figure 25-24 shows how these gauges are connected to the pumps. Always follow the gauge manufacturers instructions for proper use of the vacuum and pressure gauges.

THE NOZZLE ASSEMBLY

As the oil is discharged from the pump, it travels through a section of tubing called the **nozzle line** or **nozzle tube** (Fig. 25-25). The nozzle line connects the pump to the

Courtesy of R.W. Beckett Corp.

FIGURE 25-18 Oil pump.

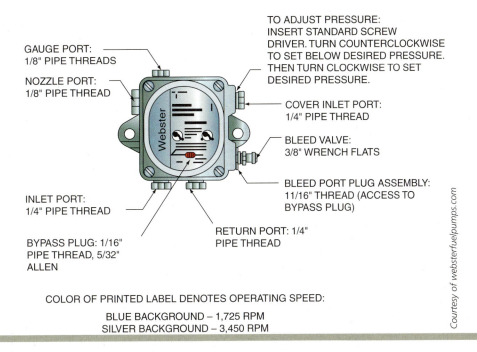

GAUGE PORT:
1/8" PIPE THREADS

NOZZLE PORT:
1/8" PIPE THREAD

TO ADJUST PRESSURE:
INSERT STANDARD SCREW
DRIVER. TURN COUNTERCLOCKWISE
TO SET BELOW DESIRED PRESSURE.
THEN TURN CLOCKWISE TO SET
DESIRED PRESSURE.

COVER INLET PORT:
1/4" PIPE THREAD

BLEED VALVE:
3/8" WRENCH FLATS

BLEED PORT PLUG ASSEMBLY:
11/16" THREAD (ACCESS TO
BYPASS PLUG)

INLET PORT:
1/4" PIPE THREAD

BYPASS PLUG: 1/16"
PIPE THREAD, 5/32"
ALLEN

RETURN PORT: 1/4"
PIPE THREAD

COLOR OF PRINTED LABEL DENOTES OPERATING SPEED:

BLUE BACKGROUND – 1,725 RPM
SILVER BACKGROUND – 3,450 RPM

Courtesy of websterfuelpumps.com

FIGURE 25-20 Diagram of a fuel oil pump.

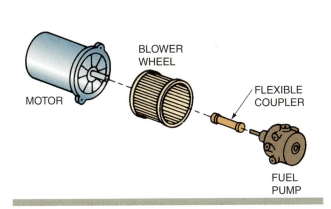

MOTOR

BLOWER
WHEEL

FLEXIBLE
COUPLER

FUEL
PUMP

FIGURE 25-21 Motor, fan, connector, and pump assembly.

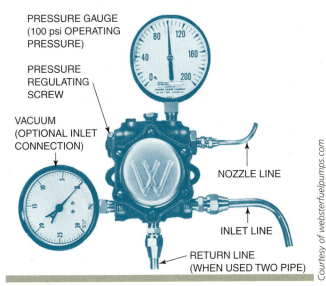

Delmar/Cengage Learning.

FIGURE 25-23 Fuel pump pressure gauge.

Delmar/Cengage Learning.

FIGURE 25-22 Vacuum gauge.

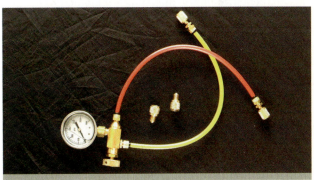

PRESSURE GAUGE
(100 psi OPERATING
PRESSURE)

PRESSURE
REGULATING
SCREW

VACUUM
(OPTIONAL INLET
CONNECTION)

NOZZLE LINE

INLET LINE

RETURN LINE
(WHEN USED TWO PIPE)

Courtesy of websterfuelpumps.com

FIGURE 25-24 Gauge locations for pump testing and servicing.

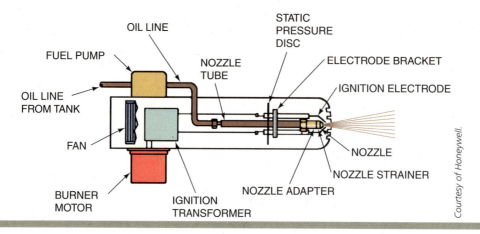

FIGURE 25-25 labels: OIL LINE, FUEL PUMP, NOZZLE TUBE, STATIC PRESSURE DISC, ELECTRODE BRACKET, IGNITION ELECTRODE, OIL LINE FROM TANK, FAN, BURNER MOTOR, IGNITION TRANSFORMER, NOZZLE ADAPTER, NOZZLE STRAINER, NOZZLE

Courtesy of Honeywell.

FIGURE 25-25 Top view of a typical oil burner.

nozzle adapter, which holds the nozzle in place. The **nozzle** is responsible for maintaining the desired flow rate and pattern of the oil as it enters the combustion chamber. The nozzle will be discussed in more detail a little later in this chapter. The nozzle line, nozzle adapter, nozzle, and electrodes comprise the **nozzle assembly**.

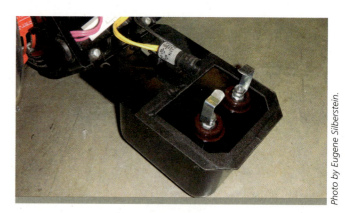

Photo by Eugene Silberstein.

FIGURE 25-26 Ignition transformer.

> ## from experience...
>
> Motors, pumps, and primary controls are often interchangeable between burners but most fans, air tubes, transformers, igniters, electrodes, air tubes, and nozzle assemblies are designed for specific burners and are usually *not* interchangeable. Always check manufacturers' specifications and recommendations when changing oil burner components.

IGNITER/TRANSFORMER

Once the fuel has been delivered to the burner, it must be ignited. The ignition of the oil can be accomplished by using either a step-up **ignition transformer** (Fig. 25-26) or a solid state ignition module to provide the spark necessary to ignite the atomized fuel. The spark voltage at the secondary of the transformer is about 10,000 volts, and the spark voltage generated by the solid state igniter is about 15,000 volts. This voltage creates

a spark between the tips of the **electrodes** (Fig. 25-27). The **electrode assembly** is held in place by the **electrode clamp**. The electrode assembly and the electrode clamp are also parts of the nozzle assembly. An exploded view of an oil burner assembly is shown in Figure 25-28.

Most new oil burners are designed to operate with **interrupted ignition**, which means that the sparking that ignites the air/fuel mix occurs only at the beginning of the cycle. Once the fuel has ignited, the ignition sparking stops. Older oil burners were designed with intermittent ignition systems with sparking as long as the burner was operating. Benefits of the interrupted ignition system include:

• Longer transformer or igniter life
• Longer electrode life
• Fewer ignition-related system failures

PRIMARY CONTROLS

The primary control on an oil-fired appliance can be thought of as the central control that performs a number of tasks including supplying power to the ignition

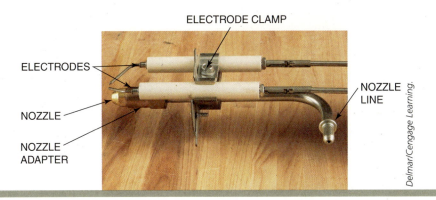

Delmar/Cengage Learning.

FIGURE 25-27 Nozzle assembly.

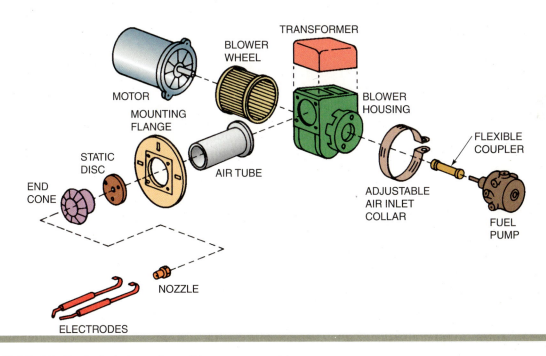

FIGURE 25-28 Exploded view of an oil burner assembly.

transformer, or igniter, and to the burner circuits. The primary control responds to the limit control and thermostat and supervises the start, run, and stop cycles of the burner. An essential function of primary controls is to respond to the presence or absence of flame. On older oil heat systems, the primary was mounted in the flue pipe and sensed temperature changes to determine if the unit was firing. This is called the stack relay. Modern oil heat systems use controls that sense the light from combustion and respond more quickly to flame failure. These are cad cell burner controls.

STACK RELAYS

The **stack relay** (Fig. 25-29) is a primary control designed to be located in the flue pipe. Constructed with a bimetal strip that extends into the flue pipe, the stack relay senses the temperature of the flue gases to determine whether or not the oil burner is firing. When the oil burner fires, the oil/air mixture is ignited, and the heat-laden flue gases, or products of combustion, begin to rise through the flue pipe. The bimetal strip in the flue senses this heat and, as it heats up, causes one set of electrical contacts to close and another set of contacts to open. The contacts that are in the closed position when the bimetal strip is cold are called the **cold contacts**; those that are in the closed position when the bimetal strip is hot are called the **hot contacts**. The hot and cold contacts are configured as shown in Figures 25-30(A) and (B). Note that the contacts are configured so that both sets cannot be in the open or closed position at the same time. When the bimetal senses high temperature, the connecting rod pushes to the right,

Delmar/Cengage Learning.

FIGURE 25-29 Stack relay.

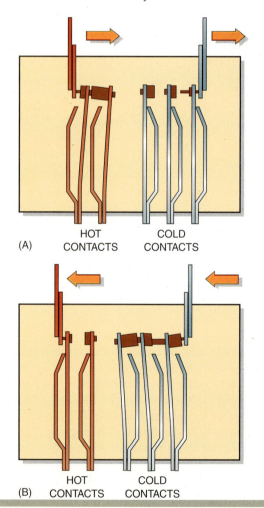

(A) HOT CONTACTS   COLD CONTACTS

(B) HOT CONTACTS   COLD CONTACTS

FIGURE 25-30 (A) Hot contacts are closed and cold contacts are open. High temperature is being sensed in the flue pipe. (B) Hot contacts are open and cold contacts are closed. Low temperature is being sensed in the flue pipe.

closing the hot contacts and opening the cold contacts [Fig. 25-30(A)]. When the bimetal senses a low temperature, the rod is pushed to the left, opening the hot contacts and closing the cold contacts [Fig. 25-30(B)]. A simplified wiring diagram for the stack relay is shown in Figure 25-31. This stack relay is shown as an interrupted ignition, but can be used as either an interrupted or intermittent primary control.

When the burner is off, the cold contacts are closed and the hot contacts are open. On a call for heat, the room thermostat closes, completing the circuit through the 1K coil, which closes the 1K contacts—1K1, 1K2, and 1K3 (Fig. 25-32). This energizes the ignition circuit as well as the circuit containing the 2K coil, which closes the 2K contacts—2K1 and 2K2 (Fig. 25-33). The 2K contacts close the burner motor circuit, thereby powering the motor that turns the fan as well as the fuel pump. The burner ignites.

As the burner operates and the flue heats up, the cold contacts will open and the hot contacts will close (Fig. 25-34). This will remove the ignition module from the circuit, allowing the burner to operate with only the burner motor energized. If the burner flame goes out for any reason during the run cycle, the system will shut down for approximately one minute and make one attempt to restart. If the flame is not re-established within approximately 90 seconds, the

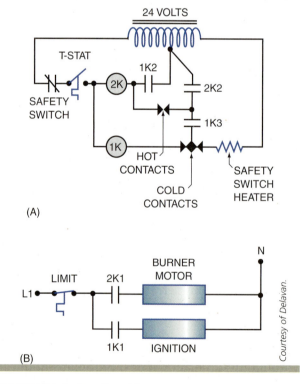

(A)

(B)

Courtesy of Delavan.

FIGURE 25-31 Simplified wiring diagram of a stack relay.

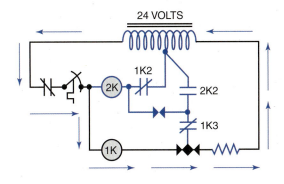

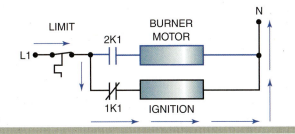

FIGURE 25-32 Circuit is complete through the 1K coil.

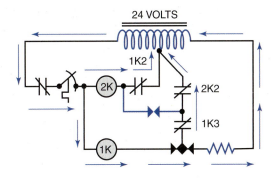

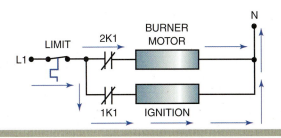

FIGURE 25-33 Circuit is complete through the 1K and 2K coils, energizing the burner motor and ignition.

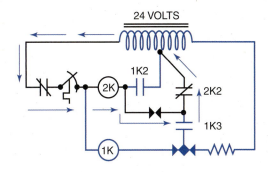

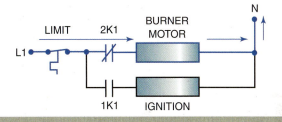

FIGURE 25-34 Cold contacts open, removing ignition from circuit as the burner continues to operate.

system will shut down on safety and will have to be manually reset.

The safety switch and the safety switch heater (Fig. 25-31) will de-energize the control circuit if the cold contacts do not open within the predetermined

Going Green

The stack relay is no longer the popular choice when selecting a primary control, although technicians will likely see them in the field when servicing older systems. More sophisticated primary controls, such as the cad cell, are now the norm. More advanced controls with self-diagnosing features are finding their way into the industry. These technological advances help to improve the operating efficiency of the oil burner.

FIGURE 25-35 Cad cell.

Photo by Eugene Silberstein.

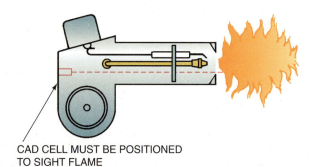

CAD CELL LOCATION CRITICAL
FACTOR IN PERFORMANCE

CAD CELL MUST BE POSITIONED
TO SIGHT FLAME

1. CELL REQUIRES A DIRECT VIEW OF FLAME.

2. ADEQUATE LIGHT FROM THE FLAME MUST REACH
   THE CELL TO LOWER ITS RESISTANCE SUFFICIENTLY.

3. CELL MUST BE PROTECTED FROM EXTERNAL LIGHT.

4. AMBIENT TEMPERATURE MUST BE UNDER 140°F.

5. LOCATION MUST PROVIDE ADEQUATE CLEARANCE.
   METAL SURFACES MUST NOT AFFECT CELL BY
   MOVEMENT, SHIELDING OR RADIATION.

FIGURE 25-36 Cad cell location.

time period, which is usually about 90 seconds. On a call for heat, the safety switch heater is energized and the current flowing through the heater generates a small amount of heat. The longer the heater is energized, the hotter the heater will get. If the heater remains in the circuit long enough, it will warp, causing the safety switch contacts to open. If the cold contacts open within the desired time frame, the heater is removed from the circuit and the safety switch contacts will remain in the closed position, allowing the burner to operate as desired.

CAD CELLS

Instead of sensing the temperature of the flue gas as does the stack relay, the **cad cell** (Fig. 25-35) senses the light that is emitted by the burning oil/air mixture. The cad cell is made from a chemical compound called cadmium sulfide, which changes its resistance in response to the amount of light it is exposed to. The resistance of the cad cell is very high, about 100,000 ohms, when it senses little or no light and very low, well under 1,000 ohms, when light, from burning fuel for example, is sensed. The cad cell must be positioned so that it can easily sense the presence or absence of the burner flame (Fig. 25-36). It is encased in glass, and a cutaway of the device is shown in Figure 25-37. The cad cell is part of the cad cell relay (Figure 25-38) that controls burner operation. The wiring is more or less similar to that of the stack relay with the major exception of the heat sensing mechanism and the introduction of a **bilateral switch** and a **triac**. These are electronic control devices that are part of the solid state circuitry in the primary control. In simple terms, working together, they start or stop current flow through the safety switch heater in response to the resistance of the cad cell.

When there is an initial call for heat, the thermostat closes, and there is a complete circuit through the 1K coil as well as the safety switch heater (Fig. 25-39). The triac is now acting as a closed switch, permitting current

FACE OF THE CAD CELL

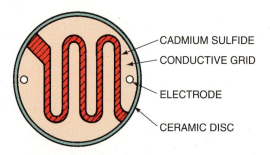

CADMIUM SULFIDE

CONDUCTIVE GRID

ELECTRODE

CERAMIC DISC

THE CAD CELL IS SEALED IN GLASS

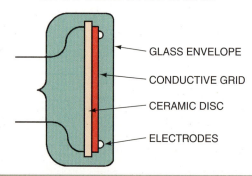

GLASS ENVELOPE

CONDUCTIVE GRID

CERAMIC DISC

ELECTRODES

FIGURE 25-37 Cad cell diagram.

FIGURE 25-38 Cad cell primary control.

Photo by Eugene Silberstein.

to flow through the safety switch heater. Completing the circuit through the 1K coil also causes the 1K1 and 1K2 contacts to close. The 1K1 contacts energize the burner motor and the ignition modules. Just as in the case of the stack relay, the safety switch heater will de-energize the burner if flame is not established.

Once the fuel has been ignited and the resistance of the cad cell has dropped, the triac opens and no longer permits current flow through the safety switch heater. This removes the heater from the circuit and prevents it from shutting down the burner. The 1K coil is still energized, as there is a complete circuit through the 1K1 contacts. The burner will continue to operate until the thermostat is satisfied (Fig. 25-40).

NOZZLES

Fuel oil will not burn in the liquid state. It must first be atomized, or changed into a vapor, and then mixed with air in order to have the oxygen needed for combustion. As you may recall from the previous chapter, fuel, oxygen, and heat are needed for combustion to take place. For combustion to continue, there must be ample amounts of all three of these components. The absence of any one of them will result in the termination of the combustion process. The nozzle (Fig. 25-41) is the system component that atomizes the liquid fuel oil, permits the correct amount of oil to enter the combustion chamber, and determines the

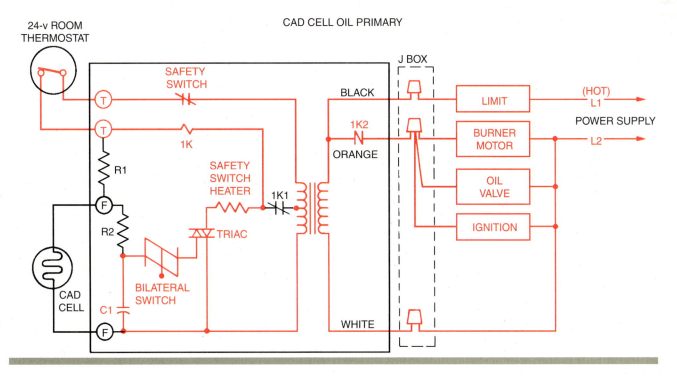

FIGURE 25-39 Cad cell primary control circuit when the cad cell senses no burner flame.

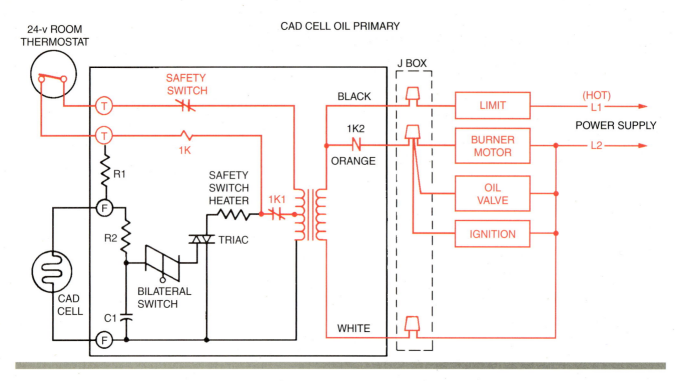

FIGURE 25-40 Cad cell primary control circuit when the cad cell senses burner flame.

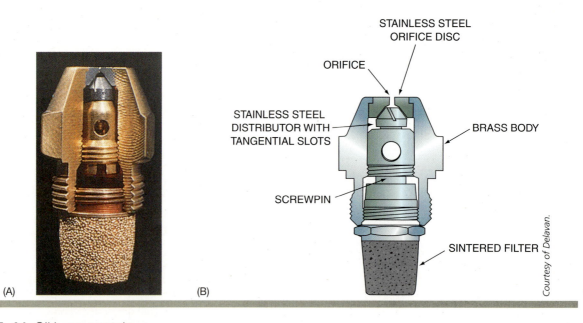

FIGURE 25-41 Oil burner nozzle.

spray pattern and angle at which the fuel oil is introduced to the chamber. By atomizing the fuel oil, its surface area is greatly increased, which in turn, increases the ignition speed. The amount of oil that a particular nozzle will permit to enter the combustion chamber is calculated by using a fixed oil pressure of 100 psig. The spray pattern and angle is determined by the manufacturer and varies

from burner to burner. Nozzles are commonly made from either stainless steel or a combination of steel and brass. They can be equated to the spud on the gas manifold that has an orifice through which the fuel enters the combustion chamber. The nozzle, however, has a number of other parts that determine the pattern and volume of the entering oil. Since the orifice in the nozzle is very small, there is a filter at its inlet to catch any particulate matter that may cause clogging.

SPRAY PATTERNS

The pattern that the oil/fuel mixture will follow as it is sprayed into the combustion chamber is determined by the nozzle. Each manufacturer has recommended guidelines for the specific type of nozzle that is to be used with a particular burner and a particular appliance. There are three common classifications for spray patterns (Fig. 25-42). The first is the hollow cone, which is used primarily for burners that have flow rates of less than 1 gallon per minute. It sprays oil around the surface of the cone, while leaving the interior area relatively oil-free. The second spray pattern is the solid cone, which forms a uniform, cone-shaped spray. It is used on burners designed to operate with long run cycles. The third type of spray pattern is the semi-solid cone, which is between the hollow and solid cone patterns. Manufacturers have codes to identify the spray patterns on the nozzles they offer (Fig. 25-43).

SPRAY ANGLES

The angle at which the atomized fuel oil enters the combustion chamber is determined by the manufacturer of the appliance on which the burner is to be used. If the combustion chamber is shallow from front to back, the angle of the spray should be high (Fig. 25-44). If the combustion chamber is very deep from front to back and short from top to bottom, the spray angle should be low (Fig. 25-45). In any event, any unburned fuel oil should not come in contact with the walls of the heat exchanger. Using a nozzle with a 30-degree angle may result in unburned fuel hitting the back of a shallow combustion chamber.

Going Green

Be sure to follow the manufacturer's recommendations when selecting the nozzle. Manufacturers test their boilers with many different types of nozzle and are able to make the best suggestion to ensure maximum system efficiency.

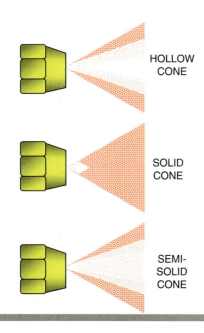

FIGURE 25-42 Oil spray patterns.

Courtesy of Hago Manufacturing Company, Inc.

FIGURE 25-43 Spray patterns.

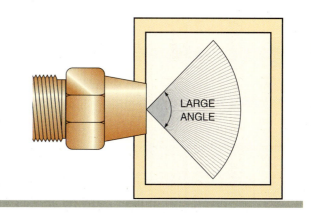

FIGURE 25-44 Large spray angle used on shallower combustion chambers.

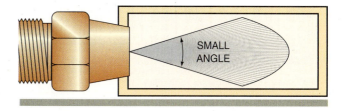

FIGURE 25-45 Low spray angle used on deeper combustion chambers.

FLOW RATE

Nozzles are rated by the amount of oil in gallons per hour (gph) that will be permitted to flow through the nozzle, assuming an oil pressure of 100 psig. If the oil pressure is higher than 100 psig, the actual flow through the nozzle will be higher than the rating on the nozzle. Figure 25-46 shows the relationship between oil pressure and the flow rate through nozzles with different ratings. One popular way to size the nozzle is to select the lowest possible firing rate that will satisfy the heating load on a structure on the coldest day of the year. A nozzle that is too small will result in improper heating, and a nozzle that is too large will cause system short-cycling and wasted fuel.

GPH	125	140	150	175	200	250	300
0.30	0.34	0.35	0.37	0.40	0.42	0.47	0.52
0.35	0.39	0.41	0.43	0.46	0.49	0.55	0.61
0.40	0.45	0.47	0.49	0.53	0.57	0.63	0.69
0.45	0.50	0.53	0.55	0.60	0.64	0.71	0.78
0.50	0.56	0.59	0.61	0.66	0.71	0.79	0.87
0.55	0.61	0.65	0.67	0.73	0.78	0.87	0.95
0.60	0.67	0.71	0.73	0.79	0.85	0.95	1.04
0.65	0.73	0.77	0.80	0.86	0.92	1.03	1.13
0.70	0.78	0.83	0.86	0.93	0.99	1.11	1.21
0.75	0.84	0.89	0.92	0.99	1.06	1.19	1.30
0.85	0.95	1.01	1.04	1.12	1.20	1.34	1.47
0.90	1.01	1.06	1.10	1.19	1.27	1.42	1.56
1.00	1.12	1.18	1.22	1.32	1.41	1.58	1.73
1.10	1.23	1.30	1.35	1.46	1.56	1.74	1.91
1.20	1.34	1.48	1.53	1.65	1.77	1.98	2.17
1.35	1.51	1.60	1.65	1.79	1.91	2.13	2.34
1.50	1.68	1.77	1.84	1.98	2.12	2.37	2.60
1.65	1.84	1.95	2.02	2.18	2.33	2.61	2.86
1.75	1.96	2.07	2.14	2.32	2.47	2.77	3.03
2.00	2.24	2.37	2.45	2.65	2.83	3.16	3.46
2.25	2.52	2.66	2.76	2.98	3.18	3.56	3.90
2.50	2.80	2.96	3.06	3.31	3.54	3.95	4.33
2.75	3.07	3.25	3.37	3.64	3.89	4.35	4.76
3.00	3.35	3.55	3.67	3.97	4.24	4.74	5.20
3.25	3.63	3.85	3.98	4.30	4.60	5.14	5.63
3.50	3.91	4.14	4.29	4.63	4.95	5.53	6.06
3.75	4.19	4.44	4.59	4.96	5.30	5.93	6.50
4.00	4.47	4.73	4.90	5.29	5.66	6.32	6.93
4.25	4.75	5.03	5.21	5.62	6.01	6.72	7.36
4.50	5.03	5.32	5.51	5.95	6.36	7.11	7.79
5.00	5.59	5.92	6.12	6.61	7.07	7.91	8.66
5.50	6.15	6.51	6.74	7.28	7.78	8.70	9.53
6.00	6.71	7.10	7.35	7.94	8.49	9.49	10.39
6.50	7.27	7.69	7.96	8.60	9.19	10.28	11.26
7.00	7.83	8.28	8.57	9.26	9.90	11.07	12.12
7.50	8.39	8.87	9.19	9.92	10.61	11.86	12.99
8.00	8.94	9.47	9.80	10.58	11.31	12.65	13.86
8.50	9.50	10.06	10.41	11.24	12.02	13.44	14.72
9.00	10.06	10.65	11.02	11.91	12.73	14.23	15.59
9.50	10.62	11.24	11.64	12.57	13.43	15.02	16.45
10.00	11.18	11.83	12.25	13.23	14.14	15.81	17.32
11.00	12.30	13.02	13.47	14.55	15.56	17.39	19.05
12.00	13.42	14.20	14.70	15.87	16.97	18.97	20.78
13.00	14.53	15.38	15.92	17.20	18.38	20.55	22.52
14.00	15.65	16.56	17.15	18.52	19.80	22.14	24.25
15.00	16.77	17.75	18.37	19.84	21.21	23.72	25.98
20.00	22.36	23.66	24.49	26.46	28.28	31.62	34.64
25.00	27.95	29.58	30.62	33.07	35.36	39.53	43.30
30.00	33.54	35.50	36.74	39.69	42.43	47.43	51.96
35.00	39.13	41.41	42.87	46.30	49.50	55.34	60.62

NOZZLE FLOW VS PRESSURE (APPROX.)
Flow Rating USGPH @100 PSI — Flow Rates in US GPH — Pressure PSI

Test Oil Standards: 35 SSU @ 100 F. Specific Gravity .825 @ 60 F.
1 US Gallon = 3.122 Kg/Gal. Test Pressure 100 PSI = 7.03 Bars.
Flow rates are estimated as equal to the square root of the pressure ratio. www.hagonozzles.com

Courtesy of Hago Manufacturing Company, Inc.

FIGURE 25-46 Nozzle flow with respect to pressure.

DRAFT AND VENTING

In the previous chapter, we discussed the venting of products of combustion from the furnace location. Removing products of combustion is important because these flue gases often contained toxic and poisonous substances such as carbon monoxide. In addition, flue gases must be removed as quickly as possible to reduce the chances of condensation, which can result in damage to the chimney. As combustion occurs, the heat-laden products of combustion are introduced to the flue pipe or chimney. Since warmer air is lighter than cooler air, the hot flue gases rise from the chimney. This creates a slight negative pressure at the bottom of the chimney that in turn helps the oil burner bring air into the combustion chamber. This negative pressure is referred to as draft.

Draft is measured in inches of water column, IWC, and is necessary to "pull" exhaust gasses out of the heating unit and up the chimney. If draft is too high, the gasses will be pulled out too fast and efficiency will be lowered. If draft is too low, the unit will smoke, develop soot, and generate odors. In most oil heat systems, recommended draft is around −.02 in. W.C. over the fire. Draft is affected by the chimney, wind, exhaust fans, and condition of the heat exchanger. Always check manufacturers' recommendations when adjusting draft. To maintain efficiency and low smoke levels, the amount of draft should be regulated, especially since winds outside the structure can have adverse effects on the amount of draft. The **draft regulator**, also known as the **barometric damper**, serves to maintain the desired draft at the unit.

DRAFT REGULATORS

The draft regulator (Fig. 25-47) is installed in a tee fitting in the flue piping (Fig. 25-48) and opens and closes as the amount of draft changes. As the amount of draft

Delmar/Cengage Learning.

FIGURE 25-47 Draft regulator or barometric damper.

Photo by Eugene Silberstein.

FIGURE 25-48 Barometric damper location.

increases, the draft regulator opens to allow air surrounding the furnace to be pulled into the flue. The introduction of the cooler, surrounding air into the flue reduces the temperature of the flue gases, thereby reducing the draft. If the draft is too low, the draft regulator remains in the closed position. In the closed position, the flue gases in the chimney can heat up, thereby increasing the draft. It is important to note, however, that the draft

---

from experience...

The draft regulator is delicate and should be handled carefully. Be sure to properly calibrate the draft regulator according to manufacturers' guidelines to ensure proper operation.

---

regulator cannot increase the draft above the level that is created naturally in the chimney.

POWER AND DIRECT VENTING

The venting method described in the previous section relied on convection currents generated by the rising heat-laden products of combustion in the chimney and natural draft created in part by wind movement across the opening of the chimney. In an effort to maintain more control over the venting of the products of combustion, mechanical means are employed to facilitate the rapid and efficient removal of the flue gases. One advantage of mechanical venting is that the termination point for the flue piping is usually the side of the structure as opposed to the chimney. There are, however, strict guidelines regarding the clearances that must be observed when installing these venting systems. For this reason, it is important to always follow the manufacturers' guidelines during the installation process.

The process of **power venting** (Fig. 25-49) uses a blower motor that pulls the products of combustion from the appliance and vents them outside the structure. The blower is located close to the flue pipe termination point, near the shell of the structure. One of the benefits of power venting is that the blower operates at the beginning of the cycle before the fuel is ignited as well as at the end of the cycle after the burner has cycled off. When the system calls for furnace operation, the blower starts and creates a pressure change in the vent piping. This results in the closing of a set of electric contacts, which initiates burner operation. At the end of the cycle, the blower will continue to operate for approximately 4 minutes until the products of combustion have been removed from the flue pipe.

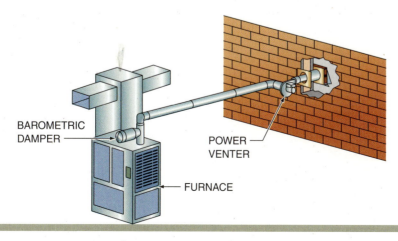

BAROMETRIC DAMPER

POWER VENTER

FURNACE

FIGURE 25-49 Power vent.

In contrast to the power vent that uses a separate blower to vent the products of combustion from the structure, the **direct vent** system uses the pressure created by the oil burner fan motor. The air introduced to the oil burner is supplied directly from outside and the products of combustion are immediately vented outside (Fig. 25-50). Some manufacturers combine the intake and discharge pipes into a single pipe-in-pipe unit (Fig. 25-51). The pipe-in-pipe configuration eliminates the need for two holes in the shell of the structure and also facilitates heating the air used for combustion prior to its entering the burner.

COMBUSTION EXHAUST SAFETY

The exhaust from any heating appliance that burns a fuel contains a combination of water vapor, carbon dioxide, nitrogen oxides, carbon monoxide and other trace gasses. All of the gasses can have a bad effect on the building and occupants so they must not be allowed to leak inside the house. Carbon monoxide is colorless, odorless and deadly. Even in low quantities, it can cause symptoms similar to a bad cold or flu (headache, nausea, vomiting and fatigue). Nitrogen oxides cause eye, nose throat and lung irritation, coughing, nausea and sometimes respiratory swelling. Water vapor raises the humidity level inside the house that leads to other indoor air quality problems like mold and biological allergens, and it can condense and cause decay inside building cavities.

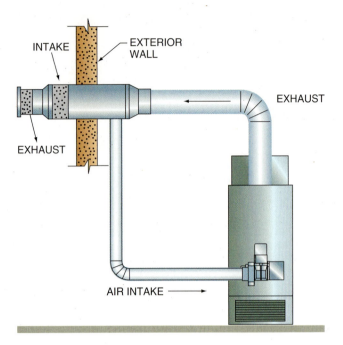

FIGURE 25-51 Pipe-in-pipe direct vent.

In order to reduce the chances of indoor air contamination, sealed-combustion appliances can be used. Sealed-combustion appliances draw in outside air for combustion and exhaust the flue gasses directly back out. There is no interaction between the combustion chamber and indoor air so there is little chance that gasses can enter the house. When the building budget does not allow for a sealed combustion model, an induced draft, power-vented model should be installed. Power-vented combustion appliances utilize blowers to efficiently remove the products of combustion from the structure.

SIZING AND INSTALLING THE FLUE PIPE

The following are general guidelines regarding the flue pipe. Always refer to the installation manual for the system being installed for specific clearances, sizes, and installation recommendations. When sizing flue pipe, keep the following in mind:

✓ The cross-sectional area of the chimney should be *at least* 1 square inch per 6,500 Btu of system input.

✓ Flue pipe should be the same size as the breech on the furnace. The breech is the opening in the appliance to which the flue pipe is connected.

✓ Flue piping should be as short as possible.

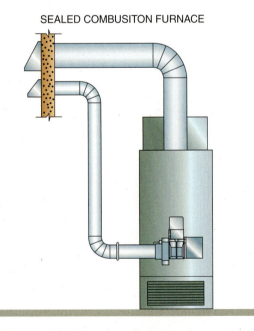

FIGURE 25-50 Two-pipe direct vent.

✓ Use as few flue pipe fittings as possible.

✓ Use 45-degree elbows whenever possible.

✓ Flue piping sections should be connected securely to each other to prevent flue gas leakage.

✓ Flue piping should be supported to prevent sagging.

✓ Flue pipe should not extend past the interior surface of the chimney.

✓ Flue pipe should be cemented in place at the chimney location.

✓ The draft regulator should be located 18 inches above the stack relay, if used.

TROUBLESHOOTING THE OIL BURNER

Just as there are a number of factors that contribute to the satisfactory operation of the oil furnace, a number of factors can contribute to system malfunction. It is important to be able to isolate and remedy the system problem as quickly and efficiently as possible to reduce customer callbacks and increase customer satisfaction. Oil-fired furnace problems can be caused by a malfunction in any one of the following areas:

• Air-distribution system
• Fuel
• Oil tank
• Oil piping
• Oil burner
• System controls

AIR-DISTRIBUTION SYSTEM

Poor airflow through a furnace and poor air distribution to the occupied space can result in overheating of the heat exchanger. This can result in the opening of system limit switches that will de-energize the system. The following is a list of items that should be checked to help ensure that there is ample airflow through the system:

✓ Check air filters—Blocked air filters result in reduced airflow through the system.

✓ Check blower motor—If the motor is not operating, check the voltage supplied to the motor and the associated contacts on the fan

relay. If the proper voltage is being supplied to the motor, disconnect the power leads to the motor and check the windings for continuity. No continuity through the windings indicates that either a winding is open or an internal thermal overload is open. If the motor is cool to the touch, the motor winding is probably open and the motor should be repaired or replaced.

✓ Check blower wheel—A dirty blower wheel reduces the amount of air that the blower can move through the duct system.

✓ Check belts and pulleys (if applicable)—Loose belts will slip, resulting in reduced airflow through the system. Belt slippage will cause the interior pulley surfaces to become polished and highly reflective. Polished pulleys should be replaced. Belts can also slip or break if the pulleys are not properly adjusted. Always make sure that the pulleys are properly aligned.

✓ Check return grill—Check the return grill for any obstructions.

✓ Check supply registers—Check for blocked or closed supply registers. Closed supply registers will result in reduced airflow through the system.

✓ Check for loose duct lining—If the acoustical duct lining should come loose, reduced airflow will also result.

FUEL AND FUEL TANK

The biggest enemy of fuel oil is water. Water in the oil tank can lead to the formation of sludge, which corrodes the tank and clogs fuel filters and nozzles. If a water problem is suspected, the tank should be checked for water content. The easiest way to do this is with **water finding paste**. Water finding paste is typically green in color when dry, but turns red when exposed to water. The paste is placed on the end of a stick that is long enough to extend to the bottom of the oil tank where it is held for approximately 30 seconds. If the paste turns color, there is water in the tank. Reasons for water in the tank include:

✓ Condensation

✓ Broken, loose, or missing fill caps

✓ Leaking tank

✓ Water-laden oil delivered to tank

OIL PIPING

Damaged or clogged oil piping between the tank and the furnace will result in reduced fuel flow to the burner. This can cause burner underfiring, which is identified by:

- ✓ Reduced heating capacity

- ✓ High oxygen readings during combustion test

- ✓ Low stack or flue temperature

- ✓ Low temperature rise (temperature differential) across the heat exchanger

Oil lines should be visually inspected for damage. Any crushed or kinked lines should be replaced. In addition, if there is sludge in the burner, there is also sludge in the oil lines. These lines can become partially blocked with sludge, so even if there is no visible restriction in the lines, flow through the lines may be lower than required. Also, it is good field practice to inspect:

- ✓ Oil filters (on periodic maintenance, filters should be changed)

- ✓ De-aerators (if installed)

- ✓ Flare fittings for cracks

- ✓ Floor under burner for oil puddles

OIL BURNER

When an oil-fired system fails to operate, attention is often turned to the burner itself. When checking the burner, keep the following in mind:

- ✓ Test the vacuum created by the pump with the vacuum gauge.

- ✓ Check the outlet pressure of the pump.

- ✓ If the pump is operating and no fuel is being sprayed from the nozzle, remove the nozzle and disconnect the ignition. Place a can under the nozzle adapter and start the pump. If oil is pumped from the nozzle adapter, the nozzle is clogged and needs to be replaced.

- ✓ If the burner fails to stay lit and there is a cad cell primary, make certain the cad cell is clean.

- ✓ If the burner lights but does not stay lit, check the primary control.

COMBUSTION

Incomplete combustion indicates that an imbalance is preventing the system from properly burning the fuel. What follows is a partial list of potential problems that can cause excessively high or excessively low readings:

- ✓ Insufficient oxygen
  - Too little combustion air. If combustion air is low, the air band can be opened, enabling the fan to pull more air into the fuel/air mixture.
  - Too much fuel delivery to nozzle. If the burner is overfiring, the nozzle may need to be changed or the pressure at the outlet of the fuel pump may need to be adjusted.
  - Low stack draft. If there is insufficient draft, the barometric damper may need adjusting. If there is no barometric damper on the system, this may be a good time to get one.

- ✓ Excessive carbon monoxide formation
  - Same as for insufficient oxygen.

- ✓ Too much oxygen
  - Too much combustion air. If combustion air is high, the air band can be closed, reducing the amount of air that the fan can pull into the fuel/air mixture.
  - Too little fuel delivery to nozzle. If the burner is underfiring, the nozzle may need to be changed or the pressure at the outlet of the fuel pump may need to be adjusted.

- ✓ High stack temperature
  - Reduced airflow across the heat exchanger. Refer to the section on blower and air-distribution problems.

- ✓ Low stack temperature
  - Burner underfiring. If the burner is underfiring, there may be reduced combustion air or reduced fuel flow to the burner.

SYSTEM CONTROLS

The main control on oil furnaces is the primary control, but there are other devices in the furnace that can become faulty. If the furnace fails to operate, be sure to do the following:

- ✓ Check the main circuit breakers.

- ✓ Check for voltage at the service disconnect on the furnace.

✓ Check for line voltage at the primary of the control transformer.

✓ Check for low voltage at the secondary of the control transformer.

✓ A reading of zero volts at any of the previous three locations indicates that there is either a supply voltage problem or a control transformer problem. If there is voltage at the primary of the control transformer but none at the secondary of the control transformer, the transformer is defective. The cause for transformer failure should be determined and corrected before replacing the component.

✓ Check the primary safety switch. If the unit operates after the switch is reset, be sure to find the reason why the primary tripped. Simply resetting the control is not repairing the problem.

✓ Check the room thermostat and make certain that the thermostat is calling for heat.

✓ At the furnace, check the voltage between the R and C terminals. If the transformer is operating properly, the voltage reading between these two points should be 24 volts.

✓ At the furnace, check the voltage between the R and W terminals. If the thermostat is calling for heat and the thermostat is operating correctly, the voltage reading between these two terminals should be 0 volts.

✓ If the burner starts but then shuts down after about 90 seconds, check the primary control. If the primary is a stack relay, there may be a problem with the bimetal strip or the hot and cold contacts in the unit. If the primary is a cad cell relay, the cad cell may be dirty or out of alignment.

SERVICE CALL
No Heat

CUSTOMER COMPLAINT

A residential oil heat customer calls her oil supplier and tells the dispatcher that her oil-fired furnace is not operating. The weather is still relatively warm as fall is just setting in. She had an oil delivery two weeks ago and, around the same time, she had her furnace started up and checked. During the startup, the technician checked the combustion of the unit and replaced the nozzle. The system has a cad cell relay primary control and the wiring of the unit is similar to that in Figure 25-39.

TECHNICIAN EVALUATION

Later that day, Michael arrives at the customer's home and listens to the home owner as she explains the problem she is having with the furnace. Since the customer presents him with the paperwork from the oil delivery, Michael is able to determine that there is indeed an ample supply of fuel oil since the delivery was just over 150 gallons. Michael first goes to the thermostat and makes certain that it is set to the heating mode and that the temperature setting is above the actual space temperature. The thermostat is properly set, so he goes to the basement where the furnace is located.

Upon arriving at the unit, Michael decides to check the line voltage feeding the unit. He turns the service disconnect off and removes the service panel from the side of the furnace. He turns the switch back on and takes a voltage reading across the wires from the power supply. He read 120 volts. He then takes a voltage reading between the orange and white wires on the primary control (Figure 25-39) and obtains a reading of 120 volts. This reading indicates to him that the 1K2 contacts are closed, given that there is power at the orange wire. Because there is a reading of 120 volts between the orange and white wires, the burner motor should be operating, but it is not.

Michael next decides to check the motor to see if the motor shaft spins freely. He disconnects the power to the unit and opens the burner to access the fan. He attempts to turn the fan by hand but the shaft of the motor is stuck. He disconnects the fan and the flexible coupler from the shaft of the motor and removes the mounting screws on the motor. He retrieves a new motor from the truck and installs it. Once the motor has been installed, he restores power to the unit and hears the blower begin to operate. A few seconds later, he hears the fuel/air mixture ignite. He allows the system to operate for about ten minutes, the time it takes to write his work ticket and gather his tools. He decides to perform a combustion test on the system. He obtains the results and compares them with the results from the combustion test that was performed a couple weeks ago. The results are similar and within acceptable limits and ranges.

SERVICE CALL DISCUSSION

Michael was able to quickly determine that the motor was defective by establishing that line voltage was being supplied to it. Once he found that the motor was stuck, it was a relatively easy task to replace the motor. While he was there, he decided to check the combustion of the system, which was good field practice, since the component responsible for supplying both air and fuel to the burner had just been replaced.

SERVICE CALL
No Heat

CUSTOMER COMPLAINT

A homeowner calls his oil company and tells the service manager that his oil furnace is not operating. He says that he can hear the motor come on, but the burner never fires. He pushed the reset button a few times but could not get the burner to fire. He does not have a service contract on his furnace and, since the weather was mild, did not have the furnace serviced last year. The system is equipped with a stack relay primary control.

TECHNICIAN EVALUATION

Later that morning, Louise arrives at the customer's house and checks the thermostat setting. The stat is set to the heating mode to maintain a temperature of 68°F, but the space temperature is 61°F. Louise goes to the furnace location and finds that the reset button has tripped. She pushes the reset and can hear the fan come on, but the burner does not fire. The system shuts down and then attempts to restart. After about 90 seconds, she hears the reset trip again. From this, she is able to conclude the following:

✓ The thermostat is calling for heating

✓ The motor on the burner is operating

Since the burner did not light, Louise determines that the problem is with one of the following:

✓ No oil in the tank

✓ Damaged piping between the oil tank and the oil filter

✓ Clogged oil filter

✓ Damaged piping between the oil filter and the pump

✓ Defective oil pump

✓ Clogged nozzle

✓ Ignition transformer failure

Since the furnace has not been serviced in more than a year, Louise decides to change the oil filter, since it obviously is in need of replacement. She closes the valve before changing the oil filter and opens it to replace the cartridge. Since the oil tank is inside and very close to the burner, Louise decides she could crack the valve open and see if oil will flow through the piping to the filter location. She places a can under the filter and opens the valve. The fuel oil flows freely. She checks the tank gauge and it registers ¾ full. From this, and assuming that the gauge is good, she determines that there is oil in the tank and that the piping between the tank and filter is good. In addition, the oil filter is good, as she just replaced the cartridge.

With the fuel valve open and the ignition disconnected, she removed the burner and reset the safety. Once again, she could hear the blower run, but no oil is being sprayed from the nozzle. She then removes the nozzle and places a can under the nozzle adapter. After setting the safety, oil begins to pour from the nozzle adapter into the can. The nozzle is clogged. She goes to the truck and gets a new nozzle, which she installs in the burner, and then replaces the burner in the furnace. She restores power to the unit and the burner fires. While the system is operating, she replaces the air filters on the furnace.

Since the system has not been serviced for quite some time, she performs a combustion test on the flue gases. All of the parameters are within acceptable ranges and, as she writes up the work ticket, she discusses the advantages of having a service contract on the furnace. The customer agrees to a service contract and thanks her for her diligence and professionalism.

SERVICE CALL DISCUSSION

Louise was very complete in her evaluation of the system and performed a number of tasks that are normally included in routine, annual contract service. She also made it a point to inform the customer about the benefits of taking out a service contract on the furnace. Even after the furnace fired and was operating, she realized that the system had not been serviced recently, so she decided to check the air filters as well.

SUMMARY

- Fuel oil is a hydrocarbon made up of carbon and hydrogen.

- Number 2 fuel oil produces approximately 139,000 Btu per gallon when burned.

- For oil to burn, it must be atomized, mixed with air, and heated above the ignition point.

- Insufficient combustion air can result in the formation of carbon monoxide.

- Combustion testing involves the evaluation of the stack temperature, carbon monoxide level, carbon dioxide level, smoke level, draft, and amount of combustion air available for combustion.

- High-efficiency furnaces use second heat exchangers to remove as much heat as possible from the flue gases.

- Oil tanks must be installed in accordance with all local codes regarding placement, clearances, and piping methods used.

- Teflon tape, cast iron pipe, and PVC piping materials are not to be used on fill and vent piping.

- Oil fill and vent lines should pitch back toward the tank to eliminate oil traps.

- Oil lines should be connected only with flare fittings.

- Oil should be filtered before it flows into the burner to avoid clogging pumps and nozzles.

- Oil de-aerators remove air from the oil before it flows into the pump.

- Most new burners deliver high-pressure oil to the combustion area and are called high-pressure atomizing oil burners.

- Flame retention burners are more efficient than older style burners.

- The movement of air and oil through the burner is accomplished by the motor.

- Pressure and vacuum gauges are used to determine the operation of the fuel pump.

- Primary controls can be of the stack relay or cad cell variety.

- The cad cell relay relies on the light of the burner flame to control burner operation.

- The stack relay relies on the temperature of the flue gases to control burner operation.

- The nozzle is responsible for metering, atomizing, and patterning the oil as it enters the combustion area.

- Flue gases can be vented by natural draft, power venting, or direct venting.

GREEN CHECKLIST

- ☐ **Combustion testing is a must!**

- ☐ **Oil burners cannot be adjusted by sight or sound. Properly calibrated test instruments are needed to properly set up an oil burner.**

- ☐ **Follow all safety precautions when installing oil tanks.**

- ☐ **Self-diagnosing primary controls are making their way into the industry.**

- ☐ **Be sure to follow the manufacturer's guidelines when selecting a nozzle to ensure maximum system efficiency.**

- ☐ **All fossil-fuel heating systems must be properly vented to the outside.**

REVIEW QUESTIONS

1. **Number 2 fuel oil is a hydrocarbon made up of approximately**

   a. 15% carbon and 85% hydrogen

   b. 50% carbon and 50% hydrogen

   c. 85% carbon and 15% hydrogen

   d. 100% hydrogen

2. **Which national agency sets the standards for production and refinement of fuel oil?**

   a. American Society for Testing and Materials

   b. Underwriters Laboratory

   c. National Fire Protection Association

   d. None of the above

3. **Which is true regarding number 2 fuel oil?**

   a. There are approximately 7.1 pounds of oil in one gallon

   b. One pound of oil will produce approximately 19,580 Btu when burned

   c. One gallon of oil will produce approximately 139,400 Btu when burned

   d. All of the above are correct

4. **All of the following are true regarding vent and fill line piping *except***

   a. Teflon tape should not be used

   b. PVC piping materials should not be used

   c. Cast iron fittings should not be used

   d. These lines should be pitched away from the oil tank

5. **The purpose of the vent alarm is to**

   a. Notify the technician that the burner is firing

   b. Notify the technician that the vent line is damaged

   c. Notify the fuel delivery person that the fuel tank is nearly full

   d. Notify the fuel company that the fuel tank is empty

6. **The oil de-aerator is located**

   a. Between the fuel filter and the oil pumps

   b. Between the oil tank and the oil filter

   c. Between the fuel pump and the nozzle

   d. Between the nozzle tube and the nozzle

7. **Flame retention oil burners**

   a. Retain the flame longer and are therefore less efficient

   b. Result in the creation of 700% more soot than conventional burners

   c. Produce a flame that is 300 to 500°F hotter than conventional burners

   d. Do not mix the air and fuel as well as conventional burners

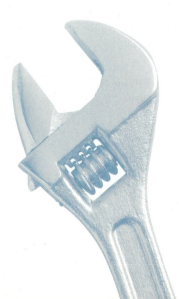

8. **The motor in the oil burner**
   a. Turns only the fan in the burner
   b. Turns only the fuel pump in the burner
   c. Turns both the fuel pump and the fan in the burner
   d. Turns neither the fuel pump nor the fan in the burner

9. **What is the vacuum requirement for a pump that is 30 feet from the tank (there is a filter in this line) and mounted 3 feet above the level of the fuel tank?**
   a. 3" of vacuum
   b. 5" of vacuum
   c. 7" of vacuum
   d. 9" of vacuum

10. **The nozzle assembly is comprised of**
    a. Nozzle line and electrodes
    b. Fuel pump and nozzle line
    c. Nozzle line, nozzle adapter, nozzle, and electrodes
    d. Fuel pump and electrodes

11. **In the stack relay primary control**
    a. The hot and cold contacts are both closed when the burner initially fires
    b. The hot contacts are open and the cold contacts are closed when the burner initially fires
    c. The hot contacts are closed and the cold contacts are open when the burner initially fires
    d. The hot and cold contacts are both open when the burner initially fires

12. **If a stack relay controlled burner is operating with a very low stack temperature**
    a. The cold contacts may remain closed and the burner will shut down
    b. The cold contacts may remain closed, de-energizing the safety switch heater
    c. The hot contacts will remain closed and the burner will shut down
    d. All of the above are correct

13. **An interrupted ignition**
    a. Sparks whenever the burner is operating
    b. Sparks only at the beginning of the cycle
    c. Sparks continually as in the case of a standing pilot
    d. Sparks whenever the burner is lit

14. **A dirty cad cell**
    a. Can result in the opening of the safety switch
    b. Can result in the overfiring of the burner
    c. Both a and b are correct
    d. Neither a nor b is correct

15. **A furnace with a very deep combustion chamber will most likely use a nozzle with**
    a. A high spray angle
    b. A hollow cone
    c. A low spray angle
    d. A solid cone

16. **The nozzle is responsible for**
   a. Metering the correct flow of fuel into the combustion area
   b. Creating the correct spray pattern and angle of the fuel oil
   c. Atomizing the fuel as it enters the combustion area
   d. All of the above are correct

17. **High winds across the chimney opening will cause**
   a. The draft to increase and the barometric damper to open
   b. The draft to increase and the barometric damper to close
   c. The draft to decrease and the barometric damper to open
   d. The draft to decrease and the barometric damper to close

18. **Which of the following is true regarding draft?**
   a. Low draft results in reduced soot formation
   b. High draft results in increases in system efficiency
   c. Both a and b are correct
   d. Neither a nor b is correct

19. **The power vent blower is ideally located**
   a. As close as possible to the furnace to increase the draft
   b. As close as possible to the shell of the structure to reduce the chances of flue gas leakage into the occupied space
   c. As close as possible to the power supply to avoid long electric power lines
   d. The location of the power vent is unimportant as there is a blower that removes the flue gases from the space

20. **A benefit of the pipe-in-pipe direct vent is that**
   a. Only one penetration is made through the exterior wall for venting purposes
   b. The air used for combustion is heated prior to being introduced to the burner
   c. Both a and b are correct
   d. Neither a nor b is correct

KNOW YOUR CODES

When installing oil-fired heating equipment, most local authorities require that the guidelines set forth by the International Code Council (ICC) or the National Fire Protection Association (NFPA) be followed. Visit the Web sites for these organizations and obtain information regarding the venting of oil-fired appliances with respect to property line clearances, proximity to electric meters, combustion sources, windows, doors, air intakes, and air exhausts.

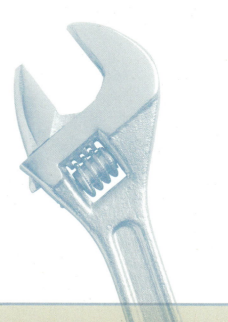

CHAPTER 26

Hydronic Heat

OBJECTIVES

Upon completion of this chapter, the student should be able to

- Explain the concept of hydronic heating.
- List the three most commonly used heat sources in boilers.
- Describe basic boiler construction.
- Identify component parts of a boiler.
- Explain the operation of a boiler.
- Describe various components that maintain the desired water temperature in a boiler.
- Explain the difference between a one-pipe and a two-pipe hot water system.
- Discuss the difference between direct-return systems and reverse-return systems.
- Describe the operation and function of centrifugal pumps.

- Explain the function of boiler controls and safety devices.
- Explain the function of an expansion tank.
- Explain the *point of no pressure change*.
- Check the pressure in an expansion tank.
- Explain primary-secondary pumping.
- Explain the concept of "zoning."
- Explain how a radiant heating system creates comfort.
- Explain how a radiant heating system operates.
- Install a boiler.
- Service boilers.

The three previous chapters dealt with forced-air heating systems—furnaces that used electricity, gas, or oil as their heat source. As heat was generated, it was used to heat air that was, in turn, transferred to the occupied space by means of fans or blowers. We also discussed three common methods of generating this heat—electric strip heaters, gas burners, and oil burners—as well as the components, controls, and heating capability of each.

In this chapter, we will concentrate not on the source of heat, but on the medium being heated. In the case of the hydronic system, this medium is water instead of the air used in furnaces. Water can be heated to generate hot water or, if heated above 212°F at atmospheric conditions, steam. Either hot water or steam boilers can provide heat, but this text will only address the issue of hot water, as steam is not commonly found in new residential installations.

Air cushion air above the semi-permeable membrane in an expansion tank

Air separator device that separates air from water in the system

Air vent fitting used to remove air, either manually or automatically, from a hydronic system

Aquastat electrical component that opens and closes its contacts to energize and de-energize electric circuits in response to the water temperature sensed by the device

Automatic air vent fitting that automatically removes air from a hydronic heating system

Balancing valve manually controlled valve used to increase resistance and reduce water flow through a given branch circuit

Baseboard sections see radiator or terminal unit

Boiler piece of heating equipment designed to heat water, using electricity, gas, or oil as a heat source, for the purpose of providing heat to an occupied space or potable water

Boiler/water feed valve valve that reduces the pressure entering the structure to the pressure required

GLOSSARY OF TERMS (CONT'D)

by the hydronic system; automatically feeds water into the system to maintain the desired water pressure

Centrifugal pump pump that moves water through a piping circuit by means of centrifugal force

Circulator see centrifugal pump

Closed loop system that is closed or isolated from the atmosphere

Compression tank see expansion tank

Diaphragm-type expansion tank see expansion tank

Direct return configuration of a hot water heating system in which the first terminal unit supplied with hot water is the first one to return to the boiler and vice versa

Diverter tee in a one-pipe hot water system, the fitting used to increase resistance to water flow in the main loop in order to direct water to the terminal units

Expansion tank system piping component that

provides additional space for expanding water to occupy

Feet of head term used to rate the pumping capacity of a pump; 1 foot of head is the equivalent of a 0.433 psig difference between the inlet and outlet of the pump (1 psig = 2.31 feet of head)

Flow check valve see flow control valve

Flow-control valve valve that prevents backward and gravity circulation through loops not requiring flow

Hydronics heating systems that circulate hot water or steam through piping arrangements located in the areas being heated

indirect water storage tank a water storage tank that has an internal heat exchanger that is serviced by the main boiler

Manifold station location of the manifold for radiant heating loops

Manual air vent fitting that, when opened manually, will remove air from a hydronic heating system

Monoflo tee see diverter tee

One-pipe hot water hydronic piping configuration that uses a main hot water loop and diverter tees to connect the terminal units to the system

Outdoor reset control used on hydronic systems that decreases the temperature of the water as the outdoor temperature increases

Pex tubing see polyethylene tubing

Point of no pressure change point in a hydronic hot water system where the expansion tank is connected to the piping system; pressure at this location cannot be affected by circulator operation

Polyethylene tubing tubing material used for buried water loops in radiant heating systems and geo-thermal heat pump systems

Pressure-reducing valve reduces the pressure of the water entering the structure to the desired pressure in the hydronic system

Pressure relief valve spring-loaded valve that opens when the pressure in a hydronic system exceeds the rating of the valve

Radiant heat system heating system that attempts to regulate the heat loss of the individual as opposed to the rate of heat loss of the structure

Radiator heat emitters or terminal units that transfer the majority of their heat to the occupied space by radiation

Relief valve see pressure relief valve

Reverse return system configuration of a hot water heating system in which the first terminal unit supplied with hot water is the last one

to return to the boiler and vice versa

Standard expansion tank see expansion tank

Steam boiler piece of heating equipment that heats water to the point of vaporization to provide heat to an occupied space

tankless hot water heating coil a heat exchanger installed in the boiler that connects to the hot water pipes of the house

Terminal unit radiator or section of baseboard

Two-pipe system hydronic piping configuration that uses one pipe as the supply and

one pipe as the return; can be configured as a direct return or a reverse return

Volume factor provides the number of gallons of water per linear foot of a piping material

Volute portion of the circulator housing that carries water from the pump

Zone valve thermostatically controlled valve that opens and closes to regulate the flow of hot water to the terminal units in the occupied space

Zoning process of dividing the structure into separate areas, each of which has its own means to regulate the temperature in the space

THEORY OF HYDRONIC HEATING SYSTEMS

Unlike furnaces that use blowers to move heated air through duct systems, hydronic systems rely on circulating water or steam to deliver heat to the remote locations where heating is desired. Instead of using large ducts, as is the case with air conditioning or forced-air systems, water is heated at one location and then pumped to the remote locations via a piping arrangement. The heat is then transferred to the occupied space by heat exchangers located in the space. These heat exchangers, typically radiators or sections of baseboard, are referred to as terminal units. Once the water has passed through the terminal units, it is returned to the heat source, where it is reheated and again pumped through the piping arrangement. This cycle continues until the occupied space reaches the desired temperature. This system is referred to as a closed loop (Fig. 26-1) because the water circuit is closed to, or separate from, the atmosphere.

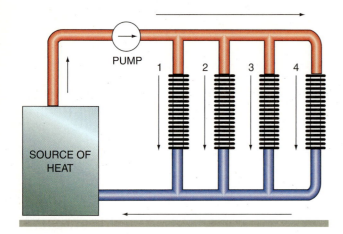

FIGURE 26-1 Hot water, two-pipe direct return hydronic system.

In hot water hydronic systems, many different possible piping configurations and controls can be used to satisfy the system requirements. A number of these components are common to all or most system configurations, but others are specific to certain configurations. As we make our way through the remainder of this chapter, we will discuss a variety of components, controls, and piping configurations designed to enhance system performance as well as meet the heating requirements for the structure.

CAUTION

CAUTION: Please note that Figure 26-1 represents only the water flow through a simple hot water hydronic system and does not contain other system components that are necessary for safe and proper system operation.

THE HEAT SOURCE

The heat used in hydronic systems can be generated by a number of different methods including burning fossil fuels; collecting solar energy from the sun; converting electrical energy directly into heat energy; and using vapor-compression, reverse-cycle refrigeration, or heat pumps. The most common method for generating the heat is by burning fuels such as natural gas, manufactured gas, or oil. The combustion processes for these fuels were discussed in earlier chapters, so please feel free to refer back to those sections for a brief review.

Heat energy is transferred from the heat source to the water in the **boiler**. The boiler facilitates the generation of the heat energy as well as the transfer of this heat energy to the water flowing through it. Figures 26-2, 3,

Courtesy of Weil-McLain.

FIGURE 26-2 Gas-fired boiler.

Courtesy of Weil-McLain.

FIGURE 26-3 Oil-fired boiler.

and 4 show gas, oil, and electric boilers. Boilers are typically constructed from either cast iron or steel. The most commonly encountered boiler in residential applications is the cast iron boiler.

FIGURE 26-4 Electric boiler.

Courtesy of Thermo 2000, Inc.

FIGURE 26-5 Cast iron boiler sections.

Courtesy of Weil-McLain.

It is often made up of individual sections (Fig. 26-5) that are bolted together to form the heat exchange surface between the water and the burning fuel. The more sections a boiler has, the higher its capacity, with all other factors remaining the same. Residential cast iron boilers typically hold between 10 and 15 gallons of water.

BOILER EFFICIENCY

Like furnaces, gas and oil boiler efficiencies are rated using the Annual Fuel Utilization Efficiency, or AFUE, rating. The AFUE is widely used to measure the efficiency of a heating appliance. It measures the amount of fuel that is actually used to provide heat to the structure compared to the amount of fuel that is supplied to the appliance. For example, a boiler that has an 80% AFUE rating converts 80% of the fuel that you supply to the appliance to usable heat, while the remaining 20% is lost from the system through the chimney. The AFUE does not take into account the electricity required to operate the boiler or the pumps that circulate the hydronic fluid. So, in addition to examining the AFUE, the HVAC installer or technician must also consider the power consumption of the systems electric components. Today's energy-efficient boilers have AFUE ratings that can range from 85% to 90%. Some very high efficiency condensing appliances can have AFUE ratings as high as 99%. Look for Energy Star-labeled models as an indicator of energy-efficient boilers.

Going Green

Boiler efficiency and system efficiency are not the same! Installing a 90% efficient boiler on an inefficiently designed and installed heating system will not produce high efficiency. It is important to keep in mind that the boiler is only a part of the system.

AQUASTAT

The **aquastat** (Fig. 26-6) is a temperature-sensing switch that is responsible for cycling the boiler on and off to keep the water in the boiler close to the desired temperature. The temperature of the water varies as the boiler cycles on and off. The temperature at which it turns on is called the cut-in temperature, and it shuts down at the cut-out temperature. The difference between the cut-in and cut-out temperatures is called the differential. If the desired water temperature is, for example, 170°F and the differential is 10°F, the boiler will cycle off once the water reaches 170°F and will cycle back on when it drops to 160°F (170°F − 10°F).

OUTDOOR RESET

The **outdoor reset** control is similar to the outdoor air, or holdback, thermostat found on heat pump systems, which will be covered in the next chapter. The reset thermostat senses the outdoor temperature and adjusts the water temperature in the boiler. When it is warmer outside, we

Photo by Eugene Silberstein.

FIGURE 26-6 Aquastat.

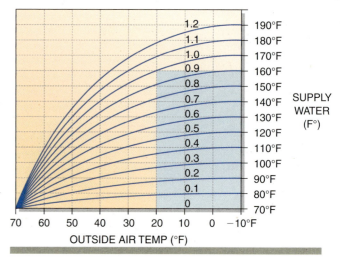

FIGURE 26-7 Representational reset control curve.

by the boiler. At the time of installation, the control must be set manually. Once set, the water temperature will change in response to the outside temperature, depending on the selected reset curve. A representational reset curve is shown in Figure 26-7. On this curve, at the 1.0 setting, the boiler will maintain a water temperature of approximately 160°F when the outside temperature is 20°F.

LOW-WATER CUTOFF

The low-water cutoff (Fig. 26-8) is responsible for de-energizing the boiler in the event the water level in the system falls below the desired level. In some states, a low-water cutoff is required by law, but it is good field practice to equip all systems with one. This is especially true for hot water boilers that serve radiant systems—covered later in this chapter—because the tubing is usually below the boiler and often buried in concrete. A leak in a buried tube can result in the draining of the boiler, which in turn, can result in the "dry firing" of the boiler.

will require less heat inside and vice versa. Therefore, as the outdoor temperature increases, the water temperature in the boiler is reduced. As it gets colder outside, the boiler maintains the water at a higher temperature.

The reset control measures two temperatures: the outside air temperature and, typically, the water supplied

Going Green

Outdoor reset devices control the water temperature that the boiler maintains based on the outdoor temperature. This way, the boiler does not operate as often nor as long when the outdoor temperature is higher. In 2012, it will be mandatory for all hydronic heating systems to have some type of reset control strategy.

Photo by Eugene Silberstein.

FIGURE 26-8 Low-water cutoff.

EXPANSION TANK

Hot water hydronic systems are closed loop systems and are, ideally, air free. As water is heated, it expands and, if the system is truly air free, the relief valve on the system will open to release the excess pressure. To prevent the relief valve from constantly opening and flooding the floor, an additional volume or space must be provided to accept the additional volume generated when the water is heated. The system component that accomplishes this is the expansion tank (Fig. 26-9). Ideally, the expansion tank is located on the supply side of the boiler near the inlet of the circulator. The location of the expansion tank with respect to the circulator is discussed in the next section.

There are two types of expansion tanks. One type of expansion tank is the standard expansion tank, which is nothing more than a large tank located above the boiler (Fig. 26-10). Initially, the air in the tank is at atmospheric pressure. As water is added to the boiler, it is pushed into the tank, which compresses the air in the tank, creating an air cushion. As the water is heated, the volume of the water increases, and more water is pushed into the tank, compressing the air even further (Fig. 26-11). The main problem with this type of expansion tank is that it eventually becomes water-logged, or completely filled with water, and the air cushion is removed. The absence of the air cushion eliminates the extra space that was

FIGURE 26-9 Expansion tank.

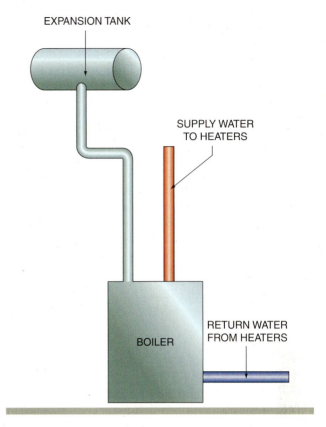

FIGURE 26-10 Steel expansion tank.

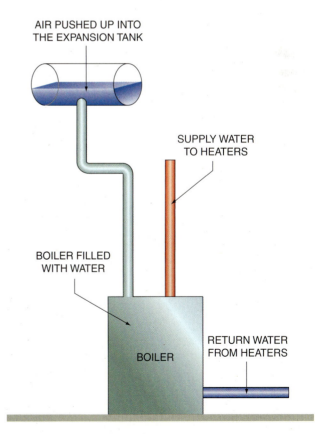

FIGURE 26-11 Pressure in the tank increases as water is added to the system.

previously available to the expanding volume of water. When this occurs, the pressure relief valve on the system will open, releasing the excess pressure.

The other type of expansion tank, which is used in nearly all new residential construction and installations, is the **diaphragm-type expansion tank**, which is divided into two sections separated by a rubber, semi-permeable membrane (Fig. 26-12). One side of the tank contains air and the other side is open to the water circuit. The air portion of the tank is pressurized by the manufacturer. Typically, for residential applications, a pressure of 12 psig is acceptable. The tank pressure at the time of manufacture is noted on the tank itself (Fig. 26-13). Note that the point at which the expansion tank is connected to the piping circuit will *always* remain at the same pressure. This is referred to as the point of no pressure change.

The **point of no pressure change** is the one place in the piping system where the pump cannot affect the pressure in the system. Consider a piping circuit that contains only the expansion tank and a pump (Fig. 26-14). Assume that the pressure in the expansion tank prior to installation is 10 psig and the pump generates a pressure difference of 15 psig between its inlet and outlet when operating. When the loop is filled, the water pressure will be 10 psig. When the pump cycles on, no water can be added or removed from the expansion tank. The water loop is filled and no additional water can be added to the loop, since water is not compressible. Water cannot leave the loop and enter the expansion tank, because this would create a vacuum in the loop, which would result in the water being pulled right back out of the expansion tank into the loop. Since water cannot be added to or removed from the expansion tank under the conditions just described, the pressure at the point where the tank is connected to the piping circuit cannot change.

Refer back to our example and the piping in Figure 26-14. If the pump is started, and it generates

from experience...

The pressure of the air in the expansion tank will increase when either the water is heated or additional water is introduced to the boiler. In either case, water will be forced into the expansion tank, increasing the pressure of the air in the tank. This will increase the pressure at the point in the water circuit where the expansion tank is connected to the piping circuit.

a pressure difference of 15 psig, a pressure of -5 psig is created at the inlet of the pump (Fig. 26-15). Since the pressure at the point of no pressure change will be 10 psig, the pump must make up its pressure difference

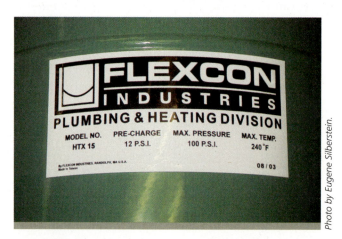

FIGURE 26-13 Expansion tank data tag.

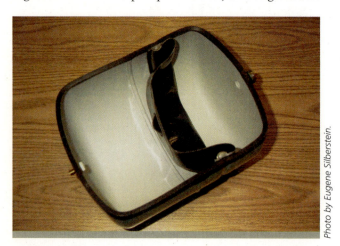

FIGURE 26-12 Cutaway of a diaphragm-type expansion tank.

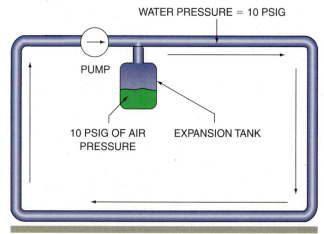

FIGURE 26-14 Simple loop with expansion tank and circulator.

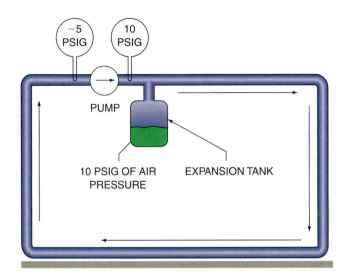

FIGURE 26-15 Pumping toward the expansion tank can result in a vacuum in the piping circuit.

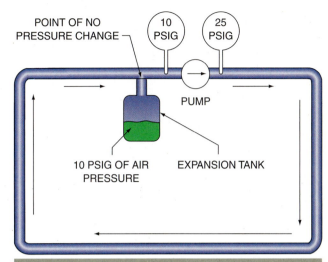

FIGURE 26-16 Point of no pressure change.

at the inlet of the pump. The vacuum created at the inlet of the pump will permit air to enter the piping circuit in the event of a leak.

Placing the pump on the other side of the expansion tank (Fig. 26-16) will eliminate this problem. The pressure at the inlet of the pump will be 10 psig and the pressure at the pump's outlet will be 25 psig. As the water circulates through the piping arrangement, friction will cause the pressure of the water to drop, and the pressure will once again be equal to 10 psig at the point of no pressure drop. Notice that, in this example, the pressure in the system has been increased to 25 psig, which prevents the piping circuit pressure from falling below atmospheric pressure. In addition, air will remain in solution at higher pressures, reducing the negative effects of air in the water circuit.

The pressure in the expansion tank will match the static fill pressure in the system. Static fill pressure is the pressure in the filled system before the boiler is fired. The actual required air-side pressure can be calculated by using the following formula:

$$P_a = (H_1 - H_2)(D_c/144) + 5$$

where,

P_a = Air side pressure in the expansion tank (psig)
H_1 = Height of the highest system pipe above the base of the boiler (ft)
H_2 = Height of the opening in the expansion tank above the base of the boiler (ft)
D_c = Density of the water at its initial, cold temperature (lb/ft^3)

Consider the following example in which cold water is introduced to the boiler at 60°F. The density of water at

> ## *from experience...*
> It is always good field practice to check the pressure in the expansion tank prior to installation. A bicycle tire gauge can be used to check the pressure and, if the pressure is too low, it can be raised by using a bicycle tire pump.

60°F is 62.4 lb/ft^3. The density of water between the temperatures of 50°F and 250°F can be found by substituting the temperature of the water for "T" in the following formula:

$$\text{Density} = 62.56 + 0.0003413\,(T) - 0.00006255\,(T^2)$$

The expansion tank opening is 6 feet above the base of the boiler, and the highest pipe is 22 feet above the base of the boiler. The required pressure in the expansion tank can be found using the following formula:

$$P_a = (22 - 6)(62.4/144) + 5 = 16(0.433) + 5$$
$$= 6.93 + 5 = 11.93 \text{ psig}$$

> ## *from experience...*
> Average-sized homes often require an expansion tank with an air side pressure of 12 psig.

Another factor that determines which expansion tank to use on a particular system is the volume of the expansion tank. The minimum required volume can be calculated as well, which requires an estimate of the volume of water contained in the system. An average-sized home with a cast iron sectional boiler often requires an expansion tank with a volume of 4 to 5 gallons.

Refer Procedures 26-1 and 26-2 on pages 728–730 for step-by-step instructions for estimating the volume of water in the system and calculating the minimum required expansion tank volume.

CENTRIFUGAL PUMPS

The pumps used to move water through hydronic systems are often called **circulators**. These circulators operate on centrifugal force and are, therefore, more accurately referred to as **centrifugal pumps**. The centrifugal pump (Fig. 26-17) is made up of a motor, a linkage, and an impeller (Fig. 26-18). The impeller slaps against the water and throws it toward the outside of the chamber or housing, called the **volute**, where it is forced from the pump assembly at a higher pressure. The pressure increase at the outlet of the pump pushes water around

FIGURE 26-17 Centrifugal pump.

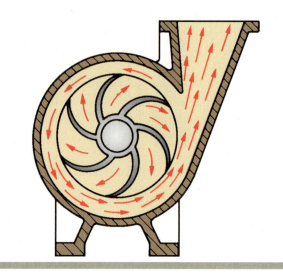

FIGURE 26-18 Impeller and volute on a circulator.

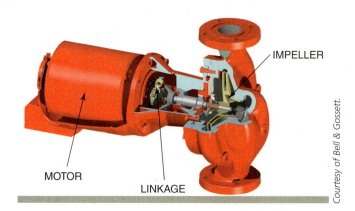

FIGURE 26-19 Cutaway of a centrifugal pump.

the piping circuit until it is eventually pushed back into the inlet of the pump by the very pressure differential created by the pump. A cutaway of a centrifugal pump is shown in Figure 26-19.

PUMPING CAPABILITY OF THE CENTRIFUGAL PUMP

The centrifugal pump is responsible for creating a pressure difference between the water at its inlet and its outlet. It is this pressure differential that facilitates water flow in the piping circuit. However, to ensure proper flow through the piping circuit, the pump must be able to overcome the resistance that exists in the piping itself. Consider a pump operating with no resistance to flow. If a pump was provided with an unlimited water supply at its inlet and was able to discharge the pumped water immediately to the atmosphere with no resistance at the pump's outlet, the volume of water moved by the pump would be the pump's maximum capacity (Fig. 26-20). Adding a section of vertical pipe to the pump's outlet would reduce the pumping capacity of the pump (Fig. 26-21). As the amount of vertical pipe at

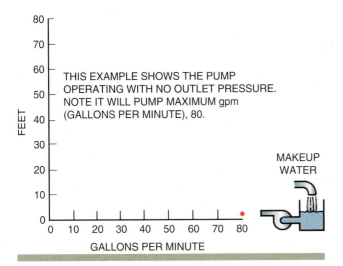

FIGURE 26-20 With no resistance at its outlet, the pump can move large volumes of water.

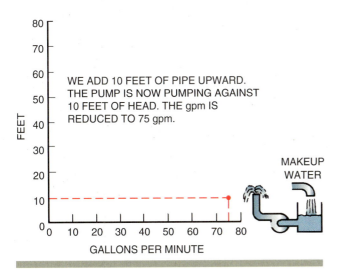

FIGURE 26-21 As the resistance at the pump's outlet is increased, the volume of water pumped will decrease.

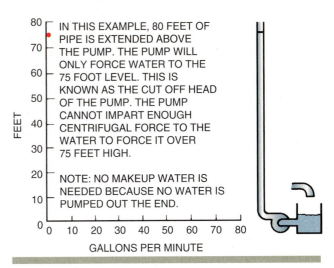

FIGURE 26-22 Point at which the pump can no longer overcome the resistance of the piping at its outlet.

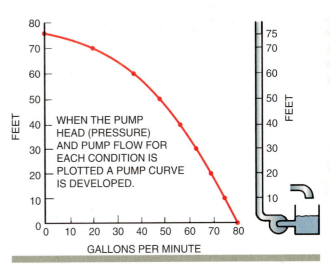

FIGURE 26-23 Sample pump curve.

the pump's outlet was increased, the capacity of the pump would continue to decrease until a point was reached at which the volume of water moved by the pump was zero (Fig. 26-22). The curve that is created when we plot the feet of pipe, or **feet of head**, against the pumping capacity of the pump in gallons per minute, is the performance curve for that particular pump (Figure 26-23).

PUMPING CAPACITY, FEET OF HEAD, AND PRESSURE DIFFERENTIAL

As a centrifugal pump operates, a pressure differential is created between the water at the inlet of the pump and the water at the outlet. A definite relationship exists between the pressure difference across the pump and the pumping capability of the pump in feet of head. In the chapter

on gas furnaces, we measured gas pressure at the furnace in units called inches of water column. It was established that 1 psig was equal to 27.7 inches of water column, which is equal to 2.31 feet (27.7 inches/12 = 2.31 feet). We can then see that 1 psig = 2.31 feet of head and that each foot of head results in a pressure change across the pump of 0.433 psig (1/2.31 = 0.433).

The pumping head of a particular pump can be determined using its performance curve if the flow rate through the pump is known. Assume that a pump has a capacity of 40 gallons per minute (gpm). From the sample performance chart in Figure 26-23, it is established that the pump is overcoming approximately 57 feet of head. We can then conclude that it will operate with a pressure differential of 24.7 psig between the pump's inlet and outlet (57/2.31 = 24.7 psig).

CENTRIFUGAL PUMP LOCATION

It has been established that the circulator should be located so that it is *pumping away* from the point of no pressure change. The point of no pressure change, discussed earlier, is the point where the expansion tank is connected to the system piping circuit (Fig. 26-16). This results in higher pressures in the system that will not only help keep air in solution, it will also reduce the chances of introducing water to the system as a result of lowering the system pressure below atmospheric pressure.

Going along with the pumping away theory, it is possible to install the circulator and the expansion tank on the return side of the boiler. The main problem with this is that once the pump is energized, the pressure at the outlet of the pump will be added to the pressure in the boiler and may cause the relief valve on the boiler to open. This is more likely to happen in larger structures but, to be on the safe side, it is good practice to install both the circulator and the expansion tank on the supply side of the boiler.

AIR VENTS AND AIR SEPARATORS

One of the biggest enemies of a hot water hydronic heating system is air. To operate properly, any air that may be present in the system must be removed. The **air separator** (Fig. 26-24), which ideally is located at the point of no pressure change, should be able to remove smaller air bubbles called microbubbles, from the system. Older air separators worked by sending the water through a straight, horizontal section of pipe, resulting in laminar or linear flow. When water flow is laminar, the air bubbles rise to the top of the pipe. These older air separators "scooped" the air out of the pipe by using a

baffle. New air separators don't depend on laminar flow. They work by a process called "collision and adhesion." Simply put, a metal mesh is placed in front of the flowing water (Fig. 26-25). The air bubbles collide with the metal and cling to it by surface tension. The air bubbles rise to the top of the air separator and leave from the top of the device.

The air then passes to the **air vents**, where it is removed from the system. The **automatic air vent** (Fig. 26-26) opens and closes automatically, but the "coin operated," or **manual air vent** (Fig. 26-27) must be opened manually.

PRESSURE-REDUCING VALVE (WATER-REGULATING VALVE)

The **pressure-reducing valve** (Fig. 26-28) automatically drops the pressure of the water entering the structure to the level at which the boiler is designed to operate. If the pressure in the boiler drops below the desired level, the valve opens to feed more water to the boiler. The **boiler/water feed valve** should be piped so that water

FIGURE 26-24 Air separator.

Courtesy of Bell & Gossett.

FIGURE 26-25 Wire screen in the air separator.

Photo by Eugene Silberstein.

FIGURE 26-26 Automatic air vent.

Courtesy of Bell & Gossett.

FIGURE 26-28 Pressure-reducing valve.

Courtesy of Bell & Gossett.

FIGURE 26-27 "Coin-operated" air vent.

Courtesy of Bell & Gossett.

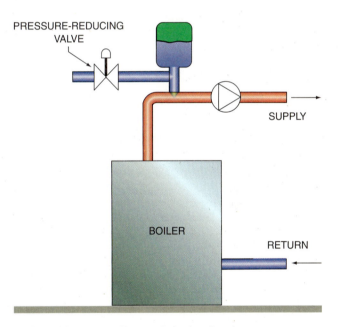

FIGURE 26-29 Proper location for the pressure-reducing valve.

is introduced to the system at the point of no pressure change (Fig. 26-29). Installing the feed valve on the return side of the boiler can result in system problems if the circulator is installed on the return side as well. Assume that the boiler piping configuration is like that shown in Figure 26-30—the air side of the expansion tank is pressurized to 12 psig, the feed valve has been factory set to feed water when the system pressure drops below 12 psig, the pump operates with a pressure drop of 6 psig, and there is a 1-psig pressure drop through the boiler.

When the pump starts, the pressure at the outlet of the pump will be 13 psig and the inlet pressure of the

pump will be 7 psig (Fig. 26-31). Since the feed valve is sensing a pressure below 12 psig, water will be fed to the boiler until the pressure is increased to 12 psig, the set point on the feed valve. This results in an increase of 5 psig to the boiler. This added water will be pushed into the expansion tank, because it has nowhere else to go, and will push against the air cushion, increasing the air pressure in the tank.

Once the circulator cycles off, the static pressure in the system will be 17 psig, which is 5 psig higher than the original fill pressure. This will cause the air pressure

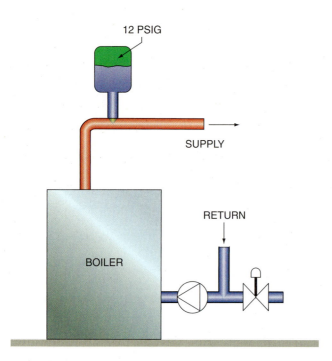

FIGURE 26-30 Potential problems can arise with this configuration.

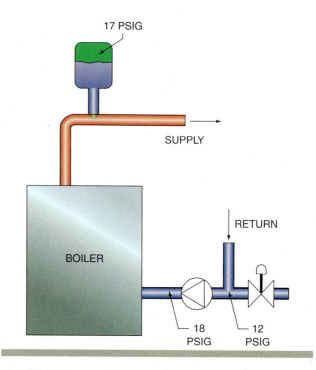

FIGURE 26-32 Static air pressure in the expansion tank is higher than desired.

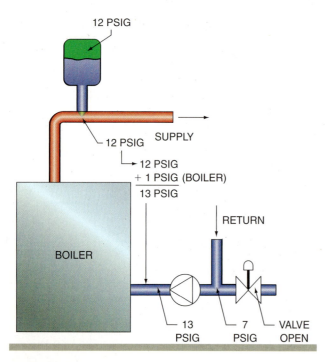

FIGURE 26-31 Feed valve will open as a result of a false reading.

Courtesy of Bell & Gossett.

FIGURE 26-33 Safety relief valve.

PRESSURE RELIEF VALVE

The **pressure relief valve** (Fig. 26-33) is designed to open if the pressure in the system reaches the set point on the valve. Once the valve opens, the pressure in the system

in the expansion tank, which was originally 12 psig, to increase to 17 psig (Fig. 26-32). In addition, the tank contains more water than originally was intended and is now undersized for the system. The increased pressure in the expansion tank puts more stress on the diaphragm and can result in premature failure of the tank.

will be relieved; when it drops to an acceptable level, the valve will close. The American Society of Mechanical Engineers (ASME) Boiler and Pressure Vessel Codes require that a pressure relief valve be installed on *every* hot water boiler. Since all hot water boilers have a pressure relief valve and most have a pressure-reducing valve, many manufacturers combine these two components into a single unit. A complete sample piping arrangement at the supply side of the boiler can be seen in Figure 26-34.

ZONE VALVES

Circulators can be used to circulate water through various loops in the structure. If water flows through all of these loops when the system is on and none of them when it is off, it is said that there is one zone in the structure. If, however, water flow can be controlled so that some

of the loops are getting heat while the others are not, it is said that the structure has multiple zones. In the case of a one-zone structure, it is likely that some parts of the structure will be too hot, some too cool, and others at the desired temperature. Zoned structures enable the occupants of each area to control the temperature of that particular area with a thermostat located there. Zoning, the process of dividing a structure into areas with separate control over the temperature in that space, is often done with zone valves (Fig. 26-35). Zone valves open and close in response to the temperature in the space. In operation, when one of the zones calls for heat, the zone valve opens and trips an end switch that starts the circulator. The circulator runs until the thermostat is satisfied. Once satisfied, the valve closes, turning off the circulator. Limitations and applications for zone valve usage are covered later in this chapter. Zone valves can also be used to control multiple zones being fed by a single circulator (Fig. 26-36).

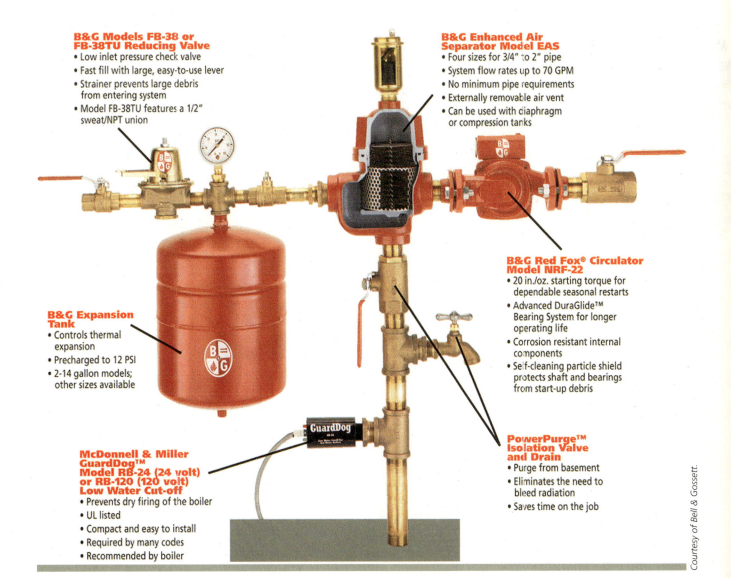

B&G Models FB-38 or FB-38TU Reducing Valve
• Low inlet pressure check valve
• Fast fill with large, easy-to-use lever
• Strainer prevents large debris from entering system
• Model FB-38TU features a 1/2" sweat/NPT union

B&G Enhanced Air Separator Model EAS
• Four sizes for 3/4" to 2" pipe
• System flow rates up to 70 GPM
• No minimum pipe requirements
• Externally removable air vent
• Can be used with diaphragm or compression tanks

B&G Red Fox® Circulator Model NRF-22
• 20 in./oz. starting torque for dependable seasonal restarts
• Advanced DuraGlide™ Bearing System for longer operating life
• Corrosion resistant internal components
• Self-cleaning particle shield protects shaft and bearings from start-up debris

B&G Expansion Tank
• Controls thermal expansion
• Precharged to 12 PSI
• 2-14 gallon models; other sizes available

McDonnell & Miller GuardDog™ Model RB-24 (24 volt) or RB-120 (120 volt) Low Water Cut-off
• Prevents dry firing of the boiler
• UL listed
• Compact and easy to install
• Required by many codes
• Recommended by boiler

PowerPurge™ Isolation Valve and Drain
• Purge from basement
• Eliminates the need to bleed radiation
• Saves time on the job

Courtesy of Bell & Gossett.

FIGURE 26-34 Hot water supply manifold.

Courtesy of Bell & Gossett.

FIGURE 26-35 Zone valve.

Courtesy of Bell & Gossett.

FIGURE 26-37 Flow control valve.

Photo by Eugene Silberstein.

FIGURE 26-36 Zone valves on a multiple zone loop.

FLOW-CONTROL VALVE

From the name, it may be concluded that the flow control valve (Fig. 26-37) controls flow in a hydronic system. However, the **flow-control valve** is actually a **flow check valve** that prevents hot water in the system from flowing through a heating loop when no flow through that particular loop is desired. If the individual loops have positive shutoffs, such as zone valves, flow control valves are not needed. If, however, water flow through the individual loops is controlled by a circulator pump, also discussed later, flow controls are needed to prevent gravity flow through the loops. Since gravity circulation can happen on either end of the secondary circuit, it is a good idea to use two flow control valves for each circuit; one on the supply and the other on the return. Always make sure the arrows on the flow-control valves face in the right direction.

from experience...

It is good field practice to install a valve after each flow-control valve. This can be either a full-port ball valve or a gate valve. This will make the future servicing of any of the components between the boiler and the flow-control valve much easier.

Photo by Eugene Silberstein.

FIGURE 26-38 Balancing valve.

BALANCING VALVES

In hot water hydronic systems, the water flow through each branch in the piping circuit must be within the desired range to help ensure that the occupied space is kept at the desired temperature. One way to help even the water flow through each branch of the circuit is to use **balancing valves** (Fig. 26-38). Balancing valves are installed in each branch of the circuit and are manually adjusted (Refer to later sections of this chapter for balancing applications). Since valves vary from manufacturer to manufacturer, review the literature supplied with the particular valve for setting procedures.

HYDRONIC HEAT EFFICIENCY

The hydronic heating system must be correctly designed beyond the boiler. Hydronic heating fluid can run through long lengths of pipe before reaching a terminal unit. Along the way, heat is lost through the walls of the pipes. When those pipes are routed through an unconditioned basement or crawlspace, the heat loss can be greater. To retain as much heat as possible, pipes need to be insulated. Building codes, specifically 2009 IRC, require that the pipes be wrapped with R-3 insulation to reduce the heat loss in pipes. And, while the building code calls for R-3 insulation, insulations with greater R values ranging from 5 to 11 are available and a better choice to improve the energy efficiency of the hydronic heating system.

Going Green

Uninsulated heating pipes result in substantial heat loss. Be sure to insulate pipes that are run through unconditioned spaces.

HYDRONIC SYSTEM PIPING CONFIGURATIONS

Having discussed the component parts of a hydronic heating system, we will now put the pieces together and examine four commonly encountered piping configurations. The piping configuration that will ultimately be used in a structure depends on a number of factors that include budgetary restrictions, the uniformity of heat loss by the structure, and the need or desire for uniform heating in the space. The four common piping configurations that will be discussed here are the

- Series Loop System
- One-Pipe System
- Two-Pipe Direct Return System
- Two-Pipe Reverse Return System

SERIES LOOP SYSTEM

In a series loop system (Fig. 26-39), all of the heaters are piped in series with each other. Similar to a series electric circuit, the water flow through any given heater will be exactly the same as through any other heater in the series loop. Remember that the current, or

from experience...

Zone valves cannot be used on a series loop system. If zone valves are installed on this type of system, hot water will flow through the loop *only if all* of the zone valves are in the open position.

flow, in a series circuit is the same at all points in the circuit. The main advantage of the series loop system is that it is very economical from a first-cost standpoint. The main drawback is that the terminal units that are fed last are often cooler than those that are fed first. This is because as heat-laden water flows through the loop, Btus are transferred from the water to the air, thereby lowering the temperature of the water in the piping circuit.

ONE-PIPE SYSTEM

In a one-pipe system (Fig. 26-40), one main piping loop extends around, or under, the occupied space and connects the supply of the boiler back to the boiler's return. Each individual **terminal unit** is connected to the main supply loop with two tees. (More on the tees later). The one-pipe system is so named because the same pipe supplies water to the terminal units and carries the return water back to the boiler.

RESISTANCE TO WATER FLOW

Consider a portion of the piping circuit that includes one terminal unit and the portion of main loop piping between the tees shown in Figure 26-41. The water

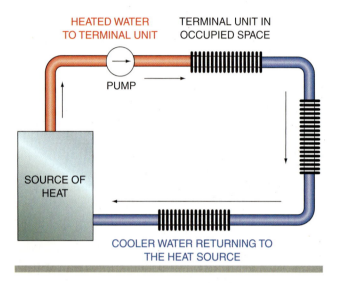

FIGURE 26-39 Series loop piping circuit.

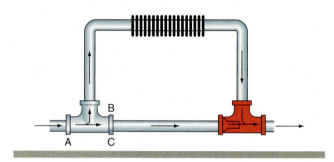

FIGURE 26-41 Common piping between the tees. Diverter tee located at the return side of the terminal unit.

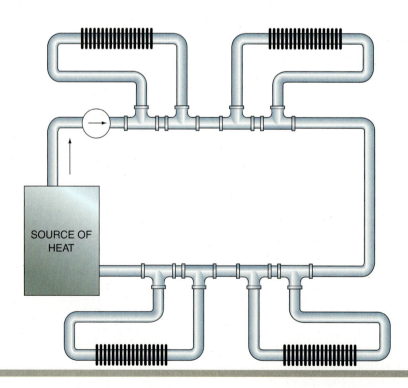

FIGURE 26-40 One-pipe system.

entering the tee at point A flows through the tee and can leave the tee through either point B, which leads to the terminal unit, or through point C, which allows the water to continue through the main water path connected between the boiler's supply and return pipes. Just as electric current takes the path of least resistance in a parallel circuit, the amount of water flow through the terminal unit is determined by the relationship between the resistance in the terminal unit piping circuit and the resistance in the section of main loop piping between the tees.

For example, if the resistance in the terminal unit branch is three times that of the main line loop section, the amount of water flow through the terminal unit branch will be one-third that through the bypass, or main loop. If the water flow through the main circuit is 4 gallons per minute, 3 gallons per minute will flow through the main loop, while only 1 gallon per minute is permitted to flow through the terminal unit (Fig. 26-42). It is, therefore, possible for all or most of the water to bypass the terminal unit branch if the resistance of that circuit is much higher than the resistance of the piping section between the two tees.

A number of factors must be considered when laying out or installing a **one-pipe hot water** system. These factors include:

- The length of the terminal unit branch circuit
- The length of the piping between the tees (the distance between the two tees)
- The size of the piping material in the branch circuit
- The size of the piping between the tees

When dealing with one-pipe hot water systems, always be aware of the following:

✓ The longer the pipe, the more resistance to flow there will be

✓ The shorter the length of pipe, the lower the resistance will be

✓ The smaller the diameter of the pipe, the more resistance to flow there will be

✓ The larger the diameter of the pipe, the less resistance there will be

✓ The further apart the tees are from each other, the more resistance to flow there will be between them

✓ The closer the tees are to each other, the lower resistance to flow there will be between them

THE DIVERTER TEE

The **diverter tee** (Fig. 26-43), also referred to as a **mono-flo tee,** is designed to increase the resistance to water flow in the main water loop. This increase in resistance will direct more water through the terminal unit branch circuit. The diverter tee is constructed with a cone inside (Fig. 26-44) that, in essence, reduces the diameter of the pipe in the main loop, increasing its resistance. This

FIGURE 26-43 Diverter tee.

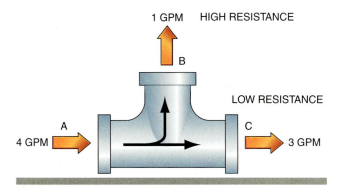

FIGURE 26-42 The higher resistance at point B results in lower flow through that branch, while the lower resistance at point C results in greater flow through that branch.

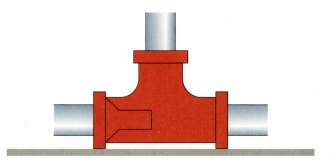

FIGURE 26-44 Cross-sectional view of a diverter tee.

pushes more water through the terminal unit loop. A typical loop using the diverter tee is shown in Figure 26-41 at the outlet of the terminal unit. When working with diverter tees, remember the following:

✓ If the terminal unit is located above the main hot water loop and the length of the piping in the terminal unit branch is not excessive (not too long with respect to the main hot water line), one diverter tee should be used on the return side of the terminal unit (Fig. 26-41).

✓ If the terminal unit is located above the main hot water loop and the length of the piping in the terminal unit branch is excessive (too long with respect to the main hot water line), two diverter tees should be used—one on the supply side of the terminal unit and one on the return side of the terminal unit (Fig. 26-45).

✓ If the terminal unit is located below the main hot water loop, two diverter tees should be used: one on the supply side of the terminal unit and one on the return side of the terminal unit (Fig. 26-46).

✓ If the terminal unit is above the main hot water loop, the tees should be no closer than 6 or 12 inches, depending on the manufacturer. (Bell and Gossett allows them to be as close as 6 inches, but Taco requires 12 inches.)

✓ If the terminal unit is below the main hot water loop, the tees should be spaced as wide as the ends of the terminal unit or radiator.

✓ If there are multiple terminal units and some are above and some are below the main hot water loop, it is best to alternate the tees. For example, from left to right: supply upper, supply lower, return upper, return lower (Fig. 26-47).

✓ Manufacturers place markings on the outside of their diverter tees so that the location of the inner cone can be identified. Be sure to install the diverter tees as recommended by the installation literature.

SIZING DIVERTER TEE SYSTEMS

When sizing the diverter tees, it is important to know the total heating load, the requirements of each **radiator** in the system, and the temperature drop from supply to return. For residential applications, a 20°F temperature drop is typical. On a hot water system with a 20°F temperature drop, each gpm of water flow will carry approximately 10,000 Btu/hr of heat.

Assume there is a system with a total heating load of 100,000 Btu/hr with a 20°F temperature drop between supply and return water temperatures. Since 1 gpm will carry about 10,000 Btu/hr, this system will have a water flow of 10 gpm. If the first radiator requires 20,000 Btu/hr, there should be 2 gpm of water flow through that branch (Fig. 26-48). The diverter tee that will facilitate the desired flow should be selected.

Once the water flows through the first radiator, the temperature of the water in the main loop after the second tee will be somewhat lower than the water in the main loop before the first tee. The temperature of the

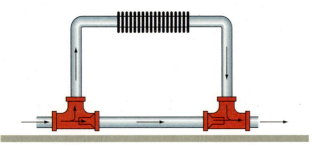

FIGURE 26-45 Two diverter tees are needed if the radiator is located above the main loop and there is significant resistance in that branch.

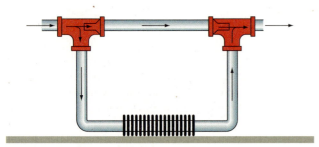

FIGURE 26-46 Two diverter tees are needed if the radiator is located below the main loop.

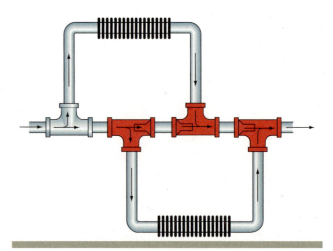

FIGURE 26-47 Alternate tees if there are terminal units both above and below the main loop.

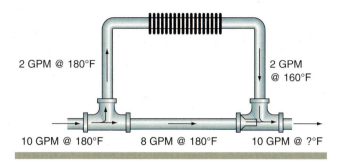

FIGURE 26-48 A flow of 2 gpm is needed through this terminal unit.

water in the main loop after the second tee can be calculated using the following formula:

$$(\text{Flow 1}) \times (\text{Temp 1}) + (\text{Flow 2}) \times (\text{Temp 2})$$
$$= (\text{Flow 3}) \times (\text{Temp 3})$$

where:

Flow 1   = the flow of liquid 1 entering the tee
Temp 1 = the temperature of liquid 1 entering the tee
Flow 2   = the flow of liquid 2 entering the tee
Temp 2 = the temperature of liquid 2 entering the tee
Flow 3   = the flow of liquid 3 leaving the tee
Temp 3 = the temperature of liquid 3 leaving the tee

Assuming that the supply water is 180°F and the water leaving the radiator is 160°F, we get the following:

$$(8 \text{ gpm}) \times (180°F) + (2 \text{ gpm}) \times (160°F)$$
$$= (10 \text{ gpm}) \times (\text{Temp 3})$$
$$\text{Temp 3} = (1440 + 320) / 10$$
$$\text{Temp 3} = 176°F$$

These results are summarized in Figure 26-49.

TWO-PIPE DIRECT RETURN

Two-pipe systems use two pipes (Figure 26-50). One pipe, the supply, carries water to the terminal units, while the other, the return, is responsible for bringing the water

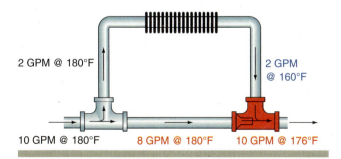

FIGURE 26-49 The temperature of the water leaving the tee is 176°F.

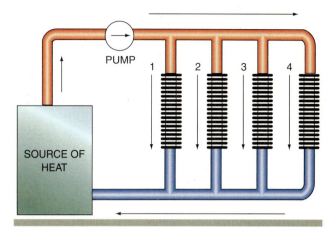

FIGURE 26-50 Two-pipe direct return system.

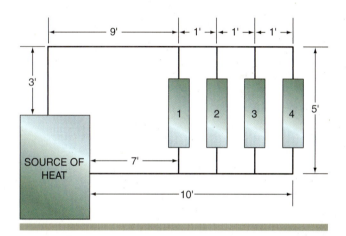

FIGURE 26-51 Water flowing through radiator #1 must travel 24 feet, while water flowing through radiator #4 must travel 30 feet.

back to the boiler. In a two-pipe **direct return** system, the terminal unit that is fed with hot water first is also the first to return its water to the boiler. Simply stated, the terminal unit closest to the boiler offers the least amount of resistance to water flow because that particular circuit or branch is the shortest. This statement assumes that all terminal units, **baseboard sections**, or radiators in the piping arrangement are identical to each other.

Consider the two-pipe direct return system in Figure 26-51. Water flowing through terminal unit 1 will have to travel a distance of 24 feet after leaving the boiler before returning to the boiler. Water traveling through terminal units 2, 3, and 4 will have to travel 26, 28, and 30 feet, respectively. Terminal unit 1 will therefore have the most water flow through it, given the lower resistance of that piping branch circuit.

To generate even and equal water flow through each terminal unit, balancing valves (Fig. 26-33) can be piped in series with each terminal unit (Fig. 26-52).

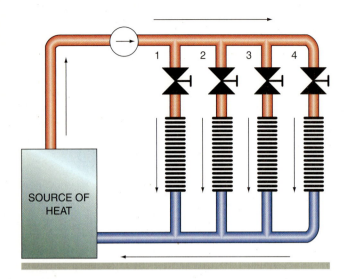

FIGURE 26-52 Balancing valves used to balance flow through the radiators.

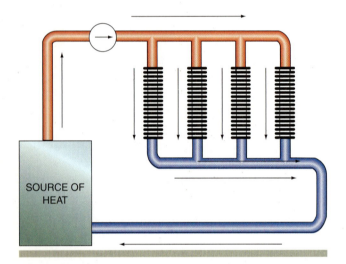

FIGURE 26-53 Two-pipe reverse return system.

These valves are manually adjustable and, when properly adjusted, should be left alone. The valve in series with heater 1 will be closed a little more than the valve in series with heater 2, which in turn, is closed a little more than the valve in series with heater 3, and so on. Closing the valves closer to the boiler will direct more water to the heaters located further from the boiler.

TWO-PIPE REVERSE RETURN

In a two-pipe **reverse return system** (Fig. 26-53), the first heater supplied with hot water is the last heater to return its water to the boiler. In other words, the heater with the shortest supply line from the boiler will have the longest return line. Configured in this manner, the distance traveled by the water through the individual heaters will be exactly the same.

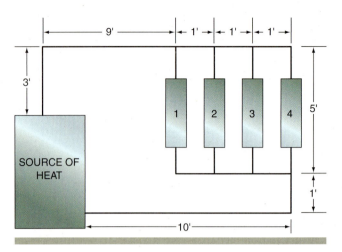

FIGURE 26-54 The length of the piping through any one branch is exactly the same as the length through any other branch.

Consider the two-pipe reverse return system in Figure 26-54. By adding the lengths of pipe between the boiler and terminal unit 1 and from terminal unit 1 back to the boiler, the water must travel a distance of 31 feet ($3' + 9' + 5' + 1' + 1' + 1' + 1' + 10' = 31'$). Similarly, the distances through terminal units 2, 3, and 4 are all 31' as well. Assuming that all terminal units in the piping arrangement are the same, the flow through each heater will be the same and no balancing valves will be needed.

PRIMARY-SECONDARY PUMPING

In primary-secondary piping, there are two distinct piping circuits just as there were two piping circuits in the one-pipe arrangement that used diverter or monoflo tees (Fig. 26-40). One main circuit flows only through the boiler, while the other flows through the boiler and one or more of the terminal units. The two circuits meet at the tees. The piping between the tees is common to both circuits. In a one-pipe system, there is just one circulator, so primary-secondary flow takes place.

The further apart the tees are placed, the more likely it is that water will flow from the boiler's circuit to the convector's circuit. The more resistance there is in the main loop between the tees, the greater the water flow through the convector branch. It is the resistance in the common piping that affects the flow through the parallel branch. The key to understanding primary-secondary

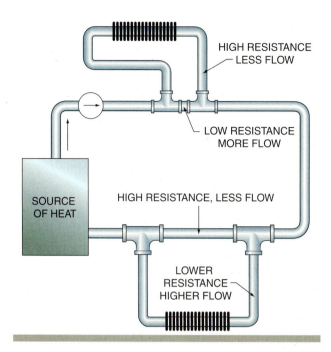

FIGURE 26-55 More resistance between the tees results in more flow through that radiator and vice versa.

pumping is understanding the resistance in the piping common to both circuits. Higher resistance results in a greater pressure drop in that circuit, just as higher resistance in an electric circuit results in a higher voltage drop across the resistance. If there is a large pressure drop along the run between the two tees, more water will flow into the branch, which is why diverter tees have the built-in restriction. Similarly, if the resistance between the tees is very low, there will be nearly no flow through the convector loop (Fig. 26-55).

PRIMARY-SECONDARY COMMON PIPING

The transition from primary-secondary flow, as in the case of the one-pipe system, to primary-secondary pumping is made by eliminating the pressure drop between the primary and the secondary circuits. To accomplish this, standard tees are used instead of diverter tees, and the tees are piped close together. The section of piping located between the tees is, as mentioned previously, the common piping, which is a part of both loops.

As a rule, the shorter the common piping, the better the primary-secondary system will work. Ideally, the common piping shouldn't be more than two feet long, but it can be as short as is physically possible. If the common piping is longer than two feet, the resistance of the common piping will increase. That's why the common piping should be as short as possible.

PRIMARY-SECONDARY CIRCUIT PIPING

Each loop in the circuit piping contains a circulator. That circulator could serve something that's either putting heat into the system, such as a boiler, or taking heat out of the system, such as a radiator. Each circulator takes care of *only* the circuit in which it finds itself. This means that even if your system is large, your circulators are probably going to be small. The short length of common piping makes this possible. As long as the common piping is kept short, each circulator will run without being aware that there are other circulators operating within that system. You can mix and match circulators of different sizes and none of them will affect any of the others. The primary circulator, however, will run only if the secondary circulator starts. The two always run together, and the primary's job is to make hot water available to the secondary. The secondary circulator usually removes only a portion of what the primary supplies. The rest of the primary flow continues on and mixes with the water that is returning from the secondary circuit. The result is that the water that flows back into the boiler is hotter than it would be if the secondary flow returned by itself. This layout gives systems designers a lot of freedom to be creative, as the following example shows.

Figure 26-56 shows a primary circulator in the main loop and a secondary circulator that feeds water to the terminal units. The secondary circulator will draw hot water from the common piping, just as it would from

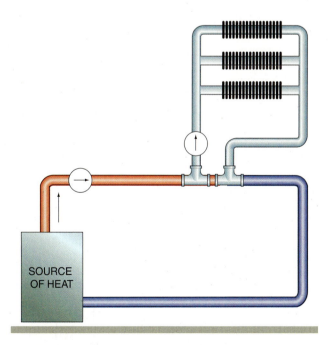

FIGURE 26-56 Primary-secondary arrangement where one circulator feeds three radiators in the secondary.

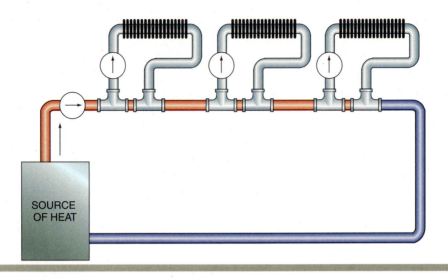

FIGURE 26-57 Primary-secondary arrangement where there are three separate secondary circuits, each with its own circulator.

a boiler, and pump that hot water through the three radiators that it serves. The secondary pump treats the main loop as the heat source or boiler. It takes hot water from the primary circuit, sends it through the radiators as needed, and returns the cooler water to the primary circuit. In addition, the secondary pump sees only the pressure drop of the piping that goes to and from those three radiators.

Another possible piping arrangement is shown in Figure 26-57. The piping in each loop is relatively simple, and the circulators in each loop are small because all they have to take care of is that particular loop. This results in lower pressure drops throughout the system and reduces the velocity of the water flowing through the loops. Even if the system were very large, the circulators would stay small.

THE CIRCULATOR PUMPS

We have established that the primary circulator is responsible for moving water through the main loop, while the pumps in the loops are responsible for moving water through the secondary loops. The pressure drop in the primary loop will be relatively low, given that the resistance between the tees is kept low. If the main loop circulator is on and the secondary loop circulators are off, water will circulate through only the primary loop. As water is pumped toward a tee, the water will bypass the loop since the resistance between the tees is much less than the resistance of the loop. The pressure drop through the common piping is practically nonexistent compared with the pressure drop through the secondary loop, so the primary flow stays in the primary circuit until the secondary circulator starts.

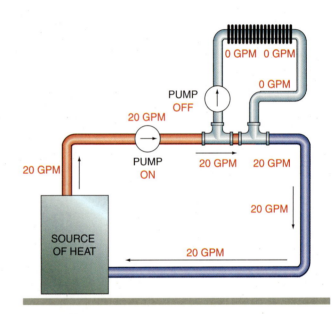

FIGURE 26-58 When the secondary pump is off, there is no flow in the secondary loop.

Once the circulator in the secondary loop starts, water will flow through that loop. How much water will flow depends on the pumping capacity of the circulators. Consider a primary circuit with a 20 gpm circulator and a secondary circuit with a 10 gpm circulator (Fig. 26-58). A short length of common piping connects them. The primary circulator is on and the secondary circulator is off. There is a water flow of 20 gpm into the first tee in the common piping. If 20 gpm of water enters the tee, then 20 gpm must come out. Since the secondary circulator is off, all of the flow will take place in the primary loop. No water is going to flow through the secondary loop because the secondary circulator is off and the resistance of that loop is much higher than the resistance of the common piping.

If the secondary circulator is turned on, too, 20 gpm is still flowing into the first tee on the common piping, but the flow splits with the secondary circulator running (Fig. 26-59). The secondary circulator moves 10 gpm of water up to the secondary circuit, while the other 10 gpm goes straight through the common piping. At the second tee, the 10 gpm from the secondary loop combines with the 10 gpm from the common piping to make 20 gpm at the outlet of the tee that returns the water to the boiler.

Now let's assume that both circulators have the same flow rate—20 gpm. When the primary circulator runs and the secondary circulator is off, the water will all flow across the common piping and continue on through the primary circuit (Fig. 26-60). Once again, the resistance of the piping circuit is responsible for this. When the secondary circulator starts, it draws all the water out of the first tee and sends it around the secondary circuit. No water flows across the common piping. The entire 20 gpm enters the second tee in the common piping and continues on its way (Fig. 26-61).

Finally, consider that the 10 gpm circulator is the primary circulator and the 20 gpm circulator is the secondary circulator. When the primary circulator is on and the secondary circulator is off, all the water will flow along the primary. No water moves through the secondary circuit, because the length of pipe the two have in common is so short. When the secondary pump starts, the secondary circulator will draw 20 gpm out of the first tee in the common piping (Fig. 26-62). The 20 gpm is made up of the 10 gpm entering the first common tee from the primary circulator and another 10 gpm from its own circuit. This portion of the 20 gpm is moving *backwards* across the common piping.

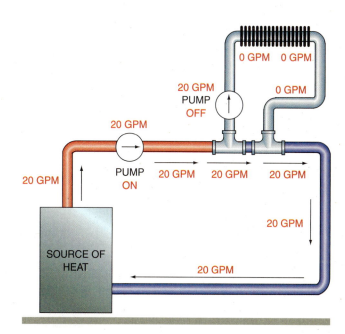

FIGURE 26-60 When the secondary pump is off, there is no flow in the secondary loop.

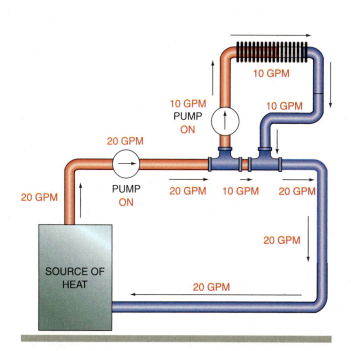

FIGURE 26-59 Water flows through the secondary loop when both pumps are on.

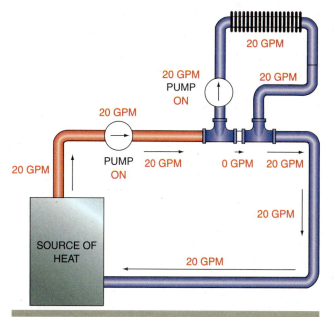

FIGURE 26-61 When both pumps are on and they have the same pumping capacity, there is no flow through the common piping.

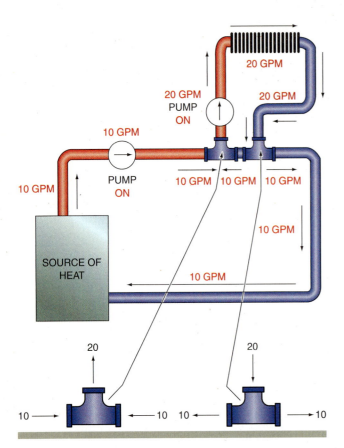

FIGURE 26-62 Water can flow through the common piping in either direction, depending on the pumping capacity of the pumps.

> *from experience...*
> Refer back to the section on one-pipe systems for the formula used to calculate the temperature of the water that leaves a tee.

MIXING VALVES IN PRIMARY-SECONDARY PUMPING

One of the many benefits of primary-secondary piping systems is that each loop can be used to supply water at different temperatures to different zones, depending on the system requirements. For example, one part of the structure may be heated with baseboard and need 180°F water, and another portion of the structure may

have radiant heat, covered later in this text, and require water at 110°F. The tees in the primary-secondary piping circuit act as mixing valves. The temperature of the water leaving a tee can be calculated in a manner similar to that used in the one-pipe system using diverter tees. Just like hot water and cold water are mixed in the kitchen sink to supply warm water, the water leaving the tee will be cooler than the hot water and warmer than the cold water. Carefully selecting the size of the pumps in both the main loop and the secondary loops will help achieve the desired water temperatures in each loop.

EXPANSION TANKS IN PRIMARY-SECONDARY SYSTEMS

It was determined earlier that the best configuration for the expansion tank and circulator was to have the circulator on the supply side of the boiler, pumping away from the point of no pressure change at the expansion tank. In primary-secondary systems the same holds true, so it may seem logical to have an expansion tank in each circuit that has a circulator. Such is not the case. The primary pump is pumping away from the **compression tank**, but the secondary circulator does not have a separate compression tank in its circuit.

When the water in the secondary circuit gets hot and expands, the "extra" water moves into the primary circuit, where the compression tank is. Because of this, the common piping will always be the secondary circulator's "point of no pressure change" or, to put it another way, its "compression tank." Therefore, whenever possible, systems should be piped so that the secondary circulator pumps *into* the secondary circuit, and away from the common piping.

DOMESTIC HOT WATER

In addition to providing the heat for a hydronic heating system, a boiler can also be used to heat the domestic hot water using either an **indirect water storage** tank or a **tankless hot water heating coil**. The indirect hot water storage tank has an internal heat exchanger that hydronic fluid from the boiler flows through on a separate zone loop (Fig. 26-63). An aquastat in the storage tank controls the boiler. A tankless hot water heating coil is a heat exchanger installed in the hydronic fluid

FIGURE 26-63 Indirect water heater. The white lines highlight the domestic heater path. The red lines highlight the boiler zone that heats the water in the tank.

Photo by Eugene Silberstein.

FIGURE 26-64 Tankless water heater circuit.

Photo by Eugene Silberstein.

jacket of the boiler that connects to the hot water pipes of the house (Fig. 26-64). The benefit of using a boiler to provide both the heat for the hydronic heating system and domestic hot water is that its one heating appliance serving two functions. The boiler eliminates the need for a separate hot water heater.

RADIANT HEATING SYSTEMS

The concept of radiant heat differs a great deal from that of forced air, which was discussed in earlier chapters. Forced air, or convective, heating systems rely on blowers to create convection currents that carry the heat-laden air to the desired location through duct systems. **Radiant heat systems**, on the other hand, concentrate on heating the shell of the structure or occupied space as opposed to the air within that space. In addition, the calculations for radiant heating are entirely different from those for convective heating. The purpose of convective heating is to determine the *rate of heat loss from the room* by conduction, convection, and radiation when maintained in the desired condition; radiant heating involves the *regulation of the heat loss per square foot from the human body.*

THE HUMAN BODY IS A RADIATOR

Under normal conditions, the human body produces approximately 500 Btuh. Each Btu is the amount of heat it takes to raise the temperature of one pound of water one degree Fahrenheit. Refer back to the beginning of the text for more on this. The body only requires 100 Btu to remain alive. The rest of this heat must be shed, making the body act as a radiator. By controlling the rate at which we shed heat, we can control our comfort level. Giving up heat too fast will make us feel cold; giving up heat slowly or not at all will make us feel hot and uncomfortable.

COLD 70

Cold 70 is the feeling one gets when there's a big difference between the temperature of the air in a room and the temperature of the surfaces in that room. This phenomenon is very common and you experience it every time you do your grocery shopping. In your neighborhood supermarket, you will find the cereal aisle, the frozen-food aisle, and the deli counter. The temperature in each of the aisles is exactly the same, but you feel cold in the frozen food aisle. That's the phenomenon of Cold 70. The air temperature in the frozen food aisle and the temperature of the surrounding surfaces in the same aisle are wildly different from each other, and your body is very quick to give up heat to the cooler surfaces in that aisle. This is what makes you cooler in the frozen food aisle. In a similar manner, you feel warmer when you get close to the chicken rotisserie at the deli counter. That's because the rotisserie is hotter than you are. It's radiating heat at you and it's also preventing your body from losing its own heat, making you uncomfortably warm. The concepts behind radiant heat are intended to provide ideal comfort.

WHAT IS IDEAL COMFORT?

We have seen that our bodies react to the temperatures of the surfaces surrounding us. If our bodies can shed and absorb heat at a controlled rate, we will be comfortable. Typically, a room that is about 68°F permits us to shed the extra 400 Btuh that our body generates. Temperatures higher than that will prevent us from releasing the heat, resulting in an uncomfortable feeling. On the other hand, temperatures colder than 68°F will cause the rapid release of heat from our body, making us feel cold. Ideally, then, the occupied space should be approximately 68°F, with the exception of the areas

closer to the floors and ceiling. Since the human body loses a great deal of heat from the feet, the floor should be somewhat warmer, about 85°F, and the space by the ceiling can be cooler. Figure 26-65 shows a cutaway of a room and the temperature changes as we go from the floor to the ceiling. This represents an ideal temperature pattern, which is very closely simulated by radiant heat. The heating patterns generated by baseboard heating and forced-air heating systems are similar to those shown in Figure 26-66.

THE RADIANT SYSTEM

Like hot water hydronic systems, radiant heating systems rely on the circulation of hot water through a series of piping arrangements located in the occupied space. There are three main differences between the two systems. First is visibility. In radiant heating systems, the piping used to carry the heat-laden water is hidden in

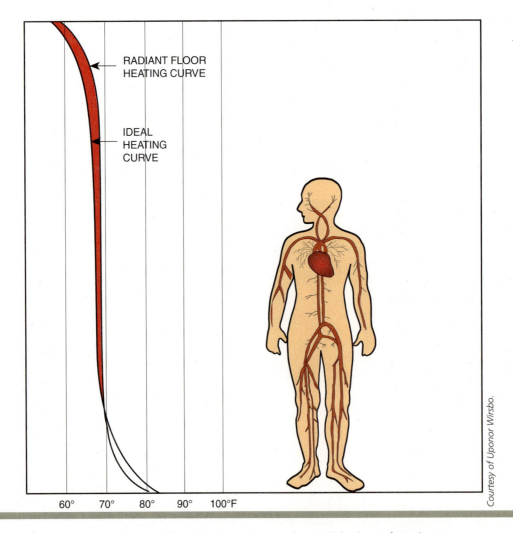

RADIANT FLOOR HEATING CURVE

IDEAL HEATING CURVE

60° 70° 80° 90° 100°F

Courtesy of Uponor Wirsbo.

FIGURE 26-65 Heating curve for a radiant system compared to an ideal comfort chart.

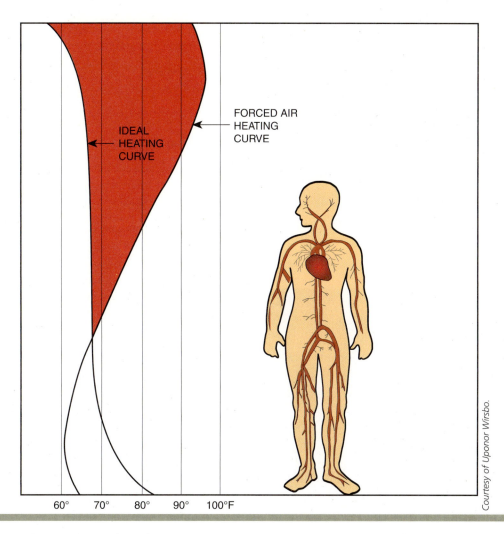

Courtesy of Uponor Wirsbo.

IDEAL
HEATING
CURVE

FORCED AIR
HEATING
CURVE

60°   70°   80°   90°   100°F

FIGURE 26-66 Heating curve for a forced-air heating system compared with an ideal comfort chart.

the floors, walls, and/or ceilings of the structure and is, therefore, not visible to the occupants. The second major difference is the piping material used. Instead of using copper pipe to carry the hot water to the occupied space, radiant heating systems often use plastic **polyethylene tubing**, more commonly known in the industry as **PEX tubing**. PEX tubing (Figure 26-67) is flexible and can be installed easily using a minimum of fittings (Fig. 26-68). The third difference is the temperature of the water required by the system. Typical hot water hydronic systems circulate water in the 180°F range. Radiant heating systems circulate water with an average temperature of 110°F.

RADIANT HEATING PIPING

Most radiant heating piping systems being installed in new residences are located in the floor, but radiant heat can be installed in walls and ceilings as well. This text

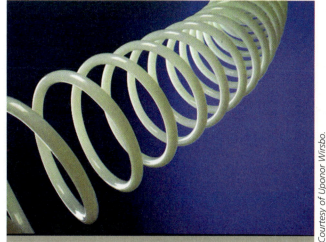

Courtesy of Uponor Wirsbo.

FIGURE 26-67 PEX tubing.

(A)

(B)

will concentrate on floor installation. Following are a few options for installing the tubing for radiant systems:

- Slab on grade
- Thin slab
- Dry

SLAB ON GRADE

The slab-on-grade configuration is the popular choice for new construction projects that will have concrete floors. Since the concrete has not yet been poured, the tubing for the radiant heat loops can be located within the concrete slab itself. The PEX tubing is secured to the steel reinforcement mesh in the floor using either wire straps or plastic clips (Fig. 26-69) to hold it in place when the concrete is poured. Generally, the tubing will be on 12-inch centers on these jobs. Near a cold surface such as a window, the distance between the tubes should be less. Insulation should be placed under the concrete slab to increase the effectiveness of the system. If possible,

from experience...

When securing PEX tubing to the steel reinforcement mesh, be sure to use clips or straps that are approved for use with the particular tubing being used. Using incorrect strapping materials can result in chemical reactions that will damage the tubing.

(C)

FIGURE 26-68 Fittings used to connect PEX tubing.

Courtesy of Uponor Wirsbo.

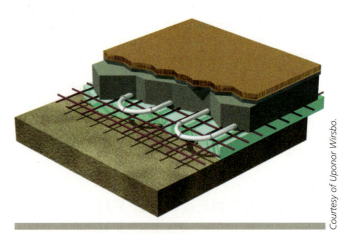

FIGURE 26-69 Cutaway of a slab-on-grade radiant piping layout.

Courtesy of Uponor Wirsbo.

at least one-inch thick polystyrene should be used. The edges of the concrete slab should have more insulation, since the heat tends to migrate toward the cooler walls. The goal is to keep the heat toward the center of the room where the people are.

In addition, there should always be a vapor barrier under the slab. It can be polyethylene and should be at least six-mil thick. The main purpose of the vapor barrier is to keep the ground water from robbing the slab of its heat. The distance between the tubing and the top of the slab is not critical, but the lower the tubing is from the top of the slab, the higher the water temperature will have to be.

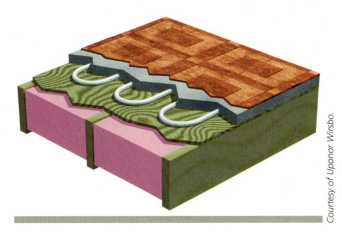

Courtesy of Uponor Wirsbo.

FIGURE 26-70 Cutaway of a thin-slab radiant piping layout.

from experience...

While pouring the concrete, it is good field practice to pressurize the tubing to at least 50 psig. This will prevent the tubing from collapsing and will also make finding leaks much easier. It is better to pressurize with air rather than water, if at all possible, since leaking air will not affect the composition of the concrete.

> **CAUTION**
>
> **CAUTION:** Consult the architect or engineer on the job when laying out a poured concrete radiant heat installation in an existing structure. An inch and a half of concrete will add about 18 pounds per square foot to the load of the floor. If the structure is not capable of handling this weight, other options should be explored.

UNDER-SLAB INSULATION

The Earth's temperature can be in the low to mid 50°F range even in moderate climates. Radiant floor hydronic fluid temperatures typically range from 90°F to 110°F, so heat energy will flow from the slab to the cooler ground. A minimum of 4″ of rigid foam insulation is needed beneath radiant-heated slab to prevent excessive heat loss and improve the efficiency of the system. The heating system installer doesn't typically install under-slab insulation but should advise the building team to have a specialist do so before the radiant tubing is set in place.

THIN SLAB

When the radiant system is to be installed in an existing structure (Fig. 26-70), the tubing can be stapled to the top of a frame floor and concrete poured to a thickness of approximately 1½″ to 2″ with a minimum of ¾″ of concrete over the tubing. If this option is being considered, make sure the framing can handle the extra weight.

from experience...

Insulate the space below the tubing to R-11 if the room below the floor is heated. Use R-19 over an unheated basement, and R-30 over a crawl space. The colder the space below, the greater the insulation should be.

If weight is a definite issue, there are alternatives to standard concrete. Gypsum concrete, a mixture of very fine sand, gypsum, cement, and bonding agents, is lighter than regular concrete. For a thin slab installation of 1½ inches concrete, the gypsum concrete will weigh about 13 pounds per square foot compared with the 18 pounds per square foot for regular concrete. One good thing about this alternative is that, because it is so thin, it levels itself. The main drawbacks are that it is more expensive than concrete and that it will leak through any holes in the floor. One other alternative to concrete is Portland-base cement mixed with a material called a super-plasticizer and a couple of other agents. This material costs about one-third of what gypsum concrete costs.

CONCRETE-FREE INSTALLATIONS

Another option for installing radiant systems is to staple the tubing under the floor instead of laying it on top in a bed of lightweight or gypsum concrete. This is a less expensive way to get the job done. Usually on a staple-up job, tubing is attached right to the bottom of the floor (Fig. 26-71) by placing a staple every six inches or so. The tubing should be in contact with the floor because the heat transfer starts as conduction and becomes radiant only when it enters the room where the people are. The heat has to conduct through the walls of the tubing and enter the floor. The main drawback to this configuration is that only a thin edge of the tubing actually touches the floor in contrast to the entire surface of the tubing being in contact with the concrete slab.

from experience...

With this type of installation, you must also use foil-faced insulation to bounce the radiant energy off the tubing and up onto the bottom of the sub-floor so the top of the finished floor heats evenly all the way across. It's also crucial that you leave several inches of airspace between the tubing and the foil, so the radiant waves of energy can diffuse down within the joist bay and bounce back up onto the underside of the floor.

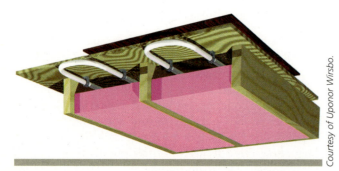

Courtesy of Uponor Wirsbo.

FIGURE 26-71 Cutaway of a staple-up radiant piping layout.

As a result, it will be necessary to circulate hotter water through the tubing arrangement to compensate for poor thermal contact between the tubing and the floor.

CAUTION

CAUTION: Be aware of the electrical wiring in the joist bays when doing a staple-up job. The temperature tolerance of older wiring is not as high as newer wiring.

When considering a staple-up radiant heating system under a wood floor, keep the following in mind:

✓ The best type of wood for hydronic radiant floor heating is laminated softwood with a layer of hardwood on top.

✓ Nonlaminated or solid wood flooring is a poor choice, because the floor will expand and contract, likely causing the wood to crack.

✓ Installing "floating floors" will reduce the chance of the floor cracking as a result of the expanding and contracting of the wood.

✓ Narrower floor boards are less likely to warp.

✓ Avoid tarpaper in the floors. It emits a very bad odor when heated.

✓ Keep the wood floor no warmer than 85°F.

TUBING

Several kinds of tubing are available to bury or staple up. PEX is cross-linked polyethylene (Fig. 26-67) with a great track record for successful installations both here and in Europe. One of the benefits of PEX tubing is that it has memory, so if it becomes kinked, it will return to its original shape when heated. The linking of the molecules in PEX tubing varies from manufacturer to manufacturer and from PEX type to PEX type. The "cross-linking" happens in the manufacturing process, and how the manufacturer chooses to do that affects the properties of the final product.

Various types of PEX tubing include PEX-A, PEX-B, and PEX-C, which have manufacturing differences, but have all been approved for use by the American Society for Testing and Materials (ATSM). Always weigh system requirements when selecting the type of PEX tubing to be used on a particular job. Other materials that can be used are PEX/Aluminum PEX, polybutylene, and rubber.

FIGURE 26-72 Radiant manifold.

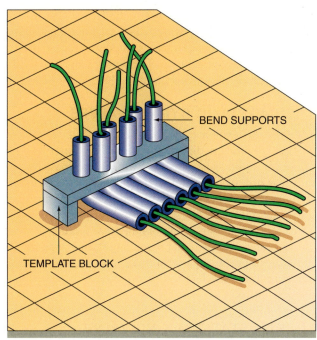

BEND SUPPORTS

TEMPLATE BLOCK

FIGURE 26-73 Template block and bend supports.

MANIFOLD STATION

The **manifold station** (Fig. 26-72) is where all the supply and return tubes come together at the manifold. In a poured concrete slab, it is important that all of the tubes penetrate the slab within the confines of future wall locations. Miscalculations at this stage of the project can result in having some or all of the tubes penetrating the floor outside the finished wall. The tubes that extend from the slab must be properly supported while the concrete is actually being poured. In addition, they should be protected from kinking at the point where they bend up into the wall. Template blocks and bend supports are used for these purposes. The template block (Fig. 26-73) holds the tubes, and the bend supports are sections of tubing into which the tubing is fed.

WATER TEMPERATURE AND DIRECT PIPING

The desired floor temperature in a radiant system is about 85°F, so the temperature of the water flowing through the piping circuit does not need to be higher than 120°F and, in many cases, as low as 100°F. If the particular boiler is being used to supply water only to the radiant loop, a direct piping system can be used. The direct piping arrangement is similar to that discussed at the beginning of this chapter, but the temperature of the supply water is much lower. In this case, the water is heated to 105°F, circulated through the radiant system, and returned to the boiler at 90°F. One major problem with this configuration is that, if the temperature of the return water is lower than the temperature at which the flue gases condense, there will be continuous condensation, which will result in damage to the chimney, flue pipe, and heat exchanger in the boiler. Since electric boilers do not have flue gases, they can be used as the heat source for radiant systems without the possibility of damage caused by condensing flue gases. Some other common piping configurations are discussed next.

BYPASS LINES

A bypass line can alleviate the problem of damage to the boiler if the water returning to it is below the temperature at which the flue gases will condense. The bypass

from experience...

Most boiler manufacturers include simple drawings of bypass piping arrangements with their installation-and-operating instruction booklets.

line takes some of the hot water that leaves the boiler and blends it with the water that's coming back from the radiant loop. This increases the temperature of the water returning to the boiler. The piping diagram of this arrangement is shown in Figure 26-74.

THERMOSTATIC BYPASS VALVES

The previous section explained the bypass line and how it is used to increase the temperature of the water returning to the boiler. The one drawback of a bypass line is that it is not automatic. The thermostatic bypass valve is an automatic device that protects the boiler against low-temperature water. It has three ports and contains a thermostat similar to the one in a car's radiator. The thermostat keeps the boiler water circulating around the boiler until the water reaches a certain minimum temperature and then lets the water out into the system (Fig. 26-75).

THERMOSTATIC MIXING VALVE

Some radiant systems can use a thermostatic mixing valve to ensure that the temperature of the water feeding the radiant loop is at the desired level. The mixing valve has two inlets—one hot and one cold—as well as one outlet. As the hot and cold water enter the valve, they mix, and the resulting temperature of the water at the outlet is somewhere between the temperature of the cold water and the temperature of the hot water. If almost no cold water is entering the valve, the temperature of the water leaving the valve will be close to the hot water temperature. Similarly, if almost no hot water is entering the valve, the temperature of the water at the outlet will be very close to the temperature of the cold water. The thermostatic mixing valve automatically opens and closes the hot and cold inlets to obtain the desired temperature at the outlet of the valve. When used to control the temperature of a radiant system, the piping arrangement might look something like Figure 26-76.

In this configuration, the boiler is supplying water at 180°F to the thermostatic mixing valve. Only a small amount of this water is actually flowing through the mixing valve; most is being returned to the boiler. The water temperature at the outlet of the radiant loop is 90°F, and the desired temperature of the water at the inlet of the radiant loop is 110°F. This will cause the hot water port on the mixing valve to open slightly so some of the 180°F water will mix with the 90°F water to reach the desired temperature of 110°F at the outlet of the mixing valve. If the temperature at the outlet of the radiant loop dropped to 80°F, the hot water port on the mixing valve would open more to keep the temperature at the inlet of the radiant loop at 110°F. The amount of water that leaves the radiant loop and flows back to the boiler is exactly the same as the amount of hot water that enters the loop through the mixing valve.

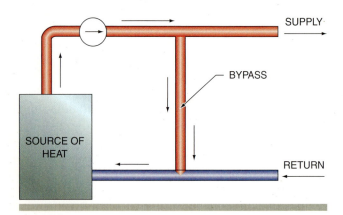

FIGURE 26-74 Bypass loop.

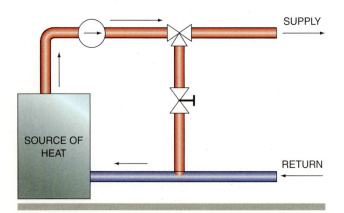

FIGURE 26-75 Thermostatic bypass valve.

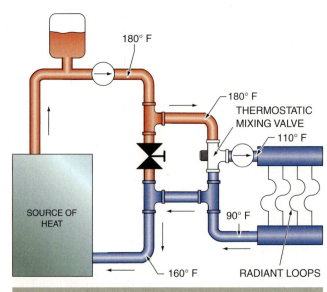

FIGURE 26-76 Thermostatic mixing valve.

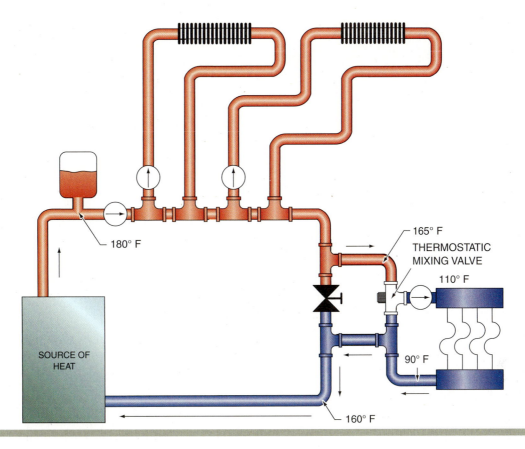

FIGURE 26-77 Primary-secondary pumping system used for both high-temperature and low-temperature water loops.

PRIMARY-SECONDARY PUMPING WITH HIGH- AND LOW-TEMPERATURE CIRCUITS

In the event that the structure has both high- and low-temperature circuits, the same boiler can be used to service both requirements by using a primary-secondary configuration. Assume that one zone in the house is heated by a radiant loop and two other areas are heated by high-temperature baseboards. By taking the secondary high-temperature loops off the primary, the operation of the radiant zone is not affected (Fig. 26-77).

In Figure 26-77, water at 180°F is being supplied to the first high-temperature baseboard loop and the temperature being supplied to the second baseboard loop is very close to 180°F as well. Each of these baseboard circuits has its own circulator, just like the primary-secondary pumping system described earlier in the chapter. The water in the main loop after the two baseboard loops is approximately 165°F, because of the heat given up by the baseboard. At the thermostatic mixing valve, the 165°F water will mix with the water at the outlet of

the radiant loop to give us water at 110°F at the inlet of the radiant loop. In addition, if the radiant portion of this system feeds multiple radiant zones, automatic valves can be used to close off individual branches of the radiant portion of the system.

INSTALLING AND STARTING THE HYDRONIC SYSTEM

Installing and putting a hot water hydronic system into operation involves a number of steps that include:

- Setting and installing the boiler
- Installing the piping
- Wiring the unit
- Filling the system
- Starting up the system

INSTALLING THE BOILER

When setting and installing the boiler, keep the following in mind:

- ✓ Make certain the boiler is level.

- ✓ Locate boiler as close as possible to the chimney.

- ✓ Make certain the flue piping is installed according to the manufacturer's literature.

- ✓ Make certain that the amount of air introduced to the area around the boiler is sufficient for combustion and dilution.

- ✓ Make certain all packing material has been removed from the boiler.

- ✓ Install the pressure relief in the tap specified by the manufacturer.

- ✓ Provide enough clearance around the boiler for future service.

- ✓ Make certain that a disconnect switch is located on or very close (within two feet) to the boiler.

- ✓ Check the combustion of the boiler on initial startup.

- ✓ Properly set the pressure-reducing valve prior to filling.

- ✓ Properly check and adjust the air pressure in the expansion tank prior to filling the system.

INSTALLING THE PIPING

When installing the piping circuit, keep the following in mind:

- ✓ Keep vertical piping vertical and horizontal piping horizontal. In addition to ensuring the proper alignment of fittings and components, customers will appreciate a neat-looking job, and so will you and your boss.

- ✓ Try to keep the number of fittings to a minimum. Excess fittings increases the cost of the job and the chance of water leakage.

- ✓ For the sake of neatness and ease in wiring, group similar components, such as circulators or zone valves, together. Keep them at the same height and have them all pointing in the same direction. Neatness counts!

- ✓ Use lots of valves. Although this increases the cost of the materials for the job, time and money will

be saved when it comes time to perform service in the future. Valves placed before and after a circulator, for example, will save the technician the trouble of draining the entire system to change a defective impeller.

- ✓ Properly support the piping and components to avoid sagging.

- ✓ Dry-fit pipe fittings prior to soldering. This avoids excessive measuring, recutting, and soldering.

- ✓ Leak check all tubing (radiant heat systems) prior to pouring concrete!

WIRING THE SYSTEM

When wiring the system, keep the following in mind:

- ✓ Color code the wires whenever possible to avoid confusion and incorrect wiring.

- ✓ Read and follow the wiring diagrams supplied with the individual components to ensure their safe and proper operation.

- ✓ Make certain that individual components will operate properly when used in conjunction with other components.

- ✓ Ensure that the controls being used are designed to perform the desired tasks prior to installation. Many electronic controls cannot be returned once they have been either removed from the package or installed.

- ✓ Follow all local electric and fire codes regarding wiring methods used.

- ✓ All wiring should be located above piping in the event of a water leak. This will prevent water from leaking onto the wires.

FILLING THE SYSTEM

Before filling the system, calculate how much water pressure you need to lift water to the highest point in the system. To do this, determine the distance between the circulator and the highest pipe in the system. Remember that 1 psig of water pressure will lift water 2.3 feet straight up, no matter what size the pipe is.

When you know how much pressure you need to fill the system, add 3-psi pressure to that so there will be some pressure at the highest point. This additional pressure will help compensate for errors in calculation as well as help remove air from the system.

Check the air pressure inside the compression tank (diaphragm-type expansion tanks). The air pressure should equal the pressure you plan to use on the water side of the diaphragm. Check the air pressure before installing the tank and filling the system with water. If you wait until there is water inside the tank you will not be able to get an accurate reading on the air side of the diaphragm.

Refer to Procedure 26-3 on page 731 step-by-step instructions for filling and purging the system.

FIRING THE SYSTEM

When starting up the system, keep the following in mind:

✓ Check the operation of all safeties on initial startup.

✓ Test the operation of all circulators, zone valve, thermostats, etc.

✓ Make certain the thermostats control the operation of the correct zone valves and circulators.

✓ Make certain that the boiler cycles on and off on the aquastat.

✓ Test the operation of all safety controls.

✓ Test the combustion efficiency of the boiler.

✓ Determine carbon dioxide and carbon monoxide levels and compare them with acceptable levels.

1 SERVICE CALL
No Heat Upstairs

CUSTOMER COMPLAINT

A customer calls her heating company and informs the service manager that her heating system seems to be working fine, except for one section of baseboard located on the second floor. The customer has a one-pipe hot water system with only one zone and this is the first time she has used the heating system this year.

TECHNICIAN EVALUATION

Upon arriving at the home, the technician, Marco, listens to the customer's complaint. She tells the technician about the problem and says that last year, toward the end of the heating season, water was coming from the vent on that section of baseboard.

Marco checks the thermostat to make certain it is set for heating and, on his way upstairs, checks the other sections of baseboard. They are all heating just fine. As Marco enters the room with the problem radiator, he can feel that the room is quite cold. He goes over to the radiator, places his hand on it, and finds that it too is very cold. Remembering what the customer told him about the water coming from the vent last year, he asks her to see the bill from that repair. The repair bill notes that there was rust in the vent and that the service tech cleaned out the vent and placed it back on the radiator.

Marco decides to remove the existing automatic vent from the radiator and replace it with a new one. He closes the hand valves on both the supply and return lines of the radiator and removes the vent. He inspects it and notes that there is indeed rust inside the vent. To confirm his suspicions, Marco slowly opens the valve on the supply side of the radiator and air begins to vent rapidly from the system. He closes the valve, installs a new vent and opens the valves on the supply and return sides of the radiator. The air vents and the radiator gets hot. On his work ticket, Marco recommends that the other vents in the home be replaced as well, given the fact that there was rust in the system. He also recommends that the boiler be drained and flushed once the weather gets a little warmer.

SERVICE CALL DISCUSSION

Marco quickly determined that there was a venting problem with the radiator and successfully solved the problem of the rusty air vent. However, Marco could have done the following things differently:

• He recommended that the rest of the vents be replaced. Because there is rust in the system, there is very likely at least some rust in all of the vents in the house. The customer herself mentioned that there was water coming from the vent last year when it had stuck open. Replacing the vents should be part of the initial job, not an option for the customer to ponder over. Chances are that at least one of the other vents will experience the same problem, possibly resulting in water damage to the home.

• For the reasons just mentioned, draining and flushing the boiler should be part of the repair, not an option. When presented with such options customers often think that this additional work is not important and, therefore, not worth the expense. How wrong they can be!

SERVICE CALL
No Heat Upstairs

CUSTOMER COMPLAINT

A customer calls his heating company and informs the service manager that there is no heat in the upstairs zone of his house. Each of the three zones in the house is controlled by a separate circulator with flow-control valves at the outlet of each. It is the middle of winter. The system had been working perfectly, and then seemed to just stop working all of a sudden. The other two zones in the house are working just fine.

TECHNICIAN EVALUATION

About an hour later, the technician, Richard, arrives at the home. He listens to the customer's complaint and checks the thermostat for the upstairs zone. It is calling for heat. He then goes to the basement and identifies the loop that services the upstairs zone. He notices that the circulator for that loop is not operating.

Richard first touches the motor on the circulator and finds it to be very hot. From this, Richard can conclude that voltage is being supplied to the motor, but there is a problem with the motor itself, the linkage or the impeller.

Richard disconnects the power to the circulator and then disconnects the linkage that connects the motor to the impeller shaft. He restores power to the pump motor and the motor starts. He then attempts to rotate the impeller shaft by hand. It is stuck and he is unable to turn the shaft even one-half of a rotation. Richard gets a bucket from his truck and places it under the circulator and closes the ball valves located before and after the circulator.

He slowly loosens the bolts, holding the circulator in place, until the water in the pump begins to drain into the bucket. When the water flow stops, he removes the pump from the system. Upon inspecting the impeller assembly, he finds that the impeller has broken and a piece of the impeller has jammed the shaft, preventing it from turning. Richard goes to his truck and gets a new pump and linkage assembly. He decides that while he has the system apart he will also replace the linkage since there was undue stress placed on it.

He replaces the assembly and furnishes new gaskets between the pump assembly and the pump flanges. He tightens all bolts and restarts the system.

SERVICE CALL DISCUSSION

Richard was very careful when he removed the pump assembly from the system. Had he removed the pump quickly without making certain that the leaking water would stop, he could very well have flooded the basement if one or both of the valves had failed to hold. He loosened the bolts slowly and, once the water began to drain, he waited until the draining stopped. He also had the foresight to replace the linkage since he already had the pump apart and knew that a lot of pressure had been put on it when the motor tried unsuccessfully to turn the impeller.

SUMMARY

- Hydronic systems circulate hot water or steam to the occupied space for heating purposes.
- Hot water is typically generated in the boiler, which can be gas-fired, oil-fired, or electric.
- Cast iron boilers are made up of bolted sections that form the heat exchange surface.
- Water temperature in the boiler can be maintained by the aquastat or reset control.
- The low-water cutoff cycles the boiler off in the event that the water level falls below a safe level.
- The expansion tank allows for the expansion of heated water.
- The point where the expansion tank is connected to the system is referred to as the *point of no pressure change.*
- The centrifugal pump is responsible for circulating the water through the system and is located so that it pumps away from the point of no pressure change.
- Air in the system is removed using air separators and air vents.
- The pressure-reducing valve opens and closes to maintain the desired water pressure in the system. It is also piped into the system at the point of no pressure change.
- The pressure relief valve is a safety device that opens to relieve system pressure if it rises above the preset limit.
- Flow control valves prevent gravity flow through hot water loops when the circulators are off.
- Zone valves are used to control water flow to different zones in the structure to maintain different temperatures in those areas.
- Common piping configurations include the series loop system, one-pipe system, two-pipe direct return, and two-pipe reverse return.

- One-pipe systems use monoflo or diverter tees to balance water flow.
- Two-pipe direct return systems often use balancing valves to evenly distribute water flow through each branch.
- Primary-secondary pumping systems use circulators in both the primary and secondary loops.
- The secondary loop uses the primary loop as the expansion tank.

- Radiant heat systems require low water temperatures in the range of 105°F to 120°F.
- Radiant heating systems often use mixing valves to maintain the desired water temperature.
- Primary-secondary systems can be used to provide both high-temperature water for baseboard terminal units and low-temperature water for use in radiant loops.

GREEN CHECKLIST

☐ Boilers are rated by their Annual Fuel Utilization Efficiency, AFUE.

☐ The higher the AFUE, the lower the equipment losses.

☐ AFUE ratings can range from 85% to 99%.

☐ Outdoor reset adjusts the boiler water temperature based on the outside ambient temperature.

☐ As the outdoor temperature rises, the boiler water temperature drops.

☐ The operating efficiency of hot water hydronic systems will be higher if air is removed from the system.

☐ Zoning allows different areas in a structure to be maintained at different temperatures, saving fuel.

☐ Hot water lines that pass through unconditioned spaces should be insulated to reduce heat loss from the system.

☐ Radiant heating systems function to heat the shell of the structure instead of the air contained within the structure.

☐ Insulation should be installed below the slab when installing radiant heating systems to reduce heat loss and improve system efficiency.

PROCEDURE 26-1

Procedure for Estimating the Volume of Water in the System (Assuming that type M copper piping is used)

- Determine the number of linear feet of each size type M copper pipe used in the piping system.

 Multiply the number of linear feet used by the appropriate **volume factor** in the figure. The volume factor provides the number of gallons of water per linear foot of the corresponding piping material.

- Add up all of the products obtained in the previous step. This will provide the number of gallons of water contained in the piping circuit, not including the radiators.

- Referring to the radiator manufacturer's literature, determine the volume of water contained in the radiators. If the manufacturer's literature is not available, estimate 1 gallon of water for each radiator. Standing cast iron radiators will require a higher estimate. Add this estimate to the total obtained in the previous step.

- Obtain the volume of the boiler from the manufacturer's literature. If this information is not available, estimate 1 to

A

Tubing Size (inch)	Volume Factor (Gallons/Foot)
1/2 Type M Copper	0.01316
3/4 Type M Copper	0.02685
1 Type M Copper	0.0454
1¼ Type M Copper	0.06804
1½ Type M Copper	0.09505
2 Type M Copper	0.1647

2 gallons for a small copper tube boiler, or 10 to 15 gallons for a cast iron sectional boiler. Add this figure to the total obtained in the previous step. This result is an estimate of the volume of water contained in the system.

Sample Calculations for Estimating Volume of Water in the System

B Consider the piping diagram in the figure.

- The total footage of 1-inch type M copper is 290 feet

- The total volume for the 1-inch type M copper is 290 × 0.0454 = **13.17 gallons**

- The total footage of 3/4-inch type M copper is 100 feet

- The total volume for the 3/4-inch type M copper is 100 × 0.02685 = **2.685 gallons**

- The total footage of 1/2-inch type M copper is 40 feet

- The total volume for the 1/2-inch type M copper is 40 × 0.01319 = **0.53 gallons**

- All radiators are the same and according to the manufacturer's literature,

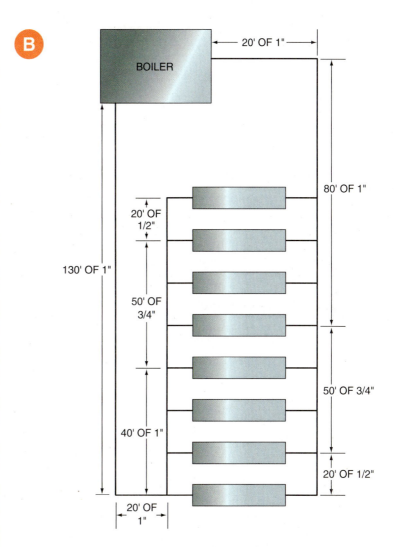

each radiator holds 0.75 gallons of water. We will also assume an extra 0.25 gallons of water for the piping that connects each radiator to the main lines. This gives us a total of 1 gallon of water for each radiator

- Eight radiators, each with 1.0 gallon of water, gives us **8.0 gallons**

- According to the boiler manufacturer, the volume of the cast iron boiler is **14 gallons.**

- Adding the values from the previous steps gives us a total estimate of the volume of water in the system (Vs) in gallons of water:

13.17 + 2.685 + 0.53 + 8.0 + 14.0 = 38.4 gallons

PROCEDURE 26-2

Calculating the Minimum Volume (Vt) for the Expansion Tank

- Determine the initial temperature of the cold water entering the boiler (Tc)

- Determine the density of the cold water (Dc) using the following formula:

$$Dc = 62.56 + 0.0003413 \, (Tc) - 0.00006255 \, (Tc^2)$$

- Determine the maximum operating temperature of the water in the boiler (Th)

- Determine the density of the hot water (Dh) using the following formula:

$$Dh = 62.56 + 0.0003413 \, (Th) - 0.00006255 \, (Th^2)$$

- Estimate the volume of water in the system (Vs) using the previous procedure.

- From the boiler's data sheet, determine the system's pressure relief valve, PRV, setting (psig)

- Determine the air side pressure setting in the expansion tank (refer back to the text for this calculation)

- Substitute the values for Dc, Dh, Vs, PRV, and Pa in the following formula to obtain the required minimum volume for the expansion tank:

$$Vt = Vs \, [(Dc/Dh) - 1] \, [(PRV + 9.7) / (PRV - Pa - 5)]$$

PROCEDURE 26-3

Filling and Purging the System

- Measure the difference between the height of the circulator and the height of the highest pipe in the system

- Divide that result by 2.3

- Add 3 psig to the new value

- This figure is the desired system pressure

- For systems with diaphragm-type expansion tanks, check the air pressure inside the compression tank

- The air pressure should equal the calculated system pressure

- Adjust the pressure in the expansion tank as needed (a bicycle tire pump works well)

A When first filling this system, close the main shutoff valve and open the boiler drain on the first supply-pipe tee to purge the air

- Keep all zones closed except one

- Use the fill valve to blow the air from that one zone. The water flows first through the system and then into the bottom of the boiler. The air will be pushed through the system piping, through the boiler, and out from the boiler drain at the top of the boiler.

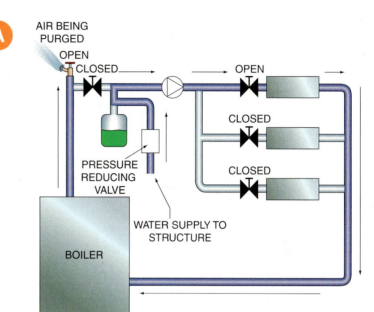

A

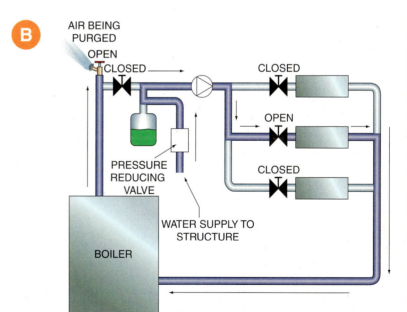

B

B Repeat the previous two steps for the next terminal unit

- Repeat the previous steps for each of the remaining zones in the system

REVIEW QUESTIONS

1. **Which of the following is true regarding cast iron boilers?**

a. The capacity of the boiler will decrease as the number of sections is increased

b. They typically hold between 10 and 15 gallons of water

c. They are desirable because they are not affected by long periods of exposure to dissolved air

d. All of the above are correct

2. **The system component that is responsible for maintaining the water in the boiler at the desired temperature is the**

a. Aquastat

b. Low-water cutoff

c. Bypass valve

d. Expansion tank

3. **If the hydronic system is equipped with a reset control and the outside temperature increases from 20°F to 50°F, which of the following is possible?**

a. The temperature of the water in the boiler will change from 170°F to 190°F

b. The temperature of the water in the boiler will change from 170°F to 150°F

c. The temperature of the water in the boiler will remain unchanged

d. Both a and b are correct

4. **When referring to the aquastat settings, which of the following correctly represents the relationship between the cut-in temperature, cut-out temperature, and differential?**

a. Cut in − Cut out = Differential

b. Cut out − Cut in = Differential

c. Cut out = Cut in − Differential

d. Cut in = Cut out + Differential

5. **Which of the following is most likely to occur if a hot water hydronic system is installed with an undersized expansion tank?**

a. The water temperature in the boiler will be higher than desired

b. The water temperature in the boiler will be lower than desired

c. The relief valve will open prematurely

d. The pressure-reducing valve will fail to feed water to the boiler

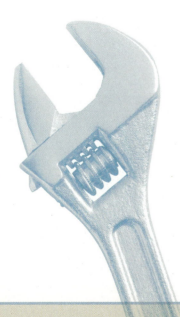

6. **As the water temperature in the boiler increases**
   a. The air pressure in the expansion tank will increase
   b. The expansion tank will contain a larger volume of air than before the water was heated
   c. Both a and b are correct
   d. Neither a nor b is correct

7. **The only point in a hot water hydronic system where the circulator *cannot* affect the pressure in the system is**
   a. The point where the return pipe enters the boiler
   b. The point where the expansion tank is connected to the piping circuit
   c. The point where the water is discharged from the circulator
   d. The point where water from the structure is introduced to the boiler

8. **Which of the following can result in a vacuum being pulled in the piping circuit if the expansion tank has an air pressure of 12 psig and the pump creates a 15-psig differential between the pump's inlet and outlet?**
   a. Both the expansion tank and the circulator are in the return line where the circulator is located between the boiler and expansion tank pumping toward the boiler
   b. Both the expansion tank and the circulator are in the supply line where the circulator is located after the expansion tank pumping away from the boiler
   c. Both the expansion tank and the circulator are in the supply line where the circulator is located between the expansion tank and the boiler pumping away from the boiler
   d. All of the above are correct

9. **A pump that has a pressure differential across it of 20 psig has a head of**
   a. 8.7 feet
   b. 20 feet
   c. 46 feet
   d. 92 feet

10. **As the height of the hot water piping in a system gets higher**
    a. The required air pressure in the expansion tank will increase
    b. The required air pressure in the expansion tank will decrease
    c. The density of the water will increase
    d. The density of the water will decrease

11. **Installing the circulator so that it pumps away from the point of no pressure change will**
    a. Result in the highest possible pressure at the outlet of the pump for that system
    b. Help keep air bubbles in solution at the higher pressure
    c. Help prevent air from being pulled into the system resulting from a vacuum condition
    d. All of the above are correct

12. It is required by law that all hot water boilers be equipped with a

a. Pressure-reducing valve

b. Pressure relief valve

c. Air separator

d. Automatic air vent

13. If the pressure in a hot water boiler rises above safe limits

a. The pressure relief valve will close

b. The pressure relief valve will open

c. The pressure reducing valve will open

d. Both b and c are correct

14. A structure that has separate areas, each with its own means to control the temperature in that area,

a. Has a series loop heating system

b. Has only one thermostat

c. Is a zoned system

d. All of the above are correct

15. Gravity flow through loops can be eliminated by using a

a. Zone valve

b. Flow-control valve

c. Both a and b are correct

d. Neither a nor b is correct

16. Two-pipe direct return systems have the advantage that

a. The flow through all of the branches is exactly the same

b. The first terminal unit being fed water is the last one to return to the boiler

c. The distance through each branch circuit is exactly the same

d. None of the above are correct

17. The benefit of the series loop systems is that

a. The flow through each terminal unit can be controlled separately

b. This system is economical to install

c. Zone valves can be located at each terminal unit

d. All of the above are correct

18. If a one-pipe system has a section of baseboard located below the main loop

a. One diverter tee should be located on the return side of the terminal unit

b. One diverter tee should be located on the supply side of the terminal unit

c. Two diverter tees should be used

d. No diverter tees are needed

19. On a one-pipe system, placing the tees very close together will

a. Increase the resistance between the tees and increase the flow through the terminal unit

b. Decrease the resistance between the tees and decrease the flow through the terminal unit

c. Decrease the resistance between the tees and increase the flow through the terminal unit

d. Increase the resistance between the tees and decrease the flow through the terminal unit

20. In a two-pipe reverse return system

a. Balancing valves are typically not needed

b. The first terminal unit supplied is the last one to return

c. Both a and b are correct

d. Neither a nor b is correct

21. **In a primary-secondary pumping configuration**

    a. The resistance between the tees should be as low as possible

    b. There is only one circulator

    c. All tees used should be diverter tees

    d. All of the above are correct

22. **Why is an expansion tank not required in the secondary loops of a primary-secondary pumping configuration?**

    a. The water in the secondary loops will not expand when heated

    b. The secondary loops will use the primary loop as the expansion tank

    c. The circulator in the secondary circuit would pump all of the water into the expansion tank

    d. Both a and b are correct

23. **In a primary-secondary pumping configuration, if a secondary pump has a higher capacity than the primary pump**

    a. There is a possibility that water will flow through the common piping in the opposite direction

    b. The system cannot work

    c. The secondary loop will be pulled into a vacuum

    d. The primary pump will turn in the opposite direction

24. **Radiant heat systems are sized to**

    a. Compensate for the heat lost by the occupied space

    b. Regulate the heat loss of the occupants of the occupied space

    c. Both a and b are correct

    d. Neither a nor b is correct

25. **Which of the following is true regarding a slab-on-grade radiant piping circuit?**

    a. The deeper the tubing is in the concrete, the higher the water temperature needs to be

    b. The PEX tubing spacing is typically 12" on center

    c. There should be a vapor barrier under the slab

    d. All of the above are correct

KNOW YOUR CODES

The requirements for installation of heating equipment vary greatly from area to area. Here is a brief list containing some of the items that need to be considered when installing a heating appliance:

- Proximity of heating appliance to combustible or flammable materials
- Insulation of pipes in unconditioned spaces
- Automatic water fill piping and valving arrangements
- Low water cutoff controls
- Material selection and usage
- Appliance venting guidelines

How many of these do you know about? Investigate the rules, laws, and guidelines that are in effect for these items as well as for other important installation considerations you might have thought of. Report your findings to your classmates and colleagues so everyone can learn!

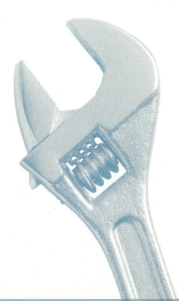

WHAT'S WRONG WITH THIS PICTURE?

Carefully study Figure 26-78 and think about what is wrong. Consider all possibilities.

✕ WRONG

Delmar/Cengage Learning.

FIGURE 26-78 When installing a relief valve on a piece of hydronic equipment, there are two things to keep in mind. The size of the relief port should not be reduced and the outlet of the relief valve should be pointing down with a section of pipe connected to it. This way, in the event the relief valve opens, the high-temperature water is directed downward and severe personal injury can be avoided.

✓ RIGHT

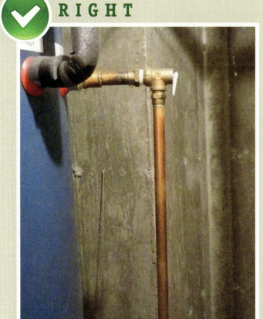

Delmar/Cengage Learning.

FIGURE 26-79 Notice that the pipe size has not been reduced and that the relief valve has a downward-pointing pipe connected at the relief valve's outlet.

Heat Pumps

OBJECTIVES   *Upon completion of this chapter, the student should be able to*

- Explain the cooling and heating cycles of a heat pump system.

- Trace refrigerant as it flows through a heat pump system in the heating and cooling modes.

- Identify the state and condition of the refrigerant as it passes through the heat pump system.

- Describe the operation of the four-way reversing valve.

- Explain the difference between a direct-acting reversing valve and a pilot-operated reversing valve.

- Explain the purpose of check valves in a heat pump system.

- Explain the importance of the suction line accumulator.

- Describe various types and classifications of heat pump systems.

- Explain how temperature, pressure, and time are used to defrost heat pump systems.

- Describe the operation of a geothermal heat pump system.

- Explain the difference between open- and closed-loop geothermal systems.

- Describe the various types of well configurations found on geothermal systems.

- Troubleshoot heat pump systems.

In this, the last chapter of the text, we will examine the heat pump system, which can also be described as reverse-cycle refrigeration. In very basic terms, the heat pump can be explained as a mechanical refrigeration or air conditioning system that provides cooling in the warmer months and heating in the cooler months.

Having reached this portion of the text, the reader has been exposed not only to the basic vapor-compression air conditioning cycle, but also to a variety of methods used to heat an occupied space. The heat pump system uses mechanical refrigeration theory to provide yet another heat source that has become very popular over the past four decades.

The heat pump system transfers heat from a fluid or substance to the air inside the occupied space. Depending on the configuration of the heat pump system, the source of heat can be a liquid, air, or the earth itself. In this chapter, we will examine the operation of various types of heat pump systems as well as installation and troubleshooting techniques.

Air source heat pumps heat pumps that use air as the source of heat while operating in the heating mode

Air-to-air heat pumps air-source heat pumps used to treat air

Air-to-liquid heat pumps air-source heat pumps used to treat liquid

Aquifer underground water formation

Ball-type check valve device that permits fluid flow in only one direction; a floating ball in the valve body prevents fluid from flowing in the wrong direction

Bidirectional liquid-line filter drier filter drier used for heat pump applications that permits refrigerant flow through the device in both directions

Bidirectional thermostatic expansion valves TEVs equipped with internal check valves to permit almost unrestricted flow through the valve in the opposite direction when the metering device is not needed in the reverse cycle

Check valve system piping component that allows fluid flow in only one direction

Closed-loop system geothermal heat pump configuration that uses buried coils of piping to transfer heat to and from the earth

GLOSSARY OF TERMS (CONT'D)

Coefficient of Performance (COP) is a measure of the amount of energy output per unit of energy input

Consolidated formation materials below the surface of the earth that contain very little water, used in geo-thermal heat pump systems; examples include granite, limestone, and sandstone

Defrost cycle heat pump mode of operation that facilitates the removal of ice from the surface of the outdoor coil by changing the system over to the cooling cycle for a period of time.

Demand defrost defrost cycle that is initiated only when needed

Disc-type check valve device that permits fluid flow in only one direction. A pivoting disc within the shell of the valve prevents flow in the wrong direction.

Dry well a gravel, sand, and rock-filled hole in the ground that accepts discharge water from the heat exchanger of a heat pump system

Earth-coupled system see closed-loop system

Energy efficiency ratio (EER) rates the efficiency of a system; calculated by dividing the Btu output of the system by the power input in watts

Four-way reversing valve device in the heat pump system that reverses the direction of refrigerant flow through the evaporator and condenser coils to allow the system to operate in either the heating or cooling mode

Geothermal heat pump heat pump system that uses the earth as the heat source in the cooler months and the heat sink in the warmer summer months

Ground coils see ground loops

Ground loops buried piping circuits used in closed-loop geothermal heat pump systems

Heat sink any substance or media that absorbs heat from another substance or media

Heat source any substance or media that supplies heat to a system

Heating Seasonal Performance Factor (HSPF) a measure of heating efficiency

Horizontal ground loop a single water loop configuration in which the ground loop is located close to the ground surface; used primarily in areas with underground rock formations that make deep trenching very costly

Impeller rotating portion of a centrifugal pump that creates the pressure differential needed to facilitate fluid flow

Indoor coil heat transfer surface in a heat pump system that absorbs heat from the occupied space in the cooling mode and rejects heat to the occupied space in the heating mode

Liquid-source heat pumps heat pumps that use liquid as the source of heat while operating in the heating mode

Liquid-to-air heat pumps liquid source heat pumps that are used to treat air

Liquid-to-liquid heat pumps liquid source heat pumps that are used to treat liquid

Open-loop system geothermal systems that obtain their supply water from an open source such as a lake, pond, well, or other underground water source.

Outdoor coil heat transfer surface in a heat pump system that acts as the condenser in the cooling mode and as the evaporator in the heating mode

Parallel ground loops in geothermal heat pump systems, the ground loop configuration that provides multiple paths for the heat exchange fluid to flow in

Pete's port self-sealing access fitting used to take temperature and pressure readings in geothermal heat pump systems

Pilot-operated reversing valves reversing valves that use system pressures to move the slide back and forth to change the system over between the heating and cooling modes of operation

Pressure switch electric switch that opens and closes its contacts depending on the pressure it senses

Pressure tanks pressurized tanks that store water

Return water in geothermal systems, the water that returns to the ground after passing through the system and the return well

Seasonal energy efficiency ratio (SEER) equipment efficiency rating determined by dividing the system's Btu output by the input power in watts, taking into account the energy used on system startup and shutdown

Standing pressure pressure inside a vessel when the vessel is at rest; for example, the refrigerant pressure inside an air conditioning system when the system is off

Suction-line accumulator a refrigeration component located in the suction line that facilitates the vaporization of liquid refrigerant before entering the compressor

Supply water in geothermal systems, the water being supplied to the system via the supply well

Timer any control device, mechanical or electronic, that opens or closes sets of electrical contacts at predetermined time intervals

Vertical ground loops used in geothermal heat pump systems, the ground loops are buried vertically in areas where ground space is limited

Wastewater system system that uses water as either a heat source or sink one time before the water is "wasted" down the drain.

Water hammer the sound generated when flowing water is suddenly stopped by a closing valve

Water loop see ground loops

Water table source of underground water used for geothermal heat pump systems; wells must be drilled to the water table, which can be 200 feet below the ground surface

HEAT PUMP THEORY

The basic purpose of an air conditioning system is to transfer heat from an occupied area where it is not desired to a location where it is not objectionable, thereby lowering the temperature of the occupied space. The heating system, on the other hand, transfers heat for the purpose of raising the temperature of a location. The basic vapor-compression refrigeration system that was discussed in the opening chapters of this text is capable of producing a refrigeration effect but cannot produce a heating effect. By adding components to the basic air conditioning system, a piece of equipment that performs both heating and cooling functions can be created. This system, the heat pump, provides comfort cooling, or air conditioning, in the warmer months and heating in the cooler months. This is accomplished by altering the path that the refrigerant takes as it passes through the system.

As we recall from basic refrigeration theory, refrigerant is discharged from the compressor and travels to the condenser to reject heat from the system. This heat is rejected to the medium passing over or through the condenser, which is typically air for residential systems and applications. As the system refrigerant transfers heat to the medium, the temperature of the medium increases. After leaving the condenser and passing through the expansion device, the refrigerant enters the evaporator, where heat from the medium that is to be cooled is absorbed by the refrigerant. In an air-cooled, residential air conditioning system, for example, the heat is removed from inside the conditioned space and transferred to the air located outside the structure. This works fine in the warmer months when the cooling effects of air conditioning are desired. But what happens in the cooler months when heating is needed?

In the colder months, heat transfer is still desired but this time the direction of the transfer is opposite to that just described. We now want heat to be transferred from outside the structure to the inside. This concept may be confusing because it is sometimes difficult to comprehend that 30 degree outside air, for example, contains enough heat to have a warming effect on the area to be heated. In reality, air that we consider cold actually contains large amounts of heat that can be used effectively for this purpose.

It would therefore be desirable for the refrigerant, upon leaving the compressor, to travel first to the conditioned space in order to reject its heat into that location, thereby increasing the temperature of the space. The refrigerant would then travel through the expansion device and evaporator to absorb heat from the air outside the structure before returning to the compressor. In essence, the locations of the evaporator and condenser have been switched. The evaporator is now located outdoors to absorb heat, while the condenser is located indoors to reject heat into the conditioned space. For this reason, the coils in a heat pump system are not referred to as the evaporator and condenser, but rather as the **indoor coil** and the **outdoor coil**, referring to the coil's physical location in the system.

When providing air conditioning, or cooling, the indoor coil of the heat pump acts as the evaporator and the outdoor coil acts as the condenser [Fig. 27-1(A)]. This is the same as a typical air conditioning system. In the heating mode, however, the indoor coil functions as the condenser, rejecting heat into the space, and the outdoor coil operates as the evaporator, [Fig. 27-1(B)].

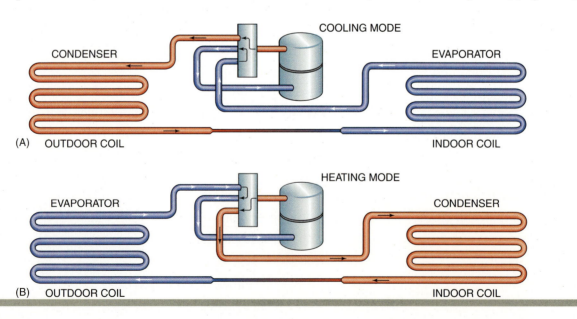

FIGURE 27-1 (A) Heat pump system operating in the cooling mode. (B) Heat pump system operating in the heating mode.

The red areas in Figures 27-1(A) and 27-1(B) represent the high-pressure side of the system, and the blue areas represent the low-pressure side of the system. Notice how the colored areas shift between the outdoor and indoor coils as the system switches over from the cooling to heating mode, and vice versa.

Using just the four major system components, achieving this switchover from heating to cooling would be impossible without having to completely re-pipe the system every time a different mode of operation was desired. However, the addition of key components has made this transition as easy as pushing a button. The two key components that make the switchover possible are the **four-way reversing valve**, (Figure 27-2) and the **check valve** (Figure 27-3).

In addition to the four-way reversing valve and the check valve, there are other components that, when

added to the heat pump system, allow it to operate more effectively and efficiently. They include:

- Suction line accumulators
- Auxiliary electric strip heaters
- Bi-directional liquid line filter driers
- Bi-directional thermostatic expansion valves

THE REVERSING VALVE

The four-way reversing valve is responsible for mechanically altering the refrigeration piping configuration to route the compressor discharge refrigerant through the:

- Outdoor coil while operating in the cooling mode
- Indoor coil while operating in the heating mode

The refrigerant is still discharged from, and returned to, the compressor in the same way. The compressor's operation is exactly the same regardless of the mode in which the system is operating. Refer to Figure 27-1(A) and Figure 27-1(B) and note that the refrigerant is still discharged from and returned to the compressor from the same ports. The four-way reversing valve makes automatic changeover between the heating and cooling modes of operation possible.

The reversing valve is controlled by a solenoid coil and changes the direction of refrigerant flow depending on whether or not the coil is energized. The valve is constructed of a shell, within which is positioned a slide that can shift back and forth (Fig. 27-4) within the shell. When the solenoid is energized, the slide is situated at one end of the valve and when de-energized, the other. Valve manufacturers have established different criteria

FIGURE 27-2 Four-way reversing valve.

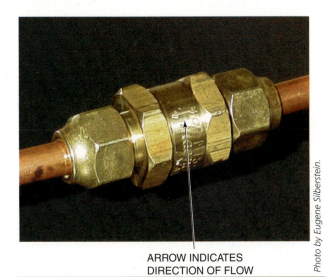

ARROW INDICATES DIRECTION OF FLOW

FIGURE 27-3 Flow check valve. Direction of flow is indicated by the arrow on the device.

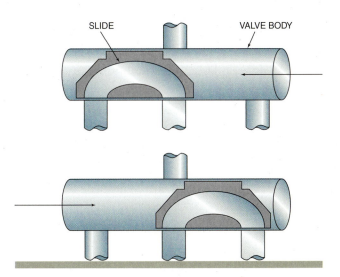

SLIDE VALVE BODY

FIGURE 27-4 The internal slide on the reversing valve moves back and forth within the shell of the valve.

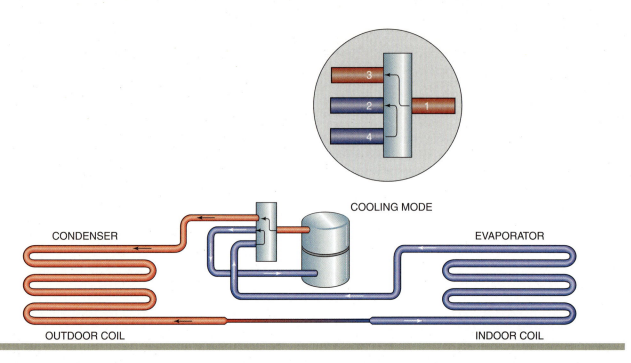

FIGURE 27-5 This valve is in the cooling mode when the solenoid coil is energized. 1. Connected to the compressor discharge line. 2. Connected to the compressor suction line. 3. Connected to the outdoor coil. 4. Connected to the indoor coil.

for determining if the system should be in the heating or cooling mode when the solenoid coil is energized. For the sake of clarity and uniformity, assume here that the system is in the cooling mode when the solenoid valve is energized (Fig. 27-5). Always check the system being worked on to determine which orientation is being used.

> ### *from experience...*
> Most manufacturers configure their heat pump systems to fail in the heating mode of operation, meaning that the system will be in the heating mode when the solenoid on the reversing valve is de-energized.

The four-way reversing valve has four ports to which the system refrigerant lines are connected. These ports are connected to:

1. The discharge line of the compressor
2. The suction line of the compressor
3. The outdoor coil (on the opposite side as the expansion device)
4. The indoor coil (on the opposite side as the expansion device)

These locations are shown in Figure 27-5. One port, usually separate from the other three, is connected to the discharge line coming from the compressor. The compressor's discharge vapor enters the reversing valve at this point. The vapor refrigerant then leaves the valve through one of the two end ports on the other side of the valve (Fig. 27-6). The refrigerant leaves the valve and travels to the condenser that is located either indoors or outdoors, depending on the desired mode of operation. In the case of Figure 27-5, the condenser is located outdoors, indicating that the system is operating in the cooling mode. Vapor refrigerant then flows back to the valve through the other end port, leaves the valve by way of the center port, and flows back to the compressor. Note that the discharge refrigerant always enters the valve at *Point 1* and returns to the compressor from *Point 2*.

COOLING MODE

When the solenoid valve is energized, putting the system in the cooling mode, the piping configuration is as shown in the top of Figure 27-7. Assume at this point that the system is equipped with a single metering device and that the same metering device is being used in both the heating and cooling modes. In this figure, note that the refrigerant leaves the compressor and first travels through the reversing valve and onto the outdoor coil, which in this case is the condenser. It then flows through the metering device and into the indoor coil. The indoor coil is acting as the evaporator and is absorbing heat from the conditioned space. This configuration is exactly the same as that of a system that operates as a cooling-only unit.

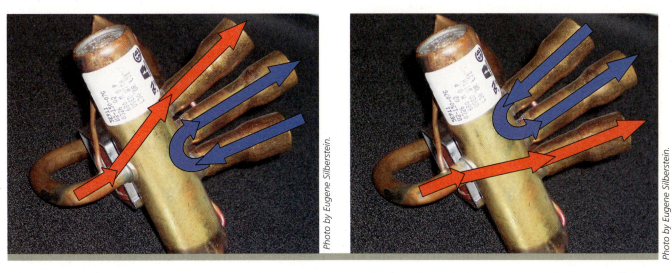

Photo by Eugene Silberstein.

FIGURE 27-6 Refrigerant flow from the compressor will be directed through the reversing valve to one of the end ports on the valve when the coil is energized and to the opposite end port when the coil is de-energized.

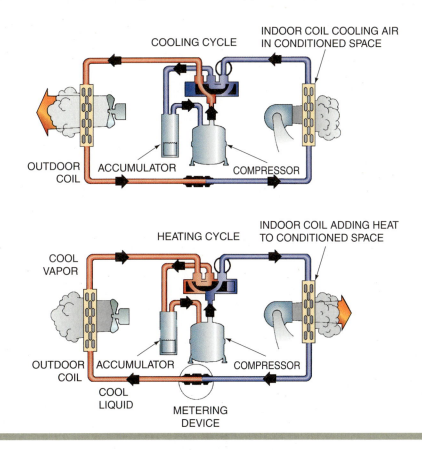

FIGURE 27-7 An air conditioning system showing the cooling (top) and heating (bottom) modes of operation.

Once the refrigerant flows through the evaporator, it then flows through the low-pressure side of the reversing valve. After leaving the reversing valve, the low-pressure refrigerant flows through the accumulator (which will be discussed shortly) and then returns to the compressor to complete the cycle.

HEATING MODE

When the solenoid is de-energized, putting the system in the heating mode, the flow through the coils is reversed, as shown at the bottom of Figure 27-7. The hot gas being discharged from the compressor still flows through the reversing valve and onto the condenser, but this time the condenser is the indoor coil. The refrigerant now gives up its heat to the occupied space, thereby heating the area. After the refrigerant leaves the condenser and flows through the metering device, it goes to the outdoor coil, which is acting as the evaporator. After flowing through the evaporator, the refrigerant travels back through the reversing valve and through the accumulator before returning to the compressor. Various key points in the system are identified in Figure 27-8; and the refrigerant flow through the system and the state of the refrigerant at these points, in both the heating and cooling modes of operation, are outlined in Figure 27-9.

DIRECT-ACTING AND PILOT-OPERATED REVERSING VALVES

On very small systems, the slide in the reversing valve can be controlled directly by the pull created by the magnetic field of the solenoid coil as described previously. When the solenoid coil is de-energized, a spring pressure pushes the slide to its normal position. This type of configuration is not found on larger valves for a number of reasons, including:

- The large distance that the slide must travel
- The large pressure differentials across the valve
- Erratic reversing-valve operation
- Premature solenoid coil failure

Larger systems are therefore equipped with **pilot-operated reversing valves**. This type of reversing valve uses the difference between the high- and low-side pressures to help push the valve slide from one position to the other. The direction of the force created by the high-side pressure is determined by the pilot ports (Fig. 27-10), which in turn are controlled by the solenoid coil. The operation of the pilot-operated reversing valve is similar to that of a direct-acting valve in that the energizing and de-energizing of a solenoid coil causes the movement of a valve slide. The main difference is this:

- The solenoid-controlled slide on the pilot-operated valve controls the refrigerant flowing through the pilot ports of the reversing valve. It does not control

the refrigerant flowing through the main body of the valve.

- By controlling the refrigerant flow through the pilot ports, the system's high pressure is directed in such a way that it pushes the main valve slide to the desired position.

In operation, when the solenoid coil is de-energized, putting the system in the heating mode, the pilot valve is in the position shown at the bottom of Figure 27-11(A). The compressor discharge gas at *D* flows into the pilot valve at *Port D1* and out of the pilot from *Port B*. This high-pressure vapor then acts on the main valve slide, pushing it to the left. The refrigerant on the left side of the main valve slide is pushed out of the valve and enters the pilot valve through *Port A*. It then leaves the pilot valve from *Port S1* and travels back to the compressor from *Port S*. This valve is in the heating position since the discharge gas from the compressor, at *Point D*, is being circulated through the indoor coil, at location *C2,* first.

When the solenoid coil is energized, and the system is calling for cooling, the pilot valve moves to the position shown at the bottom of Figure 27-11(B). The hot discharge gas from the compressor still enters the pilot valve through *Port D1* but leaves through *Port A*. This sends the high-pressure gas to the left side of the slide in the main valve, pushing the main slide to the right. The vapor at the right side of the slide is pushed out of the valve, enters the pilot valve through *Port B*, leaves through *Port S1,* and returns to the compressor from *Port S*. This system is in the cooling mode since the hot discharge gas from the compressor flows through the outdoor coil, *C1,* first. The pilot-operated reversing valves are more desirable when used on larger applications. The solenoid coil controls the movement of a small plunger which, in turn, operates the larger main slide on the valve itself. This eliminates the erratic operation often found on the direct-acting valves and also reduces the wear and tear on the solenoid coil. The purpose of both the pilot-operated and the direct-acting reversing valve is the same. They are both intended to direct the refrigerant to the correct coil to achieve the desired result. A summary of the refrigerant flow through the pilot-operated

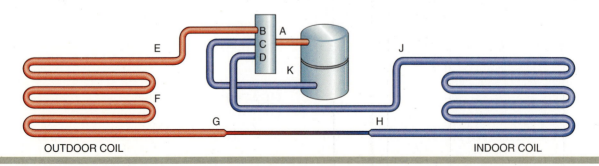

FIGURE 27-8 Heat pump system with important locations identified (see Figure 27-9).

Cooling Mode			Heating Mode		
Letter Designation	Location in the System	State of the Refrigerant	Letter Designation	Location in the System	State of the Refrigerant
A	Compressor discharge port	High-temp, high-pressure superheated vapor	A	Compressor discharge port	High-temp high-pressure superheated vapor
B	Reversing-valve port leading to the outdoor coil	High-temp, high-pressure superheated vapor	D	Reversing-valve port leading to the indoor coil	High-temp high-pressure superheated vapor
E	Inlet of the outdoor coil	High-temp, high-pressure superheated vapor	J	Inlet of the indoor coil	High-temp high-pressure superheated vapor
F	Middle of outdoor coil	High-temp, high-pressure saturated liquid	H	Outlet of the indoor coil/inlet of the metering device	High-temp high-pressure subcooled liquid
G	Outlet of outdoor coil/ Inlet of the metering device	High-temp, high-pressure subcooled liquid	G	Inlet of the outdoor coil/ Outlet of the metering device	Low-temp, low-pressure saturated liquid
H	Outlet of the metering device	Low-temp, low-pressure saturated liquid	F	Middle of the outdoor coil	Low-temp, low-pressure saturated liquid
J	Outlet of indoor coil	Low-temp, low-pressure superheated vapor	E	Outlet of the outdoor coil	Low-temp, low-pressure superheated vapor
D	Inlet of the reversing valve	Low-temp, low-pressure superheated vapor	B	Inlet of the reversing valve	Low-temp low-pressure superheated vapor
C	Outlet of the reversing valve	Low-temp, low-pressure superheated vapor	C	Outlet of the reversing valve	Low-temp, low-pressure superheated vapor
K	Compression suction port	Low-temp, low-pressure superheated vapor	K	Compressor suction port	Low-temp, low-pressure superheated vapor

FIGURE 27-9 State of refrigerant at various points in the system.

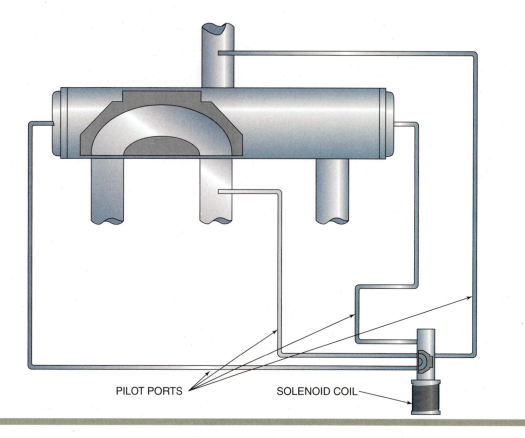

FIGURE 27-10 Pilot ports on a pilot-operated, four-way reversing valve.

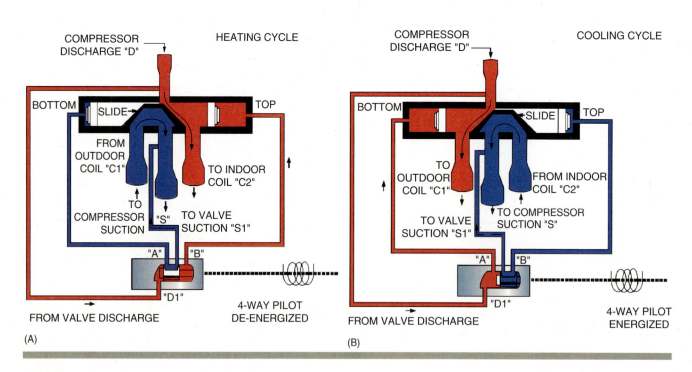

FIGURE 27-11 (A) Heating and (B) cooling positions of a pilot-operated, four-way reversing valve.

	Heating Mode	Cooling Mode
Hot Gas Leaving the Compressor	Enters port "D1"	Enters port "D1"
Hot Gas Flowing to the Condenser	Leaves port "B"	Leaves port "A"
Refrigerant Leaving the Evaporator	Enters port "A"	Enters port "B"
Refrigerant Flowing Back to the Compressor	Leaves port "S1"	Leaves port "S1"

FIGURE 27-12 Summary of pilot-operated, reversing valve port arrangement.

reversing valve, referring to Figure 27-11, is shown in Figure 27-12. Notice in Figure 27-11 that the hot gas from the compressor always enters the reversing valve at *Port D1* and the suction gas always returns to the compressor through *Port S1*. Also note in the same figure that the reversing valve simply switches the internal connections to *Ports A* and *B*.

CHECK VALVES

The purpose of the check valve is to permit refrigerant to flow in only one direction. This reduces the possibility of refrigerant trying to push its way through an expansion device or filter drier in the wrong direction, which would result in an inefficient system as well as extremely high pressure drops across the devices. Since multiple expansion devices and filter driers are often found in heat pumps, the check valve also helps ensure that only the desired components are in the active refrigerant circuit at any given time.

In Figure 27-3, it can be seen that the device is marked with an arrow that points in the direction of desired flow through the valve. As refrigerant flows through the valve in the direction indicated, there is very little resistance. If, however, refrigerant attempts to flow through the valve in the direction opposite to that indicated on the valve body, the device will close and will not permit the refrigerant to pass through. Several different types of check valves are commonly found in heat-pump systems and are differentiated by their construction and method of operation. The two common types are:

- Ball-type check valve
- Disc-type check valve

BALL-TYPE CHECK VALVE

The **ball-type check valve** is constructed of a hollow shell that is soldered or brazed into the refrigerant piping circuit at the desired location. Inside the shell rests a ball that moves freely within the device. To operate correctly, it must be installed in the vertical position. When the

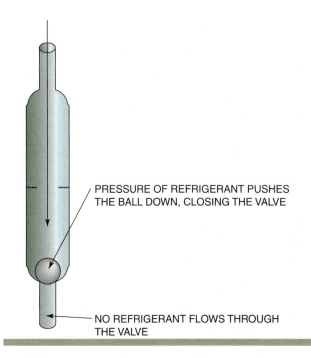

FIGURE 27-13 Ball-type check valve in the closed position.

PRESSURE OF REFRIGERANT PUSHES THE BALL DOWN, CLOSING THE VALVE

NO REFRIGERANT FLOWS THROUGH THE VALVE

system is off, the ball will fall to the bottom of the valve (Fig. 27-13). When refrigerant flows in the direction indicated by the arrow on the device, the pressure of the refrigerant lifts the ball and the valve opens (Fig. 27-14). When refrigerant attempts to flow through the valve in the wrong direction, the ball is pushed down and the valve closes, as shown in Figure 27-13.

DISC-TYPE CHECK VALVE

The **disc-type check valve** is constructed of a shell in which rests a disc that is held in place by magnetic attraction. The check valve in Figure 27-3 is a disc-type check valve. When refrigerant flows through the valve in the desired direction, the disc pivots and allows refrigerant to pass through (Fig. 27-15). If refrigerant attempts to flow through the valve in the opposite direction, the disc is pushed against the valve opening, thereby closing the valve (Fig. 27-16).

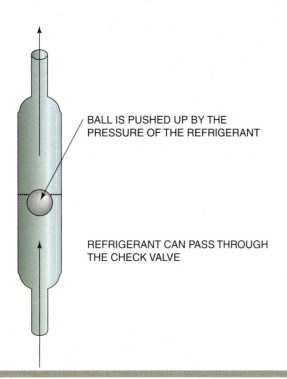

FIGURE 27-14 The ball is being pushed up, opening the check valve and allowing refrigerant to flow through it.

Inside the figure: BALL IS PUSHED UP BY THE PRESSURE OF THE REFRIGERANT

REFRIGERANT CAN PASS THROUGH THE CHECK VALVE

Photo by Eugene Silberstein.

FIGURE 27-16 The check valve is closed when the disc is perpendicular to the direction of flow.

Photo by Eugene Silberstein.

FIGURE 27-15 In the disc-type check valve, the valve is in the open position when the disc is parallel to the direction of refrigerant flow.

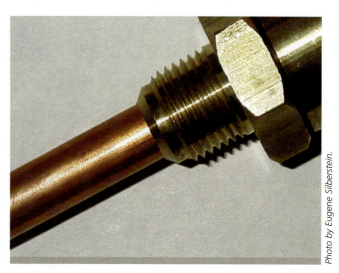

Photo by Eugene Silberstein.

FIGURE 27-17 This type of valve can be connected by either flare or soldering.

allowing the device to be soldered or brazed (Fig. 27-17). If the insert is left in place, the device is installed using flare connections (Fig. 27-18). If needed or desired, one end of the flow-check valve can be connected with a flare nut and the other end soldered into the system.

Depending on the construction of the check valve, the device can be installed in the system in a number of different ways. Two of the most common methods are soldering or brazing and using flare connections. Some flow-check valves are designed so they can be installed either way. These devices have inserts that can be removed to accommodate copper tubing, thereby

from experience...

When soldering check valves into the system, good field practice dictates wrapping the device with a wet cloth to prevent excessive heating, which could affect the operation of the valve.

FIGURE 27-18 This insert is removed when the valve is soldered or left in place when installed using a flare connection.

CHECK VALVES IN CAPILLARY TUBE SYSTEMS

At the beginning of the discussion of the components used in the heat-pump system, the assumption was made that the system used a capillary-tube metering device and that the same capillary tube was in the active circuit for both the heating and cooling cycles. Since the capillary tube is simply a fixed-bore metering device with no moving parts and no mechanical components,

the change in direction of refrigerant flow did not pose an immediate problem. However, consider the strainer or filter drier that is located at the inlet of the capillary tube.

During normal operation, the filter drier traps any particulate matter flowing in the refrigerant piping to prevent clogging of the capillary tube (Fig. 27-19). If refrigerant is permitted to flow through the capillary tube in the opposite direction, any debris or particulate matter trapped in the strainer will be pushed out of the device and back into the system (Fig. 27-20). To alleviate this problem, it is desirable to allow refrigerant to flow in only one direction through the capillary tube. Therefore, two capillary tubes are commonly used—one for the indoor coil when it functions as the evaporator and one for the outdoor coil when it functions as the evaporator (Fig. 27-21).

To introduce the proper capillary tube to the active refrigerant circuit when needed, check valves are used. In practice, the filter driers are normally installed in series with the check valve as shown in Figure 27-21. The following is true in the heating mode:

- The flow-check valve at the indoor unit is in the open position
- The refrigerant bypasses the capillary tube used in the cooling mode
- The refrigerant passes through the indoor filter drier
- The flow-check valve at the outdoor unit is in the closed position

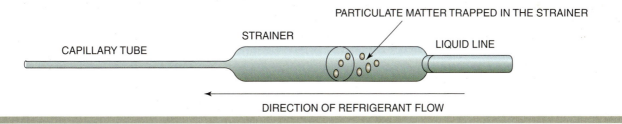

FIGURE 27-19 Particulate matter trapped in the strainer.

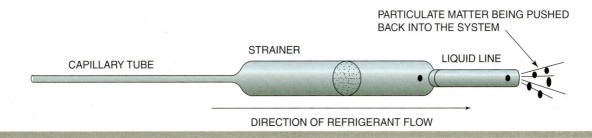

FIGURE 27-20 Particulate matter would be pushed back into the system if refrigerant flowed through the device in the opposite direction.

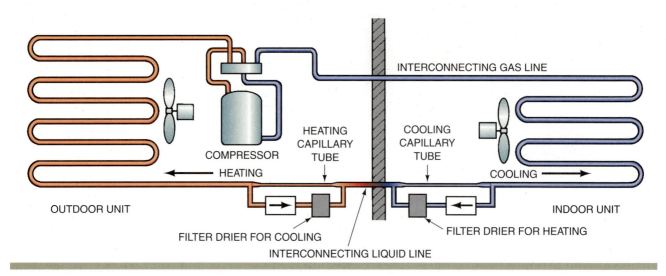

FIGURE 27-21 Filter driers are often installed in series with check valves.

- The refrigerant bypasses the outdoor filter drier
- The refrigerant passes through the outdoor capillary tube

In the cooling mode, the open check valve closes, and vice versa, as follows:

- The flow-check valve at the outdoor unit is in the open position
- The refrigerant bypasses the capillary tube used in the heating mode
- The refrigerant passes through the outdoor filter drier
- The flow-check valve at the indoor unit is in the closed position
- The refrigerant bypasses the indoor filter drier
- The refrigerant passes through the indoor capillary tube

Since the capillary tube provides a restriction to refrigerant flow, installing a check valve in parallel to the capillary tube will provide a low-resistance bypass for the refrigerant when the capillary tube is not intended to be in the active refrigerant circuit.

CHECK VALVES IN SYSTEMS WITH THERMOSTATIC EXPANSION VALVES

Since thermostatic expansion valves (TEVs) must be able to sense the temperature at the outlet of the evaporator to maintain the proper superheat in the coil, TEVs must be located at both the indoor and outdoor coils of split-type systems in order to operate effectively. Without check valves, refrigerant would be forced to travel through

one TEV in the wrong direction whenever the system was operating, as shown in Figure 27-22(A). Refrigerant would be forced through *TEV 1* in the wrong direction during the cooling mode and through *TEV 2* in the heating mode. Problems would quickly arise with a setup similar to this because of the large pressure drop created across the valve. To alleviate this situation, check valves are used in a manner similar to that described in the previous section. This type of configuration is shown in Figure 27-22(B) and Figure 27-22(C).

In the cooling mode, the refrigerant leaving the outdoor coil is able to bypass *TEV 1* because the direction of refrigerant flow is the same as the direction indicated on the check valve, shown in Figure 27-22(B). As the refrigerant reaches *TEV 2*, it is forced through the expansion device because the check valve piped in parallel with that valve is in the closed position. The shaded gray area on this piping diagram indicates the refrigerant flow pattern through the system. Figure 27-22(C) represents the piping circuit for the identical system operating in the heating mode. Some equipment manufacturers, in an attempt to simplify the piping configuration of such systems, have designed a special type of TEV with a built-in check valve. This type of expansion valve is called a **bidirectional thermostatic expansion valve**.

SUCTION LINE ACCUMULATORS

The purpose of the evaporator in an air conditioning system is to absorb heat from the medium to be cooled and, in an ideal system, completely vaporize the liquid refrigerant in the coil before allowing it to travel back to the compressor. An evaporator that functions correctly

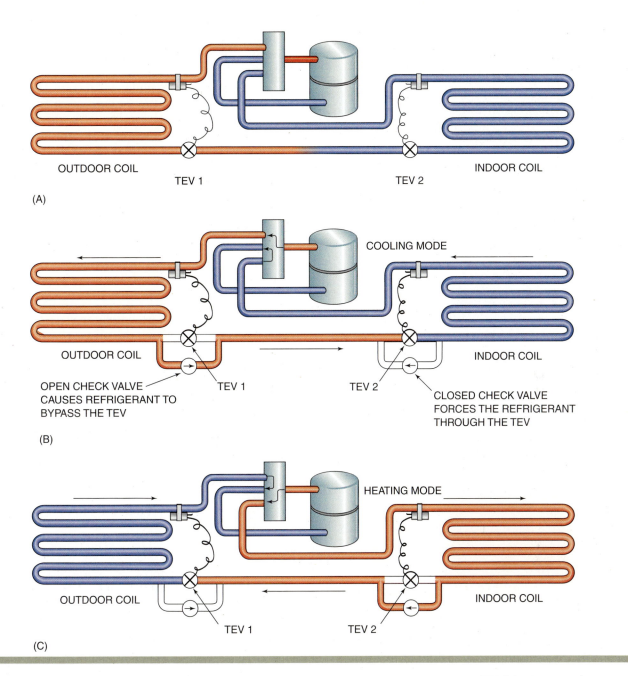

FIGURE 27-22 (A) Refrigerant would flow through the thermostatic expansion valve (TEV) in the wrong direction. (B) Refrigerant bypasses TEV1 in the cooling mode. (C) Refrigerant bypasses TEV2 in the heating mode.

therefore helps protect the compressor from liquid flood-back. However, certain system situations can result in the presence of liquid refrigerant in the suction line returning to the compressor. They include:

- Dirty or blocked evaporator coil
- Dirty or blocked air filters
- Defective evaporator fan motor
- Defective or dirty evaporator blower or fan
- Broken evaporator blower belt

Any of these situations will result in a reduction of air moving through the evaporator, thereby reducing the ability of the refrigerant to absorb heat and boil into a vapor. In addition, heat-pump systems must also contend with the fact that, when operating in the heating mode, the outdoor coil, or evaporator, has a tendency to freeze. Freezing occurs when the surface temperature of the coil drops below 32°F and the condensing moisture from the air begins to freeze on the coil. Evaporator freezing prevents the refrigerant from absorbing heat from the outside air. The ice acts as an insulator between the refrigerant in the coil and the outside air. The refrigerant in the evaporator and suction line therefore remains liquid, which can then travel back to the compressor, causing damage to the vapor pump.

To help prevent damage to the compressor, a **suction-line accumulator** (Fig. 27-23) is installed in the suction

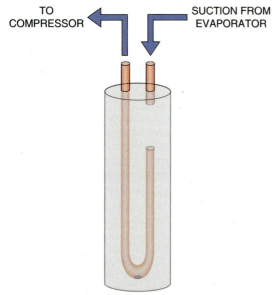

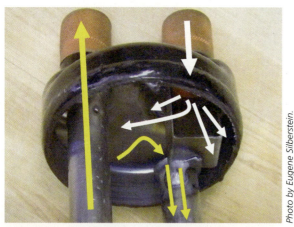

FIGURE 27-23 Suction line accumulator.

line leading back to the compressor. On heat-pump systems, the accumulator is located in the suction line between the compressor and the reversing valve (Fig. 27-24). It allows any liquid refrigerant that may be present in the suction line to boil into a vapor before returning to the compressor. The refrigerant leaving the evaporator flows into the top of the accumulator, and any liquid falls to the bottom of the shell. The outlet of the device is also at the top and is connected to a U tube that extends into the shell. The compressor's pumping action pulls the vapor refrigerant from the shell of the accumulator back to the compressor itself. Since the opening of the U tube is located near the top of the shell, any liquid refrigerant in the lower portion of the accumulator's shell must first boil into a vapor before returning to the compressor. The accumulator is usually located close to the compressor.

If the suction-line accumulator were constructed exactly as just described, the oil that normally travels with the refrigerant as it flows through the system would eventually migrate to the accumulator and cause the system to operate with less than the desired amount of oil. To prevent this oil accumulation from occurring, the U tube in the accumulator has a small hole at the bottom of the bend that allows refrigeration oil to return to the compressor crankcase. This small hole helps ensure that the oil returns to the compressor in a manner that will not damage the pump. Since the opening is not located at the very bottom of the shell, a small amount of oil is normally in the accumulator at all times.

BIDIRECTIONAL FILTER DRIERS

Just as it is common to find two expansion devices in a heat-pump system, it is also common to find two filter driers in the liquid line of the system. Such a piping configuration was shown in Figure 27-21. In another attempt to simplify the system installation and maintenance processes, specially designed filter driers allow refrigerant to flow through them in both directions, while still properly filtering and removing moisture from the refrigerant as it flows through. This device is called the **bidirectional liquid-line filter drier** (Fig. 27-25).

Driers intended for single-directional flow have an indicator arrow on the shell pointing in the direction that refrigerant should flow through it. Biflow driers have a double-headed arrow on the shell, indicating that the device can be installed in the piping circuit in either direction without affecting the operation of the device. Referring to Figure 27-25, while the system is operating in the cooling mode, the drier on the

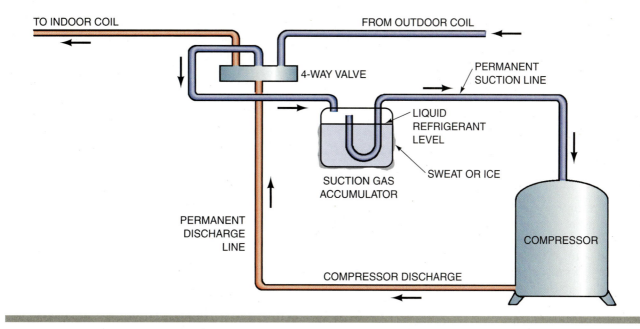

FIGURE 27-24 Accumulators are located in the suction line between the compressor and the reversing valve.

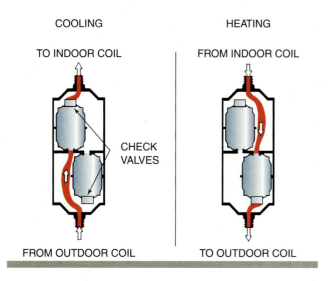

FIGURE 27-25 Bidirectional filter drier.

top is being used and, while operating in the heating mode, the drier at the bottom is in the active refrigerant circuit. This type of drier is a throwaway device, just like any other filter drier that is not equipped with replaceable cores.

CAUTION

CAUTION: When soldering or brazing a filter drier into the system, good field practice dictates wrapping a damp cloth around the shell of the drier and directing the heat from the torch away from the drier.

from experience...

The formation of frost at the outlet of the liquid-line filter drier is an indication that the filter is blocked. In this case, the drier is acting as an expansion device, creating a substantial pressure drop. A blocked filter drier must be replaced if this occurs, unless the core of the drier is replaceable. The installation of a sight glass with a moisture indicator at the outlet of the drier is also recommended. When replacing a filter drier, be sure to leave the protective seals on the device as long as possible to prevent the introduction of moisture or particulate matter into the shell of the device.

BIDIRECTIONAL THERMOSTATIC EXPANSION VALVES

When standard thermostatic expansion valves are used on heat-pump systems, check valves are often used to ensure that the refrigerant flows through the correct expansion device at the correct time, while reducing the pressure drop in the liquid line. The addition of the check valves, however, adds to the complexity of the refrigerant-piping circuit. These additional components, piping and pipe connections, increase the possibility of system malfunction and refrigerant leak. For these reasons, manufacturers have introduced specially designed bidirectional thermostatic expansion valves that have the check valve built into the valve itself. The valve looks and operates similar to other TEVs, with the exception of the internal check valve.

When refrigerant flow is desired through the expansion device, the internal check valve is in the closed position and the refrigerant is directed through the needle-and-seat assembly of the expansion valve. When refrigerant flows through the valve in the opposite direction and the bypass feature is desired, the check valve opens and the refrigerant flows around the needle-and-seat assembly, thereby reducing the pressure drop across the component.

Normally, heat-pump systems are equipped with two such expansion valves, one at the inlet of the evaporator for both the heating and cooling modes of operation. In the cooling mode, the needle-and-seat assembly on the valve installed at the outdoor coil will be bypassed; the valve at the indoor coil will be bypassed in the heating mode.

Heat Pump System Configurations

Just as in standard vapor-compression air-conditioning and refrigeration systems, the condenser can generally be either air- or water-cooled and the evaporator can remove heat from either air or liquid. Typically, in residential applications and installations, condensers are of the air-cooled variety and the evaporators are used to cool air.

Heat pumps are similar to standard air conditioning systems in that they can facilitate the transfer of heat among a number of different mediums. Common classifications of heat pumps are:

- Air-to-air heat pumps
- Air-to-liquid heat pumps
- Liquid-to-air heat pumps
- Liquid-to-liquid heat pumps

The first portion of the classification denotes the source of heat for the system while operating in the heating mode. **Air-source heat pumps—air-to-air heat pumps** and **air-to-liquid heat pumps**—transfer heat from air into the medium to be heated while operating in the heating mode. On the other hand, **liquid-source heat pumps—liquid-to-air heat pumps** and **liquid-to-liquid heat pumps**—transfer heat from a liquid to the medium being heated while operating in the heating mode.

The second portion of the classification represents the medium that is ultimately being treated by the system. In air-to-air and liquid-to-air heat pumps, air is being treated at the indoor coil. Similarly, air-to-liquid and liquid-to-liquid heat pumps treat liquid at the indoor coil. These two heat pump configurations are sometimes used to treat domestic water as well as pools and spas and are not discussed in this text.

In addition to the types of heat pumps listed here, another classification stands alone and is covered later in this chapter. This is the **geothermal heat pump**. The geothermal heat pump uses the earth as a source of heat in the colder months and as a heat sink in the warmer months. In the winter, heat is transferred from the earth to the refrigerant, which is then transferred to the medium to be treated. In the warmer months, the heat from the conditioned space is transferred to the refrigerant, which is then transferred to the earth.

AIR-TO-AIR HEAT PUMPS

As discussed earlier in the text, the air-to-air heat pump circuit uses air as its **heat source** during the heating mode of operation and as the **heat sink** during the cooling mode of operation. In the cooling mode, the hot discharge gas from the compressor is directed to the outdoor coil first so the heat absorbed by the system can be rejected from it (Fig. 27-26). The piping configuration for this mode of operation is shown in Figure 27-27.

For systems that fail in the heating mode of operation, the solenoid on the four-way reversing valve is energized, causing the valve slide to shift to the cooling position. In the cooling mode, the outdoor air is acting as the heat sink, absorbing the heat that is discharged by the system condenser.

In the heating mode, however, the outdoor air is acting as the heat source. The outdoor unit is acting as the evaporator, absorbing heat from the outside air. This heat is then transferred to the indoor unit, or the condenser. The hot discharge gas from the compressor is routed to the indoor unit, now acting as the condenser. The heat is then introduced to the occupied space (Fig. 27-28). The reversing valve, now in the de-energized position, facilitates this change in the direction of refrigerant flow.

As the outside ambient temperature drops, the efficiency and effectiveness of the heat-pump system drop as well. Notice that, in Figure 27-28, the temperature

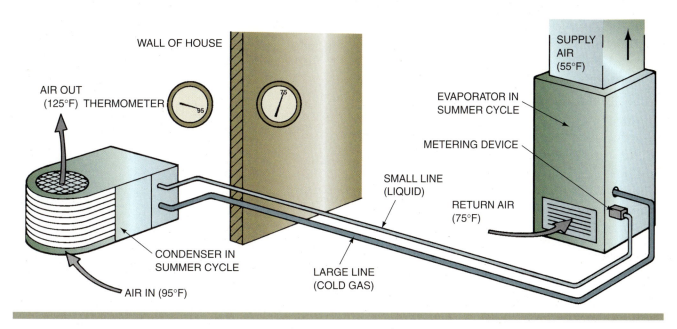

FIGURE 27-26 Air-to-air heat pump removing heat from the structure in the cooling mode.

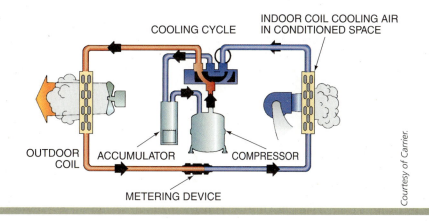

FIGURE 27-27 Piping diagram showing the direction of refrigerant flow in the cooling mode.

differential across the indoor coil is only about 10 degrees. This may not be enough to satisfy the heating requirements of the space. In this case, the second-stage heating mode, or supplementary heating strips, may be needed to help satisfy the heating requirements (Fig. 27-29). The supplementary heating strips, as mentioned earlier, are used when the first-stage heating mode cannot increase the space temperature by itself.

LIQUID-TO-AIR HEAT PUMPS

The liquid-to-air heat pump configuration uses liquid as the heat source when operating in the heating mode. Instead of using a fin-and-tube coil located outdoors as the heat source, a tube-in-tube or coaxial

heat exchanger is used (Fig. 27-30). The liquid flowing through this heat exchanger can be either water or a mixture of water and antifreeze such as polyethylene glycol. A typical liquid-to-air piping configuration is shown in Figure 27-31. The source of liquid (water) can be one of the following:

• Local municipal water supply (tap water)
• Cooling tower
• The earth
• Lake or pond

When the local municipal water supply is used, the water is normally wasted down the drain. These systems, more commonly known as **wastewater systems**, are illegal in many localities. Local zoning laws and codes must

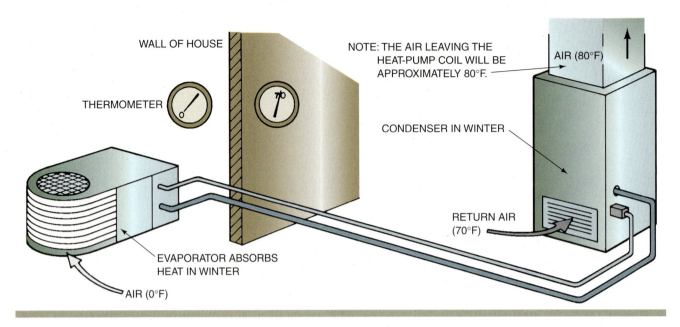

FIGURE 27-28 In the winter, the heat pump transfers heat into the structure.

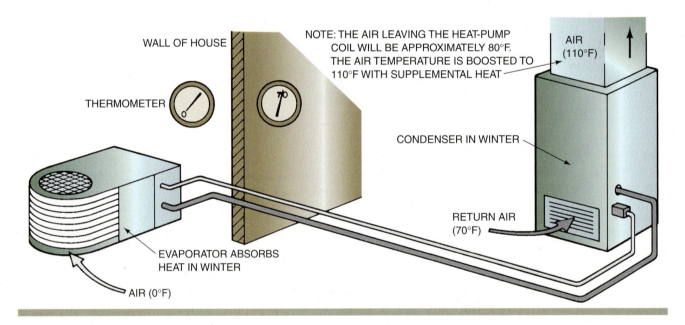

FIGURE 27-29 The use of supplementary electric strip heaters increases the temperature of the air supplied to the space.

be adhered to with regard to this type of equipment. The laws regarding wastewater systems reflect the population of the area, the availability of water, and the capacity of the system intended to be used. In urban areas, the most common source for the water is a cooling tower. Cooling towers are, however, very rarely installed as part of residential systems. In certain areas of the country, using water from the earth itself is beneficial. Refer to the geothermal heat pump sections in this chapter for more on earth-coupled systems.

In Figure 27-31, the system is operating in the heating mode. The hot discharge gas leaving the compressor at Point A flows into the four-way reversing valve at Point B. Upon leaving the reversing valve, the hot discharge gas flows to the indoor finned coil at Point C, heating the air from the conditioned space as it flows through the coil. Once the refrigerant has condensed in the indoor coil, it flows through the metering device at Point D, causing a reduction in both the temperature and the pressure of the refrigerant. The refrigerant then flows through the

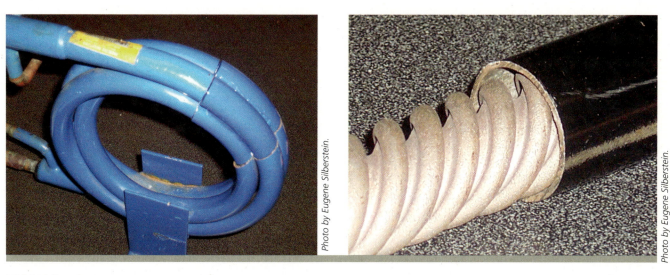

Photo by Eugene Silberstein.

Photo by Eugene Silberstein.

FIGURE 27-30 A tube-in-tube heat exchanger.

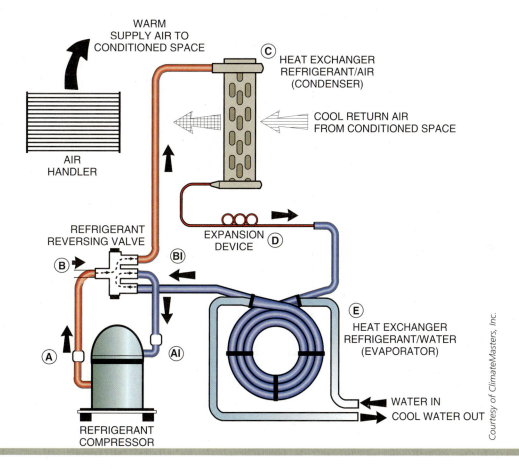

Courtesy of ClimateMasters, Inc.

FIGURE 27-31 Liquid-to-air piping diagram. Heating mode.

tube-in-tube heat exchanger at Point E, where the refrigerant absorbs heat from the liquid flowing through the exchanger. The refrigerant flows in the outer tube, and the liquid flows in the inner tube. After flowing through the heat exchanger, the low-pressure, low-temperature vapor refrigerant returns to the reversing valve at Point B1 and back to the compressor at Point A1.

The temperature of the liquid leaving the heat exchanger is cooler than the temperature of the water entering the heat exchanger, since it has transferred some of its heat energy to the refrigerant. The amount of water or liquid flowing through the heat exchanger is regulated by a water-regulating valve located on the water supply that opens as the system's low pressure drops. This water-regulating valve is set to maintain the desired low pressure on the suction side of the refrigeration system.

LIQUID-TO-AIR HEAT PUMP REFRIGERATION CIRCUIT: COOLING MODE

In the cooling mode, the tube-in-tube heat exchanger functions as the heat sink for the heat that is removed from the occupied space. The piping configuration for a liquid-to-air heat pump system operating in the cooling mode is shown in Figure 27-32.

Following the refrigerant flow through the liquid-to-air system operating in the cooling mode, the hot discharge refrigerant leaves the compressor at Point A and flows to the reversing valve at Point B. Upon leaving the reversing valve at Point B1, the refrigerant flows to the heat exchanger at Point C. Here, the refrigerant gives up its heat to the cool water that is entering the heat exchanger. After condensing in the heat exchanger, the subcooled refrigerant flows to the metering device at Point D, where the pressure and temperature of the refrigerant are reduced. The low-pressure, low-temperature refrigerant then enters the evaporator where it absorbs heat from the air passing over it. The cooled air returns

to the occupied space, thereby reducing the temperature of the area. Finally, after flowing through the evaporator, the refrigerant flows through the four-way reversing valve and then back to the compressor at Point A1.

Just as in the heating mode, a water-regulating valve controls the liquid flow through the tube-in-tube heat exchanger. In this case, though, the regulating valve senses the high pressure of the system, opening wider as the pressure increases. A potentially difficult situation then arises. How can a water-regulating valve open on a drop in pressure when operating in the heating mode and open on a pressure rise while operating in the cooling mode? This question is answered in the next section.

LIQUID-TO-AIR HEAT PUMP REFRIGERATION CIRCUIT: DUAL WATER-REGULATING VALVES

On a liquid-to-air heat pump system operating in the heating mode, the water-regulating valve tends to open as the low-side system pressure drops below acceptable levels. This is to ensure that the tube-in-tube heat exchanger does not freeze. In the cooling mode, however, the water-regulating valve is designed to open as the system's high pressure increases in an effort to keep the head

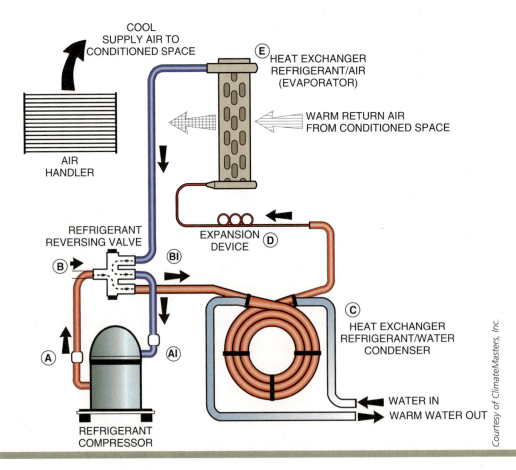

FIGURE 27-32 Liquid-to-air piping diagram. Cooling mode.

Courtesy of ClimateMasters, Inc.

pressure of the system within an acceptable range. The method often used to accommodate both scenarios is to connect two water-regulating valves in parallel with each other. These water-regulating valves sense the pressure at the four-way reversing-valve port that feeds refrigerant to the tube-in-tube heat exchanger while operating in the cooling mode (Fig. 27-33). The regulating valve arrangement can be seen at the right side of Figure 27-33.

In the cooling mode, the hot discharge gas from the compressor leaves the four-way reversing valve and flows to the tube-in-tube heat exchanger. Both water-regulating valves sense this pressure. The water-regulating valve that controls the water flow in the heating mode will be in the closed position because this valve is designed to open when the low-side pressure drops below a predetermined pressure. For an R-140A air conditioning system, this will be about 120 psig. The high-pressure refrigerant leaving the compressor will be about 335 psig, keeping the regulating valve in the closed position. Thus, all of the water flowing through the tube-in-tube heat exchanger will be directed through the other water-regulating valve. As the head pressure of the system rises above the predetermined set point, about 335 psig,

the valve will open allowing more water to flow through the exchanger, thereby reducing the head pressure.

In the heating mode (Fig. 27-34), the hot discharge gas is directed to the indoor coil where it gives up its heat to the air passing over the coil. The system is now providing heat to the occupied space. The refrigerant flowing through the tube-in-tube heat exchanger is at a low temperature and low pressure, and it picks up heat from the water flowing through the exchanger. The heat exchanger is functioning as the heat source. The refrigerant returning to the compressor is at a low temperature and low pressure, and this pressure is being sensed by both water-regulating valves. Since this pressure is low, the water-regulating valve that controls the system while operating in the cooling mode will be in the closed position. Once again, the other valve will now control all the water flowing through the tube-in-tube heat exchanger. If the suction pressure of the system falls below the predetermined set point, about 120 psig, the regulating valve will open, thereby increasing the system's suction pressure.

The operation of the two water-regulating valve configuration is outlined in Figure 27-35. This figure illustrates that, when the system is in the off position and at

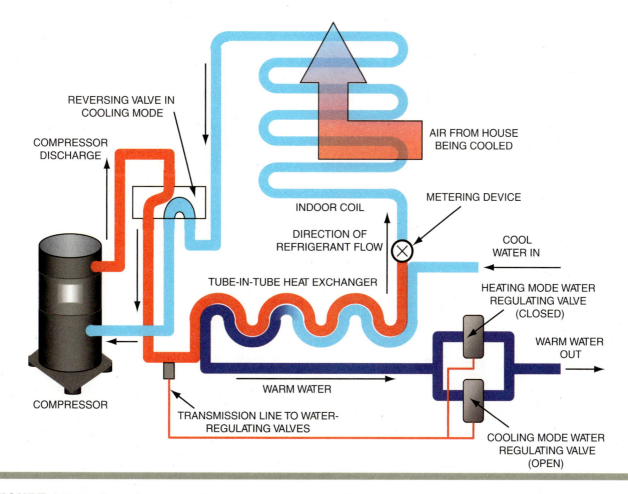

FIGURE 27-33 Dual water-regulating valve arrangement. Cooling mode.

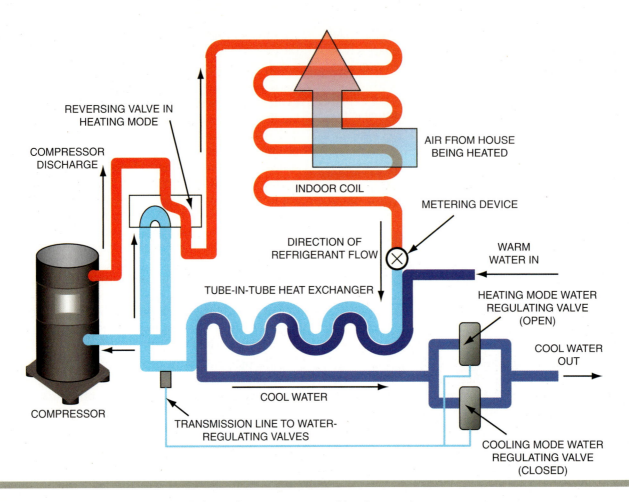

FIGURE 27-34 Dual water-regulating valve arrangement. Heating mode.

standing pressure, both valves are in the closed position. Standing pressure is present in the system when the compressor is not operating. The standing pressure at this point should correspond to the ambient temperature as indicated by the pressure/temperature chart.

from experience...

To properly evaluate the operation of the heat pump system, the technician should be able to take temperature readings of the water entering and leaving the tube-in-tube heat exchanger. For this reason, thermometer wells should be installed in both the supply and return water piping circuits. Hose bibs should also be installed in the lines to facilitate the cleaning of the water loop.

LIQUID-TO-AIR HEAT PUMP REFRIGERATION CIRCUIT: SINGLE WATER-REGULATING VALVE

The piping configuration for the liquid-to-air heat pump system using two water-regulating valves can be simplified somewhat by using a single water-regulating valve specially designed for heat pump application. This valve has a dual-adjusting screw on the top of the valve that facilitates the setting of the pressures to be maintained in both the heating and the cooling modes of operation.

This valve is closed when the system is not operating. When it is operating in the heating mode, the valve opens when the suction pressure drops below the predetermined set point. Similarly, in the cooling mode, the valve opens when the high-side pressure rises above the predetermined high-pressure set point. For an R-410A high-temperature system, assume that the low-side pressure and the high-side pressure settings are 120 psig and 335 psig, respectively. In this instance, the water-regulating valve is in the open position when it senses a pressure below 120 psig or above 335 psig. Otherwise, the valve is in the closed position.

Pressure Sensed by the Regulating Valves	Heating Mode Cooling Valve	Heating Mode Heating Valve	Cooling Mode Cooling Valve	Cooling Mode Heating Valve
80 psig	Closed	**Open**	Closed	**Open**
100 psig	Closed	**Open**	Closed	**Open**
120 psig	Closed	**Open**	Closed	**Open**
140 psig	Closed	Closed	Closed	Closed
160 psig	Closed	Closed	Closed	Closed
180 psig	Closed	Closed	Closed	Closed
200 psig	Closed	Closed	Closed	Closed
220 psig	Closed	Closed	Closed	Closed
240 psig	Closed	Closed	Closed	Closed
260 psig	Closed	Closed	Closed	Closed
280 psig	Closed	Closed	Closed	Closed
300 psig	Closed	Closed	Closed	Closed
320 psig	Closed	Closed	Closed	Closed
340 psig	**Open**	Closed	**Open**	Closed
360 psig	**Open**	Closed	**Open**	Closed
380 psig	**Open**	Closed	**Open**	Closed

FIGURE 27-35 Positions of water-regulating valves in both heating and cooling modes of operation.

HEAT PUMP SYSTEM EFFICIENCY

Air source heat pumps are rated with different measures for heating and cooling modes. **Heating Seasonal Performance Factor (HSPF)** is used to measure heating efficiency. The HSPF is a comparison of the seasonal heat output of the heat pump compared to the electricity consumed. New air source heat pumps typically have an HSPF rating that ranges from 7.7 to 10. SEER is used to measure cooling efficiency, just the same as for air conditioners. Air source heat pumps are available with SEER ratings that typically range from 13 to 19. Manufacturers will list both HSPF and SEER in the specifications of a heat pump model; 18.5 SEER − 8.8 HSPF for example.

Air source heat pumps are common in moderate and warm climates but due to historically low performance efficiencies in heating mode during cold winter conditions, heat pumps have not usually been used in northern cold climates. Advanced heat pump technologies are becoming available with improved cold temperature efficiencies that make air source heat pumps more attractive in colder climates such as the upper Midwest, the Northeast and Pacific Northwest.

DEFROST METHODS

When operating in the heating mode, the outdoor coil, the evaporator, often freezes up. Since the evaporator coil temperature is generally 20 to 25 degrees colder than the air passing over it, frost can begin to form on the surface of the coil when the outside ambient temperature is as high as 50°F. Ice buildup on the coil reduces the refrigerant's ability to absorb heat from the outside air, which in turn causes the coil temperature to drop even further, leading to even more ice buildup on the coil. Since the ice on the coil acts as an insulator between the refrigerant and the outside air, this ice must be removed for the system to operate properly. The process of removing this ice, called the **defrost cycle**, is not difficult but must be geared toward the ambient conditions that surround the outdoor unit. The following should be considered:

- Units operating in milder climates will require little, if any, defrost

- Units operating during periods of high humidity will experience more frost buildup on the outdoor coil

- Units operating during cold and wet conditions will accumulate ice at an accelerated rate

- Faster ice formation on the outdoor coil will require more frequent defrost cycles

The fact that the outdoor coil can operate as either the condenser or the evaporator, depending on the mode of operation, is the basis on which the defrosting of the outdoor coil is achieved. By switching the unit over to the cooling mode, the cold evaporator now becomes the warmer condenser. The heat-laden refrigerant vapor from the compressor flows first to the outdoor coil, transferring its heat to the ice that has formed on the coil, allowing it to melt. During the defrost cycle, the condenser fan motor is often de-energized to help concentrate the heat in the coil, helping to speed up the defrost process even when the outside ambient temperature is very low.

The defrost cycle must be initiated and terminated in a manner that will best suit the design and operating conditions of the system. Decisions about *when* to start and stop defrost, as well as *how* to start and stop the cycle, vary from one manufacturer to the next. However, all defrost systems rely on at least one of the following to initiate and/or terminate defrost:

- Time
- Temperature
- Pressure

Each is discussed in detail in the following sections; but, regardless of the methods used to initiate and terminate defrost, all are designed and intended to achieve the same end result. During the defrost cycle, the following operations are completed:

- The reversing valve switches over to the cooling mode of operation

- The outdoor fan motor is de-energized (most of the time)

- The air supplied to the occupied space is heated slightly to prevent cold air from entering the space during defrost

- The cycle is long enough to ensure proper defrosting of the outdoor coil

- The cycle is short enough to not have a great effect on the normal system operation

Some of the most common defrost methods are described in the following sections. Defrost cycles that are initiated at predetermined time intervals are not always the most efficient, as will be seen shortly. Methods that are only initiated when needed, called **demand defrost**, allow the system to operate more efficiently by reducing the defrost time, thereby allowing the system to heat the occupied space more evenly.

TIME-INITIATED, TIME-TERMINATED DEFROST

In this type of defrost, a **timer** (Fig. 27-36) is used to both initiate and terminate the defrost cycle. In operation, the time-initiated, time-terminated method of defrost places the system in defrost mode at predetermined time intervals. A typical defrost cycle might be 90 minutes of compressor run time followed by a 10-minute defrost cycle. As with other methods of defrost, this method will:

- Switch the system over to the cooling mode
- De-energize the outdoor fan motor
- Allow for tempering the air supplying the occupied space

There are, however, many drawbacks to this type of defrost method, making it a very unpopular way to

FIGURE 27-36 Electronic control board and defrost timer.

Photo by Eugene Silberstein.

remove ice accumulation on the outdoor coil. These drawbacks include the following:

- The unit may go into defrost when no ice is present on the outdoor coil
- The unit may come out of defrost before all of the ice on the coil has melted
- The unit may stay in defrost much longer than needed
- Additional defrost cycles reduce system effectiveness and efficiency

If a heat-pump system goes into defrost when no ice is on the coil or if the system stays in defrost too long, unnecessary cool air may be introduced into the occupied space, reducing the effectiveness of the system. If the system comes out of defrost before all the ice has been removed, ice will continue to form on the coil until, after a number of defrost cycles, the coil will remain frozen and the system performance will be greatly reduced. For these reasons, the time-initiated, time-terminated defrost cycle is no longer used on heat pump systems but provides food for thought in the development of newer, more-efficient defrost methods.

TIME AND TEMPERATURE INITIATION, TEMPERATURE-TERMINATION DEFROST

To help eliminate some of the problems with the time-initiated, time-terminated method of defrost, temperature was introduced as a factor in determining when the defrost cycle would be initiated and terminated. The temperature of the outdoor coil is measured by a device similar to the one in Figure 27-37. By adding the temperature factor into the mix, two conditions must be met for defrost to start:

- The proper time period must have elapsed since the previous defrost attempt. This time period is normally set by the field technician and is typically 30, 45, or 90 minutes.
- There must be frost on the coil. This is determined by the coil temperature. If the coil temperature is low enough, usually lower than 26° or 28°F, the defrost cycle can begin as long as the time condition is also met.

As seen in Figure 27-38, defrost is initiated only when both the time and temperature constraints are satisfied. Electrically speaking, these constraints can be represented by two switches wired in series with each other. The signal to defrost the outdoor coil can only pass when both switches are in the closed position. If the proper amount of time between defrost attempts has elapsed, the normally open contacts in the defrost timer will close. If, however, the outdoor coil temperature sensor establishes that the outdoor coil is too warm to warrant the initiation of a defrost cycle, no defrost will take place. Additionally, if the outdoor coil temperature sensor closes, indicating that sufficient ice has formed on the coil, and the defrost timer contacts are open, indicating that the preset time period between defrosts has not passed, no defrost will be initiated. If, however, the coil temperature sensor contacts are closed, as soon as the defrost timer contacts close, defrost will be initiated.

After these conditions are met and the defrost cycle has begun, it can be terminated solely by the temperature of the outdoor coil. A sensing bulb or element mounted on the outdoor coil determines when the system will be taken out of defrost and placed back into the heating mode. The temperature that triggers the termination of defrost is typically in the range of 55° to 60°F. Under normal conditions, a complete defrost cycle should last

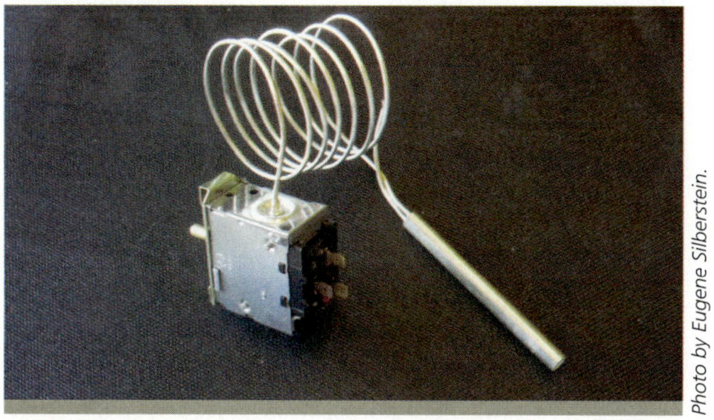

FIGURE 27-37 Defrost termination thermostat.

Photo by Eugene Silberstein.

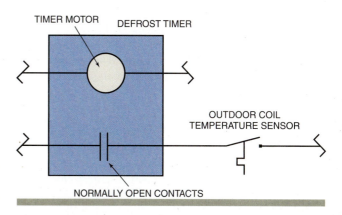

FIGURE 27-38 The timer contacts and outdoor coil temperature sensor contacts are wired in series with each other.

approximately 5 to 7 minutes with moderate frost, at which time the outdoor coil should be completely frost free. If there is excessive frost or a system malfunction, and the system fails to defrost the coil in a reasonable amount of time, usually 10 to 12 minutes, the system will automatically switch back to the heating mode. If this situation occurs, and the system is otherwise functioning properly, the time period that elapses between defrost attempts should be shortened.

TEMPERATURE AND AIR-PRESSURE INITIATION, TEMPERATURE-TERMINATION DEFROST

This method of defrosting the outdoor coil of a heat pump system is very popular, because it provides an effective and very efficient way to defrost the coil. In contrast to the method just described, time is not taken into consideration when the defrost cycle is initiated. This method relies solely on:

• The temperature of the outdoor coil
• The pressure differential across the outdoor coil

For the unit to go into defrost, the temperature of the outdoor coil must be lower than the predetermined set point, as determined by either the factory setting in the control or the field setting as adjusted by the service technician. This temperature setting is generally in the range of 26° to 28°F and is monitored by a defrost thermostat, which closes its contacts when the coil temperature drops to the desired point. In addition to the defrost thermostat, another component, the air-pressure switch, is also used to initiate the defrost cycle. The air-pressure switch can measure the pressure both inside and outside the outdoor coil casing. A bellows located within the pressure-sensing device determines the difference in pressure between the inside and outside of the condenser casing. Since the pressure outside the casing and the pressure inside the casing push against opposite sides of the diaphragm within the device, the net pressure that is sensed is the difference between the two.

For example, if the pressure inside the casing is the same as the pressure outside the casing, the net pressure difference will be 0 or, more realistically, very close to 0. These pressures are both important because the control, by taking the difference between them, can determine how much frost has formed on the coil's surface. This method of defrost is successful mainly because the pressure differential across the outdoor coil increases as the

amount of ice on the coil increases. As more ice forms on the outdoor coil, the air pressure inside the casing drops to a pressure lower than that outside the casing, increasing the difference between the two pressures. Once the pressure differential has reached the predetermined set point, the contacts of the device will close to initiate defrost, as long as the temperature requirement for defrost is also met.

SOLID-STATE DEFROST

The defrost methods already described relied on mechanical timers and other mechanical devices such as pressure switches and temperature sensors. Newer heat pump systems do not employ mechanical defrost systems but, instead, have turned to solid-state technology. Instead of bimetal temperature sensors, solid-state systems often use thermistors that sense changes in temperature and change their internal resistance according to these temperature changes. In many cases, the defrost system employs two thermistors (Fig. 27-39). One senses the temperature of the outdoor coil, while the other senses the temperature of the outside ambient air. The differential between the resistances of the two thermistors is what initiates the defrost cycle. As frost accumulates on the outdoor coil, the temperature sensed by the thermistor will drop, since the ice acts as an insulator between the coil and the air passing through it. This will cause the resistance of the thermistor to drop. A temperature differential in the range of 15 to 20 degrees will normally initiate defrost. Once the desired differential is achieved, the printed circuit board will energize the necessary contacts and holding coils to bring the system into defrost. Defrost is often terminated by a high-side pressure switch that is connected to the solid-state defrost control board.

Solid-state defrost can also be initiated by temperature; pressure; or a combination of pressure, temperature, and time. When time is used, solid-state timers are employed, which are more reliable than mechanical, cam-type clocks. The time interval between defrost attempts can often be altered in the field by repositioning a jumper wire on the defrost circuit board itself (Fig. 27-40). Many solid-state control boards give the field technician the choice of 30-, 60-, or 90-minute time intervals between defrost attempts. When pressure-initiated defrost is used, a dual-pressure-sensing switch similar to that used in the mechanical-type defrost control is connected to the circuit board and signals the system to enter defrost when the pressure differential rises to a predetermined point. Just as with the mechanical-type defrost systems described earlier, the choice of defrost method should be based on the geographic location and requirements of the system.

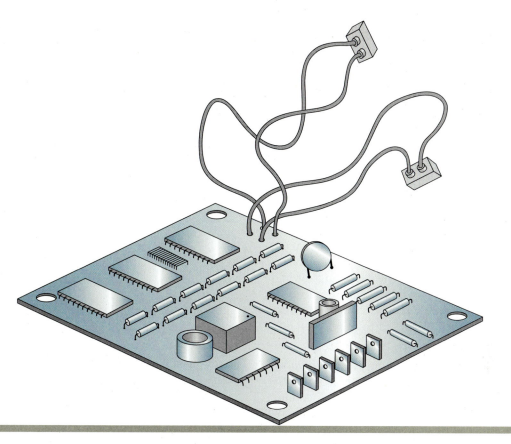

FIGURE 27-39 Solid-state defrost control board with two thermistors.

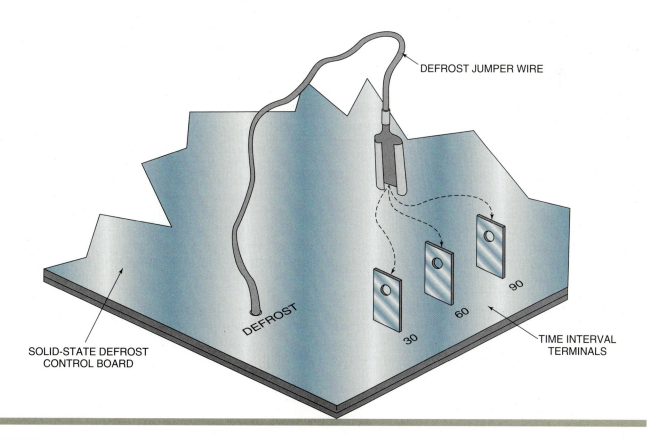

FIGURE 27-40 The time interval can be changed by repositioning the jumper wire on the circuit board.

GEOTHERMAL HEAT PUMP THEORY

When an adequate source of water or ground space is available, a geothermal heat pump system may provide a viable alternative to air- or liquid-source heat pumps. In an air-source heat pump, the efficiency of the system depends directly on the temperature of the air that passes through the outdoor coil both in the heating and cooling modes of operation. During periods of extreme temperatures—when peak efficiency is desired—the air-source heat pumps fail to operate at their full potential. Liquid-source heat-pump systems can also operate at reduced efficiency when the temperatures are out of the design range of the system. During periods of high temperature and humidity, the effectiveness of a cooling tower, for example, is greatly reduced. During the heating months, liquid must be heated to provide an adequate heat source for the system, which can be costly.

Like the other types of heat pump systems discussed earlier, the geothermal heat pump system is a reverse-acting system that provides both heating and cooling, depending on the desired mode of operation. Configured in a manner similar to that of a liquid-to-air system, the geothermal heat pump uses a liquid/liquid heat exchanger as the source of heat during the winter months and as the heat sink during the warmer summer months. The main difference between the liquid-to-air heat pump and the geothermal system is that the temperature of the liquid flowing through the heat exchanger is maintained at a relatively constant temperature by the earth itself. Although the surface temperature of the ground changes with the seasons, the temperature of both the earth and the water some distance below the surface stays constant. This constant temperature allows the geothermal system to operate consistently during a wide range of weather conditions.

This constant and unchanging operation results in an increase in the operating efficiency of the system. The **energy efficiency ratio (EER)** is used to compare its efficiency with that of other systems. The EER is calculated by dividing the Btu output of the system by the power input in watts. The higher this ratio is, the more efficient the system is. One drawback of the EER calculation is that the energy used during system startup and shutdown is not a factor in the calculation. To compensate for this, the **seasonal energy efficiency ratio (SEER)** is used. The SEER takes the energy used during startup and shutdown into consideration. Both the EER and the SEER are higher for heat pump systems, especially geo-thermal systems, than for conventional heating systems.

A good indication of the efficiency of a geothermal heat pump is by determining the system's **Coefficient of Performance (COP)**. The COP is a measure of the amount of energy output per unit of energy input. For example, a heat pump that consumes 1 kWh of energy and provides 3.5 kWh of energy to the house has a COP of 3.5 (3.5 kWh/1 kWh = 3.5). Geothermal heat pumps have a COP rating that often ranges from 3.3 to 5.5. The overall energy efficiency of a geothermal heat pump system will vary from installation to installation. Similar geothermal heat pumps installed in different parts of the country will, for the most part, operate equally efficiently.

GEOTHERMAL HEAT PUMP SYSTEM CONFIGURATION

Two common types of geothermal heat pumps are the open-loop system and the closed-loop system. The **open-loop system** is a special type of liquid-source heat pump. It uses water as a heat source, but the source of this water is the earth—often lakes or wells. The **closed-loop system** uses not the water from the earth, but the earth itself. Large coils of plastic piping, called **ground loops** or **ground coils**, are buried below the earth's surface, and the heat transfer takes place between the earth and the liquid in the piping network.

A number of factors will determine whether the open-loop or closed-loop system is used. These factors include:

- Quantity of available water
- Temperature of available water
- Quality of available water
- Local building and environmental codes
- Available ground space

For an open-loop system to operate properly, an adequate supply of water must be available. The amount of water that is needed depends on the size of the system to be installed. Larger systems will require substantially more water than smaller systems. When an open-loop system is installed, a well must be installed so that the underground water source can be accessed. The required depth of the well, as well as the amount of water that will be available, can be estimated by a contractor that specializes in well drilling in that geographic area. An inadequate water supply will reduce the efficiency and heat-transfer capability of the system. Required flow rates range from 3 to 7 gallons per minute (gpm) per ton of refrigeration. Areas that experience extreme temperatures and climates will normally require a greater flow rate. Manufacturers supply required flow-rate information based on system capacity, geographic location, and supply-water temperature.

The temperature of the groundwater varies with the geographic location. Figure 27-41 shows a map of the groundwater temperatures at depths of 50 to 150 feet.

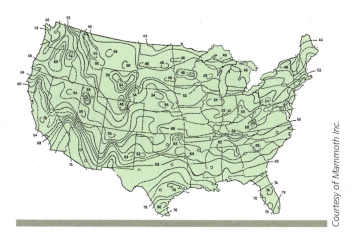

Courtesy of Mammoth Inc.

FIGURE 27-41 Groundwater temperatures (°F) in wells ranging from 50 to 150 feet deep.

The quality of the water is also a major factor that must be considered. The water that can be used in a geothermal system does not necessarily have to meet the standards set forth for human consumption, but guidelines are established, nonetheless. Factors that must be considered include:

- Calcium content
- Magnesium content
- pH level
- Hydrogen content
- Chlorine content

A water test performed by the local health department can determine if the available water meets the manufacturer's requirements. If it is safe for human consumption, it most certainly meets the requirements for heat-pump applications. Note that tying a geothermal heat pump to a potable water source will in no way affect the quality of the water. The water that is reintroduced into the water source will vary only in temperature. The quality of the water is not changed. Even in the unlikely event that a leak should develop within the water/refrigerant heat exchanger, no health hazard will result. Refrigerants and oils used in heat-pump systems are stable, nontoxic, and non-corrosive, and meet UL standards.

Two other factors that need to be considered are local codes and available ground area. Some localities do not allow the installation of an open-loop system. In these instances, a closed-loop system should be used. Closed-loop systems, however, require a sufficient amount of land space for burying the piping circuit.

CLOSED-LOOP SYSTEMS

Closed-loop systems, or **earth-coupled systems**, are designed so that the water or the antifreeze solution in the system does not come in contact with the earth or the

water in the earth. Antifreeze should be used in closed-loop systems that have a groundwater temperature of less than 60°F. The elastic piping that is buried in the ground is referred to as either a ground loop or a **water loop**. The water or antifreeze solution flows through this piping arrangement and either absorbs heat from or transfers heat to the surrounding earth. The liquid is circulated through the piping by means of a centrifugal pump. Ideally, once filled, the ground loop will remain filled. The liquid in the loop absorbs heat from the ground and transfers it to the refrigerant in the heating mode (Fig. 27-42) and absorbs heat from the refrigerant and transfers it to the ground in the cooling mode (Fig. 27-43).

Depending on the available space and the depths that can be reached in the earth, these ground loops can take on a number of different configurations, including:

- Vertical
- Horizontal
- Parallel

No matter which ground loop configuration is used, good practice requires performing a standing pressure test on the piping circuit before covering the loops with soil.

Vertical Configuration In areas where the amount of ground space is small, **vertical ground loops** are desirable. Vertical holes are drilled into the ground to a depth of approximately 150 feet. The actual number of holes that need to be drilled is determined by the capacity of the system and the earth material encountered during the drilling process. They should be drilled approximately 10 to 15 feet apart and should be about 5 to 6 times wider than the diameter of the piping material used. Generally speaking, 200 to 400 feet of piping is required for each ton of system capacity. A typical vertical ground loop configuration is shown in Figure 27-44. This arrangement is configured with three vertical U-bends connected in series with each other. During the heating season, the cool water solution enters the ground loop and absorbs heat from the soil surrounding the loop. This heat is then transferred to the refrigerant which, in turn, is transferred to the air passing over the condenser.

Horizontal Configuration In areas where drilling deep holes is not possible because of rock and other obstructions, the **horizontal ground loop** configuration (Fig. 27-45) is popular. Because these piping configurations are closer to the surface of the ground, longer piping runs are needed. In this case, roughly 300 to 500 feet of piping are required for each ton of system capacity. These systems are ideal where a lot of ground surface area is available.

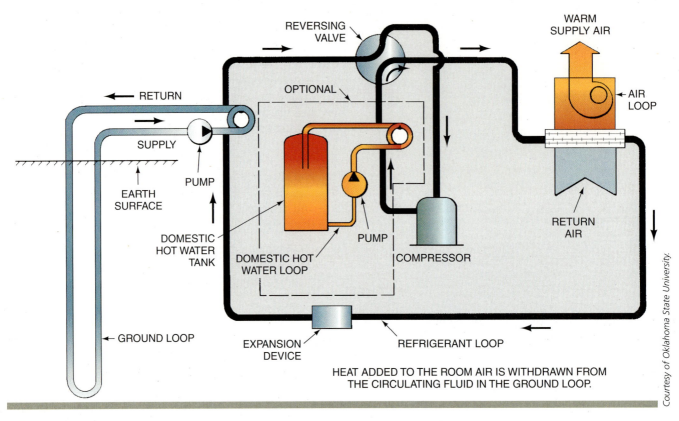

Courtesy of Oklahoma State University.

FIGURE 27-42 Closed-loop, water source heat pump in the heating mode.

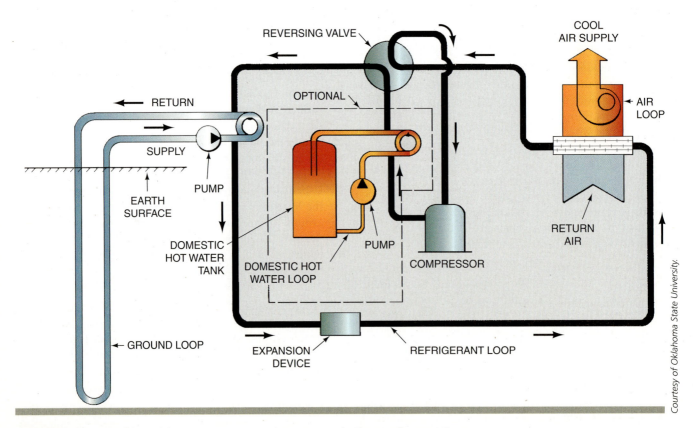

Courtesy of Oklahoma State University.

FIGURE 27-43 Closed-loop, water source heat pump in the cooling mode.

The main drawback to this type of configuration is that the length of the trench that must be dug is equal to the length of the required piping. This can become costly and is, therefore, recommended for use on smaller systems.

To reduce the cost of digging the trench and in areas where limited ground surface area is available, a modified horizontal loop configuration can be used. In this configuration, two layers of piping are installed, one on top of the other (Fig. 27-46). This cuts the amount of trenching that must be done in half. These systems require approximately 20% more piping per ton than the single horizontal loop. Ideally, the lower pipe will be at least 6 feet below grade and the upper and lower pipes should be approximately 2 feet apart.

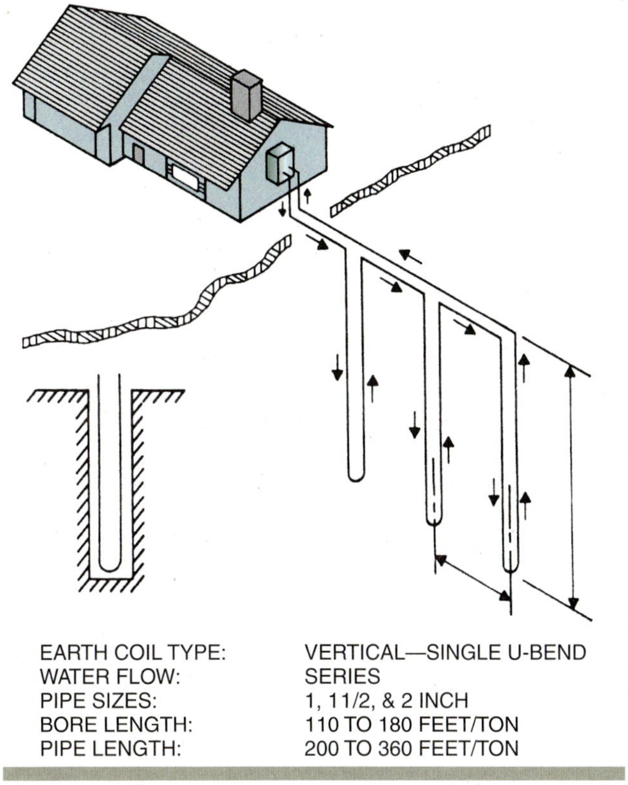

Parallel Loops The configurations described so far are all single-loop systems, in which the water flows through all of the piping material with each pass through the system. This can result in a great amount of resistance when larger systems are installed. In an effort to reduce this resis-tance, **parallel ground loops** are often used. These loops can be installed either vertically (Fig. 27-47) or horizontally (Fig. 27-48).

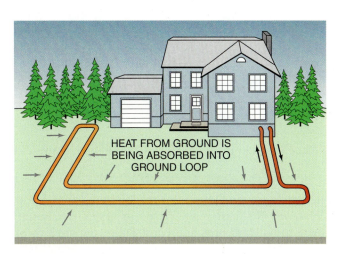

EARTH COIL TYPE: VERTICAL—SINGLE U-BEND
WATER FLOW: SERIES
PIPE SIZES: 1, 11/2, & 2 INCH
BORE LENGTH: 110 TO 180 FEET/TON
PIPE LENGTH: 200 TO 360 FEET/TON

Courtesy of Mammoth Inc.

FIGURE 27-44 Vertical, series ground loop configuration.

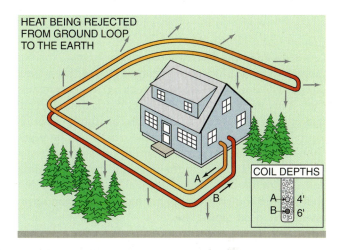

HEAT BEING REJECTED FROM GROUND LOOP TO THE EARTH

COIL DEPTHS
A 4'
B 6'

FIGURE 27-46 Two-layer, horizontal ground loop in the cooling mode.

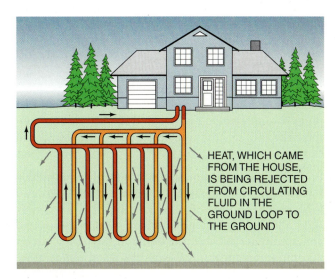

HEAT FROM GROUND IS BEING ABSORBED INTO GROUND LOOP

FIGURE 27-45 Single-layer, horizontal ground loop.

HEAT, WHICH CAME FROM THE HOUSE, IS BEING REJECTED FROM CIRCULATING FLUID IN THE GROUND LOOP TO THE GROUND

FIGURE 27-47 Vertical, parallel ground loop configuration in the cooling mode.

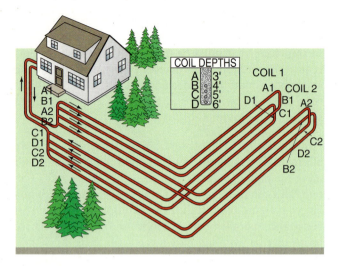

FIGURE 27-48 Four-pipe, horizontal ground loop configuration.

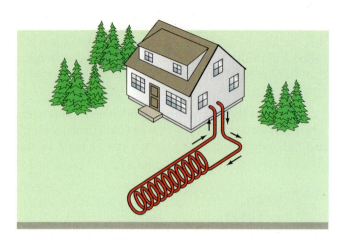

FIGURE 27-49 Spiral-type ground loop.

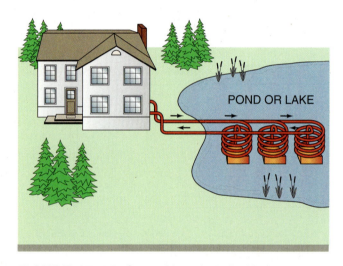

FIGURE 27-50 Ground loop installed in a pond or lake.

Because of the resulting reduced resistance, smaller pumps can be used to move the liquid through the piping circuit. The one main drawback with a parallel system is that, if one of the branches becomes blocked or clogged, it is extremely difficult to clear. On the four-layer horizontal configuration shown in Figure 27-48, the pipes should be 12″ to 18″ apart to ensure proper heat transfer between the liquid in the pipe and the surrounding soil. One minor drawback of parallel loop systems is that they require somewhat more piping material than series ground loop configurations.

Other Configurations Depending on the restrictions posed by the ground material and the availability of a nearby water source, closed-loop systems may take on a configuration different from those just discussed. For example, if limited ground area is available for a horizontal loop system, as well as limited drilling capability, a spiral loop system (Fig. 27-49) may be the answer. The spiral configuration requires that a large hole be dug—as opposed to a narrow trench. The hole should be as deep as possible to accommodate the required amount of piping. The pipes should also be as far apart as possible to maximize the rate of heat transfer between the liquid and the soil.

In areas where a nearby water source is available but open-loop systems are not permitted, a ground loop installed in the lake or pond may be the answer (Fig. 27-50). Since the ground loop is sealed, no actual contact occurs between the liquid in the piping circuit and the water source. Although the water temperature will be slightly less than with an equivalent earth loop, the reduced installation cost will compensate for the small loss in efficiency.

Going Green

Be sure to check your local laws and codes regarding the locating of a closed ground loop in a municipal (public) pond or lake. This practice is often discouraged.

Advantages and Disadvantages of Closed-Loop Systems No matter what type of system is chosen, there are always pros and cons. Following are lists of the advantages and disadvantages of choosing a closed-loop system and a comparison of different closed-loop systems.

Advantages of selecting a closed-loop system:

- Reduced scaling and corrosion occur
- Systems are not affected by drought conditions
- Fouling of the liquid piping circuit is eliminated
- Maintenance of the liquid circuit is minimal
- No chemical treatment of the water circuit is needed
- Water tests are not required to ensure compliance with manufacturer's specifications

Disadvantages of a closed-loop system:

- Water leaks are difficult to locate
- Blockages in parallel loops are difficult to clear
- Initial installations can be more time-consuming compared with open-loop systems
- Initial installation can result in more damage to finished lawns and yards compared with open-loop systems

Comparing different closed-loop systems:

- Horizontal loop systems require more trenching than vertical loop systems
- Horizontal loops require more available ground area than vertical loops
- Blockages are cleared more easily on series loop systems than on parallel loops
- Parallel loop systems require smaller pumps than series loop systems
- Parallel loop systems offer less resistance to water flow than series loop systems
- Vertical loops require deeper penetrations into the earth than horizontal loops

- Spiral loops can be installed in areas that have little ground surface area and that have a difficult earth material to penetrate
- Loops installed closer to the ground surface require more linear feet of piping per ton than those installed deeper in the ground
- Trapped air is easier to remove from a series loop system than from a parallel loop system
- The heat-transfer rate per foot of piping is greater in a series loop system
- Larger-diameter pipe is needed in series loop systems than in parallel loop systems
- The installation costs are higher on series loop systems than on parallel loop systems

OPEN-LOOP SYSTEMS

Unlike ground loop systems in which the liquid is constantly recirculated within the closed system, open-loop systems rely on a constant water source. This source is typically a well system. Just as with the ground loop system, heat is transferred from the water to the refrigerant in the heating mode, as shown in Figure 27-51(A), and from the refrigerant to the water in the cooling mode, as shown in Figure 27-51(B). As mentioned earlier, the water used in an open-loop system must meet certain requirements as indicated by the equipment manufacturer. The main factors that apply to open-loop systems are:

- Water quantity
- Water quality
- Temperature of the water
- Local governmental and environmental codes

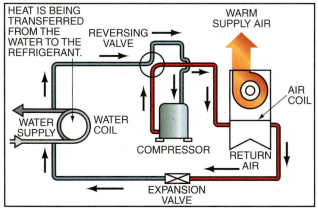

HEATING MODE

HEAT IS BEING TRANSFERRED FROM THE WATER TO THE REFRIGERANT.

IN THE HEATING MODE, HOT REFRIGERANT FLOWS THROUGH THE AIR COIL SUPPLYING WARM AIR TO THE CONDITIONED SPACE.

(A)

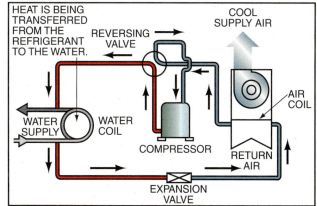

COOLING MODE

HEAT IS BEING TRANSFERRED FROM THE REFRIGERANT TO THE WATER.

IN THE COOLING MODE, COLD REFRIGERANT FLOWS THROUGH THE AIR COIL SUPPLYING COOL AIR TO THE CONDITIONED SPACE.

(B)

Courtesy of Mammoth Inc.

FIGURE 27-51 Open-loop, water source heat pump in (A) the heating mode and (B) the cooling mode.

Going Green

Be sure to check your local laws and codes regarding open loop systems. This system type is not permitted in some geographic locations.

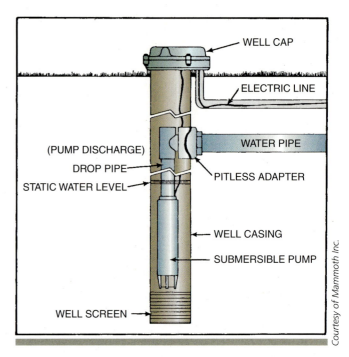

WELL CAP

ELECTRIC LINE

WATER PIPE

(PUMP DISCHARGE)

PITLESS ADAPTER

DROP PIPE

STATIC WATER LEVEL

WELL CASING

SUBMERSIBLE PUMP

WELL SCREEN

Courtesy of Mammoth Inc.

FIGURE 27-52 Basic drilled well.

These were addressed earlier, and each should be evaluated carefully before a system is installed to ensure that the system will operate correctly and that installing an open-loop system is legal. The three common types of open-loop systems are:

- Dedicated, single-well system
- Dedicated, two-well system
- Dedicated, geothermal well system

Dedicated, Single-Well Systems In a dedicated, single-well system, the source of heat for winter operation, as well as the sink for heat in summer operation, is water that is pumped to the system from a well. The well is drilled to a depth required to locate groundwater. Most wells are in the range of 125 to 150 feet, although, in some locations, groundwater can be reached at a depth of only 75 feet. Once the well is drilled, a lining must be installed to prevent it from caving in. This casing is often made of PVC pipe, but it can be made of steel. PVC is preferred in most cases due to the ease of joining sections and due to the fact that PVC is much lighter than steel. The well casing must extend below the natural level of the groundwater.

A submersible pump is located toward the bottom of the well, and the electrical connections are made at the top of the well, under the cap. Connected to the pump is a discharge pipe that carries the water being supplied to the heat-pump system. A cross-sectional view of a typical well is shown in Figure 27-52.

If the pump operated to maintain constant water flow through the heat exchanger, it would never cycle off. A pump that operated continuously would require frequent service and repair, which is not an easy task considering the location of the pump. Retrieving the pump from the location would involve the removal of the connecting discharge pipe section by section as the pump was lifted from the well. After servicing, the piping would have to be reassembled as the pump was lowered again.

In an effort to reduce the run time of the pump, **pressure tanks** are often added to well systems. A pressure tank is simply a pressurized tank that stores water and is equipped with a pressurized air bladder at the top. As the water is forced by the submersible pump up to the heat pump, it first flows into the pressure tank. As the amount of water in the tank increases, the pressure on the bladder increases (Fig. 27-53). Once the pressure in the tank reaches a predetermined set point, a **pressure switch** de-energizes the pump. The water is then stored in the tank until needed by the heat pump system. Once the pressure in the tank drops to a predetermined level, the pump is once again energized. The desired pressure level in the tank can be changed to either increase or decrease the cycle time of the pump.

The pressure tank serves another important function in the heat pump system. Since the tank is pressurized, the water in the heat exchanger is also kept under pressure. This is desirable because minerals present in the water are more soluble under higher pressures thereby reducing the amount of scaling that occurs on the interior surfaces of the heat exchanger. Keeping the water pressurized also reduces the amount of air in the heat exchanger. A slow-closing solenoid valve installed in the water circuit at the outlet of the heat exchanger keeps the pressure in the heat exchanger at the same pressure as that in the pressure tank. The slow-closing solenoid is desired because it eliminates the **water hammer**, the loud hammerlike sound generated when a water valve closes suddenly. Once the water flows through the heat

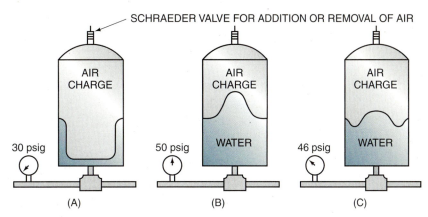

SCHRAEDER VALVE FOR ADDITION OR REMOVAL OF AIR

(A)   (B)   (C)

(A) FACTORY AIR CHARGE

(B) WELL PUMP HAS PRESSURIZED AIR CHARGE AND WATER PRESSURE IS 50 psig.
THE WELL PUMP WILL NOW SHUT OFF BECAUSE PRESSURE SWITCH HAS OPENED
ON A RISE IN PRESSURE.

(C) WHEN WATER IS USED BY THE HEAT PUMP, THE PRESSURE IN THE AIR CHARGE
PUSHES WATER INTO THE SYSTEM. THE PUMP STAYS OFF. THE WELL PUMP ONLY
COMES ON WHEN THE PRESSURE SWITCH CLOSES ON A DROP IN PRESSURE.

FIGURE 27-53 Cutaway view of a pressure tank.

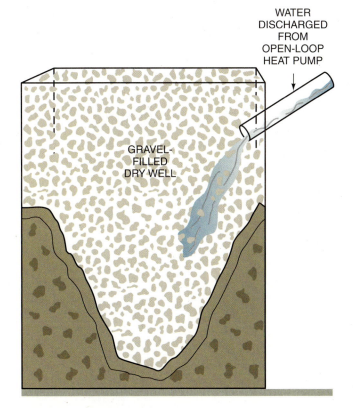

WATER DISCHARGED FROM OPEN-LOOP HEAT PUMP

GRAVEL-FILLED DRY WELL

FIGURE 27-54 A dry well.

exchanger, it is discharged into a pond, lake, stream, or **dry well**, which is simply a large gravel-, sand-, and rock-filled hole in the ground (Fig. 27-54). The discharge from the heat exchanger is pumped to the dry well where the water seeps through the sand and gravel to the **aquifer**, which is the underground water formation. A typical heat-pump setup with a dedicated single well is shown in Figure 27-55.

Dedicated, Two-Well Systems Dedicated, two-well systems are similar to the dedicated, single-well system in that they use a submersible pump to supply water to the heat exchanger. The main difference between the one- and two-well systems is in the disposal of the water after it has left the heat exchanger. In the single-well system, the water is pumped to a lake, pond, or dry well. In a two-well system, the water is pumped into another drilled well, referred to as a return well (Fig. 27-56). The return well should be at least as wide as the supply well to ensure that it can safely handle the water flowing into it. In addition, the return well and the supply well should be located as far apart as possible to prevent the **return water** from affecting the temperature of the **supply water**. The return water is the water that is returned to the ground via the return well, and the supply water is the water supplied to the system via the supply well.

Dedicated, Geothermal Well Systems Geothermal wells are commonly used when there is insufficient water to meet the system requirements. They are often used when the water source is in a **consolidated formation**, such as granite or sandstone. In a geothermal well, the return water that is discharged by the heat exchanger is sent back to the same well from which the supply water is taken. An ample supply of water should be in the well to reduce the possibility of the return water affecting the temperature of the supply water. The return water is returned to the bottom of the well, and the supply water

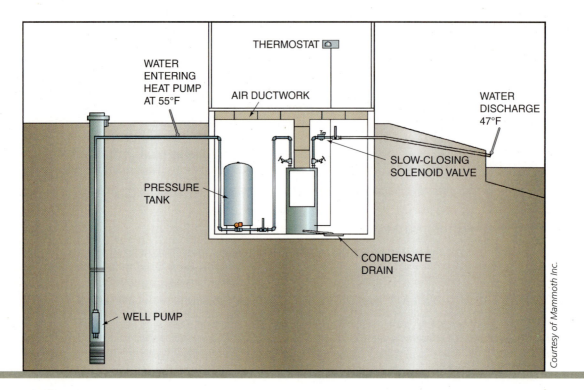

Courtesy of Mammoth Inc.

FIGURE 27-55 Dedicated, single-well system discharging return water into a pond or lake in the heating mode.

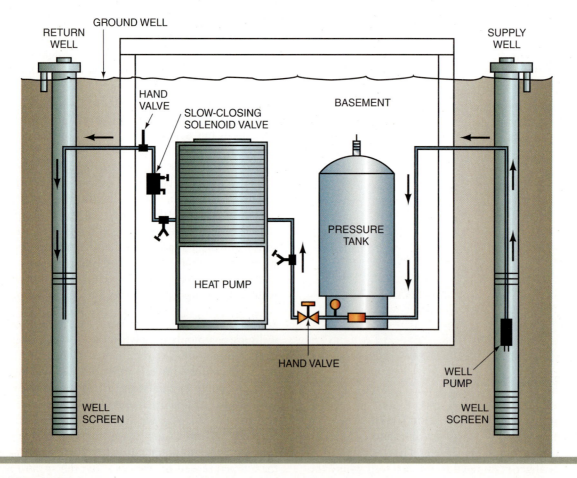

FIGURE 27-56 Dedicated, two-well system.

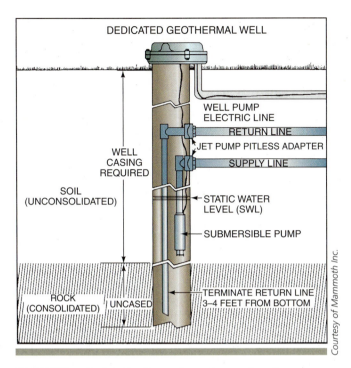

DEDICATED GEOTHERMAL WELL

WELL PUMP ELECTRIC LINE

RETURN LINE

JET PUMP PITLESS ADAPTER

SUPPLY LINE

WELL CASING REQUIRED

SOIL (UNCONSOLIDATED)

STATIC WATER LEVEL (SWL)

SUBMERSIBLE PUMP

ROCK (CONSOLIDATED) | UNCASED

TERMINATE RETURN LINE 3–4 FEET FROM BOTTOM

Courtesy of Mammoth Inc.

FIGURE 27-57 Dedicated geothermal well system.

is removed from the top portion of the aquifer but from an area still below the water table, which is the level or surface of the water. A dedicated geothermal well is shown in Figure 27-57. Because limited water is available for these types of wells, dedicated geothermal wells are typically much deeper than standard dedicated, single- and two-well systems.

TROUBLESHOOTING GEOTHERMAL SYSTEMS

A service technician who has had experience working on standard air conditioning systems or air-to-air and air-to-liquid heat pump systems may be somewhat intimidated when encountering a geothermal heat pump system for the first time. Rest assured that, even though the heat source/sink for the system may seem foreign, many of the same techniques used to troubleshoot standard air conditioning systems can be applied to geothermal systems. The refrigeration system on a geothermal system is identical to that of any other heat pump system, including air-to-air, with the simple exception of the heat exchanger configuration on the outdoor section. The air side of the geothermal system is also identical to that of any other type of system. Knowledge of

refrigerant pressures, saturation temperatures, superheat, and subcooling measurements all come into play, as well as knowledge of mechanical and electrical theory. Troubleshooting a defective motor winding requires the same skills no matter where the motor is located. Identifying a grounded or shorted compressor is a skill that goes far beyond the bounds of air-to-air systems. The majority of these techniques are directly applicable in the case of the geo-thermal heat-pump system.

Troubleshooting geothermal heat-pump systems, although very similar in most respects to handling other heat pump and conventional air conditioning systems, entails some differences of which the servicing technician must be aware. These include:

- The technician should be able to take and interpret temperature readings at both the inlet and the outlet of the water/refrigerant heat exchanger.
- The technician should be able to take and interpret pressure readings at both the inlet and the outlet of the water/refrigerant heat exchanger.
- The technician should be able to take and interpret flow meter readings in the water circuit.
- The technician should be able to interpret manufacturer's guidelines regarding proper water flow through the water/refrigerant heat exchanger.
- The technician should be able to determine approximate groundwater temperatures from a chart or table similar to Figure 27-41.

WATER CIRCUIT PROBLEMS

Like liquid-to-air heat pump systems that are connected to cooling towers, the operating effectiveness and efficiency of geothermal heat pump systems rely on a water circuit that provides an adequate flow rate through the heat exchanger. Improper water flow will result in reduced system efficiency. On all liquid-to-air heat pumps, including geothermal systems, both a liquid/refrigerant heat transfer and a refrigerant heat transfer take place. The refrigerant/air heat transfer is exactly the same, no matter what type of air conditioning system is involved. The water/refrigerant heat transfer, though, can be affected by four major factors:

- Water leaks
- Defective water pumps
- Mineral deposits
- Improper water flow

WATER LEAKS

Since both heat exchangers on geothermal heat pump systems must operate properly for the system to function correctly, the water circuit must be properly maintained. However, on occasion, water leaks do occur and the technician's job is to quickly and accurately diagnose the system problem. A wrong diagnosis on a geothermal heat pump system can result in a great deal of lost time and money. Unnecessarily digging up a ground loop, for example, would definitely not make the customer or the service company owner very happy.

Open-Loop System Water Leaks Water leaks on an open-loop system are generally not as great a problem as they are on ground loop systems. On open-loop systems, the supply water pump is submerged below the water table in the supply well. Leaks on the suction side of the pump are, therefore, almost nonexistent. For the most part, water leaks on the water circuit of an open-loop system are on the supply side of the pump. The possible locations of water leaks on the supply side of the system are:

- In the drop pipe connecting the pump to the underground water line
- In the water line leading to the pressure tank
- In the water line between the pressure tank and the heat exchanger
- In the heat exchanger itself
- In the line leaving the heat exchanger

When a water leak is present in the drop pipe or in the underground water line, chances are that the leak will never be noticed, let alone repaired. There is no way the customer would know that a leak exists, and a service technician would not look for a leak there because there are no symptoms that would indicate a leak in those locations. The submersible pump will continue to operate until the pressure in the pressure tank has reached the preset cutout pressure. If, because of a small leak on the line leading to the pressure pump, the pump takes an extra minute or two to reach the desired pressure, nobody will notice. However, if the leak becomes larger, the pump will not be able to build up the desired pressure in the tank. Then, the lines and possibly the pump will need to be evaluated.

A small leak between the pressure tank and the heat exchanger can easily be spotted. This line is generally run directly between the two pieces of equipment in full view. Routine inspection should reveal any leaks in that relatively short run of piping.

Leaks within the heat exchanger are very difficult to diagnose. The first symptom that a leak exists is the loss of system refrigerant. A refrigerant leak can be located anywhere in the system, and a heat exchanger leak should only be suspected after the rest of the system has been checked thoroughly. The best way to check the heat exchanger for leaks is to remove it from the system and pressurize it as you would a bicycle tire.

If a water leak is present in the line leaving the heat exchanger, once again it will not affect the system operation, but it might cause a flood on the floor. As with the line leading to the heat exchanger, a leak at the outlet of the heat exchanger will become evident during routine maintenance. A leak on an open-loop heat pump system will generally not affect system operation unless the leak is located in the heat exchanger. Leaks should be treated and repaired just like any other plumbing leak would be.

Closed-Loop-System Water Leaks Water leaks on closed-loop systems tend to pose much greater problems than those found on open-loop systems. The majority of the water piping circuit is buried deep underground, and a leak in the ground loop can lead to a costly repair. Luckily, though, the majority of water leaks on ground loop systems occur at the circulating pump or the flanges that connect it to the system. The following telltale signs indicate that a water leak is present in the water circuit:

- An audible crackling sound at the circulating pump
- A low temperature differential of the water across the heat exchanger in both the heating and cooling modes
- Low suction pressure in the heating mode
- High head pressure in the cooling mode
- Low pressure drop across the heat exchanger
- Low flow rate through the heat exchanger

Probably the most indicative sign of a water leak on a ground loop system is the audible crackling sound at the pump. This is a sure sign that air has entered the water circuit. A low temperature differential across the heat exchanger indicates that the heat exchange rate is lower than it should be. This can result from:

- Water flowing too fast through the heat exchanger
- High water temperature in the summer
- Low water temperature in the winter

The first possibility can be ruled out immediately unless the pump has been recently replaced with a larger, higher-volume pump. In this case, the pump should be replaced with one that meets system requirements. The second and third possibilities go hand in hand. If the ground loop is shorter than it should be, the water will not have enough time to pick up heat in the winter and reject heat in the summer. This would result in lower suction pressure while operating in the heating mode and higher head pressure when operating in the cooling mode. Assuming that the system was designed properly, the ground loop should be the correct length. The impression of an undersized ground loop can easily occur on parallel ground loop systems where one of the parallel loops has become inoperable due to an air restriction. An air restriction in the loop, which indicates that a leak is present in the system, can be remedied relatively easily if a manifold-type valve arrangement is installed on the system. Strategically placed valves and hose bibs can easily clear away an air restriction (Fig. 27-58). Under normal operation, all valves are in the open position and the hose bibs are closed. Each parallel loop can be isolated, and water can be pushed through it to alleviate any air pockets that may be present (Fig. 27-59). In Figure 27-58, all valves are closed with the exception of the one supplying water to the first loop. The hose bib on the first loop is open, and any air is pushed through the loop and out through the hose bib.

Another possible reason for high head pressure in the cooling mode and low suction pressure in the heating mode is the loss of system water. Less water in the loop means a lower heat-transfer rate in the heat exchanger. A loss of system water is indicated by reduced water flow through the exchanger and a low pressure drop across the heat exchanger. Technicians should be able to take both pressure and temperature readings at the inlet and outlet of the exchanger. The most commonly used device to obtain these readings is a **Pete's port** (Fig. 27-60). Pressure gauges or thermometers can be inserted into the port without using any tools whatsoever. Simply pushing the pressure gauge or thermometer into the port causes the rubber gasket to shift, giving access to the water circuit (Fig. 27-61). Taking pressure and temperature readings at both ends of the exchanger enables the technician to calculate pressure drops, temperature differentials, and flow rates. The piping arrangement of a typical ground loop system is shown in Figure 27-62. This arrangement gives the technician access to the ground loop by means of hand valves installed between the ground loop and the heat exchanger.

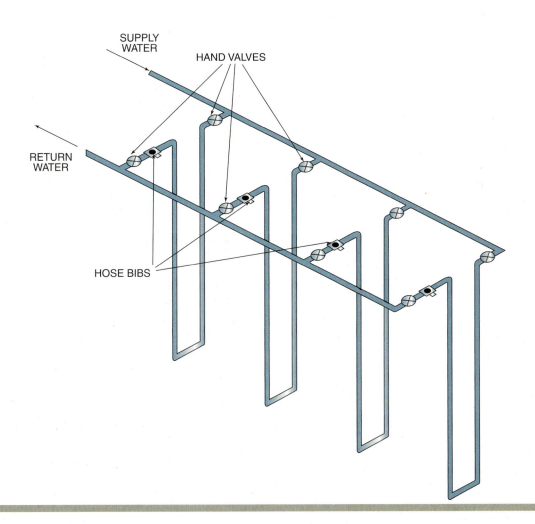

SUPPLY WATER
HAND VALVES
RETURN WATER
HOSE BIBS

FIGURE 27-58 Manifold valving arrangement on a parallel loop system.

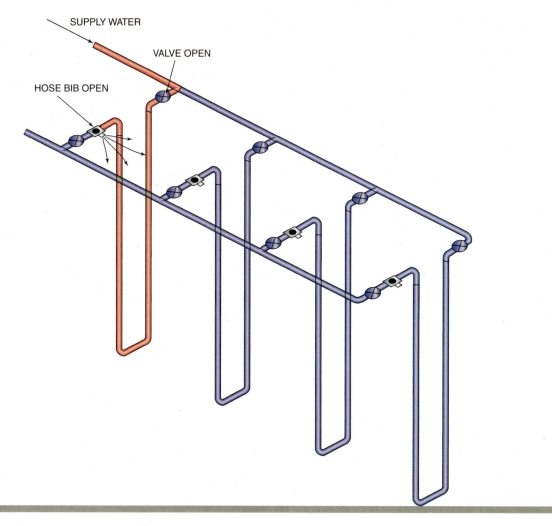

FIGURE 27-59 Water can be forced through any loop to remove air pockets.

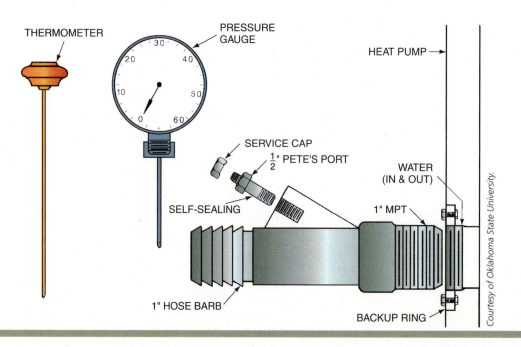

FIGURE 27-60 Pete's port used to obtain temperature and pressure readings in a ground loop.

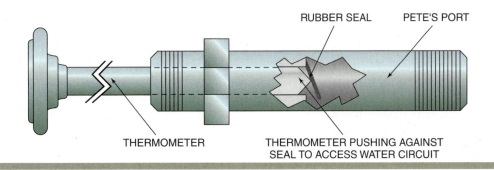

FIGURE 27-61 Cutaway of a Pete's port access fitting.

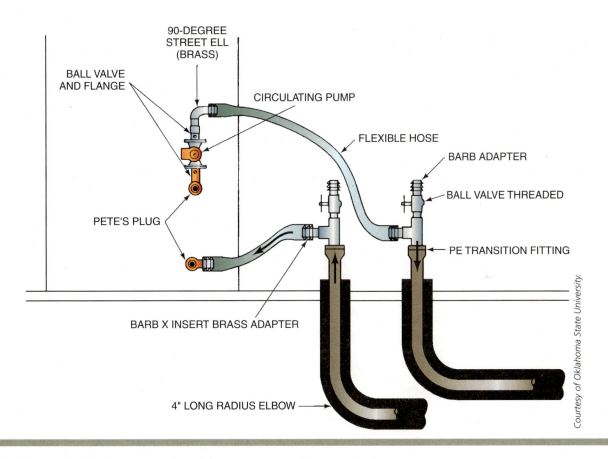

Courtesy of Oklahoma State University.

FIGURE 27-62 Piping connections between a ground loop and a heat pump system.

Defective Water Pumps Centrifugal pumps (Fig. 27-63) commonly found on ground loop systems, are much like the pumps found on domestic boilers. These pumps are comprised of the motor and the **impeller** (Fig. 27-64) connected with a spring-type linkage and a watertight seal. A defective motor does not necessarily mean that the impeller assembly is defective as well, and vice versa. The technician must be able to evaluate each of these components separately to avoid making an improper system diagnosis.

Pump Motors

Just like any other motor, pump motors can be evaluated both electrically and mechanically. Electrical problems usually involve open, grounded, and shorted motor windings. Equipped with a digital multimeter, a technician can check the motor for these conditions and diagnose accordingly. If the motor windings are found to be grounded, shorted, or open, the pump motor should be replaced. Mechanical problems include defective bearings and improper lubrication, which can lead to the motor

from experience...

Note that any of the possible results of a leaking water circuit could also be the result of other system problems. A system operating with a low suction pressure in the heating mode does not automatically mean that the system has a water leak. When a water leak exists, more than one and often all of the symptoms appear. Before concluding that a leak exists the technician must inspect the pump and make certain that it is operating properly.

jamming and not starting. When checking the motor, the technician should disconnect the impeller linkage from the motor shaft. This ensures that the motor is being evaluated and not the impeller or the linkage.

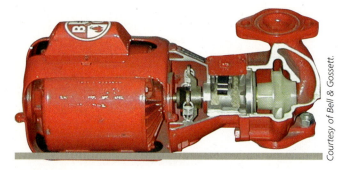

FIGURE 27-63 Centrifugal pump.

Courtesy of Bell & Gossett.

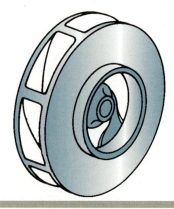

FIGURE 27-64 Pump impeller.

For example, consider the following scenario. A customer calls her service company and reports that her system is not cooling. Upon arrival, the technician notices that the circulating pump on the ground loop is humming but not turning. The technician concludes that the pump motor is defective and runs out to get a replacement. After replacing the motor and restarting the system, the new motor hums as well but does not turn. The problem is not the motor at all, but a jammed pump impeller!

Impeller Assemblies

Impeller assemblies should also be evaluated with the motor removed from the linkage. The linkage and the impeller should turn freely, but with some resistance. An impeller that can be turned only with excessive force should be a prime candidate for replacement. The impeller and the linkage can be inspected and checked without accessing the water circuit directly.

If the pump motor operates correctly and the impeller seems to spin freely, ample water flow should occur through the water loop. If a flow meter registers no water flow at all, the impeller may indeed be broken inside the casing of the pump or the flow meter may be bad. How can the technician tell? If no pressure drop occurs across the heat exchanger, no flow is occurring through it, and the pump is not moving any water through the piping circuit. If a pressure drop occurs across the heat exchanger and the flow meter indicates no flow, it is time to get a new flow meter.

Mineral Deposits Mineral deposits on the interior of water system piping are practically nonexistent on ground loop systems that are sealed. On the other hand, mineral deposits are a major concern in open-loop systems. Mineral deposits act as an insulator between the refrigerant and the water in the heat exchanger. The presence of mineral deposits can be diagnosed relatively easily. The following conditions are present when mineral deposits and scale have coated the interior surface of the water side of the heat exchanger:

- Low-temperature differential occurs across the heat exchanger in both heating and cooling modes.
- Low suction pressure occurs in the heating mode.
- High discharge pressure occurs in the cooling mode.
- The pressure drop across the heat exchanger is normal.
- The water flow rate, in gpm, through the heat exchanger is normal.

If mineral deposits are found to be present in the heat exchanger, the heat exchanger must be cleaned chemically, like the water-cooled condensers that were introduced in an earlier chapter. Before adding chemicals to the heat exchanger, the technician should check the manufacturer's recommendations for cleaning. The heat exchanger ideally should be flushed and back-flushed to remove all scale accumulation and mineral deposits within the coil.

> ## from experience...
> If mineral deposits form regularly within a heat exchanger, a water sample test may indicate that a water treatment plan is in order. Water treatment is more cost-effective than removing and cleaning a heat exchanger every few years.

Improper Water Flow Improper water flow can result from some of the conditions already discussed in this chapter. The symptoms of improper water flow are:

- Small pressure drop across the heat exchanger
- Large temperature differential across the heat exchanger in both the heating and cooling modes
- Slightly higher discharge pressure in the cooling mode
- Slightly lower suction pressure in the heating mode
- Low water flow rate through the heat exchanger

A small pressure drop across the heat exchanger is an indication that not enough water is flowing within the water circuit. This, teamed with a large temperature differential across the exchanger, indicates that the water is remaining in contact with the refrigerant for a longer period of time than normal. The water is picking up more heat from the refrigerant in the summer and giving up more heat in the winter.

Because the water is remaining in contact with the refrigerant longer, the rate of heat exchange is actually going to decrease. For example, in the cooling mode, cool water enters the heat exchanger and begins to absorb heat from the refrigerant. Because the water flow rate is low, the water remains in the coil longer and its average temperature increases. A higher water temperature results in a reduction in the amount of heat transfer between the water and the refrigerant, because these temperatures are now closer together. This will result in a higher-than-normal head pressure in the cooling season. The same argument, only in reverse, can be made to explain the lower-than-normal suction pressure during the winter season.

AIRFLOW PROBLEMS

Airflow problems are the same regardless of what type of air conditioning system is being evaluated. Insufficient airflow through a duct system can be a result of:

- A defective blower motor
- Dirty or blocked air filters
- Dirty or blocked indoor coil
- Closed supply registers
- Blocked return-air grill
- Loose duct lining
- Broken or loose belts on the blower
- Dirty squirrel-cage blower

These items are very important as far as system performance goes. A problem with the airflow through the indoor coil can result in system problems, or apparent system problems, with the other heat exchanger as well. Therefore, the technician must check all aspects of system operation before committing to a major system repair that could have been avoided. To establish that flow exists through the air side of a heat-pump system, the preceding items must be checked. Now consider two scenarios in which airflow problems can lead a technician's troubleshooting efforts astray.

COOLING EXAMPLE

Insufficient airflow through the indoor coil during the cooling mode will result in low suction pressure. This is a result of air coming in contact with the cooling coil for a longer period of time, decreasing the average temperature of the air passing over the coil. As the air temperature over the coil decreases, the suction pressure of the system decreases as well. How will this affect the rest of the system?

Consider a ground loop system that operates normally with the following conditions in the cooling mode:

- Water supplied to the heat exchanger: 68°F
- Return water from the heat exchanger: 80°F
- System refrigerant: R-140A
- Suction pressure: 115 psig
- Head pressure: 335 psig
- Water flow through the heat exchanger: 9 gpm
- Pressure drop across the heat exchanger: 2 psig

If the air filter becomes clogged with dirt, consider that the suction pressure now drops to 100 psig. This lower suction pressure will result in a lower head pressure of 320 psig. The condenser saturation temperature of the system has dropped from 105°F to 100°F. The temperature of the water entering the heat exchanger remains at 68°F. The temperature differential between the refrigerant and the water has become less, resulting in a lower heat-transfer rate. The temperature of the water leaving the heat exchanger drops from 80°F to 76°F. The normal temperature differential across the water circuit is 12 degrees (80°F − 68°F) but is now only 8 degrees (76°F − 68°F). The pressure across the heat exchanger and the flow rate through the heat exchanger remain unchanged. This may lead a technician to conclude that the reduced temperature differential is due to mineral deposits on the interior surfaces of the exchanger when, in fact, the only problem is a dirty air filter.

Service technicians must evaluate all symptoms before making a final diagnosis. In the system just examined, the head pressure of the system dropped and the temperature differential across the heat exchanger dropped as well. If, indeed, mineral deposits were present in the heat exchanger, the head pressure would have risen, not dropped!

HEATING EXAMPLE

Consider a ground loop system that operates normally with the following conditions in the heating mode:

- Water supplied to the heat exchanger: 68°F
- Return water from the heat exchanger: 58°F
- System refrigerant: R-410A
- Suction pressure: 115 psig
- Head pressure: 335 psig
- Water flow through the heat exchanger: 9 gpm
- Pressure drop across the heat exchanger: 2 psig

Once again, the air filter has become clogged with dirt. The head pressure, which was normally 335 psig, now rises to 365 psig. This higher head pressure will result in a higher suction pressure of 128 psig. The evaporator saturation temperature of the system has increased from 38°F to 44°F. The temperature of the water entering the heat exchanger remains at 68°F. The temperature differential between the refrigerant and the water has become smaller, resulting in a lower heat-transfer rate. As a result of this, the temperature of the water leaving the heat exchanger increases from 58°F to 62°F. The normal temperature differential across the water circuit is 10 degrees (68°F − 58°F) but is now only 6 degrees (68°F − 62°F). The pressure across the heat exchanger and the flow rate through the heat exchanger remain unchanged. This may lead a technician to conclude that the reduction in the temperature differential across the heat exchanger is due to mineral deposits on the interior surfaces of the exchanger when, once again, the only problem is in fact a dirty air filter.

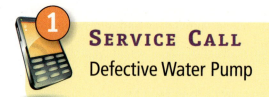

SERVICE CALL
Defective Water Pump

CUSTOMER COMPLAINT

The owner of a geothermal heat pump system calls his service company and informs the service manager that his system is not providing adequate cooling to the house. Prior to his calling the company, the home owner has noticed that the outdoor unit is not operating. The system is equipped with a single-layer, horizontal ground loop.

TECHNICIAN EVALUATION

Upon arriving at the home, the service technician, Greg, inquires about the system and is told that the unit is not cooling at all. He is also informed that the indoor fan seems to be operating but that the air coming from the supply registers is warm. Greg checks the thermostat settings and can hear the indoor fan operating. He goes to a supply register and can feel that a large quantity of air is being discharged into the room. He checks the air-return grill in the hallway and confirms that the filter is clean. Having determined that the airflow is adequate, he then makes his way to the heat pump unit in the basement.

Upon initial inspection, Greg can see that the outdoor unit is not operating. He opens the service disconnect and checks for power. He reads 218 volts. He then opens the ser-vice panel and checks the voltage at the line side of the compressor contactor. Again he reads 218 volts. A check of the voltage on the load side of the contactor yields a reading of 0 volts. Touching the compressor, Greg establishes that the compressor is not overheated, because it is cool to the touch. Since the blower motor is operating, Greg knows that the control transformer is operational. He then decides to check the voltage at the holding coil of the compressor contactor. He obtains a reading of 0 volts.

Close examination of the control circuit shows that both a low-pressure and a high-pressure switch ware wired in series with the holding coil. Greg checks voltage across the low-pressure switch and, from a reading of 0 volts, determines that the switch is in the closed position. Taking a reading across the high-pressure switch, a reading of 24 volts is obtained. The system has cut off on high head pressure. Greg now turns his attention to the water circuit.

Checking the circulator pump, he can hear a motor attempting to turn. With a flashlight, he inspects the shaft of the pump and the linkage connecting the impeller to the shaft. The motor is turning very slowly. After turning off the service disconnect at the unit, Greg removes the motor mounts and disconnects the linkage from the shaft. He then sets the motor on the floor, still connected electrically, and re-energizes the system. The motor starts immediately, so he turns the system power off again and goes to the linkage assembly.

Greg attempts to turn the linkage by hand but is unable to do so. The linkage is jammed, so the pump needs to be replaced. He leaves the customer's home to obtain a replacement pump. Upon returning, he valves off the ground loop and drains the heat exchanger into a bucket. He then mounts the new pump body, moistening the gaskets with a small amount of oil. This helps ensure a tight seal between the flanges. Once the pump body is mounted, Greg fills the heat exchanger with water and makes certain that all air is removed. He examines the flange connections for leaks and then mounts the motor back in place.

Once the motor is properly mounted, Greg opens the valves to the ground loop and restarts the system. The pump begins to operate. Knowing that the pump is now fully operational, Greg resets the high pressure control in the low-voltage compressor control circuit. The compressor starts immediately. Using the Pete's port, Greg checks the pressure drop across the exchanger. Finding everything working well, he fills out his paperwork and proceeds to his next job.

SERVICE CALL DISCUSSION

Before Greg even set eyes on the system, he had already established that proper airflow existed through the system. He was then able to narrow his search to the refrigerant and water circuits. Checking the control circuit led him to the high-pressure switch. Since the head pressure is controlled by the amount of heat transfer that takes place between the water and the refrigerant, he immediately zeroed in on the water circuit—no wasted time here, just good solid troubleshooting.

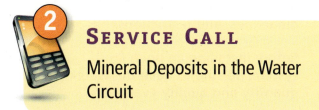

SERVICE CALL
Mineral Deposits in the Water Circuit

CUSTOMER COMPLAINT

A customer calls and informs her service company that her geothermal heat pump system is not providing adequate heat. She can hear the compressor operating, but it does not seem to be doing its job. The customer has an open-loop system connected to a dedicated, single well.

TECHNICIAN EVALUATION

Later that afternoon, Hal, the service technician, arrives at the customer's home. The home owner informs the technician that her system is operating but that it does not seem to be heating the house effectively. Hal first goes to the thermostat and checks the settings. He sets the system to operate in the heating mode and then goes downstairs to the heat pump system.

While at the unit, he can hear the blower motor operating. He checks the air filter and sees that is has been recently replaced. He checks the temperatures of the supply and return air and finds that the temperature differential across the indoor coil is only 8 degrees. Under normal conditions, Hal would expect a temperature differential of approximately 14 degrees. He installs his gauge manifold on the system and obtains a low-side pressure of 90 psig and a high-side pressure of 240 psig. For an R-410A system, these pressures seem very low. Initially, he suspects that the system might be short of refrigerant, but before adding refrigerant to the system, he checks the water circuit to confirm that the system is in need of refrigerant. He expects that, since the suction pressure is lower than normal, the heat-exchange rate between the water and the refrigerant will be higher than normal.

Using the Pete's port on the unit, Hal takes pressure and temperature readings at both the inlet and the outlet of the heat exchanger. The pressure drop through the heat exchanger is about 2 psig, which is normal for that system. The temperature differential across the coil is only 4 degrees, far below the normal 10-degree split that was expected. Hal's initial suspicion is incorrect. The system is not short of refrigerant.

Hal soon discovers that a normal pressure drop through the heat exchanger, a low temperature differential across the heat exchanger, and low operating pressures indicate that the heat exchanger has mineral and scaly buildup on its interior surfaces. An appointment is made for the coil cleaner to come to the home and clean the coil. Once the coil has been properly cleaned with the appropriate chemicals, Hal returns to check out the system. The operating pressures are well within the proper ranges, the temperature differential across the heat exchanger is back in the 10-degree range, and the system has been left in good working order.

SERVICE CALL DISCUSSION

Hal could have found himself in a real bind had he added refrigerant to the system. However, he had the insight to exhaust all possibilities before altering the refrigerant charge of the system. He had been expecting to find a large temperature differential across the heat exchanger but instead found a low temperature differential. Using his knowledge of the water and refrigerant circuits, he was able to narrow his search down to the heat exchanger itself.

SUMMARY

- The four-way reversing valve permits the automatic changeover from heating to cooling and vice versa.
- Reversing valves can be either direct-acting or pilot-operated.
- The outdoor coil functions as the condenser in the cooling mode and as the evaporator in the heating mode.
- The indoor coil functions as the evaporator in the cooling mode and as the condenser in the heating mode.
- Heat pump systems can be either air-source or liquid-source systems.
- Suction line accumulators reduce the chance of liquid refrigerant flooding back to the compressor.
- Air-source heat pumps use air as the heat source while operating in the heating mode.

- Liquid source heat pumps use liquid as the heat source while operating in the heating mode.
- Heat pump defrost is accomplished by putting the system into the cooling mode temporarily; it is controlled by a combination of time, temperature, and pressure.
- Geothermal heat pump systems provide relatively constant performance year-round.
- Geothermal systems can be open-loop or closed-loop systems.
- Closed-loop systems can be configured in either series or parallel patterns.
- Open-loop systems often require the use of a well to provide water to the heat exchanger.
- Geothermal wells have the supply and return water piped within the same well.
- Geothermal water circuit problems can be in the form of water leaks, air restrictions, defective water pumps and motors, or mineral deposits.
- A Pete's port can be used to obtain pressure and temperature readings to aid in the troubleshooting process.
- Airflow problems can result in system conditions that resemble water flow problems.

GREEN CHECKLIST

☐ **Heating Seasonal Performance Factor (HSPF) is used to measure heating efficiency of air-cooled heat pump systems.**

☐ **Seasonal Energy Efficiency Ratio (SEER) is used to measure the cooling efficiency of heat pump systems.**

☐ **SEER ratings of heat pump systems typically range from 13 to 19.**

☐ **Coefficient of Performance (COP) is a measure of the amount of energy output per unit of energy input.**

☐ **Advanced heat pump technologies make air source heat pumps more**

attractive in colder climates such as the upper Midwest, the Northeast, and Pacific Northwest.

☐ **Before installing a closed ground loop in a municipal pond or lake, be sure to check local codes, guidelines, and restrictions.**

☐ **Before installing an open loop system, be sure to have the water quantity and quality evaluated.**

☐ **Before installing an open loop system, be sure to check local codes, guidelines, and restrictions as some areas do not allow the installation of open loop heat pump systems.**

REVIEW QUESTIONS

1. **During the heating cycle of a heat pump system, the hot gas leaving the compressor will flow to the _____ immediately after leaving the reversing valve.**
   a. Evaporator coil
   b. Condenser coil
   c. Expansion device
   d. None of the above is correct

2. **A liquid-to-air heat pump system uses**
   a. Air as the heat source when operating in the heating mode
   b. Air as the heat sink when operating in the cooling mode
   c. Liquid as the heat source when operating in the heating mode
   d. Liquid as the heat source when operating in the cooling mode

3. **Which of the following can have a negative effect on the operation of a heat pump system? Excessive ice formation on the _____**
   a. Outdoor coil when operating in the heating mode
   b. Outdoor coil when operating in the cooling mode
   c. Indoor coil when operating in the heating mode
   d. Discharge line when operating in the heating mode

4. **Ideally, there will be low-pressure, low-temperature, superheated vapor at**
   a. The outlet of the outdoor coil in the heating mode
   b. The outlet of the indoor coil in the cooling mode
   c. The inlet of the compressor in the heating mode
   d. All of the above are correct

5. **A disc-type check valve is installed correctly when**
   a. It is in the horizontal position
   b. It is in the vertical position
   c. The arrow points in the direction of refrigerant flow
   d. The arrow points in the direction opposite to the direction of refrigerant flow

6. **A heat pump system that initiates its defrost cycle by temperature alone is most likely equipped with which of the following?**
   a. Thermistor
   b. Pressure-sensing switch
   c. Defrost timer
   d. Time delay relay

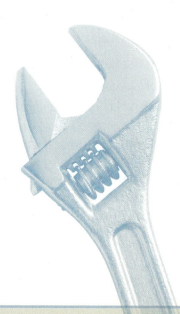

7. **Typically, a liquid-to-air heat pump system will use**

   a. One tube-in-tube heat exchanger and one fin-and-tube heat exchanger

   b. Two tube-in-tube heat exchangers

   c. Two fin-and-tube heat exchangers

   d. No heat exchange surfaces at all

8. **In a large urban city, the source of water for large liquid-to-air heat pump systems is most likely**

   a. A cooling tower

   b. The earth

   c. Local municipal water supply

   d. A lake or pond

9. **Water-regulating valves on a liquid-to-air heat pump system**

   a. Prevent the system pressure from rising above a predetermined set point

   b. Prevent the system pressure from falling below a predetermined set point

   c. Both a and b are correct

   d. Neither a nor b is correct

10. **In which of the following ground loop systems would a blockage be most difficult to clear?**

    a. Four-layer horizontal ground loop

    b. Vertical series loop

    c. Spiral loop

    d. Single-layer horizontal ground loop

11. **According to Figure 27-41, which of the following states has the warmest groundwater temperature?**

    a. Washington

    b. California

    c. New York

    d. Texas

12. **How does groundwater temperature relate to the heat transfer rate?**

    a. Low groundwater temperature results in a high heat transfer rate in the heating mode

    b. High groundwater temperature results in a high heat transfer rate in the cooling mode

    c. Low groundwater temperature results in a high heat transfer rate in the cooling mode

    d. High groundwater temperature results in a low heat transfer rate in the heating mode

13. **Which type of system gets its supply water from the top of a water column, circulates this water through the heat exchanger, and then returns the water to the same water column?**

    a. A dedicated, single-well system

    b. A dedicated, two-well system

    c. A dry well

    d. A dedicated, geothermal-well system

14. **A dry well is typically filled with**

    a. Sand

    b. Rocks

    c. Gravel

    d. All of the above are correct

15. **In the event of a severe drought, which of the following system types will be affected least?**

    a. A dedicated, geothermal well system

    b. A dedicated, two-well system

    c. A vertical series loop system

    d. None of the above systems are correct

KNOW YOUR CODES

The codes regarding air source heat pumps are typically the same as those that govern the installation of any other split-type central air conditioning system. However, codes vary greatly when dealing with earth-coupled heat pumps systems. Contact your local planning division or governing body to determine the requirements, restrictions, and regulations in place that will affect the installation of water source heat pumps.

In addition to the codes involved, water quality and quantity will have a major impact in the decision-making process about whether or not a water source heat pump is a viable option. Contact the municipal water supplier in your area and request information regarding mineral concentrations, water table level, and other pertinent information regarding your areas water supply.

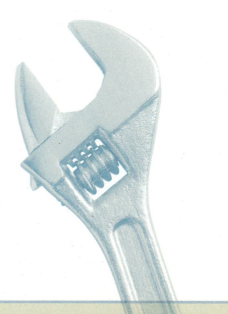

Glossary

A

absolute pressure the pressure scale that takes atmospheric pressure into account

accessible when applied to equipment such as disconnect switches, accessible means that these devices are not guarded by locked doors or other means that would prevent easy access to them. For example, you should not need a ladder to gain access to a service disconnect switch

acetylene popular gas used for soldering and brazing

ACR tubing nitrogen-charged piping used in air conditioning applications

acrylonitrile butadiene styrene (ABS) pipe rigid piping material used for drain lines

active recovery recovery method that uses a self-contained recovery unit

adjusting the process of varying fluid flow by changing the settings on fluid flow controls such as valves, dampers, and blowers

air cushion air above the semipermeable membrane in an expansion tank

air handler portion of an air conditioning system responsible for treating air from the occupied space

air separator device that separates air from water in the system

air source heat pumps heat pumps that use air as the source of heat while operating in the heating mode

air vent fitting used to remove air, either manually or automatically, from a hydronic system

air-to-air heat pumps air-source heat pumps used to treat air

air-to-liquid heat pumps airsource heat pumps used to treat liquid

alkylbenzene (AB oil) synthetic oil with HCFC and CFC refrigerants

American National Standards Institute (ANSI) organization that coordinates the voluntary formation of standards that ensures the uniformity of products, processes, and systems

American Society for Testing and Materials (ASTM) agency that established minimum specifications for fuel oils

American wire gauge (AWG) provides data, including size and characteristics, for various types of electric conductors

ampacity the current, in amperes, that a conductor can continuously carry

amperes or amps unit that indicates the amount of current that flows in an electric circuit when 1-Volt is applied to a 1-ohm resistance

analog meter piece of electrical test equipment that uses a moving needle to indicate the value of electrical characteristics

appliance branch circuit circuits that are used to supply power to outlets that will have appliances connected to them

aquastat electrical component that opens and closes its contacts to energize and de-energize electric circuits in response to the water temperature sensed by the device

aquifer underground water formation

armature portion of a relay that moves to open or close sets of electric contacts

asphyxiation loss of consciousness that is caused by a lack of oxygen or excessive carbon dioxide in the blood

atmospheric burners burners in which combustion air is supplied at atmospheric pressure

atmospheric pressure the weight of the gases that exert a force on the Earth's surface

atom the smallest quantity of a naturally occurring element

atomization process of breaking the fuel oil up into a fine mist, which facilitates combustion

atomizing the process of spraying a fine mist of water

automatic air vent fitting that automatically removes air from a hydronic heating system

automatic changeover term used to describe thermostats that switch automatically between the heating and cooling modes of operation

automatic expansion valve (AEV) metering device that maintains a constant evaporator pressure

automatic reset feature that allows an overload or

safety device to reset itself automatically when the dangerous or undesirable condition no longer exists

auxiliary drain pan installed under the air handler in the event the condensate pan overflows

azeotropes blended refrigerants that behave as a single refrigerant with one saturation temperature for a given pressure

B

balancing the process of matching the actual fluid flow rates to the desired fluid flow rates within an acceptable range

balancing valve manually controlled valve used to increase resistance and reduce water flow through a given branch circuit

ball-type check valve device that permits fluid flow in only one direction; a floating ball in the valve body prevents fluid from flowing in the wrong direction

bare conductor term used to describe a conductor that has no covering at all.

barometric damper see draft regulator

baseboard sections see radiator or terminal unit

belt drive the blower arrangement where the blower is connected to the motor by means of belts and pulleys

bidirectional liquid-line filter drier filter drier used

for heat pump applications that permits refrigerant flow through the device in both directions

bilateral switch used in cad cell primary controls, the electronic control switching device that works in conjunction with the triac to control burner operation

bimetal or bimetal overloads see bimetal strips

bimetal strips metal strips made up of two dissimilar metals that expand and contract at different rates at different temperatures

bimetal thermostats thermostats that open and close electrical contacts as a result of the flexing of a bimetal strip

blended refrigerants refrigerants that are created by mixing two or more other refrigerants

boiler piece of heating equipment designed to heat water, using electricity, gas, or oil as a heat source, for the purpose of providing heat to an occupied space or potable water

boiler/water feed valve valve that reduces the pressure entering the structure to the pressure required by the hydronic system; automatically feeds water into the system to maintain the desired water pressure

boot duct section that connects the branch duct to the supply register

branch circuit the wiring that is installed between an

overcurrent device and an outlet branch termination points

brazing process of joining metallic piping sections using filler materials with melting points in the range of 1500°F

British thermal unit (Btu) the amount of heat required to raise the temperature of 1 pound of water 1°F

bubble point for blended refrigerants, the temperature used to calculate condenser subcooling

building setback the required distance between a property line and structures on the property, or between two structures on a piece of property

bulb pressure pressure that facilitates the opening of the thermostatic expansion valve

bulldog snip sheet metal snip with a short nose. Used for cutting multiple layers of sheet metal

bypass humidifiers humidifiers that use the difference in pressure between the supply and return ducts to move air across the humidifier media

C

cad cell cadmium sulfide device that changes its resistance as the amount of light it senses changes

canvas connector a flexible connection between the air handler and main trunk line to lessen vibration and noise transmission in the duct system

capacitor energy-storing device that can help increase

the starting torque or running efficiency of a motor

capacitor-start-capacitor-run (CSCR) motor single-phase motor that starts and operates with the aid of start and run capacitors

capacitor-start-induction-run (CSIR) motor single-phase motor that uses only a start capacitor that is removed from the circuit after the motor has started

capillary tube a fixed bore metering device

carbon dioxide (CO_2) extinguishers fire extinguisher that uses vaporizing liquid carbon dioxide to remove heat from a fire

carbon monoxide tester (CO tester) instrument used to measure the percentage of carbon monoxide in flue gas

cardiopulmonary resuscitation (CPR) emergency first aid procedure to maintain circulation of blood to the brain

centrifugal pump pump that moves water through a piping circuit by means of centrifugal force

centrifugal switch starting device that opens and closes its contacts based on the speed of the motor

CFM ft³ per min

chassis ground common connection point in a piece of equipment that is not actually driven into the ground

cheek the side portion of a hammer head

circuit breakers circuit protection devices that have the ability to be reset

circulator see centrifugal pump
closed loop system that is closed or isolated from the atmosphere

clamp-on ammeter test instrument used to measure the amperage in an electric circuit

class A fire extinguishers fire extinguishers intended for use on fires that result from burning wood, paper, or other ordinary combustibles

class B fire extinguishers fire extinguishers intended for use on fires that involve flammable liquids such as grease, gasoline, or oil

class C fire extinguishers fire extinguishers intended for use on electrically energized fires

class D fire extinguishers fire extinguisher typically used on flammable metals

cloud point temperature at which wax crystals form in fuel oil, typically 20 degrees higher than the pour point

cold anticipator device on a cooling thermostat that causes the cooling contacts to close shortly before the space reaches the cut-in temperature to facilitate even cooling of the space

cold contacts contacts in the stack relay that are in the closed position when the flue pipe is cold

cold junction in a thermocouple, the loose ends of the dissimilar wires

color pure, fresh heating oil is normally champagne colored but is dyed red to differentiate it from taxable on-road diesel fuel; fuel tends to darken when exposed to light, so color is not a good indication of fuel quality.

combination duct system a duct system constructed of more than one material

combustion analyzer test instrument used to measure operating conditions on a fossil fuel heating appliance

combustion reaction that results when fuel, oxygen, and heat are combined in the proper proportions

complete combustion achieved when only nontoxic and nonpoisonous products of combustion, such as carbon dioxide, water vapor, and nitrogen, are released during the combustion process

compression tank see expansion tank

compression the portion of the compression process in which the refrigerant is compressed

compressor the component of an air conditioning system that pumps refrigerant through the system by increasing the pressure of the vapor refrigerant

condensate drain line piping arrangement that carries condensate from the structure

condensate drain trap "U" bend piping arrangement at the outlet of the condensate drain that facilitates proper condensate removal

condensate pump a mechanical pump that carries condensate from the structure

condensate the moisture removed from the air by the air conditioning system

condenser a heat transfer surface in an air conditioning system that rejects heat

condenser saturation temperature the temperature at which system refrigerant will condense in the condenser; this temperature corresponds to the high side pressure in the system on a pressure/temperature chart

condensing furnaces high-efficiency furnaces that remove large amounts of heat from the flue gases, resulting in the condensing of the flue gases

condensing medium the medium, usually air or water, that absorbs the heat that is rejected by the system condenser

condensing unit portion of an air conditioning system made up of the compressor, the condenser coil, and condenser fan motor

conduction the method of heat transfer by which heat is transferred from molecule to molecule within a substance

conductors materials that have one or two electrons in their outermost shell and facilitate the free flow of electrons

contaminants unwanted particulate matter in air that may or may not lead to health problems

continuity the term used to indicate that a complete current path exists

continuous run timer defrost timer wiring that determines defrost cycles by the time the system is operating

control circuit circuit that is made up of control devices such as thermostats, relay coils, pressure controls, and safety switches

control transformer device that creates the voltage supply for the control circuit

convection the method of heat transfer that is facilitated by the flow of a fluid—typically air or water

covered conductor term used to describe a conductor that has a covering *not* recognized by the NEC as electrical insulation

critically charged system a system that requires an exact quantity of refrigerant; all of the refrigerant is moving through the system at all times

cumulative compressor run timer defrost timer wiring that determines defrost cycles by the actual run time of the compressor

current magnetic relay (CMR) starting device that opens and closes its contacts based on the current flow through the run winding of a split-phase motor

current the amount of electron flow through a conductor

cut-in pressure pressure at which a pressure switch or control closes its contacts

cut-out pressure pressure at which a pressure switch or control opens its contacts

cylinder the component part of a reciprocating compressor that houses the piston

cylindrical heat exchanger heat exchanger with only one combustion chamber

D

de-aerator oil system accessory that facilitates the removal of air from the oil before it passes into the fuel pump

deep vacuum the term used to describe a vacuum that is in the 50–250 micron range

degassing the removal of non-condensable gases from a system during the evacuation process

dehumidifying the process of removing humidity from the air

dehydrating the removal of water vapor and moisture from a system during the evacuation process

dehydration the process of removing water vapor from a system

delta configuration three-phase wiring configuration used primarily for applications that require a large number of three-phase circuits; found primarily in commercial and industrial structures

desuperheating the process by which the discharge refrigerant from the compressor is cooled down to the condenser saturation temperature

dew point the temperature at which air reaches 100% humidity. For blended refrigerants, the temperature used to calculate evaporator superheat.

diaphragm-type expansion tank see expansion tank

differential mathematical difference between the cut-in pressure and the cut-out pressure

digital meters pieces of electrical test equipment that use internal electronic circuits to take voltage, current, and resistance readings; these readings are shown on a liquid crystal display

dilution air air that is mixed with the flue gases as they enter the flue pipe to dilute the concentration of products of combustion

direct drive the blower arrangement where the blower is connected directly to the motor shaft

direct return configuration of a hot water heating system in which the first terminal unit supplied with hot water is the

first one to return to the boiler and vice versa

direct spark ignition (DSI) ignition method that does not use a pilot; an electric spark is used to ignite main gas directly

direct vent venting method that uses the blower on the burner to pull outside air in for combustion and to push flue gases from the structure

discharge line the refrigerant line that carries the discharge refrigerant from the compressor to the condenser

discharge pressure the pressure of the refrigerant in the high-pressure side of the air conditioning system; Also referred to as the high-side or head pressure

discharge the portion of the compression process when the refrigerant is discharged from the compressor

discharge valve the component part of a reciprocating compressor that opens to discharge refrigerant from the compressor into the discharge line

distribution panel panel that distributes incoming power to protected circuits throughout the structure

diverter tee in a one-pipe hot water system, the fitting used to increase resistance to water flow in the main loop in order to direct water to the terminal units

DOT Department of Transportation

DOT-approved cylinders vessels used to contain recovered refrigerant

draft gauge instrument used to measure the pressure over the flame in a heating appliance

draft regulator automatic damper that opens and closes in response to the amount of draft in the flue pipe; high draft will cause the draft regulator to open, and low draft will cause the draft regulator to close

drive pulley the pulley that is connected directly to the shaft of the motor

driven pulley the pulley connected to the blower shaft

drop-out voltage voltage at which the potential relay returns its contacts to their normally closed position

dry chemical extinguisher fire extinguisher that contains an extinguishing agent and a compressed, nonflammable gas, which is used as the propellant

dry-bulb temperature temperature measured with a standard thermometer

dry-type evaporator an evaporator that is designed to have all of the refrigerant boil off into a vapor before leaving the coil

E

earth ground the point at which a physical grounding rod or pipe is driven into the Earth

electrode assembly portion of the oil burner comprised of the electrodes, electrode clamp, and ceramic insulators

electrode clamp strap that holds the electrodes in place

electrodes component part of the oil burner that generates the spark needed for ignition

electromotive force (EMF) see voltage

electronic air cleaners see electronic air filters

electronic air filters electrically operated filters that offer multiple stage filtering as well as odor removal

electronic manometer test equipment that reads gas pressure in inches of water column

electrons the negatively charged components of an atom

electrostatic air filters filters that capture particulates with electric charges opposite to that of the filter media

element see media

endplay the side-to-side motion of a motor shaft that indicates that the motor bearings are damaged

energy efficiency ratio (EER) rates the efficiency of a system; calculated by dividing the Btu output of the system by the power input in watts

energy the ability to do work

equipment ground see chassis ground

ethane a hydrocarbon made up of two carbon atoms surrounded by six hydrogen atoms

ethanol yellow liquid commonly used in levels

evacuation the process of removing moisture and noncondensable gases from an air conditioning system prior to startup and charging

evaporator pressure the pressure that corresponds to the temperature at which refrigerant vaporizes in the evaporator on a pressure/temperature chart

evaporator temperature differential the difference between the temperature of the supply air and the temperature of the return air

evaporator the component part of an air conditioning system that is responsible for absorbing heat from the space to be cooled

expansion tank system piping component that provides additional space for expanding water to occupy

extended surface filters filters that have pleated filter media to increase the surface area of the filter

extended-plenum duct system duct arrangement that facilitates shorter branch duct runs by extending the plenum closer to the branch termination points

externally equalized TEV a thermostatic expansion valve that measures the evaporator pressure at the outlet of the coil

F

factory wiring wiring installed at the factory at the time of system manufacture

fan switch device that energizes and de-energizes the blower circuit in response to temperature conditions within the system

fan-limit switch control device that incorporates both the fan switch and the limit switch

feet of head term used to rate the pumping capacity of a pump; 1 foot of head is the equivalent of a 0.433 psig difference between the inlet and outlet of the pump (1 psig _ 2.31 feet of head)

fiberboard rigid fiberglass material used to fabricate duct sections

fiberglass sheeting see fiberboard

field wiring system wiring that is installed by field technicians or installers

filter-backed return grill return grill in the occupied space that supports an air filter

filters means by which particulate matter is captured from the air

filtration process of removing particulate matter from the air

fire point see ignition point

fittings preformed piping components which make connections and turns in rigid piping arrangements possible

fixed-bore metering device a metering device that does not open or close in response to changes in the load on the system. Examples of the fixed bore metering device are the capillary tube and the piston

flame retention burner type of oil burner that concentrates the air flow pattern to create a better fuel/air mixture at the time of combustion

flaring process by which a 45-degree angle is formed on the end of a soft-drawn tubing section

flaring yoke portion of the flaring tool used to create the 45-degree angle on the cut portion of the tubing section

flash gas the process by which some of the liquid refrigerant instantly vaporizes upon entering the evaporator

flash point temperature at which fuel oil will momentarily flash and immediately extinguish

flexible duct fiberglass duct material that is used to make connections between the main trunk line and the

floodback the condition that exists when liquid refrigerant leaves the evaporator and flows back to the compressor

flooded see overfed evaporator

flow check valve see flow control valve

flow-control valve valve that prevents backward and gravity circulation through loops not requiring flow

flow-through humidifiers humidifiers that have a constant flow of water through them

flue gases substances that are created during the combustion process

flue pipe piping arrangement used to vent products of combustion from the structure

flux material applied to piping materials prior to soldering to help reduce oxidation and remove particulate matter from the joint

fossil fuel fuel made up of the remains (fossils) of prehistoric plants and animals

four-valve manifold refrigeration gauge manifold that is equipped with four hoses and four valves

four-way reversing valve device in the heat pump system that reverses the direction of refrigerant flow through the evaporator and condenser coils to allow the system to operate in either the heating or cooling mode

fractionation the process by which blended refrigerants separate into their component refrigerants from flowing through the winding

frontseating in the case of the liquid line service valve, the process of closing off the liquid line, leaving the gauge port open to the condensing unit side of the valve

frostbite injury to the skin resulting from prolonged exposure to freezing temperatures

frostnip the first stage of frostbite

ft³ per Min (CFM) term used to describe the volume of airflow

fuel pump component part of an oil burner that is responsible for feeding pressurized oil through the nozzle

fuel pump pressure gauge instrument used to measure the pressure at the outlet of the fuel pump on an oil burner

fully halogenated a hydrocarbon-based refrigerant that has all of its hydrogen atoms replaced with chlorine or fluorine atoms

fuses a one-time circuit protection device

fusible link one-time control device that will melt to open its contacts if the temperature in a furnace reaches an unsafe level; used primarily as a backup should the limit switch fail to open

G

galvanized sheet metal sturdy material used to fabricate duct systems

galvanized steel pipe steel pipe that is coated with zinc to resist rusting

Gas Appliance Manufacturers Association (GAMA) organization that provides venting tables for properly sizing and installing vents for gas appliances

gas laws laws of physics that govern the behavior of gases or vapors

gas valve furnace component that modulates (starts, stops, or adjusts) the flow of gas in response to system and safety requirements

gauge pressure the pressure scale that does not take atmospheric pressure into account. At sea level, the gauge pressure will be 0 psig

general purpose branch circuit circuit that supplies power to two or more outlets for lighting and appliance use

geothermal heat pump heat pump system that uses the earth as the heat source in the cooler months and the heat sink in the warmer summer months

global warming potential (GWP) the index used to measure a chemical substance's effect on global warming as compared with the effects of carbon dioxide

global warming the result when the atmosphere traps the heat radiated from the earth

grains weight measurement where 7,000 grains equals 1 pound

gravity indication of the heat value of fuel; ASTM standards indicate fuel oil's density to

be 7.1 pounds per gallon and 139,400 Btus per gallon

greenhouse effect see global warming

ground coils see ground loops

ground fault circuit interrupter (GFCI) electrical device designed to sense small current leaks to ground and de-energize the circuit before injury can result

ground loops buried piping circuits used in closed-loop geothermal heat pump systems

ground term used to describe an electrical connection between the casing of equipment or tools and the earth

grounded connected to ground or to a conductive body that is connected to ground

H

halogen chemical elements such as chlorine and fluorine that replace hydrogen atoms to create halogenated hydrocarbons. Other halogens include iodine and bromine

halon extinguisher fire extinguisher that contains a gas that interrupts the chemical reaction when fuel burns

hard-drawn pipe see rigid pipe

hard-drawn tubing term used to describe rigid pipe

heat anticipator device on a heating thermostat that causes the contacts to open shortly before the space reaches the desired temperature to facilitate even heating of the space

heat energy that increases molecular movement within a substance

heat exchanger in heating, the component that keeps the air from the occupied space separate from gases and by-products of combustion resulting from burning fossil fuels

heat recovery ventilators mechanical ventilation systems that exchange air between the inside and outside of a structure while recovering heat from the discharged air.

heat sink any substance or media that absorbs heat from another substance or media

heat source any substance or media that supplies heat to a system

heaters component part of a motor starter that senses the heat generated by current flow, facilitating overload protection for the motor

HEPA filters (high efficiency particulate arresters) filters capable of capturing a minimum of 99.7% of all particulate matter at least 0.3 microns in size

hex heads six-sided screw or bolt-head shape

high efficiency particulate arresters see HEPA filters

high-pressure atomizing oil burner oil burner that delivers atomized fuel oil through the nozzle at pressures above 100 psig

high-voltage term used to describe voltage levels greater than 600 Volts

holding charge initial amount of refrigerant added to the system before it is initially energized

hollow cone spray pattern where the oil is concentrated on the outer edge of the spray

horizontal configuration of air handlers that causes air to move through the unit from side to side

horizontal ground loop a single water loop configuration in which the ground loop is located close to the ground surface; used primarily in areas with underground rock formations that make deep trenching very costly

horsepower power rating of motors that is equal to 746 watts

horsepower unit of power equal to 33,000 ft lb/min

hot contacts contacts in the stack relay that are in the closed position when the flue pipe is hot

hot junction in a thermocouple, the point at which the dissimilar wires are connected

hot-surface igniter silicon carbide or silicon nitride strip used to ignite fuel in gas heating appliances; see hot-surface ignition

hot-surface ignition ignition method by which electric current is passed through a high-resistance material to ignite gas as it passes through the gas valve; see hotsurface igniter

humidification process of adding moisture to the air

humidifiers system components that facilitate the addition of moisture to an air sample

humidity amount of water vapor in an air sample

hydrocarbons molecules that are comprised of only hydrogen and carbon atoms. Methane and ethane are examples of hydrocarbons

hydronics heating systems that circulate hot water or steam through piping arrangements located in the areas being heated

I

ignition point lowest temperature at which the fuel will rapidly burn in the presence of air; the standard for fuel oil is 637°F

ignition transformer step-up transformer that provides a secondary voltage of approximately 10,000 Volts to vaporize and ignite the fuel

impeller rotating portion of a centrifugal pump that creates the pressure differential needed to facilitate fluid flow

inches of mercury vacuum when reading gauge pressure,

a reading below atmospheric pressure

inches of water column (in. W.C.) units used to measure pressures that are too low to read in psi; pressure readings are typically obtained with a manometer in which one psig is equal to 27.7 in. W.C.

incomplete combustion combustion that results in the formation and release of carbon monoxide as well as other toxic and poisonous substances

individual branch circuit circuit that supplies power to only one piece of equipment

individual section heat exchanger heat exchange configuration with common openings at the top for common venting and at the bottom for common fuel ignition

indoor coil heat transfer surface in a heat pump system that absorbs heat from the occupied space in the cooling mode and rejects heat to the occupied space in the heating mode

induction-start-induction-run (ISIR) motor single-phase motor that starts and operates solely on the imbalance in magnetic field generated by current flow through the windings

inductive loads electric devices that primarily generate a magnetic field

inductive reactance the additional resistance in an ac circuit that results from

the constant building up and collapsing of magnetic fields

infrared thermometer instrument that measures the infrared energy emitted by an object to determine its temperature

in-line fuse low amperage fuse in the control circuit that protects the control transformer from excessive current

insulated conductor term used to describe a conductor that has a covering that is recognized by the NEC as electrical insulation

insulators materials that have a stable atomic structure and do not permit the free flow of electrons from atom to atom

interlock any condition or set of conditions that must be met before an unrelated or separate electric circuit can be energized

intermittent pilot ignition method that lights the pilot only during a call for heat

internally equalized TEV a thermostatic expansion valve that measures the evaporator pressure at the inlet of the coil

interrupted ignition ignition method that initiates spark only at the beginning of the cycle

K

kilowatt power measurement equal to 1,000 watts

knockout plug commonly used to close unused openings in electrical boxes

L

ladder diagram see line diagram

latent heat heat energy that results in a change in state of a substance while maintaining a constant temperature

left-hand threads threaded connection that is tightened by turning the connection to the left

legend the portion of a wiring diagram that identifies the abbreviations used in the diagram

level term used to indicate a perfectly horizontal line or plane

limit switch safety device that de-energizes heating equipment if unsafe temperature conditions exist

limit switch typically used in furnaces, the control device that opens its contacts if an unsafe or undesirable high temperature is sensed

line diagram wiring diagram that is configured with each circuit on a separate line; used to facilitate effective circuit troubleshooting

line set term used to describe the bundle of refrigerant lines that connects the indoor and outdoor portions of an air conditioning system

lines of force fields of force generated by magnets and magnetism

line-voltage term used to describe voltage levels between 50 Volts and 600 Volts

liquefied petroleum liquefied propane, liquefied butane, or a mixture of the two

liquid line the refrigerant-carrying line that connects the condenser to the metering device

liquid-source heat pumps heat pumps that use liquid as the source of heat while operating in the heating mode

liquid-to-air heat pumps liquid source heat pumps that are used to treat air

liquid-to-liquid heat pumps liquid source heat pumps that are used to treat liquid

load any electric component that uses or consumes electric power

locked-rotor amperage (LRA) amperage that a motor draws on initial startup

low loss fittings fittings attached to refrigerant hoses to prevent the release of refrigerants from them when removed from a system

low-voltage term used to describe voltage levels lower than 50 Volts

M

magnetism the force exerted by magnetic field; see lines of force

mandatory rules practices that are either allowed or not allowed

manifold station location of the manifold for radiant heating loops

manometer instrument that measures very low pressures in air conditioning and heating systems

manual air vent fitting that, when opened manually, will remove air from a hydronic heating system

manual reset device feature that requires an individual to physically reset the device

manufactured gas gas mixture made from coal, oil, and hydrocarbon gases

MAPP™ gas composite gas that burns at temperatures higher than that of propane (methyl acetylene-propadiene)

mastic sticky putty used to seal air-leaks in metallic duct systems

Material Safety Data Sheets (MSDS) forms that provide storage, transport, and first aid information regarding chemicals used in the field

matter any substance that has weight and mass and occupies space

media humidifier component that holds water and extends the surface area of the water

metering device the component of an air conditioning system that controls the flow of refrigerant to the evaporator

meters see multimeters

methane a hydrocarbon made up of one carbon atom surrounded by four hydrogen atoms

micron a unit of linear measurement equal to 1/25,400 of an inch

micron gauge the test instrument used to measure the level of vacuum in an air conditioning system

millivolt quantity of voltage equal to 1/1000 of a volt

mineral oil lubricant used with CFC and HCFC refrigerants

mold microscopic organism that can cause allergic symptoms similar to those caused by plant pollen

monoflo tee see diverter tee

Montreal Protocol Act 1987 legislation resulting from a meeting of 23 countries that jump started the program that slowed the production of ozone depleting substances

motor controllers devices that are normally used to start and stop a motor by making or breaking electrical contacts

motor single-phase motor that has low starting torque but good running efficiency. This motor uses a run capacitor and operates with both the run and start windings in the circuit at all times

motor starter component used to start three-phase motors and protect them from single-phasing

motor-off-delay the process of keeping the burner blower energized for a period of time at the end of the heating cycle; also referred to as post-purge

multimeters pieces of electrical test equipment that measure a combination of voltage, amperage, and resistance

multistage thermostat thermostat that can control more than one heating or cooling device at any given time

multitap transformer control transformer that is wired for use with multiple primary voltages

mushroomed head term used to describe the misshapen, damaged head of a chisel or other struck tool

N

nanometer distance equal to 1×10^{-9} power, or 0.000000001 meter

national Electric Code, (NEC) codebook that provides specific safety guidelines for installation and service personnel working in our industry

National Electrical Code (NEC) a publication that sets the standards for electrical installations

National Fire Protection Agency (NFPA) agency that provides codes, standards, research, training, and education regarding safety and fire-related issues

natural gas mixture of hydro-carbon gases consisting mostly of methane, with smaller amounts of ethane, propane, and butane

natural return process by which air is returned to the air handler without a physical connection between the return grill and the air handler

near-azeotropes blended refrigerants that behave similar to azeotropes, but that operate with temperature glides and are subject to fractionation

negative temperature coefficient (NTC) thermistor that experiences a decrease in resistance as its temperature increases

neutrons the neutrally charged components of an atom

nichrome wire made up of nickel and chromium that is typically used in the fabrication of resistive heating elements

noncondensable gas any gas that is unable to condense at the pressures and temperatures normally encountered in air conditioning and refrigeration systems

noncondensable gases gases that will not change from a vapor to a liquid within the normal pressure ranges commonly found in air conditioning and refrigeration systems

noncondensables term used to describe gases that will not condense within the normal operating pressure ranges encountered in an air conditioning system

normally closed term used to describe the closed or connected position of a set of electrical contacts when the device is either exposed to atmospheric pressure, de-energized or at room temperature

normally open term used to describe the open or disconnected position of a set of electrical contacts when the device is exposed to atmospheric pressure, de-energized, or at room temperature

nozzle adapter pipe fitting that connects the nozzle to the nozzle line

nozzle assembly unit comprised of the nozzle line, nozzle adapter, nozzle, and electrodes

nozzle component part of an oil burner that atomizes, meters, and sprays the oil into the combustion area

nozzle line section of tubing that connects the outlet of the pump to the nozzle adapter

nozzle tube see nozzle line

number 2 fuel oil most common type of fuel oil used in residential applications

nut driver tools that are similar in appearance to screwdrivers but are configured to tighten and loosen screws and bolts with hex heads

O

offset tool configuration that facilitates the reaching of tight or awkward spaces

Occupational Safety and Health Administration (OSHA) branch of the U.S. Department of

Labor that strives to reduce injuries and deaths in the work place

offset duct section a duct section that permits a rigid duct to be redirected around an obstacle

one-pipe hot water hydronic piping configuration that uses a main hot water loop and diverter tees to connect the terminal units to the system

one-pipe system fuel oil piping arrangement that uses one supply pipe connected between the oil tank and the oil burner

open circuit the term used when there is no available path for electric current to take

open motor winding condition in which an internal motor winding experiences a break that prevents current

open-loop system geothermal systems that obtain their supply water from an open source such as a lake, pond, well, or other underground water source.

operational controls switches that open and close their contacts during the normal operation of the equipment

orifice opening in the spud that delivers the fuel to the burners

outdoor coil heat transfer surface in a heat pump system that acts as the condenser in the cooling mode and as the evaporator in the heating mode

outdoor reset control used on hydronic systems that decreases the temperature of the water as the outdoor temperature increases

outside ambient temperature the temperature of the air that surrounds the outdoor coil of an air conditioning system

overfed evaporator an evaporator that is being fed too much refrigerant; these evaporators operate with very low, or zero, superheat

overload protector internal thermal device used to protect motors from overheating

ozone depletion potential (ODP) the index used to measure a chemical substance's effect on ozone depletion as compared to the effects of CFC-11. The ODP ranges from 0–1.

ozone layer refers to the layer of good ozone that exists in the stratosphere, which is 7–30 miles above the Earth's surface

P

packaged units systems that have all components in one cabinet

parallel ground loops in geothermal heat pump systems, the ground loop configuration that provides multiple paths for the heat exchange fluid to flow in

partially halogenated a hydrocarbon based refrigerant that has some of its hydrogen atoms replaced with chlorine or fluorine atoms

part-wind start three-phase motor designed to have only part of the windings in the circuit upon initial startup

PASS acronym used to describe fire extinguisher use (Pull, Aim, Squeeze, Sweep)

passive recovery recovery method that uses the system compressor to remove refrigerants from the system

peen the rounded back portion of the head of a ball peen hammer

perimeter duct system duct system configuration in which the duct system runs around the perimeter of the occupied space

permanently lubricated term used to describe motors that typically will not require lubrication during their expected useful lives

permanent-split-capacitor (PSC) single-phase motor that has low starting torque but good running efficiency. This motor uses a run capacitor and operates with both the run and start windings in the circuit at all times

permissive rules practices that are allowed, but not required

personal protection equipment (PPE) any equipment that will provide protection from potential injury

Pete's Plug a self-sealing access port by which system temperature and pressure readings can be obtained

Pete's port self-sealing access fitting used to take temperature and pressure readings in geothermal heat pump systems

PEX tubing see polyethylene tubing

pick-up voltage voltage at which the potential relay's contacts open

pigtail splice an electrical wire connection made by twisting two or more wires together

pilostat coil of wire that generates a magnetic field to hold the gas valve open

pilot-operated reversing valves reversing valves that use system pressures to move the slide back and forth to change the system over between the heating and cooling modes of operation

piston the component part of the reciprocating compressor that moves back and forth within the cylinder to compress or expand the refrigerant in the cylinder

pitting condition in which the surfaces of relay or contactor contacts become worn

planning division local agency that oversees the zoning issues for a particular area

plenum-type duct system duct system configuration commonly used when the air handler is located close to the branch termination points

pleated filters see extended-surface filters

plumb term used to indicate a perfectly vertical line or plane

point of no pressure change point in a hydronic hot water system where the expansion tank is connected to the piping system; pressure at this location cannot be affected by circulator operation

pole term used to describe the number of sets of contacts that are opened or closed when the position of a switch is changed

pollutants unwanted particulate matter in air that often leads to health-related problems

polyethylene tubing tubing material used for buried water loops in radiant heating systems and geothermal heat pump systems

Polyol ester oil (POE) lubricant used with HFC refrigerants

polyvinyl chloride (PVC) pipe rigid plastic material commonly used for drain lines

positive temperature coefficient (PTC) a resistor that reacts to changes in temperature; as the sensed temperature increases, the resistance of the PTC will increase as well

positive temperature coefficient (PTC) thermistor that experiences an increase in resistance as its temperature increases

post-purge the process of keeping the burner blower energized for a period of time at the end of the heating cycle; also referred to as motor-off-delay

potential relay starting device that opens and closes its contacts depending on the induced voltage across the start winding of a motor

pour point minimum temperature at which the fuel will flow; the standard is 20°F

power circuit circuit that carries power to main system components, such as compressors and fan motors

power the rate at which work is done; work per unit time

power venting venting method that uses a powerful blower to pull flue gases from the furnace

pre-purge the process of energizing the burner blower for a period of time before allowing fuel to enter the combustion chamber; also referred to as valve-on-delay

pressure force per unit area; common units are pounds per square inch (psi)

pressure relief valve springloaded valve that opens when the pressure in a hydronic system exceeds the rating of the valve

pressure switch electric switch that opens and closes its contacts depending on the pressure it senses

pressure tanks pressurized tanks that store water

pressure/temperature chart shows the relationship that exists between the temperatures and pressures of saturated refrigerants

pressure-reducing valve reduces the pressure of the water entering the structure to the desired pressure in the hydronic system

preventive maintenance tasks, such as oiling motors and changing air filters, that are performed on a system to help reduce the chances of future system failure

process tubes the lines through which hermetically sealed systems are pressurized, leak tested and charged

proportional balancing the balancing process which ensures that all outlets supply the same percentage of design air to the occupied space

protons the positively charged components of an atom

psia pounds per square inch absolute; takes into account the pressure of the atmosphere and is approximately equal to the gauge pressure plus 15

psig pounds per square inch gauge; ignores the pressure of the atmosphere. Used to measure the pressure in sealed vessels such as car tires and air conditioning systems

pumping down process of storing the system refrigerant in the condenser by frontseating the liquid valve at the outlet of the condenser

R

radiant heat system heating system that attempts to regulate the heat loss of the individual as opposed to the rate of heat loss of the structure

radiation method of heat transfer by which heat travels through the air and heats the first object the rays come in contact with

radiator heat emitters or terminal units that transfer the majority of their heat to the occupied space by radiation

reactor plate the copper disc in a halide leak detector

reamer tool used to remove burrs from both the inside and outside of a pipe section

recovery process of removing refrigerant from a system and storing it in a cylinder without testing, cleaning, or reprocessing it

reduced-voltage start three-phase motor starting technique that adds resistance to the circuit to reduce the amount of voltage supplied to the motor during startup

reducing-extended-plenum duct system duct system configuration used to maintain a constant air velocity in the trunk line as air moves through it; not commonly found in residential applications

reexpansion in a reciprocating compressor, the process by which the refrigerant trapped in the cylinder at the end of a cycle is expanded to reduce the pressure in the cylinder

refrigerant trap device located at the outlet of the evaporator when the air handler is located below the compressor; used to facilitate oil return to the compressor

regulator torch kit component that regulates the pressure at the outlet of the vessel

relative humidity ratio that compares the amount of moisture in an air sample to the maximum amount of moisture the same sample can hold

relative humidity relationship between the amount of moisture in an air sample and the maximum amount of moisture the air sample can hold

relative humidity the amount of moisture in an air sample compared to the maximum amount of moisture the air sample can hold

relief valve see pressure relief valve

repeating cycle a process that can repeat itself indefinitely without depleting resources

resistance opposition to current flow

resistive loads any electric devices that primarily generate light or heat

return air air from the conditioned space that is returning to the air handler

return duct duct that facilitates the return of air to the air handler from the conditioned space

return grill decorative grill located in the occupied space that connects to the return duct or provides access for air to return to the air handler

return water in geothermal systems, the water that returns to the ground after passing through the system and the return well

reverse return system configuration of a hot water heating system in which the first terminal unit supplied with hot water is the last one to return to the boiler and vice versa

right-hand threads threaded connection that is tightened by turning the connection to the right

rigid pipe piping material that is not intended to be bent

rotor portion of a motor that rotates within the motor shell

run capacitor energy storage device that provides additional starting torque and helps improve running efficiency of single-phase motors; it is oil-filled and is intended to be in the active electric circuit whenever the motor is energized

run winding low-resistance winding of a single-phase motor

running-load amperage (RLA) amount of current a motor draws after it has been energized and permitted to reach its normal operating speed

S

saddle valves two-piece valves that clamp around a domestic water line to provide supply water to the humidifier

safety device a switch that opens or closes its contacts only when an unsafe condition is present.

saturated refrigerant that is a mixture of liquid and vapor; saturated refrigerants follow a pressure/temperature relationship

schematic diagram wiring diagram that provides a representation of every wire in the circuit as well as wire color-coding and electric component location

schrader valve a self-sealing valve similar to those found on car or bicycle tires

Schrader-valve stems the portion of the Schrader valve that creates the seal

seasonal energy efficiency ratio (SEER) equipment efficiency rating determined by dividing the system's Btu output by the input power in watts, taking into account the energy used on system startup and shutdown

self-tapping screws sheet metal screws with drill-like tips; eliminates the need to pre-drill holes in the sheet metal

semi-solid cone spray pattern where most of the oil is concentrated on the outer edge of the spray, while a lesser concentration is found within the spray area

sensible heat heat energy that results in the change of temperature of a substance

sequencers control devices used to energize and de-energize electric heating elements at regular time intervals

service wrench tool designed to turn square stems on valves

shank the portion of a screwdriver-like tool that connects the tip of the tool to the handle

short circuit a circuit that offers zero resistance to current flow

short circuit to ground condition in which an internal motor conducting surface comes in contact with the shell of the component

shorted motor winding condition in which damage to a motor winding exists, creating a reduction in the resistance of the winding

sleeve bearings type of motor bearing often used on residential systems because of their quiet operation

slips and drives materials used to join sections of sheet metal duct

sludge organic solids made up of moisture, acid, and refrigerant oil

smoke tester instrument used to measure the presence of unburned fuel in flue gases

soft metal tubing term used to describe tubing and pipe made from soft metals such as copper and aluminum

soft start term used to describe the gradual increase in motor speed on start up

soft stop term used to describe the gradual decrease in motor speed on shut down

soft-drawn copper tubing piping material that has the ability to be bent into the desired configuration, reducing the number of fittings required to connect a system

soft-drawn tubing refers to piping materials that are typically flexible and easily bent

soldering process of joining metallic piping sections using filler materials with melting points in the range of 500°F

solderless connectors used to create a nonsoldered, mechanical electrical connection between two wires or between a wire and an electrical component or device

solid cone spray pattern where the oil is evenly dispersed within the entire spray pattern

split-phase motor single-phase motor in which the power is split between the start and run windings of the motor

split-type systems systems that have the major system components in two different locations

spring pressure one of the closing pressures on the thermostatic expansion valve;

also the opening pressure on the automatic expansion valve

stack relay primary control that controls the operation of the burner by sensing the temperature of the flue gases

standard expansion tank see expansion tank

standing pressure pressure inside a vessel when the vessel is at rest; for example, the refrigerant pressure inside an air conditioning system when the system is off

Standing-pressure test leak detection method that involves the pressurization of system and the evaluation of the system pressure

start capacitor energy-storing device intended to increase the starting torque of a single-phase motor

start winding high-resistance winding of a single-phase motor

starters see motor starters

starting torque twisting force used to initiate the motion of an electric motor

starved the condition that exists when an evaporator is fed less refrigerant than needed to provide adequate cooling; also known as an underfed evaporator

static pressure the pressure that pushes against the sides of the duct sections

stator portion of a motor that does not rotate and houses the start and run windings

steam boiler piece of heating equipment that heats water to the point of vaporization to provide heat to an occupied space

steel filters permanent, cleanable air filters

stepwise balancing method of air balancing that involves stepping down the airflow to the outlets with the greatest amount of excess air

stratosphere the atmospheric shell around the earth that is between 7 and 30 miles above the surface of the earth

striking face the portion of a hammer head that comes in contact with the object being struck

stubby-type term used to describe tools with very short shanks that are used to get into very tight areas

subcooling the process by which the refrigerant in the condenser is cooled below the condenser saturation temperature

suction in a reciprocating compressor, the process by which refrigerant in the suction line is pulled into the compression chamber prior to being compressed

suction line the refrigerant-carrying line that connects the outlet of the evaporator to the compressor

suction pressure the pressure of the low side of the system; also called low-side or back pressure

suction valve the component part of a reciprocating compressor that opens to pull refrigerant from the evaporator into the compressor for compression

suction-line accumulator a refrigeration component located in the suction line that facilitates the vaporization of liquid refrigerant before entering the compressor

sulfur chemical element found in fuel oil; the maximum amount of sulfur in fuel oil is limited to 0.5%

superheat the process by which vapor refrigerant is heated above its evaporator saturation temperature

supply air the air that is returned to the conditioned space after it has passed through the evaporator

supply water in geothermal systems, the water being supplied to the system via the supply well

swaging process of expanding the end of a piping section to accept another section of same-size piping material

swaging process of expanding the end of a tubing section so that another section of same size tubing can be inserted into it

switches component parts of a motor starter that open to protect three-phase motors from single-phasing; electric devices that open or close electric circuits, starting and stopping current flow

system charge the total amount of refrigerant that an air conditioning system holds

T

takeoff duct section connected to the main trunk line, providing a connection point for the branch duct

temperature differential term used to describe the difference between two temperatures

temperature glide the range of temperatures in which a blended refrigerant will vaporize or condense at a given pressure

temperature term used to describe the level of heat intensity

terminal unit radiator or section of baseboard

testing the process of using specialized, properly calibrated instruments to measure characteristics of fluid flow and electricity

thermal bulb TEV component that senses the temperature at the outlet of the evaporator and converts this temperature to the valve's opening pressure

thermal kinetic energy see sensible heat

thermal potential energy see latent heat

thermistor an electronic device that changes its resistance as its temperature changes

thermostatic expansion valve (TEV) metering device designed to maintain a constant evaporator superheat

three-phase motor motor designed to operate when supplied with three-phase power

throw term used to describe the number of active (on) positions on a switch

timer any control device, mechanical or electronic, that opens or closes sets of electrical contacts at predetermined time intervals

torque the twisting force that is exerted by a rotating object. In the case of a screwdriver, the torque is the force put on the screw, causing it to turn

total pressure the sum of static and velocity pressures

total system charge the amount of refrigerant a system contains

transition duct section duct section that either reduces, enlarges, or changes the shape of the existing duct section

transmission line the small diameter tube that connects the thermal bulb to the body of a thermostatic expansion valve

triac used in cad cell primary controls; the electronic control switching device that works in conjunction with the bilateral switch to control burner operation

troposphere the atmospheric shell around the earth that is between 0 and 7 miles above the surface of the earth

trunk line main run of a duct section coming off the air handler

two-pipe oil delivery system fuel oil piping arrangement that uses two pipes connected between the oil tank and the oil burner; one serves as the supply, and the other serves as the return for any unburned fuel

two-pipe hydronic system hydronic piping configuration that uses one pipe as the supply and one pipe as the return; can be configured as a direct return or a reverse return

two-stage rotary vacuum pump vacuum pump that is made up of two rotary pumps where the outlet of one pump is the inlet of the second

two-valve manifold refrigeration gauge manifold that is equipped with three hoses and two valves

U

ultraviolet (UV) solution dye additive that is introduced to air conditioning systems for leak detection purposes.

underfed evaporator an evaporator coil that has insufficient liquid refrigerant being fed into it

V

vacuum gauge device used to measure pressures below atmospheric pressure. Used on both air conditioning and heating system

vacuum pump piece of equipment used to reduce the pressure in an air conditioning system during the evacuation process

vacuum pump oil oil specially designed for use in vacuum pumps

vacuum the condition that is present when the pressure in a vessel is below atmospheric pressure, 0 psig

valence electrons electrons in the outermost shell of an atom

valve-on-delay the process of energizing the burner blower for a period of time before allowing fuel to enter the combustion chamber; also referred to as pre-purge

velocity pressure the pressure that results from the speed, or velocity, and the weight of the air

velocity term used to describe speed. In the case of airflow, the unit "feet per min" is commonly used

vent alarm device that whistles as an oil tank is being filled, but stops when the tank is nearly full

vertical downflow air handler configuration where the air enters the unit from the top and is discharged from the bottom

vertical ground loops used in geothermal heat pump systems, the ground loops are buried vertically in areas where ground space is limited

vertical upflow air handler configuration where the air enters the unit from the bottom and is discharged from the top

viscosity the ability of a fluid to flow—higher viscosity results

in a lower ability to flow; the viscosity of fuel oil increases as its temperature decreases

voltage drop the voltage reading between two points in an electric circuit

voltage the amount of electrical pressure in an electric circuit that causes electron flow

voltmeter piece of electrical test equipment that measures the voltage or potential difference between two points in an electric circuit

Volt-ohm-milliammeter (VOM) electrical test instrument used to measure voltage, resistance, and small currents in electric circuits

volume factor provides the number of gallons of water per linear foot of a piping material

volute portion of the circulator housing that carries water from the pump

W

wastewater system system that uses water as either a heat source or sink one time before the water is "wasted" down the drain.

water finding paste substance that changes color when exposed to water; used to determine if water is present in an oil tank

water hammer the sound generated when flowing water is suddenly stopped by a closing valve

water loop see ground loops

water table source of underground water used for geothermal heat pump systems; wells must be drilled to the water table, which can be 200 feet below the ground surface

watt unit of electrical power that causes 1 amp to flow in a circuit that is powered by 1 Volt DC

wet-bulb temperature temperature measured with a thermometer that has a wet sensing bulb

wick component part of a humidifier that increases the surface area of the water, increasing the efficiency of the humidifier

wire gauge tool or scale used to determine the gauge or thickness of a wire

within sight term used to describe the relative location of two pieces of equipment that implies visibility of one from the other and a distance that is no further than 50 feet

work force exerted on an object times the distance the object is moved, measured in foot-pounds (ft lb)

Z

zeotropes blended refrigerants that behave similar to azeotropes, but operate with temperature glides and are subject to fractionation; these refrigerants experience larger temperature glides than near-azeotropes refrigerant blends

zone valve thermostatically controlled valve that opens and closes to regulate the flow of hot water to the terminal units in the occupied space

zoning process of dividing the structure into separate areas, each of which has its own means to regulate the temperature in the space

Index

Note: Items in bold indicate table or figure entry.